# ECOLOGICAL GEOGRAPHY OF THE SEA

Second Edition

# ECOLOGICAL GEOGRAPHY OF THE SEA

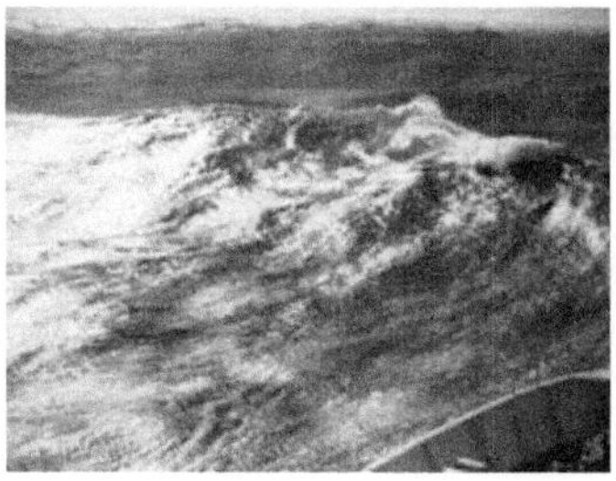

Alan R. Longhurst

ELSEVIER

AMSTERDAM • BOSTON • HEIDELBERG • LONDON
NEW YORK • OXFORD • PARIS • SAN DIEGO
SAN FRANCISCO • SINGAPORE • SYDNEY • TOKYO

Academic Press is an imprint of Elsevier

Academic Press is an imprint of Elsevier
30 Corporate Drive, Suite 400, Burlington, MA 01803, USA
525 B Street, Suite 1900, San Diego, California 92101-4495, USA
84 Theobald's Road, London WC1X 8RR, UK

This book is printed on acid-free paper. ♾



**Library of Congress Cataloging-in-Publication Data**
Longhurst, Alan R.
Ecological geography of the sea/Alan R. Longhurst. –2nd ed.
p. cm.
Includes index.
ISBN 0-12-455521-7 (hard cover : alk. paper) 1. Marine ecology
2. Oceanography. I. Title.
QH541.5.S3L65 2006
577.7–dc22

2006006124

**British Library Cataloguing in Publication Data**
A catalogue record for this book is available from the British Library

ISBN-13: 978-0-1245-5521-1
ISBN-10: 0-12-455521-7

For all information on all Elsevier Academic Press publications
visit our Web site at www.books.elsevier.com

Printed and bound in the United Kingdom
Transferred to Digital Printing, 2011

*"Distinguishing between myth and science is subtle, for both seek to understand the things around us. The characteristic style of mythic thinking is to place special emphasis on a selctive conjecture, which is thereafter given privileged status over alternative suggestions. Concepts in geoscience are quite often mythic in that sense."*

William R. Dickinson (2003)

# Contents

# Preface

How do pelagic ecosystems respond to regional oceanography? Why does each region have a characteristic pelagic ecosystem, differing from all others in the kind and relative abundance of its biota? Why does the production and consumption of organic matter not proceed similarly everywhere in the ocean? These were some of the questions discussed in the first edition of this book, but because progress in biological oceanography has been so rapid, a revision of that discussion is already appropriate.

The first edition dealt only with the planktonic ecosystem and ignored the strong coupling between benthic and pelagic processes that occurs over continental shelves, and between plankton and larger pelagic organisms. Here, I have tried to remedy both these omissions, although the main emphasis remains on the pelagic realm so that abyssal and deep-water organisms are not discussed, because although they respond surprisingly promptly to intermittent sedimentation of organic material from the surface, several kilometers above them, this is a one-way street. Then, the littoral zone is sufficiently isolated ecologically from the oceanic and shelf ecosystems that it is perhaps better discussed separately, so, as in the first edition, you will find reference to estuaries, coastal lagoons, the sublittoral fringe, or coral reefs only where it is essential to understand what happens offshore.

Setting spatial bounds to ecosystems is an essential step toward their quantitative study, and this is best done according to a natural logic, rather than arbitrarily. Partition, therefore, is central to the argument of this book and is set by reference to features of the physical circulation of the oceans rather than to the distribution of individual species. This approach is intended to suggest that regional oceanographic processes are paramount in molding the characteristics of regional ecosystems. Once again, emphasis in the second half of the book is placed on providing a simple account of the characteristic ecology of each region of the oceans, and of the physical factors that shape it.

So, in the same spirit that Tomczak and Godfrey (1994) published their innovative text on regional oceanography for physicists who lacked a modern geography of the oceans, this book is offered as a regional oceanography for ecologists. There is a key phrase in Tomczak and Godfrey's preface: "We have been surprised to learn how much of the ocean's behavior as a component of climate . . . can be understood by combining an understanding of simple physics with knowledge of the ocean's geographical features." I have been equally surprised to find how far the distribution of characteristic marine ecosystems is predictable from a few ecological principles, a simple understanding of regional oceanography, and an elementary "knowledge of the ocean's geographical features." Central to my argument is how ecological processes in the ocean are forced by the geography of the continents to depart from the ideal state that would be observed on a world entirely without land.

One of the problems of writing about marine ecosystems, or of proposing a series of formulations for their structure and dynamics, is the rapidity with which our knowledge has evolved in recent years. In the Preface to the first edition of this book, I noted that during the previous decade there had been a major cognitive shift from a simple "nitrate-diatoms-copepods-fish" model of the pelagic food chain, to more complex constructs incorporating picoautotrophs, mixotrophic protists, and the role of bacteria in regenerating nitrogen. Mills (1989) suggests why it took so long to incorporate in our models of the pelagos the observations of Lohman, made at the end of the 19th century, of the relative abundance of nanoplankton cells. I shall devote a section of this second edition to a discussion of the consequences of our new knowledge about the kinds of organisms that populate the oceans and how they interact. Much has been discovered since 1995–96, when I was working on the first edition.

Inevitably, many of the region-specific studies I have consulted were done when the simpler model was the accepted norm, and the new understanding about the structure and functioning of pelagic ecosystems has come out of studies at just a few locations. For some regions I have been able to extrapolate from modern work, to suggest how the older studies should probably now be interpreted, whereas for too many others the best that I can do is to describe what I think is known and leave you to guess the rest.

The regional scientific literature is far from even-handed in how it deals with different parts of the oceans, as becomes rapidly evident when one tries to analyze any process, or describe the regional expression of any phenomenon, in all seas and oceans. Where significant resources have been expended in oceanographic research in recent years the riches may be embarrassing, but in other seas the embarrassment has nothing to do with riches. The new observations and new understanding about the structure and functioning of marine ecosystems that have accumulated in the past 10 years ago or so are, quite literally, voluminous and I have made extensive use of the results of recently completed or ongoing global research programs—WOCE/WHP, HOT, BATS, JGOFS, GOOS, GLOBEC, and so on—in this edition.

Nevertheless, you will quickly come to appreciate that I have been unable to locate comprehensive ecological studies for several regions (usually distant from Europe or North America) even if their physical oceanography is quite well known. This says something about the priorities of oceanographic agencies, and also about how physical and biological oceanographers have been trained look differently at the world. But once again, the literature of ocean science is evolving and I have the impression that a larger number of regional oceanographic studies are now available than was the case 10 years ago. These, such as the remarkable volume on regional coastal oceanographic regimes recently edited by Robinson and Brink (1998), have been of great assistance in the present task.

In this edition, I have not avoided controversy and, in one chapter, I have indeed rather sought it. This is a reaction to the fact that generous research funding and perhaps, just perhaps, in-group peer reviewing have encouraged the rapid development of novelties that appear to overturn previously held concepts. The quotation from an essay by William Dickinson, a geologist, on the flyleaf is a reminder that we should take care when pursuing the latest scientific quarry. I expand on this issue in Chapter 5, using as my illustration the fashionable hunt for ecological consequences in the open oceans of "iron from the skies."

Once again, my ex-colleagues at what remains today of the Bedford Institute of Oceanography have given me very willing support. In particular, George White obtained and processed the SeaWiFS sea surface chlorophyll data (1997–2002) that are used for the seasonal, regional environmental cycles illustrating Chapters 9–12; I have been greatly assisted in the task of interpreting these data by Trevor Platt and Glen Harrison. I am also grateful to Frederic Mélin of the Joint Research Institute, Ispra, for the provision of a comparable series of regional data from the MODIS sensors that, unfortunately, it was not possible to include. Once again, Gene Feldman of NASA helped me greatly in obtaining images for use in this book: in fact, I wish to emphasize once again the indebtedness of all oceanographers to those U.S. and European agencies that obtain and freely distribute satellite data and images to the rest of us. Their skill and generosity has revolutionized oceanography.

I also remain acutely conscious of the debt that I owe to the many people who taught me enough about the oceans to risk undertaking this book in the first place, some of whom I listed in the preface to the first edition of 1998. I am also very grateful to those people whose warm reception of that work has encouraged me to undertake this rewriting and expansion of it; you were very generous in your comments and very patient in replying to my often naïve enquiries about your special fields: you know who you are, and I thank you one and all.

# Acknowledgments

The following sources of material for figures new to this edition are also gratefully acknowledged: S. Alvain, unpublished thesis of the University of Paris (Fig. 1.1); John Wiley Inc. (Fig. 4.3); National Research Council, Canada (Fig. 8.4); Elsevier (Figs. 3.5, 9.24, and 11.16). Sources of significantly redrawn figures are acknowledged in the legends. SeaWiFS images (Figs. 8.5, 9.14, 9.29, and 11.8) were provided by GeoEye for which processing was provided by NASA. SeaWiFS data for the time-series that illustrate Chapters 9–12 were provided by NASA.

# Chapter 1

# Toward an Ecological Geography of the Sea

Ideally, marine biogeography should have three components. First, it should describe how, and suggest why, individual species from bacterioplankton to whales are distributed in all oceans and seas. Second, it should tell us how those species form characteristic ecosystems, sustaining optimum biomass under characteristic regional conditions of turbulence, temperature, irradiance, and nutrients. Third, and most important for some purposes, it should document the areas within which each characteristic ecosystem may be expected to occur: such a partition of the ocean is the principal focus of this work.

Unfortunately, marine biogeographers have almost exclusively occupied themselves with the first of these tasks, the distribution of individual species and genera, and even in this they have made only limited progress in describing the complexity that evidently exists. Meanwhile, biological oceanographers or marine ecologists, occupied with the analysis of marine ecosystem structure and function, have—on the whole—not been very interested in how ecosystems are constrained spatially.

This, then, has become the objective of ecological geographers, who are interested in the regional distribution of characteristic types of ecosystem. Their studies go a little beyond what has come to be termed *macroecology* (Brown, 1995a), which is the analysis of pattern in the characteristics and relative abundance of biota in complex ecosystems, often using statistical techniques. Macroecology can have a spatial element and may seek geographic patterns in ecosystem structure and function, but it is largely constrained to the identification of statistical pattern in large data sets, over large scales of space and time, in which local details are subsumed. In ecological geography, all relevant data and indicators are used to interpret ecosystem characteristics and spatial distribution: all is grist that comes to its mill.

To be useful at the global scale, ecological geography requires that sufficient information be available from all oceans and seas, a requirement that has been met only in the last few years. All that could be done even 20 years ago (e.g., Longhurst, 1981) was the analysis of functions within some kinds of ecosystem: coastal upwelling, low-latitude gyres, subarctic gyres, abyssal depths, and so on. It was not possible at that time to map the characteristics of these ecosystems globally or even regionally, because we lacked the tools with which to gather the required information.

But, since then, there has been a steep increase in our ability both to acquire and manage knowledge about oceanic ecosystems, and so the task has now become feasible. The tools have come to hand not only to describe how marine ecosystems differ from one region to another, and to set bounds to each, but also to understand how the biological functions within them are forced by physical processes in ocean and atmosphere. Finally, we can go beyond merely descriptive geography toward its functional explanation.

It will be useful, before proceeding further, to consider how oceanographic research at the end of the 20th century changed our understanding of the pelagic ecosystem, and to what extent the new observing systems that have matured in recent years have changed our knowledge of how oceanographic and ecological processes may best be partitioned. Many things have evolved, both in fact and in perception, since the first edition of this work was being drafted. The task set for this first chapter is briefly to review some aspects of these changes.

## The Progressive Exploration of Oceanic and Shelf Ecosystems

The simple, descriptive phase of biological oceanography occupied almost a full century, beginning with the straightforward exploration of the marine biota of all oceans by the *Challenger* in 1872–1876. Though some of the very early voyages, notably the Plankton Expedition of 1889, also laid the foundations of our understanding of the biology of oceanic plankton, post-*Challenger* research on ecological processes was principally pursued in coastal laboratories—especially at Naples, Kiel, Plymouth, and Woods Hole. Quite rapidly, people in these laboratories obtained a basic understanding of the cycle of pelagic production and consumption, a learning process well described by Mills (1989).

Of course, everything changed rapidly in the mid-20th century. It may be hard now to appreciate to what extent the Second World War altered the face of oceanography; governments suddenly understood that this science could produce answers relevant to their strategic concerns. In the United States, this was the period when coastal marine laboratories, such as Scripps in California, grew into the great ocean-going institutions. In those years, modern oceanography was born from the hunger of the U.S. and Soviet navies for information about the ocean: it seemed that the flood of new posts, new ships, and new institutes was unstoppable, a situation that continued on into the 1960s. Pelagic ecologists began seriously to explore the open ocean, perhaps mainly because of naval interest in deep sonic scattering layers.

It was at this time that the first modern regional surveys were mounted to begin to understand the distribution of properties in the open ocean. These were multiship, seasonal explorations of large oceanic regions of which some, like EASTROPAC and EQUALANT, responded to the interest of the U.S. fishing industry in the exploration of the then-unknown resources of the high seas. Others, like the International Indian Ocean Expedition (IIOE), represented the beginnings of international cooperation between oceanographers at sea. The results of these surveys lie in the "Atlas" section of your library and have not yet been integrated into any modern database.

During the same period, large Russian research ships were also exploring the oceans, led by scientists from the Academy of Science in Moscow; a major focus of their work was the biogeography of the pelagos, leading to the studies of K. Beklemishev, M. Vinogradov, and others, to which I shall refer later.

As Karl Banse reminded us, a major catalyst for rapid progress during this period was the introduction into biological oceanography during the 1950s of bulk methods for chlorophyll, seston, proteins, and carbon uptake rates. These enabled us to take samples that matched the bottle data of the physical and chemical oceanographers, and it is to them that we owe the first basin-scale maps of biological processes, such as the primary production of plant biomass. It was by using these methods that we could finally associate biological rate processes with oceanographic processes at regional scale—but always in hindsight, not yet in real time.

There followed a period of maturation and digestion of observations, marked by the steady increase in the capability of the instruments used for measurements at sea, and the progressive sophistication of shipboard experimental methods. Rather than exploring species distributions, the objective of large-scale and multiship expeditions during this period was to quantify the rate constants for physiological processes across a wide range of oceanographic conditions. The catalyst for this change of emphasis was surely the development of solid-state electronics for underwater sensors, for shipboard laboratory equipment, and for data processing.

During the last 25 years of the century, and parallel with equivalent progress in the other branches of oceanography, these instruments delivered nothing less than a revolution in our understanding of ecological dynamics in the oceans. Almost as important was the development of a novel ability to archive and process vary large quantities of numerical data, especially concerning the physical environment of the pelagic biota. Those who did not go to sea prior to the arrival of the personal computers that now dominate shipboard laboratories must find it hard to realize how little information we could take ashore with us. Instruments were read by eye, and data were recorded into deck logbooks.

One technological innovation that was progressively developed during the 1990s is, I believe, so important and brought such revolutionary new possibilities to biological oceanography, and the analysis of the functioning of marine ecosystems, that I must discuss it in a separate section later. I refer, of course, to the availability of images of various properties of the sea surface obtained by earth-orbiting satellites, after the brief proof-of-concept SeaSat mission flown in 1978.

But instrumentation and technical methods are not science, they only enable it to be done. In fact, the most fertile developments in biological oceanography toward the end of the century arose once the observations of the '60s and '70s were sufficiently digested as to permit the formulation of important questions that could be answered only by carefully planned work at sea. We are only now reaping the full benefits of this revolution in oceanography that was forced by (or fed upon, depending on your point of view) three public concerns: environmental pollution, the depletion of fish stocks, and climate change.

All this has been accomplished by an alphabet soup of national and international agencies and research programs, with which it is difficult to keep up. Many, indeed most, of the programs were process-oriented, but regionally based, so that we now have a series of studies that are of great assistance in assembling an ecological geography of the oceans at the global scale. You will, however, certainly have noticed that this has been almost entirely a revolution in our understanding of the pelagic ecosystem because most of the initiatives had to do with the open ocean, where the epipelagic ecosystems lie above and interact with, not a benthic ecosystem, but the bathypelagic biome of the interior of the ocean. There have been few cooperative investigations of the dynamics of the benthic ecosystem that are intellectually comparable to (or as innovative as) those concerned with the pelagos. For the benthos, we shall just have to do the best we can with what we have.

Two initiatives in physical oceanography have provided us with a hitherto-unmatched library of observations, from which to generate the regional analyses of circulation patterns required to interpret regional ecological processes. These initiatives were, of course, the World Ocean Circulation Experiment (WOCE) and the Tropical Ocean/Global Atmosphere (TOGA) program. Oceanographers from 30 nations participated in WOCE, a component of the World Climate Research Programme, and worked at sea from 1990 to 1998 on a logically spaced grid of sections that very adequately covered all four ocean basins. These were arranged as a "One-time Survey" to give a snapshot of the entire ocean circulation, and as a series of "Repeated Hydrography" sections to achieve seasonal and decadal variability. Together with additional Time Series stations, and the deployment of expendable bathythermographs (XBTs), of acoustic doppler current profilers (ADCPs), of surface drifters and of subsurface floats, these data form a set of observations that will remain the primary description of ocean physics for a long time to come. WOCE data are, of course, intended primarily to support climate modeling, but they are accessible to us all on DVD or at various Web sites.

The principal interest in TOGA (1984–1994) for ecologists may be the array of moorings arranged on 11 sections (10°N–10°S) across the entire equatorial Pacific that permitted the first long-term description of the response of the ocean to the atmospheric Southern Oscillation that will be discussed later. Permanent monitoring of ocean conditions is promised by the GOOS (Global Ocean Observing System), currently organized by the IOC of UNESCO. This is intended to match, and mesh with, the atmospheric World Weather Watch to facilitate regional modeling of the coupled atmosphere–ocean system.

Concern over the effects of the increasing content of carbon dioxide in the atmosphere on future climate states has, of course, led to attempts to model the fluxes of carbon between atmosphere and ocean, and the cycling of biologically active elements in the oceanic biogeochemical cycle. But, as McCarthy *et al.* (1986) commented, it had been hard to make the numbers add up and, in particular, to compute with any conviction the role of water column photosynthesis and the subsequent sedimentation to the interior of the ocean of biogenic carbon detritus. Arising from this uncertainty, a Joint Global Ocean Flux Study (JGOFS) was planned in the mid-1980s to clarify the workings of this system at the scale of ocean basins, and to quantify the links among atmospheric, oceanic, and sedimentary carbon pools. This, it was understood, would require careful and sophisticated investigation of the production and consumption of organic material in the upper ocean, and it was toward these processes that JGOFS was specifically directed. Oceanographers from at least 20 nations participated in the work at sea, as JGOFS studies were undertaken in the equatorial Pacific, the North Atlantic, the Arabian Sea, the Subarctic Pacific, and the Southern Ocean, as well as in the marginal seas of the Gulf of St. Lawrence and the Bay of Biscay. From each of these representative regions we now have excellent information about the relations between organisms from microbiota to zooplankton, and concerning the production, consumption, excretion, and sedimentation of organic aggregates. Our knowledge of the functioning of the planktonic component of oceanic ecosystems became immeasurably greater as a result of JGOFS, which has been a critical catalyst in the unfolding "microbial revolution" that is the subject of the next section.

JGOFS also gave birth to two long sets of time-series observations, both starting in 1988, the Bermuda Atlantic Time Series (BATS) and the Hawaii Ocean Time-series (HOT). The data from these programs provide an unprecedented description of seasonal changes in open ocean physics and chemistry, but also in some ecological processes, including rates of phytoplankton production and organic sedimentation. These data have already been used to suggest that a response of the oceanic inorganic carbon system to the rising levels of atmospheric carbon dioxide is already observable.

Finally, it will be useful to mention one collective response of biological oceanographers to the global fishery crisis of the end of the century and to the failure of many fish stocks to recruit successfully. GLOBEC (Global Oceans Ecosystems Dynamics) is a study of the response of top predators in marine ecosystems to variability in physical oceanographic conditions, and—like the other international programs I have mentioned—was implemented as a series of regional studies, starting in 1992. In each, the variability of the physical regime was analyzed in terms of the response of a small series of "target organisms," essentially the regionally important commercial fish and the planktonic or nektonic organisms forming their main food. Studies of fisheries ecosystems, predominantly in continental shelf regions, have been undertaken in both the North Atlantic and the North Pacific, as well as in the Southern Ocean. Cod, salmon, and antarctic euphausiids have been especially targeted for study.

The published studies and data sets arising from these programs are, of course, only a fraction of the new material available from which to craft an ecological geography of the sea. I have reviewed them briefly simply to support my contention that biological oceanography has evolved in the last 20 years to the extent that what was not possible in 1985 can at least be attempted today.

## The Availability of Timely Global Oceanographic Data from Satellites

Our new ability to specify, using satellite-derived data, a few critical ecological processes under a wide range of ocean conditions, at sites selected to represent characteristic regions, has far outstripped our ability to describe the distribution, abundance, and biomass of the biota themselves. Although our accumulated biogeographic data clearly demonstrate that individual taxa are indeed distributed discontinuously, as will be discussed in Chapter 2, these data are inadequate to partition ecological models among a series of compartments to represent this discontinuity across the entire surface of the ocean. Lacking this possibility, global biogeochemical models usually integrate ecological processes as a parameterized continuum.

However, an ongoing technological revolution now provides the information required to partition such models realistically, by locating at least some of the partitions in the ocean to which pelagic ecology is sensitive. During the same period in which WOCE and JGOFS delivered new understanding in their respective spheres, great advances were made in the ability of satellite remote sensing systems to deliver high-precision information at the global scale, describing aspects of the physical and biological state of the oceans. Time-sensitive mapping of significant relationships between physics and biology in the pelagic ecosystem is now possible and we can—for the first time—speak seriously about dynamic observation of ecological processes at global and regional scale. Sensors carried aboard environmental satellites now routinely obtain data representing conditions at the surface of the oceans at very short time intervals and specify these at very small spatial scales. To be sure, nothing is perfect, and data from high latitudes and from regions with much cloud cover may be inadequate at some temporal scales, but what we have from the rest cries out for innovative approaches to ecology.

Weather satellites operated by several national agencies deliver data on several important parameters of the sea surface; their scatterometers, altimeters, synthetic aperture radars, and microwave radiometers deliver information on such characteristics as surface roughness, elevation, topography, temperature, and upwelling radiation. The Topex-Poseidon series of precision radar altimeters deliver sea surface topography, while spectroradiometers from OCI to MODIS measure upwelling irradiance at several wavelengths from the sea surface and hence "ocean color." Although only limited information is obtained at each data point on the ocean surface, the flow of simple data at a resolution of only a few kilometers, over the whole surface of the ocean, has been revolutionary. These global products are supported by modeled output representing (for instance) mixed-layer depth at short time intervals, though this is among the products now unfortunately hidden behind a security fence.

Sea surface temperature (SST) is obtained with a precision of $\pm 0.3°K$ by the Advanced Very High Resolution Radiometer (AVHRR) instruments carried by the polar-orbiting US-NOAA environmental satellites, starting with TIROS-N in 1978. The AVHRR is a scanning spectral radiometer, sensing in the visible, the near-infrared, and the thermal infrared portions of the electromagnetic spectrum. The infrared sensor enables solar radiation at the sea surface to be measured to $\pm 10\,Wm^{-2}$. Apart from High Resolution Picture transmission as full-resolution data in real time to dedicated ground stations, AVHRR data are rendered in two formats. Local Area Coverage images are obtained at 1.1 km resolution, but are available only for limited periods and places, whereas Global Area Coverage data have a resolution of 4 km. Although the primary mission of AVHRR is to study the solid earth and terrestrial vegetation, SST data are accumulated as weekly and monthly global integrations that may be consulted or downloaded from several Web sites. Ten-day global integrations are available for the decade of the 1990s on the WOCE CD-ROMs.

One of the immediate and (relatively) unsophisticated benefits of the data from the new sensors has been an unprecedented ability to locate and map thermal fronts at sea down to the kilometer scale. In this way, there has already been a rapid increase in our knowledge of the locations of individual fronts, their evolution, and the physical processes that maintain them. Such information has been particularly valuable in understanding the nature of fronts at the shelf edge and also those associated with mesoscale eddies in the open ocean. By inference, and by comparison with chlorophyll images, the ecological significance of these features is now much better understood. However, it is as well to bear in mind that SST is not a conservative property and that global, monthly integrations can provide little more detail than is shown in "old-fashioned" seasonal maps of SST. Integrated at 15-day intervals, however, the LAC images are striking and contain rich information on the regional distribution of surface water masses.

The use of active altimetry from space to observe the topography of the sea surface (SSH) was initiated in 1973 from Skylab and was progressively developed during the 1980s by a series of U.S. Navy missions, including SeaSat and GEOSAT during which precision and sustainability in space were progressively improved to fully operational status in the early 1990s. TOPEX-POSEIDON is a U.S.–France observing mission that is currently in orbit at a height of 1336 km and uses active radar altimetry to measure the elevation of the surface of the ocean in relation to the geoid. More than 90% of the ice-free ocean is observed in each 10-day period, during which 127 orbits are completed. Sea truth is obtained at 10-day intervals at two sites, one off Corsica and the other off California, to calibrate the altimeter so that it maintains the desired accuracy of $\pm 3$ cm, compared with 50 cm for SeaSat in 1982. Data have flowed from TOPEX-POSEIDON since the instrument was launched in 1992 and will be continued by the planned follow-up mission, Jason 1.

From the details of the shape of the returned radar pulses, useful information is extracted concerning wind speed (to $\pm 2\,m\,sec^{-1}$) and wave height (to $\pm 0.3$ m). From the slope of the regional sea surface, geostrophic velocity vectors can be quantified. From the topography represented by anomalies in regional sea surface height in relation to the geoid, the mesoscale eddy field can be mapped. Maps of each of these products may be obtained from browse files representing shorter or longer periods at several Web sites. Personally, I find the University of Colorado global near-real-time SSH data viewer to be extremely effective; here, the user easily specifies an area and period for viewing as a color-coded and/or contoured image, invaluable for the interpretation of features in the sea-surface chlorophyll field.

Perhaps even more importantly for ecological analysis, sea-level anomalies (SLAs) serve as an inverse proxy for anomalies in the topography of the thermocline: small changes in sea surface elevation require much larger, inverse changes in the mixed layer which is relatively deep below an anticyclonic elevation in the sea surface compared with surrounding areas, and vice versa. This is important because all changes in the pycnocline depth, even when wind mixing is not the cause, may have consequences for nutrient availability within the euphotic zone. A recent study in the equatorial Pacific used anomalies in the TOPEX-POSEIDON sea surface elevation to predict anomalies in the depth of the 20°C isotherm, as a proxy for mixed-layer depth, and so to model changes in new production rate using a relationship established by observation (Turk *et al.*, 2001). This particular analysis went further by using satellite-derived zonal wind stress to demonstrate the primacy of nonlocal dynamics of the tropical ocean in determining local production rates.

Finally, although it is not directly of relevance to us here because we are not concerned directly with bathyal ecosystems I cannot refrain from noting the amazing—to those not involved—recent advances in mapping the sea floor by satellite radar altimetry that requires precision at the millimeter scale. Even today, of course, the solid surface of much of our planet is mapped with a horizontal resolution of only about 15 km, and about 250 m in the vertical, whereas parts of the surface of Mars are mapped to resolutions of 1 km in the horizontal and 1 m in the vertical.

The global measurement of phytoplankton chlorophyll biomass in the ocean depends on analysis of the small fraction of incident radiation that is not scattered or absorbed at the surface or deeper and consequently may be observed as water-leaving irradiance by a passive spectroradiometer. Both pure seawater and chlorophyll have, of course, wavelength-specific characteristics: seawater absorbs red-green and reflects blue light, while chlorophyll absorbs the deeply penetrating blue light and preferentially reflects red-green wavelengths. Thus, reflected red-green light seen from space is proportional to chlorophyll integrated over a large fraction of the first optical attenuation depth for the relevant wavelengths, biased toward the surface when the chlorophyll profile is not uniform. The sensed depth is $Z_e/4.6$, where $Z_e$ is the euphotic depth (Morel and Berthon, 1989). Using the attenuation analysis of Smith (1981b), we can infer that sensed depths range from about 25 m in clear oligotrophic water (0.1 mg chlorophyll $m^{-3}$) to about 5 m in eutrophic ocean water (10 mg chlorophyll $m^{-3}$).

Multispectral radiometers have the further capability of discriminating between some taxon-specific combinations of individual chlorophyll pigments and of distinguishing coccolithophore blooms by their white reflectance (Brown and Yoder, 1994). Global maps of coccolith blooms show that these occur preferentially in the open NE Atlantic and NE Pacific and also in some large shelf regions: the North Sea, off eastern Canada, on the Falklands plateau, and off northern Australia. Other taxonomic discriminations are now becoming possible; one such is the ability to classify the individual pixels comprising a SeaWiFS or MODIS image as representing either a "diatom-dominated" or a "mixed" population of generally smaller autotrophic cells (Sathyendranath *et al.*, 2004). The algorithm used to achieve this is based on the fact that the absorptive properties of diatoms differ from those of other phytoplankton taxa, and also requires the use of a theoretical reflectance model that relates water-leaving radiance to chlorophyll concentration. We may then prepare maps to show the probability on a pixel-by-pixel basis that the phytoplankton represented was diatom-dominated; such maps describe the real dimensions of episodic diatom-dominated blooms. More recently, Alvain *et al.* (2005) have analyzed a large set of relationships between different phytoplankton taxa and the spectral characteristics of their water-leaving radiance based on the spectral signatures derived from taxon-specific inventories of seven phytoplankton pigments. Using these relationships, Alvain *et al.* are able to separate four major phytoplankton groups in sea surface chlorophyll data from satellite sensors: haptophytes, *Prochlorococcus, Synechococcus*-like bacteria, and diatoms. Experimental plots of relative distributions of the four taxonomic groups thus obtained (e.g., Color plate 1) conform to our general understanding of how phytoplankton taxa are distributed, obtained from surveys done at sea. Diatoms really do dominate in spring blooms and coastal upwellings, whereas *Prochlorococcus*- and *Synechococcus*-like bacteria dominate open-ocean, low-latitude gyral situations and also the high-pigment signal from the South Subtropical Convergence zone around the Southern Ocean, except briefly in summer (see Chapter 12). Haptophytes form a sort of universal background to higher concentrations of the other groups. Taxonomic mapping of this kind should become routine as experience grows with new sensors, but I venture to suggest that Alvain *et al.*'s illustrations are destined to become textbook cases.

A lengthening series of radiometers has already been flown, starting with the Coastal Zone Colour Scanner (CZCS). This was launched in October 1978 aboard NIMBUS-7 and remained operational until 1986; because it was a "proof-of-concept" mission, the sensors were activated only intermittently so that a comprehensive, repetitive coverage of the oceans was not obtained, and the data were most useful for developing regional, seasonal climatologies. The CZCS radiometer operated at five spectral bands, and some difficulty was encountered in correcting for the atmospheric scattering that was responsible for 95–99% of radiance sensed by the instrument (Feldman *et al.*, 1989). Nevertheless, the images were a revelation to biological oceanographers and were used in many studies, including the first regional and global estimates of primary production in the ocean: the regional analyses presented in the first edition of this work were derived from CZCS output. Subsequent evolution of sensors and their deployment has been rapid, and an introduction such as this is out of date before going to press; to keep up with developments, I recommend reading the bulletins issued regularly by the International Ocean-Color Coordinating Group (IOCCG).

After CZCS failed in 1986, two other sea-surface color missions obtained data for relatively brief periods before spacecraft failure—the Japanese ADEOS and the CNES-France POLDER instruments; the IOCCG lists a couple of dozen sensors with ocean color capabilities, either launched or due to be launched in the next few years by the United States, France, China, India, Australia, Japan, Korea, and so on. Here, I shall use images only from SeaWiFS and MODIS, both currently operational.

In 1997, the SeaStar satellite, owned by the GeoEye Corporation, placed the SeaWiFS sensors in sun-synchronous orbit at a nominal altitude of 705 km, so as to cover a swath that varied between 1502 and 2801 km, giving a revisit time of 24 hours; at the time of writing it remains operational. The radiometer is sensitive to eight spectral bands from the violet (412 nm) through blue-green and green to the near infrared (865 nm). Twelve binned geophysical parameters are obtained, including the chlorophyll-like pigments (band 1 for colored dissolved organic material (CDOM) and bands 2–5 at 443 to 555 nm for chlorophyll-*a*), the diffuse attenuation coefficient, the ratio of chlorophyll-*a* to the diffuse attenuation coefficient at band 3, the epsilon value for the aerosol correction of bands 6 and 8 (660–785 nm), and the aerosol optical thickness at band 8 (845–885 nm). For each of these parameter values and for each bin, the mean, median, mode, and standard deviation may be derived.

SeaWiFS images, like those of CZCS, are available in two formats, Local (1 km resolution) and General Area Coverage (4 km resolution), the former being available for only a few specific regions where ground stations able to receive data directly from the satellite have been installed. GAC data, processed at NASA Goddard, are available from a variety of sources and at a variety of processing levels, from raw irradiance measured at the satellite (Level 1) to global data integrated over weekly, monthly, seasonal, or annual periods (Level 3). For NASA, SeaWiFS was a "data-buy" project, and although the NASA-generated images are freely downloadable on-line, they remain the intellectual property of GeoEye; their use in a book such as this is restricted to license holders. Most of the images I shall present, therefore, will be from other sensors.

From the beginning there has been some uncertainty about the validity of radiometer images of optically complex seawater typical of coastal regions where bed stress keeps fine particulate material, both organic and inorganic, in suspension. Here, a wide range of factors influences the optical properties of seawater (Lavender *et al.*, 2005), and such complexity requires algorithms "soundly based on theoretical considerations" according to the IOCCG. Although both sensors and algorithms for processing their output are currently evolving with great rapidity, the evaluation of data from optically complex seawater ideally requires both special algorithms and also sensors able to distinguish the relative contributions to an optical property of suspended material, of CDOM, and of phytoplankton.

Although SeaWiFS images are still being obtained and made available, global chlorophyll images are also obtained routinely from the MODIS (Moderate-resolution Imaging Spectroradiometer) sensors that are operated by NASA's Earth Observing System for the U.S. Global Change Program. These are carried aboard the NASA Terra and Aqua satellites: these began work in tandem so that data from every area of the planet were returned twice daily—during local morning and afternoon periods—but, because of technical problems, we now obtain these data only from the Aqua satellite.

After the application of cloud and atmospheric correction algorithms, these data deliver 36 wavelength-specific parameters that describe water-leaving irradiance and total absorption coefficients; these comprise computations of chlorophyll-*a* to first optic depth, fluorescence efficiency, CDOM, and calcite concentrations. Otherwise, four parameters are used to compute SST and eight for the computation of primary production while yet other channels are used for data quality control and environmental conditions: sea surface temperature, opacity of cloud cover, aerosol properties, and photosynthetically available radiation.

Unlike SeaWiFS, MODIS measures chlorophyll not only by absorption, but also by reference to fluorescence (F), the emitted light energy that is not used for photosynthesis. Estimates of F will be especially useful in estimation of chlorophyll where other materials that scatter and absorb light, such as sediments and CDOM, are abundant. Finally, progress has already been made in deploying two algorithms for computation of primary productivity from MODIS data: parameter P1 is based on the vertically generalized model of Behrenfeld and Falkowski (1997b), integrated over the photic zone, and parameter P2 on a Howard, Yoder, and Ryan version of the Epply/Peterson polynomial, integrated over the mixed layer depth. You can obtain further details of the processing and output of MODIS data from the MODIS Users Guide available on the Web.

At first order, we may assume that the indicated chlorophyll values from SeaWiFS or MODIS represent mixed-layer pigments and that subsurface chlorophyll maxima deeper than ~25 m are not detected directly. For this reason, if uplift and illumination of the nutricline lead to a subsurface bloom, this may not be observed in water-leaving radiation; however, we can expect that such a bloom will enhance mixed-layer "background" chlorophyll values by the same vertical mixing process that maintains the mixed layer itself. This enhanced background may then be regionally detectable in the chlorophyll field, though this may not represent the true intensity of the subsurface bloom. However, as will be discussed later, we do possess a sufficient archive of observations of chlorophyll profiles to make a reasonable prediction of the seasonally variable subsurface chlorophyll field in all parts of the oceans.

It is difficult now to remember how ignorant we were of the extent, variability, and seasonality of algal blooms prior to their global visualization by the CZCS (Banse and McClain, 1986; Brock *et al.*, 1993; Muller-Karger *et al.*, 1989; Longhurst, 1993). CZCS data were at first available simply as a brochure of a few striking global and regional images, but the files soon became available as a climatology of monthly means for 1978–1986 on a 1° grid covering all oceans and seas, comprising 42,732 data points, of which many were blank because sensors were often switched off and because of cloud cover. This was the archive used by Banse and English (1994) in their groundbreaking study of the seasonality of near-surface chlorophyll in the ocean, but we are now immeasurably richer, having already access on the NASA Goddard site to 8 full years of global SeaWiFS and 4 years of MODIS images, including SST. These data have all been used to compute primary production of ocean basins and smaller regions and already permit suggestive analysis of decadal-scale trends in the global sea surface chlorophyll field; surface chlorophyll in oligotrophic midocean gyres has significantly declined during the SeaWiFS era while over some of the major shelf regions it has progressively increased (Gregg *et al.*, 2003, 2005).

The analytical model that was used in the first ocean basin and global estimates of oceanic primary production from satellite data required (i) surface chlorophyll, (ii) an assumed photosynthesis/light relationship, (iii) surface irradiance, from sun angle and a cloud cover climatology, and (iv) parameters descriptive of the chlorophyll profile. The last item was obtained by reference to a global data base of 26,232 profiles each of which was parameterized by its fit to a shifted Gaussian distribution, as described by Platt and Sathyendranath (1988). This procedure delivered a unique definition of the shape of each profile: the depth of the chlorophyll maximum, the standard deviation around the peak value, the total pigment within the peak, the ratio of peak height to total pigment, and a background chlorophyll value. Specifying that a minimum of six depths were required for a successful fit of the model, 21,872 sets of profile parameters were used in a partitioned global estimate of autotrophic production in the sea (Longhurst *et al.*, 1995).

More recently, semiempirical methods have been developed that compute the same rate from the surface chlorophyll field using a global equation that estimates the form of the chlorophyll profile by means of an irradiance-independent, vertically integrated model (e.g., Morel and Berthon, 1989). Semianalytical models, including that of Behrenfeld and Falkowski (1997b), are routinely used by NASA at the Goddard Space Flight Center to deliver global ocean production indices, binned weekly, monthly, and annually. An image-based method has been devised for separating production fuelled by nitrate from what is fuelled by regenerated ammonium and urea, a vital distinction in biological oceanography (Sathyendranath *et al.*, 1991).

The seasonal cycles of chlorophyll biomass and primary production for each province that illustrate this book, from SeaWiFS sensors, were computed by George White with a version of the Platt and Sathyendranath analytical model. Because some uncertainty remains concerning the specification of photic depth from satellite images, integrated pigment (Chl mg $m^{-2}$) was derived from the relationships obtained by Morel and Berthon (1989) from a global set of chlorophyll profiles:

$$\mathrm{Chl_{tot}} = 40.6\ \mathrm{Chl_{sat}}^{0.425}\,(r^2 = 0.686),\ \text{where } \mathrm{Chl_{sat}} = < 1.0\,\mathrm{mg}^{-3}),$$

$$\text{and } \mathrm{Chl_{tot}} = 40.2\ \mathrm{Chl_{sat}}^{0.507}\,(r^2 = 0.776),\ \text{where } \mathrm{Chl_{sat}} = > 1.0\,\mathrm{mg}^{-3})$$

But this uncertainty will have only minor consequences for the computation of column-integrated productivity, and the available model will serve us well enough here where absolute values are not important; for the majority of the provinces subsequently to be described, we have mean monthly values (1998–2005) for SeaWiFS and/or MODIS sea-surface chlorophyll and computed primary production rate.

The graphs describing this seasonality offered in Chapters 9–12 are intended to be an extension of the descriptions of regional ecology given for each biogeochemical province. Another key difference between these and the CZCS-derived versions offered in the first edition is that the boundaries between provinces are here dynamic and sensitive to the real conditions observed each month: this was performed for this book by George White, using an early version of his "MakeShift" routine. For each nominal province, for each time period, this routine compares the mean values for each variable (surface chlorophyll, temperature, cloud cover, and depth) for adjacent 1° rectangles on either side of a boundary with that of its neighbors, and reassigns it to that which it most closely resembles. Because of the very different scales of variables, these are first converted to rank scale so that the observed differences are essentially the different number of pixels with individual values in each of the two rectangles being compared. This routine is especially useful where province boundaries can be induced to respond to the month-to-month variability of regions such as GFST or ARAB, where mesoscale, chlorophyll-enhanced eddies dominate the pattern. In other regions, little change in location of boundaries is induced. Note that this routine for obtaining dynamic, rather than static, boundaries differs from that described by Platt *et al.* (2005) that is discussed in Chapter 7.

## Internal Dynamics of Satellite-Observed Algal Blooms

Because one of the most useful kinds of observations now made by satellite-borne sensors is changes in the location and concentration of regions of relatively high chlorophyll concentration, it will be useful briefly to review what we may reasonably infer from these data. Because the images risk being interpreted by those not entirely familiar with the complex dynamics of phytoplankton, it may be useful to emphasize once again that there is not necessarily a simple relationship between changes in chlorophyll concentration and changes in phytoplankton growth rate.

What do we really mean by "algal blooms"? This is one of the most frequently used terms in biological oceanography and one that is often given rather little thought although it always refers to an increase in the standing stock of chlorophyll. As biological oceanographers, our thinking about algal blooms is probably still colored by the title of Sverdrup's (1953) classical paper "On the Conditions for Vernal Blooming of the Phytoplankton" and by terms such as the "spring outburst" or "vernal flowering" of diatoms commonly used not so long ago. To examine this phenomenon, Sverdrup returned to Gran and Braarud's concept of a critical depth of mixing that retains algal cells within sufficient irradiance as to permit net growth (see Chapter 4). We should remember that Sverdrup discussed the critical depth concept by analysis of those oceanographic conditions that would permit "an increase in the phytoplankton population" to occur—adding only, and somewhat as an afterthought, that "if grazers are present . . . the phytoplankton population may remain small in spite of heavy production." Despite this caution, it has been all too easy to interpret an increase or decrease in chlorophyll as having some proportionality to production rate. It is also very easy to interpret a change in the sign of biomass change as necessarily reflecting a similar change in sign of the production rate. None of which, as we should know, is necessarily so.

One of the earliest findings in the study of plankton dynamics was that even large diatoms can be consumed almost as rapidly as they are produced. In 1935, Harvey and others thought that 98% of the production in the English Channel of the spring bloom was grazed down by copepods in a few weeks, and later workers, who included Fleming, Riley, Gauld, and Cushing, reached similar conclusions for coastal waters. By 1942, Hart had extended such calculations to the Southern Ocean, where he found that the standing stock of phytoplankton represented only about 2% of its daily production. From these early observations, the dynamic balance between production and disappearance of phytoplankton cells should subsequently have been the central theme of biological oceanography but, as Banse (1992) pointed out, this has, most disconcertingly in retrospect, not been the case. Perhaps, he suggests, because of the difficulty of research on consumers (great functional diversity and dimensional range) and the easier access to simple demonstration of biological principles in phytoplankton research, teachers have given priority to the latter: consequently, research on production and consumption has been only very loosely coupled. So it may be instructive to ask what satellite imagery can tell us about the dynamics of algal blooms.

For each province subsequently to be described and for each month, we have mean values that specify surface chlorophyll, integrated chlorophyll, and integrated primary production rate. By comparing rates of actual and potential increase of phytoplankton from these, it is simple to propose a first-order plankton calendar for each province that integrates both production and loss of cells. This can be done by computing a value for the difference between the algal biomass in each province between successive months. The sign of this value indicates loss or gain. From the rate of change of algal biomass and the rate of primary production, both on a monthly basis, it is then possible calculate how much of the production of algal biomass does not accumulate but must be lost to consumption, sinking, advection, or entrainment.

The result of this simple investigation is shown in Figure 1.1, where I have computed the relationship in as simple terms as I can devise: monthly accumulation of biomass in terms of the daily production rate and standing stock in terms of the monthly production rate. In each case, you may be surprised by the result if you have never thought seriously about the relationship. Look first at the open ocean areas lying under the trades and westerlies—monthly increase of biomass usually represents less than a quarter of one day's primary production, and the standing stock represents less than one-tenth of the production of a whole month. These values are higher in the polar regions, but not by very much. If we examine individual months, rather than their climatological mean values, we consistently get the same result. The bloom induced by the southwest monsoon in the Arabian Sea in June and July accumulates no more than 3.0% and 2.3%, respectively, of the production that occurs during each of these months. The spring bloom in the temperate North Atlantic in April and May accumulates even less: 0.26% and 0.58%, respectively. These percentages are almost matched in May and June in the Atlantic subarctic regions.

The series of seasonal graphs for production rate and chlorophyll standing stock presented in Chapters 9–12 illustrates the lack of confidence we should have in interpreting underlying ecological dynamics from indications of plant biomass alone. Phytoplankton standing stock may track seasonal trends in primary production rate quite closely, or may significantly diverge from them. Therefore, the graphs show that chlorophyll may accumulate when the production rate is decreasing or may decline when the primary production rate is increasing. Such deviations should be welcomed as implying something about the dynamic biology that forces the observed pattern. Where possible, I have suggested in Chapters 9–12 what the reasons may be for each important example shown. Associating $Chl_{sat}$ data with global maps that relate the critical depth to the mixed layer, Obata *et al.* (1996) assume that a doubling of the surface chlorophyll concentration must indicate a bloom. However, as I shall suggest later, perhaps this idea may not be quite as useful as it seems to be.

What are we to make of all this? Obviously, these observations support the concept that the production and loss rates for plant material in the ocean are fiercely coupled and rarely resemble those pertaining to the spring outburst of terrestrial plants, where caterpillars in no way keep up with the growth of oak leaves. Perhaps it is only in some circumstances, as in the North Atlantic spring bloom south of Iceland, that the latter model would be appropriate. Loss of observed chlorophyll cannot be accurately partitioned between sinking and grazing, but the former term is usually accepted to be the smaller. Diatoms, after all, unless aggregated, have sinking rates of no more than $1\,m\,d^{-1}$, whereas the small cells that inhabit the viscous realm, and that comprise the largest biomass component over much of the surface of the ocean, sink even more slowly—if at all. Further, the measured precipitation of organic material from the photic zone is usually observed to comprise carbon that has already been consumed; diatom frustules obtained from the sediments are mostly broken as if by copepod grazing, and sediment traps capture mostly fecal pellets. The sinking flux (the j-flux of Berger *et al.*,

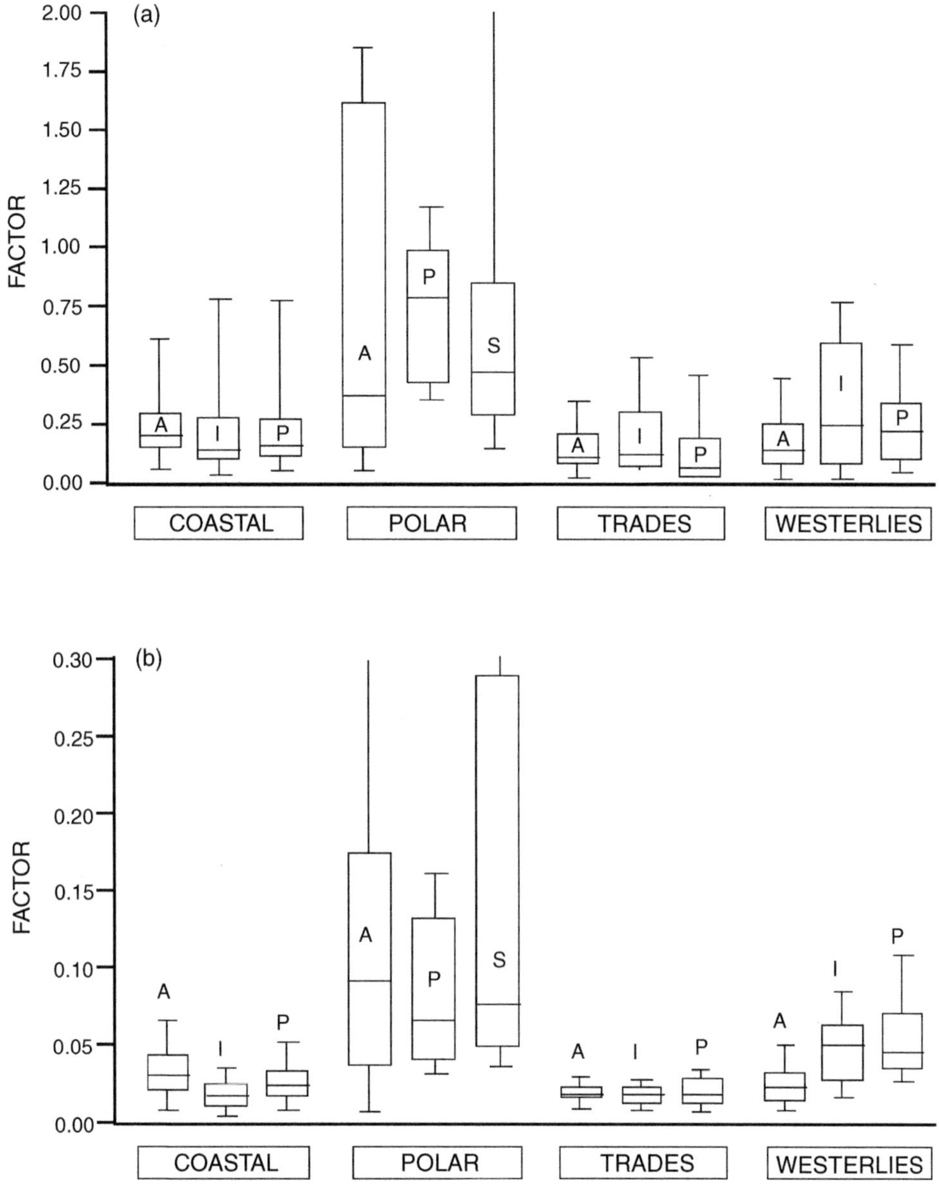


**Fig. 1.1** (a) Monthly accumulation of phytoplankton carbon is generally about one-half of one day's primary production. (b) Standing stock is equivalent to about one-half to three days' productivity. The analysis, based on CZCS data, is split between oceans and biomes.

1987) is generally believed to comprise only around 5% of total primary production in oligotrophic areas, 10–15% in productive oceanic situations, and 25–30% in coastal waters. Given the already recycled nature of much of this material, and the degradation of chlorophyll in it, these figures are not out of line with what is suggested by the satellite observations of chlorophyll.

Demonstration of the control of algal growth by herbivores must be based on observed or modeled budgets for the flow of material through this part of the pelagic food chain. This is a difficult thing to do satisfactorily, given the complexities of the pelagic food web, not only taxonomically but also spatially. It might seem strange that herbivore control of the uptake of nitrate by algal cells should so perfectly balance the externally forced flux of nitrate across the pycnocline as to result in a predictable and characteristic concentration of mixed-layer nitrate and chlorophyll observed in data fields. The only stable and predictable balance would appear to derive from the case in which nitrate or another limiting nutrient is driven to undetectable levels, as indeed over most of the ocean.

The answer to this riddle must lie in the dynamics of grazing organisms, including protists. For copepods, there is a wide literature on the response of grazing rate to the concentration and size of available phytoplankton cells and much attention has been given to grazing thresholds and the general response of grazers to increasingly dense or increasingly attenuated concentrations of food particles. The questions, somewhat simplified, have been the following: do they eat until they burst, or do they stop when satisfied and, on the other hand, can food be so scarce that they quit and go on a fast? It is the lower threshold that is of particular interest here. Plankton ecosystem modelers have found it necessary to introduce such a threshold to stabilize their models, and one has been demonstrated experimentally for copepods on some occasions (Conover, 1981). I have been informed (Lessard, personal communication) that protists in an oligotrophic sea have a threshold for successful feeding that is close to the average chlorophyll value outside the brief winter bloom. If this is so, writes Karl Banse (1992), perhaps small cells are kept near this concentration by protistan grazers, whereas the macronutrients that are measurable in the oligotrophic phase do not control plant growth but are—in his words—"left over." In fact, Karl Banse has returned profitably to this same theme on a number of occasions, most recently in 2002, in a reminder that Steeman Neilsen—he of the $^{14}C$ technique—had, already in the 1930s, commented that the abundance of algal cells that we observe is a product of the balance between production, consumption, and sinking. We now have the tools to tell us how tight that balance is in most seas, most of the time.

One of the conclusions that I draw from these musings is that it is no longer very useful to examine only one of the two terms in the production/consumption equation as, for example, was done in the study by Obata *et al.* (1996). Their analysis, based on the global CZCS data and the Levitus mixed-layer depth climatology, appears to confirm some aspects of Sverdrup's model—that the North Atlantic and northwest Pacific spring blooms occur after "euphotic conditioning" (i.e., after the mixed layer becomes shoaler than the critical depth). They also suggest that some blooms are terminated by the opposite condition. However, I suggest that this can tell us very little about what causes the observed chlorophyll accumulation: simple inferences that we may make from maps of change of biomass can turn out to be entirely erroneous. We should all have understood that a long time ago.

## Our New Understanding of the Role of Very Small Organisms

During the very recent past, one emergent novelty is so fundamental to our understanding of the characteristic functioning of the pelagic ecosystem in different regions of the oceans that it is imperative to discuss it, at least superficially, before proceeding to a synthesis of oceanic and shelf phytoplankton. This novelty is the role of extremely small organisms in all pelagic ecosystems.

We may visualize two realms in the pelagic ecosystem, and both are intimately connected—so that the one cannot get along without the other—but, until very recently, almost everything you read about, and most of what you cared about, belonged to only one of the two. These realms are those in which, respectively, viscous and inertial forces dominate. What has been learned about the ecosystem of the viscous realm in the last 20–25 years has turned our understanding of ecological processes in the ocean upside down. But the new knowledge has not yet completely penetrated our collective consciousness, even though we know that the classical model of a "microbial loop" inhabiting this realm no longer suffices.

In the *viscous realm* water is a viscous medium and gravity a negligible force; it is largely populated by extremely small, very ancient forms of life that are ubiquitous and abundant in all seas, and at all seasons. In the inertial realm water is a fluid, gravity is a significant force, and inertia is an important consequence of motion; this realm is inhabited by the marine organisms that most people eat or study, such as fish or zooplankton, of linear dimensions one to several hundred millimeters. According to Stokes' law, the relative significance of inertial and viscous forces for a moving particle is a function of its size and velocity, together with appropriate values for the density and viscosity of the medium. From this relationship is derived a simple ratio, Reynolds' number, which takes small values when viscous forces dominate. As Purcell (1977) put it, a swimming man takes values around $R = 10^4$, a small fish about $10^2$, and a motile bacterium about $10^{-4}$ or $10^{-5}$. The last can move relatively very much more rapidly than you can swim, at about $>10\,\mu m\ sec^{-1}$, using a very small fraction of its metabolic energy. But when it stops, it stops. And does not sink.

At the dimensions of bacteria, diffusion of nutrient molecules toward and waste molecules away from the organism occurs very rapidly and indeed, as Purcell pointed out, local "stirring" of the medium cannot increase its nutrient supply. At nutrient concentrations typical of oligotrophic water, $NO_3$ molecules are about $1.0\,\mu m$ apart so that a bacterium-sized cell would have only a few such molecules adjacent to its surface at any one time. Motion is therefore critical to diffusion-limited nutrient uptake, and small cells are favored in low-nutrient water. Most biota of the viscous realm are either rod-shaped or approximately spherical, unlike those of the inertial realm where form has evolved so as to reduce sinking rates, to sense food particles, and to serve various other functions.

But what kinds of organisms inhabit the curious viscous realm? Certainly, very many more kinds than there were thought to be only 20 years ago although, before getting into that, I note that there is also an abundant population of nonliving particles to be thought about. In clear coastal seawater, particles of dimension $0.5–1.0\,\mu m$ ESD occur at abundances of about $1 \times 10^7\,ml^{-1}$, together with about three orders of magnitude higher numbers of colloidal particles of around $0.01\,\mu m$. Such numbers suggest a blurring of the concept of dissolved and particulate organic material (Longhurst *et al.*, 1992). Viral particles are also very numerous everywhere in seawater, at $10^7–10^9\,ml^{-1}$, are of equivalent dimension to the small nonliving particles, and are more abundant by an order of magnitude than the picoplankters having dimension $<2.0\,\mu m$. Not much larger than viral particles is the bacterial clade SAR11 in coastal seawater, which is the smallest living cell yet cultured (Rappé *et al.*, 2002). The lunate cells of these bacteria are only $\sim 0.1\,\mu m$ long.

Also among the picoplankton ($<2.0\,\mu m$), we now know that many oceanic bacteria are aerobic, anoxygenic photoheterotrophs, containing bacteriochlorophyll. These $\alpha$-proteobacteria, such as *Erythrobacter*, obtain energy from sunlight and use it not only to metabolize DOM molecules obtained from seawater, but also to enable them to function as photoautotrophs. Such cells are globally distributed in the oceanic photic zone, preferentially in oceanic gyres where DOM concentrations are low and where they may form $<10\%$ of the total oceanic microbial community (Kolber *et al.*, 2000, 2001). Finally, in the picoplankton we should note the newly found Archaeobacteria, which are an independently evolved domain differing in its ribosomal DNA structure from both Prokaryotes and Eukaryotes. Though these are very abundant below the oceanic photic zone, forming 40% of the microbial biomass at 1000 m in mid-Pacific, there may be surprises yet in store for us. Archaeobacteria have very recently been found abundantly in the Southern Ocean at relatively shallow depths.

The most important novelty of recent decades has been the discovery of two groups of prokaryotic autotrophs among the oceanic picoplankton: the gram-negative cyanobacterial *Synechococcus* (Johnson and Sieburth, 1979) and the slightly larger prochlorophyte *Prochlorococcus* (Chisholm *et al.*, 1988). The latter is furnished with divinyl chlorophylls-*a* and -*b*, has no phycoerythrin, and has an accessory chlorophyll-*c*-like pigment. These cyanobacteria are of order 1.0 μm in diameter, and occur at densities of $<10^6\,ml^{-1}$ form 30–40% of bacterial numbers in most parts of the surface oceans. These are now known to be the most abundant phytoplankters in the ocean and are the clearest indication that the concept of the classical "microbial loop" must be replaced.

Exploration of the ecology of *Synechococcus* and *Prochlorococcus* has proceeded very rapidly, and already many solid generalizations may be made. Both are warm-water forms and, at <14°C, numbers of cells of both genera scale with temperature; in warm seas, nutrients determine abundance. *Prochlorococcus* is the more abundant, has a greater depth range, and may perform 10–80% of water column primary production, the higher values being in oligotrophic waters between the 40° parallels (Partensky *et al.*, 1999). Thus, in the mid–North Atlantic in summer the two genera form about 88% of all phytoplankton numerically, about 39% by its carbon biomass, and generate about 30% of all primary production (Li, 1995). *Prochlorococcus* can grow over an irradiance range of four orders of magnitude and distinct populations are adjusted to each light environment and depth. In the Arabian Sea and the eastern tropical Pacific, clades of *Prochlorococcus* occur at <2% surface illumination, below the oxygen minimum layer, and almost as monocultures. *Synechococcus*, on the other hand, is most abundant around the margins of the subtropical gyres rather than toward the more central regions (Li, in press).

We should not forget that eukaryotic cells also contribute to the pico fraction of the autotrophic cell population. Green flagellates (<2 μm), such as *Micromonas* and *Pedinomonas* and very small, scaly prasinophytes (0.5–1.0 μm) from the North Atlantic, are examples of these organisms. They are certainly more diverse than was revealed even by the devoted exploration in the mid-20th century by Mary Parke and Irene Manton. Oceanic 18S rDNA sequences from 75 m in mid-Pacific have very recently been used to demonstrate unexpectedly high diversity among eukaryotic cells of <3 μm in the pico fraction of the cell population. Most are assignable to prasinophytes, haptophytes, dinoflagellates, choanoflagellates, and acantharians, but some appear to belong to novel lineages (Moon-van der Staay *et al.*, 2001).

The borderland between the viscous and inertial realms is occupied by ultraplankton (2–5 μm) and nanoplankton (5–20 μm), some of which exert the primary control over numbers of <2.0-μm picoplankton, whether these are heterotrophic bacteria or *Synechococcus* and *Prochlorococcus*. These heterotrophic organisms represent, in general, the smallest eukaryotes in the sea and comprise a great range of mastigophoran flagellates and aplastidic dinoflagellates, ciliates (both tintinnids and oligotrichs), and sarcodines (both radiolarians and heliozoans). They are responsible for the removal of up to 35% of cyanobacteria and prochlorophytes daily in ocean water and a greater proportion in coastal water, and are capable of controlling blooms of single-celled phytoflagellates like *Phaeocystis*.

These size fractions of the microplankton also include a photosynthetic capability because symbiosis by plastid retention, more or less permanently, occurs in a wide variety of mixotrophic genera, especially in late summer. Ciliate biomass in coastal water may equal that of larger zooplankton, while 40% of the ciliate cells may contain active, photosynthesizing plastids (Stoecker, 1987). Wholly photoautotrophic genera of microplankton range in size from μ-flagellates to the larger dinoflagellates that border on the dimensions of large diatoms.

Chlorophyll biomass seems more tightly regulated in smaller than in larger cell fractions because the rapid reaction of ultraplankton consumers to an increase in numbers of their picoplanktonic food induces greater population stability than is characteristic of the realm of larger organisms. Although, as we shall see, large cells characteristically dominate algal blooms, Li (in press) points out that microbial size spectra conform in shape and slope with those characteristic of larger plankton: partial size overlap between components results in smooth ensembles. Li suggests that such observations tend to support previous speculation that plankton ecosystems tend to self-organize into holistic states characteristic of each oceanic environment.

Li *et al.* (2002, 2004) have examined some of the basic relations between small planktonic organisms and between them and their environment, using macroecological analysis of very large data sets, some obtained pixel-by-pixel from satellite images. These studies show that we may no longer use a straight-line model for the ecological coupling between heterotrophic bacterioplankton and phytoplankton biomass. Li *et al.* point out that the relationship between these two trophic groups is different at high and low levels of phytoplankton biomass, suggesting a difference between bottom-up resource control and top-down mortality control. This is seen when all data for the North Atlantic are consolidated, but when a region-by-region approach is taken, region-specific differences in slopes and intercepts in the relationship become apparent.

The differential distribution of heterotrophic bacteria and autotrophic cells within the viscous realm between regions or in relation to (say) temperature is complex; for one thing, bacterial biomass and numbers are known to be relatively constant everywhere, while the biomass of autotrophs varies strongly: how, then, ask Li and Harrison (2001), is balance achieved? Where phytoplankton productivity is low, they suggest, bacteria "are respiring a great deal and not growing very much," while where it is high "bacteria channel their resources more equitably between respiration and growth." Of course, as they also remind us, the proportion of metabolically active bacteria in the heterotrophic population is quite small in oligotrophic waters but increases as phytoplankton productivity increases; such a mechanism appears capable of accounting for the apparent invariance in total heterotrophic bacterial numbers in all surface waters.

It may be useful to remind ourselves very briefly at this point that benthic ecology has recently undergone at least one sea change in its acknowledgement of the diversity of the very small biota that inhabit the interstices between sand grains and other particles in the substrate. This was progressively exposed to science during the second half of the last century and (as Tom Fenchel commented in 1978) "next to the astonishing morphological adaptations . . . the next most remarkable thing about the meio- and micro-fauna is perhaps its richness in species." He refers to the extraordinary manner in which organisms like holothurians, polychaetes, or mollusks manage to fit their complex anatomy into dimensions of only 0.5–5.0 mm. For instance, the tiny medusa *Halamohydra* is $<1.0$ mm long and lives between sand grains. These interstitial organisms comprise both protists and metazoans and their interest for us is the complexity of their ecological relations, and how these change with the nature of the substrate. Although mostly investigated along the shore and in shallow water, the interstitial fauna is known to extend to great depths and even into anoxic, reducing sediments of the shelf where ciliates and zooflagellates dominate.

For such reasons, the interstitial fauna has become a paradigm for how species within ecosystems partition the available resources, and for the existence of specialized food niches. Rather than the generalized type of food gathering that we might anticipate, specialists abound for both food particle size and type. We shall have to bear the existence of this fauna in mind when considering the dynamics of benthic–pelagic coupling in continental shelf regions.

Chapter 2

# Biogeographic Partition of the Ocean

As suggested in the previous chapter, ecological geography is a special case of the wider subject of biogeography—the latter being a discipline that, ideally, should tell us not only how species are distributed but also how they aggregate to form characteristic ecosystems, sustaining optimum biomass under characteristic regional conditions of turbulence, temperature, irradiance, and nutrients. Unfortunately, classical biogeography has made only limited progress in even the first of these tasks and since ecological geography cannot proceed without some understanding of how and why individual species are distributed as we observe them to be, it will be useful—before proceeding any further—to consider the present status of the biogeography of the sea. How far can our present knowledge of species distributions, or biogeography, satisfy the needs of ecological geographers? In this chapter, I propose to explore the problems associated with the use of biogeographic data for our purpose and to suggest how we may profit from some earlier analyses of patterns of distribution of organisms at the regional and ocean-basin scale.

Of course, oceanic biogeography has inherent difficulties compared with terrestrial biogeography, as was discussed long ago by de Beaufort (1951). The relatively high cost of collecting samples at sea, the high levels of expatriation of plankton species, the relative lack of isolation between natural regions, and the problem of observing three-dimensional distributions that vary in both space and time are among the most serious difficulties. From the start, therefore, progress was slow compared with terrestrial biogeography, which (for example) by about 1870 had already accurately located "Wallace's line" in the Indo-Pacific archipelago, separating the Australian from the Oriental faunas. But, at sea, progress was slower and only the biogeography of the benthic fauna of shallow seas was equally well developed by mid-20th century, as described in the classical work of Sven Ekman (1953).

Modern marine biogeography may be said to have taken its origin in the suggestion by Mary Somerville, in the 1862–70 editions of her *Physical Geography*, that the global distribution of the marine fauna was best described by a series of nine latitudinal "homozoic" zones. These were "neither parallel with one another nor do they coincide with lines of latitude . . . but (respond) to the effect of warm and cold currents." In the North Atlantic, then, the Arctic zone has its "greatest breadth between the pole and the Gulf of St Lawrence, and its least extension is (to) the extreme north of Scandinavia." This is far more sensitive to reality than some partitions of the global ocean that have been proposed in recent years. She also described a progressive faunistic change from the intertidal zone to the shelf edge and inferred the tropical submergence of high latitude, cold-water fish.

Although only a small part of the total effort of oceanography has ever been devoted to biogeography, even as late as the 1960s strong teams, expressly devoted to this task, were being recruited at some major oceanographic institutions. But despite all the effort expended, even now—150 years after the *Challenger* voyage—the total number of species in each major group of pelagic organisms is not even approximately agreed upon and we have descriptions of the seasonal distribution of no more than a very small proportion of them. Although a quantitative comparison cannot be made, the number of stations at which marine plankton have been collected, identified, and enumerated must be many orders of magnitude smaller than it is for the terrestrial invertebrates, although the area of the oceans is more than twice that of the continents.

Sven Ekman (1953) remarked that biogeography should not confine itself simply to describing the occurrence of living forms and arranging them regionally. It must also proceed historically, and an important objective of biogeographers has long been to understand the progressive dissemination of species or higher taxonomic groups during the evolution of the ocean basins and seas. For this reason, a criterion very often used in even relatively recent global biogeographic systems for marine biota (e.g., Briggs, 1974) is the degree of endemism of the fauna of each region. Although ecologists must be alert to the consequences of the opening and closing of connections between oceans during geological time, this will not be a serious concern to them: ecologists will be more concerned with the functional differences between the warm-water benthic fauna of the western coasts of Atlantic and Pacific Oceans than with the wonderful occurrence on each coast of a remnant of the Jurassic fauna, the horseshoe crabs *Limulus polyphemus* and *L. moluccensis*.

## Taxonomic Diversity: The Shifting Baseline of Biogeography

The basis of biogeography is said to be the distribution of individual species, but this simple adage is questionable today when the term *species* remains imprecise. Integration of accumulated biogeographic data is hampered by the continual evolution of taxonomy, so that biogeographers must work with a shifting taxonomic baseline.

Indeed, one of the more enduring discussions in biology has been about the nature of "species." Many biologists and the general public have become convinced in the couple of centuries since Carolus Linnaeus that the diversity of animals and plants is necessarily partitioned into this fundamental unit, although it has long been recognized that the Linnaean binomial does not, in practice, represent a single level of biological organization.

During 50 years or more of modern discussion, biogeographers have settled down to use the concept of the biological species, following the magisterial examination of the "species problem" by Huxley (1942) that bears rereading today. We should remember his admonition that taxonomy exists for human convenience and not in the interest of "some Platonic eidos stored up in Heaven," as he put it. He concluded that for everyday biological purposes the binomial species, even if broadly interpreted, must remain the central entity, although the trinomial geographic race or subspecies may also be required. Since, in this view, the basic unit of taxonomy is loosely defined, closure of the global or regional listing of species must remain an elusive objective.

The taxonomy of marine biota is almost entirely based on biological species and, to classify the geographically separated populations of individual species, we have come to use the very satisfactory trinomial nomenclature associated with this concept. Taxonomic compilations, such as FishBase, are conformable with this arrangement and are useful in some activities—as, for example, the management of marine fisheries. But, as I shall discuss later, the new on-line taxonomic data archives may not be as useful as hoped, both because they risk being undermined by a new "phylogenetic" taxonomy and also because we cannot yet determine how far we are from a final listing of species, however we decide to define that unit.

In fact, how many "species" may there be in the sea? It is certainly unhelpful to speak in terms of billions—the most extreme case I have seen is a claim for $1 \times 10^{13}$ species. The extraordinarily high numbers projected for some groups of organisms have been arrived at by what has been called statistical legerdemain, also practiced in the enumeration of terrestrial insects. Extrapolation from the fauna of 19 forest trees in Panama gave an estimate of 30 million insects globally and, similarly, by extrapolating from a few box-core samples yielding 800 species of small metazoa, Grassle and Maciolek (1992) derived an estimate of about 10 million marine species worldwide. I prefer the "expert opinion" reasoning of John Briggs (1994), who, by extrapolating from the taxonomic literature, arrived at a total of fewer than 200,000 potential species of marine eukaryotes: Porifera 9000, Cnidaria 9000, Nematoda 35,000, Annelida 15,000, Arthropoda 37,000, Mollusca 29,000, Bryozoa 15,000, Chordata 15,000, remainder 14,000. I have suggested that there are perhaps 3000 Linnaean species of metazooplankton (Longhurst, 2001), an estimate that sits well with the ~4000 species each of planktonic protists and of phytoplankton proposed by others.

Unfortunately, the lack of closure in taxonomic procedures has important consequences for marine biogeographers. A typical example is what lies behind a statement of John McGowan, who wrote in 1971 that "holoplanktonic zooplankton are taxonomically well-known at the species level." This was surely an overoptimistic statement that is worth brief reexamination. The taxonomic status of copepods of the genus *Calanus*, as it has developed since McGowan's comment, is probably typical of copepod genera. Two North Atlantic species are surely the best known of all zooplankters, but *C. finmarchicus* (Gunnerus, 1770) and *C. helgolandicus* (Claus, 1863) were not clearly distinguished until the work of Fleminger and Hulseman (1977). They were, for a long time, thought to be geographical races (or subspecies) of a single cosmopolitan species. Bucklin *et al.* (1995) suggested that we should recognize 14 species of *Calanus*, of which 3 (*C. hyperboreus*, *C. simillimus*, and *C. propinquus*) are morphologically distinct from the remainder, which are themselves distinguishable only by expert analysis of fine differences in secondary sexual characters of the exoskeleton. These comprise two species groups: a small arcto-boreal group (*C. finmarchicus*, *C. glacialis*, and *C. marshallae*) and a larger group of the mid-latitudes (*C. helgolandicus*, *C. pacificus*, *C. australis*, *C. orientalis*, *C. euxinus*, *C. aguihensis*, *C. chilensis*, and *C. sinicus*).

But there are also significant differences between individuals of *C. pacificus* from the California coast, from the North Pacific gyre, and from Puget Sound, so that subspecies rank is accorded to the populations of these three regions. This arrangement recalls the work of Fleminger and Hulseman (1977), who referred to 10 morphologically distinguishable forms of *C. helgolandicus* from eastern and western Atlantic regions. In neither case do we know if the characters—morphological or genetic—lie along gradients or are discontinuous, and we know very little of the extent to which species of *Calanus* may have sympatric distributions. We should not be surprised that authors of papers on the ecology of *Calanus* still feel constrained to explain in detail the taxonomic status of their material. If this is the unhappy situation for what is perhaps the most extensively studied genus of marine plankton, you may imagine the status of less-studied genera. And, don't forget, there are almost 200 other genera of calanoid copepods to be specified, not to speak of all those cyclopoids and harpacticoids.

Of course, much can be done once a group has been critically revised in the classical manner, as was done for chaetognaths and siphonophores by Alvarino (1965, 1971) or for *Clausocalanus* by Frost and Fleminger (1968). Alvarino reduced the 106 species of chaetognaths listed in 1935, and the 70–80 in a 1961 listing, to only 52 species. To follow the trail of a chaetognath species quoted in the literature so as to integrate all information about its distribution, it would be necessary carefully to consult the synonymies proposed by Alvarino. Unfortunately, comprehensive and scholarly reviews like these (S. J. Gould called them "works of genius") cover but a small fraction of the total marine biota. The present, start-up phase of electronic taxonomic data archives seems not to understand that unless critical taxonomy of this kind has been done, the simple listing of species that have been described in a genus is at best useless. There appears to be inadequate recognition of how widely the views of taxonomists may vary: Boltovskoy long ago discussed a case in which identical sets of specimens of benthic foraminifera, each apparently containing 200 species, were submitted to four taxonomists who were specialists in that group. The resulting species lists shared only 10 generic names, and but one species. Only those who have hands-on experience of trying to establish regional species lists can appreciate how general is the problem of the unconsolidated nomenclature of marine organisms.

But everything concerning the species concept according to Huxley and Mayr has now potentially been turned on its head. The molecular characterization of individual populations by sequence analysis of subunits of mitochondrial RNA and DNA has become routine and is widely used in some fields. This technique, supported by statistical techniques for analyzing relationships between clades, results in neighbor-joining or maximum-parsimony trees and other presentations that give taxonomists a new ability to quantify relationships between populations. This school cleaves to a very restrictive species concept, suggesting that "the smallest diagnosable cluster of individual organisms within which there is a parental pattern of ancestry and descent" represents a "phylogenetic species" (Cracraft, 1983). For Nixon and Wheeler (1990), these are "the smallest aggregation of populations (sexual) or lineages (asexual) diagnosable by a unique combination of character states in comparable individuals." Phylogenetic species are thus quite distinct from the morphological or nominate species with which we have been familiar in the past.

A typical example of the kind of problem that phylogenetic taxonomy will induce in our taxonomic understanding is given by a recent ribosomal RNA analysis of 41 specimens of the small mesopelagic fish *Cyclothone alba*, which is ubiquitous and abundant in all subtropical and tropical oceans (Miya and Nishida, 1997). This analysis identified five monophyletic populations with low levels of mutual gene flow under conditions in which there appears to be no discernible barriers to prevent complete dispersion and intermingling of stocks. The central North Pacific population is genetically closer to those of the North Atlantic than to the three populations of the Indo-Pacific Ocean. I note, without comment, that the Web-based Oceanic Biogeographic Information System (OBIS) associated with the multinational Group on Earth Observations (GEO) offers a map to represent the global distribution of *Cyclothone alba*: this shows only 25 North Atlantic locations, being based on the data and specimen holdings in national and institutional data banks—there is, apparently, no reference to the scientific literature that, of course, comprises the majority of our knowledge about this fish.

Consider also the problems that would arise following the application of the phylogenetic species concept to the genus *Calanus*, whose taxonomy—as I suggested earlier—we thought was beginning to settle down. Recently, molecular systematics using variations in the DNA base sequence of the mitochondrial 16S rRNA gene (Bucklin *et al.*, 1995) have been used to explore the systematics of *Calanus*, as of other organisms. This approach has confirmed the reality of the relationship *Calanus finmarchicus* + *glacialis* + *marshallae* and also of the more genetically diverse *C. helgolandicus* group, though genetic information on *C. orientalis* was lacking. Further, the same technique distinguishes between populations of *Calanus finmarchicus* from different regions of the North Atlantic: fifteen samples of *C. finmarchicus* were thus grouped into four populations, one each in the Norwegian Sea, in the Gulf of Maine, on Georges Bank, and in the Gulf of St. Lawrence. Given the oceanic distribution of *Calanus*, of which these samples were marginal, one may conclude that we shall require several dozen Linnaean binomials to describe the Atlantic population structure. And I shall need to revise my opinion, noted earlier, concerning the number of species that coexist in today's oceans.

It has even been suggested that all existing species designations that are not based on DNA analysis should be relegated to a category to be called the LITU or the "least inclusive taxonomic unit," to emphasize our present lack of understanding of their lines of descent. Such suggestions, it seems to me, come from a cloud-cuckoo-land inhabited by those who choose to ignore the practical uses for which a classification of the living world is required. It is very regrettable that the phylogenetic taxonomists did not accept earlier suggestions that the basic units identifiable by their technique should become "clades" or "syngens," named according to a new code. That way, we could have had the best of both worlds, by accommodating the trickle of new cladistic information into the general body of our existing taxonomy of the living world. Of course, for microbiota—archaea, bacteria, cyanobacteria, and prochlorophytes—a taxonomy that mingles numbered or named clades with Linnaean genera and species has proved essential where cell form is of little or no assistance in classification. I see no reason why the same approach should not have been taken with macrobiota.

Whichever way the current fashion for "phylogenetic taxonomy" goes, we ought now to be aware that two very different categories of "species" are already accommodated within the existing taxonomy of both macroinvertebrates and vertebrates. Fortunately for biological oceanographers, the problem is not yet acute and I am still very comfortable with the use of trinomials to designate the individual, geographically isolated populations of marine organisms discussed in works that I shall cite as we go along.

This discussion of taxonomic diversity may be a suitable point to introduce a brief polemic concerning another new concept, that of "biodiversity." The changes wrought by us in natural ecosystems have started to capture public attention, and this process has fed back into science in a manner that is not entirely without problems. To bring complex issues to the general public requires the reduction of complexity to simplicity, and the generation of shorthand expressions to stand for complex and improperly understood phenomena. The newly minted term *biodiversity* with its derivative phrase "loss of biodiversity" has become a mantra for much of what presently ails the biosphere. It is also used in some scientific writing as a marketing ploy, as in "Biodiversity of North Atlantic and North Sea calanoid copepods," the title of a paper published recently that contains nothing that might not have been published under the more correct, but less arresting, title "Diversity of . . . ." This novel usage merits some attention here, because biotic diversity is an aspect of ecological geography that will concern us in later chapters.

In fact, the omnipresence only 10 years later of a term so "loosely introduced" in 1988 attracted the critical attention (as you might expect it to have done) of the late Ramón Margalef (1997). He commented that he was "not altogether happy with the success of biodiversity" and its confusion with the more usual term and supposed that if it has any specific meaning, it must be the "total specific, taxonomic or genetic richness contained in nature or in any local or taxonomic part of it." This, he suggested, is very different from the classical use of *diversity* to mean the processes of change and succession, analyzed by reference to the actual numbers of individuals, species and of their relative biomasses, so he suggested that we may have to rename the older concept *ecodiversity*. Thus, biodiversity would concern the repository of genotypes—as Margalef put it "the actual richness of nature's dictionary"—and would complement the concept of diversity as the study of ecosystems that are evolving in response to external conditions. Ecodiversity, on the other hand, is in continual flux: an active algal bloom pushes the index down as a small number of algal taxa comes to dominate the whole spectrum present in a water body. In one of his inimitable evocations, he suggested that there is a reservoir of ecodiversity in the sea near the lower boundary of the photic zone, and adjacent to the nutricline, comparable to the seed banks of forests. Here, many taxa of algae and heterotrophs exist, each taking small numbers, but each available to populate a bloom under appropriate and different conditions.

With all this, one cannot but agree but unfortunately it is difficult to know what is in the minds of those who use the term *biodiversity*: is it political rectitude, a marketing ploy, or an interesting analysis of the processes of microevolution? Unfortunately for Margalef's useful suggestion, I believe that the word has now lost the possibility of having any special meaning. It will be better to leave it where it is now most useful, as a label for well-meaning efforts to stem the loss of species from the biosphere. Further, we shall then not need to consider such odd notions as the "levels of diversity that are useful for Nature and for Society" (van der Spoel, 1994).

As Steele (1991) demonstrated, the most significant diversity in marine ecosystems is functional diversity, or the variety of responses that any marine ecosystem can make to environmental change. This concept has led to that of the regime shift in marine ecosystems that we shall meet frequently in the following chapters. These shifts may be caused either by a change in top-down forcing of ecosystem structure induced by the removal of a large organism, as in the "trophic cascade" theory of lake ecosystems, or in bottom-up forcing, induced by changes in the oceanographic regime. In neither case can we advance our understanding as rapidly by simply concentrating our attention on "biodiversity" or genetic richness, as we can by analysis of the structural diversity of the ecosystems under question. Temporal shifts in ecosystem structure, contingent on shifts in climatic regime, have been investigated principally in the eastern Pacific and to a lesser extent on the European continental shelf, but the process certainly occurs everywhere. In the chapters that follow, I shall want on many occasions to discuss the various and complex manner in which the diversity of marine ecosystems varies in space, time and function, but I shall not expect to use the term *biodiversity* again.

## The Useful Results from 150 Years of Marine Biogeography

It will be useful to remind ourselves at this juncture exactly what we are looking for from the accumulated biogeographic literature, to know what will be useful to our purpose.

If we want to find a partition of the ecology of the oceans into biotopes comparable with characteristic vegetation types ashore (forest, steppe, tundra, desert, and so on) we need to know if changes in the distribution of the characteristic marine species assemblages are spatially continuous, or are discontinuous? And if there are discontinuities between them, how are they forced and are they predictable in space and time? All this shall be discussed in the following chapter, but it will be useful to examine how the concept of discontinuity emerged progressively in biogeographic studies.

Even a superficial examination of the subject reveals some simple principles. Perhaps the most important is the consequence of the different motility of adult benthic and pelagic organisms, even if both usually resort to planktonic larvae. Pelagic and planktonic organisms are highly cosmopolitan and many species occur in all three nonpolar oceans, while bipolarity is rather common. Benthic organisms of the continental shelves are, on the other hand, rarely cosmopolitan so that very few species, although many genera, occur in both Indo-Pacific and Atlantic faunas. Bipolarity in benthos is unusual—demersal fish, for example, of boreal and austral polar seas have a very high degree of endemism. Ekman suggests that 90% of species and 65% of genera of demersal fish of the austral polar shelf are endemic.

A further obvious difference between patterns of distribution of benthic and pelagic organisms is the degree to which benthic distribution patterns are locked to physical features of the coastal regions: both the pattern of the distribution of sediment types on the continental shelf, and the association of the circulation pattern of water masses over the shelf with coastal features such as gulfs, bays, capes and river mouths. However, some of the major discontinuities in the distribution of benthic communities match those that form the basis of pelagic biogeography. We shall return to this in detail later, but it is striking to what extent the boundaries between the benthic faunistic subregions of Ekman (1953) and Briggs (1974) match those that have been proposed by several authors for the major faunistic subdivisions of the pelagos. These locations tend to be where major frontal regions between oceanic gyres, or coastal boundary currents, intersect with the coastline.

## Biogeographic Regions of the Pelagos

The boundaries between physical oceanographic regimes are coincident with the primary discontinuities in pelagic biogeography. So, in thinking about partitions in ecological geography, we should perhaps first consider how far taxonomic biogeographers agree on how to partition the oceans to reflect discontinuities in their data. Agreement, we shall find, is quite good between individual suggestions, and most of them support the thesis to be reviewed in the next chapter that links oceanic frontal regions to partitions in ecological geography.

Not surprisingly, oceanic phytogeography, essentially based on the distributions of diatoms (e.g., Semina, 1997), does not yield a partition as specific as marine zoogeography. The generalized series of phytoplankton ranges suggested by Margalef in 1961 can be sustained in more recent data, except that we now know that few species are entirely excluded from one or other ocean basin: cosmopolitan distributions within climatic zones appears to be the dominant pattern. The resultant partitions are not mutually exclusive, presumably because of the extensive passive drift of cells, and do not form a useful categorization of the surface ocean. Even less useful for our present purpose is the recognition of "range bases" in the five subtropical gyres, and "expatriation areas" elsewhere, by Semina. We have a much more comprehensive phytogeography of the North Atlantic than for anywhere else, thanks to the many years of deployment of towed Continuous Plankton Recorders from the Plymouth laboratory. In the Atlas of those data, repetitive pattern is clear: arctic, Atlantic, and subtropical oceanic species of diatoms and dinoflagellates are distinct, as are shelf species. Expatriation is also evident from these areas. Had we equivalent observations for the whole ocean, the required partition would perhaps become clear. But we do not have such observations, and consequently I shall not expect to evoke a global partition of phytoplankton as for other organisms.

Reviews of historical progress in oceanic zoogeography generally start with the map of Steuer (1933), based on copepod distributions. He recognized circumpolar arctic and antarctic regions and then divided each ocean basin (where appropriate) into subpolar, subtropical (north and south), and tropical regions. This is not very different from Sven Ekman's proposal at approximately the same time for a "pelagic warm-water fauna, northern and southern cold-water plankton, and a neritic plankton." In those early years, the emphasis was on sea surface temperature as the primary determinant of pelagic species distributions, so summer and winter boundaries between biogeographic zones were often defined. Progressively, interest came to be given rather to features in the global circulation rather than to temperature itself. Whatever the basis of the classification, all authors were agreed on one thing: that the pelagic realm should be divided into zonal features stretching across each ocean at approximately the same latitude. These zones were annular around the Southern Ocean, and displaced in the North Atlantic because of the open connection to the Arctic Ocean in its northeastern quadrant.

There was also agreement, I believe, that these zones were defined by the distributions of widely different groups of organisms, from algae to fish, but especially by their more abundant and widely distributed taxa. Rarer species, of course, confuse the general pattern and perhaps for no better reason than poor representation in the data, so that a survey by routine tow-net hauls across a grid may render their distribution incorrectly: the proliferation of smaller biogeographic regions by some authors may be an artifact of this problem. Beklemishev's 1969 text on oceanic biogeography is one such, dividing the surface of the ocean into 21 zonal regions, subdivided into very many more—often overlapping—subregions.

Apart from the Russian biogeographic work reviewed by Beklemishev, the other major sets of observations from which useful patterns emerge are those accumulated by the biogeography group at Scripps in California, covering the whole Pacific Ocean. By 1978, led by Joe Reid, this group was able to review ocean circulation and marine life comprehensively, extending their Pacific experience to the global ocean: the resultant maps of euphauiid distributions by Brinton (1962) or of chaetognaths by Alvarino (1965) are central to the corpus of knowledge concerning pelagic biogeography. The Pacific biogeography of McGowan (1971) offers an impressive documentation of the distribution envelopes in both hemispheres of many species of copepods, euphausiids, chaetognaths, and fish and their relation to water masses (Fig. 2.1). The congruence of species distributions is very convincing and demonstrates the reality of subpolar, transitional, central, equatorial, and eastern tropical groupings, together with species inhabiting the transition zones. McGowan discusses useful nonparametric statistical techniques useful in sorting species of several phyla into recurrent groups, and to associate the distribution of these with the superficial water masses.

In a later study (Reid *et al.*, 1978), an approach was taken to biogeographic pattern that is close to the partition used in this book. The gross pattern of the ocean circulation and of the wind-driven convergence and divergence of surface water of the Pacific Ocean, including vertical structure of the upper kilometer, was used to locate regions that should be ecologically unique. Inferences were drawn as to relative biological activity at ocean-basin scale from the overall distribution of dissolved phosphate. Reid *et al.* then showed how overall biomass and the distribution envelopes of species of copepods, pelagic mollusks, and euphausiids could be made to match the physical features. Their

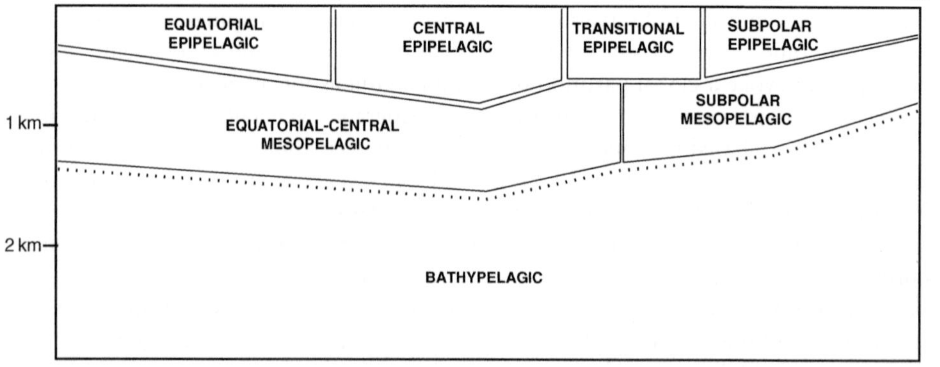


**Fig. 2.1** The distribution of groups of pelagic species, with latitudinal and depth along a meridional section in an idealized ocean.

*Source: Redrawn from Brinton, 1962.*

study, appropriately, was entitled "Ocean circulation and marine life," words that might have been used as a subtitle for this book.

The critical conclusion to be drawn from the now-classical studies of the Scripps biogeographers is that the distribution envelopes of species of epizooplankton represent neither a random pattern at the ocean surface nor a simple gradient from poles to equator, but fall into a small number of discrete classes that are related to distance from coasts and to latitude. Since we must assume that the distribution of each species represents its response to the living conditions, then we have direct confirmation from biogeography that the oceanic habitat may reasonably be partitioned into ecologically consistent regions rather then being treated as a continuum. That some species exist in several adjacent regions does not nullify the argument: Brinton lists 6 species of epipelagic Pacific euphausiids that occupy both his equatorial and central regions, compared with 11 and 7 species, respectively, that are restricted to these regions. Mesopelagic and bathypelagic species are progressively more ubiquitous as, indeed, we should expect them to be.

The integration that best, to my mind, summarizes the accumulated information got by almost 150 years of investigations of the pelagic fauna is the brief sketch of the problem by Backus (1986). This is especially useful because he formally (for I think the first time) relates the boundaries of biogeographic zones to major frontal systems in the global ocean circulation. The thrust of his argument was that there was, by 1985, sufficient agreement in the work of many authors that a synthesis was within reach, and he was clearly correct in this. Indeed, his proposals for zonal patterns of organisms as divergent as oceanic phytoplankton and epipelagic fishes converge very closely and are inapplicable only to organisms whose distribution, like that of some siphonophores, cannot be understood by reference only to the upper few hundred meters of the water column.

Backus argues for the retention of the classical nine-zone system: Arctic and Antarctic zones, Sub-Arctic and Sub-Antarctic zones, North and South Temperate (or Transition) zones, North and South Subtropical zones, and a Tropical zone. He suggests that the Temperate zones are bounded equatorward by the Subtropical Convergence of each hemisphere and that, at least in the south, the oceanic Polar front is the poleward limit of the Sub-Antarctic zone. He allows that further subdivision is both inevitable and desirable, to reflect the reality of species distributions. As examples, he suggests that the Gulf of Mexico, the Mauritanian upwelling, the eastern and western Mediterranean basins, and other regions of this dimension can profitably be treated as entities in any biogeographical partition of the oceans. Though he did not expressly make the point, I note that the examples he gives are peripheral to the major ocean basins, or else comprise semienclosed marginal seas. This, as we shall discuss later, closely resembles the partition proposed here.

If we accept that this arrangement is a reasonable model for partitioning the biogeography of the open ocean, our attention is then directed toward discontinuity in the ocean circulation, and so toward the convergent and divergent frontal systems associated with discontinuity in the global wind field. Between the Subtropical and Subpolar gyres at 40–60° latitude lie the convergent Polar fronts, convergent because their western root is at the meeting place of the opposing western boundary currents of each gyre. Vertical Ekman transport of opposite sign in the two gyres results in the subduction and equatorward transport of Intermediate Water masses into the interior of the ocean. At yet higher latitudes, forced by the change in strength of the resultant Ekman transport with increasing latitude, a Polar divergence occurs, bounding the characteristic density stratification of polar oceans.

At the Subtropical Convergence (STC) fronts, usually near 30° latitude, Central water masses are formed by subduction, passing down and equatorward within the deep permanent thermocline. Poleward of each STC, the seasonal thermocline is destroyed each autumn by the increase of westerly wind stress so that the mixed layer (as defined by a density criterion, sigma-$t = 0.125$) is significantly deepened. Equatorward of each STC the relative lack of seasonality in trade wind stress and irradiance induces the formation of a relatively shallow permanent tropical thermocline; here, seasonal changes occur in mixed-layer depth principally due to geostrophic response to the regional wind field. In an ideal equatorial current system, as in the eastern Pacific, the South Equatorial Current lies above the equator; here, because Ekman transport takes opposite sign in the two hemispheres, transport is poleward on each side of the equator. Evidently, within the upper 200 m or so, this must create an equatorial divergence zone readily observable both at sea and in satellite images. This may have an extraordinarily sharp boundary between cool upwelled water and warmer surface water: it has been described as "a line in the sea" (Yoder *et al.*, 1993).

We shall find that the evidence that these oceanographic frontal zones act as biogeographic boundaries is equivocal. For some authors, for some organisms and especially those confined to the epipelagic zone shoaler than the seasonal or tropical thermoclines, they clearly separate different faunas. For other authors, for other organisms and especially those with representatives in mid-depths, there is no such evidence.

A final problem presents itself in the deep expatriation of species at convergent fronts. The surface layers (in which we are interested) lie above deeper water masses that may have been subducted from the surface at convergent fronts even thousands of kilometers distant, together with some of the planktonic organisms that inhabit that water mass. In this way, the surface organisms of a subtropical water mass may lie only a few hundred meters above organisms that were expatriated from a subarctic environment. Oblique plankton net samples may deliver organisms from two entirely separate environments. Species characteristic of both may appear with great regularity in the same sample, and one such example puzzled me a long time ago now. About 500 m below the tropical copepods in the EASTROPAC zooplankton profiles, I often found quite large numbers of *Eucalanus bungii*, a copepod of the Subarctic North Pacific that had apparently been subducted at the convergent Polar Front; the specimens were translucent, sluggish, nonreproductive, and existed—one might think—only to confuse the biogeographer.

## Geographic Component of Benthic–Pelagic Coupling

To what extent can the logic of an open ocean partition be relevant also to the biogeography of continental shelf organisms? This problem is not so divorced from the ecology of the pelagic realm as it might appear to be, because the nutrient regime experienced by phytoplankton over the continental shelves is modified both by the biotic regeneration of dissolved nutrients from sedimented pelagic organic material and by the consumption of planktonic biota by filter-feeding benthic invertebrates. It may be useful at this point, then, very briefly to review the biogeography of shelf benthos communities. We shall find that some of the early generalizations concerning the composition and differential distribution of communities of benthic invertebrates are highly relevant to the present study.

Ekman (1953) partitioned the biogeography of the continental shelf benthos simply by water temperature, thus: warm-water, temperate, boreal, and polar faunas. He suggested that the principal geographic distinction was between the warm- and cool-water faunas, and he placed the discontinuity between these near the latitudes of the oceanic subtropical divergences. The separation, for instance, of his North Atlantic boreal from temperate benthic faunas lies near Cape Cod in the west, and off the western English Channel in the east. This arrangement reflects the penetration of Labrador Current water south to Nova Scotia, and the bifurcation of North Atlantic Drift flow off Western Europe. Ekman's next level of partition derives from coastal topography rather than from any features of ocean circulation.

Ekman commented that the global distribution of benthos reflects the geological history of the oceans more strongly than does the biogeography of plankton. The benthos of the warm-water shelf in the western Atlantic has considerable affinities with that of the eastern Pacific; although the Isthmus of Panama in our days is a barrier (or was until the Canal was constructed) between Atlantic and Pacific, in the not-so-distant Pliocene past there was a sea-level connection, testified to by the fact that many genera and species of invertebrates and fish are common to both oceans, but occur nowhere else.

Nor can ecological consequences of events at even longer geological time scales, such as the relatively late opening of the Atlantic itself, be ignored: the entire Atlantic warm-water fauna of benthos and demersal fish is significantly less diverse than in the Indo-Pacific. Consider the tropical coral reefs of the Atlantic and of the Indo-West Pacific, as discussed by Wells (1957): although some of the dominant coral species are identical in the two oceans, the Atlantic reefs are less diverse and comprise only 35 species of 26 coral genera. The Indo-West Pacific supports 700 species of 80 genera. *Acropora* and *Porites* each comprise only 3 Atlantic species, but include 150 and 30, respectively, in the Indo-West Pacific. Important structural differences in the coral communities occur so that, for instance, there are relatively very few Atlantic calcareous algae or coral gall crabs, while there is a relative lack of alcyonarians on Pacific reefs.

More generally, regional characteristics of the benthic fauna are very largely dictated by the nature of the sediments: the animals living in muddy sediments must differ from those of, say, a shelly-sand ground. This was the basis of C. J. Petersen's dictum in his classical 1918 paper, that it is the nature of the physical biotope that determines the nature of the benthic biocoenosis: what follows from this is of much greater importance to ecological geography, and to the task at hand, than is the biogeography of Ekman and later students.

The importance of Petersen's contribution was that it led him to propose that structurally similar and recognizable assemblages of benthic macroorganisms—mollusks, echinoderms, polychaetes, and so on—were associated with each characteristic type of sediment on continental shelves around the world at all latitudes. This concept has startling implications, not often recognized; it means that organic material originating in phytoplankton carbon fixation or in terrestrial runoff is processed in the same manner, by similar benthic organisms inhabiting each characteristic sediment type, no matter where these occur. The simple observation that similar genera occur in the benthic communities of both polar and tropical seas has yet to be properly assimilated by marine ecologists. Only recently I read (in a paper that I shall not cite) that the benthos of the Gulf of Guinea is not "truly tropical" because many "nontropical" genera occur there.

Petersen's synthesis, of course, led to bitter arguments concerning the nature of these communities, for it was pointed out—quite correctly—that it had yet to be demonstrated that they react collectively to changes in the environment. His approach to benthic community ecology has also been heavily criticized on statistical grounds, and because interaction has not been demonstrated to be essential to the formation of each species association: but, as Mills pointed out in 1989, this has been a sterile and emotionally charged debate.

Fortunately, Gunnar Thorson and others continued to exploit the concept, and out of their geographically more extensive work was erected the very useful structure of parallel benthic isocommunities occurring on all continental shelves. This global classification of benthic communities is frank natural history, but no worse for that, and it works observationally. I shall very frequently have recourse to it in my descriptions of regional benthic ecology: Thorson's chapter in Joel Hedgepeth's 1957 *Marine Ecology* is not the only gem in that volume, but it is the one to which I refer back most frequently.

These isocommunities form a very useful preliminary description of the benthic community found in sandy, muddy, and sandy-mud deposits. Thorson's system of about a dozen series of parallel communities has stood the test of time (Rosenberg, 2001): these are grouped into *Macoma* communities on inshore soft, muddy deposits; *Amphiura*, *Amphipoda*, or *Amphioplus* communities in slightly deeper muddy sands; *Tellina* communities in clean sands in shallow water; and *Venus* communities in shelly sand deeper on the continental shelf. Each, you will note, is characterized by one or several dominant genera and each forms the dominant species association in deposits of a certain geological grade and organic content.

Thorson recognized *Venus* isocommunities in the White Sea and Scoresby Sound, from the Baltic to the Adriatic, in the Persian Gulf and off Madras, and also off New England. I found that clean shell-sand grounds on the continental shelf off Sierra Leone (8N) in the Gulf of Guinea were occupied by a species grouping that closely matched those that had been described elsewhere as *Venus* isocommunities. During a multistation benthos survey of this community a total of 47 bivalve mollusks were taken, while a *Venus* community sampled with similar equipment in the Celtic Sea comprised 57 macrobenthic species, many congeneric with those I took off Sierra Leone (Longhurst, 1958; Warwick *et al.*, 1978). In fact, the Sierra Leone genera would be familiar to any North Sea biologist: *Abra*, *Cardium*, *Nucula*, *Venus*, *Macoma*, *Mactra*, *Spisula*, *Tellina*, *Pecten*, *Ensis*, and so on.

*Macoma* isocommunities were recognized by Thorson on muddy, estuarine-influenced deposits in the White Sea, off East Greenland, along the Danish Baltic coast, off Vancouver-Seattle, on the Japanese coast, and in the English Channel. One was subsequently located in the Bristol Channel and I found yet another example off Sierra Leone.

These observations were later supported by a formal study of k-dominance in three counterpart benthic communities on the continental shelves of northern Java and western Britain (Warwick and Ruswahyuni, 1987). Diversity indices were computed to represent benthic communities occupying similar habitat and composed of similar (or even the same) genera in the two regions. Off Java, three isocommunities were identified on appropriate deposits: an *Amphioplus/Lovenia*, a *Tellinoides*, and a *Lacionella* community. These parallel, respectively, European *Amphiura/Brissopsis*, a *Tellina*, and a *Macoma* community. Strict comparison of the first two pairs of isocommunities from Java and Britain showed that within-habitat diversity is similar in the two regions. In each comparison, k-dominance was similar even though total community biomass differed. The third comparison was less satisfactory because at their shallower depth of occurrence, physical disturbance was found to be maintaining the communities in various subclimax stages.

I am confident that these results may be generalized, although you will find many contrary statements in the literature, such as those of Gray (2002), who demonstrated, using data from 6 intertidal, 13 shelf, and 7 deep-sea locations, that species richness increases into deep water and toward the tropics. But the sites were selected only for their representation of depth zones, so I am not surprised at the result. The significant comparison is between sites carefully chosen to represent similar habitats, differing only in the latitudes at which they occur.

Perhaps equally interesting is the generalization, also attributable to Thorson, that the larval development of benthic invertebrates in cold and warm seas is quite different. He remarked that in benthic species at very high latitudes, the planktonic larva is abandoned in favor of viviparity, or of yolk-supported external development. Thorson's results were based largely on data from prosobranch mollusks in which up to 90% of species in tropical seas have planktonic larvae, compared with 35–65% in temperate seas and 5–27% in cold seas. Later compilations showed essentially the same thing for lamellibranch mollusks, echinoderms, and some decapod crustaceans. Subsequent investigations (Mileikovsky, 1971) supported these findings but emphasized that it is necessary to distinguish several kinds of planktonic larval existence: planktotrophic, lecicotrophic, and those larvae that do not venture more than a few centimeters above the deposits.

An exception that proves Thorson's rule has been provided recently by Gallardo and Pencheszardeh (2001) in their study of the response of hatching mode to latitude in South American littoral gastropods. On the Pacific coast, the rule is observed impeccably, the proportion of prosobranch species having pelagic larvae decreasing poleward. But along the Atlantic coast, benthic development predominates even into subtropical latitudes. This difference, they say, is attributable to the different composition of the gastropod faunas on the two coasts: the scarcity of pelagic development on the Atlantic shelf reflects the near-continuous soft-bottom habitat there, whereas on the Pacific coast the benthic habitat at all latitudes is more diverse. Thus, an exception to Thorson's rule may be stated: that, other things being equal, very soft muddy habitats favor the *in situ* development of gastropod larvae at the expense of pelagic development.

## From Pristine to Modified Ecosystems

I have suggested elsewhere, not entirely facetiously, that perhaps Intelligence is a deadly pathogen that infects all habitable planets. Observations here on Earth show how rapidly it learns to obtain energy by reversing the geochemical fluxes that hitherto maintained the surface conditions of the planet hospitable to life. Intelligence appears to be capable of sufficiently modifying the surface environment of a planet as to put its own survival in doubt.

Indeed, the consequences of "intelligent" exploitation and management of marine resources over the last century are such that they cannot now be ignored in any discussion of the regional characteristics of marine ecosystems, even if this is not done in the spirit of seeking solutions. I should state my position at the outset, so that you should not expect any helpful suggestions in this section: these problems are without scientific solution.

Studies of the modification of the natural world have been a growth industry for some decades even as the process continues, very little abated by the acquisition of an understanding of what is happening. For the marine environment, I suppose that most attention has been given to the problems of chemical contamination and its effects both insidious and spectacular. Rather than adding to the torrent of material about marine pollution, I propose instead to discuss—very briefly—two other aspects of our unintentional modification of marine ecosystems: first, the insidious corruption of biogeography by the careless transport of organisms from one region to another and, second, the much more spectacular modification of the structure of marine ecosystems by industrial fishing.

The unintentional transport of living organisms has increased significantly in recent decades, principally because of recent change in the design of ships and shipping patterns. Commercial ships are much larger and somewhat faster than they were 50 years ago, and on every voyage may carry unintended passengers, either attached to their hulls or in ballast water to be discharged when the ship takes on cargo. Examples of what has been called the "globalization of marine ecosystems" abound in any appropriate text and although we cannot know the rate at which nonindigenous species are transported, it has been suggested that the rate at which we are discovering them in the Mediterranean is very high: during the last 5 years, a new exotic species there has been recorded each month in the scientific literature!

Most of these travelers are coastal biota, and we might think that this problem is unlikely to occur—or that should it occur, it would have little significance—in the open ocean. But this is not the case, and we must be prepared for significant changes to occur in the entire pelagic ecosystem structure, as has occurred recently in the Black Sea (q.v.).

Diatoms, all older textbooks tell us, lie near the base of the open-ocean ecosystem. Although, as we shall see in the next section, this is no longer seen to be a correct model, diatoms nevertheless remain very important constituents in some regions and at some seasons. If a species is transported unintentionally in ballast water and implants itself as an exotic in a new region, consequences may ensue that are unpredictable in their details, but are perhaps not insignificant. Take the case of *Coscinodiscus wailesii*.

*Coscinodiscus* is, of course, a well-known genus of discoid centric diatoms, represented in most regions where phytoplankton blooms occur by several species having a range of cell diameters from 50 to 350 μm. Relative taxonomic abundance changes seasonally, and the small to moderate-sized species, at least, are important in the diet of herbivorous copepods. Larger species exist, principally oceanic in distribution. One such is *C. wailesii* of the northwest Pacific, having rather wide salinity and temperature tolerances (Nishikawa *et al.*, 2000); its cells of 160–450 μm are dominant in the spring bloom in the Seto Inland Sea, reaching cell concentrations sufficiently high as to interfere with the raft culture of the macroalga *Porphyrea* by competing for light and nutrients. Mass sedimentation of dead cells after a bloom may create oxygen deficiency in bottom water.

In the mid-1970s, *C. wailesi* was observed in the English Channel for the first time, whence it spread rapidly into the North Sea, the German Bight, and more widely in the northeast Atlantic (Edwards *et al.*, 2001). It became firmly established as an invasive species and continues to be dominant in the spring diatom bloom of these regions. More recently, it has been observed on the continental shelf off southern Brazil, where it now forms seasonal blooms in the Gulf of Parana, and also off the coast of Nova Scotia in the NW Atlantic. All that might not matter very much if the cells were not so large and could be consumed by the indigenous copepods and other herbivores. One author has described this as an inedible diatom in the seas that it has invaded, although that is not strictly the case for "sloppy" feeding by copepods has been recorded. Perhaps only the largest cell sizes of *C. wailesi* are not readily available to copepods—recall that cell volume progressively diminishes as diatoms reproduce during the season of growth. In warm oceans, of course, centric diatoms as large as *C. wailesi* are the central item in the diet of some filter-feeding fish (i.e., *Ethmalosa, Brevoortia*), rather than of copepods.

We may expect that such invasions will have measurable impacts on the structure and flow of material through pelagic food chains and it has been suggested (Sommer, 1998) that the effects may be comparable to those expected from eutrophication of semienclosed seas by waste discharge. This is deficient in Si, compared to N and P, and so leads not to diatom blooms, but to a more complex phytoplankton community and less efficient transfer of energy up the food chain. Tunicates and coelenterates may be favored at the expense of diatoms and fish. In this case, the fear is that the diatoms will not be consumed.

The second consequence of the role of "intelligent" biota that I want to discuss briefly is far more urgent and is perhaps—just perhaps—controllable. A century or more of industrial fishing in large stretches of the coastal ocean has created havoc with their pristine ecosystems. Indeed, as Jackson *et al.* (2001) comment "Ecological extinction caused by overfishing precedes all other pervasive human disturbance of coastal ecosystems, including pollution, degradation of water quality and anthropogenic climate change." One must agree with their assessment.

Unlike the introduction of exotic species of plankton, the impacts of fishing on the marine food chain are (in the ecological sense) top-down, rather than bottom-up, and are more immediate. Some regional food webs have already been entirely transformed, to the common detriment of fisherfolk and fish. Recent studies of the destruction of benthic invertebrates caused by the passage of a commercial trawl over the seabed have given alarming results: individual scours left by otter boards may be visible for more than a year. The mortality of gastropods, echinoderms, crustaceans, and annelids caused by the passage of a single beam trawl is 5–40%, and for bivalve mollusks it is 20–65%. Taken as a whole, population mortality of macrobenthos due to fishing in the eastern North

Sea ranges from 5% to 39%, but for about half the species is in the vicinity of 20%. It would be unreasonable to expect long-term occupation of habitat by species where such mortality is additional to natural causes of death.

The extent to which the ecosystems of continental shelves can be modified by industrial fishing has taken fisheries and ecological science by surprise. Indeed, the imposed modifications can go far beyond the fish species and the benthic invertebrates of the seabed. The existence of an abundant large herbivorous mammal, Steller's sea cow, must have significantly structured the Gulf of Alaska shelf ecosystem, but they are all gone now; the ecological consequences of the greatly reduced Southern Ocean biomass (by perhaps 40 million tons) of large whales toward the end of the whaling period must have had a significant effect on the biomass of krill; other examples are not hard to find.

The evidence is somewhat circumstantial, but there is general agreement that prior to industrial fishing the stocks of demersal fish on temperate shelves were very much larger than the stocks we became familiar with during the 20th century. Steele and Schumacher (2000) suggest that the pristine cod and haddock stocks on the NE Atlantic shelves may have been as much as an order of magnitude more abundant, while pristine pelagic fish stocks were probably less abundant.

The progressive decline in the abundance of demersal fish has everywhere been associated both with excessive mortality of the larger species, and with a progressive reduction in the average age of the populations of both large and moderate-sized species: a reduction in the average individual weight of commercially exploited fish off eastern Canada of about 50% occurred in the period 1970–1990, whereas a few Icelandic kitchen middens from the mediaeval period were found to contain larger absolute numbers of 100-cm cod than the Icelandic trawl surveys took during entire seasons in recent years.

Stergiou (2002) refers to this process as the "tropicalisation" of fish species of temperate seas. By this he means that the fish have now come to possess growth rates and age-at-maturity appropriate to warm, rather than cold, water fish. They come to have, as I have suggested, life history traits unsuitable for the habitat in which they must continue to try to survive (Longhurst, 2002). Typically, a series of years with conditions likely to result in poor recruitment is associated with the final collapse of a fish stock already stressed by fishing pressure. It would have seemed literally incredible only a few decades ago that cod should be fished to ecological extinction on the Grand Banks of Newfoundland and that their ecological and economic role should be replaced by lobsters and other crustacea. But so it is. And, even as I am writing this section, the headlines proclaim the imminent probability of a similar regime shift in the North Sea, because of tardy action by fishing regulators, themselves caught in a morass of eurobureaucratic niceties. The "regime shifts" that Steele and Schumacher (2000) suggested would occur as a consequence of heavy fishing seem not, on balance, to have been radical enough to encompass everything that has happened in the NE Atlantic.

If the landings of commercial fisheries reflect, in general, the populations of fish currently occupying the continental shelves then, as has been shown repeatedly (e.g., Pauly *et al.*, 1998), the mean trophic level of fish has progressively decreased since 1950. This change reflects a gradual transition from species assemblages containing a due proportion of high-trophic-level piscivores to short-lived, low-trophic-level planktivores. All this we shall have to bear in mind when discussing the regional ecology of every region, in both high and low latitudes. Though the extent of the process is very poorly evaluated in most places, the effects of fisheries can be found wherever you look.

It is slowly coming to be acknowledged that long-lived fish have life history traits that render them peculiarly vulnerable to the additional adult mortality imposed by any industrial fishery. The age structure of the population is rapidly truncated, reproduction and recruitment begin to fail, and a terminal population decline may be induced if the fishery persists. Tropical groupers such as *Epinephalus* and *Mycteropercus* are such fish and have almost everywhere suffered the same fate: widely, their populations have now collapsed under very moderate fishing intensity and have become so modified in their age structure that it is no longer possible to compute the parameters of their pristine populations—such as natural rates of mortality-at-age—that are essential for the management of population recovery. Fisheries may be considered as producing "new species" having life history traits sufficiently modified that they are no longer appropriate to their habitat, to which the characteristics of the pristine population were closely adapted.

The conclusion is inevitable, and difficult to handle in the context of this book, and in this concerned age. Some people may consider it futile, and even counterproductive, to try to describe the characteristics of marine ecosystems as if they remained in their natural state. Certainly, it is not very politically correct. But the simple fact remains that it is critical to understand this state because it is the naturally evolved biological response to the physical characteristics of each region of the oceans. Human intervention in the biosphere obscures this natural relationship and makes the task of understanding it much more difficult.

Chapter 3

# Fronts and Pycnoclines: Ecological Discontinuities

Boundaries of biogeographical or ecological regions will be least ambiguous where discontinuities in the physical environment are strongest and these, in the open ocean, will be located along major fronts and frontal systems; for our purposes as ecologists, however, the treatment given these features in most oceanographic texts does not really tell us what we need to know. Fortunately, what we can now observe in high-resolution satellite images of sea-surface chlorophyll and sea-surface elevation brings us closer to reality, for where the texts may lead us to expect featureless central gyres, we see instead much spatial nonuniformity; we also see the real complexity of fronts and frontal systems that cannot be observed at sea level. So TOPEX-POSEIDON and MODIS have given us new confidence in extrapolating from studies of ecological dynamics done at sea along necessarily very short sections of a frontal system.

Although, as discussed in the previous chapter, fronts are the locations of greatest ecological discontinuity, they can really represent no more than a leaky boundary between different regimes. This is because the physical dynamics of fronts require that parcels of water should pass from one side to the other, a mechanism that may be observed even in those cases where tidally mixed shelf water meets stratified water at a very sharp discontinuity. This exchange of water has an important consequence that again is clearly observed in satellite imagery: frontal zones are frequently areas of biological enhancement. This occurs at all scales, from tidal fronts in the North Sea to the globe-encircling, convergent frontal zone in the Southern Ocean at which subantarctic water passes below the subtropical surface water mass. As Margalef (1997) has emphasized, fronts in the ocean are not only boundaries but also habitats having the attributes of ecotones, in the sense of Shelford (1963) or Odum (1971).

The ecotone, or transition zone between two ecological communities, became an established concept in terrestrial ecology but has been very little discussed in biological oceanography. Ecotones are, by definition, linear and less extensive than the communities they separate; they are associated with a gradient either in the physical environment or in an external stress, such as might be imposed if herbivore biomass differs on either side. Ecotones may exhibit special ecological characteristics that differ from either of the separated communities, and they may be the habitat of specialized "edge-effect" species. For all these reasons, Odum notes, "we would not be surprised to find the variety and density of life greater in the ecotone." Because the biota in convergent oceanic fronts may have access to resources supplied from each of the adjacent water masses, and because of physical aggregation there, a greater biomass may indeed build up within the frontal zone than on either side. Generally, we may expect that ecotones at sea shall be associated (i) with conjunctions, principally convergent but also divergent, between two surface water masses, (ii) with mesoscale eddies and filaments in the open ocean, but especially associated with the edges of continental shelves, and (iii) in shallower seas, associated with the effect of the semidiurnal tide.

# FRONTS AND FRONTAL SYSTEMS

## OCEANIC FRONTS AND EDDY STREETS

In the open oceans, we should distinguish between two kinds of frontal systems, although the one is, in a sense, included in the other. These are (i) basin-scale zones of convergence or divergence at the confluence between major current systems, and (ii) mesoscale filaments around, or shed from, mesoscale eddies that may occur as eddy streets lying across almost entire ocean basins. Observations of enhanced levels of biomass are commonplace in both categories, but it is at oceanic convergent fronts that the accumulation of biota is greatest. These are the "siomes" of Japanese oceanographers and fishermen, where turbulence at the confluence may disturb the wind-wave system and where the passive aggregations of biota attracts an active concentration of pelagic fish, birds, and fisherfolk (Sournia, 1994). At such fronts, accumulation of biota may depend on passive flotation/sinking rates, or on their individual swimming ability so as to avoid submergence in the descending water mass (e.g., Owen, 1981).

The most important oceanic fronts for partitioning both biogeographic and ecological processes are the Polar, Subtropical, and Equatorial frontal systems. Where the alignment of the coasts permits it, examples of each occur in each hemisphere of each ocean, appearing in basin-scale images of sea-surface elevation (Fig. 3.1) as linear fields of

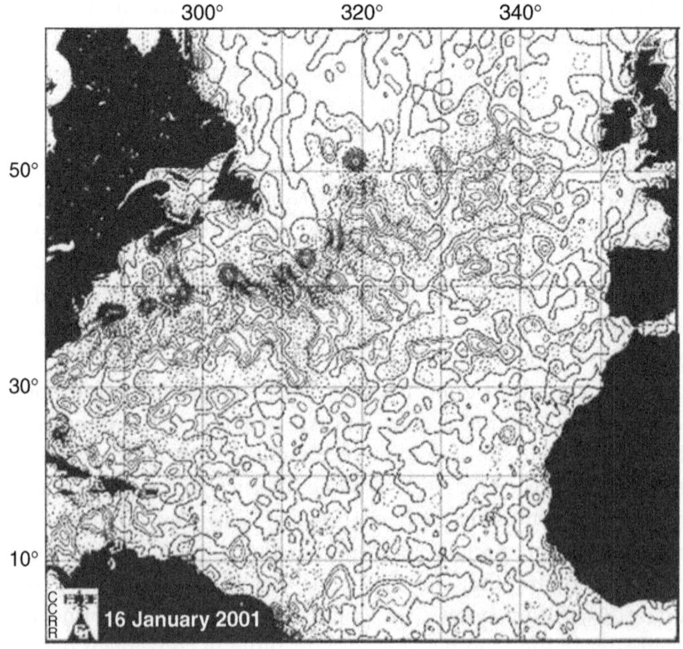


**Fig. 3.1** Sea surface elevation (±30 cms) of the North Atlantic Ocean, to demonstrate where mesoscale eddy activity is most intense, and where the major frontal regions occur in this ocean: the data are from TOPEX/POSEIDON radar altimeter data.

*Source: University of Colorado.*

eddies, stronger than background; at the local scale they include fronts that may be so sharply defined at the ocean surface as to surprise oceanographers who come across them at sea. For those who like their geography to be tidy, the TOPEX-POSEIDON images must have been a difficult awakening to the real world. The images also explain why I have suggested that some oceanic frontal regions should be regarded not as boundaries between biogeographic regions, but as regions having unique characteristics.

The group of fronts associated with the equatorial currents differs strongly from other frontal systems. The system of parallel **Equatorial fronts** is most clearly developed in the eastern tropical Pacific Ocean, which will act as an exemplary case from which to interpret what we observe in the other two oceans. The parallel ridges and troughs at the sea surface, indicated by TOPEX-POSEIDON, are aligned with the equator and match a diagrammatic section (Fig. 3.2) of the equatorial current system obtained at 150–160°W

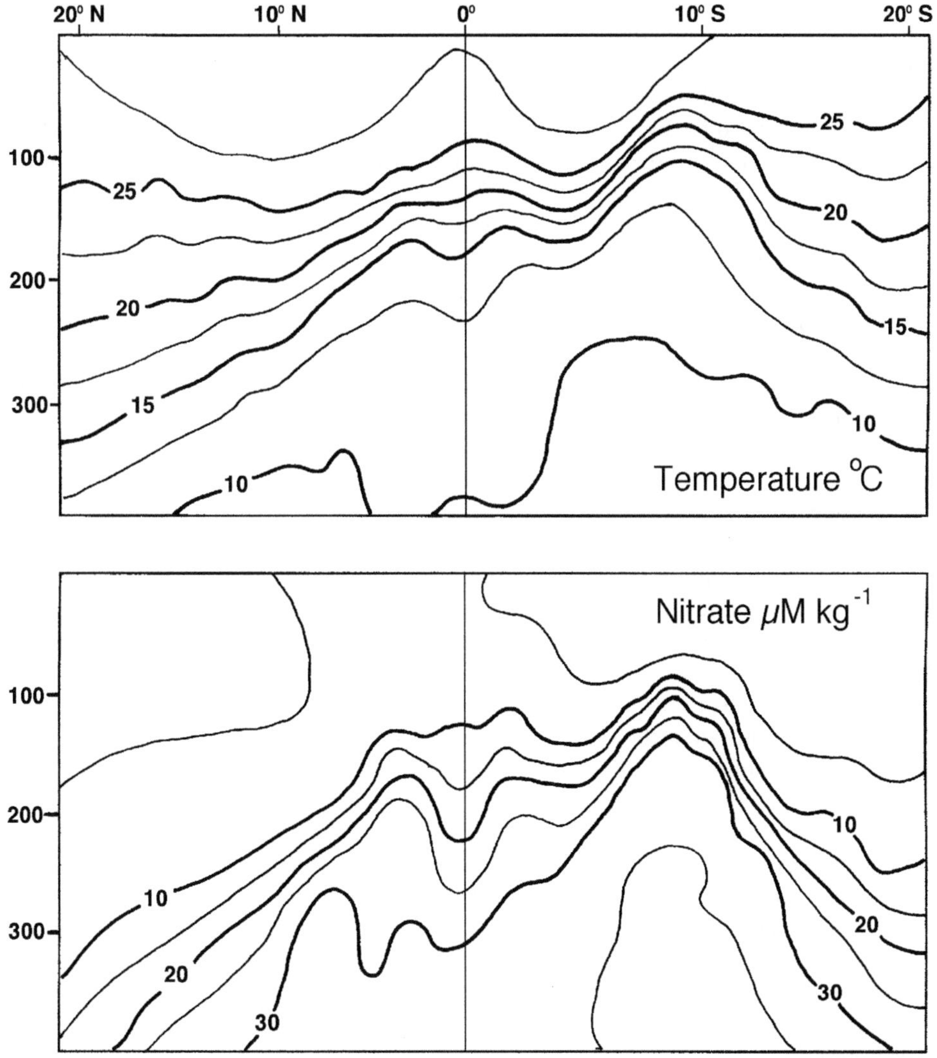


**Fig. 3.2** Diagrammatic meridional section of temperature and nitrate in the eastern tropical Pacific at 150–160°W; compare these with the surface distribution of nitrate-replete and nitrate-deficient water (Fig. 11.13), and with the relations between surface chlorophyll and elevation in the same region (Fig. 11.15).

*Source: Redrawn from Wyrtki and Kilonsky (1984).*

(Wytki and Kilonsky, 1984). Between the 20°N ridge and the 10°N trough the NEC flows westward. From the 10°N trough to the 5°N ridge the NECC flows eastward, while south of this ridge the SEC flows westward. The near-equatorial trough represents a narrow linear zone of cooler water, seen also clearly as a band of high chlorophyll concentration in the relevant SeaWiFs image and of low temperatures in AVHRR images. Here, evidently, we can use physical features unequivocally in our partition of the tropical ocean, in a manner sensitive to the seasonal, zonal displacement of the equatorial current system, because we can readily follow them in routine TOPEX-POSEIDON images.

The convergent **Subtropical fronts** are dominated by meandering flow along an eddy street in which cyclonic and anticyclonic eddies alternate, once again readily observable in the sea-level anomaly signature obtained by orbiting radar altimeters at around 40° latitude. These eddying flows lie along the poleward limbs of the subtropical gyres, so that in the southern hemisphere the individual subtropical fronts of the Indian, Pacific, and Atlantic Oceans form a quasicontinuous feature around the Southern Ocean. Although it is, as we shall see, interrupted as it approaches the western coasts of South America and Africa, passing equatorward, the Subtropical front or convergence is treated by most authorities as a continuous feature. Indeed, it is so prominent in any global analysis of sea surface features that I am constrained to consider it a province of the ecological geographic system to be discussed in later chapters. As we shall see later, the geographies of the Subtropical Convergence Fronts of the North Pacific and North Atlantic differ strongly because of the strikingly different flow system in the subtropical gyres of the two oceans. Recall that in the Atlantic most of the transport of the western boundary current is lost to the Arctic Ocean, while in the Pacific essentially all flow passes directly eastward across the ocean. The Azores Front, then, which is the Subtropical Convergence Front of the North Atlantic, is a weak version of that of the North Pacific. There, it is often referred to as the Transition Zone between the subtropical and subarctic gyres and is an important feature in the biogeography of the North Pacific, selectively harboring some species of, for instance, euphausiids.

Further poleward, the **Subarctic and Polar Fronts** must be crucial delimiters in any ecological geography of the sea, but to map these we get less help from satellite imagery because of cloudiness above 45–50° in winter, and because the radar altimeters of TOPEX-POSEIDON cannot deliver images from above latitude 60°. In the Southern Ocean, the location of these fronts is a relatively simple annular arrangement (see Chapter 12), but in the northern hemisphere there are significant differences between the Pacific and Atlantic Oceans, once again because of the open passage between the Atlantic and the Arctic Ocean.

The high-latitude fronts of the North Atlantic are both complex and confusingly named. The north wall of the Gulf Stream in the northwest Atlantic forms a continuous feature with a frontal system that lies SW-NE across the ocean toward the north of Scotland. This is the Oceanic Polar Front of many authors but is the homologue of the subpolar fronts of other oceans. Farther into the cold northern waters, the frontal system that lies above the Iceland-Faeroes Ridge separates surface water originating in the subtropical gyre from surface waters of the Barents Sea: it is this front that is most usually termed the Polar Front in the North Atlantic. North of Iceland it is almost continuous with another frontal system that can be traced continuously from the coast of Spitzbergen to Greenland and around into the Labrador Sea. Truly polar water lies to the north of this front, so I am proposing to treat this entire linear feature as the Polar Front of the North Atlantic.

Finally, it will be well to note at this juncture that frontogenesis occurs at a variety of scales in the ocean, in a manner that threatens to confuse us as we attempt to use major frontal systems in biogeographical or ecological analysis. Consider the atmospheric convergence that lies between the westerlies of the temperate zone and the easterly Trade

Winds of the tropics. This lies athwart the subtropical oceanic gyres and may be associated with the formation of ephemeral, but biologically significant fronts: such is the case, for example, in the subtropical North Atlantic to the southwest of Bermuda at 22–32 °N and at comparable latitudes in the North Pacific. Here, the convergent wind regime generates convergent surface transport. These surface fluxes in turn generate relatively strong horizontal gradients in surface properties so that, early in the year, frontogenesis is particularly strong at the 100-km scale and some eastward geostrophic flow is induced (Voorhuis, 1969; Weller, 1991).

Similarly, in the southern hemisphere parts of each ocean, sea-level anomaly (SLA) images reveal unambiguous mesoscale eddying in regions where oligotrophic conditions almost always prevail, below the atmospheric convergence. Lying across each subtropical gyre is an arc-shaped field of large mesoscale eddies, having strong sea-level signatures that lie below the atmospheric boundary between the westerlies and the trade winds. These eddies are of diameter 100–250 km, have SLA signatures of 15–25 cm, and in the Indian Ocean tend to be strongest off Australia and near the termination of the field off Southeast Madagascar. The cyclonic eddies frequently lie equatorward of the anticyclonic eddies, so forming a ridge-trough system in the SLA field. This description could be adjusted to fit the other two southern subtropical gyres. Especially in the Indian Ocean, this eddy field is associated with episodic biological enhancement, apparently forced by appropriate vertical motion associated with eddy dynamics. The analogous eddy field in the Pacific and Atlantic Oceans is associated only more weakly with biological enhancement, although "dendritic" blooms may be forced in the SW Atlantic at the same seasons as those in the SW Indian Ocean. This phenomenon shall be discussed in more detail in Chapters 9 and 10.

## Shelf-Edge and Upwelling Fronts

Water masses of shelf seas are generally more buoyant than the oceanic water offshore, because freshwater (and hence buoyancy) is supplied by rivers, or in some locations by the equatorward coastal flow of light polar water. Shelf sea hydrography is therefore often characterized by buoyancy-driven, longshore currents in which the sea surface slopes downward offshore. Separation between such coastwise flow and the oceanic water mass offshore then occurs across a retrograde front in which the density isolines slope downward inshore. In some cases, as in those parts of the Gulf of Guinea where the shelf is unusually wide, a series of fronts may lie approximately parallel to the coast and at varying distances offshore, isolating different water masses. In such cases, we often observe enhanced nutrients along the outer shelf, perhaps attributable to nutrient pumping as the nutricline is periodically lifted by the passage of crests of internal waves interacting with the shelf break. This process has the same frequency as the semidiurnal tides and may explain the permanence of the increased phytoplankton cell numbers often observed along the break of slope; it is not to be confused with the consequences of geostrophic upwelling.

Of course, the discontinuity between oceanic and shelf water that lies above the upper slope and shelf break off western Europe was one of the first of these systems to be investigated: one of the earliest suggestions was simply that westerly gales might overturn (Leslie Cooper actually used the more colorful term *capsize*) the water column in autumn. There is not yet complete agreement on the mechanism by which deeper, cooler, and nutrient-rich water is brought to the surface although the most widely accepted explanation involves the generation of internal standing waves on the thermocline at the shelf edge where tidal streams encounter rough topography. These are thought to originate principally during the period of maximal velocity in the seaward, offshelf tidal stream and the existence of long standing waves is now confirmed by observation of sun-glint pattern at the sea surface.

The enriched phytoplankton biomass that occurs in these frontal zones may arise also from other mechanisms; Horne and Petrie (1986) described "shear-flow" dispersion by which nutrient-rich subsurface water passes into shallower areas with the incoming tide, to retreat a half-tidal cycle later with a diminished nutrient content. Whatever physical mechanisms may dominate at each location, rates of cross-frontal transfer may be quite high and apparently sufficient to maintain the observed enhanced phytoplankton biomass. It should also be noted that where the shelf is sufficiently wide, a shelf edge front may also occur seaward of coastal upwelling cells, in which case it will appear in the chlorophyll field as a secondary, outer zone of biological enhancement.

Shelf-break fronts occur almost everywhere but, in some special circumstances, they risk being confused with major oceanic fronts that lie parallel to the coast but further seaward. Such is the case in places where western boundary currents, representing poleward flow around the subtropical gyres, pass offshore. Thus, the North Wall of the Gulf Stream separates offshore poleward flow of subtropical water from inshore equatorward flow of arctic water from the Labrador Sea (Fig. 3.3). Satellite imagery clearly shows that slope water is separated from shelf water over the Grand Bank and the Nova Scotian shelf by a well-marked front that follows the break of slope and can be traced, at least seasonally, as far south as Cape Hatteras.

Where offshore forcing by Ekman divergence at the coastline dominates the circulation pattern as it does along upwelling, western coasts in lower latitudes, offshore motion is bounded by a prograde front, the so-called upwelling front: this usually lies near the shelf edge. Here, the upwelling of cooler subsurface water is translated into offshore motion, usually in the form of cold jets associated with features of coastal morphology. This transport forces the isopycnals of the permanent thermocline to slope upward toward the coast and, if upwelling is sufficiently vigorous, these may reach the surface and an upwelling frontal zone is formed. Water that is upwelled closer to the coast then passes seaward at shallow depths causing convergence to occur on the inshore side of the front as some of this water passes back down again within a coastal circulation cell. Divergence occurs toward the open ocean at the front, carrying water that has upwelled on its seaward

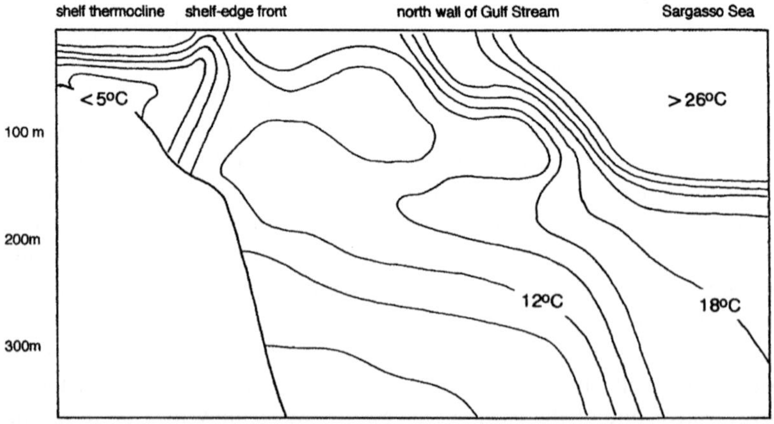


**Fig. 3.3** Temperature section across the shelf and slope in the NW Atlantic at about 40°N in summer, to show the relationship between the convergent front associated with the break of slope and the oceanic front at the north wall of the Gulf Stream. Note the cell of shelf water (<10°C) calving at the shelf edge; these are the "*bourrelets froids*" of French oceanographers, where the stratification parameter takes high values and bottom water is thus shielded from solar heating.

*Source: Redrawn from Bowman and Esaias, 1978.*

side. So, not only is physical accumulation of plankton and nekton to be anticipated at the front, but the relatively stable, shallow pycnocline region just seaward of the front generally supports the most intense production of phytoplankton.

## Tidal Fronts and River Plumes of the Shelf Seas

Stratification of coastal seas is forced not only by solar heating at the surface, but also by coastal freshwater run-off or by an excess of precipitation over evaporation. Several processes contribute to the development of fronts between stratified and nonstratified regions of the shelf: most important, perhaps, is vertical shear within the tidal streams, together with the effect of baroclinic eddies of semidiurnal frequency.

Where a river enters the sea, a plume of relatively fresh, often silt-laden water flows from the estuary. If the coast is insufficiently low-lying to permit the development of a delta that will disperse the effluent, the effluent will pass seaward as a buoyant plume above the coastal water mass. In the absence of entrainment at the interface, the buoyant plume would spread indefinitely, but this does not occur, of course, and it takes dimensions appropriate to the seasonal pattern of discharge. As the horizontal interface between the river water above and shelf water below progressively shoals and breaks the surface, a rather distinct frontal boundary forms around the edges of the plume. Although rivers entering the sea may be so small as to be trivial, or so large as to dominate the regional hydrography, all will be found to have plumes that follow these principles. Off very small rivers, a plume may form only at certain states of the tide.

The injection of land-based nutrients by river plumes into coastal seas may lead to biological enhancement there, although this is not always the case and biological enrichment may have more to do with divergence in the zone of interaction between the two water types than with riverborne nutrients. At the boundary between regions where tidal stress fully mixes the water column and where the river plume enforces stability, estuarine fronts may be formed between stratified and mixed water. These, in general, lie parallel to the axis of the estuary. But the most important effect of tidal streams is to break down the inherent stratification of shelf seas, temporally or permanently, and over relatively large areas.

When the semidiurnal tide encounters shoaling water, as over the continental slope, the amplitude of the tidal wave and its horizontal velocity progressively increase. At some depth, usually shoaler than the continental edge, vertical turbulence produced by friction between the tidal stream and the seabed is sufficiently enhanced (when added to turbulence produced by wind stress at the sea surface) as to overturn seasonal thermal stratification of the water column. As an aside, John Simpson likened the frictional effect of tidal streams to "hurricane force winds blowing regularly twice each day": no wonder, then, that stratification is so regularly broken down on continental shelves. Where this occurs, the tidally mixed and stratified regions of the shelf are separated by a frontal region that migrates semidiurnally and also seasonally because the area of vertically mixed water over the shelf is usually larger during seasons of higher-than-average wind stress. Although there exist more complex formulations, it is convenient to use maps of the parameter $\log_{10} h/u^3$ (where $h$ is water depth and $u^3$ is the tidal current at the surface) to locate the transition between mixing and stratification. For most shelf areas, assuming that the surface heating rate is spatially uniform, the critical value of this parameter is between 1.8 and 2.0. More simply, a tidal stream of 1 m sec$^{-1}$ in 100 m of water falls within this range and therefore mixes (Fig. 3.4).

Phytoplankton observed at shelf sea fronts may exceptionally result from accumulation, but growth *in situ* is the more normal case: Franks and Chen (1996) show that, at the tidal front which encircles Georges Bank, off New England, the distribution of nutrients mirrors that of productivity. They show that the strong tidal mixing over the central bank

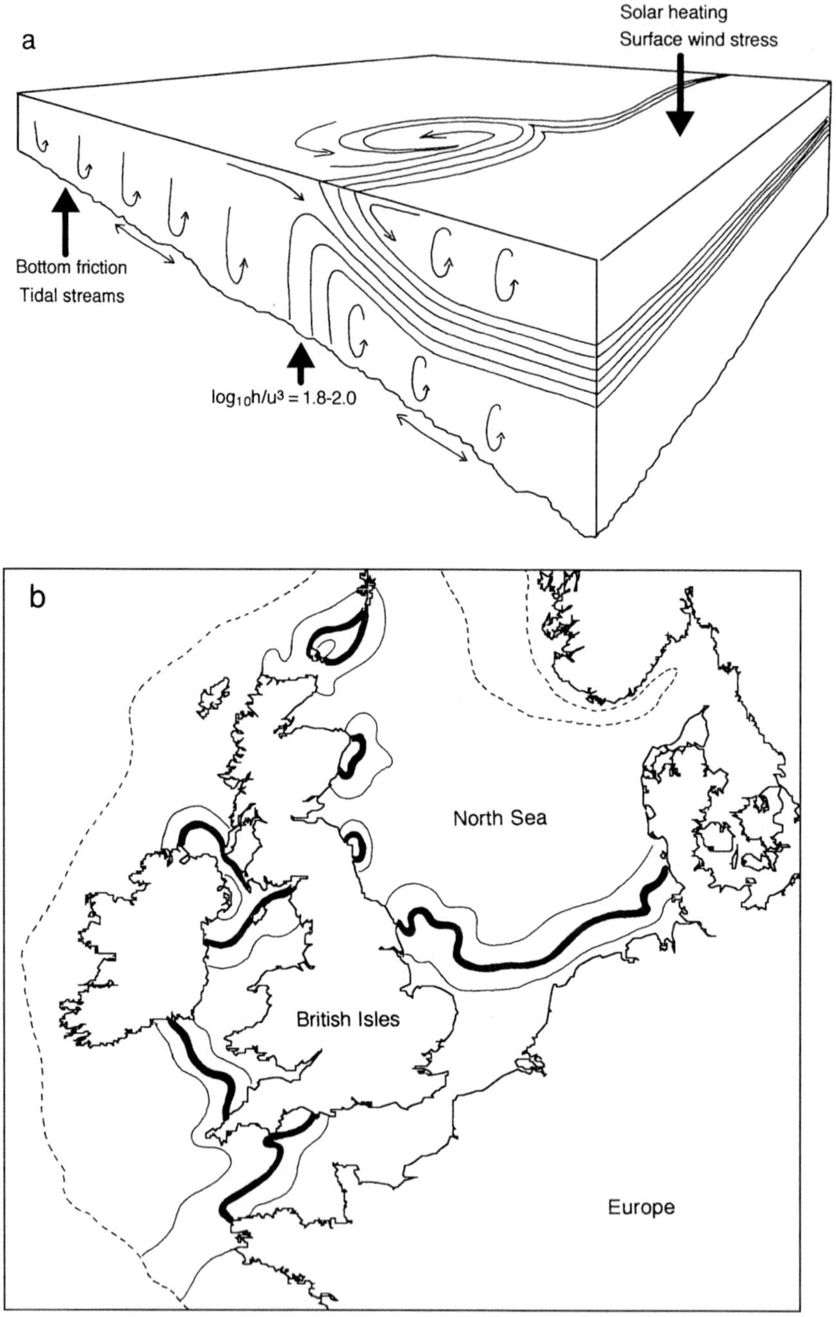


**Fig. 3.4** (a) The structure and circulation within a convergent, prograde tidal front showing how bottom friction in tidal streams may overturn the pycnocline of offshore water to form a front between tidally mixed and stratified water where the critical value of the $h/u^3$ parameter occurs. (b) The European continental shelf, showing the average location of tidal fronts; mean and extreme positions of the critical value are shown as heavy and thin lines respectively.

*Source: Redrawn from several sources.*

decouples population biomasses of autotrophic cells and their zooplankton herbivores by dispersing the latter and so decreasing their local concentration. There is no doubt, then, that localized, enhanced population growth of phytoplankton is responsible for the observed patches of high chlorophyll concentration; this conclusion is supported by regional analysis of f-ratios and by the observations of Horne *et al.* (1989) that nitrate uptake in the front was 0.36 nM $m^{-2}$ $sec^{-1}$ compared with 0.09 on the mixed side and 0.18 on the stratified side of the front; in nearby oceanic water, the demand was only 0.02 nM $m^{-2}$ $sec^{-1}$. Phytoplankton demand at the front was partitioned between nitrate and ammonium, showing that 61% of the daily production within the front was new production, based on nitrate introduced into the front by the processes discussed previously, compared with only 41% and 27% in stratified and mixed water, respectively. We can therefore have reasonable confidence that the remarkably precise coincidence that has been observed between shelf sea thermal fronts and chlorophyll, as observed on the Plymouth-Roscoff section in the western English Channel using towed, undulating instrumentation (Aiken, 1981), does represent *in situ* growth rather than accumulation. This conclusion supports the concept of fronts as ecotones, though at the scale of shelf-sea fronts the exchange between adjacent waters will be sufficiently dynamic that we will not expect that edge-effect species could maintain self-sustaining populations within the front, except perhaps for species at high trophic levels. Thus, there is also some explanation of the common observations that shelf sea fronts are attractive feeding zones for fish and seabirds.

Several general mechanisms have been advanced to account for the enhanced availability of nutrients in shelf sea fronts, always on the assumption that here phytoplankton growth is nutrient limited. The simplest explanation is that as the lunar month advances, tidal friction increases and the front advances toward its stratified side, progressively incorporating nutrients from below the stratification. Two other, more general models were proposed independently, both in 1981. Holligan (1981) suggested that on the stratified side of fronts, phytoplankton growth is released from constraints of light limitation that would occur below the stratified layer and from nutrient limitation that would occur on the mixed side of the front. Tett (1981), on the other hand, suggested that vertical eddy diffusion from bottom friction should be stronger in the front than on either side, thus constantly supplying nutrients from the near-bottom layer. The model of Franks and Chen (1996) of processes at the Georges Bank fronts emphasizes, as noted previously, the decoupling of phytoplankton and herbivores in the mixed zone alongside the front (compared with the spatial balance achieved on the stratified side of the front), and the consequent release from grazing pressure on the autotrophs. This is likely to prove to be a general phenomenon.

Other potential mechanisms exist to explain linear zones of enhanced plankton biomass in shelf seas. For example, in the North Sea, where the velocity of tidal streams falls below a critical value, organic material sinks from suspension and forms linear zones of soft mud at midshelf depths, of which that below the Friesian Front (Baars *et al.*, 1991) is a good example. In summer, above such benthic fronts the flux of remineralized nitrogen from the organic sediments is thought to be responsible for a zone of enhanced phytoplankton production in the water column. However, in practice it may be difficult to separate this effect from that of tidal fronts that may be aligned with the linear benthic remineralization zone.

## The Ubiquitous "Horizontal Front" at the Shallow Pycnocline

As biogeographers have long been aware, the most significant environmental gradient and discontinuity in the ocean is horizontal, between shallow and deeper layers, rather than in the vertical plane at the frontal systems discussed earlier. This gradient lies at the seasonal or tropical pycnocline and is globally associated with the change from epipelagic to deeper ecosystems, so it is the most significant feature in the three-dimensional ecological geography of the oceans. It may be useful to remind ourselves of some of its main characteristics: most importantly, what determines its depth and the strength of its density gradient and whether or not it occurs within the lighted zone.

The epipelagic zone, whose ecological geography is the subject of this book, is a thin layer of light, lighted, and wind-mixed water lying atop the cooler mass of the interior of the ocean; the change in physical and chemical properties between the two depth zones is more frequently abrupt than gradual. When abrupt, the ecological changes that occur across a few tens of meters are greater than across most vertical fronts that intersect the sea surface. This phenomenon is most striking in the eastern parts of the tropical oceans: here, at around 35–40 m below the surface, the temperature drops from 28 °C to about 16 °C. Although the organisms of the epipelagic zone differ fundamentally between the tropical and temperate regions, the deeper-living biota are much more similar. Consequently, as has been noted by many authors, a diel migrant copepod or euphausiid moving across this boundary in the tropics at dawn and dusk makes an environmental adjustment as great as traveling several thousand kilometers equatorward or poleward.

Though the depth at which the pycnocline occurs is determined principally by baroclinicity associated with the ocean circulation, local processes also intervene, at least seasonally: turbulence induced by wind stress at the sea surface, shear-stress turbulence due to inertial oscillations of the mixed layer water mass that are induced by impulsive changes in wind strength, local heating at the surface by short-wave solar radiation, and the local supply of fresh or brackish water. All these factors will be discussed in the next chapter, so here it is sufficient to note that the depth of the surface wind-mixed layer varies from 25 m in the eastern tropical oceans to 250 m in the center of the subtropical gyres. It may also be noted, parenthetically, that the baroclinic upsloping of nitrate isopleths toward the edges of the anticyclonic subtropical gyres is a necessary consequence of the bowl-shaped mixed layer of the gyres and is reflected in the surface chlorophyll field. For this reason, chlorophyll takes consistently higher values around the margins, and lowest values in the center, of the subtropical gyres of each ocean.

At higher latitudes, where seasonality of wind stress and solar heating is stronger, there is a clear discontinuity between the mixed layer of one year and that of the next. Winter mixing and surface cooling progressively deepen the surface mixed layer down to the depth of the deep permanent pycnocline until, with the return of surface heating and reduced wind stress in spring, a new shallow seasonal mixed layer develops. This progressively deepens and strengthens during the summer, only to be eroded again by wind mixing during the subsequent winter. Peak development of the summer mixed layer is typically several months after midsummer, and greatest penetration of deep convection occurs several months after midwinter. In some regions, a halocline may somewhat obscure the simple model.

You should be aware that seasonal graphs of mixed layer depth, which are based on archived time series of monthly mean values (such as those used in this book), may not capture the establishment of near-surface stratification in spring. This occurs *de novo*, far above the winter pycnocline, by solar warming and the consequent induction of buoyancy in the near-surface layer. Instead, seasonal graphs may seem to show a progressive shoaling from a very deep to a very shallow mixed layer depth—a process that simply cannot happen, or at least only as a consequence of geostrophic adjustment. This, of course, would violate Dodimead's first rule for thermoclines: that they can deepen by mixing warmer surface with cold deeper waters but they cannot shoal by unmixing the same! Confusion appears to have arisen between the diel depth of wind mixing, which does of course follow the pattern often used by modelers, and the depth to the seasonal pycnocline. It is this latter, of course, that has the greater biological significance because it carries the seasonally variable nutricline with it. The confusion may also arise by the averaging of shoal and deep mixed layer values during spring, thus indicating a false mean value somewhere between the two.

At all latitudes, vertical velocity (Ekman's WE) is imparted to the water column by the curl of the local wind stress at the sea surface, and this motion shoals or deepens the pycnocline. Cyclonic stress imparts positive values; anticyclonic stress imparts negative values. In this book I shall follow the notation of Isemer and Hasse (1987). Therefore, positive values of WE (upward motion) shall be labeled "Ekman suction" and negative values (downward motion) shall be labeled "Ekman pumping." You should be aware that in many papers the latter term is loosely used without definition and often in the opposite sense to that used by Isemer and Hasse, or by Tomczak and Godfrey (1994).

The pycnocline, whether seasonal or permanent, lies at a depth generally predictable for each region and has a gradient that is predictable from its depth: generally, the shoaler the pycnocline, the sharper the gradient. From its depth (and some knowledge of regional oceanography) the vertical gradients in other properties of ecological significance that are associated with it may also be predicted, especially nutrients, light, and the biomass of both phytoplankton and zooplankton. The interactions between nutrients, light, turbulence, and the biota in the mixed layer and pycnocline have been a central theme of biological oceanography in recent decades: for a good discussion of all this, see Mann and Lazier (2006) or Banse and English (1994).

The simplest expression of the feature is the typical tropical profile (TTP) of Herbland and Voituriez (1977); this is a description of an oligotrophic profile that is either the end member of plankton succession in midlatitudes from spring to summer, or the permanent condition in tropical seas. Under these conditions, the depths of nutricline, pycnocline, deep chlorophyll maximum (DCM), and productivity maximum do not differ by more than a few meters, and the vertical distribution of zooplankton and micronekton also conforms in a predictable manner (Longhurst and Harrison, 1989). Below the pycnocline, in regions where the TTP is a permanent regional feature of the water column, bacterial oxidation of sinking organic material commonly leads to the development of an oxygen minimum zone and a characteristic anomaly in the usual vertical distribution of plankton profiles (Longhurst, 1967; Saltzman and Wishner, 1997a,b).

It is now generally accepted that in many situations the appropriate model for an oligotrophic profile has a two-layered euphotic zone. Where the mixed layer water is sufficiently clear (few algal cells and little suspended particulates) that light attenuation is dominated by seawater absorption, the 1% isolume often lies within the pycnocline. In this case, the upper (mixed layer) zone is well lit and nitrate poor, while the deeper (pycnocline) zone is poorly lit and nutrient rich. Here we have a sufficient model for the steady-state DCM in which algae are larger, shade-adapted, and receive a constant vertical flux of fresh nitrate so that new production is high relative to production based on regenerated ammonium. On the contrary, the well-lit and nitrate-poor upper mixed layer has low rates of new production relative to ammonium-based, regenerated production. Production in the DCM relative to the production in the mixed layer is complex to compute but usually lies within the range 5–50%. Though the shade-adapted cells of the DCM may have access to adequate nitrate, their $P_{max}$ (photosynthetic rate per unit of chlorophyll at light saturation) may be 10 times lower than for near-surface cells. The plots of seasonal cycles for individual provinces (see Chapters 9–12) show that whereas an illuminated pycnocline is the normal condition in low latitudes, it is ephemeral in polar seas. These graphs thus illustrate the potential for increase in the absolute rate of primary production in the deep chlorophyll maximum during seasons when the pycnocline lies shoaler than the photic depth.

In short, all this describes an ecosystem vertically ordered about a discontinuity in the density, nutrient, and biotic gradients, as shall be discussed later. Nitrate diminishes to very low values upward across the pycnocline as both chlorophyll biomass and the rate of primary production increase. Zooplankton herbivores are often concentrated as dense layers at the depths of maximum rate of algal growth, though the total vertical distribution of zooplankton is complicated by the vertical migrations of some species, usually with diel or seasonal frequency, across the pycnocline. In this way, the migrant species utilize to their advantage the contrasting ecological conditions of both euphotic and bathypelagic zones.

Ocean basin-scale baroclinicity defines the topography of pycnocline troughs, ridges, bowls, and domes that are associated with the geostrophic flow while, everywhere, the pycnocline itself and the other gradients associated with it remain in the same depth sequence. Generally, the deeper the pycnocline, the greater the depth interval over which the features of the oligotrophic profile are spread. In Chapters 9–12, I shall discuss many regional variations on this general theme, forced by regional topography and climate, but this introduction is sufficient to demonstrate the ubiquity of this most important ecological boundary in the ocean, excepting only the sea surface, the sea floor, and the shoreline.

All this tells us that this boundary layer is also an ecotone, comparable to the examples discussed previously at vertical frontal zones, and has all the characteristics of one; that is, not only are ecological conditions different above and below it, but there are very special ecological conditions within it. If a characteristic flora could be recognized at the pycnocline, as distinct from occasional shade adaptation of phytoplankton also occurring in the upper euphotic zone, we could regard it as the shade flora of the ocean and comparable with that of the forest floor. In fact, for each major group of photosynthetic cells, there is indeed evidence that an oceanic shade flora must be recognized (Longhurst and Harrison, 1989). In discussing this evidence, let us begin with the smallest cells.

Cyanobacteria and prochlorophytes are most abundant in the mixed layer across the entire North Atlantic from the Gulf Stream to Morocco, whereas peak abundance of small eukaryotes, mostly chlorophytes, prymnesiophytes, and chrysophytes, occurs at the DCM (Li and Wood, 1988; Li, 1995). In the Northwest Pacific, very small eukaryotes, especially *Micromonas* (1–3 μm), also dominate the DCM. Thus, there is some evidence for the existence of a specialized shade flora within the smallest photosynthetic cells.

But for the larger cells, especially dinoflagellates and diatoms, we have much better evidence. Early in the season in temperate or subtropical regimes, during periods of relatively high turbulence and diapycnal mixing, the taxa at the DCM resemble those in the mixed layer above (Venrick *et al.*, 1973; Taniguchi and Kawamura, 1972). However, at 26°N in the Pacific in summer, after pycnocline and DCM have developed, two distinct diatom assemblages meet at the top of the nutricline: the shallower assemblage is nutrient limited while the deeper is light limited (Venrick, 1988). Each assemblage has the characteristics of a mature, predation-controlled assemblage, and community diversity increases to a maximum near the DCM.

From the more extensive general literature on phytogeography, conclusions may be drawn that seem to support the few available floristic profiles. For example, the widespread existence has been noted of a diverse "shade flora" of four Bacillariophycae, ten Dinophycae, one Prasinophycae, and three Prymnesiophycae. Some of these shade species are very large organisms, such as the diatom *Plantoniella sol* (Furuya and Marumo, 1983) and the widespread prasinophyte *Halosphaera viridis.* In the Kuroshio region another 11 species, in addition to *P. sol*, are shade species (including the diatoms *Asteromphalus sarcophagus, Oolithotus fragilis, Thorosphaera flabellata*, and *Thalassionema* spp.). Thus, the hypothesis that the ordered ecosystem of the euphotic zone is partitioned among two different assemblages of algal cells, one of which constitutes a shade flora, appears to be supported.

The protistan consumers of the many size classes of phytoplankton are themselves, not unexpectedly, likewise distributed in a vertically ordered manner in the upper part of the water column. Unpublished pumped profiles that I obtained at oligotrophic stations in the North Atlantic (31° and 34°N) in 1987 showed that protist taxa (both genera and species) occurred preferentially across restricted depths within the mixed layer and the thermocline, in the same way as the larger zooplankton. Unlike these, however, protist cells have negligible powers of locomotion. The depth-differential distribution of protists is best illustrated by the tintinnids, a group whose taxa are relatively simple to identify: at Station PURPLE, south of the Convergence, 62 species of 31 genera were tallied at 10-m intervals from 0 to 110 m. The vertical distribution of these showed that the simple hypothesis—that within the mixed layer, all species should be distributed randomly by wind-induced turbulence—must be rejected. Some genera (*Salpingia*, *Tintinnoposis*) occurred preferentially within the subsurface chlorophyll maximum, others (*Xystonella*, *Eutintinnus*) within the mixed layer. Of three species of *Dictyocystis*, one occurred preferentially above, and two within, the subsurface chlorophyll maximum. The mechanism by which these taxon-specific layers of single-celled consumers are maintained is yet to be explained, but one presumes that depth-differential growth rates must be invoked in some way.

For the larger zooplankton, the pycnocline represents a special depth zone that not only has unique characteristics but also lies close to the separation between the two principal life zones of the ocean: lighted and dark (Longhurst and Harrison, 1989). The increase of zooplankton biomass (five or six orders of magnitude) over the vertical distance (5 or 6 km) from the ocean floor to the sea surface would represent an unprecedented degree of variability if translated into horizontal change within the mixed layer; it would also have an unprecedented predictability (Longhurst, 1985a). In the interior of the ocean, the vertical rate of change of biomass is very small, and the gradient is greatest over a few tens of meters of the pycnocline, where the sparse bathypelagic plankton is separated across a planktocline from the much more abundant epiplankton above. The epiplankton and the deeper acoustic scattering layers of diel migrants are the most prominent features in full-depth profiles of pelagic biota and, like the DCM and the pycnocline, they may be traced across ocean basins.

At some depth within the epiplankton, and most often also within the pycnocline, a maximum of zooplankton abundance ($Z_{max}$) usually occurs. Where the water column has stabilized, $Z_{max}$ lies somewhat shallower than the DCM, especially at night, and closer to the depth of the productivity maximum (Pt). Where there is a very shallow mixed layer, as in upwelling regions or at the start of a spring bloom, $Z_{max}$ occurs very close to both PM and DCM, which are coincident in these circumstances. The depth difference between $Z_{max}$ and DCM is positively correlated with the absolute depth of DCM. Such observations lead one to enquire whether all taxa aggregate at the $Z_{max}$, or is it only certain species that are specialized for life in the ecotone we are considering? It is convenient to discuss this question by reference to special investigations of vertical distribution of zooplankton species made at the BIOSTAT station in the eastern tropical Pacific (Longhurst, 1985b), where there was a shoal pycnocline together with all the features of a TTP. Groups of species could be identified that feed similarly and that occur within common depth horizons, though some rearrangement of the vertical pattern occurs at dawn and dusk, because several of these groups are diel migrants (Fig. 3.5).

The following characteristic groups of copepods were identified by species-specific depths of maximum abundance ("preferred depths") and depth ranges ("layer depths") of the central 50% of the populations of the 72 most abundant species:

*Small herbivores* (<*2.0 mm*): All species (genera: *Calanus, Clausocalanus, Calocalanus, Undinula, Nanocalanus, Acrocalanus, Paracalanus, Ischnocalanus, Acartia, Eucalanus, Oncaea*, and *Corycaeus*) had preferred day depths on the upper shoulder of the DCM and thus close to the Pt. Some of these shifted a few meters upward into the lower mixed layer at night. *Oithona* lay deeper than the other genera—closer to

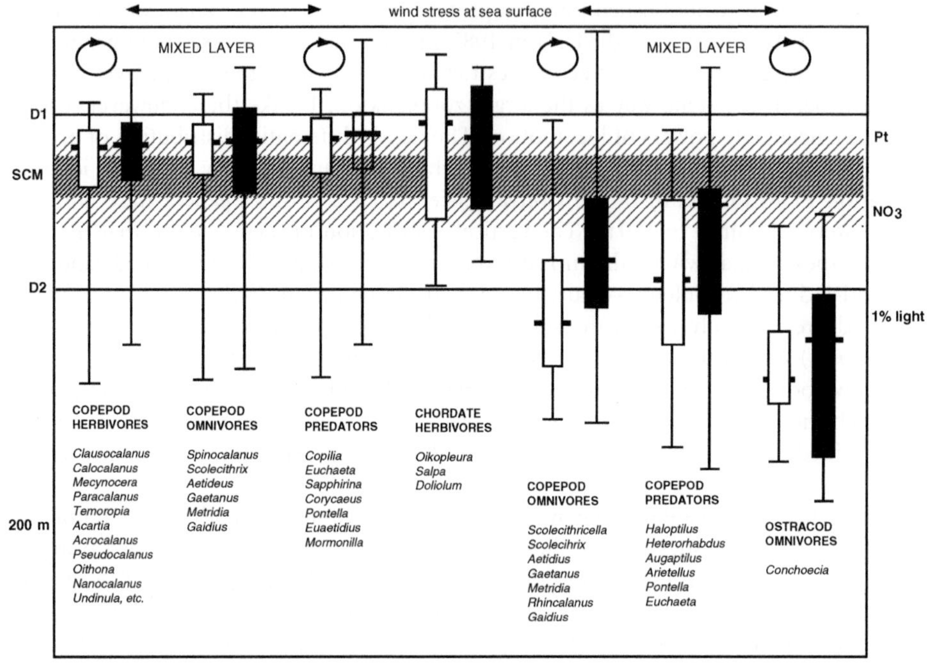


**Fig. 3.5** Some ecological relationships in a "typical tropical situation" in the eastern tropical Pacific at BIOSTAT in day and night LHPR profiles of groups of plankton that have similar feeding strategies. Shown are the depth of maximum numbers, the depth range of 50% of the population, and the depth range that excludes only extreme outliers. The top and bottom of the thermocline is D1 and D2, respectively; the DCM lies within the depth zone of relatively high chlorophyll while P is the depth of maximum production, $NO_3$ is the depth at which nitrate become undetectable, and 1% is the depth of this level of illumination.

the DCM. However, this is the most narrowly specialized group in its selection of preferred depths.

*Larger herbivores (2.5–6.5 mm) and most omnivores (1.5–4.6 mm)*: These occupied a wider layer depth than the small herbivores. Most exploit the lower shoulder of the DCM down to and even below the bottom of the thermocline, with some extending as deep as 250 m. These are species of *Eucalanus, Rhincalanus, Scaphocalanus, Lucicutia, Gaetanus*, and *Neocalanus.*

*Interzonal migrant omnivores*: Mostly *Pleuromamma*, these occurred at night both in the mixed layer and at the DCM; diel migrant *Euphausia*, small fish, and siphonophores were similarly distributed. By day, all diel migrants occurred well below the pycnocline.

*Predatory species*: Preferred depths included both the mixed layer and depths down to 250 m, although there was much individual specialization. For example, the two smallest (2.5–3.3 mm) species of *Euchaeta* preferentially occurred with the small herbivorous copepods on the upper shoulder of the DCM.

Obviously, this analysis supports the hypothesis of a specialized fauna of the ecotone associated with the pycnocline and emphasizes how herbivorous copepod taxa cluster in the upper part of the feature. Other taxa were similarly ordered down these profiles. Ostracods, which are all detrital feeders, occurred preferentially just below the pycnocline. Noncrustacean herbivores (e.g., *Oikopleura*) tended to aggregate like herbivorous copepods on the upper shoulder of the DCM, though doliolids exploited a wider depth range; predatory chaetognaths, like predatory copepods, had overlapping depth ranges covering the whole mixed layer. This vertical distribution of feeding groups is recognizable elsewhere in the Pacific Ocean, where the mixed layer is both extremely shallow (Costa Rica Dome) or very deep (at the CLIMAX station in the North Pacific central gyre).

We should not expect to find such a consistent pattern of vertical distribution in latitudes with deep winter mixing, where plankton profiles are driven alternately by diapausing and feeding strategies. However, available summer profiles from all oceans suggest that, as the summer progresses, plankton profiles approach the vertical pattern of the tropical climax community described previously.

The North Atlantic from 44° to 62°N in August exemplifies this situation; at this time, many genera have depth distributions remarkably similar to their occurrence in tropical profiles (Longhurst and Williams, 1979). Small herbivorous copepods (*Acartia, Paracalanus, Pseudocalanus, Oithona*, and *Clausocalanus*) and cladocerans occur most abundantly in the upper thermocline, near the DCM, with only scattered individuals deeper. Already in this season, ontogenetic migrations had carried diapausing individuals of the large herbivores to 1000 m. At night, the interzonal copepods and diel migrant euphausiids occurred at the same depths as the epiplanktonic filter-feeding copepods.

Therefore, there seems to be abundant evidence from observations of plankton organisms from the smallest autotrophic cells to herbivorous mesozooplankton to support the original statement. We should, indeed, consider the horizontal front represented by the pycnocline not only as a way station where sinking particles may aggregate, but also as an ecotone that is the preferred living space of a characteristic group of plankton organisms.

# Chapter 4

# Physical Control of Ecological Processes

A suggestion of Yentsch and Garside, made to a congress of pelagic biogeographers in 1986, might serve as a synopsis of the argument presented in this book:

> Are the large-scale biogeographic distributions of oceanic biology a response to seasonal and spatial patterns of primary production? Geostrophic currents dictate the shape of the density structure and what we term the degree of baroclinicity is a global mirror of primary production. The fact is that for at least fifty years we have recognized that . . . light, wind and temperature . . . regulate phytoplankton growth. We recognize that seasonality in the primary processes is controlled by stirring of the upper layers of the ocean, and the nutrient and density fields. Why, then, are we so hesitant to tie the climatological primary process together with concepts concerning biogeography? One answer might be that we do not have enough knowledge of primary productivity over large areas of the world's ocean.

Now, of course, we do have enough knowledge from satellite imagery to begin the integration we were encouraged to undertake. A little later, Yentsch gave one of the earliest demonstrations that the CZCS chlorophyll field could be used to demonstrate the control of algal growth by irradiance and nutrients. He showed that the front of the migrating North Atlantic spring bloom, as revealed by CZCS, closely matched predictions obtained from an independent analysis of the seasonal poleward march of the shoaling of the mixed layer.

How such thinking had much earlier led to the formulation of Sverdrup's critical depth theory is well known. Had Sverdrup not left Scripps, as he did in 1948 to return to Norway, he might not have become involved in thinking about the Ocean Weather Station "M" observations, and his formulation would now bear another's name. It is so fundamental that with or without his initiative it would have become central to biological oceanographic theory.

Nevertheless, we tend to forget that Sverdrup's important contribution was to encapsulate, in a formally stated theorem, several concepts that had already been discussed more generally. Nathansohn (1909), Gran (1931), Riley (1942), and others made major advances as the old "nitrification" theory of plankton production came to be replaced by modern understanding of the nitrogen cycle. Prior to Sverdrup, the concept of a compensation illuminance ($I_c$) at which photosynthesis of individual cells would exactly balance cellular respiration had already been evoked from studies done in the Gulf of Maine (Gran and Braarud, 1935).

Sverdrup (1953), using time-series observations of daily mixing depths and cell counts made at OWS "M" (66°N2°E) analyzed the interaction between vertical mixing and $I_c$ and introduced the concept of a critical depth ($Z_{cr}$) above which phytoplankton respiratory loss exactly balances growth. He showed that if mixing penetrates deeper than this, then cells will be light limited because they will experience a time-integrated mean irradiance lower than $I_c$ and their time-integrated growth will take a negative sign. If, on the other hand, mixed-layer depth is shallower than $Z_{cr}$, then cells will be light-sufficient, their time-integrated growth will have a positive sign, and a bloom may ensue. Unstated by Sverdrup, though well recognized by Riley, is the necessary auxiliary assumption that deepening of the wind-mixed layer in winter recharges surface layers with inorganic nutrients.

Sverdrup, of course, was also well aware that his model requires that the loss term should include not only algal respiration but also excretion, grazing, and sedimentation: the balance is, as Sathyendranath and Platt (2000) reminded us, between total growth and total loss of phytoplankton in the superjacent water column. This balance may be discussed in general terms, but the exact expression offered by Platt, Caverhill, and Sathyendranath (1991a) leads to a characteristic time scale for the development of a bloom. Sverdrup's original formulation invoked very simple assumptions to calculate the irradiance at the critical depth, and it now understood to be necessary to recognize how this responds to changes in the optical parameter K. This parameter, of course, itself responds to the biomass of phytoplankton in the surface mixed layer, so that critical depth shoals during the growth of a bloom; perhaps the most significant modification of the original concept is that self-shading by autotrophic cells is now known to account for a smaller fraction of total shading than had been previously expected. Because individual taxa of autotrophs have individual characteristic values of $I_c$, this is usually stated as a community mean.

The validity of the Sverdrup model is not vitiated by observations that blooms can occur in the absence of significant stratification under special circumstances. This occasionally occurs in the Gulf of Maine (and surely, then, also elsewhere?) when deep penetration of light in early spring, accompanied by very little wind mixing, permits exceptionally buoyant phytoplankters to accumulate biomass (Townsend *et al.*, 1992). This situation has been revisited recently by Ebert *et al.* (2001) in the context of a "critical turbulence" rate for buoyant, for slow-sinking, and for fast-sinking phytoplankton cells: the conclusions reached by these models for the initiation of blooms are intuitive. Further, since a chlorophyll-rich layer in the water column must induce local warming by absorption of incident radiation at that depth, this may directly induce local stratification (e.g., Sathyendranath and Platt, 1994). The consequent changes in the photosynthetic parameters for algal growth will then add new complexity to the overall biological response to irradiance. The Sverdrup model, therefore, is seen to be but one component of a complex and coupled biophysical system.

For this reason, more general effects must also be accommodated if the small set of critical factors for algal ecology required by Sverdrup are to be useful in global analysis, so as to be relevant to regions where other processes may override the effects of local winds and sunshine in determining the depth of the surface mixed layer. Two principal effects must be accommodated as anomalies to Sverdrup, of which we have already discussed the first: in low latitudes the seasonal changes in mixed-layer depth may be a response to distant wind forcing rather than to changes in locally induced wind mixing. The second effect is that of a surface layer of low-salinity water, not always induced by local rainfall, but which may come to dominate near-surface stratification.

A simple relationship between seasonal values for $I_c$ and $Z_{cr}$ results in characteristic changes in the strength of seasonal blooms at different latitudes (Follows and Dutkiewicz, 2003). The relationship between these two depths may be expressed as the dimensionless parameter $h_c/h_m$ that takes values near unity in the subtropics and as low as 0.05 in polar seas where winter mixing penetrates very deep under dark skies. Increased winter mixing beyond a threshold depth in high latitudes may result in reduced strength of the subsequent spring bloom, because cells, though nutrient-replete, continue to be intermittently mixed sufficiently deep as to become light-limited. Such a threshold does not occur in lower latitudes where, in general, unusual seasonal deepening of the mixed layer leads subsequently to enhanced strength of the seasonal bloom, by increasing the nutrient supply. By this model, Follows and Dutkiewicz suggest mechanisms for the control of the timing of the high latitude spring bloom and for coherent regional changes in bloom strength observed in the subtropics.

## Ecological Consequences of Mesoscale Eddies and Planetary Waves

General maps of ocean circulation suggest that flow is linear and that free-drifting particles, such as planktonic biota, will be transported parallel to the apparent axis. This assumption is inherent in discussions of the biogeography of plankton and nekton, where retention within oceanic gyres is generally assumed.

Satellite sensors now reveal that flow in the real ocean is far from linear and occurs within a complex and constantly varying field of mesoscale eddies, both cyclonic and anticyclonic, associated with a global field of planetary waves (reviewed by Chelton and Schlax, 1996), previously observed only with difficulty because of their small signature at the sea surface: a sea-level anomaly (SLA) of a only a few centimeters is associated with the displacement of the thermocline of several tens of meters. Rossby waves are the dynamic response at large scale to wind and buoyancy forcing on the eastern boundaries of ocean basins; the velocity of their westward propagation is an inverse function of latitude. Waves at the first baroclinic mode propagate slowly across the ocean basin, requiring several months to cross the Pacific in low latitudes and several years at high latitudes. Very close to the equator, most of the observed sea-level variability is associated with eastward-propagating Kelvin waves, traveling along the equatorial wave guides. These equatorially trapped waves may originate in wind events in the central part of ocean basins or by the reflection of Rossby waves at the western boundary. The global field of planetary waves may now, of course, be visualized in the SLA field obtained by TOPEX-POSEIDON and is the template for an important component in the sea-surface pattern of chlorophyll. Physicists have commented on the unexpected nature of this relationship: ecologists, if they had thought about the matter, would have predicted it. I shall have frequent occasion to refer to it here.

It is now clear that flow in the ocean, as in the atmosphere, is everywhere dominated by synoptic mesoscale eddies that extend deep into the water column, have lifetimes of order 100 days and maximal rotational velocities of order $10\,cm\ s^{-1}$. These are, it is trite to repeat, the "weather systems" of the ocean and contain an order of magnitude more energy than do the mean currents. Eddies are generated preferentially in regions of strong horizontal gradients of properties, as, for instance, where coastal boundary currents turn seaward. Flow around mesoscale eddies may be an order of magnitude faster than that of the mean current.

Mesoscale eddies are, of course, both cyclonic and anticyclonic. In bowl-shaped, cyclonic eddies, pressure is low below the sea surface so that the thermocline shallows centrally, hence these are "cold-core" eddies. In anticyclonic features associated with doming of the sea surface, pressure is high below the surface so that the thermocline deepens centrally, resulting in a "warm-core" eddy. It is important to note the relative scale of vertical motions likely to be observed centrally in an eddy in response to motion: a few centimeters at the sea surface, a few meters at the thermocline, and a few tens of meters at the main pycnocline. Thus, the very small SLAs observed by TOPEX-POSEIDON indicate much larger vertical displacement, deeper in the water column. It is a trite experiment to match images of sea-surface chlorophyll with sea-level anomaly maps from TOPEX-POSEIDON and to observe the match between chlorophyll patches and the centers of cyclonic eddies; indeed, a University of Colorado site now matches the two data sets for you, within rectangles and for dates of your specification.

Motion of an eddy depends on where it lies in relation to the axis of major current. Those intimately associated with the flow of the Gulf Stream near its origin, for example, travel eastward while those lying farther from the axis travel west and north as a result of their interaction with the general field of potential vorticity (Flierl and McGillicuddy, 2002). In regions especially active in eddy formation, such as along the Gulf Stream and Kuroshio fronts, meanders of the main jet may become pinched off, to enclose a parcel of the adjacent water mass. These are the cold-core and warm-core "rings," having a larger dimension than simple mesoscale eddies, of order 200–300 km. Cold-core rings, enclosing slope water surrounded by a jet of warm Gulf Stream water, form southeast of the jet current five to eight times a year and may persist for up to 2 years; warm-core rings are more rapidly wrecked against the shelf edge and have shorter life spans (Wiebe *et al.*, 1976). The subtropical front of the North Atlantic (see Chapter 3) is probably typical of midocean frontal systems (Pingree, 1997; Pingree and Sinha, 2000); here, westward motion of Rossby waves can be observed in TOPEX-POSEIDON imagery at about 32–34°N, carrying eddies having SLA signatures above background. The larger features may be followed westward, crossing the mid-Atlantic Ridge, over periods of several months.

Because eddies create vertical motion, a different response of phytoplankton to irradiance and nutrients is to be anticipated in these features and in the surrounding ocean. Early studies of SLA and temperature in the Sargasso Sea showed a correspondence between cold (warm) temperature anomalies and higher (lower) pigment anomalies. Phytoplankton in the oligotrophic ocean that are not growing at their maximum specific growth rate may be stimulated by such "eddy-pumping" (Falkowski *et al.*, 1991). Especially in low latitudes, the lifting of nutrients from below into the photic zone in cold-core eddies stimulates an instantaneous response of the phytoplankton that is at first linear; later, the accumulating nutrients decrease rapidly as the phytoplankton enters—at least theoretically—a phase of exponential growth. McGillicuddy and Robinson (1997) suggested that in the subtropical gyre of the North Atlantic the contribution of the fertilizing effects of mesoscale eddies is significantly larger than the combined entrainment of nutrients into the mixed layer during winter convection, and by thermocline mixing and wind-driven transport. Eddy pumping may therefore be the dominant nutrient flux in such regions.

In anticyclonic eddies the process is more complex and marginal, and enhanced chlorophyll concentrations occur preferentially in the high-velocity region encircling the eddy, where elevation of the pycnocline (and hence of the nutricline in most situations) is associated with a strongly baroclinic structure (Yentsch and Phinney, 1985; Lohrenz *et al.*, 1993). Vortex contraction may strengthen the upwelling of nutrients, especially on the anticyclonic side of the transient jets (Woods, 1988). This effect is usually limited to patches on the 10-km scale, which serves to explain the beaded string of high chlorophyll often observed around anticyclonic eddies in chlorophyll images. The same effect is observed on the filaments shed by mesoscale eddies, especially on their curved rather than their straight sections (Tranter *et al.*, 1983). Production within warm-core rings in the Agulhas eddy field may be limited by convective instability, but around the edge of such rings stability and enhanced productivity is conferred by the warm water of the ring overlying the cooler water in which it is embedded (Dower and Lucas, 1993). The same, of course, may occur elsewhere.

Planetary waves present major anomalies in the global field of surface chlorophyll, and Killworth *et al.* (2004) recently reviewed the possible causes of the observed mesoscale chlorophyll anomalies in such eddying flow. Three possible mechanisms were examined: (i) horizontal advection of phytoplankton cells and their concentration within a frontal structure, (ii) vertical advection of phytoplankton cells from below to the surface, and (iii) upgrowth of cells following vertical transport of nutrients into the photic zone. The latter, the *rototiller effect* (Siegel, 2001), differs significantly from what occurs ideally in a cyclonic eddy, which only upwells nutrients when it forms or intensifies; during their entire westward propagation across the ocean, planetary waves will be expected to upwell nutrients continuously on their leading edge. Nevertheless, cross-spectral analysis of TOPEX-POSEIDON SLA and SeaWiFS chlorophyll suggested to Killworth *et al.* that this process is less significant than the meridional transport of chlorophyll against the background field. This is sufficient to account for most of the observed wave propagation seen in the SeaWiFS chlorophyll field, although other mechanisms must contribute. One of these is the effect of increased wind stress seasonally when Rossby-wave upwelling may cause simple vertical entrainment of cells that then dominate the regional surface chlorophyll pattern; this can only occur when the mixed-layer depth lies close to that of the chlorophyll maximum (Kawamiya and Oschlies, 2001) and is important in establishing the surface chlorophyll pattern at low latitudes in the Indian Ocean.

However, an even simpler "cell-lifting mechanism" explanation of mesoscale pattern within the basin-scale chlorophyll field is proposed both by Kawamiya and Oschlies (2001) and Charria *et al.* (2003). If the passage of a Rossby wave induces no significant input of nutrients into the mixed layer and if the deep chlorophyll maximum (DCM) lies at or just below the thermocline, it is sufficient for the passage of a Rossby wave to raise the DCM into contact with the mixed layer; this will erode the DCM and increase the observed cell count within the mixed layer. Thus, although Rossby waves may have an appreciable effect on variation in primary and export production, the surface pattern of chlorophyll observable by satellite sensors "is controlled by the difference between MLD and the depth of the DCM." This may be, suggest Kawamiya and Oschlies, yet another complication for algorithms used in the computation of regional productivity from the sea surface chlorophyll field. For Charria *et al.* this mechanism is sufficient explanation of the observed regional chlorophyll field in the Subtropical Convergence Zone of the South Atlantic.

## Stratification and Irradiance: The Consequences of Latitude

Any ecological geography of the ocean must be sensitive to the physical processes that determine when and where phytoplankton growth will occur, and which taxa will participate. From Sverdrup, we may assume that what we need to analyze are seasonal changes in (i) the light field and (ii) stratification within the water column. Apparently, these are the consequences of solar irradiance and of wind stress at the sea surface; however, both are strongly modified by continental geography and the distribution of land masses, and each also responds characteristically to latitude. It will be useful at this point to review the consequences of latitude for irradiance at the sea surface and on the processes that are induced by wind stress at the sea surface. The first are easily stated, the second are much more complex.

The regional, seasonal characteristics of irradiance at the sea surface are readily encapsulated in two relatively simple propositions: (i) in low latitudes the seasonal cycle has a smaller amplitude than the diel cycle, whereas the annual cycle exceeds the diel at high latitudes (Hastenrath, 1985) and (ii) the irradiance field is dominated by a simple and predictable meridional gradient, which migrates with the seasons and is modified by regionally variable cloudiness and atmospheric clarity that respond to processes above adjacent continents. At this point in the discussion, that is all that needs to be said.

Prominent in any textbook on biological oceanography are the ecological consequences of the balance between turbulence and mixing caused by wind stress at the surface and stratification induced by solar heating. That this balance is relevant only at higher latitudes is rarely mentioned, and the student may be left to discover that the physics required by the Sverdrup model is quite different at low latitudes. There, seasonal changes in depth of the surface mixed layer are due not to seasonal changes in turbulence induced by varying local wind stress but rather, through geostrophic adjustment, to changes in circulation that may occur in distant regions (Philander, 1985; Katz, 1987; Hastenrath and Merle, 1987). To understand this it is necessary to start with the essential, but often overlooked, difference between the effects of wind stress at the sea surface at high and low latitudes.

Simply stated, wind stress is translated mostly into potential energy at high latitudes, but into kinetic energy at low latitudes. This is not the place for a formal discussion of this effect, for which a good modern source is Tomczak and Godfrey (1994), but, simply stated, flow is induced by the pressure-gradient force at the sea surface, and because geostrophic balance is maintained between this gradient and the Coriolis parameter induced by Earth's rotation (f, with dimensions of 1/sec), the slope of the sea surface required to produce motion diminishes equatorward as a direct function of latitude, and without discontinuity (Lighthill, 1969; Philander, 1985). It is perhaps characteristic of the evolution of oceanographic theory that this was not understood until the 1960s.

Consequently, seasonal moderation of wind stress over the North Atlantic in summer has a negligible effect on the flow of the Gulf Stream, whereas seasonal reversal of monsoon winds in the Arabian Sea reverses the Somali Current to almost 1000 m depth in just a few weeks. Only after reversal of the North Atlantic westerly winds for about 10 years would the same effect be produced on the Gulf Stream. At the extreme, geostrophic balance ceases to function at less than about 2° from the equator, where motion is determined by nonrotational fluid dynamics; the most striking manifestation of this effect is the undercurrent flowing eastward in the pycnocline below the equator.

That seasonal changes in wind stress should quickly modify the flow in major barotropic currents at low latitudes, such as that along the coast of Somalia and Arabia, is a fundamental characteristic of tropical oceans; moreover, since the spatial scale of the adjustment of barotropic currents is similar at all latitudes, seasonal geostrophic adjustments in the mixed-layer depths can be identified across whole ocean basins and have important consequences when considering the Sverdrup model. This, of course, is usually invoked where it is changes in local wind mixing and irradiance that force changes in mixed-layer depth, although it has also been used to interpret phytoplankton seasonality in lower latitudes (Obata *et al.*, 1996). This investigation of the interaction between relative changes in mixed-layer depth and $Z_{cr}$ and the concomitant increase or decrease of surface chlorophyll used Levitus mixed-layer depth ($Z_m$), CZCS surface chlorophyll, and the surface light field. The results suggested that whenever and wherever $Z_m$ shoals up through $Z_{cr}$, whether by changes in wind stress or by geostrophic adjustment, a bloom follows. The reverse effect was also simulated: whenever and wherever $Z_m$ became deeper than $Z_{cr}$, a decrease in surface chlorophyll was produced in the following month.

Also consequent upon the equatorward vanishing of the Coriolis parameter are changes in the scale taken by internal wave trains, implicated in wind mixing of the surface layers of the ocean. The period of these waves changes from only 12 hours at the poles to much longer at the equator (Garrett, 2003). Their propagation into the interior of the ocean induces a cascade to smaller scales and to ever-smaller eddying and, hence, mixing. This process becomes less energetic as a function of latitude, with a slower cascade toward the equator. Thus, wind-induced turbulence and mixing is—other things being equal—a direct function of latitude and another reason for the physical stability of tropical oceans.

The final consequence of latitudinal change in the Coriolis parameter is that the values taken by the Rossby radii of deformation are latitude-dependent (Emery *et al.*, 1984; Houry *et al.*, 1987). The external radius (Re) determines the length scales for barotropic (geostrophically balanced) phenomena; it is the length scale over which gravitational phenomena balance the tendency of f to deform a surface and may be computed as (gH/f), where H is bottom depth and g is acceleration due to gravity. The internal radius (Ri) also depends on stratification and determines the horizontal dimensions of quasigeostrophic features—mesoscale eddies and Rossby waves. It may be approximated as (1/first eigenvalue N) × f and, because f vanishes at the equator, Ri here takes a value of infinity. At 50° latitude, Ri takes values of only 25 km whereas at 5° latitude, values are around 300 km. For this reason, we may expect larger and fewer mesoscale eddies as we approach the equator.

There are also latitudinal differences in the density gradients (and thus in their resistance to wind mixing) between the summer pycnocline of midlatitudes and the permanent shallow pycnocline of low latitudes; the latter is steeper and has greater resistance to mixing, thus enhancing the effects discussed earlier. Furthermore, the heat balance at the surface of low-latitude oceans is such that there is a mean downward heat flux across the sea surface that is balanced by horizontal transport to higher latitudes; seasonal changes in incoming irradiance here are sufficiently small that at no season is there sufficient loss of heat to produce convective deepening of the permanent tropical thermocline. Finally, there is often an excess of precipitation over evaporation at the surface of the tropical ocean, so mixed-layer water at low latitudes is characteristically not only warmer but also a little fresher than deeper water. The tropical Ekman layer, therefore, lies above a pycnocline having greater stability and resistance to mixing than elsewhere; this is expressed as a relatively high subsurface maximum of the Brunt-Väisälä buoyancy frequency (N, in cycles $hr^{-1}$; see later discussion).

Finally, consider also the effect of the seasonal increase in the westward stress of the trade winds across the tropical ocean that occurs within about 15–20° latitude of the equator. It is this stress that maintains an upward slope of the sea surface and downward slope of the pycnocline to the west in each ocean. The adjustment time of an ocean basin to wind stress is related to the time taken for planetary waves to propagate across the ocean, so an equilibrium seasonal response cannot occur within a single season across the great width of the tropical Pacific (15,000 km). Additionally, over this great distance, wind stress in the west is not in phase with that in the east, so the response of the ocean cannot be simple; consequently, it is only when wind stress changes for longer periods, on the interannual El Niño–Southern Oscillation scale, that an equilibrium response in Pacific circulation can occur. However, the tropical Atlantic (just 5000 km wide at the equator) responds to the seasonal cycle of trade wind stress with a seasonal basin-wide geostrophic adjustment of mixed-layer depth that maintains equilibrium.

The aspects of ocean physics discussed in this section serve to distinguish very clearly the characteristics of low- and high-latitude oceans as biological habitats. But they do not suggest the discontinuities we might like to have in order to partition the apparent continuum forced by the continuous vanishing of the Coriolis parameter equatorward. However, if we look more closely at one feature from the previous discussion—relative pycnocline stability, the property that most closely controls resistance of the water column to vertical mixing—then I believe that the required discontinuities can be demonstrated.

## Regional and Latitudinal Resistance to Mixing in the Open Oceans

In this section I shall attempt to locate such discontinuities, starting with those that may occur between the strong pycnocline of low latitudes and the weaker pycnoclines of higher latitudes. To do this the distribution of the buoyancy or Brunt-Väisälä frequency ($N$) was examined in all oceans, with Java OceanAtlas (Osborne *et al.*, 1992). In this way, a maximum value for $N$ (and hence of maximum resistance of the pycnocline to turbulent mixing), and the depth at which it occurred in the water column, was obtained at 1° intervals along each of 19 long sections (11 meridional and 8 zonal) from oceanographic voyages and from 10 meridional sections synthesized from the accumulated profile data at the U.S. National Oceanographic Data Center in Washington, DC (Levitus, 1982; Levitus *et al.*, 1994; Levitus and Boyer, 1994a,b).

The 25°W Atlantic section that was worked by *Oceanus* and *Melville* in 1988–1989, from Iceland to the Southern Ocean (Fig. 4.1), will serve to illustrate this approach to the problem of locating discontinuities in the resistance of the pycnocline to mixing. Obtained during summer in each hemisphere, it shows clearly the distinction between the permanent tropical pycnocline, where $N$ takes values of about 12–14 cycles $hr^{-1}$, and the shoaler summer thermocline (7 or 8 cycles $hr^{-1}$) of the subtropical gyres. The transition occurs across the equatorward limbs of the subtropical gyres in each hemisphere; the North Equatorial Countercurrent and the main flow of the South Equatorial Current both lie above the tropical pycnocline, which thus lies somewhat asymmetrically to the north in relation to the equator. The actual depth and slope of the maximum value of $N$ reflects (as does the depth of the pycnocline itself) the flow of the zonal currents within the Ekman layer.

The same features are readily identifiable in climatological sections derived from archived data along 30°W, though the seasonal summer thermocline is weaker (as it should be in archived data that represents all seasons), taking a value of only about 5 cycles $hr^{-1}$. The strong tropical pycnocline from 20°N to 20°S is clearer (as it should be) in the archived data and takes the same values of $N$ as in the *Oceanus–Melville* section. Furthermore, it lies in the same relationship to the zonal tropical current systems, inferred from the depth of the pycnocline. Because the archived data extend further poleward they better show the effects of the low-salinity polar surface layer. At 60°N, as the section passes over the Irminger Basin, and at 40°S across the South Subtropical Convergence Zone, the value of $N$ increases progressively poleward as the pycnocline approaches the surface toward the ice edge.

This comparison gives us confidence that the Levitus sections may be used to explore systematically the strength and depth of $N$ at 30° latitude intervals in all oceans. Such an investigation confirms that, by reason of its stability and depth, the tropical pycnocline is a unique feature in all oceans and, moreover, that the transition to this feature tends to occur at about 20° of latitude (Longhurst, 1995).

What else can be discerned about the tropical pycnocline? Certainly, the data suggest that the relatively abrupt transition between the weak subtropical and the strong tropical pycnocline (which is a prominent feature in most meridional sections) does not occur always at the edge of the equatorial zonal currents, as one might expect it to do. Consider again the *Oceanus–Melville* section: in the North Atlantic, the transition occurs clearly at about 22–23°N, coincident with the northern margin of the North Equatorial Current (NEC) flowing west around the equatorward limb of the subtropical gyre. However, in the South Atlantic no such relationship exists, perhaps because the South Equatorial Current (SEC) is wider and more diffuse than the NEC.

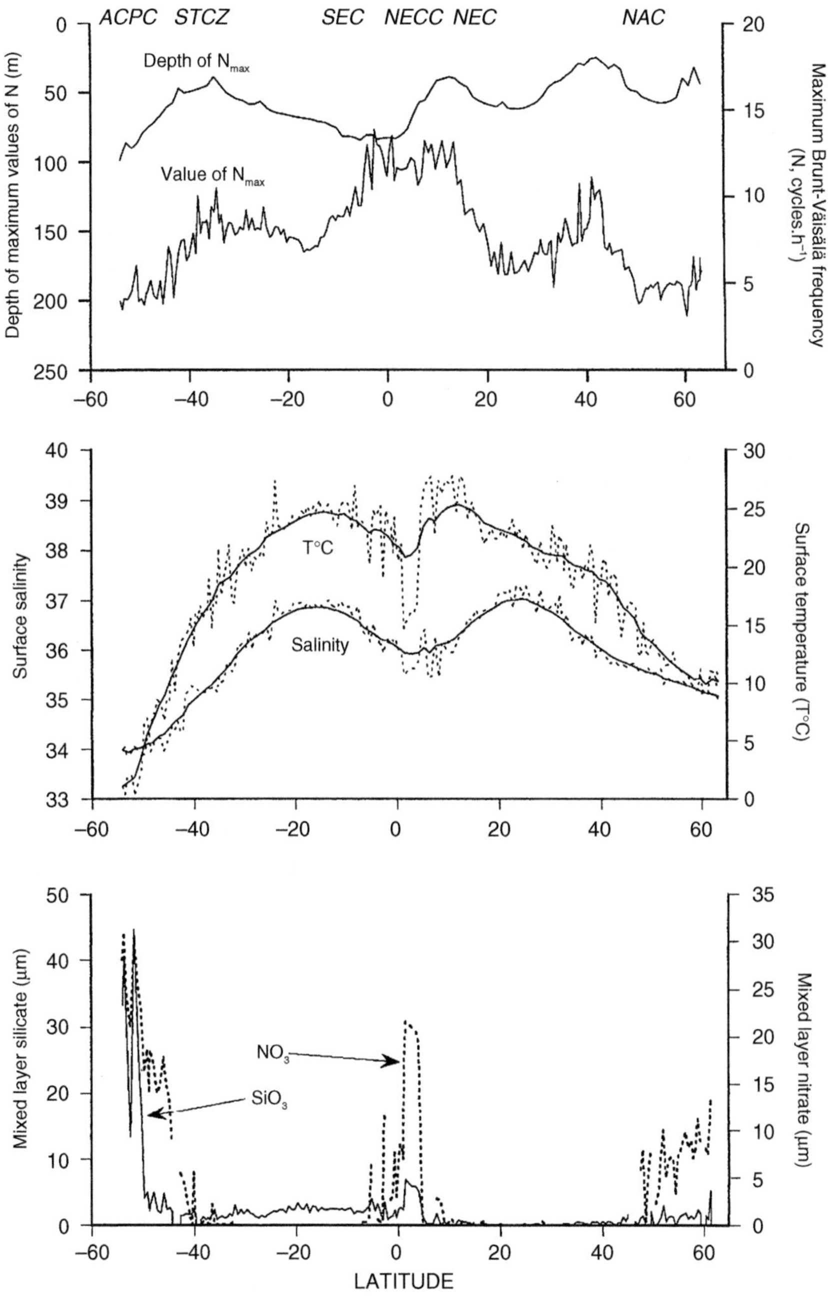


**Fig. 4.1** Meridional *Oceanus/Melville* section down 25°W in the Atlantic for summer in each hemisphere, showing maximum values for stability (buoyancy frequency, $N$) in the pycnocline and the depth at which $N_{max}$ occurs. The relative slope of the pycnocline reflects the locations of the zonal current systems. The lower panel shows the relatively continuous meridional fields of mixed layer temperature and salinity and the highly discontinuous fields of available macronutrients.

*Source: Computed and drawn using OceanAtlas 2, Osborne* et al.*, 1992.*

Clearly, the area of the tropical pycnocline with high values of $N$ follows the distribution of low-salinity tropical surface water more closely than the margins of the zonal currents. This effect is clearer in the southern than the northern hemisphere and is seen not only in the *Oceanus–Melville* section at 25°W but also in *Knorr, Melville*, and *Meteor* sections, which run down the Greenwich meridian, and in the "Atlantic Western Basins" sections. In all these, part of the SEC (4°N to 8°S) and the entire South Equatorial Countercurrent (8–12°S) lie equatorward of the abrupt transition from weak subtropical to strong tropical pycnoclines; the poleward part of the SEC (from 12°S to 23°S) exhibits a relatively weak pycnocline.

The Brunt-Väisälä frequency takes higher values in the east than in the west of each ocean (Fig. 4.2). Thus, the evolution of the values for $N_{max}$ along zonal sections shows that the pycnocline is most stable ($N = \sim 15$ cycles $hr^{-1}$) in the eastern tropical Atlantic and Pacific, illustrating the well-known westward deepening of the tropical thermocline in response to westward trade wind stress at the sea surface. This slope will be about two orders of magnitude greater than the associated wind-driven upsloping of the sea surface. The discontinuity between the relative strengths of the tropical and subtropical pycnoclines is also strongest in the east. It is in the western Indo-Pacific (180°W to 90°E) that the value of N between 20°N and 20°S is lowest (<10 cycles $hr^{-1}$), and transition to the southern subtropical condition is most continuous. Along the Greenwich meridian, south of Ghana, typical values for $N_{max}$ are the following: South Equatorial Current 12–20 cycles $hr^{-1}$; equatorial zone 10–17 cycles $hr^{-1}$; and in the Guinea Current, 12–14 cycles $hr^{-1}$. In the western tropical Atlantic, in comparable regimes, the Western Basins Section shows typical values in the range of 8–12 cycles $hr^{-1}$.

Because it forms an exception to arguments just made, it will be useful at this juncture to introduce the special case of the western part of the tropical Pacific, the "warm pool" of high surface temperatures. This lies below the heavy cloud cover of the low-pressure cell at the conjunction of the Intertropical and South Pacific atmospheric convergence zones (Tomczak and Godfrey, 1994). In this region there is a near-surface halocline within the deeper isothermal mixed layer, the two features being separated by a "barrier layer" (Yan *et al.*, 1992; Lukas and Lindstrom, 1991) between thermo- and haloclines. Meridional sections along 130° and 137°E show that the strongest inflection in the density profile occurs at about 50 m, near the thermocline. The highest values of $N$ (10–14 cycles $hr^{-1}$), and therefore the greatest resistance to mixing, occur rather deep, and well into the thermocline, at 100–150 m. These conditions extend almost to the date line, across a zone of rapidly changing values for $N$ in the tropical and subtropical pycnoclines. This is an extreme case of the situation in the tropical Atlantic and Bay of Bengal, where the

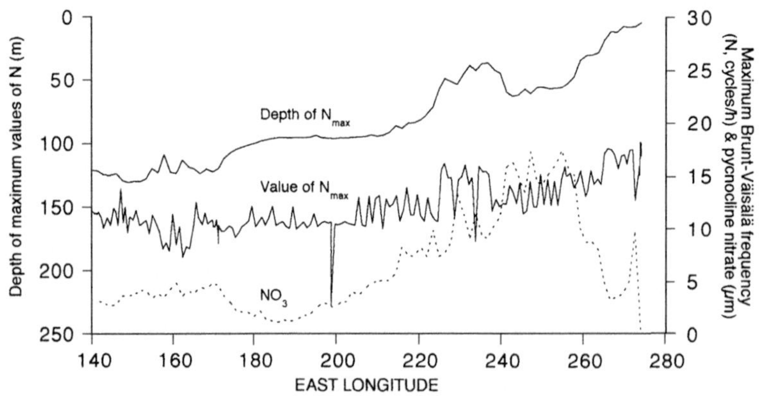


**Fig. 4.2** The zonal *Moana Wave* section along 10°N from Costa Rica to the Philippines showing the deepening and weakening (smaller values of $N_{max}$) of the pycnocline towards the west and the relative lack of discontinuities compared with the meridional section shown in Fig. 4.1.

*Source: Computed and drawn using OceanAtlas 2, Osborne* et al.*, 1992.*

halocline does lie shallower than the thermocline, but only by about 10 m (Sprintall and Tomczak, 1992), so that both contribute to the high stability associated with the density gradient. I shall return to the ecological consequences of the barrier layer later.

The discontinuities discussed in this section are perhaps sufficiently solid to suggest that the warmer parts of the ocean may not be treated as an ecological continuum, but what of the cooler regions where wind stress in winter may deeply mix the water column? I propose to examine just one example of those frontal zones, discussed in the previous chapter, where sufficient discontinuity for our purpose can be demonstrated—in this case, at the boundary of the polar regions. These, defined by the Polar Fronts, are relatively small and comprise only about 6% of the whole ocean. Because it is the critical defining feature of the polar seas, we should first ask if there are definable limits to the imposition of resistance to mixing by the polar brackish surface layer. If no discontinuity can be located, the effect may not be useful for our present purpose and it may be difficult to define a boundary for polar seas. We will first look for evidence of this in some relevant oceanographic sections.

A hydrographic section worked by *Hudson* into the Norwegian-Greenland Sea from the regime of Atlantic water into that of Polar water serves to illustrate the winter situation in the open waters of these polar regions. This section crossed the Iceland Gap Front and the Oceanic Polar Front (Johannessen, 1986) encountering (i) arctic water, north of the polar front, which is a cold (<2°C), brackish (34.6) surface water mass with a weak $N_{max}$ (1 or 2 cycles $hr^{-1}$) at about 75 m; (ii) subarctic water, between the two fronts, which is isothermal (~3°C) and isohaline (~35), with a weak pycnocline at 200 m with $N_{max}$ of 0.5–2.0 cycles $hr^{-1}$; and (iii) Atlantic water south of the Iceland Front that is warmer (7°C), saltier (35.2), and lies above a deep pycnocline having its upper inflection at about 250–275 m and an $N_{max}$ of <2.0 cycles $hr^{-1}$. Here, in winter, the deep mixing typical of the northern part of the North Atlantic subtropical gyre extends into the open, stormy waters of the polar seas. The Western Basins Section (25°W) shows the transition to the surface low-salinity layer in the Denmark Strait.

In the Pacific, the *Washington* (150°W) and *Hakuho Maru* (170°W) sections clearly show the Polar Frontal Zone (32–45°N, depending on longitude) to be the southern boundary of the surface low-salinity water of the subarctic gyre. In the Southern Ocean, the near-surface salinity gradient across the South Subtropical Convergence Zone is especially well seen in the *Knorr* section along the Greenwich meridian and the *Oceanus–Melville* 25°W section; south of this feature, shallow low-salinity layers are frequently encountered in the sections. A discontinuity occurs in the southern hemisphere within the South Subtropical Convergence Zone and in the northern hemisphere at the polar fronts of each ocean. Here, an abrupt increase in the values of $N_{max}$ and a shoaling of the pycnocline occur at the boundaries of the polar biome.

However, specification of the geographical limits of the polar seas is complicated by seasonal changes. Even beyond the polar fronts, in winters when sea ice cover is not established, the surface mixed layer may be overturned by thermal convection and by wind mixing. It is only in spring that a shoal pycnocline becomes reestablished below the polar layer of low-salinity water, and it is only during that season that we may expect to see a distinction between the strong polar and the weak gyral pycnoclines.

It might be protested that a surface brackish layer, constraining vertical mixing, is not a unique characteristic of high-latitude seas. After all, as has been noted previously, the western Pacific warm pool also has a near-surface halocline in a deeper isothermal mixed layer over an area equivalent to 81% of the polar oceans as defined previously. However, this is really a special case because the nutricline occurs as much as 100 m deeper than the halocline. The surface brackish layer is thus ineffective in controlling the exchange of water across the nutricline.

## Rule-Based Models of Ecological Response to External Forcing

Those who originally coined the biosphere concept during the 19th century believed that pattern within it was created principally by reaction between its component animals and plants (Westbrook, 2002). But central to the argument of this book is the observation that the structure and functioning of both the marine and terrestrial parts of the biosphere result principally from the effects of their external physical environment: both terrestrial and marine ecologists have used very similar constructs to represent how the external forcing of ecosystems selects dominance among plant assemblages. Margalef (see later discussion) suggests that phytoplankton assemblages respond principally to turbulence and nutrient concentration, and he is remarkably close to the logic of Holdridge (1947), or Atjay *et al.* (1979), who showed how the occurrence of vegetation types or biomes can be predicted from just a few indices: Holdridge requires only annual precipitation, annual mean temperature, and the potential evapo-transpiration ratio to map tropical plant biomes. To predict the distribution of terrestrial biomes, Atjay *et al.* require information on only latitude, altitude, exposure, rainfall, and geological substrate.

Rule-based modeling of this kind can be carried sufficiently far to reproduce the observed pattern of natural terrestrial vegetation types, except (of course) where these have been replaced by intensive agriculture. By coupling these routines to atmospheric General Circulation Models the dynamic response of terrestrial ecosystems to changes in the global climate may be predicted.

The same could be done at sea if we could identify a comparable set of factors in marine ecology that control the growth of plant cells in the open ocean. To the extent that this approach is successful, and yielded a partition defined by characteristic modes of phytoplankton ecology, it would also suggest which organisms of higher trophic levels are likely to occur in each biome because phytoplankton-herbivore-predator relationships in characteristic pelagic habitats are well known at least in outline, differ significantly between habitats, and should be generally predictable for each partition or biome defined by a unique pattern of environmental forcing of algal growth. Here, *biome* is used in the sense of Odum (1971): "The largest community unit that it is convenient to recognize. In a given biome the life form of the climatic climax vegetation is uniform. Thus, the climax vegetation of the grassland biome is grass." Here, it is assumed that knowledge of the characteristic seasonal phytoplankton cycle for any region is equivalent to a description of its "climax vegetation" and that much can be inferred about characteristic regional ecology from just this knowledge.

Although the principles involved should be common to both fields, such models have been taken much farther in terrestrial than marine ecology, perhaps because of the relative ease of observing and mapping terrestrial plant communities for verification. Production of a sufficiently detailed field of the forcing factors required by the simulation process is surely now possible, using the synoptic global fields of satellite data discussed in the previous chapter. Marine ecologists can surely now hope to match the achievements of their dry-land colleagues.

To do so, a useful starting point might be Margalef's diagram, which has been the basis of many fruitful discussions of the relations between physical processes and pelagic ecosystems. Partly because it was drafted before the existence of autotrophic microbiota was known, and emphasized the reaction of larger phytoplankton to their environment, Cullen *et al.* (2002) revisited Margalef's diagram that suggests that four types of phytoplankton community are determined by differential interaction between nutrients

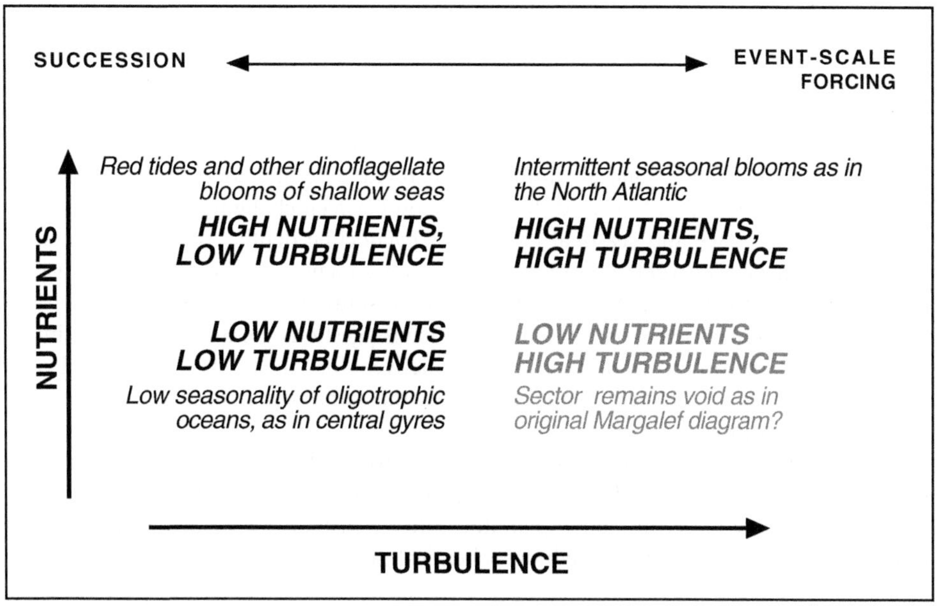


**Fig. 4.3** This figure is a restatement of Margalef's analysis of the varying response of phytoplankton communities to a range of conditions of available nutrients and of ambient turbulence.

*Source: Redrawn from Cullen* et al., *2001.*

with turbulence. A simplified version of Cullen *et al.*'s version is shown in Fig. 4.3, in which the four quadrants derived from Margalef's model are characterized as follows:

1. *Low nutrients and low turbulence*—In the ideal case, this represents the phytoplankton of central gyres dominated by picophytoplankton, including autotrophic bacteria; here, retention of nutrients by recycling in the microbial loop (by protists and specialized microherbivores) maintains a successional community with relatively low seasonality.
2. *Low nutrients and high turbulence*—This quadrant either is void or is represented by the case of the so-called HNLC conditions in the Southern Ocean. The model of Cullen *et al.* assumes this to be the case and assigns characteristics of austral high-latitude phytoplankton to this quadrant: selection for efficient use of light and nutrients, low biomass, and slow turnover. As I shall discuss in detail in the following chapter, I now very much doubt if this assumption is valid.
3. *High nutrients and low turbulence*—The ideal case to represent this quadrant are the "red tides" that occur in permanently stratified coastal waters where nutrient input is both continuous and strong. Some dinoflagellates specially adapted to such conditions (e.g. *Gymnodinium splendens* and *Gonyaulax polyhedra*) release toxins that kill or exclude zooplankton and fish. Without herbivores, such blooms may be quasi-permanent.
4. *High nutrients and high turbulence*—Once again, in the ideal case, this represents situations similar to the North Atlantic spring bloom in which diatoms are important if not dominant and where a seasonal bloom is associated with rapid accumulation of biomass and exhaustion of nutrients. Note, of course, that the bloom occurs only when seasonal turbulence is minimal; it would be stretching the model too far to suggest that intermittent upwelling blooms could also be accommodated in this quadrant.

The problem with Margalef's diagram, and with Cullen *et al.*'s restatement of it, is that it does not tell us very much about the complexity that is characteristic of the real ocean. From it, one may infer only four states of phytoplankton ecology: red tides, spring blooms, oligotrophic central gyres, and (just possibly) austral high latitudes. This, it seems to me, is an inadequate characterization of the ecological states that we can observe at sea.

Furthermore, I believe that a partition based solely on interaction between varying nutrients and turbulence conceals more than it reveals. Perhaps one could construct a "Margalef diagram" that interpreted the interaction between changing seasonal illumination at the sea surface and seasonal changes of pycnocline depth. Another informative diagram might be constructed to represent the consequences of the fact that nutrients known to be limiting to phytoplankton have three sources: atmospheric, riverine, and from subpycnocline water. I am not sure that such suggestions are useful, but one simple Margalef-type relationship cannot perform predictively for the variety of ecological conditions that we now recognize in the oceans, in the same way that such diagrams can apparently perform in the terrestrial realm.

So, if this what we want (and I think it is), then we shall have to proceed in a slightly less reductionist manner. To this end, I have proposed (Longhurst, 1985b) that we might classify the ecology of oceanic phytoplankton into six primary cases, or models, defined below, to be partitioned among the biogeochemical provinces that will be discussed later:

## Case 1—Polar Irradiance-Mediated Production Peak

This represents the ideal production cycle beyond the Polar Front in all oceans. A halocline defines the surface mixed layer within a relatively isothermal water mass, especially at very high boreal latitudes. The pycnocline mixes deepest in winter in the open ocean regions and in the absence of ice cover. The shallow polar halocline may induce stability earlier in the spring than at lower latitude.

The extreme range of irradiance in polar regions forces a unique seasonal cycle of primary production rate, having a single, light-limited maximum at the summer solstice that may be independent of the concentration of nitrate within the photic zone. Once sustained algal growth is initiated, a shallow subsurface chlorophyll maximum develops at the halocline and summer oligotrophic conditions with irradiance of the pycnocline are very briefly established. Chlorophyll accumulates during the period when productivity is increasing and tracks its initial decline. A secondary accumulation of chlorophyll during the late summer period of declining primary production rate is consistent with reduced consumption as herbivores descend out of the photic zone to their overwintering depths.

This sequence is not necessarily observed at all individual locations; blooms occur locally when winter ice cover breaks up, when irradiance within the photic zone is at a local annual maximum, and the date of breakup is determined not only by latitude but also by topography and circulation.

## Case 2—Nutrient-Limited Spring Production Peak

This model represents the cycle of productivity within the regions of maximum westerly wind stress, where winter deepening of the mixed layer may reach depths >300 m. It may be convenient to specify the mixed layer either by a temperature discontinuity (as in the Atlantic) or by a salinity discontinuity (as in the Pacific Subarctic gyre).

The rate of primary production is minimal in winter, and about 1–2 months after the winter solstice the rate begins to increase toward a unimodal maximum about 2–3 months later. This spring bloom is usually thought to conform closely to the Sverdrup critical depth model, but there is increasing evidence that near-surface spring blooms may be initiated in unstratified (winter-mixed) water columns if the mixing rate is below a critical threshold and there is sufficient light. Once stratification is established, evolution occurs toward a summer oligotrophic situation. The summer pycnocline and nitracline lie within the photic zone for 4–6 months and during this period a deep chlorophyll maximum is formed. Production there varies from about 5% to 25% of total production.

Phytoplankton biomass tracks increasing production rate in early spring, but the rate of biomass increase is suppressed when hibernating migrant herbivores (copepods, euphausiids) return to the surface. When the main biomass of these organisms descends again toward the end of summer, a secondary chlorophyll accumulation is initiated that reaches a weak peak a few weeks later. The subsequent progressive recharging of the mixed layer with nitrate at the end of summer may be accompanied by a brief surge of primary production before being dissipated by autumnal mixing and declining light.

## Case 3—Winter–Spring Production with Nutrient Limitation

This model is applicable to the open ocean at midlatitudes where winter winds are relatively light, primary production rates are always low, and seasonality in phytoplankton biomass is weak. Such regions cover almost half the area of the ocean, so there are significant regional differences that cannot be ignored. However, most of these regions share the common attribute that from low values in midsummer, the rate of primary production begins an increase that is sustained through the autumn and culminates in a spring maximum after a second rate increase in late winter.

Characteristically, the winter increase in production rate is accompanied by an increase in chlorophyll values that exhibit a broad peak in midwinter, but fail to track the spring surge in primary production rate. The nutricline is illuminated for a longer period each summer than in Case 2 situations. A Case 3 seasonal cycle seems not to be associated with accumulation of chlorophyll at any season when the rate of primary production is declining, as occurs in Case 1 and 2 situations. This suggests a greater degree of biological coupling between growth and loss rates than occurs in higher latitudes.

Because of latitudinal and between-ocean effects it will useful to consider the Case 3 regions in three groups: (i) the North Atlantic and Pacific Oceans and the Tasman Sea and, diverging somewhat from this type, (ii) the southern gyres of the three ocean basins, and (iii) the Southern Ocean. These groups differ in their characteristic depths of winter mixing, in relative productivity, and in the timing of the winter increase in production rate.

## Case 4—Small-Amplitude Response to Trade Wind Seasonality

This model is appropriate to that part of the ocean at low latitudes (forming 22% of the whole ocean) that lies under the influence of the trade winds, where seasonality is imposed principally by geostrophic adjustment of the zonal equatorial current systems. There is weak seasonality in mixed-layer depth, while primary production rate and phytoplankton biomass take characteristically low values; a deep chlorophyll maximum is almost always present.

The nutricline is perennially shoaler than the photic depth, except during exceptional events and, consequently, the vertically integrated production rate is not light limited, although nutrients or trace elements may be limiting. Minor rate increases are forced by geostrophic response of the pycnocline to seasonality in trade wind stress, and to open-ocean Ekman suction and divergence, particularly at the equator. Rates of phytoplankton chlorophyll accumulation and consumption are closely balanced at all seasons.

The photic depth lies deeper, and in some cases much deeper, than the mixed layer. This situation obtains essentially year-round, and the pattern breaks down with deepening of the mixed layer only briefly and only in some regions. Nowhere is there any evidence of accumulation of chlorophyll unaccompanied by a simultaneous increase in the rate of primary production: growth and loss terms are closely coupled, with the biomass values varying over a relatively small dynamic range.

It is not simple to devise a poleward boundary for the region where Case 3 is appropriate that excludes the effects of some increased wind stress at the surface from winter westerlies. For this reason, in some zones that we might want to think of otherwise as Case 3 models, winter winds may force the mixed layer deeper than the photic zone for several months seasonally. The equatorial regions of the Pacific Ocean have less seasonality in mixed-layer depth, rate of primary production, and chlorophyll accumulation than comparable Atlantic regions because of the differences in east-west dimension of the two oceans.

## Case 5—Large-Amplitude Response to Monsoon-like Reversal of Trade Winds

This model describes those parts of the trade-wind zone of the oceans where a small-amplitude Case 4 response is modified by strong seasonality in local trade-wind forcing. These are the monsoon regions where, in the extreme case of the northwest Indian Ocean, seasonal reversal of the trade winds results in seasonal reversal of the monsoon currents and seasonal alternation between eutrophic and oligotrophic biological systems. The small zonal dimension of the tropical Atlantic permits a strong seasonal response to the surge of southerly trades across the equator that is analogous to the effect of monsoon reversal in the Arabian Sea.

Though the seasonal variance in frictional wind stress at the sea surface in these regions is equivalent to that in the latitudes of winter westerlies, the consequence is quite different. In the northwest Indian Ocean the onset of both monsoons is accompanied by increases in mixed-layer depth only about one-tenth of that achieved in the North Atlantic by the onset of similar seasonal winds. But the mixed layer does deepen sufficiently with the onset of each monsoon that the pycnocline is alternately within, or below, the photic zone. Four seasons may therefore be distinguished.

As in Case 3 waters, productivity is not light limited and responds rapidly to nutrient entrainment into the photic zone by coastal upwelling, by offshore Ekman suction, and by geostrophic adjustment of the pycnocline associated with reversal of the monsoon winds. The pycnocline is illuminated during the oligotrophic season, but the mixed layer shoals above the photic zone during upwelling episodes. Changes in phytoplankton biomass match the seasonal changes in production rate, consistent with consumption and production rates being closely matched.

## Case 6—Intermittent Production at Coastal Divergences

This model is appropriate to the four canonical eastern boundary current upwelling areas, and a few other places where seasonal coastal upwelling occurs in low to moderate latitudes. The mixed layer shoals, and primary production rate takes high values when coastal winds are appropriate for upwelling, usually in summer, so that deep nitrate-rich water is entrained into the photic zone. Chlorophyll accumulation coincides with duration of upwelling and accumulation of chlorophyll is balanced by advection and consumption loss terms. Herbivores, as in high latitudes, enter a resting phase rather deep off upwelling regions in periods of stability. The coastward upwelling of deep water therefore brings not only nutrient-rich water into the photic zone, but also seed populations of large calanoid copepods. In all eastern boundary currents there is a seasonal, latitudinal march of upwelling, poleward in summer, equatorward in winter, with a central region where upwelling is essentially nonseasonal. The mean seasonal changes in, for instance, chlorophyll values for each region may obscure these facts and present a pattern that is not as helpful as it might be.

You may have noticed that I have not identified the so-called HNLC regions as an individual case, as was done by Cullen *et al.* (*op. cit.*). This is because it is the stratification and irradiance cycles that determine which kind of pelagic ecosystem shall develop, not whether the Liebigian limiting element is a trace metal, such as Fe, or one of the major nutrients, such as $NO_3$. As I shall discuss in the next chapter, only very recently has anybody commented on the simple fact that, in all so-called HNLC situations, the nutrient that limits phytoplankton growth is delivered to the phytoplankton upward from deeper layers by Ekman upwelling or by entrainment, but not from the skies. We can therefore include equatorial iron-limited regions in Case 4 and those at high latitudes in Cases 1 or 2. This argument is expanded in Chapter 5.

It is more difficult to extend the foregoing analysis into the coastal seas, so as to fit Margalef's "low turbulence and high nutrients" quadrant, appropriate to uncontrolled blooms in the littoral zone. Next, I shall review some of the physical processes to which phytoplankton must respond in shallow seas over the continental shelves; if such processes did not exist, then the six Cases discussed above would be appropriate over the shelves, right in to shoal water. The role of the physical forcing that actually does occur in shelf regions strongly modifies the pelagic regime that is characteristic of the adjacent open ocean. Unfortunately, we can place much less confidence in the seasonal cycle of chlorophyll biomass in shallow water derived from satellite imagery because of the interference of pigmented dissolved organic material (CDOM) in river water: some components of CDOM absorb light preferentially at wavelengths very similar to those of chlorophyll absorption. A routine filter for these has yet to be devised and applied to the available images.

Even as we argue that ecosystem types are determined primarily by the action of the external environment on plant growth, we must also recognize the moderating effects of consumer organisms. In the jargon of general ecology, we must evaluate both top-down and bottom-up forcing of the ecosystem. For terrestrial ecologists it is evident that dry forest, savanna, and grassland country is maintained in a certain state by the action of large herbivores and the observed vegetation represents a dynamic balance between browsing or grazing and the response of plants. In the pelagic ecosystem, the same dynamic balance exists between production of algal cells and their consumption by appropriate herbivores—diatoms by copepods, nanoplankton by tunicates. But it is neither copepods nor tunicates, but rather their physical-chemical environment, that determines whether nanoplankton or diatoms shall dominate and whether a Case 2 or a Case 4 model shall characterize a region. Similarly, it is neither kelp-eating sea urchins nor sea-grass-eating manatees that determine whether macroalgae or corals shall dominate the littoral fringe.

## Coastal Asymmetry, Geomorphology, and Tidal Forcing

So far, we have been concerned only with the consequences of ocean physics for the development of pelagic ecosystems characteristic of the open ocean. We should also note, very briefly, the fact that in the shallow seas, whether enclosed or over continental shelves fronting the open ocean, ideal physical processes are dominated by the consequences of land form. This brief excursion into the asymmetries of coastal oceans, and their consequences for the forcing of ecological processes, will touch only on two aspects—tides and tidal streams, and the nature and distribution of terrigenous sedimentary material. You have to know something about the special regional characteristics of each of these if you want to understand the ecological processes occurring in any region of the coastal seas.

This is not the place for a treatise on tides and tidal streams, nor am I equipped to write one, but it will be useful to bear in mind the extraordinary complexity of the processes that determine regional tidal characteristics when considering the ecology of coastal seas. Complexity derives from the fact that tides are driven by an equally complex suite of gravitational forces, each of which produces an individual effect on sea level so that the observed tides represent the interaction between several tidal elements. Consider then the fact that the ideal oceanic tides are modified by the location and the shape of the continents and are further molded by the shape of the open coastline, islands, and adjacent enclosed seas as the tides run up over the continental shelf into shallow water. No wonder that the understanding of tidal phenomena dominated research on ocean physics from classical times until the mid-19th century.

It is no more than common knowledge that tides are raised in the ocean by the gravitational effect of moon and sun. Because the gravitational effect of the moon is rather more than twice that of the sun, we are accustomed to thinking of tides as related exclusively to pull of the moon, but that is far from the case. Sun, moon, and Earth dance a complex ballet for which you will find a description in any good text; here, we shall need only to recall that the consequence of this ballet, in which the relative positions of the three spheres, and the relative distances between them, are changing constantly, but in a repetitive manner. These changing orbital relationships produce 4 semidiurnal tides, of which 2—the lunar and solar M2 and S2 tides—command most attention, together with 3 diurnal and 3 long-period tides. The observed tides are raised, then, by a series of harmonic oscillations (or "partial tides") having the periods of the changing orbital relationships.

The interactions between these forces are expressed differently in each region. In the Atlantic, with the minor exception of the Gulf of Mexico, semidiurnal tides prevail: that is, there are two high and two low waters each day. In the NW and W Pacific, tides are predominantly diurnal, with only one high and one low water daily. Some enclosed seas, such as the Baltic and most of the Mediterranean, have extremely small tidal ranges and hence weak tidal streams. Further regional complications arise because the ideal tidal sequence and the ideal relative heights of tides of various periods is under topographic control and differs strongly from place to place on each coast. The orientation of the coastline, and the arrangement of its headlands and bays, will tend to favor one or another of the various tidal components. A single explanation cannot account for the great tides of places like the Bristol Channel and the Bay of Fundy: in the Bay of St. Malo, for example, the tidal range is greater than can be accounted for simply by the narrowing and shoaling of the bay. This is due to the fact that the tide, advancing up the English Channel, takes the form of a Kelvin wave with small magnitudes on its left side, on the English coast, but great magnitude on its right, French side.

But we are not concerned with coastal processes *per se*, and it will be the consequences of coastal morphology for the velocity of tidal streams and the consequential overturning of stratification that will be of greater significance to us in thinking about the pelagic ecosystem over continental shelves. Tidal velocities are, of course, modified by coastal form and the placement of headlands, and the most direct effect will be the acceleration of flow in shoal water or through straits, or wherever else flow is constricted. On great continental shelves that are relatively flat, such as parts of the coasts of the Arctic Ocean, friction between the rough sea bed and the superjacent tidal stream may almost dissipate the flow. The consequences of all this for the location of shallow-water fronts between tidally mixed and stratified water was discussed, as you will recall, in Chapter 3.

The form of continental terraces reflects the movement of the continents themselves: trailing-edge terraces along the eastern coast of the Americas differ strongly in profile from the "collision" or leading-edge terraces along the western coastline (see, for example, Eisma, 1988). Because mountain building occurs preferentially along leading-edge coasts, rivers opening there are short and carry little organic material; in contrast, great rivers tend to flow to trailing-edge coasts and carry larger sediment loads, richer in humus. Note, however, that the present-day sediment loads carried to the sea may be very different from the pristine state: the Amazon carries only about half as much sediment as the smaller Ganges/Brahmaputra system, because of the relative intensity of human development in the two drainage basins.

We shall expect these facts to have major consequences for the ecology of each type of continental terrace. Thus, "upwelling" coasts associated with eastern boundary currents have narrow, steep-to continental terraces, whereas coasts associated with western boundary currents tend to have broader terraces with a less steep slope. On geologically active, collision coasts the continental crust is thinly covered with sediments, whereas on trailing-edge shelves the sediment cover may become very deep. In extreme cases, on prograding shelves along trailing-edge coasts, the crustal topography becomes covered with a thick blanket of sediments within which bottom currents, themselves heavily loaded with sediments, may cut deep canyons. Slumping and folding of sediments may also occur by local tectonic activity. It is also helpful to consider the relative ages, or maturity, of continental shelves, depending on the era in which the large continental blocks separated. Young shelves are characteristic of the eastern Asian region, from New Zealand to the Sea of Okhotsk, with the exception of those around the South China Sea. More mature shelves are characteristic of much of the Atlantic coastline.

The relative strength of tidal streams on the shelf has, of course, a second significance for ecological geography. This has to do with the transport, sorting, and redistribution of terrigenous sediments, which largely determine the location of characteristic faunas of benthic invertebrates and demersal fish. It may therefore be useful to use a little space in a discussion of how shelf deposits were constituted and came to be distributed as we observe them to be.

Recall that the continental shelves were largely dry land during the recent glaciations and that their superficial geology is no more than an extension of that of the continents, partially overlain by terrigenous and pelagic sediments, and calcareous biogenic material of benthic origin. Therefore, present-day bottom types range from rocky ridges, cobbles, and water-worn gravel, through sand and shell sand, to organic muds, in addition to biogenic reefs of calcareous skeletal material. During glacial periods, coarse sediments were preferentially deposited near the actual shelf edge, and this pattern still dominates the pattern of sedimentation on many shelves. Along eastern coasts of both the North American and Asian continents, relic sediments from glacial epochs still dominate much of the shelf, and modern sand and silt lies preferentially shallower, closer to the coast, dominating in bights and gulfs where sedimentation from suspension is induced.

It was originally supposed that there would be a simple particle-size gradient toward the shelf edge and that a "mud line" would be found everywhere at about 50 m, shoaler than which wave action would prevent the deposition of the smallest organic particles. This approximately describes the situation off Western Europe, where the earliest investigations were made, but it is not a useful generalization.

Sedimentation is not, as it might appear to be from the resultant arrangement of sediments, a continuous process, but is rather episodic and the arrangement of sediments is evolutionary. Episodic settlement of fine-grained material occurs when tidal streams (or other sources of turbulence) weaken. The deposit at first tends to form patches, reflecting the topography of the bottom terrain, filling hollows and bowls, until a plane surface is formed. Within a few hours, the new sediments become dewatered and their surface becomes rippled by subsequent water movement, the dimension of the ripples depending on current speed and sediment particle size. In mature sediments, under rapid water movement, ripples may become parallel sand waves or ridges several meters high, arranged at right angles to mean flow. Resuspension occurs continuously, and sorting of particles by their density follows, sand being deposited as ripples, and mud mainly in the troughs between. Bedding of sediments of different density occurs also at tidal and event scales. Bioturbation by burrowing benthic organisms powerfully modifies the ideal, physically driven arrangement.

Terrigenous material delivered by rivers to the shelf comprises mineral sand particles and organic material derived from the decay of terrestrial and estuarine vegetation. The latter tends to remain incompletely oxidized in estuarine regions and close inshore, whence the black, organic-rich, reducing and acidic muds that may dominate here where tidal streams weaken. Pelagic organic material sinking to the shelf deposits comprising fecal pellets, phytoplankton cells, and zooplankton carcasses is more completely oxidized and forms blue or gray muds, often with high calcium content from the skeletons of foraminifera. Sedimentary material of benthic origin comprises fragments of the shells of mollusks, and of the exoskeletons of echinoderms and crustacea. These various material are sorted and transported by coastal currents and tidal streams according to their relative density and particle size to form the mosaic of bottom types we observe on continental shelves. Where, as off California, the topography of the shelf is characterized by the existence of deep basins and offshore islands and the input of terrigenous material is negligible, the grain size and organic content of sediments follows the general pattern of depth. Close inshore and in the shoal water around islands, grain size is large and organic content low. Progressively into the deep basins and troughs, grain size decreases and organic content increases.

Where, as in the many tropical regions, wave action at the coast is relatively slight and input from rivers is significant, terrigenous mud deposits may occur close inshore, while offshore muds are principally of pelagic origin. The distribution of sediments and water types on the shelf determines, quite precisely, the composition of the benthic invertebrate and demersal fish faunas on the continental shelves.

Thus, ecological processes on continental shelves are fragmented spatially, and predictable only from a specific knowledge of the forces that drive this fragmentation. It is this problem that makes it so much more difficult to generalize satisfactorily about the geography of ecological processes in coastal seas than in the open ocean. Research into benthic ecology has, in recent decades, turned very largely toward an understanding of fluxes of energy within the invertebrate and microbial community, and between this and the pelagic ecosystem. Consequently, the exploratory surveys characteristic of benthic ecology of the first half of the 20th century are no longer done, and we are left with adequate descriptions of the distribution of benthic species associations covering only a very small fraction of the continental shelves of the world. From these, we must extrapolate to the unknown regions.

Chapter 5

# Nutrient Limitation: The Example of Iron

Physical forcing is not a sufficient explanation of the seasonal and spatial pattern of production in the ocean and, as noted in the previous chapter, a necessary auxiliary assumption is that the surface mixed layer should be recharged with inorganic nutrients. It will be useful, then, at this point to discuss the role of both macro- and micro-nutrients in enabling and terminating phytoplankton growth. Because iron limitation has been so widely discussed in recent years, this chapter will be focused primarily on that element.

Although nitrate and other molecules have long been known also to reach the sea surface in rainwater and dry aerosols, this flux has seemed unlikely to modify significantly the assumptions of either Margalef or Sverdrup, both of whom assumed that flux across a nutricline was dominant. Further, most biological oceanographers thought only about the macronutrients nitrogen, phosphorus, and silica and ignored the early experiments of Menzel and Ryther (1961), who clearly demonstrated that iron was the initial limiting element in the Sargasso Sea off Bermuda and that iron limitation was due to differential supply rates from below of iron and the macronutrients: winter mixing was found to control the seasonal production cycle here. These experiments were not done with the ultra-clean rigor of the subsequent studies performed by the Moss Bay team, but that is no reason for their subsequent neglect.

Martin's contrary suggestion, made in 1991, was extraordinarily influential: he proposed that in "open ocean upwelling regions far from Fe-rich continental margins, the only way phytoplankton can obtain Fe is via long-range, wind-blown transport of Fe-rich atmospheric dust derived from arid regions." Iron-fertilization experiments at sea in the tropical Pacific (IronEx I and II), the Southern Ocean (SOIREE), and in the NE Pacific (SEEDS and SERIES) soon appeared to confirm his insight and were widely interpreted as demonstrating that areas with residual nitrate are causally related to low rates of deposition of terrestrial dust (Martin and Fitzwater, 1988). It came to be widely accepted that such High Nutrient–Low Chlorophyll, or HNLC, regimes are characteristic of the subarctic Pacific, much of the eastern tropical Pacific, and parts of the Southern Ocean. But, as I hope to demonstrate, little evidence was presented to support the concept of growth limitation in these regions by insufficiency of subaerial input.

Martin's intervention was so uncritically accepted as to become mythic in the sense of William Dickinson, quoted on the fly-leaf of this book. In the years that followed, many authors turned to the assumption that in at least some parts of the ocean, the regional pattern of productivity and chlorophyll accumulation was forced by the regional pattern of subaerial deposition of Fe in terrestrial dust. Menzel and Ryther were soon forgotten, and the Bermuda area was omitted from the paradigmatic "HNLC" regions.

The clearest statement of belief in Martin's suggestion that I have yet seen was contained in a Web-published report of an open-ocean iron-fertilization experiment. It read:

> ... as water is the primary factor determining how much life can be sustained in a particular terrestrial environment ... so the supply of iron is the primary determinant of how much life can be sustained in the ocean. Dust whipped off the continents by winds supplies iron to the open ocean, and the productive regions receive more "iron rain" than the unproductive regions.

From there, it is but a short step to the astonishing evocation of a relationship between seasonality of chlorophyll biomass and of modeled dust deposition in the Southern Ocean (Erickson *et al.*, 2003).

Such statements are, I believe, mythical, although similar assumptions had already led Banse and English (1993) to go so far as to suggest that the HNLC areas should be regarded as one of only three characteristic natural domains of an ocean partitioned according to the forcing and seasonality of phytoplankton growth. To confront the tension between myth and reality will require special consideration of these areas in what follows.

In the years when this concept was being developed, the global chlorophyll fields obtained by the CZCS sensors were not so familiar to most biological oceanographers as such products are today. Now, we can appreciate that the HNLC areas are not distinguishable in global images of sea-surface chlorophyll as anomalies inexplicable by ocean physics and by reference to Sverdrup. We can readily see that the distribution of surface chlorophyll (and the pattern of primary production derived from it) in no way matches the subaerial supply of Fe, but rather reflects those physical processes that induce (i) vertical motion of deeper water into the photic zone, and (ii) stability in the upper part of the water column. This relationship, foreseen by Yentsch and Garside (1986), is valid down to the mesoscale, and it is reasonable to believe that the observed correspondence between phytoplankton growth and physical processes, at all scales, represents cause and effect.

These images also show us that the term HNLC itself is unhelpful because the "LC" part of it is downright misleading: chlorophyll accumulates in these regions in the same pattern as in other places having comparable physical conditions, but where macronutrients are taken down to very low levels. It is certainly time to abandon the term: where we observe persistent unused nitrate in a sufficiently lit, sufficiently stratified euphotic zone we may assume that the vertical flux of this element exceeds the requirements of phytoplankton, whose population growth is limited by some other element. For this relationship, the dimensionless $S$ proposed by Platt *et al.* (2003), where $S =$ (nitrate supply rate)/(nitrate equivalent of new production), is appropriate. In so-called HNLC regions, $S$ must take a value higher than unity, and I shall subsequently refer to these simply as the high-$S$ regions.

In the first edition of this book I suggested that these areas were, by virtue of their nutrient regimes, "exceptional regions" that required to be, to some degree, treated independently of the four major biomes of the seas discussed in the next chapter. This was not a good idea, and now I much prefer to emphasize how the pattern of productivity—both spatial and temporal—within each high-$S$ area is forced primarily by regional physics; these areas, in fact, function very similarly to comparable areas elsewhere in the biomes within which they are located.

## Nutrient Distribution and the Consequences of Differing Supply Ratios

So, why should nitrate characteristically remain in excess in surface waters in some regions even after the season of phytoplankton growth? Consider the relative distribution of just two of the many essential plant nutrient molecules in the ocean: dissolved iron ($Fe_{td}$ that passes a 0.4-μm filter), and nitrogen as $NO_3$. Each potentially limits phytoplankton growth although at very different concentrations. A stoichiometric ratio for N:Fe cannot be stated precisely, but it is thought to range from 0.06 to $2.0 \times 10^4$ for cells at optimal and minimal growth rates respectively (Martin and Gordon, 1988). Where required, I shall use the intermediate value of $1.5 \times 10^4$ proposed by Geider and LaRoche (1994).

Although nitrogen is abundant and readily accessible to plant cells in our present era, the supply of iron is much more problematical, being an element whose abundant reservoirs on Earth have been profoundly modified by biota. Oxygen, evolved by the earliest photoautotrophic cyanobacteria, could accumulate in the atmosphere only after the complete oxidation of free iron and its deposition as $Fe_2O_3$ in the Banded Iron Formations 2.5–3.0 billion years ago. The cellular chemistry of present-day microbiota (and the cells of higher plants) retains the mark of their origin in an iron-rich environment; Fe is required in the elaboration of cytochromes and of the redox proteins involved in photosynthesis, respiration, and nitrate reduction, for example.

A substantial proportion of dissolved iron now occurs in forms that are largely unavailable to phytoplankton, principally together with strong Fe-complexing ligands (Wells *et al.*, 1994). Nevertheless, mechanisms exist to ensure that sufficient Fe can be obtained by plant cells in today's iron-poor environment, although cellular Fe:C ratios may respond both to floristics and to dissolved iron concentration, unlike the relatively constant C:N:P ratio of Redfield (Fung *et al.*, 2000). Iron-limitation for oceanic phytoplankton typically occurs around 0.1 nM $kg^{-1}$. Picoplanktonic *Synechococcus* are not strongly Fe-limited, perhaps only because of their extremely small dimension, and hence large surface area/volume ratio, even in equatorial high-N regimes. These cells also produce Fe-binding siderophores that enable Fe acquisition even where dissolved iron takes very low values indeed (Wells *et al.*, 1994; Tortell *et al.*, 1999). Some photosynthetic cells (e.g., *Ochromonas*) are able to obtain Fe through ingestion of bacteria (Maranger *et al.*, 1998).

We should also perhaps recall here that Fe is recycled, like nitrogen, within the microbial ecosystem of the mixed layer in the same manner, and with about the same efficiency, as nitrogen (Bruland *et al.*, 1991; Maranger *et al.*, 1998). Indeed, for populations of phytoplankton cells to be sustained in the mixed layer of an oligotrophic ocean, both Fe and N (as well as other elements) must be recycled *in situ*. Indeed, some pre-Martin observations of Fe excretion by mesozooplankton off India were used to support Cooper's 1935 suggestion that regeneration of useful forms of oligonutrients in this way might be critical for phytoplankton (Ramadhas and Venugopalam, 1977). Deeper than the mixed layer, as shall be discussed later, Fe is regenerated by a mechanism that differs markedly from that appropriate to the major nutrient molecules.

Iron presently enters the ocean from two sources: by wet or dry deposition of terrestrial dust everywhere at the sea surface, and by transport from shelf deposits or river effluents into the adjacent water mass. The regional deposition of terrestrial dust at the sea surface reflects aerosol trajectories from sources in arid lands, mostly from major topographic depressions with dry alluvial soils. The mesoscale pattern within air masses, and hence of dust deposition at the sea surface, is clearly different from the familiar mesoscale, eddying circulation pattern in the ocean (see Color plate 2). The global pattern of dust deposition at the surface of the ocean is known only rather generally from observations, so that many studies of the process have been based on the output from models; these are perhaps regarded with too much confidence, for the uncertainties are very great.

Nevertheless, simulation of sources and aerosol transport of Fe can now successfully reproduce the seasonal pattern of accumulated haze observations at sea discussed by Prospero (1981). The GOCART model represents the state of this art, but significantly overestimates deposition in regions far from sources. The chemical composition of aerosol deposition requires attention, because the N:P ratio of airborne dust particles is rather high relative to Redfield. Thus, new production induced by aerosol nutrient input will tend to drive a previously Fe-deficient phytoplankton toward P limitation (Baker *et al.*, 2003).

It was thought until recently that 40–50% of subaerial Fe went rapidly into solution in seawater provided this contained relatively little already in solution (Zhuang and Kester, 1990), but a much lower solubility is now assumed for geochemical models: 1–10% by Fung *et al.* (2000) and 2% by Jickels and Spokes (in press). Many aerosol particles become incorporated in rainwater or in snow, and this is an important mode of delivery of trace metals and other essential molecules to the surface water mass (Jickels, 1995). A significant fraction of this material is directly available to biota after deposition. In squalls over the equatorial Pacific, rainwater may contain as much as 3 nM Fe $kg^{-1}$ that may then be retained in a lens of lighter water at the surface (Hansen *et al.*, 2001). Aerosol fallout during the intermittent passage of Gobi dust across the North Pacific may induce a doubling of mixed-layer phytoplankton biomass in the subsequent 2 weeks (Bishop *et al.*, 2002). In all these reports, Fe fertilization was assumed, but not observed, to have occurred. However, since rainwater also delivers nitrogen and other macronutrients in forms usable by plant cells, this fashionable assumption is in fact unproven; we can anticipate that mesoscale blooms will occur after rainfall in a wide range of circumstances. In fact, it has been suggested that as much as 10–15% of global primary production above the thermocline may be induced by atmospheric deposition of useful nitrogen (Prospero, 2002).

The available data on the distribution of $Fe_{td}$ within the ocean are sparse, but two sources are sufficient to characterize global distribution in a preliminary manner. These are (i) the 30 stations of the Moss Landing data set (Johnson *et al.*, 1997) representing North and Central Pacific, North Atlantic, Southern Ocean, and Arabian Sea, and (ii) the Atlantic Meridional Sections (AMT) discussed by Bowie *et al.* (2002). These show that the high concentrations observed near continental masses decay rapidly seaward, and also that low-latitude Atlantic profiles are exceptional.

Over deep water, Fe profiles resembles those of other biologically active metals such as Ni, Zn, Ge, Cd, and Y, being nutrient-like with a discontinuity at the pycnocline; this is clear in the Moss Landing data set and also (for the North Pacific alone) in the vertical profiles of the periodic table elements offered by Nozaki (2001). The discontinuity in the generalized Fe profile, resulting from depletion in the photic zone, is designated the ferricline. The 354 profiles in the Moss Landing archive are dominated by northeastern Pacific data (2°S to 60°N) but are consistent with profiles from the North Atlantic, the Southern Ocean, and the Arabian Sea, also shown by Johnson *et al.* Western equatorial Pacific profiles of Mackey *et al.* (2002b) are similar, with very low mixed-layer values above a strong ferricline at 150 m, near the pycnocline. The AMT sections show that under the African dust veil (25°N–10°S) a secondary, near-surface maximum of $Fe_{td}$ is imposed on profiles otherwise similar to the Pacific at similar latitudes (Bowie *et al.*, 2002). In the SE Atlantic, close to the American continent at >40°S, a surface layer is again observed. Chester (2000) offers profiles with a secondary surface layer of higher concentration from the central North Pacific gyre, presumably under the Asian dust veil that, as he remarks, are "typical of a scavenged-type trace metal having a significant external source." The 6°W transect of the Antarctic Circumpolar Current (Löscher *et al.*, 1997b) showed that here, too, $Fe_{td}$ increases with depth and it is only in the region of the Polar Front attributable to input from shelf water.

Although the relatively uniform $Fe_{td}$ field below the ferricline may be perturbed where a water mass passes across shallows, or is adjacent to a continental mass, this does not negate the generalizations about the deep Fe field of Johnson *et al.* Such an anomaly originates in the Bismarck Sea, where a coastal undercurrent accumulates high concentrations of Fe (2–5 nmol $kg^{-1}$ at 300 m) from iron-rich shelf sediments originating in the Sepik

River (Mackey *et al.*, 2002b). Although Wells, Vallis, and Silver (1999) suggested that this enrichment might be sustained eastward in the EUC to induce plant growth when upwelled in the eastern Pacific, the equatorial FeLine profiles at 140°W showed that this was not the case: here, Fe profiles across the EUC are not anomalous, but nutrient-like (Coale *et al.*, 1996a).

In the Scotia Sea, the anomalously high iron content of near-surface Polar Front water has been attributed to its passage along shoal topography in the Drake Passage (see later discussion), and close to the Antarctic continent, Fe profiles may have surface maxima caused by the presence of iron in melt water from icebergs and at the ice front. However, recent observations made both across the Southern Ocean (Measures and Vink, 2001) and in the eastern Pacific (Coale *et al.*, 1996a) require no revision of the generalizations discussed earlier.

Below the oceanic ferricline, the concentration of $Fe_{td}$ is maintained by the balance between solution from sinking particles and scavenging by readsorption onto other organic particles so that values cluster narrowly around 0.8 (0.4–1.4) nmol $Fe_{td}$ $kg^{-1}$ at >500 m in each ocean. For this reason alone, we may expect that the remineralization schedule of Fe from sinking organic particles will differ markedly from that of $NO_3$; despite this difference, models of Fe dynamics continue to ignore this distinction (e.g., Fung *et al.*, 2000). The JGOFS-NABE profiles at 47° and 56°N in the North Atlantic closely resemble North Pacific VERTEX profiles, but those from the AMT in mid-latitudes differ, because they lie either below the Saharan dust field or within the influence of a continental borderland. In the former case, $Fe_{td}$ profiles have a shallow near-surface maximum (1–5 nM), a subsurface minimum at 75–100 m, and then a deep profile resembling the Moss Landing profiles from other oceans. Where the AMT stations are adjacent to the continental slope, the $Fe_{td}$ maximum occurs at 80–100 m.

Johnson *et al.* remarked that the residence time of dissolved Fe in the ocean (100–200 y) is much shorter than the thermohaline circulation time (1000 y) or than the residence time of other biologically reactive elements, such as N or P ($<10^{-5}$ y). They conclude, therefore, that the processes controlling $Fe_{td}$ concentration are unique: nutrient-like profiles and the relative constancy of deep-water concentrations are maintained because the rate of removal of Fe from solution by organic ligands decreases substantially below concentrations of about 0.6 nM $kg^{-1}$. Further, remineralization of Fe from sinking particles must be essentially continuous throughout the water column. This mechanism differentiates the dynamics of Fe in the deep ocean from those of the $NO_3/NH_4$ system and explains the relative uniformity of the distribution of iron in the ocean interior.

The distribution of nitrate in the oceans is much better described than that of Fe and reflects the formation of the nitrate molecule within the deep ocean. The relatively very long residence time of nitrogen means that interocean transport in the deep circulation intervenes in the global distribution of the molecule; nitrate concentrations are twice as high in the interior of the Pacific as they are in the Atlantic. Thus, the relatively young Atlantic deep water at 1–3 km contains less nitrate than does older Pacific water at the same depth, although concentrations are remarkably uniform within each ocean. However, shoaler than 500 m, the distribution of nitrate becomes progressively less uniform and around the periphery of each subtropical gyre, nitrate concentrations are an order of magnitude greater than in mid-gyre at the same depth (Fig. 5.1). These "textbook" high-nitrate conditions exist as an equatorial feature in both Atlantic and Pacific Oceans, as a zonal band across the high-latitude regions and as a meridional feature along the eastern and western margins. The North Atlantic, open poleward so that the subarctic gyre is not enclosed as it is in the Pacific, has a relatively very weak high-latitude subsurface zone of high nitrate.

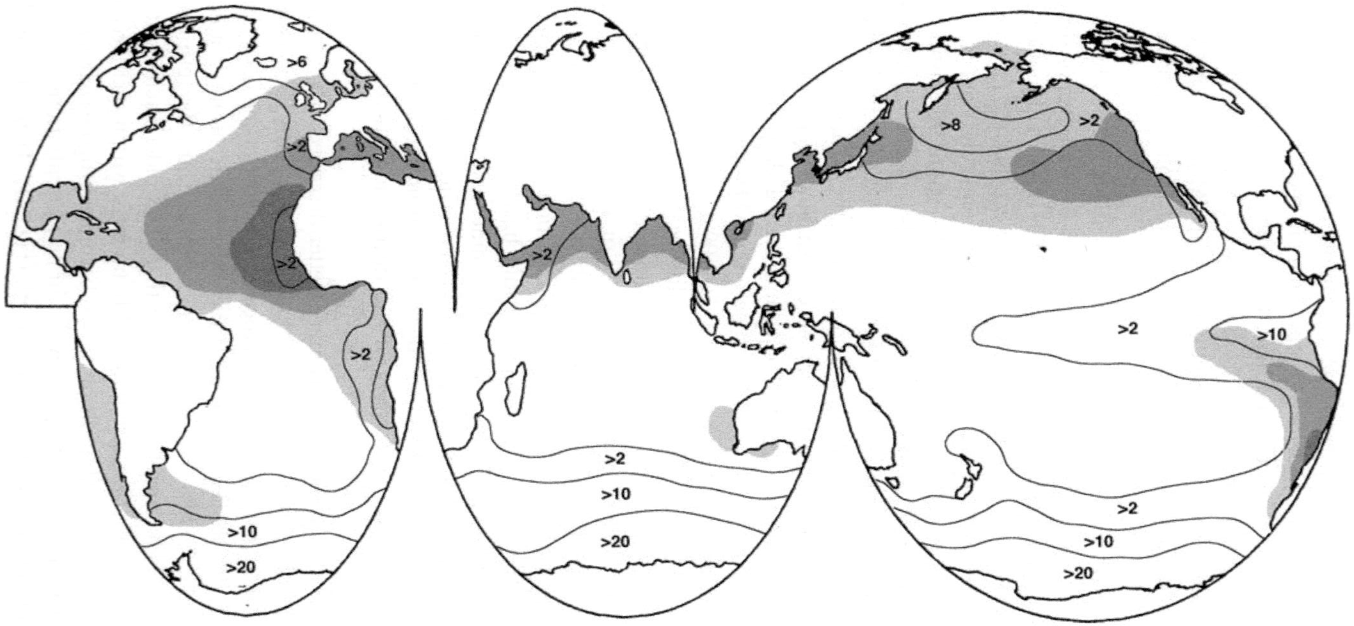


**Fig. 5.1** The lack of coherence between patterns of mean annual surface nitrate, shown as isolines from 2 to 10 μM, and of the deposition of dust at the sea surface rendered as shading at three levels of intensity. It will be observed that nitrate is a superior indicator of regional phytoplankton activity compared to dust-fall.

*Source: Redrawn from Ginoux* et al. *(2001).*

These are, of course, the "geostrophic nitrate ridges" of Yentsch (1990), and analogous features exist in the silicate and phosphate fields. Yentsch appears to have been the first to have noticed the consequences of the fact that (in his words) the surfaces "of equal nitrate are bowed sharply upwards between latitudes 10–20°N and 40–50°N" in the North Atlantic. He went on to remark that these baroclinic features reflect the geostrophic, westward flow of the equatorial currents at lower latitudes and of the west wind drift at higher latitudes. For a more comprehensive description of this phenomenon, consult the WOA 1991, or Gruber and Sarmiento (2002).

It will not have escaped your notice that each of the three "classical" high-*S* regions lies above one of Yentsch's nitrate ridges: in the subarctic Pacific, in the eastern tropical Pacific, and around the Southern Ocean. What you may not have noticed is that above the nitrate ridge of the northern Atlantic, which is somewhat atypical because of the open northern border of this basin, there is another (but largely ignored) high-*S* region. Here, nitrate is rarely drawn down to limiting concentrations: thus, at OWS I at 60°N 20°W, throughout the summer, at least 2.0 $\mu$M kg$^{-1}$ nitrate remains within the upper mixed layer and is reduced to nearly undetectable levels only within the intermittent and brief near-surface stratifications that may form at any time from April until August (Fig. 5.2). The explanation for the Bermuda region not being included in the paradigmatic high-*S* regions is simple: it does not lie above a nitrate ridge, so excess nitrate in surface waters (the "HN" condition) is not induced by winter mixing.

Where the data permit the construction of long meridional sections of dissolved iron, geostrophic ridges seem not to occur, and would not be expected to occur. Consequently, there are significant regional differences in the vertical flux ratio of nitrate and dissolved iron across the oceanic nutricline that are forced by geostrophic effects on the nitrate field. The consequence of this variability in the Fe:$NO_3$ ratio in the water masses apt to supply the euphotic zone with macronutrients may be computed quite simply, provided we don't take the results too seriously. By using the stoichiometric ratio for Fe:N of $15 \times 10^3$ (see earlier discussion), and typical values for subeuphotic zone nitrate, some indicative calculations may be made. The mean concentration of $Fe_{td}$ at such depths being 0.76 nM, we can compute quite simply that this quantity would support the uptake by

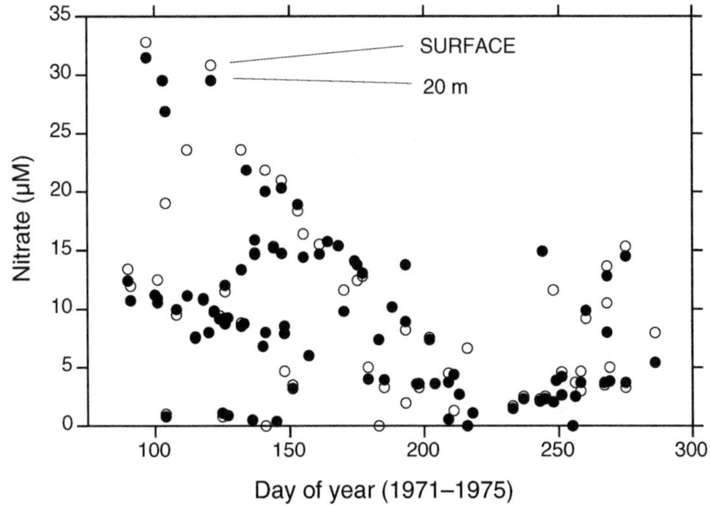


**Fig. 5.2** The seasonal cycle of available dissolved nitrate at the surface and at 20 m at OWS "India" at 59°N, just to the south of Iceland, 1971–1974. Observe how the sequence differs between years: the string of high values that occurred between year-days 100–150 represents 1971 data alone.

*Source: Data courtesy of Bob Williams.*

new phytoplankton production of about 11.4 μM $kg^{-1}$ of nitrate. Further, it seems likely that the observed range of $Fe_{td}$ concentrations (0.34–1.3 nM $kg^{-1}$) would allow uptake of nitrate within the range of 6.0–19.5 μM $kg^{-1}$.

Allowing a modest input (modest, that is, over short time scale) from dry deposition, it is reasonable to infer from this computation that the variable quantity of Fe relative to N transported in vertical flux is likely to result in both iron-limited (high–*S*) and nitrate-limited (low-*S*) regions in the surface waters of the oceans. This is, of course, just what we observe and I venture to suggest that the variable $NO_3$/Fe ratio just below the euphotic zone must be regarded as the simplest possible explanation of the observations. Even where a different macronutrient limits phytoplankton growth, the same argument may be made; the preferential uptake of silicate by diatoms in the Subantarctic Zone of the Southern Ocean causes the water mass that is formed there in winter to be highly silica-deficient.

In fact, the relative flux of $Fe_{td}$ and other nutrients to the euphotic zone, across either the nutricline or the sea surface, is rarely in the ratio required by plant cells, with consequences that are both physiological and ecological. The study of the ecology of high-*S* regions has greatly advanced understanding of the consequences for phytoplankton of a nutrient supply in which Fe is relatively lacking. I shall argue that, while relative Fe shortage in high-*S* regions does induce some characteristic features in their phytoplankton ecology, it does not modify the regional distribution of chlorophyll accumulation: this is forced principally by physics and by herbivory, both in high-*S* regions and elsewhere.

## Regional Anomalies in Nutrient Limitation

Let us now examine the consequences of the distribution of iron and nitrate in some characteristic regions. Of course, each of these regions will be discussed in more detail in Chapters 9–12, but it will be useful here to deal specifically with the consequences of differing subaerial flux of Fe in, especially, the high-*S* regions. The most accessible source for defining the high-*S* regions is any convenient nutrient atlas; for the mean distribution the 2.0 μM $kg^{-1}$ contour for $NO_3$ at (say) 10 m may be taken as defining such areas.

The equatorial Pacific is a good entry into the characterization of high-*S* conditions. However, although this region is often characterized as comprising only those areas directly influenced by the equatorial upwelling itself, the area where $NO_3$ is >2.0 μM $kg^{-1}$ at the surface is actually larger than that extending from 15°N to 15°S in the east and tapering westward to mid-ocean, so that significant parts of the subtropical gyres exhibit some aspects of Fe limitation (Behrenfeld and Kolber, 1999). Most biological oceanographers would now agree that the familiar pattern of productivity and chlorophyll accumulation in the eastern tropical Pacific very largely represents the response of phytoplankton to physical processes: however, in the early years of investigations of high-*S* regions this was not the case, and it was supposed that the gyre field downstream from the Galapagos was the one region in the HNLC waters of the eastern Pacific where phytoplankton do bloom. This opinion was based on the analysis of Barber and Chavez (1991), who suggested that rarefaction of the aerosol flux of Fe from Asiatic sources eastward across the tropical Pacific at low latitudes had consequences for regional productivity. Barber and Chavez suggested that productivity is a simple function of 60 m nitrate beyond 115°W, but is independent of nitrate in the eastern equatorial Pacific, and thus limited to ~36 mg C(mg $chl^{-1}$)$d^{-1}$. Downstream of the Galapagos they postulated that Fe input from slope sediments might maintain higher productivity. The only pattern of haze and aerosol deposition over Pacific low latitudes then available (Duce and Tindale, 1991) for this region rather poorly matches the pattern obtained from modern satellite AVHRR data (e.g., Husar *et al.*, 1997), or delivered by the GOCART model (see Fig. 5.2) of Ginoux *et al.* (2001). There is now ample satellite evidence of phytoplankton blooms in this part of the ocean other than those we might attribute to the island wake of the Galapagos.

Fiedler, Philbrick, and Chavez (1991) showed that both production rate and productivity in the eastern Pacific respond to the seasonal pattern of divergence (equatorial and countercurrent) and coastal upwelling, forced by the consequent relative depths of mixing and of light penetration. Now that we can routinely access 7- and 30-day images of this region, both as surface chlorophyll and translated into productivity, it is very clear that the observed pattern is a direct response to physical forcing. We are now quite familiar with the consequences of the range of circulation processes that occur in this region, responding to both seasonal and Niño/Niña cycles; the consequences of reduced upwelling at the equator in the Niño situation are plain for all to see. Of course, we anticipate that strong vertical motion will be induced in the Galapagos Islands wake; this is now seen to be just one feature in the larger pattern of chlorophyll accumulation characteristic of the eastern tropical Pacific.

A nutrient budget of the **eastern Pacific high-*S* region**, at 140°W and from 3°S to 6°N, suggests that flux of iron and nitrate to the equatorial photic zone occurs in sufficiently differing ratios that most of the nitrate remains unused. Coale *et al.* (1996a) assume a molar C:Fe ratio of 167,000, a sustained upwelling velocity of 1.23 m $d^{-1}$, and a subaerial Fe supply rate of 5–25 nM $m^{-2}$ $d^{-1}$. Since the observed new production rate here is 5–30 μM C $m^{-2}$ $d^{-1}$, iron must limit growth to <25% of the potential offered by nitrate flux at this location, of which >80% remains unused. However, at 9°N, where upward $NO_3$ flux is much lower, nitrogen was found to be the limiting element and is assimilated down to very low concentrations indeed, as it is at the ADIOS station at about the same longitude, but centrally in the gyral circulation at 26°N.

I have seen no discussion of the fact that, although a different nutrient is limiting in each, both tropical Pacific and Atlantic have rather similar geographic patterns of chlorophyll accumulation. In both oceans, just below the nutricline, nitrate and phosphate take similar values, respectively 32–36 μM $kg^{-1}$ $NO_3$ and 1.5–2.5 μM $kg^{-1}$ $PO_4$. However, residual nitrate values in the euphotic zone of the two oceans are very different (Atlantic, about 0.5 μM $kg^{-1}$; Pacific, 2–10 μM $kg^{-1}$ $NO_3$). Differences in euphotic zone phosphate between the two oceans are also significant, but weaker than for nitrate. The simplest, and perhaps correct, explanation for this difference in nutrient availability lies in the greater subaerial Fe deposition in Atlantic low latitudes where euphotic zone nitrate is reduced to near-limiting values by phytoplankton uptake. Thus, I suggest that in each ocean it is the supply of an element by vertical flux across a nutricline that limits phytoplankton growth. This is a very different matter from the paradigmatic and unthinking "It is established that the high-nitrate, low-chlorophyll character of these waters is attributable to the limited availability of iron in the photic zone" that we read too often.

Despite all this, it would be simplistic to suggest that the greater accumulation of phytoplankton observed in satellite chlorophyll fields in Atlantic low latitudes is Fe-driven. I have seen this suggestion made, but it ignores many other factors, of which the principal are the different orientation of coastal margins in the two oceans and the fact that two great rivers—Amazon and Congo—open into the Atlantic at near-equatorial latitudes. These issues will be discussed in Chapter 9, so here it is only necessary to point out that each represents a major input at very shallow depth of a mixed bag not only of nutrients, but also of CDOM (see Chapter 1). Hu *et al.*'s (2004) analysis of SeaWiFS images revealed a dominance of CDOM over chlorophyll in the low-salinity Amazon and Orinoco plumes up to 2000–3000 km from the river mouths, both coastwise and across the ocean in the NECC. We may assume that the same phenomenon occurs in the Congo plume because this is the root of the chlorophyll enhancement that lies seasonally across the equatorial currents. Of course, no rivers of the magnitude of these three enter the eastern tropical Pacific. We must also remember the Niña-like thermocline uplift that occurs seasonally in the Atlantic (Longhurst, 1995), but only interannually in the Pacific. For all these reasons, I believe it would be very rash to attribute the apparent difference in chlorophyll accumulation in the tropical zones of the two oceans simply to their differing subaerial Fe supply.

The high-*S* region of the **subarctic Pacific** is enigmatic and has engendered much discussion. Why here, under dusty skies, where dry deposition rates are somewhat higher than in the North Atlantic, should significant residual nitrate ($\leq$18 μM kg$^{-1}$) remain in the euphotic zone at the end of summer? Why should there normally be no spring bloom in the open ocean? This high-*S* region both is larger and has an unusually high residual concentration of nitrate just below the photic zone; in the subarctic Atlantic, nitrate concentration at 150 m is only ~14 μM, but in the subarctic Pacific is ~30 μM. This is the highest value on this depth horizon than anywhere else in the ocean except at the Antarctic Divergence.

The lack of a spring bloom here was originally attributed to the presence of large calanoids (the pre-Martin "Major Grazer Hypothesis"; see Chapter 11) but then later to Fe limitation of the growth of larger cells. The synthesis drawn from the results of the SUPER investigations (Miller *et al.*, 1991) remains valid today: that Fe limitation prevents the development of a diatom bloom but does not restrain the high growth rate of small phytoplankton cells that dominate the autotrophic biomass. These preferentially utilize $NH_4$, which is rapidly recycled in the mixed layer, while their cell numbers (and those of bacteria) are held down by protozoa that have doubling times shorter than their food organisms, and by small copepods. This "mixing and micrograzer" (or SUPER) hypothesis requires that winter mixing be insufficiently deep as to sweep clear the micro- and picophytoplankton and their protozoan herbivores. This synthesis forces our attention, as it should, on the permanent stratification imposed by the low salinity of the surface layer of the Subarctic Pacific. Restraint of winter mixing enables the survival through winter of a community of small-celled phytoplankters and their protist consumers that is not mixed down and eliminated—as it would be in the North Atlantic—thus enabling a rapid resumption of active growth whenever sufficient irradiance is available.

Episodic diatom blooms that occur preferentially in ENSO years draw mixed-layer silicate down to limiting levels while nitrate remains in excess (Wong and Matear, 1999). These blooms have been attributed to episodes (and between-year differences) of Fe deposition at the surface and "not through vertical mixing from a concentrated supply at depth" (e.g., Miller, 1993). It was suggested that a dust-deposition event in the North Pacific induced an observed doubling of phytoplankton biomass over a 14-day period and at an appropriate spatial scale (Bishop *et al.*, 2002). In this context, it will be recalled that the SUPER hypothesis suggests that the relative lack of diatoms in the Subarctic Pacific is due to deficiency in the subaerial Fe supply. However, we must surely note (as the authors of the hypothesis apparently did not) that the North Atlantic lies under less dusty skies than the North Pacific at similar latitudes: yet diatoms are prominent in the Atlantic spring bloom. It seems inescapable, therefore, that it is rather the permanent stratification of the subarctic Pacific than the surface flux of Fe that forces the fundamental difference between the pelagic ecosystems of this region and those of the North Atlantic.

In fact, the SUPER hypothesis was crafted when understanding of Fe distribution in the oceans was very weak, and prior to the publication of the Moss Landing data archive. Now, of course, we know that the Vertex section along 140°W revealed nutrient-like profiles with the anticipated additional enrichment below the ferricline close to Alaska at 48–49°N; in retrospective analyses, where data are lacking for mixing or deposition rates, or a nutrient budget is unavailable, vertical flux in either direction provides equally plausible mechanisms.

Despite this, the episodes of high diatom flux that have been observed in sediment cores here were ascribed by McDonald, Pedersen, and Crusius (1999) to the influx of Fe-rich meltwater from Alaska or to increased transport of dust from Asian sources. Episodic breakdown of stratification seems not to have been considered, but would produce the same effect. Given the conditions now understood to characterize the subarctic Pacific, it should be no surprise that Fe-fertilization of the mixed layer here should produce a diatom bloom, as has been observed.

The **Southern Ocean** is the most talked-about high-*S* region. Of course, much of the recent attention is due to the suggestion that the artificial addition of Fe in tanker quantities to surface waters might perhaps promote sufficient algal growth to sequester significant amounts of $CO_2$ from the atmosphere. This idea has been enthusiastically taken up by some oceanographers and many "ecologists"—in the political sense of that term—and has perhaps assisted in obtaining support for extensive investigation of the role of iron in supporting phytoplankton growth. These entrepreneurial suggestions have drawn substantial criticism from Miller (2004) and others.

Because the Southern Ocean is the focus of Chapter 12, what follows serves only to put this ocean in the context of the present discussion: a wider analysis of how phytoplankton production is controlled—and consumed—is presented in that chapter. So here, we need deal with only a very few salient points of the arguments in favor of iron limitation of phytoplankton growth. In particular, it may be useful to use this region to illustrate the complex relationships between irradiance and iron as limiting factors for the growth of phytoplankton—or, more strictly, their production of carbohydrates. A useful primer to this complex problem is that of Tim van Oijen (2004): I recommend it highly.

During the initiation of blooms, phytoplankton must utilize $NO_3$ as their principal nitrogen source, because of the relative absence of $NH_4$ at this stage of the production cycle. Because Fe is required in the cellular processes involved in the reduction of $NO_3^-$ to $NH_4^+$ their requirement for Fe is higher (by as much as a factor of 1.6) than during later stages of a bloom (Maldonado *et al.*, 1999). Then, the cellular Fe requirement of phytoplankton varies inversely with irradiance, because the photosynthetic apparatus has a higher Fe demand at low light levels. Thus, phytoplankton in regions of shallow mixing are capable of growth at ambient Fe levels that would preclude growth in deeper mixed layers where the irradiance experienced during each diurnal cycle would be lower. Then, as noted earlier, small cells can obtain sufficient Fe from lower concentrations than can large cells such as diatoms: the half-saturation constant for cell growth of phytoplankton responds to cell size across several orders of magnitude, ranging from $<0.001$ to $1.20\,\mathrm{nM\,kg^{-1}}$ Fe (Blain *et al.*, 2001).

It is customary to restrict analyses of the role of iron in the ecology of this ocean to regions poleward of the Sub-Antarctic Front where, it is said, the "real" Southern Ocean begins. However, it will be more instructive to consider the whole ocean south of 35°S, as is done in Chapter 12. This includes the entire region having remnant nitrate at the surface, as well as both of the annular zones of high chlorophyll that dominate the open Southern Ocean (see Color plate 20); it is highly probable that these two linear regions of seasonally high chlorophyll have some functional commonality. Each is a complex frontal zone that encompasses several degrees of latitude, and each has characteristic conditions that result in higher chlorophyll biomass within the front than in the intervening zones.

The Southern Ocean presents us with an apparently simple question, posed by the observation that it habitually exhibits regions of clear, blue water where phytoplankton biomass and primary production must take low values. In fact, satellite chlorophyll fields suggest that the widely held view that most of the Southern Ocean does not accumulate chlorophyll is an overstatement: episodic seasonal blooms are widespread and accumulate chlorophyll biomass to levels characteristic of other high-latitude regions. Differential cellular demand for ambient Fe under varying conditions, discussed earlier, generally support the observation that small cells dominate the Fe-depleted areas of the Southern Ocean and that diatoms are restricted largely to the frontal zones where Fe is more available and deep mixing is restrained. Nevertheless, we must address the general question of whether the constraints on phytoplankton growth in this ocean are due to Fe deficiency, given the characteristic excess availability of macronutrients, or due to Sverdrup dynamics appropriate to unusually deep wind-mixing and low solar irradiance. Or, do both factors intervene—as in the case where Fe limitation responds to ambient irradiance?

Several generic solutions have been proposed. Regional differences in the regulation of algal growth by iron in the annular zones were invoked by de Baar *et al.* (1995) to explain the observations: Fe-deficient water south of the STF, Fe-replete water in the PFZ, and Fe-limiting water to the south of the PF. Fe-replete water in the coastwise East Wind Drift completed the logic that "iron availability was the critical factor in allowing blooms to occur." Versions of this model are still quoted as the basic explanation of the observation of higher chlorophyll within the PF than elsewhere. It responds well to the observations that in the Southern Ocean, where episodic aeolian dust deposition is as low as anywhere on earth (Watson, 2001), the supply of Fe to surface water must be almost entirely from below, forced by autumn-winter mixing or by convergence and upwelling in the frontal zones. Here, it should be remembered, Fe concentrations in deep water do not diverge from what is normal at similar depths in other oceans. Surface deposition is seasonal and restricted to the marginal ice zone, and near drifting fields of melting icebergs.

The recent proposal of Erickson *et al.* (2003) appears to ignore the consensus that the dominant supply of iron to the euphotic zone of the Southern Ocean must by vertical flux across the pycnocline (Watson, 2001; de Baar *et al.*, 1995) and is based rather on what is called the *Patagonian plume* of atmospheric dust. Statistical techniques are used to suggest that the zonal band of high chlorophyll accumulation along about 50°S from the SW Atlantic around to the Pacific is causally correlated with dust deposition rate, spatially and (perhaps) seasonally. I shall return to the modeling aspects of this suggestion later, but it is important to understand that the term *plume* that has been applied to this phenomenon is inexact. Global surveys of aerosol optical thickness by NOAA AVHRR (e.g., Husar *et al.*, 1997) show that the distribution of haze over the Southern Ocean is very diffuse, not at all plumelike, and—relative to such haze in the northern hemisphere—very light. Further, it is far more diffuse than the circumpolar, meandering bandlike signature of chlorophyll enhancement in the ocean below. Moreover, deposition rates based on pelagic sediment cores in the Southern Ocean (Rea, 1994) are lower by factors of 5–10 than those estimated by Duce *et al.* (1991) from haze observations.

In each of the zonal partitions of the Southern Ocean, to be discussed in Chapter 12, the seasonal cycle of chlorophyll accumulation is unimodal. Biomass accumulation follows the seasonal irradiance cycle, as does the mean value for $Z_m$ so that chlorophyll biomass is maximal in midsummer when $Z_m$ is most shoal. A return to net growth in response to reduced mixing depth and increased irradiance is entirely in accordance with Sverdrup. An argument will be made in Chapter 12 that the phytoplankton ecology of the Southern Ocean can only be understood if its physical characteristics are fully considered: unusually deep wind mixing, associated with characteristically low sun angles, requires that we consider the probability that irradiance may be an important factor limiting phytoplankton growth here. Mixing is much deeper between frontal zones than within them and, at least in late summer, phytoplankton growth in the Antarctic Circumpolar Current (ACC) south of the Polar Front is light-limited (van Oijen, 2004). In spring and summer, however, Fe limitation is important here.

As I have already noted, the frontal zones appear generally to be enhanced in Fe compared with the deeper-mixed ACC water between them, although de Baar's observations of Fe enhancement in the Polar Front were made just east of the Drake Passage. This is not a typical region and even if this condition appears to obtain around the entire

Southern Ocean, it seems improbable that enhanced Fe arising in this manner could be conserved within frontal jets over long distances. It appears more likely that the vertical motion and convergence associated with frontal dynamics will ensure a continual supply of Fe from the deeper water masses below the ferricline. In this case, the rate of supply will respond to relative strength of meandering of the fronts; perhaps the strengthening of chlorophyll biomass downstream of shallow topography is due to this mechanism rather than to supply from shoal water. It is also surely significant that meandering fronts are rich in sloping, relatively shallow stratification that may serve to sustain blooms once initiated. All this must be integrated with the observations that, between the annular frontal zones, each Fe fertilization experiment (though all done in the Atlantic sector) has succeeded in generating an ephemeral diatom bloom.

I have not dealt here with the situation in the coastal biome of the Southern Ocean and will note in passing that irradiance is very frequently limiting here even in summer, because of much cloudiness and low sun angle, yet near-surface stratification is widely induced by meltwater, itself relatively high in dissolved Fe. Though not well observed by satellite sensors, surface blooms are, not unexpectedly, relatively widespread here.

How then should we evaluate the balance between the several mechanisms of bloom inception and limitation in the Southern Ocean? Certainly, it will be essential to examine the conflicting possibilities in the light of William Dickinson's evocation of scientific myths. Perhaps the general model of Boyd (2002) is the best guide yet available: this suggests a seasonal progression of limitation and colimitation of Fe, irradiance, and silicate so that irradiance limits in all months except mid-October to mid-March and Fe is limiting only in that period alone, while silicate may be limiting, in some regions and for some cells, in October and November.

## Models of Regional Nutrient Flux and Limitation

So far, I have made little or no reference to global or regional models of production processes that involve Fe inputs from aerosol or upward mixing, largely because no consensus has been reached by modeling studies. They have not greatly advanced our understanding of the problem, although their indications cannot be ignored.

Because it directly supports the concept that I believe has become mythic, it will be useful to address at once the model of Erickson *et al.* (2003). These authors merge SeaWiFS data with the output from the global model of dust-deposition of Ginoux *et al.* (2001) that, incidentally, they describe as a "state-of-the-art geophysical data set." They obtain fields of anomaly correlation between surface chlorophyll and dust deposition at $2° \times 2.5°$ resolution over the Southern Ocean, where the anomalies are the time-series mean subtracted from each monthly value. Correlations of $>0.6$ are obtained for a zonal swath from 40 to 55°S across the entire ocean, but principally in the Atlantic and western Indian Ocean. The location of high correlation is commonly displaced by 2–5° of latitude from locations of chlorophyll maxima, and so Ekman transport is invoked to explain the resultant bloom. Their analysis, they suggest, elucidates "the spatial response of chlorophyll to iron flux" and identifies "those regions where ocean biology is possibly tightly coupled with atmospheric Fe deposition."

Really, of course, it does neither of these things. Statistical correlation between two factors that have been selected from among the very many that are involved in a complex ecological process cannot, of itself, demonstrate causation: that much we learn in elementary statistics courses. So I remain unimpressed by apparent correlation between dust deposition and those places where phytoplankton growth and accumulation is most active, because it is quite certain that many other factors, not considered, are involved in the complicated processes that lead to an algal bloom. In the Southern Ocean blooms are largely restricted to meandering frontal zones and are possibly enhanced downstream from shallow topography, as I have already discussed.

Then, the authors find it informative that both dust deposition and surface chlorophyll have seasonal maxima in the Southern Ocean in midsummer. They did not, apparently, consider that each of these cycles is an independent response to sun angle. The algal response is obvious at high latitudes, but it is not so well known that the dust-carrying winds from Patagonia are "monsoonal," so that dust flies NW in winter and SE in summer. Nevertheless, the Erickson *et al.* model has been influential.

The wide range of assumptions used for the input field of Fe at the sea surface indicates the level of uncertainty that is still inherent in geochemical nutrient models. To obtain this input field in models, extrapolation has been made from observations of haze at sea (Duce and Tindale, 1991), from global vegetation cover data (Mahowald *et al.*, 1999), from vegetation, soil texture, and land surface modification data (Tegen and Fung, 1994), and from the location of topographic depressions (Ginoux *et al.*, 2001). If Fe is 3.5% of deposited dust (the usual assumption), the estimates of Fe deposition on the ocean ranges sixfold from 15 to 100 Tg $y^{-1}$ so the global pattern of deposition flux cannot yet be narrowly constrained. Yet it is probable that only between 5 and 20°N in the eastern Atlantic, and seasonally in the extreme NW Arabian Sea, is flux greater than 1–5 g $m^{-2}$ $yr^{-1}$. In the southern hemisphere, it seems to exceed 0.25 $m^{-2}$ $yr^{-1}$ only patchily, in some small regions, and seasonally.

We remain uncertain even of the relative magnitude of the fluxes of labile iron across the sea surface and across the nutricline, although this must be critical to any quantitative understanding of Fe limitation in the ocean. For instance, the suggestion of de Baar *et al.* (1995) that an aerosol flux of 30 mg $m^{-2}$ $y^1$ dominates over vertical flux across the nutricline in the North Pacific is based on what they called an "order of magnitude assessment" of Duce and Tindale's classical map. Modeling of the atmospheric flux has given equally diverse results: Fung *et al.* (2000) simulated a deposition rate almost twice that estimated by the model of Mahowald *et al.* (1999) and almost four times what was suggested by Duce and Tindale's data (263 and $132 \times 10^9$ mol $y^{-1}$, respectively).

Fung *et al.* remarked that uncertainty concerning aeolian flux to the photic zone is as large as a factor of 5–10, with an order of magnitude uncertainty in determining the soluble fraction. Despite this, they asserted that entrainment and upwelling deliver only $0.7 \times 10^9$ mol Fe/$y^{-1}$ across the nutricline, as against $96.0 \times 10^9$ mol Fe/$y^{-1}$ across the surface: the consequences of entrainment and upwelling flux of Fe are trivial for these authors. On the other hand, Archer and Johnson (2000) compare aeolian flux from the sources quoted earlier and conclude that 70–80% of global carbon export production by phytoplankton can be supported by the "upwelling of iron in seawater rather than by atmospheric deposition." These authors go on to propose that "ocean recycling of Fe appears to play a major role in determining the strength of the biological pump in the ocean and the $pCO_2$ of the atmosphere."

Three recent intermediate-complexity ecosystem models appear to support, inadvertently, the primacy of physical processes in determining the pattern of productivity in the ocean. Each is coupled to an OGCM and each comprises several classes of phytoplankton with appropriate herbivore classes and sinking terms. The models are distinguished principally by their nutrient assumptions. The first has three macronutrients with no Fe input, the second by the same team (Gregg *et al.*, 2003) is similar, but has Fe input from the GOCART model of Ginoux *et al.* (2001), and the third (Moore *et al.*, 2002) has a full range of nutrients ($NO_3$, $NH_4$, $SiO_3$, $PO_4$, and Fe). In this case, Fe input is from the models of Tegen and Fung (1994) and Mahowald *et al.* (1999), with solubility set at 2%. A fourth model (Christian *et al.*, 2002) is structured like that of Gregg *et al.* (2003), but is applied only to the tropical Pacific domain and uses Tegen and Fung (1994) for Fe input.

What is remarkable is that each of this wide range of models tracks rather well the CZCS or SeaWIFS chlorophyll fields, with believable interyear variability. Gregg and Conright (2001) remark that the model lacking Fe input "was able to represent the seasonal distribution of chlorophyll during the SeaWIFS era and was capable of distinguishing the widely different processes" that occur globally in Niño and Niña years. The model presented global monthly maps of sea surface chlorophyll that adequately matched SeaWIFS maps for both pattern and values except near coasts, and SeaWIFS monthly means in 12 domains were tracked quite closely, excepting only the equatorial domains where the lack of river input (Amazon) and coastal influences (upwelling coasts) produced values that were overall too low. A similar model, with Fe input and a more complex representation of phytoplankton functional groups (Gregg *et al.*, 2003), performed very similarly, with very similar deviation from observations in the equatorial domains.

The model of Moore *et al.* (2002) likewise presents monthly global patterns of surface chlorophyll that closely match SeaWIFS observations, and global patterns of derived properties. Standard runs were compared with runs having no Fe input or having saturating Fe input globally; both produce global maps of primary production that are realistic and represent all the major features seen in maps of production rates computed directly from SeaWIFS surface chlorophyll. Yentsch's effect of geostrophy is remarkably well preserved. The Fe-saturated model notably retains the high chlorophyll in the frontal zones of the Southern Ocean, strongly isolated from the more oligotrophic conditions between; the run lacking Fe input produced unrealistically low chlorophyll accumulation, unlike the model of Gregg (2002) that also lacks an Fe input but simulates reasonable chlorophyll biomass fields. Depending on the solubility assumption, the standard runs of the Moore *et al.* model indicated annual global primary production of 42–47 Gt C $y^{-1}$. This range was extended to 35–70 Gt C $y^{-1}$ by including the runs with no Fe and with saturating Fe input; direct computations from satellite data all fall within this range.

Now, what would happen if dry deposition patterns of Fe were significantly to change in some way, so that deposition rates over the eastern Pacific resembled those over the eastern Atlantic? If indeed skies everywhere became as dusty as under West African *harmattan* winds then, after a brief period of adjustment, each regional ecosystem would adopt a pattern of primary production and cell accumulation rather closely resembling what we see today. Further, there would be very little change in the specific composition of the characteristic phytoplankton communities, although the greatest uncertainty might be in the distribution of N-fixing organisms. The eastern Pacific would continue to accumulate chlorophyll in a manner typical of equatorial regions but quite different from what we expect in higher latitudes, or in low-latitude oligotrophic regions. I doubt if the sea-surface chlorophyll field of a dusty Southern Ocean would differ greatly, once equilibrium was established, from that of the present day. The new field would continue to reflect the simple fact that where Sverdrup does not permit growth of cell populations, none will occur, no matter what the ratio and concentration of available nutrients.

Obviously, where Sverdrup does permit cell population growth, but this does not occur because of lack of an essential element, then its addition—as in the iron-enrichment experiments—must induce a bloom. That such events occur naturally is not in doubt and there are now many observations that confirm this. For example, a response to dry deposition of Fe-rich terrestrial dust particles was observed in the ADIOS project in the oligotrophic, high-*S* North Pacific at 26°N (Young *et al.*, 1991). Here, in an oligotrophic ocean, when mixed-layer $NO_3$ was in very low concentration, each deposition event created a brief, descending pulse (as the aerosol particles sank) of increased production rates of about 50% over ambient. Both iron and nitrogen were delivered in the dust, so that the near-surface concentration of $NO_3$ increased from 0.2 to 0.7 nM $kg^{-1}$ after one episode. Nevertheless, it was thought that the dust-fall released the autotrophic community from Fe limitation, rather than from $NO_3$ limitation. Similarly, on the West Florida Shelf, deposition pulses of African dust may provide an explanation of blooms of *Trichodesmium* whose nitrogenase enzyme system has a high iron demand; after a dust event, background iron levels increase from <0.5 to 16 nM $kg^{-1}$, and *Trichodesmium* colonies increase a hundredfold (Lenes *et al.*, 2001).

To what extent these observations can be generalized is unknown, but such events can be expected only in regions where the prevailing winds carry heavy dust loads: this occurs commonly under the Atlantic trade winds and the North Pacific westerlies, but not elsewhere. Their frequency is unknown but probably follows the relative rate of dry deposition in each region. It has also been suggested that the flux of both iron and nitrogen at Midway (180°W, 30°N) and Bermuda (60°W, 32°N) may represent a major part of nutrient flux to the photic zone, and sufficient to induce a significant growth response (Donaghey *et al.*, 1991). Suggestions have also been made that the strong increases in deposition of both desert dust and pollutant aerosols after the mid-20th century are currently modifying the structure of the base of the pelagic ecosystem by inducing anomalous population growth both of pathogenic microbes and of diazotrophic cyanobacteria (e.g., Hayes *et al.*, 2001).

Observations such as these support the view that there must be regional consequences for phytoplankton growth of the deposition of aerosols at the sea surface and that we should expect that the most significant deposition occurs at the sea surface of aerosol particles, whether of desert dust or industrial haze. Accepting this view, we may conclude that the high productivity of parts of the Atlantic, especially in low latitudes, is one result of this process. Unfortunately, this is very difficult to demonstrate and, indeed, has yet to be done: nevertheless, I have already seen statements making this connection. The problem, of course, as discussed earlier is that it is difficult to isolate the deposition effect from the many other factors that control phytoplankton growth and accumulation. The physical forcing processes that are involved differ strongly between oceans, and between comparable biogeochemical provinces in each ocean: this makes meaningful comparison of aerosol deposition effects very difficult.

Recourse may be had to an estimate of phytoplankton productivity, partitioned among 50-odd biogeochemical provinces (Longhurst *et al.*, 1995), that has been revised for this volume, using the same algorithms, with data both from SeaWiFS and from MODIS for 2002–2005. In only 8 provinces does total phytoplankton production exceed 400 gC $m^{-2}$ $y^{-1}$ and, of these, 3 are in Atlantic low latitudes: CNRY (Canary Current upwelling), GUIA (Amazon shelf and plume), and GUIN (tropical West African coast). The other high-productivity provinces are the coast of China, the NE, SE and NW Atlantic shelf regions, the Alaska shelf, and the NW Arabian Sea. In each of these 8 provinces, as in the remainder, serial surface chlorophyll images clearly suggest that the pattern of productivity matches that of physical processes, rather than the more diffuse pattern of aerosol deposition.

It may be significant to the question of Fe deposition that the Canary Current upwelling province, lying directly below the Saharan dust plume, has almost twice the productivity of each of the other three eastern boundary current upwelling provinces: 710 gC $m^{-2}$ $y^{-1}$ compared with 269–396 $m^{-2}$ $y^{-1}$. However, these four upwelling provinces are not directly comparable, and the Canary Current region is unique in the great width of its shelf. Then, tropical West Africa (GUIN) has the only coastline anywhere that is aligned close to, and parallel with, an equatorial current system. The Amazon shelf (GUIA) receives the total nutrient flux of the largest river on any continent. So, it is altogether too early to ascribe the relatively high production of some Atlantic regions to the effect of African dust with any confidence. For what it is worth, which isn't much, statistical correlation between regional production in all biogeochemical provinces and a "dust factor" (obtained from the interaction between contours in the Ginoux model and province boundaries) does not suggest a global relationship ($y = 2.9x - 0.0003$, $R = 0.04$) between deposition and productivity. Obviously, although many assume that such a relationship must exist, it cannot yet be demonstrated: isolated studies of only deposition processes and phytoplankton nutrient dynamics are insufficient for such a demonstration, because integration of holistic regional ecology and ocean physics is required.

Preoccupation with surface flux of a micronutrient was one part of our response to the possibility that the global $CO_2$ flux budget might be manipulated by artificial inputs of particulate Fe, repeating what may have occurred naturally during glacial epochs. Of course, this possibility aroused intense interest and major funding, so it is not surprising that studies tended to place more emphasis on atmospheric flux than on more classical problems of nutrient flux into the photic zone from below. Nor would it be very surprising that results tending to demonstrate Fe limitation should have been emphasized, as has perhaps happened in recent years (e.g., Behrenfeld and Kolber, 1999).

So, if myths in science are "selective conjectures . . . thereafter given privileged status" lacking an "alternative suggestion" (Dickinson, 2003), then I suggest that when we think about nutrient limitation in high-*S* regions, we should remember the "alternative suggestion" of complex vertical flux of several nutrients across the nutricline. We should not think only about fluxes of iron across the sea surface, just because this happens to have been the fashionable topic of the recent past. However, in complex fields of natural science like geology and biological oceanography, where neither observations nor models can capture the true complexity of the dynamic system, it is true that myths—in the sense of geologist Dickinson—do have a role to play in arriving at scientific truth.

Chapter 6

# Biomes: The Primary Partition

Wherever the wind blows and the sun shines over the ocean, turbulent diffusion and thermal stratification may be imposed locally and may interact locally, as in all one-dimensional models of the spring bloom. Wherever distantly forced adjustment of mixed-layer depth occurs, the elements of such algal growth models will be shoaled or deepened.

That these two statements apply uniquely to fundamentally different and characteristic latitudinal zones in the oceans, as was discussed in Chapter 4, has only recently become clear. That both local and distant forcing must be accommodated in analyses of local algal dynamics is perhaps still not generally understood, though the relative importance of local and distant forcing is what principally distinguishes algal dynamics at low and high latitudes. This is the principal distinction between the two most extensive biomes of the pelagos: those regions generally lying below the high-latitude westerly winds and those below the easterly trades characteristic of low latitudes.

Wherever the winter sun lies so low on the horizon that the ocean freezes, brine is extracted from the freezing seawater. When sea ice melts with the climbing sun in spring, fresh water is released to form a brackish surface layer above the deeper, more saline layers so that the two water bodies are separated by a sharp density gradient. Because this pycnocline constrains wind mixing, the polar brackish-water layer retains its integrity while spreading in the surface circulation. The oceanic polar fronts of each hemisphere define the extent of its dispersal and in this way serve to define a third primary biome of the pelagos.

Circulation and stratification of surface water masses is modified where the continental topography and coastal wind patterns are encountered. These changes, for Mittelstaedt (1991), define a coastal habitat within a shelf-break front that lies above the upper part of the continental slope. This coastal zone is here considered to be the fourth biome, characteristically more diverse than the other three. As already noted, we exclude from this biome the narrow alongshore zone that is very difficult to generalize except in a few exceptional regions, such as those coasts dominated by the effluents of great rivers where the turbid littoral zone is relatively wide in relation to the entire shelf.

Because it is a principal argument of this book that the processes that force stratification of the surface layers thereby also determine characteristically different phytoplankton regimes, recognized as the primary partitions of pelagic ecology, it will be useful to consider in greater detail the mechanisms and processes that specify the four primary biomes before proceeding to consider the secondary division of these biomes into the provinces of a biogeographic system.

## The Four Primary Biomes of the Upper Ocean

The partition discussed here is of the upper ocean and rests principally on observed or inferred regional discontinuities in physical processes, particularly those that affect the stability of the upper kilometer of the ocean; regional differences in other ecologically significant variables, such as irradiance, are also considered. In short, the partition is based largely on the factors invoked by the Sverdrup model and proceeds from the suggestion made in Chapter 4 that six simple models, or cases, are sufficient to accommodate the observed range of pelagic production mechanisms. You will remember that I suggested that this might be more useful in partitioning oceanic phytoplankton ecology than the simpler analysis of Margalef, who defined only four quadrants in the nutrient/turbulence relationship.

Little rearrangement and only modest consolidation is required to apply these six cases to my earlier suggestions to recognize four biomes, or basic vegetation types, within the pelagial realm of the oceans; these are very similar in concept to those of Beklemishev (1969) and to the eco-regions of Bailey (1983). With each biome is associated one or more of the six models already discussed, as in the following arrangement:

**Polar biome**: where the mixed-layer depth is constrained by a surface brackish layer that forms each spring in the marginal ice zone:
*Case 1—Polar irradiance-mediated production peak.*
**Westerlies biome**: where the mixed-layer depth is forced largely by local winds and by local irradiance:
*Case 2—Nutrient-limited spring production peak.*
*Case 3—Winter-spring production with nutrient limitation.*
**Trades biome**: where the mixed-layer depth is forced by geostrophic adjustment on an ocean-basin scale to local or distant wind forcing:
*Case 4—Small-amplitude response to trade wind seasonality.*
*Case 5—Large-amplitude response to monsoon reversal of trade winds.*
**Coastal biome**: where many diverse coastal processes modify the mixed-layer depth and nutrient inputs:
*Case 6—Intermittent production at coastal divergences and upwellings.*

While reviewing the general properties and boundaries of these biomes it will be useful to bear in mind that only the trades biome represents a continuous body of water in each ocean. The polar and westerlies biomes each exist as two separated boreal and austral units and, furthermore, because land masses are not uniformly distributed in each hemisphere, their boreal and austral expressions have individual characteristics: thus, the boreal polar biome comprises a mediterranean sea containing a major archipelago, whereas the austral polar biome comprises an annular open ocean surrounding a central continent. Despite such asymmetries, the degree of ecological commonality is sufficient to support the biome concept presented again here.

In putting the case for this partition, I want to emphasize that the definition offered refers to an ideal ocean on a landless globe. In applying it to the real ocean, pragmatism must be used in setting limits to the individual partitions, this being especially true for the Coastal Biome, where the distribution of shallow continental shelves and deep basins may be too fractal for the dogmatic application of the definitions. Similarly, partially isolated basins such as the Mediterranean and Caribbean, or archipelagic regions such as the SW Pacific, also require to be treated pragmatically.

There is, obviously, a wider set of ecological factors that, if they were available and if we could identify discontinuities in their global fields, could be used to define more precisely the ecological characteristics and boundaries of each biome. For instance, it would be useful to quantify how plankton diversity (and hence the complexity of trophic networks) changes regionally, because this is clearly an attribute likely to take characteristic values for individual ecosystems (or biomes): unfortunately, for very few water bodies is there a complete enumeration of the species present, let alone any comprehensive measure of characteristic ecosystem diversity. A good illustration of this problem is provided by one of the most complete and recent enumerations of pelagic biota, that of Margalef (1994), which compares species lists of phytoplankton from small flagellates to very large dinoflagellates and diatoms taken at two stations, one each in the Caribbean and in the western Mediterranean. These yielded lists of 353 and 257 named species, respectively, with about 50 more unidentified taxa at each. To complete an inventory of autotrophic cells, to these counts would have to be added the flagellate flora of the nano- and ultraplankton and the cells of the photosynthetic picoplankton (cyanobacteria and prochlorophytes). It is very unusual to be able to locate even single samples of the heterotrophs (micronekton, meso- and micro-metazoans, and the protists) that have been analyzed taxonomically as well as Margalef's phytoplankton.

To obtain regional fields of diversity within which we might hope to discern discontinuities, we would need observations such as those described by Margalef (but completed by inclusion of all taxonomic groups) on a suitable grid across the ocean, repeated seasonally. Obviously, as was discussed in Chapter 2, such information exists for no group of organisms, and probably for no location in all the oceans is there a complete listing of the kind required.

Another way of proceeding would be to seek information on the global distribution of biota at higher taxonomic levels than species, hoping to aggregate these into ecologically meaningful groups. It is reasonable to hope that such data might be accessible in a more uniform format than the almost hopelessly diverse descriptions of species lists at individual stations that can be found in the biogeographic literature. Fortunately, the Smithsonian Institution has for many years used a standardized first-order sorting technique for the plankton samples it archives, and these protocols have been adopted uniformly by other sorting centers. An archive of these sheets from several plankton sorting centers thus enables a first-order analysis of the composition of zooplankton globally; in one case, this was done with data from 4166 stations from all oceans where sampling nets had been worked between the surface and 250 m (Longhurst, 1985b).

The counts were allocated to six functional groups (gelatinous predators, raptorial predators, micro- and macroparticle herbivores, omnivores, and detritivores) and also among taxonomic groups (medusae, siphonophores, chaetognaths, polychaetes, ostracods, copepods, mysids, euphausiids, and pteropods). These counts were then stratified regionally and seasonally to represent first-order differences between oceans and continental shelf faunas of the polar, temperate, and tropical zones. The general results are shown in Table 6.1, which enables the quantification of some latitudinal trends and seasonal changes in plankton composition that were well known for only a few study sites, although also suspected to occur more generally.

To illustrate the possibilities offered by such data, note how they quantify the change in dominance of copepods from 67% of zooplankton biomass in polar seas to 33% in the tropics, and how predators (both gelatinous and raptorial) increase from 22% in high latitudes to 47% at low latitudes. A striking aspect of these zonal differences is the increasing contribution of gelatinous predators equatorward: from 0.9% of total zooplankton biomass in polar latitudes to 14.3% in the tropics. The more equal distribution of relative biomass among both the taxonomic and trophic groups in the tropical, compared with the polar, seas is also clearly recorded.

Other data can also be consulted for confirmation of regional boundaries established otherwise: consider the case of calanoid diversity in the samples from the Continuous Plankton Recorder (CPR) survey of the North Atlantic. Long-term taxonomic richness of calanoid copepods in several thousand transects (Beaugrand *et al.*, 2000a) shows a zonal

**Table 6.1.** Relative composition of zooplankton in coastal and oceanic biomes derived from data on >4000 samples held in the Smithsonian and Kuroshio Plankton Sorting Centers. Gelatinous predators are medusae and siphonophores; raptorial predators are chaetognaths and some genera of cladocerans and copepods, with all amphipods, annelids, and gymnosomes; macro-filtering herbivores are some genera of cladocerans, copepods, and thecosomatous mollusks; gelatinous herbivores are tunicates, appendiculaians, and doliolids with mucous-net filtration; omnivores are some genera of copepods, some euphausiids, all mysids, and penaeids; detritivores are all ostracods.

| | | *Medusae var.* | *Siphonophora* | *Chaetognatha* | *Polychaeta* | *Cladocera* | *Ostracoda* | *Copepoda* | *Amphipoda* | *Mysidae* | *Euphausiidae* | *Penaeidae* | *Pteropoda* | *Appendiculariae* | *Salpida* | *Doliolida* |
|---|---|---|---|---|---|---|---|---|---|---|---|---|---|---|---|---|
| **N individuals (%)** | | | | | | | | | | | | | | | | |
| OCEANIC | Polar | 0.06 | 0.15 | 3.88 | 0.23 | 0.00 | 2.45 | 86.32 | 0.56 | 0.00 | 0.52 | 0.02 | 2.44 | 3.32 | 0.02 | 0.00 |
| | Westerlies | 1.39 | 0.45 | 3.99 | 0.54 | 0.47 | 3.12 | 80.45 | 0.81 | 0.02 | 2.77 | 0.33 | 1.76 | 2.79 | 0.52 | 0.12 |
| | Trades | 0.85 | 2.61 | 6.38 | 0.85 | 0.14 | 3.39 | 72.31 | 0.54 | 0.15 | 3.02 | 0.30 | 1.70 | 3.85 | 2.10 | 0.66 |
| COASTAL | Westerlies | 0.22 | 0.59 | 3.32 | 0.39 | 7.07 | 0.31 | 80.10 | 0.78 | 0.07 | 1.02 | 0.27 | 1.50 | 3.51 | 0.46 | 0.05 |
| | Trades | 0.67 | 2.11 | 5.83 | 0.43 | 6.46 | 5.72 | 62.95 | 0.82 | 0.02 | 1.17 | 3.15 | 0.60 | 5.45 | 5.15 | 0.10 |
| **Carbon biomass (%)** | | | | | | | | | | | | | | | | |
| OCEANIC | Polar | *0.23* | *0.75* | *8.62* | *0.47* | *0.00* | *3.05* | *67.58* | *4.57* | *0.00* | *7.66* | *0.13* | *6.73* | *0.18* | *0.03* | *0.00* |
| | Westerlies | *3.28* | *1.78* | *8.62* | *0.47* | *0.00* | *3.05* | *67.58* | *4.57* | *0.00* | *7.66* | *0.13* | *6.73* | *0.18* | *0.03* | *0.00* |
| | Trades | *1.98* | *8.51* | *9.00* | *1.10* | *0.01* | *2.61* | *33.13* | *2.58* | *0.00* | *30.21* | *5.02* | *2.66* | *0.10* | *2.89* | *0.01* |
| COASTAL | Westerlies | *0.62* | *2.66* | *6.51* | *0.79* | *0.78* | *0.28* | *64.66* | *5.22* | *0.00* | *11.79* | *1.11* | *4.83* | *0.17* | *0.80* | *0.00* |
| | Trades | *1.39* | *7.88* | *9.81* | *0.64* | *0.48* | *4.98* | *37.51* | *4.25* | *0.00* | *10.67* | *15.26* | *1.15* | *0.18* | *5.80* | *0.00* |

| **Carbon biomass (%)** | | *Entrapping predators* | *Raptorial predators* | *Micrivorous herbivores* | *Macrivorous herbivores* | *Omnivores* | *Detritivores* |
|---|---|---|---|---|---|---|---|
| OCEANIC | Polar | 0.88 | 21.72 | 0.21 | 56.13 | 18.08 | 2.98 |
| | Westerlies | 4.08 | 33.65 | 0.68 | 19.80 | 39.80 | 1.98 |
| | Trades | 14.30 | 33.05 | 4.22 | 20.48 | 24.64 | 3.44 |
| COASTAL | Westerlies | 3.62 | 14.01 | 1.07 | 43.08 | 37.90 | 0.31 |
| | Trades | 9.84 | 28.14 | 6.28 | 7.84 | 42.58 | 5.31 |

*Source: Longhurst, 1985.*

discontinuity across the ocean at between 40 and 50°N. Taxonomic richness, supported by five separate diversity indices, followed very closely, as the authors commented, the boundaries between the partition between biomes discussed here. The authors note that the poleward decrease of taxonomic diversity anticipated in the pelagic ecosystem is discontinuous, and that the gradient is interrupted by local variability where different surface water bodies are interleaved.

Such information, although very restricted in scope, does gives some confidence that the primary partition discussed here is based on reality.

## Polar Biome

Beyond the polar fronts, strong near-surface stratification is induced by the effects of the freeze-thaw cycle of sea ice: thus, brine rejection occurs when surface water freezes so that the dense water sinks, and fresh water is released when sea ice melts in the spring.

Deep winter mixing occurs only in the absence of ice cover and, after the spring thaw, the superficial low-salinity layer stabilizes the upper water column so that seasonal change in mixed-layer depth is relatively small during the open-water season. Despite these constraints, sufficient mixing occurs to recharge the surface layers seasonally with inorganic nutrients, and this is critical because the stability imposed by the near-surface halocline may be sufficient to allow algal growth to occur as soon as there is sufficient sunlight, even though values of $N_{max}$ (commonly 2–8 cycles $hr^{-1}$) within the shallow polar pycnocline are not as high in the tropical pycnocline. In this connection, note that because of its low salinity, equivalent changes in the density of polar water are forced by salinity changes of 0.1/ml or temperature changes of 5°C.

The extreme range of irradiance in polar regions forces a unique seasonal cycle of primary production rate, represented by the simple Case 1 model and having a single, light-limited maximum at the summer solstice. Poleward of 66° latitude, after the vernal equinox, day length increases exponentially with latitude so that there is insignificant effect of latitude on the timing of a spring bloom in open water. A shallow chlorophyll maximum develops at the halocline in open water, and summer oligotrophic conditions with irradiance of the pycnocline may be very briefly established. Paradoxically, one might think, early spring thaws produce less strong ice-edge blooms than late thaws when the sun is high in the sky. This is because the accumulation of an ice-edge bloom requires sufficient irradiance that algal growth will exceed its local sinking rate, given that it may occur prior to the establishment of stratification in the water column.

The polar biome is characterized by low taxonomic diversity at all trophic levels; during the short pulse of primary production that occurs during the brief period of open water and high sun angle, phytoplankton biomass is dominated by large cells (>90% diatoms and coccolithophores) although the smaller fractions of pico- and nano-phytoplankton are nevertheless responsible for 10–25% of production. Diatom frustules form a diatom ooze on the deep ocean floor, especially in the Southern Ocean where the northern limit of ooze corresponds very closely with the location of the Antarctic Divergence, itself defining the equatorward boundary of the austral polar biome. When a brief oligotrophic phase follows the spring bloom, as it may do in open water in Baffin Bay, smaller cells dominate as everywhere when nitrate is no longer available and biologically regenerated ammonium is utilized.

A second accumulation of chlorophyll that occurs in some regions, in some years, during the late summer period of declining primary production rate is consistent with the effect of reduced herbivore consumption as these organisms descend to overwintering depths. Herbivores are dominated by large copepods in both hemispheres; if you contribute to the popular belief that the polar seas are unique because euphausiids dominate the plankton and provide the abundant and easily strained food required to sustain baleen whales, you may be surprised by Table 6.1. Contrary to tradition, the polar oceans are not dominated by euphausiids ("krill"), although obviously these occur in great abundance in some restricted regions. In fact, in both northern and southern hemispheres, euphausiids form <1% numerically and <8% by biomass of all zooplankton in regional surveys of polar oceans. In tropical latitudes, euphausiids comprise 30% of zooplankton biomass overall.

The herbivore response to the availability of phytoplankton may not be as direct as in lower latitudes because of relatively long generation times; some of the larger species of copepods require more than a single season to complete their growth and reproductive cycles, passing the winter in water as deep as 1000 m. In their first season, having reached the fifth-stage copepodite stage (the pre-adult instar) they cease to feed and migrate down. Apart from the consequences for the standing stock of phytoplankton cells, the annual cycle of the large copepods has another, often neglected, consequence. Where water depth is insufficient for the seasonal ontogenetic migration to occur, large copepods do not complete their life cycles successfully and exist principally as expatriate populations. A difference between northern and southern hemisphere polar zones is that in the latter there are relatively more important populations of herbivorous euphausiids that, unlike the large calanoid copepods of both north and south, do not perform seasonal migrations: they remain relatively near the surface during winter, usually associated with sea ice.

At the upper trophic levels, the marine mammal fauna is more diverse than in lower latitudes and includes several whale species that feed directly on plankton and micronekton. Less than 1% of the approximately 20,000 species of marine fish occur south of the antarctic convergence, and there are no antarctic species of densely schooling pelagic fish. Small shoaling fish occur only in the southerly parts of the boreal polar biome: for instance, capelin (*Mallotus villosus*) is a significant component of the ecosystem of the Labrador and Barents Seas. In the high arctic, the mesopelagic polar cod *Boreogadus saida* is a key species in the transfer of material from planktonic production to birds and mammals and occupies a niche similar to that of schooling pelagic euphausiids *Euphausia superba* and *E. crystalophorius* in the antarctic, together with another small gadoid, the silverfish *Pleurogramma antarcticum*.

## WESTERLIES BIOME

Here, either the Case 2 or Case 3 model describes the seasonal production cycle, depending on latitude and on the relative depth of winter mixing by the westerly winds. These, of course, extend equatorward to the Subtropical Convergence zones between the opposing hemispheric Trades and Westerly wind systems; because the westerlies are globe-encircling, stronger, and more consistent in location in the southern hemisphere, the subtropical convergence (STC) there is more readily located. It also occurs at much higher latitudes than in the northern hemisphere, being constrained by the geography of South America and Australia-New Zealand to encircle the globe at around 60°S. In the northern hemisphere, the STC crosses the oceans SW-NE at between 30 and 40° of latitude.

Wind stress (see Color plate 3) is most strongly seasonal across the North Atlantic (winter <30 and summer $<5 \times 10^{-1}$dyn cm$^{-1}$) and least seasonal in the Southern Ocean ($<20$–$40 \times 10^{-1}$dyn cm$^{-1}$ at all seasons). Westerly storm tracks in the North Atlantic track across the ocean from Florida to the Europe and, in the North Pacific, more zonally from southern Japan to the west coast of Canada. In the Southern Ocean in winter, the track is circumpolar, crossing Patagonia at 45–50°S.

Mixed-layer depth is strongly seasonal, deepening in winter with increased wind stress and convective cooling forced by the seasonal changes in sun angle, cloudiness, and strength of the zonal westerly winds. In the North Pacific, and especially in the Alaska gyre, winter mixing is to some extent constrained by a shallow halocline, though convective cooling extends much deeper.

Increasing irradiance in spring and the onset of stratification may induce an algal spring bloom. One of the first revelations of the global CZCS images was the degree to which the North Atlantic spring bloom is a singular feature of the oceans. Nevertheless, the discipline of biological oceanography is based historically on studies of this bloom, and what we know about its induction and evolution serves us well as a starting point for discussing the other, mostly less well researched, regions under the midlatitude westerlies.

The classical vernal sequence, described by the Case 2 model, is too well known to require more than very brief mention here. Initial pre-bloom conditions for nutrients at the end of winter are related to the depth of winter mixing and the baroclinicity of the subsurface nitrate field. There is increasing evidence that near-surface spring blooms may be initiated in unstratified (winter-mixed) water columns if the mixing rate is below a critical threshold and if there is sufficient light (e.g., Dandonneau and Gohin, 1984). Once stratification is established, evolution occurs toward a summer oligotrophic situation when the summer pycnocline and nitracline lie within the photic zone for 4–6 months and during this period a deep chlorophyll maximum is formed. The bloom becomes nutrient-limited when the initial charge of inorganic nitrate (or, in some cases, silicate) in the mixed layer is exhausted; the summer growth that follows is nutrient-limited, fueled largely by biological regeneration of nitrogen as $NH_3$, with some minor contribution of $NO_3$ from entrainment (wind and eddy events) and weak Ekman suction (curl of wind-stress field). An autumn bloom (in the sense of biomass accumulation), usually weaker than the spring bloom, may follow when the mixed layer deepens so that vertical nutrient flux increases, driven by increasing wind stress and reduced surface warming; this may also result from the reduction of grazing pressure when deep-hibernating herbivorous copepods descend to overwintering depths.

At lower latitudes where westerly winter winds are relatively light, a Case 3 model is appropriate. Here, primary production rate is always low, and seasonality in phytoplankton biomass relatively weak, depending on the depth of winter mixing; this is here rather variable, so that in some years it is sufficiently weak that the nutricline may not be significantly eroded. From low values in midsummer, the rate of primary production begins an increase that is sustained through the autumn and culminates in a spring maximum after a second rate increase in late winter. At these latitudes, the smallest photosynthetic cells dominate phytoplankton biomass during those long periods when stratification is strong, nutrient flux is suppressed, and blooms of larger algal cells do not occur. The picoplankton fraction, dominated by cyanobacteria and prochlorophytes, may comprise 70–90% of chlorophyll and contribute 80–90% of primary production at 20–30° latitude. Likewise, the herbivore fraction of the zooplankton is more diverse than in Case 2 situations and comprises many genera of small copepods, together with tunicates.

Herbivore ecology is complex in the Westerlies biome, because both the pattern of seasonal ontogenetic migration typical of the polar seas and the diel vertical migration pattern typical of trade-wind seas occur here. As at higher latitudes, the timing of the fall migration to depth appears to be ontogenetic, when a certain instar has been achieved, rather than a response to lack of food. At all latitudes within this biome, during at least some part of the oligotrophic summer phase, some large calanoids perform diel vertical migrations, rising to feed at night from daytime residence depths of around 200–500 m, in this way presumably escaping predation in the clear water during daylight hours.

Pelagic fish here are not as diverse as they are in the Trades biome. Some migratory stocks of herring and other shoaling clupeids spend much of the year in the open ocean and return to the coastal zone to spawn. Others, such as the saury of the North Pacific, pass the entire year over deep water. Both North Atlantic and North Pacific salmon, though spawning in freshwater, spend almost their whole lives in the open ocean. Some species of tuna perform meridional migrations that enable them to exploit the westerlies biome even to quite high latitudes during summer in all oceans. Here, too, we encounter the poleward limits of the small, vertically migrant bathypelagic fish that are more characteristic of the Trades biome.

## Trades Biome

The Case 4 model is appropriate to that part of the ocean at low latitudes (forming 22% of the whole ocean) that lies under the influence of the trade winds, between the STC zones that lie across the subtropical gyres of each hemisphere, where seasonality is weakest and where changes in depth of discontinuity layers is imposed principally by geostrophic adjustment of the zonal equatorial current systems. Mixed-layer depth is therefore relatively constant, so that primary production rate and phytoplankton biomass take characteristically low values; a deep chlorophyll maximum is almost always present, the "Typical Tropical Profile" being a paradigm for the vertical arrangement of the planktonic food web.

The nutricline is perennially shoaler than the photic depth, except during exceptional events, so vertically integrated production rate is not light-limited, although productivity is nutrient or trace element limited. This situation obtains essentially year-round, and the pattern breaks down with deepening of the mixed layer only briefly and only in some regions. Minor rate increases are forced by geostrophic response of the pycnocline to seasonality in trade wind stress, and by open-ocean Ekman suction, or by divergence, particularly at the equator.

Because of the different shape of ocean basins, and the different arrangement of high terrain on the continents, the pattern of trades and westerly winds is not identical over each central gyre. As noted earlier, the STC zones of the southern hemisphere lie much farther poleward than in the north, so that the partition between their characteristic biomes differs significantly. Thus, there is an important distinction between the proposal for an ecological partition discussed here, and previous proposals in which conditions within central gyres were assumed to be essentially uniform (e.g., Miller, 2004). On the contrary, here it is thought to be significant that almost the whole South Atlantic gyre lies below trade winds, while the demarcation between trades and westerlies lies across the North Atlantic central gyre at about the latitude of Bermuda: a similar partition of wind regimes occurs in the North Pacific.

The Case 5 model describes the dynamics of those parts of the trade-wind zone where strong seasonality is a feature of local wind forcing; these are the monsoon regions where, in the extreme case of the northwest Indian Ocean, seasonal wind reversal results in seasonal reversal of the monsoon currents, and seasonal alternation between eutrophic and oligotrophic biological systems. The small zonal dimension of the tropical Atlantic permits a strong seasonal response to the surge of southerly trades across the equator that is analogous to the effect of monsoon reversal in the Arabian Sea.

Though the seasonal variance in frictional wind stress at the sea surface here is equivalent to that in the latitudes of winter westerlies, the consequences are quite different, as was discussed in Chapter 4. It is, of course, the seasonal meridional migration of the north and south trade-wind belts and of the intertropical convergence zone (ITCZ), marking the doldrum region of calm winds between the trades, that is the source of the variability in wind stress that forces seasonality in the tropical ocean. The migration of the ITCZ modifies the sign and intensity of Ekman vertical motion over large regions (Isemer and Hasse, 1987).

Productivity is rarely light-limited and responds rapidly to nutrient entrainment into the photic zone by coastal upwelling, offshore Ekman suction, and geostrophic adjustment of pycnocline associated with reversal of the monsoon winds. The pycnocline is illuminated during the oligotrophic season, but the mixed layer shoals above the photic zone during upwelling episodes. Changes in phytoplankton biomass match the seasonal changes in production rate, consistent with consumption and production rates being closely matched. Seasonal algal blooms are diatom-dominated, with picocyanobacteria and other prochlorophytes contributing only 30–40% of phytoplankton carbon, compared to >90% in the oligotrophic regions/seasons.

The trade-winds regime covers about 45% of the total area of the ocean, and for that reason alone it will be useful here to review very briefly its principal physical characteristics that determine its singular ecology:

- *The radiation balance* is such that there is a mean annual positive downward heat flux across the surface. The area so defined has its widest latitudinal extent in the Atlantic.
- *The seasonal radiation flux* is such that surface mixed layer is maintained continuously; at no season is there sufficient loss of heat to produce convective deepening of the permanent tropical thermocline.
- *The level of solar radiation* is such that autotrophs are less commonly light-limited than they are at higher latitudes.
- *The Rossby radius of internal deformation* increases into the tropical zone as a consequence of diminishing Coriolis force; therefore, eddies are increasingly larger but fewer toward the equator.
- *Baroclinic time scales* are weeks in the tropics rather than years at higher latitudes. The equatorward-diminishing Coriolis parameter ($f$) means that the slope of the sea surface required to force horizontal motion diminishes toward the equator.
- *The pycnocline* is coincident with permanent tropical thermocline and has very high stability ($N$; the Brunt-Väisälä frequency takes high values) equatorward of latitude at which winter convective mixing becomes trivial.
- *The tropical Ekman layer* [$Z_{ekman}$ = friction of imposed stress/($Nf$)] lies above a very sharp density discontinuity where turbulence is weak. Most energy from wind friction becomes kinetic and little is used to further deepen Ekman layer.

For all these reasons, the locally forced $Z_m$ model of Sverdrup is insufficient to account for observed algal blooms in the tropical ocean, although it is still often invoked for this purpose (e.g., Wroblewski *et al.*, 1988; Yentsch, 1990; Obata *et al.*, 1996). The seasonal changes in the depth of the mixed layer that occur in responses to distant geostrophic forcing are relatively smaller than those caused by deep winter convection in the westerlies biome. Nevertheless, some of these distantly forced changes do result directly in a bloom, as in the case of the spin-up of geostrophic domes and where a basin-scale thermocline tilt shoals the nutricline into the photic zone. Vertical Ekman velocities in the tropical ocean are similar to rates at high latitudes (Isemer and Hasse, 1987) but may act over larger areas and may be effective in sustaining relatively high chlorophyll at the surface for extended periods, as occurs in the Arabian Sea (Brock et al., 1991).

An increase in the rate of primary production is normally accompanied by a simultaneous accumulation of chlorophyll because growth and loss terms are closely coupled, so that standing stock or biomass values vary over only a relatively small dynamic range. It here that the very small cell fraction of the phytoplankton is most important, and even the eukaryotic cells themselves are of relatively small size, while the submicron picoplankton fraction dominates: <90% of all plant biomass and <80% of its production may be attributable to this fraction. The perennially relatively low nutrient concentrations have induced the evolution of special strategies, such as the symbiotic relationships of heliozoans and their autotrophic plastids, and some large cells have evolved special forms that enhance the rate of uptake of scarce nutrient molecules across their cell walls, and minimize their sinking rate.

Although nitrogen fixation occurs in the pelagos of the Westerlies biome, especially by epiphytic cells on *Sargassum* of the North Atlantic and as endosymbionts within diatoms, the blooms of nitrogen-fixing cyanophytes of this biome have become the paradigm for $N_2$ fixation among phytoplankton cells. The apparent coincidence of these blooms with trajectories of aeolian dust may be related to the high demand for cellular iron during $N_2$ fixation (de Baar *et al.*, 1997). Four pelagic species of the cyanobacterium *Trichodesmium* (= *Oscillatoria*) contribute macroscopic cellular mats to these blooms, which occur seasonally during the most windless period of the year and may color the sea surface red as far as the eye can see; the global distribution of actively growing *Trichodesmium* populations is limited by the seasonal 20°C sea-surface isotherm (Carpenter, 1989). It has a remarkable ability to fix gaseous nitrogen and release oxygen under aerobic conditions as well as to offer a substrate for a wide range of consorting organisms within its loose colonial mats (Capone *et al.*, 1997).

Though diel vertical migration of planktonic organisms is a global phenomenon, occurring in all kinds of waters from small freshwater ponds to the deep ocean, it is in the Trades biome that it extends over greater depth intervals and is most ubiquitous. At all seasons and in all areas, we may expect that a substantial fraction of the zooplankton and nekton, especially copepods (*Pleuromamma Metridia*, and *Euchaeta*), euphausiids, myctophid fish, and squids, will arrive at or close under the surface soon after dusk and to descend again to 200–500 m at dawn. Watching the swarming of deep red bathypelagic squid and black or silvery myctophid fish under the lights of a research ship on station at night, apparently feeding on migrant euphausiids, is one of the pleasures of tropical oceanography.

In this biome, the pelagic ecosystem is at its most taxonomically diverse and represents the climax community of the pelagos. All other pelagic ecosystems can be viewed as derivatives from it, constrained or stressed in different ways that suppress some of the complexities of the tropical pelagos. As shown in Table 6.1, it is in the tropical oceans that the overall dominance of copepods, both numerically and as relative plankton biomass, is weakest; copepods are reduced to only 33% of total biomass. The difference between this and the 68% of polar oceans is accounted for by relatively greater biomass of several groups in tropical seas.

It is also in this biome that pelagic fish reach their greatest development, and another pleasure of tropical oceanography is to stand on the foredeck under way and watch the scatter of flying fish from under the bows. A great variety of shoaling clupeids, schooling tuna and other scombroids, and solitary sharks also inhabit the tropical pelagos so that there are multiple food chains, each more complex and longer than those in high latitudes. This is reflected in the temporal stability of the tropical biome and the climax nature of the ecosystem. Populations are more sustainable over the long term in tropical seas, recruitment in fish species between years usually having a variance of less than 3 whereas, for high-latitude fish, recruitment may vary by a factor of 30—or even more.

## Coastal Biome

The physical forcing of water motion and stratification over continental shelves is very similar at all latitudes equatorward of the Polar Fronts, beyond which the presence of sea ice induces unique conditions. I suggest that this commonality of physical processes on shelves is more significant than the consequences of latitude; for this reason, I suggest that all continental shelf regions, save only within the Polar biome, should comprise a single unit, the Coastal biome.

Here, tidal streams and water depth control in the same manner where tidally mixed and stratified shelf regions should occur, and the consequences of the flow of silty, nutrient-rich river water out across the shelf are very similar everywhere. With some exceptions that shall be noted, sedimentary regimes and the nature of deposits is rather uniform and the ecological interaction between benthic and pelagic biota follows rather similar logic everywhere. The strong interaction between the oceanography of any continental shelf and that of the adjacent oceanic biome that undoubtedly exists everywhere is, nevertheless, of secondary importance to regional ecology compared with the physical processes over the shelf.

The concept of a *coastal boundary zone* in the ocean was introduced by Mittelstaedt (1991) and defines what many oceanographers regard as the critical regional distinction between processes over continental shelves and those in the open oceans. The width of the coastal boundary will tend to be a function of latitude, if topography permits, because (as discussed in Chapter 4) the Rossby radius of deformation is a function of the relative strength of the Coriolis acceleration, itself a function of latitude.

Continental shelves, the fundamental habitat of the Coastal biome, reflect the geomorphology of the continent of which they form the margin so that not only the width, depth, orientation, and topography of the continental shelf are significant, but also the nature of the coastline itself. Rias and fjordlands, shallow sedimentary regions (with or without fringing mangroves), and the existence of major rivers are some of the characteristics that will determine the nature of the regional coastal ecosystem. Finally, the discharge of silt by rivers onto the shelf is much greater in low than high latitudes. Three-quarters of all silt delivered by all rivers comes from the Amazon/Orinoco (11%) and from the rivers that enter the Bay of Bengal, the South China Sea, and the basins of the Indo-Pacific archipelago (65%). In these regions, then, we may expect to find greater areas of continental shelf dominated by silty, organic-rich deposits, and so it is. Shallow seas in regions such as the Gulf of Thailand have almost entirely soft, muddy deposits. Nevertheless, there is significant commonality in the ecology of muddy bottoms at all latitudes, and whether the facies is small or is extensive.

The concept of shelf and slope water regimes, originally applied only to wide midlatitude shelves, is probably relevant at all latitudes. Shelf-break fronts are not only a useful defining feature for the boundaries of this biome but, as discussed in Chapter 2, they also have consequences for algal growth that are likely to be of importance everywhere, and they will require special attention.

Coastal upwelling, and nutrient enrichment, occurs off all coasts and at all latitudes, seasonally forced by interaction among local topography, coastal currents, wind, and Kelvin waves induced by distant wind stress. However, with the exception of the Somali-Omani upwelling, all these are minor compared with upwelling in the California, Humboldt, Canary, and Benguela current systems, and it is toward these that the attention of the biological oceanography community has been largely directed. The wide range of conditions in these sites invites study by the comparative method, and this has been a feature of research here: their individual characteristics will be discussed in later chapters.

Over the eastern boundary currents, the equatorward trade wind field has two maxima. One of these, at 100–300 km from the coast, is also the line of zero wind curl so that cyclonic curl (upward Ekman motion) occurs landward of the wind maximum, and anticyclonic curl (downward Ekman motion) seaward. The offshore current velocity maximum is also coincident with this offshore maximum in the wind field. A second equatorward wind velocity maximum often occurs at the coast and is the boundary-layer response of the eastward component of the oceanic wind field to its encounter with coastal topography. Coastal winds also have a greater diel component than the oceanic wind field, forced by the opposition of daytime solar heating and evening wind mixing, and this phenomenon may give rise to diel changes in near-shore water column stability.

Plankton, and especially the zooplankton, of the coastal biome at all latitudes differs consistently from the plankton of the open ocean because it comprises a higher proportion of meroplankton—the planktonic larvae of benthic and littoral invertebrates. But because benthic invertebrates in higher latitudes tend toward direct development, so the relative contribution of meroplankton to the coastal plankton is another variable that is modified by latitude. For the rest, species comprising the coastal holoplankton are most frequently congeneric with species of the oceanic plankton; only a few higher taxonomic groups of holoplankton, such as the cladocerans, are coastal specialists. These crustacea, most species of which occur in freshwater, occur as seasonal swarms, especially after phytoplankton blooms, in coastal regions at all latitudes (Longhurst, 1985b).

It is characteristic of both major and minor upwelling sites that the specialized zooplankton herbivore should be a seasonally migrant calanoid copepod: in low latitudes this is frequently *Calanoides carinatus* or a closely related taxon. The life history strategy of these organisms is a modification of that of oceanic calanoids of high latitudes: a descent to depths of 500–1000 m during the non-upwelling season and a rapid ascent when upwelling commences. Unlike their boreal analogues, however, their generation time is rapid and several generations may be achieved during a single upwelling season. Nevertheless, in all sites investigated, the resting population comprises only copepodites of the final instar before sexual maturity. In this way, reproduction occurs rapidly after the rise to the surface. In the coastal biome at higher latitudes, adjacent to oceanic regions in which some of the large copepods perform seasonal vertical migrations, parts of these populations are advected over the shelf during summer and are unable to descend to their full overwintering depths at the end of the season. These individuals may gather in deep basins on the shelf in some regions, where they may form apparently permanent expatriate populations.

Although the existence of coral formations in only the tropical seas is a very important source of nonuniformity within the coastal biome, we should not make too much of it in this analysis. Since the littoral zone itself is to be excluded from this study, we shall be concerned only with the consequences of offshore barrier and fringing reefs lying deeper than the atoll and platform reefs of shoal water. I have already discussed in Chapter 3 the coherence of the benthic invertebrate fauna of continental shelves at all latitudes, so that ecologically similar organisms, forming associations of similar diversity, occupy similar kinds of deposits wherever they occur.

We shall have to be aware when discussing biological production and consumption of organic material in the pelagic ecosystem over continental shelves that the trophic coupling of pelagic and benthic ecosystems introduces a factor with which we do not need to concern ourselves when considering the ecology of the oceanic biomes. We shall encounter numerous examples of shelf regions where only quite a small fraction of the phytoplankton biomass is consumed within the plankton ecosystem, and we shall have to follow the fate of these cells as they sink unconsumed to the shallow sea floor. There, they are metabolized within the benthic ecosystem and sustain a bewildering array of pathways through the macro, the meio-, and the micro-benthic communities. It is no more than a statement of the obvious, but remineralized nutrient molecules in shallow water are of course much more directly accessible to the photoautotrophic plankton than in the open seas, even if (as a generality) 80% of the available nutrients on shelves originates in the deep sea. Not only do such molecules "leak" continuously from sediment porewater into the overlying water mass, but any transient mixing process will tend to resuspend superficial organic matter from the surface of the sediments and so accelerate the process of renewing mixed-layer nutrients.

We must also be aware that even in the absence of macroalgal beds as occur along temperate-zone upwelling coasts, there is significant autotrophic production on many continental shelves by benthic diatoms on the surface of sediments down to as deep as 50 m. Where irradiance at the sediment surface is >5% of surface irradiance, we may find that the rate of growth of benthic diatoms is equivalent to phytoplankton production in the overlying water mass.

To understand the biological processes by which pelagic organic matter is consumed and transformed in the benthic ecosystem, and how nutrients arising from this process are reinjected into the planktonic ecosystem, we shall have to have to understand the outlines of active material transfer within the benthos. This is not simple to generalize, because specialized feeding strategies are greatly more numerous and greatly more specialized than within the pelagic ecosystem. Moreover, their study has not been central to the thrust of biological oceanography in recent years and the information we require is widely scattered. However, some generalizations can be made concerning relative rates of these processes on continental shelves. With the use of benthic respiration chambers, it has been shown that activity of macro-benthos and meio-benthos are mutually exclusive spatially, depending on the nature of the sediments, principally their relative content of soft organic material. Compartmentalized, macrofauna accounts for as little as 25% of all respiration, whereas microbiota account for as much as 45% of total community respiration. Further, relative activity diminishes into deep water, other things being equal, and this trend continues into the deep sea. From 10 m to 100 m, the rate of benthic respiration generally falls by as much as an order of magnitude.

The nature and amount of settling organic material on continental shelves determines the response of the benthic ecosystem, which under most circumstances very rapidly remineralizes it and ensures its burial, sometimes its deep burial by the activity of the constituent macrofauna. The C/N ratio of settling material varies seasonally and is higher than in the open ocean, but always much lower than the ratio of riverborne debris from vascular plants (where C/N < 100) so the material itself is much more labile. A settling phytoplankton bloom has a C/N ratio of around 7, which is close to the ratio of living phytoplankton, but the slower rate of settlement in post-bloom, stable conditions is of order C/N = 10. So not only does the supply of organic material to the benthos pulse, following the phytoplankton calendar, but it differs in its composition seasonally. This pulsed pattern is enhanced by the well-known general and probably near-linear relationship between export production (or the percentage of cells produced that sink) and total production rate.

Although in most models of benthic-pelagic coupling this simple vertical flux is invoked as the principal input, in reality the energetic horizontal water movements over shallow water induces a much greater horizontal flux, and resuspension. Thus, one cannot assume that there is a more than general relationship between water column production of organic material and the response of the benthic community spatially below. In addition, the existence of a permanent or summer thermocline may act to isolate the two ecological systems. Such hydrographic factors have significant influence on the manner in which benthic organisms are distributed on the seabed and their response to production rates in the overlying water column. Where conditions are unusually favorable for the local supply of organic material by some combination of physical processes, benthic biomass may be highly specialized and extraordinarily abundant: the slope off western Sweden may support aggregations of the brittlestar *Amphiura filiformis* at densities of around 3000 $m^{-2}$. This organism is facultatively both a suspension and a deposit feeder, and it functions particularly well where physical conditions of turbulence, current speed, and siltation rate are highly variable.

Chapter 7

# Provinces: The Secondary Compartments

A partition into four biomes clearly does not fully satisfy the requirement for a regional geography of the ocean because it fails to capture the fine detail observed in satellite images of the surface chlorophyll field, and so fails to reflect the regional structure of the pelagic ecosystem. Equally obviously, the two polar regions function differently, though following similar general rules, and the midlatitude and tropical regions of the Pacific, Atlantic, and Indian Oceans each have individual characteristics that we wish to understand. And each semienclosed sea and shelf region differs from every other in its characteristic ecology.

The primary biomes of the pelagic realm were characterized rather simply in Chapter 6, but if we are interested in a smaller-scale partition of the ocean, then these four biomes must each be rationally subdivided. To do this, we must have recourse to a wider set of factors, especially those apt to define interfaces between physically, and therefore ecologically, distinct regions. We should examine the factors that determine the characteristics of regional circulation and stratification at all scales, of which the resistance of the mixed layer to deepening (the primary determinant of biomes) is only a single example. Bathymetry, river discharges, characteristic coastal wind systems, location of islands, and the distribution of land masses must all be reviewed. The procedures adopted to perform this finer partition are discussed in this chapter.

It is important to accept, as I have reiterated at several points in this book, that establishing a partition of the ocean's surface is a fractal problem, because physical phenomena occur across such a wide range of dimension. Though some phenomena have a characteristic scale, many others that are critical to this task do not. An example that is frequently used to illustrate the fractal concept is vorticity in the atmosphere, which occurs across a wide range of horizontal dimension: dust devils (10 m), tornadoes (1000 m), cyclonic storms (100 km), and weather systems (1000 km). Similarly, vorticity in ocean circulation exists from small swirls, to mesoscale eddies, and up to entire ocean gyres. Perturbation of flow, and fronts between opposing flow, in water bodies may be observed alongside a jetty or across an ocean basin.

We cannot expect, therefore, that an objective review of the processes that are likely to be relevant to pelagic ecology shall constrain us to converge on an objective number of compartments. Rather, it will be better to pursue our investigations with a subjective ideal in mind. It has already been suggested that about 50 compartments globally would be convenient for the computation of global primary production (Sathyendranath *et al.*, 1995), a suggestion that has been allowed to guide these proposals for a global partition of the pelagic ecosystem. This is a matter not only of practical convenience but also of necessity because we are constrained by the available number of observations of any ecological variable or parameter. Where too few observations are binned among too many compartments, we must expect to have insufficient numbers in each to judge whether the characteristics of adjacent compartments are significantly different, and therefore if the boundaries between them are real.

Obviously, the idea of partitioning the ocean into provinces is not new, and many schemes to do this already exist beyond those from strictly descriptive study of the geographical distribution of biota discussed in Chapter 2. Why, you might ask, do we need yet another scheme? This is not the place to review the validity for our purpose of existing schemes, such as the Large Marine Ecosystem concept introduced for fishery management purposes by Ken Sherman that—as I have suggested elsewhere—is perhaps principally useful as a symbol and rallying point for funds for research on "ecosystem-sensitive" fisheries management. The compartments of this scheme, as of other such schemes, have only a limited basis in regional oceanography. Consider the partitioning of global fishery statistics among subareas of the ocean by the United Nations Food and Agricultural Organisation (FAO) for the past many decades. These compartments have been discussed (Gulland, 1971) as if they represented natural areas of the ocean, though a glance at the FAO map will show that they conform to no possible oceanographic reality. One of the most extreme examples is FAO area "H2," which runs from the Bay of Bengal to Tasmania and includes the whole eastern part of the Indian Ocean. To put a part of the Southern Ocean together with the Bay of Bengal for fishery purposes is a breathtaking denial of the natural order of the ocean and also (dare I say?) of political reality.

All this suggests that biological oceanographers face the same difficulty as did Tomczak and Godfrey (1994), who commented that previous regional systems fail to match our present understanding of ocean circulation. They pointed out that the widely used subdivision of the ocean basins adopted by the International Hydrographic Bureau (IHB) is not optimal for scientific description of natural processes. The IHB map assumes that the Atlantic, Pacific, and Indian oceans extend poleward down to Antarctica and was drafted before the true nature of the circulation of the Southern Ocean was understood. In fact, the polar ocean is a good starting point to introduce the present suggestion for a system of oceanic biogeographic provinces.

Since the days of the seminal investigations of Alister Hardy and his colleagues in the *Discovery* expeditions to the Southern Ocean during the 1920s, oceanographers, including biologists, have found overwhelming evidence for the ecological, chemical, and physical unity of the Southern Ocean. This ocean is unique because of the annular currents that flow continuously around the globe and because of the importance of convergence and divergence within these flows. Therefore, contrary to the view of the International Hydrographic Bureau, an oceanographer must place the southern boundaries of the three principal oceans—Atlantic, Pacific, and Indian—at the Subtropical Front. This lies between Patagonia and Tasmania and just to the south of Australia and South Africa. Further poleward is the Southern Ocean.

In the other hemisphere, the Arctic Ocean must be treated differently. This is a relatively small basin, almost land-locked and with narrow connections to the Atlantic and Pacific Oceans, and is almost perpetually ice-covered, except marginally along the Asiatic coast. It will be appropriate to follow the physical oceanographers and treat this as a land-locked sea such as the Baltic, Mediterranean, Caribbean, Okhotsk, and Red Seas. Such seas could either be discussed collectively or allocated to the adjoining ocean as a more-or-less open "marginal sea." The latter option is greatly more informative and will be followed. The Arctic shall be considered a marginal sea of the Atlantic, to which it is connected through Fram Straits and the Norwegian-Greenland Sea. Tomczak and Goodall consider the Arctic Ocean, the Barents Sea, and the Norwegian-Greenland Sea to be a useful unit (the "Arctic Mediterranean Sea") when describing regional circulation, but we shall find it more useful to treat them separately.

The coastal biome, as we have defined it in Chapter 6, shall therefore comprise not only open and reasonably linear continental coastlines, but also marginal seas and archipelagos where the islands are sufficiently large and numerous to modify the ocean circulation, and important areas of shallow water. For instance, we shall include in this category the Arctic, Indonesian, and Philippine archipelagos but not (for example) the oceanic Marquesas or Cook Islands.

Our analysis will profit from careful comparisons between the three major oceans and their marginal seas. We shall use the powerful comparative method of geographical analysis and utilize the effects of differing dimension and boundary conditions between different oceans and seas as natural experiments to understand regional oceanography and ecology. Such apparently simple matters as the difference between how far Patagonia and South Africa extend poleward have profound effects on the ecology of the tropical Atlantic and Indian oceans.

## Ecological Provinces in the Open Ocean

It was partly the availability of the global CZCS chlorophyll field that reactivated the search for a satisfactory way of defining ecological provinces. Platt and Sathyendranath (1988) suggested that calculations of primary production should be partitioned between biogeochemical provinces (BGCPs) within which photosynthetic parameters, and the form of the chlorophyll profile, might be seasonably predictable. Their proposal rests on the fact that regional differences exist in the photosynthetic response of phytoplankton to changes in environmental conditions, probably associated with changes in the composition of the phytoplankton. Following this proposal, but using different bio-optical criteria, Mueller and Lang (1989) suggested how the northeast Pacific might be partitioned objectively into provinces compatible with the concept of the BGCP.

At the same time, not every biological oceanographer thinks that ecological models must be partitioned to achieve satisfactory global integration. Morel and Berthon (1989), for instance, suggested that surface chlorophyll specifies subsurface chlorophyll distribution sufficiently well that a one-dimensional, vertically integrated model of primary production may be scaled to the surface chlorophyll field and applied globally, as is now done routinely by Behrenfeld and Falkowski. This is the approach now used in a routine compilation of global maps of primary production using a vertically integrated mode and is not so far from the concept of ecological continuity that is often assumed when one-dimensional models of biological processes are incorporated into general circulation models. Ecological continuity requires either that the parameters of a biological model remain constant everywhere or that they always respond in the same way to changes in the environmental covariables such as temperature and nutrients. However, biological responses to changes in environmental conditions are often species dependent and, in nature, further complicated by species succession. Therefore, they may be highly nonlinear.

In fact, Platt and Sathyendranath (1999) suggested that because the parameters descriptive of the phytoplankton profile respond to regional ocean physics they are "neither globally uniform, nor invariant with season." These, and physiological parameters (they comment), are distributed in such a manner that the "functional structure of the ecosystem is determined everywhere by physical forcing, the variation in the physical forcing from place to place, and by the associated biological response." Consequently, "In the ocean, these parameters are believed to be distributed in a manner that is not smoothly continuous. Rather, they seem to have a piece-wise continuous distribution."

Ideally, global fields of relevant data would carry information enabling us to set objective boundaries between provinces or would demonstrate that a set of provinces that had been postulated subjectively could not be supported objectively. Available data archives, for both biological rate processes and the distribution of biota, are inadequate for this task and even the most elementary physical observations, such as mixed-layer depths, are adequate only for some parts of the ocean. Elsewhere, as in the South Atlantic and South Pacific, there are too few data even to contour seasonal mean mixed-layer depth fields (e.g., Levitus, 1982).

However, some attempts have been made to devise an objective methodology for deriving boundaries between meaningful provinces in oceanographic data. Using sub- and near-surface temperature and salinity from a repeated meridional transect of the Atlantic, Hooker *et al.* (2000) used the pattern of local extrema in the first spatial derivative of density ( $_t$) to locate boundaries between zonal hydrographic provinces for the Atlantic Ocean. The result was intuitive, and most of the indicated boundaries were located at the edges of flow patterns. And, according to the authors, "if judicious choices are made in grouping the 14 indicated provinces" the basic arrangement of the provinces used in this book can be reproduced.

But the latter were not derived in this objective manner. Because, to be frank, it was rather the result than the method that was paramount, I took a more classical approach—following, for example, Colborn (1975), who derived operational partitions of the stratification of the Indian Ocean, from the Bay of Bengal to Antarctica. He assembled all available hydrographic data and subjectively partitioned this region into 34 provinces by what he called their "Distinct Thermal Structure." This was derived from the annual cycle of stability in the upper half kilometer of the water column, and the provinces were grouped into "Types," distinguished by the following criteria. Is the permanent thermocline shallow or deep? Is a summer thermocline formed in the mixed layer or not? Does winter mixing—if it occurs—reach shallow or deep? If one were judiciously to group those adjacent provinces having the same type of seasonal cycle, one would arrive at a total of about eight or nine "mega-provinces" for the noncoastal parts of this ocean. These would be the Red Sea and Gulf, the Arabian Sea, the Bay of Bengal, the equatorial regions from 5°N to 15°S, the southern gyre down to the subtropical convergence, and some subdivision of the Southern Ocean. As we shall see, such an arrangement would conform quite closely to the system used in the later chapters of this book.

Colborn's provinces, Sverdrup tells us, must also have seasonal cycles of phytoplankton production and of everything that stems therefrom, appropriate to the observed seasonal water column stability, transparency, and local irradiance. This is precisely what Banse (1987) found when he merged Colborn's compilation with CZCS surface chlorophyll to obtain three types of "Area" in the northern Arabian Sea, each having similar phytoplankton calendars. Had Colborn performed his critical analysis for the whole globe, we should probably have what we want. Unfortunately, he did not look beyond the Indian Ocean.

Also unfortunately for an enterprise such as this, the planetary wind systems, and the driving forces they exert on the ocean circulation change not only from season to season, but from year to year and also at longer time scales. Consequently, the global circulation pattern is always changing, sometimes more and sometimes less, from the long-term mean seasonal pattern, a problem that shall be discussed in the following chapter. Although we know that the features we choose to delineate the provinces are in a constant state of change, it may be necessary to draw the boundaries as if fixed in space and time as a matter of convenience when describing the distinctive characteristics of each, as discussed below. As Platt and Sathyendranath (1999) remarked, "When we name a province, we give it a nominal boundary only for convenience . . . on the other hand, the instantaneous boundary of the province has practical value in real applications." It should not be necessary to emphasize that dynamic province boundaries must be the ideal and that static boundaries can never perfectly resemble reality, which must be the ideal.

But, before we get to that, this problem raises another one that we must deal with here. The partition of the ocean required by biogeochemists must be useful for the deployment of models of (say) primary production of phytoplankton to analyze a field of (say) satellite observations of chlorophyll. For such tasks, it may be more important to locate the areas and periods over which a certain parameterization of a model is appropriate, than to enable coherent regional statements about the scientific natural history of the sea to be made. But it is really the latter that is my object in this book so that, despite the use of the edge-adjustment routine "MakeShift" of George White (see Chapter 1), I have somewhat distanced the provinces used here from the concept of the dynamic biogeochemical province.

The partition of the Arabian Sea of Brock *et al.* (1998) done to support the deployment of a production model illustrates how others have established provinces within which to partition computations of ocean productivity. These authors merged climatologies of incident light, mixed-layer depth, and chlorophyll so as to identify three fundamental types of "biohydro-optical class" defined by the form of the chlorophyll profile: (i) Typical Tropical Profile and (ii) Mixed Layer Bloom, plus (iii) a Transitional type between the first two. These classes were discussed as if they could serve as the biogeochemical provinces in a partition of this region even though, at different seasons, different profile types might occur at any given location. The seasonal output from interaction between the three climatologies, computed pixel by pixel across the whole Arabian Sea, indeed showed a seasonal, meridional march of the three classes that was imposed by changing irradiance and wind speeds.

This partition was later used to model nitrate-based production in the same region that was partitioned by the use of a decision tree for water depth, water temperature, and rate of primary production to allocate each pixel to one of six provinces (Watts *et al.* 1999); thus, the concept of a "province" was equated with the location of a particular phase in a seasonal plankton calendar. This is very different from the concept of bounded geographical entities that are modified by seasonal or between-year anomalies to their average hydrographic situation. Such, of course, is the concept of an ecological province defined and used in this book, and that I have described elsewhere; this concept can perhaps best be explained by a description of how the partition was actually achieved.

So, what follows is an account of the steps that were taken toward defining a global set of ecological provinces and their boundaries. Things are often tidier in the telling than in the doing, and the following description implies a logical progression that is rather different from what actually occurred: most of the activities described in this section were undertaken more or less simultaneously, and it was only the testing of boundaries that logically had to follow the other work, and did so. Furthermore, the work was performed in the early 1990s by reference to CZCS data, prior to the availability of the much-superior SeaWiFS and MODIS images, and prior to the general availability of sea-surface elevation images at the basin scale. Modern data have been used to make minor modifications to, and clarifications of, the boundaries originally proposed.

The first step was an examination of all available regional and seasonal images of the surface CZCS chlorophyll field in a variety of formats for characteristic, observable, and repetitive regional patterns, both spatial and temporal. Where necessary, the individual scenes for critical regions were scanned to clarify the nature of blooms observed in the monthly and seasonal composites. This subjective technique of interrogating the images to locate boundaries was a proxy for the objective and numerical technique that would be the method of choice.

The second step was to examine the regional oceanography of all parts of the ocean not only by bibliographic search but also by consulting data archives. The physical oceanographic literature for each ocean basin was reviewed extensively to compare the seasonal and regional distribution of chlorophyll values found in the SeaWiFS images with current concepts of surface circulation, together with the distribution of oceanic frontal zones. Original data compilations for several properties were obtained and examined:

*Global climatology of mixed-layer depth*: Obtained from NOAA-NODC (updated 1994 archive) on-line data, based on temperature (0.5°C) and density ($0.125_t$) criteria as appropriate for each province. These data were arranged as monthly mean depths on a 1° latitude and longitude grid.

*Brunt-Väisälä frequency*: Profiles were computed from OceanAtlas data (Osborne *et al.*, 1992) along many archived zonal and meridional oceanographic sections, and also along synthetic sections derived from archived and WOCE density profiles at standard intervals of 30° latitude and longitude.

*Rossby internal radius of deformation*: Obtained along meridional sections for the main oceans from tables and maps of Emery *et al.* (1984) and Houry *et al.* (1987).

*Photic depth (m)*: Calculated from the light field and the chlorophyll profile thought to be typical of the season and region derived from the profile archive. The estimate is therefore valid only for Jerlov oceanic type I-III waters (Jerlov, 1964), where phytoplankton comprise the principal source of turbidity. The photic depth is expressed as the monthly mean depth of the 1% isolume (i.e., 1% of integrated surface light, itself based on sun angle and regional cloudiness archives) at each 1° grid point (Sathyendranath *et al.*, 1995). An independent measure of water clarity was afforded by an archive of Secchi disc depths obtained from the NODC, enabling a comparison to be made between regional, seasonal cycles of Secchi depth and of surface chlorophyll (Fig. 7.1).

*Surface nutrient fields*: Obtained from seasonal multidepth climatologies (NOAA-NODC 1994 archives) for all regions of the ocean, issued originally in atlas format but available also as data files for manipulation (Conkright *et al.*, 1994).

Even these data sets failed to describe all parts of the ocean uniformly and adequately: in particular, the South Atlantic and the South Pacific continue relatively data-poor and are also relatively poorly understood or described. This is undoubtedly reflected in how well the northern and southern hemisphere subtropical regions have been partitioned in this review and will be borne in mind by the cautious reader. As we noted in Chapter 4, in relation to the concept of "ecotones," important regions of chlorophyll enhancement cannot be accommodated within provinces that are defined either by circulation features or by water mass types because the bloom is itself induced by boundary conditions between components of the near-surface circulation. The extreme case of this situation is the linear zone of high chlorophyll that is a prominent feature in chlorophyll images of the southern hemisphere, coincident with the austral Subtropical Convergence. Logically, if circulation features are allowed to dominate the setting of boundaries between provinces, this zonal band of high chlorophyll values must either be split between two adjacent provinces meeting at the convergence or assigned arbitrarily to one of them. Either solution would be unsatisfactory, so in this case it is better to allow the chlorophyll field to determine the boundary. The Convergence should be a province, following this view, rather than a boundary: and so it is here.

These data and a detailed consideration of many previous proposals for partitioning the oceans into a global set of provinces led, by a process of trial and error, to a proposal for a series of 51 provinces. The notional boundaries of these provinces were forced onto a rectangular grid at a scale chosen to facilitate the task of assigning data (such as SeaWiFS surface chlorophyll values) to individual provinces: it is this grid that has become the familiar map of provinces shown in Color plate 22.

But a static grid is clearly unreal in the sense that we have already some knowledge of the extent to which the field of features chosen to characterize, or to bound, each province is variable seasonally and between years. Techniques are emerging to assign dynamic

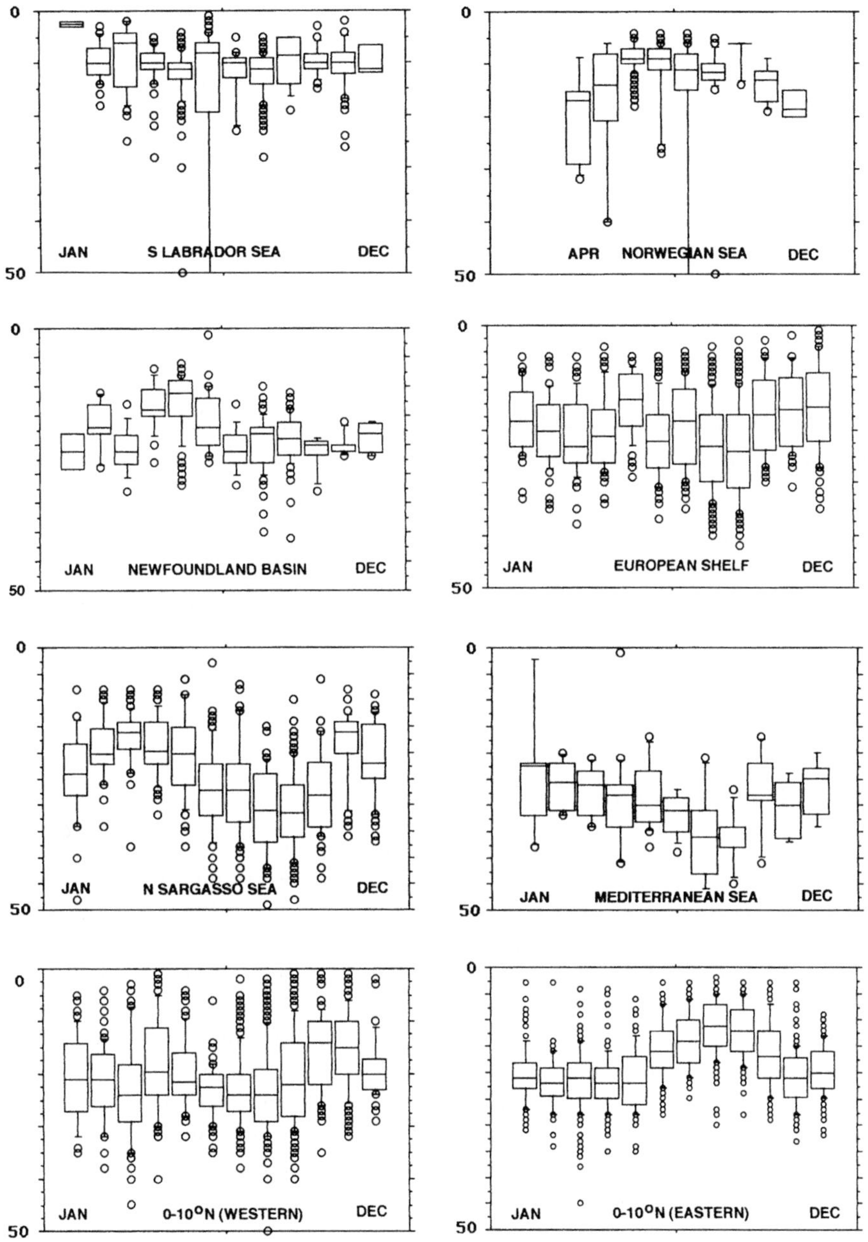


**Fig. 7.1** Water transparency in regions of the North Atlantic to support the interpretation of satellite radiometer images of changes in the seasonal, regional surface chlorophyll concentrations. The European shelf data confirm that it is not only chlorophyll that dominates transparency in coastal regions.

*Source: The analysis was based on 6720 Secchi disc observations archived at the NODC, Washington DC.*

boundaries to provinces in an ocean whose characteristics are defined by time-specific data fields obtained by satellite sensors. An operational protocol for the dynamic assignment of boundaries to provinces is described by Platt *et al.* (2005) based on surface fields SST and chlorophyll as primary indicators, and using bathymetry as a secondary indicator. For each of the seven static provinces as used in this book for the NW Atlantic, frequency distributions of SST are plotted: it is assumed that multiple peaks indicate multiple water-mass conditions and hence nonuniformity within a province. The boundaries are then readjusted using an iterative procedure until, ideally, a single SST peak appears in each province; chlorophyll concentration is taken as an indicator of water mass boundaries in the same way where SST cannot be used. Applied to the North Atlantic provinces, and for spring and autumn of 2003, this enabled their boundaries to be refined to match the distribution of properties in the ocean in those periods, and suggested that another entity ("Slope Water") might usefully be integrated into the partition. Finally, the use of an algorithm that isolates diatom-dominated regional phytoplankton biomass from other populations shows that the dynamic boundaries derived from the temperature field in the NW Atlantic were a good match for boundaries between diatom-dominated and other phytoplankton populations.

## Ways of Testing Static Province Boundaries in the Open Ocean

It would be useful to have some confirmation that the boundaries selected intuitively by analysis of climatological data, and the number of provinces so identified, were not entirely subjective. This can be done in several different ways: (i) to test statistically that conditions differ in adjacent provinces, (ii) to compare data on the distribution of individual biota with boundaries between provinces, and (iii) to use analytical techniques to partition a relevant global data set.

### A Statistical Test

Confirmation of the reality of boundaries was obtained by distributing the individual data from a suitable global archive into compartments representing provinces, and then investigating statistically the similarities and differences between the individual data sets that then represented the attributes of individual provinces.

This was done with the 21,872 sets of parameters descriptive of the chlorophyll profile obtained from the global archive described previously. These values were aggregated seasonally, centered on the 15th day of January, April, July, and October to represent the boreal (austral) winter (summer), spring (fall), summer (winter), and fall (spring) in each province. Of the resulting 204 possible cases (four seasons in each of 51 provinces), for 145 there were >25 profiles, for 38 there were <25 profiles, and for 19 cases there were no profiles at all, though some of these represented polar provinces during winter darkness. This seemed a much better result than might be expected.

Because the depth and relative strength of the deep chlorophyll maximum layer is an integral of the nutrient, light, and density environments of phytoplankton and because it is the most consistent feature of chlorophyll profiles, it seemed to be a good criterion for testing the proposed boundaries. The depth of the chlorophyll maximum is described by the parameter $Z_m$. Where this is a positive integer, it represents the subsurface depth of the layer; where $Z_m$ is a negative integer, the chlorophyll maximum lies at the surface. It is not surprising, therefore, that of the parameters describing the chlorophyll profiles, $Z_m$ consistently showed (as I shall discuss) the greatest differences between regions and seasons when partitioned by the proposed set of boundaries.

Some of the resulting seasonal frequency distributions of $Z_m$ were bimodal, suggesting nonuniform conditions for that season with a province. The reason for this was resolvable by identifying the data creating the minor peak. In most cases, this showed that the anomalous data were a discrete set of observations, concentrated in one part of the province, often over a restricted period. In such cases, it was usually possible to obtain a unimodal set of data to represent a province by a minor adjustment of its boundaries. Obviously, if the historical record of oceanographic expeditions and experiments had been different, the final solution would also have been somewhat different.

Using this process, a set of unimodal distributions of parameter values describing the chlorophyll profiles was obtained that represented each case (province and season) and could be used to check that the boundaries between provinces were realistic. This was done by investigation of the differences between seasonal mean values (for each parameter) across the boundaries between each pair of adjacent provinces using analysis of variance and the Bonferroni-Dunn *post hoc* test. This investigation showed that measures of chlorophyll biomass and depth of its maximum value ($Z_p$, B(0), $B_{int}$, and $Z_m$) differ between adjacent provinces more significantly than the parameters describing the form of the chlorophyll peak itself (s, h, and R), which show weaker regional differentiation; this is evident for both annual and seasonal means for each province.

For all parameters, the most significant boundaries were found to be between provinces of the coastal and oceanic biomes (Sathyendranath *et al.*, 1995), whereas within the oceanic provinces the most distinct boundaries lie to the north and south of the subtropical gyres and thus between the provinces of the westerlies and trades biomes. Most boundaries between oceanic provinces proved to have some significance for the more variable measures ($Z_p$ and $Z_m$), though even these do not discriminate convincingly between boundaries within the most northerly and southerly groups of provinces. These same parameters confirm the reality of the east-west partition of both the subtropical gyre and the tropical regions of the Atlantic.

## Analytical Tests

Such a test was performed by Kawamiya, Kishi, and Suginohara (2000), who embedded an ecosystem model in a general circulation model of the North Pacific Ocean. The output represented chlorophyll biomass, obtained by interaction between the ecosystem model and the relevant physics of the general circulation model. The data used for the ecological partition of the North Pacific (between 60°N and 10°S) represented the monthly average integrated 0- to 150-m chlorophyll (strictly speaking, the "phytoplankton" component in the model) obtained in the sixth year of model integration, when stability had been achieved. Partition of the data set was obtained by empirical orthogonal function (EOF) analysis. EOF 1 showed that the amplitude of seasonal variation of chlorophyll was higher at middle and high latitudes with maxima in April, whereas the peaks in EOF 2 are located in regions of deep mixed layers in spring. Based on the EOF analysis and annual mean abundance, boundaries may be located by fronts either in the EOF modes or in the annual mean abundance. This analysis achieves a delimitation and definition of seven provinces, for each of which a seasonal cycle of phytoplankton abundance is available.

The authors find a "remarkable coincidence" between this result and the arrangement of North Pacific provinces proposed here, several being directly comparable in area and in their seasonal chlorophyll cycle. Several others proposed by them equate, as they remark, with adjacent pairs of the provinces reviewed in this book. The provinces whose location and properties are thus confirmed are BERS, PSAG(E) + PSAG(W), KURO, NPST(W) NPST(E) + NPTG, PNEC + PEQD. Further, by direct comparison with sea surface chlorophyll obtained from several satellite-borne sensors, Kawamiya and his coauthors subjectively confirm that their own proposals are valid—thus closing, as it were, the circle.

Another such test, using the observations made along the repeated Atlantic Meridional Transect (AMT 1–7, 1995–1998) from the UK to Antarctica, was made by Hooker *et al.* (2000). These authors devised an objective methodology for identifying oceanic provinces from multiparameter data taken along the meridional cruise tracks that covered

110° of latitude. In this case, shipboard observations of water properties to 200 m were used to identify geographic discontinuities by means of derivative analysis. The authors suggest that the results validate the concept of biogeochemical provinces discussed here. An independent test using the same techniques from derivative analysis were run with remotely sensed sea-surface bio-optical data obtained from SeaWiFS gave the same result and identified the same province boundaries. The results obtained in this investigation were almost entirely supportive of the partition suggested here for the open Atlantic Ocean—NADR, NAST, NATR, WTRA, and SATL.

## BIOGEOGRAPHIC TESTS

Alain Fonteneau investigated the distribution of all species of tuna in the records maintained by the Inter-American Tropical Tuna Commission in La Jolla, California, and elsewhere, in order to model trophic chains on the high seas. He found a singular concordance between distribution of different species and the boundaries of the provinces proposed here. He remarks, for instance, that there is an excellent concordance between the distribution of yellowfin (*Thunnus albacares*) and skipjack (*Katsuwonus pelamis*) distributions and the equatorial provinces—PNEC and PEQD in the Pacific and ETRA and WTRA in the Atlantic Ocean. It is all the more remarkable that this should be so, given the extraordinary mobility of oceanic tuna, their very extensive migrations, and the independence of the various integrated data sets. I shall make repeated reference to the ORSTOM tuna atlas that embodies his analysis (Fontenau, 1997).

Accumulated data on the distribution of euphausiids in the South Atlantic were used by Gibbons (1997) to test the congruence between the partition of classical biogeography discussed in Chapter 2 into climatic zones (Tropical, Subtropical, Temperate, and so on) and those proposed in early versions of the biogeographical partition discussed here. The test was performed both with all species and also with epipelagic species alone. Using Bray-Curtis clustering techniques, six groups were obtained for all species (Tropical, Warm and Cold Temperate, Subantarctic, and Antarctic) together with and two local groups of epipelagic species (Benguelan and offshore Falkland regions). Gibbons comments that congruence is good between the distributions of these groups and biogeochemical provinces in the south, but poor in the north and coastal regions. Given (i) the static location of Gibbons data points and the dynamic location of province boundaries and (ii) the stronger and more predictable pattern of province boundaries in the Southern Ocean than farther north, this is a predictable result.

The largest biogeographical data base that has been applied to verification of the provinces discussed here is, of course, that of Hardy's Continuous Plankton Recorder (CPR) by Beaugrand and his colleagues at Plymouth (e.g., Beaugrand *et al.*, 2002a,b). Although the available data points are extremely numerous ($>18 \times 10^{-6}$ to date) they are not ideally disposed in the open ocean for the purpose, being mostly along "great circle" routes between the UK and North American ports; over the shelf and adjacent regions of the NE Atlantic, they are closer to ideal. The CPR data for calanoid copepods were processed for taxonomic richness and for the occurrence of some 96 taxa, partitioned between day and night samples; Bray-Curtis clustering is followed by a complete linkage clustering technique that offers several cut-off levels of indicator values: using the first eight cutoff levels, a dendrogram comprising 15 species clusters was obtained, each of which mapped coherently into a specific region of the North Atlantic. Of these, several appear to represent transitions between other clusters and are interpreted as ecotones—either between two oceanic clusters or between shelf and oceanic regions.

The first cutoff level (0.5) separated the Arctic biome from the Westerly Winds biome and suggested a boundary between them that was consistent with that used here; also consistent was the relative diversity—4 indicator species in the former, 58 in the latter.

A cutoff level of 0.4 isolated the Coastal biome from the remainder. Separation of additional groups was achieved at progressed cutoff levels; the principal entities recognized may be equated (at least informally) to the GFST, NADR, NAST and NECS provinces.

However that may be, it is not very useful to attempt to verify a partition that is usually presented as a series of rectangular spaces, fixed in location, by reference to the ungridded observed distribution of biological properties; difficulty of verification by this method becomes the more acute the longer the period over which the observations are accumulated: in this case over more than 50 years, during which time we know that species distribution patterns have changed.

I believe we can go little further at the moment in assuring ourselves that the boundaries proposed in this book for the open ocean have some objective reality. The data fields needed for a purely objective ecological partitioning of the ocean neither exist, nor are likely to exist in the foreseeable future. The proof of the pudding must be in the eating. If the provinces proposed here, or something like them, prove to be useful for practical purposes, such as the partitioning of fishery statistics to investigate the biological basis of fisheries production, then they probably resemble those that might have been objectively drafted had the needed data been available. If not, they will rapidly join the heap of discarded biogeographic systems.

## Practicable and Useful Partitions in Coastal Seas

For a partition of the coastal zone, one cannot proceed so methodically as for the open ocean, and it is here that the greatest difficulty exists in proposing a logical and objective system. It is here that the fractal nature of spatial diversity is most pronounced. It is here that one is faced with the greatest number of choices in how to make a partition and no single solution is evident. I have, therefore, adopted a very flexible approach to the selection of unit areas within which it seems useful to consider the regional ecology in an integrated manner. A series of provinces within the Coastal biome has been established that match the characteristic dimension of the oceanic provinces: there are 22 of these, bringing the number of provinces close to the total of 50 that was the original target. Their boundaries are, seaward, the shelf-break or upwelling front present along many continental slopes. Longshore boundaries of coastal provinces are set to match the natural features of the coastline: at prominent capes, especially at those where a major coastal current passes offshore into the open ocean. The suggested boundaries also match the major partitions between the Trades, Westerlies, and Polar biomes so that coastal provinces may be associated with each for some purposes. Several of the provinces proposed in this system correspond at least approximately with one of the well-known "Large Marine Ecosystems" proposed by Ken Sherman as suitable units within which to integrate fishery management research. The boundaries of these "ecosystems" are not, however, entirely based on natural features, and LMEs are not considered further here. For those who may be interested in making a comparison between these and biogeochemical provinces, Pauly *et al.* (2000) provide an analysis of the compatibility of the two systems.

Of course, almost all of the 22 coastal provinces used here are too large to be entirely satisfactory for some purposes, but it is simple for users of the system to establish their own smaller partition. For instance, the Northeast Atlantic Continental Shelf province runs from Cape Finisterre to southern Norway, a region that may be subdivided quite naturally and easily. The following would represent a sensible second-level partition: (i) North Sea from the Straits of Dover to the Shetlands, (ii) English Channel west to Ushant, (iii) Bay of Biscay from Ushant to Cape Finisterre, (iv) northern outer shelf, from the Celtic Sea to the Hebrides, (v) Irish Sea, (vi) the central Baltic, and finally (vii) Gulfs of Bothnia and Finland.

The fractal nature of the coastal zone (remember that a coastline is a common paradigm for fractal geometry) readily leads to an apparently logical, yet very fine regional subdivision, which, for the purpose of devising a global partition, must be resisted. Here, the Gulf of California is treated as one potential subunit of the California Current Province. But Santamaria-del-Angel *et al.* (1994) divided this narrow gulf into 14 biogeographic regions on the basis of seasonal cycles of chlorophyll at 33 locations obtained from the CZCS files—a methodology very close to that used in this book. However, because of the fractal coastline geography, it would be just as logical (if more complete chlorophyll data were available) for a further subdivision of each of these 14 regions to have been made.

I have found it convenient to treat this level of partition rather informally and use it only where it seems essential to increase clarity of the regional description. There will certainly be some who find that this arrangement does not do justice to the smaller shelf region that especially interests them; to these people, I can only say that books, like lives, have their natural limitations.

I hope it will have become obvious to any user of the partition reviewed in this book that I have avoided, so far as I can, being dogmatic concerning the principles on which it is based. Any attempt to partition the surface of our planet immediately encounters tension between the theoretical symmetry of the seasonal irradiance pattern, or the wind systems characteristic of each latitude, with the consequences of nonsymmetric continents. I have suggested a partition of pelagic ecology that seems to be ideal in an ideal world, and have attempted to apply it to a world in which the ideal nature of ocean basins is disrupted by the consequences of plate tectonics. Locations of adjacent mountain ranges, of river drainage basins, of archipelagos, and even the different shapes and sizes of the ocean basins all result from this process. In the following chapters, you will encounter many examples of provinces for which I have had to bend my original guidelines in order to accommodate the facts of geography. The most glaring, perhaps, is the solution found to accommodate the Indonesian archipelago—a great scattering of large and small islands standing on the largest shelf seas anywhere, but within which several deep ocean basins—of the magnitude of small seas, and named as such—exist. I am not proud of the solution presented, but it's the best I can make of a bad job. You will discover others, also, though perhaps not so obviously messy as this one.

# Chapter 8

# Longer Term Responses: From Seasons to Centuries

So far, I have discussed a partition of marine ecosystems into biomes and provinces as if these were permanent entities, fixed in space and invariant in their characteristics at time scales beyond the seasonal, but—as we all know—that is very far from their real condition. This is not the place for a full review of how life in the ocean responds to changes in Earth's physical systems, but it will be useful briefly to consider the problems raised for us by the impermanence of conditions in the sea. For an ecological geographer, the core of the problem arising from this impermanence is the extent to which the boundaries set between natural regions of the ocean shift in response to changing conditions; for an ecologist, it is rather the changing conditions within each region that are more important.

Marine ecosystems certainly have less permanence than terrestrial ecosystems. Ashore, ecologists are not confronted with shifting ecological discontinuities, or with changes in the characteristic conditions of individual ecosystems, because, unless man intervenes, the tree line on a mountain or the passage between grassland and savannah remains approximately static over a human lifetime. It is only on the millennial scale that such boundaries migrate significantly, or that characteristic regional ecosystems disappear. Urban sprawl, deforestation, overgrazing, and intensive agriculture are accomplishing in a few decades what nature can only do in centuries, but that sad fact does not alter the argument. Although the human population explosion can produce pressures that rapidly shift ecological boundaries and modify ecosystems ashore, it is paradoxically more difficult directly to modify the average locations of the ephemeral and shifting ecological boundaries of the seas that are the subject of this study. We can accomplish this only indirectly by atmospheric modification, resulting in a changed global climate and a shifted ocean circulation.

Indeed, if we are agreed that the regional characteristics of marine ecosystems are consequent on the characteristics of the physical environment, then we must assume that ecological conditions are as impermanent as the physical conditions themselves. And these, it is now well understood, are in continual flux and state of change at all scales of variability, both spatial and temporal. Although for most practical purposes we regard the circulation of ocean and atmosphere as having a "normal" behavior (we call it the "average climate"), we really know that this is a moving average and capable of important excursions. At least some of these must be accommodated in any model of how biota are distributed and interact in the oceans. And so they must be accommodated in any account of how marine ecosystems are distributed and function in today's oceans.

# Scales of External Forcing

We need not concern ourselves directly with events at the millennial scale, but it will be useful to remember what has occurred in the not-so-distant past, and the consequences of these events for what we observe at the present time. Two lessons stand out from retrospective studies: the first is how fundamentally and rapidly the ocean circulation may change, and the second is the extent to which it is in constant flux—so that during no two consecutive years are ocean conditions identical.

Self-evidently, the hydrosphere and atmosphere together form a closely coupled system, the thermodynamic state of each being determined by transfers of heat, moisture, and momentum between these two "fluids," which must remain in equilibrium. This balance is, of course, modified by secondary effects induced by the cryosphere and the biosphere: these four systems act out their play on the stage set by the arrangement and topography of the land masses of the geosphere. Energy is transferred within the two "fluids" by the planetary wind systems and by the wind-driven surface currents of the ocean, each of which transfers sensible heat from low to high latitudes. Exchange of latent heat between ocean and atmosphere occurs by means of rainfall and evaporation, balancing the heat capacities of each. Vertical exchange of latent heat between lower atmosphere and stratosphere occurs by convective processes, and between surface and deep oceans by the density-driven flux of the global thermohaline circulation (THC).

We should briefly consider the impermanence of the THC. This major oceanic flux was understood only much later than the wind-driven current system (Stommel and Arons, 1960; Gordon, 1966). The THC is forced by the formation of an extremely dense (cold, highly saline) surface water mass by evaporation during winter in small regions of the Norwegian-Greenland and Labrador Seas, and also in the Weddell Sea. Very recently, it has been shown that it is also formed in the Irminger Sea, beneath the atmospheric "tip-jet" that forms intermittently in winter off southern Greenland (Dickson, 2003). These dense surface waters sink by deep convective flow, in the Atlantic forming the North Atlantic Deep Water, which can be traced south through the Indian and Pacific Oceans, to be progressively and eventually returned surfaceward. Were the THC to be interrupted, the distribution of heat within the ocean, and its exchange with the atmosphere, would be strongly modified and there must be major consequences for global climate.

And, of course, it can be interrupted—as has happened frequently in the past—at the origin of its descending branch. All that is required is an accelerated release of freshwater from glaciers or from sea ice, sufficient to increase the stability of the surface layer in small, critical areas of polar seas and so to cap the deep convective process. More fundamentally, the same may be accomplished by an extension of sea-ice cover, which may completely shut down the process of deep convection. There is evidence that each of these processes occurred in the (climatically) recent past, during the Holocene glaciations. But, closer to our times, it is postulated that the rapid cooling of the Younger Dryas event (c.11 ky BP) was forced by a shutdown of the THC, which entered an alternative steady state. Such a process would result in a complete rearrangement and relocation of the oceanographic features that I have suggested may be useful to indicate the limits of characteristic marine ecosystems in the present ocean. The THC appears to exist in three modes (Clark *et al.*, 2002). The Normal mode is what we see today, whereas during the Glacial mode deep convection reaches only to <2500 m. During the Heinrich mode, northern deep convection is absent and the Atlantic basin is filled, to within 1000 m of the surface, with very cold water originating in the Antarctic deep convective zone. Dansgaard-Oeschger events in the paleoceanographic record represent switches from one state to another, occasioned by rapid (years to decades) warming, followed by a very slow cooling trend. These events have a characteristic return period of $1500 \pm 500$ years. Note our habitual use of the word *normal* for an Atlantic circulation that is no more normal than any other, but just happens to be our own.

I have discussed the THC only as an example of the responsiveness of the ocean to changes in the external forcing agents that together determine the quantity and distribution of solar radiation received by Earth. These are, of course, the factors required by the Milankevitch theory of climate variability: the obliquity of Earth's axis and the eccentricity and precession of our orbit around the sun. To these must (possibly) be added variability in solar output, having periodicity at 2400, 200, 80–90 years, and on down to the 22- and 11-year cycles. Noncyclic, episodic changes in volcanic activity also modify the radiative balance of Earth through changes in atmospheric clarity. Episodes of low activity (1100–1250 A.D.) and of strong vulcanism (1550–1700 A.D.) have been linked to the Mediaeval Warm Period of the 11th to 14th centuries and to the Little Ice Age of the 17th and 18th centuries, respectively. The recovery of species distributions from this last cold event was very slow: the progressive northward advance of "warm," Atlantic water benthic organisms (sponges, decapods, and echinoderms) along the continental slopes of the Barents Sea continued on into the middle of the 20th century (Blacker, 1965).

Even single major eruptions, like that of Mount Pinatubo in 1991, may produce measurable atmospheric cooling. Vulcanism and solar output may also interact: one notes the coincidence of the Maunder and the Spörer minima of sunspot activity during the recent cold centuries. The failure of the trade winds of the tropical Pacific, to be discussed later, had a return interval that varied on a millennial time scale during the Holocene, becoming more frequent until about 1200 years ago, and then declining in frequency toward the present.

Because the integrated response of the atmosphere, cryosphere, and ocean to changes in the radiative balance of Earth includes complex feedback mechanisms and nonlinearities, it is not surprising that a steady state is not achieved. Rather, what takes our attention is a series of major changes of state, at various time scales, but these occur against a white-noise background of continual and apparently random change. It is this variability, or weather, around the present-day climatic mean state that renders the analysis and prediction of change so very complex and unsatisfactory. Yet the consequences of neither the white-noise background nor the larger, discrete events (for which explanation is relatively simple) can be ignored by marine ecologists: for just one example, consider the lack of predictability in annual recruitment success to marine fish stocks. This is a serious problem, both for ecosystem analysis and for the stability of marine fisheries, and it is forced importantly by unpredictable between-year differences in ocean conditions.

The recurrent major changes that take our attention are generated through interaction between ocean and atmospheric circulation. Because it is the latter that appears to lead the dance, we are accustomed to using an index derived from the global atmospheric pressure systems as an index of change in the ocean. As a matter of practical convenience, of course, we have a much better capability for monitoring, daily or weekly, the state of the atmosphere than that of the ocean. The low-frequency (or long time-scale) variability of the circulation of the atmosphere exhibits recurrent and persistent patterns that may persist for several years, or even several decades, and may be of ocean-basin or planetary scale. These patterns involve shifts in atmospheric wave and jet-stream locations, of centers of high and low atmospheric pressure, and are associated with anomalously low or high sea surface temperatures. A dozen or so major teleconnections of this kind are recognized by meteorologists: the North Atlantic Oscillation (NAO, winter months), East Atlantic Pattern (winter), East Atlantic Jet (summer), North Pacific Pattern (spring), Southern Oscillation (all months), and so on.

For each, we have recourse to an index, or comparison of atmospheric pressure at two distant points, which provides a simple description of the general state of the wind pattern over a major ocean basin. However, we should note that such indices may mislead us in two ways. First, their use suggests greater independence of each indexed phenomenon

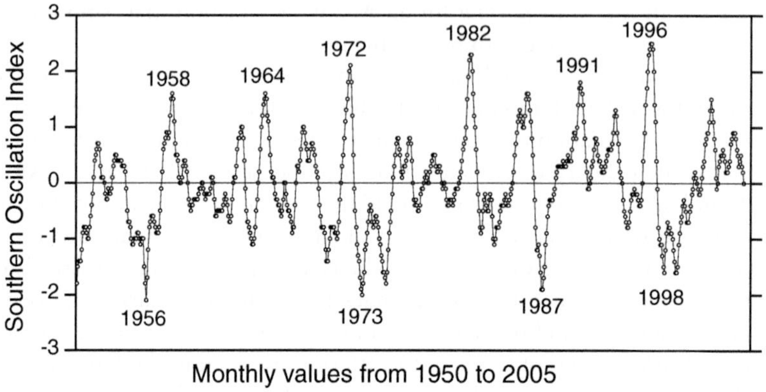


**Fig. 8.1** A half-century of monthly values for the Southern Oscillation Index, 1950–2005, plotted to emphasize the pattern of returns of extreme values, both positive and negative; the 1957–58, the 1973, and subsequent major "Niños" are clearly identifiable in the relevant low values.

than is really appropriate, for in truth none can be isolated from the remainder of the interactive atmosphere-ocean system. And, second, what they are thought to indicate is sometimes oversimplified. Consider the Southern Oscillation Index (SOI, discussed later), used to indicate two, or sometimes three, states of the circulation of the tropical Pacific Ocean. The standardized numerical value of the SOI does not fall easily into three populations (Fig. 8.1), but is closer to white noise on which is imposed excursions held to coincide with changes of state of the ocean. So, although it is convenient to recognize only three states (see later discussion), and to study the consequences of their recurrence, the reality is that the Pacific Ocean exists in a constantly varying set of states: sometimes more like one ideal, sometimes more like another.

Somewhat different is the recently described Pacific Decadal (or North Pacific) Oscillation (PDO) that also merits some discussion. This occurs at similar spatial scale as the SOI or NAO, but at much longer temporal scales. The PDO is thus far characterized only by its effects on the ocean, not by its atmospheric driving factors that remain to be analyzed satisfactorily.

In what follows, I shall discuss important and recurrent changes in conditions at the present day, ocean by ocean, both as a matter of convenience and because this is, after all, a regional ecology of the oceans. These changes recur not cyclically, but with a characteristic return period that includes variance of, usually, a factor of about 2. We shall deal with these processes under two headings or scales of variability: the so-called El Niño–Southern Oscillation (ENSO) scale having a return period usually of less than one decade (2–7 years is usually quoted), and the more obscure decadal changes in global weather patterns and in ocean conditions. Although climatologists are moving very rapidly to integrate disparate phenomena within this complex whole, we cannot hope to achieve a satisfactory integration here of all environmental changes we want to discuss. Many well-observed and regionally important recurrent changes must be part of a greater whole, even though the distant teleconnections have often not been identified.

## Recurrent, ENSO-Scale Changes of State

The tropical trade winds and monsoon rains are reliable from year to year but are notoriously fickle in the longer term, and this unreliability is associated with the significant interannual variability that occurs in ocean circulation in tropical seas. Soon after the

Great Famine of 1877 on the Indian subcontinent, meteorological observatories were established there and undertook to monitor the strength and timing of the monsoon. In the early years of the 20th century, searching for an explanation of monsoon variability, Gilbert Walker, then directing the Indian observatories, examined surface pressure records across the Indo-Pacific. He uncovered a repetitive "Southern Oscillation" in the Pacific-wide pressure field, and so gave us the critical tool to understand interannual change in global weather patterns and ocean variability. Walker's Southern Oscillation (SO) refers to the observation that when atmospheric pressure is high over the Pacific Ocean (the southeast Pacific subtropical high pressure cell) it is low over the Indian Ocean (the Australian-Indonesian low-pressure trough) and vice versa. The state of the SO is quantified by the SO Index (SOI), which, in its simplest form, is the differential sea-level pressure between Darwin and Tahiti. It usually is computed as [dP(Tahiti) – dP(Darwin)]/s.d., where dP represents the monthly pressure anomaly, as the monthly mean minus the 1882–1997 mean.

The seesaw effect of the SOI has attracted a major research effort in recent decades because it involves a great deal more of the global weather pattern than was at first apparent. Low values of SOI, associated with a generalized weakening of the trade winds, foreshadow events in the tropical Pacific that merit more than passing attention, though for a deeper understanding of the physics involved the reader is referred to the now-abundant literature. My preferred starting point would be Tomczak and Godfrey (1994). It is now customary to quantify the SOI into three phases keyed to sea surface temperature in the eastern Pacific: (i) *cool*, with strong coastal upwelling off Peru, off California, and at the equator, (ii) *warm*, when this upwelling weakens, and (iii) *neutral*, or the intermediate condition. Traditionally, the warm phase was regarded as anomalous and greeted in South America as El Niño, the Christ Child who came in winter. Somewhat unnecessarily, oceanographers and meteorologists have coined the term La Niña for the cool, upwelling periods when trade winds are at their most robust and sustained. Some have attempted to classify El Niño events according to their duration of strength, but as Philander (1990) reminds us, no two events are alike: there is no statutory Niño.

Because Niño events are associated with other major climate anomalies than the rains that it brings to the South American coastal deserts (Fig. 8.2), direct historical records exist of the return interval of the stronger events since 1525 and of all events since 1803 (Enfield and Cid, 1990). The period of the SO is determined by the rate of propagation

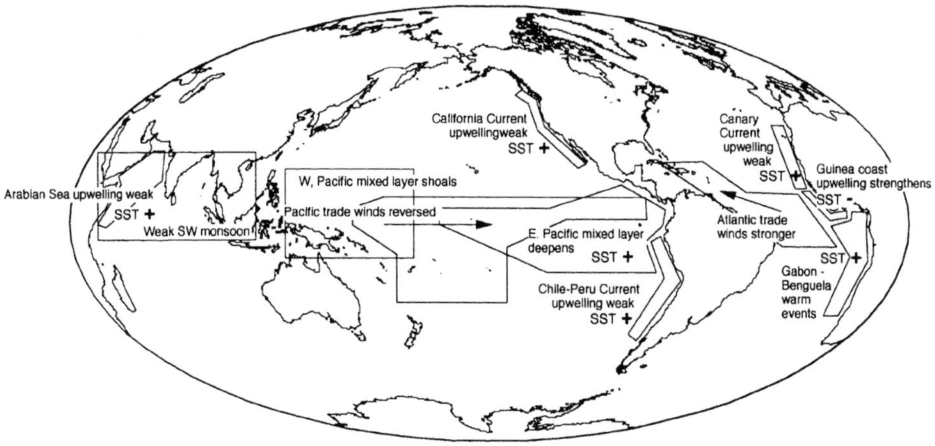


**Fig. 8.2** Responses at low latitudes to a canonical Niño event, from various sources. Note that these events vary significantly and not all the responses indicated will occur during every event, or even simultaneously. In particular, the connection between events in the Atlantic and Pacific is far from secure.

of planetary waves across the Pacific Ocean, modified by orbital and solar constants. From 1803 to 1987, El Niño had a mean return time of 3.7 years during periods of high solar constant and 3.2 years when the solar constant was low. Extreme return intervals ranged from 1 to 8 years. As I shall discuss later, in recent decades the return interval has shortened significantly. The SO is, in truth, the dominant interannual climate signal on the global scale, as Philander and Rasmusson pointed out in 1985. If we assume that the relaxed state of the tropical weather patterns is the existence of a well-developed trade wind belt in both hemispheres, then a strong zonal pressure gradient across the tropical Indo-Pacific, associated with a high value of the SOI, is also the norm. A Niño event is foreshadowed by a period of anomalous southwesterly winds in the Tasman Sea and by a weakening of the trades in the western Indo-Pacific. It is important to understand how the reinforcement of this process is induced by feedback mechanisms: for instance, the warming of the surface waters of the western Pacific caused by the previous weakening of the trades acts to enhance trade-wind anomalies even further. This reinforces, rather than weakens, the contingent surface warming. The event develops by the progressive extension eastward of outbursts of rainy, westerly winds initiated between a succession of pairs of tropical cyclonic cells, north and south of the equator in the western Pacific.

These bursts of eastward wind stress transport surface water to the east and induce the propagation of Kelvin waves along the equatorial wave guide. Consequently, the thermocline tilts across the entire Pacific basin, deepening in the east even as it shallows in the west. There are also smaller scale changes in thermocline depth along the equator associated with the passage of each Kelvin wave: deepening equatorward and shoaling poleward. Initially, flow of the Equatorial Undercurrent accelerates as Kelvin waves pass along it, but then it weakens in response to reduced zonal pressure gradient across the Pacific basin. When they encounter the continent, the Kelvin waves propagate poleward as coastally trapped waves along the western coast of the Americas. A fully developed ENSO event is characterized by weakening of coastal upwelling on the western coasts of the Americas and along the equator so that the surface waters of the eastern tropical ocean are warm, oligotrophic, and overlie a deep thermocline. Coastal upwelling does continue in these circumstances, but the unusually warm water brought to the surface will not be nitrate-rich. The biological consequences of all this are well described by Arntz (1986). In contrast, in the mid-ocean portion of the subtropical gyres (at 20–25°N), an increase in overall productivity may occur, together with significant changes in the functioning of the pelagic ecosystem (Table 8.1) that are readily observed in regional satellite images. Ryan *et al.* (2002) comment that the 1998 transition from El Niño to La Niña conditions was the first to have been observed by modern satellite imagery and offer striking examples of the extent of the large-scale blooms that are induced by the rapid and extreme shoaling of the thermocline that occurs during the transition period (see Color plate 4, discussed later). This is an important phenomenon to which I shall refer again in this and in following chapters.

Anomalous conditions may also occur in the trade-wind zone of the Atlantic during some Niño events, as noted previously. This was observed in both 1968 and 1984 (Hisard, 1980; Tomczak and Godfrey, 1994). In both instances, the expected coastal upwelling did not occur during boreal summer in the Gulf of Guinea, and heavy coastal rainfall occurred in the winter dry season. The whole of the eastern part of the ocean had an anomalous deep, warm mixed layer. In the same year, cold anomalies occurred in the northwest Atlantic near Newfoundland. Recently, the response of the Atlantic to El Niño events has been more satisfactorily analyzed (Bakun, 1996; Binet, 1997). During these events, the anomalous low-pressure region (warm, wet, rising air) over central America strengthens the Atlantic trade-wind system, and the westward slope of the sea surface is enhanced, as is thermocline uplift in the east, leading to anomalous cool sea surface temperature (SST) in the eastern tropical Atlantic. When the event relaxes, high pressure

**Table 8.1. Effect of 1991–92 El Niño on Pacific marine ecosystems.**

| Region | La Niña conditions | El Niño conditions |
|---|---|---|
| Eastern Tropical Pacific Ocean | Coastal upwelling strong | Coastal upwelling weak |
| | Mixed layer <40 m | Mixed layer 80 m |
| | Large nutrient flux | Small nutrient flux |
| | High primary production | Low primary production |
| | High fish production | Low fish production |
| N. Pacific Subtropical Gyre | Mixed layer 40–100 m | Near surface stratification |
| | Vigorous surface flows | Slack surface flows |
| | *Trichodesmium* present | *Trichodesmium* abundant |
| | Eucaryotic cells dominant | *Prochlorococcus* dominant |
| | Metazoan herbivores dominate | Protistan herbivores important |
| | Moderate primary production | Increased primary production |
| | Photoautrophic biomass low | Photoautotrophic biomass high |
| | New production $NO_3$-based | New production N-based |
| | New production N-limited | New production P-limited |

*Source: Modified from Karl* et al., *1995.*

(cool, dry, sinking air) over central America causes a relaxation of the Atlantic trades, a slumping back of the sea-surface slope, and a deepening of the mixed layer in the eastern part of the ocean. The seasonal oceanic-scale tilt of the thermocline (see Chapter 9, Western Atlantic Tropical Province) in response to seasonally varying trade wind strength is then reestablished. El Niño is thus marked by a strengthening of the NECC (and the Guinea Current), by unusually high saline deep water surfacing in the Gulf of Guinea, and by anomalous southward flow along the Namibian coast. This latter leads to the occurrence of a "Benguela Niño" such as was observed in 1934, 1950, and 1963 and perhaps subsequently.

More generally, beyond the trade-wind zone, where the most important shifts in weather patterns occur, El Niño events are associated with a general strengthening of the midlatitude westerlies of the Northern Hemisphere. Although there is much variability between events, a stronger Aleutian atmospheric low pressure often develops over the North Pacific and winter weather systems over northern Canada are modified so that the spring breakup of ice in Hudson's Bay occurs up to 3 weeks earlier in strong Niño years. Associated with these far-field effects, a warming of the whole northeast Pacific and a northward shift of the subarctic boundary, defined as the zoogeographic transition between subarctic and central water masses and biota, may also occur in ENSO years. This effect is especially marked at the eastward margin of the ocean along the coasts of Oregon, British Columbia, and Alaska and is caused partly by poleward Kelvin waves passing along the continental margin and partly by changed atmospheric forcing at the sea surface in the northeast Pacific. In the northwest Pacific, the seas to the east of Japan (25–45°N) experience a negative temperature anomaly of about 1°C largely caused by the changed weather patterns. Anomalous poleward extension of the ranges of individual species of pelagic fish and nekton (Longhurst, 1966) and changes in bulk biomass of plankton both in the California Current and the oceanic North Pacific in ENSO years (Chelton *et al.*, 1982) are related to these oceanwide changes in near-surface circulation.

Binet (1997) has reviewed fisheries responses in the Canary Current. Here, two periods (in the early 1970s and the late 1980s) when coastwise wind stress was anomalously strong resulted in strengthening of the upwelling cells and a very significant extension southward of the temperate sardine *Sardina pilchardus*. This led to a threefold increase in its abundance, at the expense of the home range of the tropical sardines *S. aurita* and *S. maderensis*. During these wind-stress anomalies, *S. pilchardus* extended its range even south of Senegal, onto the Arguin Bank (see GUIN province). We may reasonably suppose that changes in organisms at such relatively high trophic levels as sardines must reflect very profound changes in the planktonic ecosystem on which their growth depends.

Obviously, the seasonal evolution of the atmospheric circulation pattern is not identical from year to year, associated with changes in oceanic circulation. It may be useful to identify some of the characteristic and recurrent aspects of these changes. Consider, for example, the consequences of variability in the seasonal march of the atmospheric Intertropical Convergence Zone (ITCZ), which follows the apparent movement of the sun, seasonally, from about 2°S to about 9°N in the Atlantic Ocean and between appropriate latitudes in the other oceans. The date on which its poleward movement begins is variable between years, as is the northing attained during each boreal summer. In this ocean, in some years, the ITCZ reaches only to 8°N, whereas in others it reaches to 10°N and the annual displacement is matched by the northing achieved by the ITCZ in other oceans. The amplitude of the between-year differences may appear to be minor, but are associated with distant changes in weather patterns that are very significant—for example, the continental wind and rainfall fields over sub-Saharan Africa, and trade wind stress on the sea surface (Delacluse *et al.*, 1994). The effect of the location of the ITCZ on the surface temperature anomalies in the eastern tropical Atlantic is complex but observable. The anomalously warm summer of 1984 in the Gulf of Guinea was caused by unusually intense trade-wind stress during the previous summer and fall, leading to an unusually deep mixed layer of warm water in the western basin. On the seasonal relaxation of the trades, this mass of warm water surged eastward to form an unusually deep and warm mixed layer in the Gulf of Guiana that persisted for many months (Citeau *et al.*, 1988; Carton and Huang, 1994). Between-year variability in the location of the ITCZ, the principal convergence zone in the planetary atmospheric circulation, is associated with major anomalies in SST, of scale 3000 to 4000 km, and observable for periods of up to 12 months. Such anomalies may be coherent over even longer periods, spanning more than a decade, and are characteristic of the variability of planetary weather systems as discussed, for example, by Cushing and Dickson (1966) or by Cushing (1982).

We may expect to observe significant effects of these anomalies on pelagic ecology over large areas of the warm oceans. The curl of the wind stress in the ITCZ is associated with Ekman suction in the North Equatorial Counter Currents of both the Pacific and the Atlantic Oceans and, hence, with a zonal band of enhanced chlorophyll in each ocean. Two of the provinces of the Trade Wind biome are partially defined in relation to this process so that their actual locations will be sensitive to annual variation in the meridional position of the ITCZ. Strong El Niño events, like that of 1982–1983, may generate sufficiently energetic, poleward Kelvin waves along the continental margin that these establish conditions for the radiation westward of Rossby waves at mid-latitudes (35–40°N). These then advance very slowly across the North Pacific and eventually (a decade after the original ENSO event) modify the latitude of the axial flow of the Kuroshio extension. The consequent advection of anomalously warm water to the northeast Pacific has the same amplitude and extent as that which occurs by atmospheric forcing during an ENSO event (Jacobs *et al.*, 1994).

# Multidecadal Trends and Changes

El Niño–Southern Oscillation events occur against a background of longer, decadal and secular-scale change. While emphasizing that they are integral components of the planetary weather system, I shall discuss two examples: the Pacific Decadal Oscillation (PDO), and the North Atlantic Oscillation (NAO). Both involve major and sustained changes in conditions across whole ocean basins of the Northern Hemisphere. We must assume that comparable changes are sustained in the Southern Hemisphere, but these are yet to be identified. The NAO is better understood than the PDO, and a simple meteorological index suffices to indicate its state, whereas the more recently recognized PDO is usually indexed by the resultant sea surface temperature changes. The two phases of the PDO are characterized by anomalously warm and cool sea surface temperatures along the western coast of North America, although these are accompanied by sea-level pressure and wind anomalies that are large-scale, low-frequency, and occur rapidly. These changes are, in turn, reflected in ocean properties and circulation pattern and, not surprisingly, the oceanic ecosystems respond on similar time and space scales (Francis *et al.*, 1998). During the 20th century, the warm phase dominated during the periods 1900–1915, 1922–1945, 1958–1961, and 1976–1998.

Thus, during the winter of 1976–1977 the entire North Pacific weather system underwent a change of state. The Aleutian low-pressure cell intensified, thus shifting storm tracks further south than normal and from then until the 1990s the tropical ocean remained in a quasi-permanent warm mode. Full development of El Niño events was more frequent during this period so that until 1995 the mean return interval was less than 2 years, and in only a single year (1988–1989) was there full development of the trade winds (Graham, 1994). During this period, upwelling in the California Current was often constrained by a cap of light, warm water, and winter zooplankton biomass there progressively declined by about 80% from the 1970s to the 1990s (Roemmich and McGowan, 1995). In the central North Pacific, the carrying capacity of the ecosystem significantly increased, with integrated chlorophyll almost doubling from 1975 to 1985 (Venrick *et al.*, 1994). This signaled that the central North Pacific was undergoing a major ecological change, close to what general ecologists have called "catastrophic ecosystem shifts" (e.g., Scheffer *et al.*, 2001), representing a very rapid step change from one set of dominant organisms to another, or from one dominant process to another. In this case, both of these changes occurred. The mixed layer of the central gyre became more strongly stratified than previously, and the vertical flux of nutrients from below the nutricline was accordingly constrained.

These changes represent a phylogenetic, or domain, shift from a photoautotroph community dominated by larger eukaryotes to one dominated by very numerous, very small cells of the domain Bacteria, and by bundles of filaments of nitrogen-fixing cyanobacteria, inedible to most herbivores. Such domain shifts, as Karl *et al.* (2001) remark, will echo up the food chain. Selection for protistan "herbivores" over metazoan herbivores must have occurred and will have involved another fundamental shift in ecosystem structure and function, toward a more complex food web. Carbon will have tended to be retained within the system, rather than being exported via herbivorous crustaceans and their fish predators.

But this is just the most recent example of a phase shift of the PDO. During the 20th century, several such events were sustained, each for 20–30 years; a "cool" phase dominated until 1924 and from 1947 to 1976, and a "warm" phase in the intervening years. Within these periods were embedded many El Niño events, each of 6–18 months duration, whose effects are somewhat spatially distinct: PDO phase changes are most clearly expressed in high latitudes, whereas those of ENSO events are centered in the tropical ocean. Both ENSO and the PDO are associated with rather similar climate anomalies across North America, warm and cool phases of each mutually reinforcing their effects. There are large-scale, low-frequency, and sometimes very rapid changes in the distribution of atmospheric pressure over the North Pacific that are, in turn, reflected in ocean properties and circulation. Oceanic ecosystems respond on similar time and space scales to variations in physical conditions. Perhaps the best known response is that of the fish and birds of the eastern boundary currents of the Pacific, reviewed recently by Chavez *et al.* (2003). I shall discuss some of these phenomena in detail in Chapter 11, so here it is only necessary to note that the relative abundance of sardines and anchovies, and the relative productivity of the planktonic ecosystem, changed in our instrumental and sedimentary records throughout the 20th century. In the historical and archaeological records this phenomenon has been traced back several centuries and, in sediment cores off California, for two millennia.

A general relationship between long-term changes in SST and pelagic productivity has been proposed by Kamykowski and Zentara (2005). The very long SST time series now available was made nitrate-sensitive by subtracting "nitrate-depletion temperatures" associated with SST in water of salinity >28 ppt; this enabled estimates to be made of effects of decadal-scale SST changes on the relative nitrate supply to the surface. In periods of anomalously warmer SST, some regions of the ocean are predicted to have suffered nitrate loss in a manner that is consistent with long-term fishery production data and of blooms of dinoflagellates, better able to access subsurface nitrate supplies than nonmotile phytoplankters.

The North Atlantic Oscillation (NAO) Index is quantified by the departure from the mean pressure gradient between the Azores high-pressure and the Iceland low-pressure cells during winter. Winter westerly winds intensify and storm tracks shift northward during positive phases of the NAO, leading to milder conditions in the eastern part of the ocean and heavier weather toward the north. The converse occurs during negative phases.

Using sea level and meteorological records, it has been possible to reconstruct a record for the Gulf Stream transport and the strength of the midlatitude westerly winds extending back to 1850 from which long-term trends of the NAO can been identified. Planque and Taylor (1998) review the connections between the two phases of the NOA, the position of the Gulf Stream, changes in stratification and hence in the timing of the spring bloom and in the subsequent zooplankton response. Because the north wall of the Gulf Stream is constrained to lie below the line of zero Ekman pumping, where no wind-driven divergence or convergence occurs at the sea surface, its position may be predicted from the value of the NAO index (Taylor *et al.*, 1998). When this takes low values, the north wall of the Gulf Stream shifts to the south, and southward flux of the Labrador Current is increased; in the Norwegian Sea, under these conditions of wind forcing, flow of arctic water along the eastern coast of Greenland is enhanced.

A trend toward increasingly positive NAO values was sustained from 1920 to about 1950 and was renewed after 1964, so that in the early 1990s the NAO took stronger positive values than at any time in the previous 175 years (Fig. 8.3 and Dickson, 2003). This trend represented a major and sustained climate change that is imposed on the year-to-year consequences of variability in the value of the NAO that may have significant regional consequences (e.g., Color plate 10).

From 1950 to 1964, mean sea surface temperatures increased progressively in parts of the North Atlantic, and this induced major shifts in the distribution of fish species: for example, the penetration of the boreal seas by cod (*Gadus atlantica*), which built up stocks progressively further north along the coast of western Greenland (Cushing and Dickson, 1966). Northerly winds were progressively strengthened along the seaboard of western Europe (Dickson *et al.*, 1988), so that upwelling on the Portuguese coast became unusually strong. Northerly winter winds created a high-salinity anomaly east of

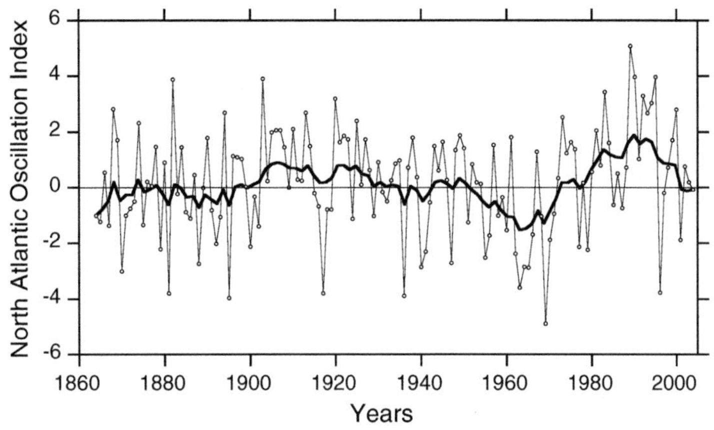


**Fig. 8.3** A long time series, with running mean, of the winter index for the North Atlantic Oscillation, 1864–2005, to emphasize the unusual pattern of values during recent decades. This index covaries positively with sea temperature in the Barents Sea and with winter air temperatures over Iceland.

*Source: Online data from NCAR, Boulder, Colorado.*

Greenland that could be traced for the following 10 years during its advection around the North Atlantic circulation. After 1950, there was a period of decreasing overflow from the Nordic Seas through the Faeroe channel southward into the Atlantic that, if not compensated by increased flow from other sources, could have led to a weakening of the global thermohaline circulation (Hansen *et al.*, 2001). Such periods may also have local consequences for the regional ecology of plankton, for one of the main overwintering sites of the *Calanus finmarchicus* population that summers on the NE European shelves is deep in the Faeroe-Shetland Channel. Changes of state of the NAO, therefore, will modify the transport of these diapausing individuals and hence the relative availability of individuals to recruit to shelf populations in spring. It has already been observed that important trends in plankton abundance around Britain may occur in relation to changes in values of the NAO (Colebrook, 1986; Heath *et al.*, 1999; Beare and McKenzie, 1999b; Greene and Pershing, 2000).

Abundance of *Calanus finmarchicus* on both European and North American coasts responds to changes in the NAO, but with opposite sign (Greene *et al.*, 2003): abundance in the North Sea is negatively correlated with the index, whereas abundance in the Gulf of Maine and western Scotia Shelf has a positive correlation with the value of the NAO. Since shelf populations of *C. finmarchicus* everywhere are expatriates from the deep ocean, sustained by regular recruitment of individuals from offshore, it is no surprise to find that abundance on the shelf should be sensitive to changes in oceanic circulation forced by the NAO. Positive NAO values and increased westerly winds shift the Atlantic storm track to the north and are associated with a northerly shift of the Gulf Stream axis and increased deep convection in the Labrador Sea; conditions on the European shelf are milder. Negative values bring a southerly Gulf Stream and increased southward transport in the Labrador Current and milder conditions in the NW Atlantic but colder ones on the European coasts. Years of positive NAO bring flow onto European shelves dominantly from the south, so that relatively few *Calanus* recruits arrive, whereas with the opposite sign, flow onto the North Sea is from the Norwegian Sea and transports abundant *Calanus* onto the shelf. The supply of recruits on the American shelf is more complex and depends on relative strength of transport of the Labrador Current south along the shelf edge off Nova Scotia: weak with positive NAO and strong with negative NAO. Weak transport of Labrador Sea water under positive NAO conditions permits Atlantic Temperate Slope

Water to lie adjacent to the shelf edge, and cross-shelf incursions of this carry abundant *Calanus* recruits and also render water in the basins and Gulf of Maine warmer and saltier than during negative NAO conditions.

After 1965, the long-term trend toward increasing positive values of the NAO was reestablished. The poleward extension of cod populations was reversed, their reproductive success faltered, and the populations moved progressively southward, even as winters became harsher, with increased seasonal ice cover and nor'westerly wind stress. The spring bloom and zooplankton population maxima now occurred progressively later each year. By the end of the period, cod had been so reduced by environmental change and continued fishing pressure that the stocks on the Grand Bank had collapsed and the great ancestral fishery off Newfoundland was closed. The failure of cod to maintain a population, as they were able to do during earlier sieges of strongly positive NAO conditions earlier in the century, has been attributed to the fishing-induced truncation of their population age structure. If insufficient fish are permitted to survive sufficiently long to bridge each period before conditions became once more favorable for reproduction, then the stock must collapse, as indeed happened in 1992 (Longhurst, 2002).

In the Trade Wind biome, where reversing monsoon winds and monsoon currents dominate the regional environment, we are used to strong interdecadal variability in the strength and reliability of the monsoon regime. These are some of the factors that force drought, famine, and warfare. Although, as I have suggested, it was the fickleness of the Indian monsoons that led to the uncovering of the Southern Oscillation Index, the SOI is now mostly applied to studies of the ENSO phenomenon. But monsoon variability in the Indian Ocean is expressed principally at much longer time scales that can be examined in the excellent records that are available for most of the 20th century. These deliver information not only about the variability of monsoon winds but also about their oceanographic and biological consequences (Longhurst and Wooster, 1990). As shall be discussed in Chapter 10 (see INDW), the southwest monsoon forces upwelling on the west coast of India and this process, in turn, produces diatom blooms and supports a large population of *Sardinella longiceps*, the oil sardine, which feeds on them. Because *S. longiceps* is a very short-lived species whose abundance responds rapidly to year-to-year changes in ocean conditions, fishery records can tell us much about long-term changes in circulation and upwelling.

However, we also have direct information because, used with appropriate caution for steric effects, sea level is an indicator of upwelling along the Indian coast, as it is elsewhere. Long time series can be constructed for several stations, and those from Cochin and Mangalore are particularly useful. Average seasonal variation of the sea level at Cochin was calculated for the period 1939–1987; we can compare sea level for the monsoon period of locally forced Ekman divergence (May–September) with sea level for months when remotely forced, baroclinic upwelling occurs (March–April). The two kinds of upwelling seem to be somewhat independent because the strongest decadal trends occur in the April data, which show a trend of increasing sea levels beginning about 1942 and reaching its maximum around 1960. After this, a slightly lower level was sustained until the end of the data set. In contrast, for the monsoon months (exemplified by June) decadal trends are less conspicuous, but there is an important change of state from weak to strong upwelling over a period of 2 or 3 years culminating in 1960. Somewhat similar trends occur in a longer data set (1878–1982) for Bombay, which shows rising sea levels from about 1930 until the mid-1950s: it is unlikely that the occurrence of highest sea levels along western India in the 1950s at two such distant stations could be a matter of chance.

Comparison of the trends in Cochin sea level and oil sardine catches confirms the anticipated relationship. A long period of relatively low but variable abundance from 1900 until 1959 was transformed by a massive population increase in 1960, coincident with the

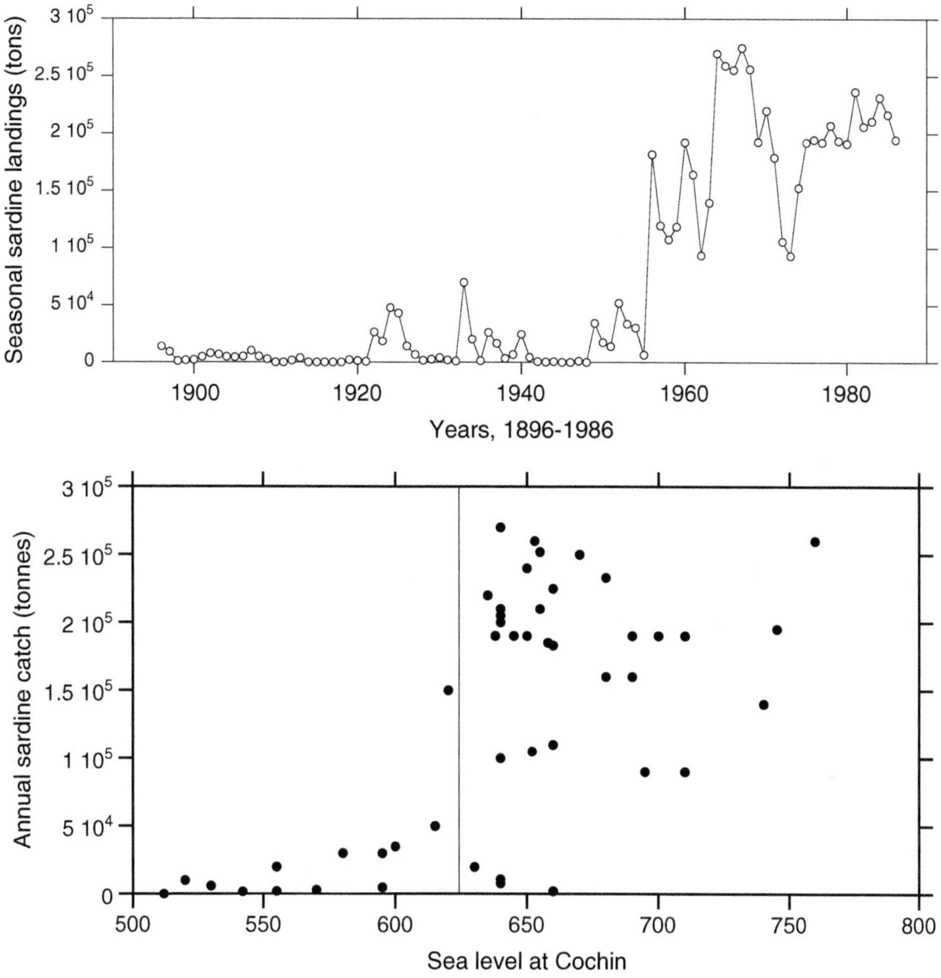


**Fig. 8.4** Relationship between *Sardinella longiceps* landings on the west coast of India, and sea level at Cochin; the vertical line is intended to suggest the existence of two response regimes.

*Source: Redrawn from Longhurst and Wooster, 1990.*

brief period of exceptionally strong upwelling noted earlier. The catch is dominated by O-group fish, so that recruitment variability is reflected almost instantaneously in catch rates so that the statistical data (Fig. 8.4) confirm more anecdotal information concerning the relationship between diatom blooms and oil sardine response and are a powerful illustration of the ecological effect of decade-scale climatic changes in the wind patterns over the ocean (Longhurst and Wooster, 1990). A similar increase in catch rates had been recorded also much earlier: in 1855–56 stocks were <1000 tons annually, but in the three subsequent years catches increased very rapidly, reaching 70,000 tons in 1859. In such preindustrial fisheries we may be sure that rapid change of this nature is resource-based, not due to an evolution in catching methods as it might be in an industrial fishery.

In the Atlantic Ocean, along the coast of tropical West Africa and especially off Ghana and the Ivory Coast, the sustained threefold increase in sardine abundance that was initiated in the late 1970s and apparently continued until at least the mid-1990s must be explained. This biological event is matched by increased zonal wind stress but not, as might be expected on this coast (see GUIN), in enhanced upwelling strength. This is strange, as this is obviously the first type of biological enhancement we should look for in order to explain any massive natural increase in sardine abundance. In fact, the explanation is quite unexpected and is surely explicable in terms of the "larval retention area/adult stock size" hypothesis of Sinclair (1988), as argued convincingly by Binet (1997). The argument runs that the Atlantic Ocean experienced anomalously strong flow of eastward zonal currents during this period. Downstream of both Cape Palmas (Ivory Coast) and Cape Three Points (Ghana), these unusual flows supported larger than usual turbulent gyres. These gyres serve as the retention areas for the larvae of *Sardinella aurita*, the sardine of this upwelling region, so that their enlargement is likely sufficient reason for the observed increase in stock size.

These case studies in different oceans are directed at two sardines having quite different ecologies. *S. longiceps* is one of the few species of clupeids to graze directly on blooms of large diatoms, often *Coscinodiscus* spp., whereas *S. aurita* mainly eats copepods and depends principally on a population of *Calanoides carinatus* for its nutrition. Both fish provide evidence, in their time-dependent variability, of the extent to which the whole pelagic ecosystem undergoes long-term variability in the tropical seas.

## Conclusion: Stable Partitions in a Varying Ocean?

You may have wondered at my emphasis in this chapter on the impermanence of marine ecosystems (or at least of ecological conditions) after earlier having proposed that it is feasible, and would be useful, to partition the ocean into regions, each having characteristic conditions. You may have concluded that the constant change in planetary weather systems likely induce so much (and such constant) change in circulation patterns, and hence in the response of ecosystems, that any attempt to partition the ocean must be fallacious. I suggest, on the contrary, that some degree of formalism and partition may, in fact, assist us in comprehending changes in such a vastly complex, interacting whole. It may, at least, help us to keep our thoughts in order.

We shall have to be alert to the fact that at least some of the key features of ocean circulation that we use to define our partition may themselves change location with changes in sign of the SOI, the NAO, or another index. We shall have to be alert also to the fact that within many of our compartments the ecosystem may exist in more than a single state. Finally, we shall have to accept that in some cases, it is possible that partitions between adjacent provinces can under some conditions no longer be identified.

An instructive case is the observation that the circulation of the NE Pacific exists in two states, Type A when winter atmospheric sea-level pressure is anomalously high over the Gulf of Alaska, and Type B when it is relatively low (Francis *et al.*, 1998). Under Type A, the flux of the West Wind Drift, lying across the ocean at 45–50°N, splits strongly to the south and weakly to the north on encountering the west coast of North America. During Type B conditions, the reverse obtains, so that flow down the California Current is weak, and flow north around the Alaska coastal current is strong. But note that the general oceanographic features characteristic of each region remain approximately static. Conditions within regions change, but boundaries between them do not.

It is perhaps only in the Trade Wind biome and perhaps the equatorward parts of the Westerlies biome that the boundaries of provinces risk being modified significantly during, for example, an El Niño event. It is, of course, the provinces of the Indo-Pacific Trade Wind biome that have the potential to show the greatest modification, and especially the western Pacific Warm Pool Province (WARM), whose eastern boundary might become difficult to define. In extremely strong events, the conditions across the whole Pacific basin, from the Indo-Pacific archipelago to the coastal boundary of the Americas, might perhaps exhibit sufficiently uniform oligotrophic conditions as to be thought of as a single ecological entity. But even in such cases, although seasonal production cycles will be strongly modified within them, the coastal boundary provinces (California Current, Central American Coastal, and Humboldt Current Coastal Provinces) will retain their identity and unique characteristics as part of the Coastal Boundary biome of the ocean. There may be some change in the effective coordinates of the eastern part of the zonal province representing the transition between subarctic and subtropical regimes (North Pacific Transition Zone Province). In the Atlantic Ocean, the distinction between the two zonal trade wind provinces (WTRA and ETRA) might become so slight that we could, perhaps, recognize only a single entity having the usual poleward boundaries.

But, in practice, a simple comparison of pairs of global SeaWiFS images, one for each extreme situation, is comforting to the idea that the oceanic provinces discussed here are relatively permanent, and may be identified usefully at either state of the Southern or Pacific Decadal Oscillations. I illustrate this (see Color plate 4) for two cases, January 1998, during an El Niño event, and January 2001, when cool water and strong trade wind (or La Niña) conditions obtained. You will see, I think, that it is only in the eastern Pacific that significant anomalies occurred in 1998. Elsewhere, at all latitudes, the pattern of sea surface chlorophyll was essentially normal, indicating that the processes that induce the transport of nutrients into the photic zone were performing normally. Everywhere in the southern hemisphere the chlorophyll field in the two years is essentially similar, and all the features that we expect to observe are readily identifiable. The chlorophyll field of the central Atlantic Ocean shows very little difference between these two situations, and although the surface chlorophyll values are admittedly somewhat lower during the El Niño year, their pattern is perfectly normal. The same is true for the Indian Ocean, except that there the zonal band of chlorophyll along the equator was rather stronger than in some other years. In the eastern Pacific itself, it is the lack of equatorial upwelling that is most striking. But although characterized by weaker chlorophyll than normal, the locations of upwelling cells in the eastern boundary currents are perfectly clear, as also are the locations of the minor, Central American coastal upwellings in the province characterized by that process.

Despite the significant lack of chlorophyll biomass along the equator, the location of the equatorial currents can still be identified, and so can—by inference—the locations of the equatorial provinces. Evidently, the Equatorial Divergence province (PEQD) is defined northward by a zonal *increase* in surface chlorophyll that corresponds to the processes we expect to meet within the Equatorial Countercurrent Province (PNEC). Although the conditions within PEQD are highly anomalous along the equator itself, nevertheless conditions with that province are distinguished from the much more oligotrophic conditions within the Subtropical Gyral Province (SPSG) to the south. An appropriate and familiar discontinuity along about 15°S serves, as usual, to locate a suitable boundary between PEQD and SPSG.

Chapter 9

# The Atlantic Ocean

The symmetrical planetary forces that drive circulation and mixing can be modeled with great precision for a landless world, but in reality they operate within irregular basins shaped by the asymmetric drift of continental masses during geological time: what we observe in the real ocean, therefore, departs very significantly from the ideal state. So, the following chapters, each devoted to a single ocean basin, open with a brief introduction to its basic geography because it is this that determines its water circulation patterns and the distribution of turbulent energy.

It is remarkable how little emphasis is placed on the individual geography of each ocean basin in most texts on oceanographic processes, whether these are physical, chemical, or biological. Some knowledge of the geography of each ocean must be one of the most important items in any oceanographer's tool kit. With each ocean it is logical to include discussion of its marginal seas so that, in the case of the well-endowed Atlantic, the Arctic Ocean together with the Caribbean, Mediterranean, Black, and Baltic Seas will be included. As Tomczak and Godfrey (1994) point out, this gives it by far the longest latitudinal extent of any ocean basin—21,000 km from the Bering Straits, over the pole and down to Antarctica.

The significant features of the geography of the Atlantic Ocean that influence its circulation are self-evident. South America extends much farther poleward into the Southern Ocean than does Africa, and so creates a unique asymmetry in the surface temperature fields of the South Atlantic that has no homologue in other oceans. Interruption of the strong westerly flow of the Circumpolar Current induces a northward loop along the continental edge east of Tierra del Fuego and of the Falklands plateau that continues until it meets the southward flow of the warm western boundary current off Mar del Plata. If the Falkland Islands arose from deep water, instead of from a shoal plateau, the circulation pattern of the whole southern Atlantic would be different. Again, if Cape Hatteras did not exist, the Gulf Stream would separate from the American continent at another latitude and the ecology of the whole North Atlantic would be different from what it is today. Perhaps most important, if the triangular protuberance of Brazil and the Guianas lay farther north, the flow of the westward trade wind currents would be more evenly divided between North and South Atlantic basins. Then, equatorial surface water from both hemispheres would not, as it does now, flow almost entirely into the Gulf Stream of the North Atlantic and, consequently, 10°C water would not penetrate to 60°N off Iceland.

In winter, between the Iceland low and the Azores high, wind stress ($>250 \times 10^{-2}$ dyn $cm^{-2}$) and heat flux ($-40$ to $-60\,W\,m^{-2}\,y^{-1}$) both take very high values (Hellerman, 1967) and drive winter mixing to 750–900 m, especially at 50–60°N between Ireland and Newfoundland. The spring bloom takes nitrate to very low levels in the "nutrient hole" of the North Atlantic, which is a striking anomaly on the global nutrient field. This deficit, which extends to 1000 m, is presumably maintained because export of organic material to the sea floor is faster than horizontal advection of nutrients into the region of deep mixing. The result is a nutrient front at 1000 m across the ocean at about 30°N for nitrate, phosphate, and silicate that has no counterpart in other ocean basins.

In the North Atlantic, the bifurcation of the Gulf Stream over the Newfoundland Basin adds complexity to the gyral system, which is reflected in its dynamic biogeography, and requires that we recognize compartments within the gyral circulation, though they may not be readily identifiable in surface features. At about 40–45°W the Gulf Stream feeds a southeasterly flow (the Azores Current) directly into the gyral circulation and a northeasterly flow (the North Atlantic Current), of which part recirculates into the gyre only when it encounters the European continent. Midocean ridge topography restricts the main gyral recirculation to the western basin, so there are significant biological differences between the eastern and western parts of the gyre. The passage of Gulf Stream water across the ridge is often marked by a southerly loop associated with the topography.

# ATLANTIC POLAR BIOME

The coastal geography of the polar region of the Atlantic is rather complex, especially in the Canadian archipelago, and the marginal ice zone (MIZ) that migrates through these regions during the summer has many of the characteristics of a coastal zone. It is therefore not useful to recognize a coastal boundary zone poleward of the North Atlantic continental shelf provinces.

## BOREAL POLAR PROVINCE (BPLR)

### Extent of the Province

It is ironic that for BPLR, the first province to be discussed in detail, it is better not to follow the logic of the four biomes discussed in Chapter 6. This is because the complexity of the arctic archipelago of northern Canada and the difficulty of access of the northern Asiatic coastline make it very difficult to sustain an adequate description of the coastal biome here. This province, therefore, comprises the Arctic Ocean, lying between North America, Greenland, and Asia, together with all but one of its marginal seas: the Chukchi, East Siberian, Laptev, Kara, and Beaufort Seas. Only the northern part of Barents Sea is included in BPLR; the southern part is assigned to SARC, because it is strongly influenced by Atlantic Current water and does not freeze over in winter. The boundary between BPLR and SARC in the Barents Sea is the well-marked thermal front between Atlantic and Arctic water that passes around Bear Island and meanders eastward from Hope Island at about 74–78°N. Hudson Bay is included in BPLR, as is the entire coastline of the Arctic basin except that of the Barents Sea. The coasts of both eastern and western Greenland, and of Labrador south to Hamilton Bank, are included: all of these have extreme polar conditions.

### Continental Shelf Topography and Tidal and Shelf-Edge Fronts

The Arctic Ocean ($3 \times 10^6$ km$^2$) has wide continental shelves reaching 700–800 km along much of the north coast of Russia: the Eastern Siberian, Lapta, and Kara Seas have extensive shelf regions although these seas, as defined by geographers, include both shelf and adjacent deep water. Tidal streams are very strong in some embayments of the Laptev shelf.

Off northern Alaska the shelf narrows to only $<250$ km and is even narrower to the east, across the Beaufort Sea toward the Mackenzie delta. The shelves of the north coasts of the Canadian archipelago and Greenland are narrower still, though the archipelago itself extends over 15° of latitude. The continental shelf off Labrador is remarkable for its anomalous average depth, close to 200 m, with a shelf break occurring at around 400 m. The Labrador shelf is also anomalous for the deep troughs that lie about 70 km offshore and parallel to the coast, for about 550 km north and south of the Strait of Belle Isle.

Hudson Bay requires special comment for it is really a very shallow inland sea, of more than one million square kilometers in area, having a mean depth of around 150 m. It is also remarkable that the Hudson Bay region is still rising because of eustatic rebound in our postglacial period, by as much as 1 m $y^{-1}$, resulting in a continually evolving, young coastline.

Along the eastern coast of Greenland, the shelf is unusually deep, so that the 200-m contour lies close to the steep-to coastline, with a shelf break at 400 m offshore. To the west of Cape Farewell (60°N), the shelf is narrower but shoaler, having a break of slope at 200 m. This section of the Greenland shelf occupies a significant part of the entire width of Davis Strait (65°N).

## Defining Characteristics of Regional Oceanography

Here, at the top of the world, seasonal cycles of irradiance are extreme and the perpetually low temperatures induce almost complete ice coverage of the Arctic Ocean year-round, and of its marginal basins in winter. Another, often overlooked, consequence of such high latitudes is the extreme value taken by the Coriolis parameter, and the effect of this on water motion and especially on shelf-break tidal regimes.

Arctic continental shelves generally have a complex density circulation, behaving as salt-wedge estuaries in summer and, by the deep export of salt-rejection brine produced as the sea freezes, as negative estuaries in winter. Thus, the surface water of the continental shelves is more dilute and shows greater seasonal salinity variation than that of the central part of the Arctic Ocean (Carmack, 1990). The Laptev and East Siberian Seas along the northern coast of Asia carry particularly heavy effluents from the Lena, Kolyma, and Indigirka Rivers that drain the north Siberian plains. Hudson Bay has a cyclonic circulation, driven by buoyancy and wind forcing; near-surface low-salinity water passes out of the embayment to be exchanged with deeper water of higher salinity.

The Arctic Ocean is dominated by the Circumpolar Boundary Current, within which major water mass transformation occurs. This circulation is initiated by a flow of warm North Atlantic water that enters the Arctic Ocean through the eastern Greenland Sea (see SARC Province) to pass around northern Norway and through the Bear Island Channel. Most recirculates within the Greenland Sea before entering the Barents Sea, but some passes on eastward and so enters the anticlockwise circulation of the boundary current that circles the basin along the edge of the wide shelf of northern Russia and the narrower shelf of north America and Greenland. The source for Arctic Ocean surface water is mainly the warm Atlantic Current, which is progressively modified at high latitudes by heat flux to the atmosphere, river runoff, and meltwater in summer and by salt rejection during freezing in winter (Jones *et al.*, 1990). These processes induce both a buoyant surface layer and a dense and highly saline bottom layer over the shelf regions. The 1994 section by Canadian and U.S. icebreakers confirmed the continuity right across the Arctic Ocean of the thermocline lying at about 110 m between surface water of −1.2° to −1.7°C, with warmer water of 0.2 to 1.0°C below. With this is associated a nitracline above which both nitrate and silicate are reduced to very low levels in summer.

The main flow of the boundary current leaves the Arctic Ocean through Fram Strait and this is the source of the narrow, buoyancy-forced East Greenland Current, whose seaward boundary is the strongly marked East Greenland Polar Front, a temperature discontinuity that defines the outer boundary of BPLR in the warmer Nordic Sea water (Muench, 1990). The southerly East Greenland Current, locked to topography, is continuous along the coast as far as Cape Farewell, which turns the flow up the west coast. Here, the province is less well defined by surface temperature, because the surface water of the West Greenland Current has by this time been significantly modified by Atlantic water entrained across the Polar Front. Icebergs are glacier-fed into the Greenland coastal currents and many survive the passage around Baffin Bay to reach Newfoundland.

The permanent ice cover of the Arctic Ocean opens significantly in summer only in the coastal seas; in the central Arctic Ocean, even in summer, ice is from 1.0 to 3.5 m thick, covering from 80% to 100% of the sea surface, while a snow-pack above the sea ice ranges from 5 to 30 cm thick. In some places, open-water polynyas occur during seasons when one would otherwise expect complete ice cover. During the summer, these polynyas become the centers of the seasonally open water regions surrounded by complete ice cover. They occur at locations now predictable from topography and currents and are usually accounted for by one of two models. The simplest case is that of the "latent-heat polynyas" that exist because of the balance between rates of ice formation and loss; in these polynyas the latent heat occasioned by ice formation is removed by wind, often offshore in these cases. If wind flux becomes insufficient to support this process, the polynya will freeze over. "Sensible-heat polynyas" result from heat flux caused by vertical motion of warmer, deep water either by wind mixing or by wind-induced upwelling; in these, there is less modification of the density regime of the water column than in latent-heat polynyas.

The North Water, between Greenland and Ellesmere Island, is more complex than either simple model (Ingram *et al.*, 2002). Here, not only is there sensible heat flux up into surface layers from the West Greenland Coastal Current, but the southerly transport of pack ice is interrupted by an ice bridge that forms across the narrows of Nares Strait. Ice that is wind-drifted southward from the northern part of Baffin Bay is therefore not replaced. However, despite the interest that it has attracted in polynya studies, the North Water conforms to the definition of a polynya for only a few months in spring.

Around the margins of the polar ocean, the summer retreat of the ice forms an ecologically significant MIZ, associated with an ice-edge front in the subsurface density field where the warmer, less saline open water meets colder, more saline water along and below the ice edge (Fig. 9.1). This frontal region may either be simple, or more

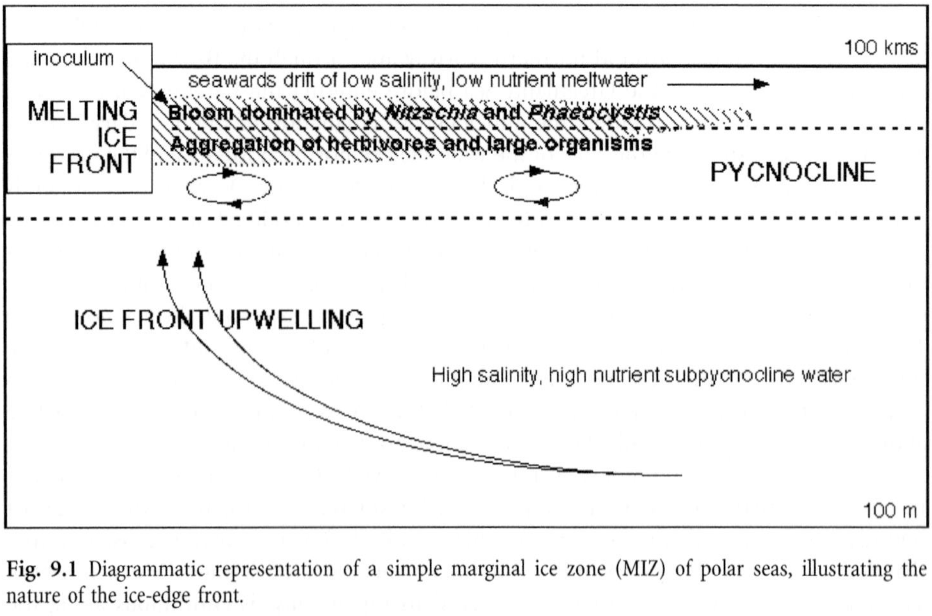


**Fig. 9.1** Diagrammatic representation of a simple marginal ice zone (MIZ) of polar seas, illustrating the nature of the ice-edge front.

*Source: Redrawn from Walton Smith, 1987.*

complex where lateral interleaving of water masses occurs along a convoluted ice front. Embayments along the receding ice edge correspond to the flow of warm water associated with topographic troughs on the Chukchi shelf and elsewhere.

Little eddy activity occurs below the permanent ice cover of the Arctic Ocean, and mesoscale features there are thought to be very long-lived (Muench, 1990), but ice-edge fronts and shelf-edge fronts are significant elsewhere. The position of the ice edge of the Beaufort and Chukchi Seas during the summer is constrained by the position of the shelf break. Warm, northward currents are steered by deep troughs on the Chukchi shelf, and the ice-edge frontal system therefore follows this topography (Muench, 1990).

You will perhaps have noted that I have not commented on the concern now widely expressed that the permanent ice-cover of the Arctic ocean is yielding to atmospheric $CO_2$-induced climate change; throughout this work, such issues are avoided: it is enough here to try to characterize the pristine regional ecosystem. But, if the breakup is progressing as is currently reported, then much of the foregoing account will require modification rather soon.

## Response of the Pelagic Ecosystems

The general ecology and physiology of arctic organisms is now reasonably well explored, perhaps especially because their diversity is as low as anywhere in the oceans, so that relationships between them are simpler to understand than in warm seas. Arctic heterotrophic bacteria, for example, act as a uniform psychrophilic assemblage and actively metabolize glucose and amino acids down to subzero temperatures, ensuring turnover rates for glucose and amino acids of 1–50% daily in the upper 100 m of the North Water (Li and Dickie, 1984).

While recognizing that they represent a continuum, it may help us to understand the ecology of the BPLR province if we recognize three different ecological zones: (i) those areas that are permanently ice covered, (ii) the marginal ice zone during retreat in spring or formation in autumn, which may or may not have "polynya" status, and (iii) the more extensive open-water areas of summer. As noted earlier, each of these zones has been extensively investigated in the Canadian archipelago and northern Baffin Bay.

***Permanently Ice-covered Regions*** The Arctic Ocean will serve as a model for all parts of BPLR that remain ice-fast in summer, with the caution that in ice-covered passages within the archipelago strong currents must advect biota from adjacent open-water areas. Biota found below the ice may not have developed there. In fact, it has been calculated that during their larval period of $< 90$ days, boreal cod may drift as far as 2500 km through the Canadian archipelago from their point of origin.

The strong nitracline at the halocline of the Arctic Ocean is not mediated by local biology. Rather, it is due to the fact that the halocline water originates on the surrounding continental shelves and carries from there the nutrient signatures of Pacific and Atlantic Ocean water. Though some biological uptake of nitrogen does occur in the surface water, levels remain relatively high compared with those of other oceans; 4 or 5 $\mu$M is normal in the central Arctic Ocean (Jones *et al.*, 1990). For silicate and phosphate, concentrations above and below the halocline are similar.

In computing both the sub-ice irradiance field and the total plant production in the Arctic Ocean in summer, it will be necessary to recognize the existence of a subaerial algal community, the so-called nivalis cryobionts dominated by *Chlamydomonas nivalis*, on the surface of the snowpack. These cells form highly visible patches ranging in color from green to red in which the cells occur either singly or as mucilaginous aggregates associated with wind-blown dust particles: cell concentrations may reach $1.5 \times 10^5$ cells $ml^{-1}$ in melted snow from such patches. The 1991 *Polastern* voyage across the central basin of the Arctic Ocean found these colored patches along most of the track from 83°N to the pole. In some places in the central basin, >10% of the snow surface was discolored. Estimates of their potential gross production vary greatly, but are significant in relation to the equally variable estimates for total primary production below the ice of the Arctic Ocean (Gradinger and Nürnberg, 1996).

It is also necessary to recognize that the ice itself or, rather, cavities within the ice along its lower surface are an important habitat for many species of planktonic algae. These epontic cells, although they occur in a low-light environment, are not an obligate shade flora and their population growth is limited—as in the water column—by both light and inorganic nutrient supply: *Melosira arctica* and loosely attached mats of both centric and pennate diatoms are characteristic of this flora. As Smith *et al.* (1987) commented, ice algae are a case study in photoautotrophic growth and of metabolism under chronically low temperature and irradiance: they display marked shifts in metabolism that are consistent with changes in light and nutrient supply. The growth of epontic algae below sea ice begins as soon as sunlight returns in spring. On the underside of ice 1–2 m thick the light regime resembles that of the lower photic zone in open water, receiving <2% of incident illumination. The photosynthetic efficiency of epontic cells is unusually great so that they are able to photosynthesize down to 0.01% of surface illumination. Nutrient supply and demand are more complex than in the water column and involve nutrients dissolved in brine rejected from the ice, whereas external supply depends on hydrodynamic forcing of wind and tide, and the turbulent water flow along the lower surface of the ice (Cota *et al.*, 1990). Some local heterotrophic regeneration occurs, just as in the water column. The epontic ice flora is intolerant of low salinities and collapses rapidly when ice melts during the arctic summer.

Phytoplankton blooms under the ice may occur at times and places where the sea ice becomes snow free, albeit at rates of about one-third of the water column over the surrounding continental shelves and MIZs (Legendre *et al.*, 1992; English, 1961).

The water column below the Arctic Ocean ice is inhabited by a sparse but permanent zooplankton community, higher biomass being located in the central basin, where ice cover is not so thick as around the margins; on the 1994 polar transect, highest biomass occurred at 87°N, in the Amundsen Basin, and here it represented—because of very high individual lipid content—40% of the total POC in the upper 100 m of the water column. These organisms, during the daylight period, were ingesting <30% of their body carbon daily as phytoplankton (Thibault *et al.*, 1999).

Zooplankton biomass (0–200 m) below ice cover is dominated, as elsewhere in the Arctic, by calanoid copepods: *Calanus glacialis, C. hyperboreus*, and *Metridia longa* and larger numbers, but smaller biomass, of *Pseudocalanus, Oithona, Microcalanus*, and *Oncaea*. In the Nansen Basin, *Oncaea borealis* forms 40–80% of the individuals. It may be noted that the geographical distribution of *C. glacialis* approximately matches the boundaries of the BPLR province (Conover, 1988) and this species is often associated with shallow water over very high latitude shelf regions, having a preferred water temperature of about −0.5°C (Longhurst *et al.*, 1984). In the Labrador Sea region, it is restricted essentially to this habitat, being replaced by *C. hyperboreus* and *C. finmarchicus* over deep water (Head *et al.*, 2003). The small *Pseudocalanus minutus* is distributed similarly in this province and the even smaller *Microcalanus pygmaeus* is the most abundant copepod of the Arctic Ocean proper. *Metridia longa* is also distributed in such a manner as to be consistent with having an Arctic source, compared with the Atlantic *M. lucens*.

*Calanus hyperboreus* requires 3 years to complete its life cycle in the Arctic Ocean. In the first year it achieves growth to C2, in the second year it reaches C4 or C5, and in the third year becomes adult and reproduces (Conover, 1988; Smith and Schnack-Schiel, 1990). Of course, there is some effect of latitude on generation time, so that in the North Water, many individuals become adult during their second year of life. Long generation times are also characteristic of other arctic biota: chaetognaths (e.g., *Sagitta elegans*) have a 2-year life cycle and reach an unusually large size in BPLR. Other polar organisms, such as the predatory medusa *Aglanthe digitale*, have an entirely different pattern, reproducing at depth essentially year-round, even in the Arctic Ocean.

Vertically stratified tows across the Lomonosov Ridge show that *Calanus hyperboreus* in summer has a distribution that extends very deep, essentially to the bottom, while maintaining a zone of maximum abundance in the upper 25 m, just below the ice cover. Somewhat deeper, with maximum biomass at 200–300 m, we may expect to encounter *Gaetanus tenuispinus*, whereas forms like *Lucicutia* spp. occupy the deepest water masses (Kosobokova and Hirsche, 2000). Diel migrations are generally thought to be largely absent in summer, though Groendahl and Hernroth (1986) did find both *O. borealis* and *M. longa* performing short-range diel migrations in the Nansen Basin, to the north of Spitzbergen, when the daily irradiance cycle became sufficiently strong to induce such behavior; nevertheless, ontogenetic migrations are the principal determinant of vertical distribution of herbivores. Conover (1988) described the seasonal succession of copepods below fast ice as an ontogenetic escalator on which waves of the F0, F1, and F2 generations (successively, and in that sequence) of *C. hyperboreus* and *C. glacialis* rise toward the surface. *Pseudocalanus*, in contrast, remains close below the ice surface. With the erosion of sheets of diatoms from the undersurface of the ice as summer advances, all copepods come to be concentrated in the upper 20 m of the water column. At this season, depth partitioning of the near-surface layer occurs between species and their growth stages, which tend to be differentially distributed with depth.

In Resolute Sound in the Canadian archipelago, under full ice cover in February–March, hyperiidid amphipods (*Themisto libellula*), small calanoid copepods (*Pseudocalanus acuspes, Acartia* spp., *Oithona* spp.), and naupliar stages of larger species aggregate at very high concentrations just below the bottom surface of the ice. At this season the epontic flora is starting to be dispersed, largely in response to tidal currents, and it was observed experimentally that the copepods could fill their guts in just a few moments from appropriate *in situ* suspended cell concentrations (Conover *et al.*, 1991). This fauna, in the period after light returns but while ice cover is still complete, perform a routine short-range diel migration, mostly within the upper 10 m below the ice, rising at dusk to feed and descending later in the night, replete (Hattori and Saito, 1995).

***Marginal Ice Zones*** As Smith (1987) emphasized, ice edges or marginal ice zones occur in a variety of geographic locations in this province over both deep and shallow water. These zones are influenced by both physical conditions and biota and either may form an abrupt edge, or else are represented by zones as much as several hundred kilometers wide. For these reasons, it is not easy to generalize their ecology. Nevertheless, we must try, and it may help if we concentrate our attention on the properties of a receding ice edge in spring.

Seasonal ice cover in the open ocean is necessarily accompanied by a front where pack ice meets open water. These seasonally migrating zones may extend laterally across hundreds of kilometers, within which mesoscale variability is imposed by the interaction of wind, current, and the ice edge (Smith, 1987). Retreat of the ice edge may occur rapidly (10 km per day in the Bering Sea), and in 3 months the edge moves several hundred kilometers. Rapid algal growth is induced in a shallow (often <25 m) mixed layer. Retreating MIZs are then the focus of biological activity at all trophic levels; local "biological spring" may, of course, occur on almost any date throughout the summer months. Available nitrate is progressively depleted, though renewal may occur by ice-edge upwelling (either density or wind driven); without such input, the algal bloom at a rapidly retreating ice edge may be sustained for as little as 10 days (Smith and Sakshaug, 1990). There is an important ecological connection between epontic algal communities and the phytoplankton. As frazil ice forms in the fall, crystals floating upward scavenge algal cells and incorporate them in the forming ice pack, and algae released from the melting ice in spring may either sink as large flocs to the sediments or seed the burgeoning phytoplankton community.

It is principally the effects of the freeze-thaw cycle of seawater that set the ecological stage in the MIZ by determining the strength of near-surface stratification. Thawing sea ice in spring releases freshwater to create strong density gradients, whereas freezing in the autumn partitions seawater between freshwater ice and strong brine that is rejected and sinks, causing instability and deep mixing. In late winter (March–April), as daylight returns, the surface water column is deeply mixed, nutrient levels are high, and net carbon fixation is not yet established (Smith and Brightman, 1991). It is precisely the effect of a thawing, receding ice edge in spring that stabilizes the water column and gives conditions likely to lead to phytoplankton bloom, especially where (as expected) cell growth will be principally light limited. Where a spring bloom in an MIZ reduces nutrient levels below optimal, further enhancement of algal growth may occur by ice-edge upwelling forced by wind-driven transport of surface water away from the ice and its replacement from below. For all these reasons, we may expect that MIZs should be a focus for biological activity. Indeed, this is seen in the satellite imagery for the Greenland and the Labrador Currents, both characterized by a receding ice edge in summer, and each of which tends to be located in satellite chlorophyll fields as an area of enhanced chlorophyll. The Kara Sea, east of Novaya Zemlya, has higher surface chlorophyll in summer than the Barents Sea to the west, which has little winter ice cover (see the later section on the Atlantic Subarctic Province).

Marginal ice zones develop from west Spitzbergen through the northern and eastern Barents Sea with the vernal retreat of the ice edge. These induce upwelling, which may be brief, wind-induced events, or longer density-induced events (Johannessen, 1986). The low-salinity surface layer is on the order of 10–30 m deep at its most shoal position near the ice edge when density-driven or wind-forced upwelling occurs. The MIZ of the Chukchi Sea appears only in June, but only a small part of the sea becomes ice-free; extremely strong stratification occurs at the ice edge, and reported rates of primary production seem improbably high ($<7\,\text{gC m}^{-2}\ \text{d}^{-1}$). At any rate, a strong chlorophyll maximum occurs at the base of the pycnocline, and nitrate levels are reduced to undetectable. In the Barents Sea, the MIZ retreats very rapidly northward across shallow water in spring, whereas in the deep Fram Strait it remains close to the boundary between the northward flow of Atlantic water and the southward flow of polar water. In both regions, nitrate is reduced to low values after the summer bloom occurs.

In the shelf polynyas of the Arctic Ocean, as in the Beaufort Sea, a second phytoplankton bloom may occur toward the end of summer; this may be more intense than the initial bloom and appears after some summer stratification of surface water has occurred (Arrigo and van Dijken, 2004). The timing and intensity of blooms varies strongly between years, responding to anomalous early warming and stratification (as in 1998) or a later summer ice melt and resulting stratification (as in 2002). In general, it is found that nutrient limitation occurs over the deeper parts of the polynya, whereas light limitation obtains near the coast below land-fast ice that delays the bloom by about 1 month. Somewhat similar regional characteristics force earlier and more intense algal blooms on the Greenland side of the North Water in the early part of the season than on the west (Odate *et al.*, 2002).

A special case of MIZ occurs at glacier fronts in the Canadian archipelago and around Greenland where diatom blooms occur, often an order of magnitude denser than elsewhere in the region; these blooms utilize the increased nutrient levels associated with upwelling on the underwater ice cliffs. This process is forced within meltwater-driven convection cells at the face of the glacier accompanied by caballing (mixing of waters of equal density, but of different T and S to produce water of greater density). Upwelling at ice fronts by this mechanism occurs very commonly in BPLR, even alongside large free-floating icebergs.

***More Extensive Open Water Situations*** Where extensive open water develops during summer, as in Baffin Bay and the Chukchi Sea, a "spring" bloom occurs in the open water, although it is not long sustained because nutrients are rapidly stripped from the surface brackish layer. Strong blooms, sustained by topographic upwelling, occur in many places in the Canadian archipelago, notably in Lancaster Sound. Here, the chlorophyll maximum occurs consistently during summer between the 1% and 10% light levels, usually near the base of the pycnocline, at which depth a nitracline also occurs. The timing of the bloom is dependent, naturally, on the schedule of the breakup of fast ice and hence of local "spring"; in Barrow Straits, for example, this occurs only in July. It is largely by virtue of its relatively early opening that the North Water "polynya" is exceptional in the Canadian region. On this timing, of course, depends the instantaneous state of development of the entire planktonic ecosystem; the recruitment of the first cohort of copepodites of *Calanus* and *Pseudocalanus* occurs 1.5–3.0 months earlier in the North Water than in the Barrow Strait (74°N) (Ringuette *et al.*, 2002).

A "typical arctic structure" for phytoplankton has been described from the open North Water region in summer (Harrison *et al.*, 1987): chlorophyll maxima consistently occur between the 10% light level and the bottom of the photic zone, itself usually deeper than the mixed layer. These maxima are usually six times the surface values of around 1.25 mg chl $m^3$. Below the photic zone, enhanced microbial activity associated with aggregated sinking cells reaches maxima usually below 500 m. $NO_3$ values are often below the limit of detection in the mixed layer and a nutricline is coincident with the pycnocline. Reduced nitrogen is relatively available throughout the summer and is importantly utilized by autotrophs, a fact apparently missed by earlier investigators (Harrison *et al.*, 1982). The photosynthetic index (PI), as gC $(g\ chl)^{-1}\ d^{-1}$, of arctic open-water phytoplankton takes "normal" values (giving a mean of $PI = 6.7$ in a large data set), but very low values obtain below the Arctic Ocean ice cover, where values of $PI < 0.5$ are more typical.

Diatoms and coccolithophores were long assumed to dominate polar phytoplankton under these open-water conditions, at concentrations reaching 750 and $925 \times 10^3$ cells $liter^{-1}$, respectively, but recent work has revealed that a substantial proportion of chlorophyll, cell numbers, and RuBPC activity are actually contributed by the pico fraction, just as in other seas. These small cells perform 10–25% of all carbon fixation in polar seas compared with 20–30% in mid-latitudes and >50% in the tropical ocean (Trotte, 1985). Generally, 60% of polar algal biomass is contributed by >35-mm cells, whereas >50% of respiration is contributed by <1-mm cells (bacteria and other microheterotrophs). Limited by low temperatures are the prokaryotic cyanobacteria and prochlorophytes, which occur only in relatively low abundance and serve as a biological marker for the transport of southern water in summer. No novel physiological mechanisms need be invoked to explain the success of arctic phytoplankton in extremely cold water and it would, in fact, have been possible to predict their performance by extrapolation from what is known of the physiology of temperate-zone organisms (Li and Dickie, 1984).

The general ecology of arctic zooplankton (especially pteropods, copepods, and euphausiids) has been well understood for many years. Growth rates are very slow, individuals are large compared with congeners in warm seas, seasonal ontogenetic migration dominates vertical distribution, and copepods dominate the total biomass. Copepods form >85% numerically of all mesoplankton and (as carbon biomass) polar zooplankton have the following composition: 70% copepods, 11% pteropods, 10% amphipods, and lesser amounts of ostracods, coelenterates, and appendicularians among the major taxa.

Euphausiids form no more than 0.1–7% of individuals. Recently, it has been found that protozoan microplankton comprise only about 10–20% of total biomass.

Lipid storage is very important for copepods, especially in diapausing copepodites in winter, whose ammonium excretion is reduced to very low levels (Head and Harris, 1985), this being an especially important process in those taxa having multiyear generation times. *Calanus hyperboreus* accumulates lipid especially during the C5 stage in preparation for the adult molt, and for egg production at depth during the last winter of each multiyear generation cycle (Conover, 1988). Wax ester levels rise from 0.4 to 2.1 mg $ind^{-1}$ between June and September of their penultimate summer.

In the open-water summer situation there is a strong differentiation in depth selection by zooplankton genera. Species respond to the three-layer partition of the water column into (i) warmed surface water to 30–50 m, (ii) cold arctic water from there to 250 m, and (iii) warmer Atlantic water below that. Zooplankton respond also to the depth of the maximum aggregation of plant cells, usually at the junction between (ii) and (iii). In summer, in northern Baffin Bay, most individuals of *Gaidius tenuispinus* and *Metridia longa* occur in the upper part of the Atlantic water, whereas *Oithona* spp., *Parathemisto libellula*, and *Pseudocalanus minutus* occur preferentially in the upper part of the cold arctic water mass and just below the depth of maximal algal cell numbers. *Limacina helicina*, for its part, is restricted to the warm surface water (Longhurst *et al.*, 1984). Stage-specific depth selection occurs during development and we must understand the reproductive status of a population in order sensibly to interpret its vertical distribution. C1–C4 of *Calanus glacialis* are restricted almost exclusively to the upper warm layer, whereas stages 5 and adults are widely distributed within the depth zone of the cold arctic water at 75–300 m.

As in warmer seas, detailed depth distributions (as are obtained with BIONESS or LHPR sampling systems) indicate that herbivores tend to select either the depth of maximum production rate or the slightly deeper chlorophyll maximum. Intrageneric specialization may occur, because *C. glacialis* and *C. finmarchicus* follow the depth of maximal production rate, whereas layers of *C. hyperboreus* occur deeper, near the chlorophyll maximum (Herman, 1983). One is not surprised, therefore, by such observations as that in high summer, at the deep chlorophyll maximum in the MIZ northeast of Spitzbergen, *C. hyperboreus* and *C. glacialis* consume from 65% to 90% of the daily primary production. At the same time, they support about 35% of the local cell growth by their own excretion of ammonium (Eilertsen *et al.*, 1989a).

Nevertheless, the primary production/herbivore consumption ratio may takes a positive value in the photic zone (Longhurst and Head, 1989): in late summer, the remnant herbivore grazers remove less than 1% of daily production in Baffin Bay and the fate of most algal cells must be to sink through the pycnocline as soon as winter mixing destroys this feature. Rapid, mass sinking of diatom-dominated algal floc is suspected to be a feature of the final phase of blooms in this province.

## Regional Benthic and Demersal Ecology

The extensive shelf sediments of the continental shelf encircling the Arctic Ocean are relatively inaccessible to study, but we have sufficient information from parts of it for a preliminary interpretation. One characteristic that is mentioned by several authors is the massive end-of-summer sedimentation of phytoplankton cells, resulting in soft organic-rich deposits on the floor of the continental shelf where its topography encourages settlement. The relationships among benthic diversity, benthic biomass, and sediment type (indicated by the relative content of organic matter and the C/N ratio) have been investigated in the Chukchi Sea (Feder *et al.*, 1994). There is some indication that although hydrographic fronts encouraged the sedimentation of organic material and the development of high benthic biomass, relative diversity is not related, as one might expect, to gravel-sand-mud ratios, or the water contents of deposits. Instead, it is related to disturbance to the sediments by ice scour and by the feeding of gray whales and walrus.

The benthic macrofaunal communities of the Arctic shelves, below the shallow ice-scoured zone of the littoral, are dominated by variations of the polar *Macoma* and *Astarte* communities, having high biomass down to about 50 m—that is to say, within the influence of the surface water mass. On muddy sediments in this depth zone, *Macoma calcarea* is associated with other bivalve mollusks, such as *Mya* and *Leda*, whereas on sandy, coarser sediments this "boreo-arctic *Macoma*" association is replaced by one dominated by *Cardium ciliatum* and *Venus fluctuosa*, the "Arctic *Venus* community" (Thorson, 1957; Ellis, 1959). Deeper on the shelves, benthos of lower overall biomass is characterized by foraminifera (*Rhabdomina, Axinopsis*).

A curious benthic association of considerable ecological significance is the boreo-arctic *Ampelisca* community of Thorson (1957), which is characterized by amphipods of the families Ampheliscidae, Isaeidae, and Phoxocephalidae that together may comprise <85% of individuals, together with some bivalve mollusks (*Cardium, Tellina*), polychaetes (*Maldane, Nephthys*), and various ophiuroids. These tubicolous amphipods and bivalve mollusks are suspension feeders, utilizing the near-bottom layer of high-turbidity water, so that benthic associations of this kind are very probably widely distributed around the Arctic Ocean shelves where this layer is characteristic. The same association of tubicolous amphipods has recently been encountered on the slope in northern Baffin Bay at about 500 m depth, where the density of *Haploops tubicola* is around 3500 ind $m^{-2}$.

The earlier observations of benthic communities and associations of species by Thorson and others have been confirmed in their objective reality by factor analysis of 204 quantitative benthic samples from the shelves of the northern Baffin Bay region (Thomson, 1982). By this analysis, 21 factors (each comprising one or several species) were grouped into three species assemblages that bore sufficient resemblance to the "communities" previously described as to suggest that these are realistic entities, comprising groups of species having mutual interactions and requirements. Factor 2, for instance, associated six of the nine species listed by Thorson as being characteristic of the boreo-arctic *Macoma* community. We now have observations of this association from the northern Barents Sea, both coasts of Greenland and the Canadian Arctic: it (like the other associations) is probably circumpolar in distribution.

Unlike the rich arctic benthic communities, the associated fish are extremely reduced in diversity compared to those associated with similar benthic communities in warm seas. As I have already discussed in Chapter 4, this is perhaps related to the greater uncertainty of successful recruitment from planktonic larvae in cold than in warmer seas. The planktonic eggs of *Boreogadus saida*, the arctic cod, require a 45 to 90-day period of "incubation," whereas occlusion occurs overnight in tropical clupeids; it is chiefly *Boreogadus* that should take our attention in this province, along with the Greenland shark, halibut, and several small liparids. *Liparis koefoedi* is typical of the latter and is probably circumpolar, as is *Eumicropterus derjugini*, a small lumpsucker occurring on the shoaler parts of the shelves. Along the southern Greenland shelves, the bentho-pelagic cod, *Gadus morhua*, was at one time very abundant but is now reduced by industrial fishing to a small remnant.

*Boreogadus saida* is a small (<32 cm), slim, largely pelagic gadoid that is the key species in the transfer of organic material from planktonic production in Arctic seas to marine mammals and birds, for which it forms a vital source of food. It is consumed by murres, guillemots, and kittiwakes, by harp and bearded seals, and by narwhal and beluga. It is largely confined to cold circumpolar water, where it is widely distributed, but its spawning concentrations in shallow water are not yet well known. It occurs at a range of depths, from cavities in sea ice down to at least 300 m. It is entirely nektonic for its first year and subsists throughout life very largely on planktonic crustaceans.

Benthos-feeding walrus (*Odobenus rosmarus*) and gray whales (*Eschrichtius robustus*) are restricted to open water, the former occurring in polynyas in winter, the latter migrating south to the California coast. It is the abundance of bivalve mollusks, which they dig from the sediments with their downwardly directed tusks, in the benthic communities described earlier that enable walrus to maintain populations at such high latitudes as that of the Chukchi Sea. Walrus must have access to dry land for hauling out, which they do in very large herds.

Gray whales are, of course, now long extinct in the Atlantic, but a population of 15,000 or so still migrates between the Chukchi Sea and California (to the great profit of the tour-boat operators there). These curious whales occupy the whole shelf habitat of the Chukchi Sea in summer and fatten by feeding on the benthic invertebrate infauna. These they dredge—it is the only word to use—from the bottom, and then filter the organisms from the soft deposits with their baleen plates. Nevertheless, their principal item of food is not so different from that of the baleen whales that feed on krill in the Southern Ocean, for example. Recall that one of the main benthic species associations on the Chukchi shelf is that of the *Ampelisca* community, which is characterized by crowded masses of small tubicolous, filter-feeding amphipods, and it is these that form the principal diet of the whales. Feeding heavily but moving little, they are able to fuel themselves not only for reproduction, but also for the long southern migration to their wintering grounds. They take little food there, or on the journeys to and fro.

### Synopsis

*Case 1—Polar irradiance-mediated production peak.* Because some parts of the province are south of the Arctic Circle, the mean seasonal cycle does not include winter darkness; midwinter data that are incorporated into the seasonal cycle represent regions where sunlight occurs year-round. Nevertheless, mean regional photic depth is strongly affected by ice cover and the winter darkness that does occur over most of this province. $Z_m$ is shallow at all seasons in open water, and $Z_{eu}$ is usually deeper (at least in the sunlit regions), so that the response of productivity to the seasonal light field is very rapid (Fig. 9.2) and the seasonal evolution of P rate is symmetrical about midsummer peak rate. Chlorophyll tracks the linear increase of P rate to a midsummer peak, then declines (with P) but subsequently accumulates again to reach the annual chlorophyll maximum in October. Thus, consumption and sinking do not constrain accumulation during the vernal increase in P and the late summer renewal of plant biomass accumulation is consistent with relaxation of the loss term as the large herbivorous copepods descend to overwintering depths in July–September.

## Atlantic Arctic Province (ARCT)

### Extent of the Province

This is an entirely oceanic province that lies between the edge of the Greenland coastal currents and the Oceanic Polar or Subarctic Front (Dietrich, 1964) that crosses the ocean diagonally from Flemish Cap to Spitzbergen. The province thus excludes the East Greenland Coastal Current but includes the central part of the Labrador Sea and a broad zone between eastern Greenland and Spitzbergen including the Iceland and Greenland Seas. The limits of the province are rather variable, because set by oceanic and shelf-edge fronts, and its southern boundary probably cannot often be traced in satellite imagery.

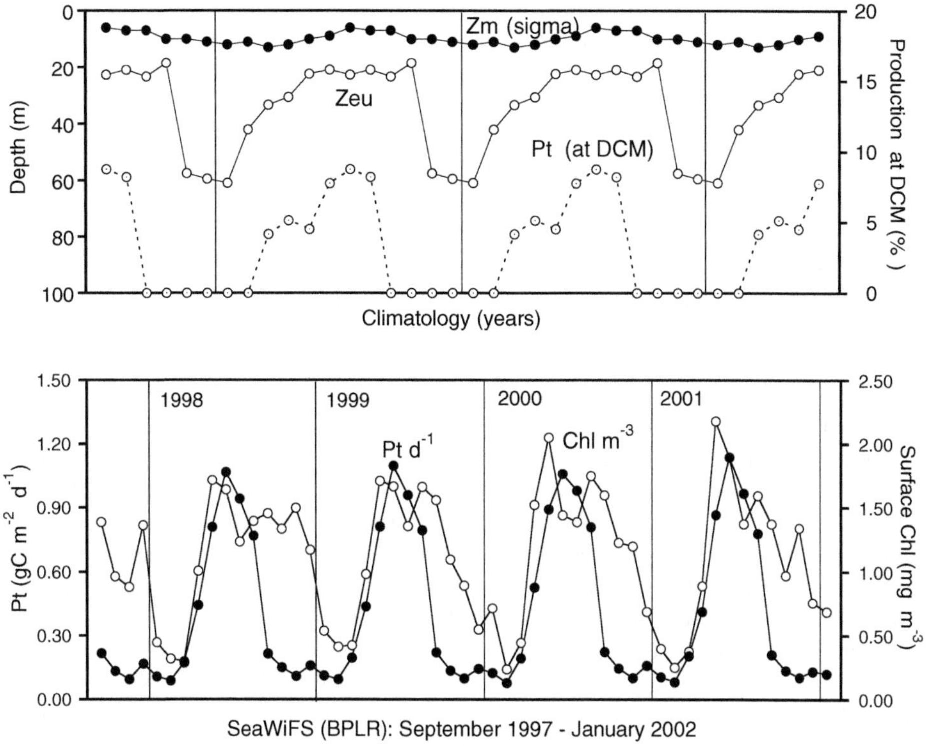


**Fig. 9.2** BPLR: seasonal cycles of monthly surface chlorophyll and depth-integrated autotrophic production for the years 1997–2002 from SeaWiFS data together with characteristic seasonal cycles of mixed-layer depths from Levitus climatological data and photic depths computed from characteristic irradiance and the archive of chlorophyll profiles discussed in Chapter 1.

## Defining Characteristics of Regional Oceanography

The Arctic province, as defined earlier, comprises the two cyclonic Subpolar Gyres of the North Atlantic Ocean. The larger of these lies south of Greenland, occupying the Labrador and Irminger Basins, while the smaller lies to the northeast of Iceland occupying the northern part of the Norwegian Basin. These gyres are characterized by strong eddy fields, associated both with the Greenland Coastal Front that lies just above the break of slope, and southward by the Oceanic Polar Front that lies across the North Atlantic and its extension into the Nordic seas.

James Swift (1986) defines Arctic waters in these seas as a "hydrographic middle ground lying between the domains of the polar and Atlantic waters . . . relatively cold (0–4°C) and saline (34.6–34.9 ppt)." The Arctic water entrained around the Labrador-Irminger gyre is modified by the influence of Atlantic water, eddied across the Oceanic Polar Front (see later discussion), and transported north around the western part of the gyre in the Irminger Current. The water of the cyclonic gyre of the Norwegian Sea is less modified, although an arm of the Irminger Current does pass north around the Icelandic shelf and enters the western boundary current of the Norwegian cyclonic gyre. The classic cartoon showing the circulation of this region according to Dietrich is confirmed by more recent flow analysis.

The province is bounded to the northwest by the front at the edge of the Greenland coastal current and to the south and east by the frontal region at the edge of the flow of Atlantic water. There is some confusion over how these fronts should be named, but I shall use Greenland Coastal Front for the edge of the polar water flowing south from

Fram Strait (see BPLR) and Oceanic Polar Front (OPF), in the classical sense, for the front that lies between Iceland and Bear Island.

Though the definition of the OPF is simple, its location at the sea surface is complex. It is the conjunction between warm, salty, subtropical gyral water and cold, less saline subpolar water and is thus the extension of the North Wall of the Gulf Stream. This feature retains its coherence, meandering eastward as far as the Mid-Atlantic Ridge at about 30°W (Dietrich, 1964). Thence, it passes north around Iceland, then meanders strongly again as it continues northeast along the Faeroe-Iceland Ridge. It then turns more northerly along the Jan Mayen section of the Mid-Atlantic Ridge (Johannessen, 1986; van Aken *et al.*, 1991; Wassman *et al.*, 1991). Along the cold side of the OPF there is strong flow of relatively cold, low-salinity water from the west, recirculating within the Subpolar Gyre of the Labrador-Irminger Basin (Krause, 1986).

Ice formed in the polar water of Baffin Bay and in the Fram Strait extends patchily and irregularly into the Arctic Province during some winters, and broken pack ice is frequent in the regions bordering on BPLR; complete ice cover occurs in the Odden region between northeast Greenland and Iceland, and in the northern Barents Sea west of Spitzbergen. The inner part of the Labrador Current, shoreward of the break of shelf, remains largely ice-covered well into the summer period while the more offshore segment, comprising (see later discussion) the western boundary current of the cyclonic gyre in the Labrador Sea, is relatively ice-free. Some pack ice is drifted southward from the Fram Strait region in the flow along eastern Greenland.

North of Iceland, winter mixed-layer depths are deepest in the central Norwegian Sea (around 500 m), whereas a thermocline dome (<100 m) appears to be associated with the northward-turning flow of Arctic water east of Iceland (Levitus, 1982). Not only is winter mixing deepest at 50–60°N but also, because the subsurface nitrate field slopes with the baroclinicity of the subtropical gyre, initial mixed-layer nitrate concentration in spring are progressively higher toward the northern edge of the gyre (Glover and Brewer, 1988; Yentsch, 1990).

The surface waters of the Greenland and Labrador Seas have a relatively lower stability than adjacent areas so that cooling and wind stress at the surface cause unusually deep regional winter mixing. In fact, the Subpolar Gyres are two of only three global centers of deep convection of surface water into the global thermohaline circulation, the Weddell Sea off Antarctica being the third. Consequently, at the end of winter a very deep trough in the mixed-layer topography extends in an arc around the northern flanks of the subpolar gyres, from the southern Labrador Sea to the Norwegian. It occupies a somewhat different area according to the data and criteria used to display it (e.g., Levitus, 1982; Robinson *et al.*, 1979; Glover and Brewer, 1988; Woods, 1984). This process is associated with high winter nitrate values in the mixed layer, reaching 16 μM.

The actual depth of deep winter convection is very variable. Exceptionally intense convection down to 2300 m occurred during 4 years in the early 1990s and created a new deep pool of Labrador Sea Water but, in response to warmer winter weather in the later years of the decade, convection reached to only half that depth: stratification was therefore reestablished at between 150 and 1000 m (Lazier *et al.*, 2002).

## Response of the Pelagic Ecosystems

The chlorophyll images available since 1997 suffice to build a climatology of the seasonal cycle of chlorophyll accumulation of this province, and it is now clear that our earlier assumptions about the seasonality of the Arctic spring bloom were not satisfactory. It has long been evident that the progression of the bloom does not represent a simple poleward, irradiance-forced progression, although the pattern has not been easy to read. Now, both the individual years and the climatology make it clear how blooms really evolve. In March, the entire province exhibits low surface chlorophyll values, and it is only during April, and only in the northeast quadrant of the Labrador Sea over deep water, that a bloom is initiated. This pattern is concordant with the progression of regional stratification, as we should expect it to be, although in the Labrador Sea there appear to be regional consequences of buoyancy induced by ice-melt water. The Levitus data show that it is in the northeast quadrant of the Labrador Sea that near-surface stratification first develops. Furthermore, in April, mixed-layer depths shoaler than 40–50 m appear here, while the remainder of the province remains deeply mixed.

Subsequently, in May, bloom conditions are much more extensive. The NE Labrador Sea deep-water bloom is now continuous with chlorophyll accumulation along the West Greenland Current, around Cape Farewell, and north along the coast to the Greenland-Iceland Ridge. Again, this matches the evolution of regions having shallow mixed depths. Similar conditions now also occur in the northern part of the Norwegian Basin, where the mixed layer may exceed 30 m. So it is especially around the northern limbs of the two subpolar cyclonic gyres, poleward of the trough of very deep winter mixed layers, that chlorophyll accumulation is initiated. Also in May, the northward flow of the Irminger Current around the western gyre supports a bloom.

In June, almost the entire province has shallow mixed-layer depths and while the Labrador Sea bloom regresses into the southern limb of that gyre, the bloom in the Greenland Sea occupies the entire eastern gyre. The Irminger Current east of Iceland and its eastward extension within the Iceland Gap Front are now both prominent in the chlorophyll field. From July onward, the regional features in the chlorophyll field become more diffuse, and the chlorophyll concentration progressively diminishes. Finally, in October, the Iceland Gap Front is the most prominent feature, though higher latitudes are progressively less available to satellite imagery. Thus, later in the season, chlorophyll accumulates preferentially in relation to active eddying along oceanic fronts, rather than in relation to the start of regional stratification.

The spring bloom is dominated by diatoms (*Chaetoceros, Nitzschia*, chains of *Melosira*, and pennate forms such as *Navicula* and *Pleurosigma*) and abundant athecate dinoflagellates. The colonial prymnesiophyte *Phaeocystis pouchetti* may be the dominant organism in very early spring blooms, which are supported by extremely high rates of primary production of $<2$ g C $m^{-2}$ $d^{-1}$, or about the same rate as occurs in blooms at polar ice edges (Smith *et al.*, 1991a). The *Phaeocystis* blooms that occur in the Norwegian basin in early May are contemporaneous with the spring bloom in the North Atlantic Drift Province (NADR), initiated by thermal stratification approximately 30° farther south. Recent investigations in the Labrador Sea (Lutz *et al.*, 2003) confirm that the pico fraction of autotrophic cells is relatively less important here than in adjacent lower latitudes; this is now becoming a routine observation as is, perhaps, the universality of the haptophyte fraction at both low and high latitudes.

The basic distribution and ecology of macrozooplankton of this province was established during the 1960s during Russian fishery investigations, although it should be noted that they used the term "Davis Strait" for what we now call the entire Labrador Sea; properly, of course, Davis Strait is the passage between Baffin Bay and the Labrador Sea. It is also appropriate to recall the early time-series work at OWS "B" in the central Labrador Sea, where daily BTs and weekly net tows and bottle casts were obtained throughout 1950 and into 1951 (Kielhorn, 1952). Ongoing Canadian investigations since 1994, along the WOCE transects in the Labrador Sea, have emphasized the ecology and physiology of the dominant taxa (e.g., Head *et al.*, 2000). The partition of the accumulated CPR data by Beaugrand *et al.* (2002a, b), already discussed in Chapter 7, offers further insights for this region.

Herbivore ecology has the same characteristics as in BPLR, and the same organisms are dominant. The ARCT province is clearly distinguished from more southerly provinces by the "mean community body size" for copepods computed by Hays (1996). A line from Iceland to the Flemish Cap, approximately coinciding with the province boundary proposed here, divides small copepods (adults, mostly about 0.5 mg wet weight) to the south in NADR from large copepods to the north in ARCT (mostly about 2.0 mg wet weight). The same line divides two characteristic patterns of diel vertical migration (DVM). Mean community percentage biomass translocation by DVM was computed by Hays from about 130,000 Hardy Continuous Plankton Recorder samples archived for this region, and these show clearly that in ARCT this index takes values of about 80%, whereas in NADR values of 10–50% are more usual. DVM is strongest in summer, weakest in winter in both provinces, and not only intensity but also timing differs on either side of the line defined earlier: in ARCT, maximum values for DVM occur at midsummer, whereas in NADR there are spring and autumn peaks.

Seasonal ontogenetic vertical migration follows the same pattern as in BPLR although *C. hyperboreus* descends to depths of as much as 1000–1500 m in the Greenland Sea (Hirsche, 1991). Though some populations of this copepod may require only a single year to complete their life cycle, the probable duration for most populations in ARCT is 2 years. When deep convection events occur, they disrupt the arrangement of the deeper water masses within which overwintering populations of copepods reside. Richter (1994) proposed that this process causes the sporadic but massive recruitment failures of copepods that are known to occur here.

Although the ARCT province is a meeting place for polar, subpolar, and temperate faunas, and although their points of entry to the province are evident, the Russian studies here emphasized that recirculation and mesoscale activity ensures that the various faunistic elements remain partitioned only in the most general way. There is some fidelity of polar forms such as *Calanus hyperboreus, Metridia longa, Themisto libellula*, and *Limacina helicina* to water entrained from the Labrador coastal current and the East Greenland Current, but individuals are more widely distributed than that. Similarly, *Calanus finmarchicus, Oithona similis, Limacina retroversa*, and *Physophora hydrostatica* are characteristically more southerly species and are preferentially distributed where Irminger water, carrying some Atlantic signature, dominates—that is, around the northern limb of the subpolar gyre in the Labrador and Irminger basins. Thus *C. hyperboreus* is a dominant copepod in the extreme NE parts of the Norwegian Basin gyre and in the western parts of the Labrador Sea gyre, but not elsewhere. Although the fidelity of the copepod faunas to Atlantic and Arctic water masses seems very clear, the effect of depth cannot be excluded. In spring 2000, the Labrador water that normally occupies the shelf spread out over the slope region, but this did not displace the slope fauna that, with a strong Atlantic biogeographic element, remained *in situ* (Head *et al.*, 2003). The CPR analysis confirms that the characterizing species in the two subpolar gyres are different: *Calanus hyperboreus, C. glacialis*, and *Metridia longa* occupies the Labrador Sea gyre and *C. finmarchicus, Heterorhabdus norvegica, Euchaeta norvegica*, and *Scolecithricella* spp. that of the Norwegian Sea.

Overwintering strategies differ between species (Hirsche, 1991), partly by depth selection and partly by differential timing of their reproductive periods. *C. hyperboreus* overwinters as C3s to adults and *C. finmarchicus* as C4s to adults but mostly C5s, whereas *C. glacialis* overwinters mostly as CIVs. *Metridia longa* overwinters in a more advanced stage, >50% of all individuals in fall already being adult. *C. hyperboreus* overwinters deeper (1000–1500 m) than the other species. Experiments in the Labrador Sea suggest that egg production in spring by *C. finmarchicus* is fueled importantly by stored lipid reserves conserved during winter (Cabal *et al.*, 1997).

The life cycle of *Calanus finmarchicus* will be reviewed in the next section, devoted to the SARC province, so here it is sufficient only to note some characteristics of its behavior in the ARCT province. In the Labrador Sea, *C. finmarchicus* produces a single generation yearly, and the timing of the algal bloom in relation to egg laying appears to determine recruitment success of the subsequent generation (Head *et al.*, 2000, 2003). When the bloom is early and intense, maturation of overwintered adults is rapid and eggs are produced when phytoplankton concentrations are high. If the bloom is late, or weak, maturation is delayed so that eggs are produced when phytoplankton concentrations are low. In the latter case, not only will fewer eggs survive, but larvae will have a shorter period to develop prior to entering the overwintering phase. In the later copepodite stages, survival appears to be less dependent on phytoplankton availability.

The pelagic fish of most significance in this province is the capelin (*Mallotus villosus*), a small (15–20 cm) osmerid smelt that is significant in the support of the extensive sea-bird populations of the region. It is extremely abundant but very little appears to be known of its pelagic ecology, save that its diet is dominated by copepods; the relations between capelin and demersal fish will be touched on in the next section.

## Synopsis

*Case 1—Polar irradiance-mediated production peak.* Deep winter mixing ensures that $Z_{eu}$ is shoaler than $Z_m$ except very briefly in summer months (July–September), when thermocline may be illuminated. The seasonal cycle shown in Fig. 9.3 indicates that the rapid near-surface thermal stabilization in May is not readily captured by archived data. Seasonal evolution of P is symmetrical about a midsummer peak rate that is significantly higher than in high austral latitudes. Vernal increase in P tracks both the changes in

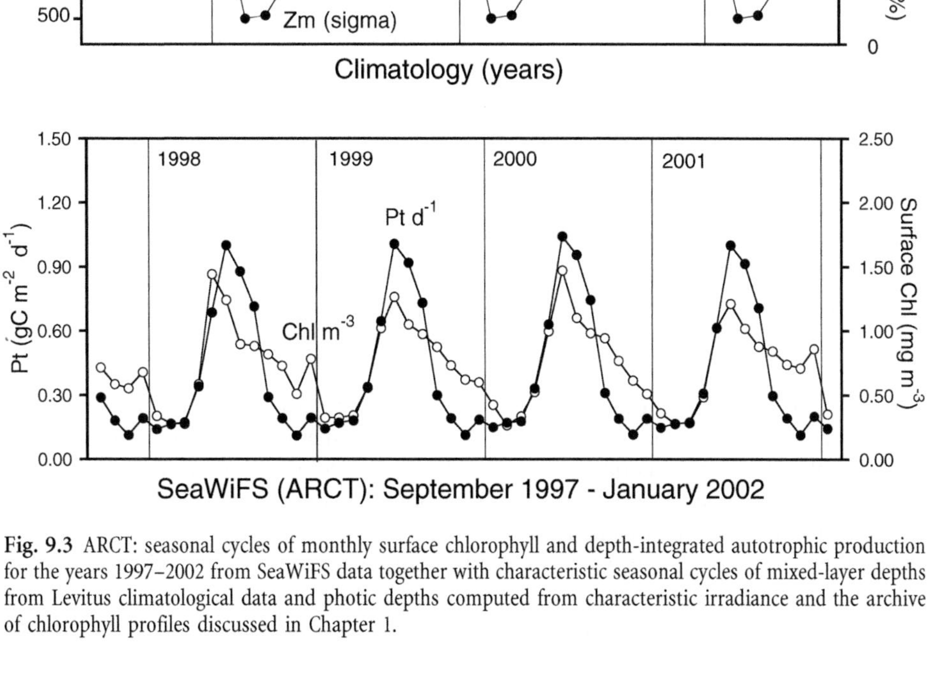

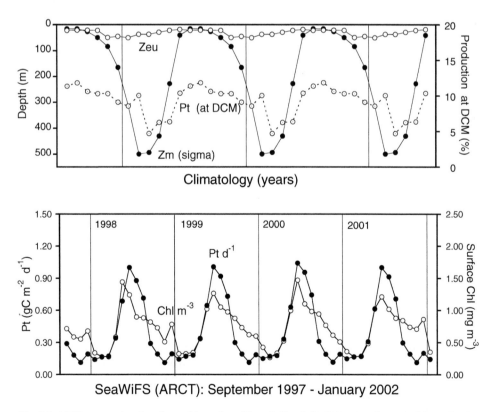


**Fig. 9.3** ARCT: seasonal cycles of monthly surface chlorophyll and depth-integrated autotrophic production for the years 1997–2002 from SeaWiFS data together with characteristic seasonal cycles of mixed-layer depths from Levitus climatological data and photic depths computed from characteristic irradiance and the archive of chlorophyll profiles discussed in Chapter 1.

the light field and the shoaling of $Z_m$ but, after June, P tracks the declining light field. Accumulation of chlorophyll tracks changes in rate of P until late summer, after which a secondary accumulation of biomass occurs in September–October and is interpreted as a response to the descent of herbivores to overwintering depths. Balance between production and consumption may be closer than in BPLR.

## ATLANTIC SUBARCTIC PROVINCE (SARC)

### Extent of the Province

From above the Rejkanes Ridge, southeast of Iceland, this province extends to the northwest, including the entire Norwegian Sea and the southern Barents Sea as far as Novaya Zemlya. The province is bounded to the north by the Arctic Front (or OPF), beyond which is the ARCT province. In the Barents Sea the adjacent province is BPLR, from which SARC is divided by the well-marked thermal front between Arctic and Atlantic water that passes close around Bear Island and meanders eastward from Hope Island at about 74–78°N. Between SARC and NECS an effective boundary is the break of slope between the Shetlands and the Faeroes. In the south, an arbitrary line at 60°N separates the SARC portion of the North Atlantic Drift from that lying in NADR.

### Continental Shelf Topography and Tidal and Shelf-edge Fronts

For the same reason that the coast of Greenland and the shelves of the Arctic Ocean were included in the BPLR province, the coast of Norway, from Stavanger to the North Cape, is included in this province; this coast is very steep-to with a very narrow terrace across which many troughs extend the topography of fjords. Only at the Lofoten Bank (68°N) is there any significant shelf width. The Barents Sea is nowhere deeper than 500 m; the eastern half, associated with Hope and Bear Islands and Spitzbergen, is <200 m deep and is one of the wider shelf regions in any sea. The Faeroes and Iceland lie on small 200-m platforms arising from oceanic depths.

The Faeroe platform carries a coastal water mass, which is confined within a persistent tidal front that surrounds the archipelago at a depth of about 150 m. A Norwegian Coastal Front lies outside the northward flow of the permanently stabilized water mass of low salinity of the coastal current (see later discussion), following the break of slope; it thus has the character of a shelf-edge front. Tidal amplitude increases northward along the Norwegian coast from a few centimeters to >1 m at North Cape; in fjords and narrow straits in the more polar regions, then, very strong tidal streams may be expected, with all that they entail.

### Defining Characteristics of Regional Oceanography

This is a complex oceanographic region, especially in the north, where warm Atlantic water penetrates into very high latitudes indeed: it has no equivalent in the Pacific Ocean. This flow is highly variable between years, forcing strong variability in both physical and ecological processes.

South of Iceland, part of the North Atlantic Current begins to flow toward the Nordic seas rather than continuing eastward in the gyral circulation. It is here that Atlantic subtropical water begins its odyssey toward the Arctic Ocean, penetrating even into the Kara Sea (Dickey *et al.*, 1994). This flow exists both as a broad west wind drift, and also as a jet current along the Oceanic Polar Front (OPF) whose characteristics were described in the previous section, and which forms the northwest boundary of SARC, separating Atlantic water from Arctic water. The eastward zonal flow to the south of the OPF passes across the Iceland-Faeroes Ridge to the south of the Iceland Gap Front (Johannessen,

1986). It is joined by Atlantic water that forms a rather variable slope current north of the Shetlands, and large offshore eddies are persistently generated as this flow enters the Shetland-Faeroe Channel (Sherwin *et al.*, 1999).

Low-salinity water from the Baltic and eastern North Sea passes around the south-western coast of Norway into the Norwegian Coastal Current. This is bounded by a salinity-dominated coastal front, with which is associated a highly unstable eddy field (Johannessen, 1986), modifying both its Atlantic and coastal components as they pass northward. Off North Cape, most water from the combined flow passes eastward into the southern Barents Sea, while some continues poleward above the continental slope west of Spitzbergen and so on to Fram Strait as the West Spitzbergen Current. The water passing into the southern Barents Sea feeds a gyral circulation, which returns westward to the south of Spitzbergen. A meandering convergence extends toward Bear Island from the central Barents Sea, and this front is both the southern limit of winter ice cover and the boundary between SARC and the adjacent BPLR region.

In the oceanic parts of this province, winter mixing is deep, especially along the OPF and to the south of Iceland, although not so deep as in the southernmost parts of the ARCT province. There has been a significant change of physical conditions in these seas during the past 30 years, several time series suggesting a progressive decrease in both salinity and temperature of the surface waters of the Norwegian Sea. Decreasing overflow of deep cold water across the Faeroe-Iceland Ridge since 1950 is associated with decreasing flow of warm Atlantic water into the Nordic seas (Hansen *et al.*, 2001). Increased wind-induced eastward advection of Arctic water in recent decades has resulted in an extension of an intermediate layer of water of Arctic origin over the whole Norwegian Sea, and to a freshening of the Atlantic water. The OPF has moved eastward in response by up to 300 km, and there is a close correlation between the location of this front and the winter index of the North Atlantic Oscillation (Blindheim *et al.*, 2000).

## Response of the Pelagic Ecosystems

We have good information on the cycles of pelagic production and consumption for several regions in the SARC province. It was, of course, from his studies of data from Ocean Weather Station (OWS) M at 64°N in the Norwegian Sea that Sverdrup (1953) formulated his classical model of a spring bloom, establishing firmly the relations between mixed-layer depth and critical depth. His results were closely matched by the longer series of observations by Bob Williams at OWS I at 59°N just south of Iceland, also in this province. Here, serial observations were made from April to October in each year from 1971 to 1975, including weekly multidepth profiles of density, nutrients, chlorophyll, and zooplankton (e.g., Williams and Robinson, 1973; Williams and Hopkins, 1976). Further useful work near the same position was undertaken in the summer of 1980 during the JGOFS North Atlantic Bloom Experiment (NABE). The weather ships observations have been confirmed by daily data obtained more recently from a moored optical profiling array (Dickey *et al.*, 1994) near OWS I.

Because of the presence of low-salinity surface water of Arctic origin, the spring bloom at OWS I occurs earlier than would be predicted by some models of mixed-layer evolution and initiation of the algal bloom for the North Atlantic (Wolf and Woods, 1988: Strass and Woods, 1988). Chlorophyll and $^{14}C$ data for 1972–1975 show that phytoplankton growth rapidly follows the reestablishment of a shallow mixed layer (e.g., Williams and Robinson, 1973). Ephemeral blooms associated with temporary near-surface pycnoclines after periods of calm weather occur even earlier in the year, and continuous spring bloom conditions are usually established by mid-April, with chlorophyll biomass reaching 3–4 mg $m^{-3}$ by mid-May. But there is much between-year variability: in 1972, nitrate was reduced from $>10.0$ to $<2.0\,\mu M$ in only 7 days in April, whereas in 1973 the equivalent uptake took the entire months of June.

In the early stages of the bloom, during April and May, chlorophyll concentrations of <0.25 mg $m^{-3}$ extend well beneath the mixed layer to 150–200 m, suggesting that ungrazed cells at that time sink quickly through the weak pycnocline. As discussed in Chapter 5, nitrate is only occasionally exhausted in the euphotic zone and, as Fig. 5.2 shows, minimal values are not reached until mid-July. This is, of course, the consequence of the geostrophic nitrate ridge that lies across the periphery of the North Atlantic gyre and of small wind-mixing events. Chlorophyll values remain relatively high throughout the summer, usually >2.0 mg chl $m^{-3}$ in the core of the near-surface, high-chlorophyll layer. Late summer blooms of coccoliths, probably *Emiliania huxleyi*, are consistently observed as backscattered light in surface color images in the southern part of SARC (Brown, 1995b).

For the SARC province we have almost unrivalled information about the three-dimensional, quantitative evolution of zooplankton communities. In 1948–48, at OWS M, weekly and throughout the year, depth-stratified Nansen net tows (0–2000 m) were obtained and complete analysis made of zooplankton taxa on almost all profiles (Østvedt, 1955). At OWS I, the dynamics of the zooplankton were investigated from 1971 to 1974, again almost weekly, from late March to mid-October, using the Longhurst-Hardy Plankton Recorder (LHPR). The 111 profiles obtained comprise about 4700 individual, metered plankton samples, each representing depth intervals of approximately 10 m from the surface to 500 m, and in some cases deeper. One hundred and twenty species or species groups of copepods were identified, of which only seven comprised about 90% of the total numbers of individuals: *Oithona* spp. 29%, *Acartia tonsa* 19%, *Calanus finmarchicus* 16%, *Metridia lucens* 14%, *Euchaeta norvegica* 5%, *Oncaea* spp. 4%, and *Pleuromamma robusta* 3% (Williams, 1973). These *C. finmarchicus* data were revisited recently by Irigoien *et al.* (2000).

The profiles at OWS M and OWS I confirm Gran's earlier suggestions that the zooplankton species associations in the Norwegian Sea comprise (i) a surface epiplankton, (ii) a plankton of intermediate depths corresponding to the subsurface layers of Atlantic water, and (iii) a deeper plankton association. The surface plankton is by far the most diverse, and not only because some individuals of all seasonal vertical migrants are found there in all months. The Atlantic water zone is characterized especially by the occurrence of *Paraeuchaeta norvegica* and (during daytime) of diel migrants such as *Pleuromamma* and *Metridia*. The deeper, colder zone is characterized by *Calanus hyperboreus* (which does not rise seasonally in this province) and the large amphipod *Cyclocaris guilelmi*.

It is, of course, *Calanus finmarchicus* and its seasonal life cycle that has attracted most attention in studies of the North Atlantic plankton, although in this province, as elsewhere, the smaller epipelagic copepods are numerically dominant, so that the total population of copepods is always biased surfaceward, numbers being an order of magnitude higher in the upper 30 m than at all depths below 100 m. This should be remembered when assessing the food available to—for instance—fish larvae (see later discussion) and is also reflected in the distribution of zooplankton predators, such as chaetognaths. These, likewise, have a distribution in the LHPR profiles biased surfaceward at all seasons. A similar argument can be made for other predators. Coelenterates are dominated at OWS I by *Aglanthe digitale* (>80% dry weight biomass) and individual distribution is once again biased surfaceward with most individuals in the upper 100 m. Only in late July were some larger individuals found to the bottom of the profiles in 800 m.

But to return to *C. finmarchicus* itself, which does nevertheless maintain the largest species-specific copepod biomass at OWS I. Four distinct phases can be identified in the seasonal cycle in the profiles there from 1971 to 1974:

1. *Rise of the overwintered population*: From the middle of March until the end of April, the overwintered population of C5s and adults dominates the biomass between 0 and 800 m. During this period, and prior to the spring bloom in some years, but coincident

with it in others, C5 individuals and adults rise from overwintering depths that are in excess of 500 m. By mid-April, essentially the whole population lies shallower than 100 m, and from about April 10, the first copepodites of the new generation begin to appear.

2. *Production of first generation*: Throughout May and during the first 10 days of June, population biomass increases very fast with the growth of copepodites, and in the second half of this period adults of the first new generation appear. The population remains almost entirely within the upper 150 m, with many profiles showing crowding into the upper 50 m. Diel migration, if it occurs, is shallow.
3. *Multigeneration period, some individuals descending*: From mid-June until mid-September, population biomass in the upper 500 m declines progressively, even as the late copepodites and adults of later generations appear. Although there must be continual loss of biomass to predation, the decline during the late summer is mainly caused by the progressive migration of cohorts to depths >500 m. After mid-June, most profiles show bimodal populations with layers of high abundance in the upper 50–100 m and also layers deeper than 250–350 m. The upper population contains early and late copepodites, whereas the deep populations are almost entirely C4 and C5.
4. *Main population at overwintering depths*: Between mid-September and mid-October, there are few early copepodites (and no C1 at all). The bimodal vertical distribution is progressively replaced by profiles in which most of the biomass is layered deep, usually below 300–350 m, and in such a way as to suggest that only the upper parts of the deep layers are being sampled.

At OWS M, the same seasonality occurs, *Calanus finmarchicus* biomass being concentrated in midwinter at 600–1000 m. In May, the surviving population aggregates shallower than 100 m. The population of *C. finmarchicus* in the southern part of the Norwegian coastal current starts its reproductive period earlier in the year than the oceanic population and may attain three generations in a single summer period, south of the Lofoten islands, beyond which only a single generation is achieved; there is, of course, a continual advection of individuals coastwise into the Barents Sea (Pedersen *et al.*, 2001). In the East Icelandic Current, spring warming of surface layers starts in May, and peaks in August; below 75 m, temperatures remain subzero during the summer (Astthorsson and Gislayson, 2003). Here, *C. finmarchicus* has a 1-year life cycle while the *C. hyperboreus* cycle is of at least 2 years. The former species peaks in abundance in July (c. 16,000 ind $m^{-2}$), whereas *C. hyperboreus* and *C. glacialis* numbers peak a little earlier in June (c. 370 and c.7700 ind $m^{-2}$, respectively), apparently more coordinated with the timing of the spring bloom (<1 mg chl $m^{-3}$ in late May–early June).

Østvedt observed that the numbers of *C. finmarchicus* declined progressively at OWS M during winter so that only about 15% survived until spring, and the same was observed at OWS I, where 70–80% of the initial overwintering biomass fails to return to the surface in spring (Longhurst and Williams, 1992). The same progressive decline in numbers of an overwintering population can be observed in the data for *Pseudocalanus minutus*, another seasonal vertical migrant.

*Pareuchaeta norvegica*, the second largest component of copepod biomass at OWS I, also rises from overwintering depths below 500 m soon after mid-March. Reproduction occurs at depth in late winter so the rising population is structured quite differently from that of *C. finmarchicus*. Early (C1–C3) copepodites already numerically dominate the population. During the summer the biomass of *P. norvegica* is maintained at a more consistent level than that of *C. finmarchicus*, and there is less evidence of the progressive establishment of deep layers and little indication that the population descends to overwintering depths before mid-October. Thus, *P. norvegica* spends a shorter period at depth during the winter than *C. finmarchicus*. The next ranking species in terms of absolute biomass, *M. lucens* and *P. robusta*, appear not to undertake seasonal migrations unless they descend after sampling at OWS I ended in October. These genera are, in any event, strong diel migrant genera wherever they occur.

Individuals of *Thysanoessa longicaudata* comprised >99% of the eight euphausiid species that occurred in the OWS I profiles and were found to have a vertical migration and reproduction strategy markedly different from that of the copepods just discussed. The main cohorts of eggs are produced in March–April at from 100 to 800 m, with >90% of nauplii subsequently occurring at >500 m. The first-feeding stage of euphausiids is the calyptopa, and these are strongly biased surfaceward, with highest concentrations in the upper 200 m in the period May–June, though with remnant individuals down to 500 m. In June, furcilia larvae are restricted to the upper 200 m, and most occur in the upper 50 m. This, as Williams and Lindley (1982) suggested, is an ontogenetic migration that represents the consequence of the progressive maturation of the larvae. Adults appear in significant numbers in the profiles only from July until October, during which period their distribution is, again, biased surfaceward, almost entirely within the upper 100 m. Juveniles of the hyperiid amphipod *Parathemisto gaudichaudi* occur only within the upper 50 m, but the adults make significant diel migrations within the upper 200 m.

At OWS I, the near-weekly sampling schedule gave the somewhat surprising information that for many other planktonic organisms the generation time was relatively short: 6–8 weeks for amphipods and only 6 weeks for coelenterates.

Turning now to the southern Barents Sea, three areas having special characteristics take our attention: (i) adjacent to the MIZ bordering BPLR to the north and east; (ii) adjacent to the northern coast of Scandinavia, with its narrow shelf region; and (iii) the main central area. Processes affecting stability of the water column are not identical and, hence, the timing of the spring bloom differs in each area (Loeng, 1991) and, as already noted, there is significant between-year variability in the location of the ice edge, which is more consistent in the western than in the eastern Barents Sea. Thus, down the western coast of Spitzbergen and around the south of Bear Island, this variability is on the order of only a few tens of kilometers, but in the eastern end of the Barents Sea it increases to about 500 km. In 1984, the ice front lay zonally along 75°N to encounter the northwest coast of Novaya Zemyla; in 1979, it lay southeasterly from Hope Island to the White Sea, a full 5° of latitude further south. Adjacent to the MIZ, a surface layer of low-salinity water floods south and west above the more saline Atlantic water. This creates sufficient near-surface stratification to sustain a very early spring bloom as soon as sun angle increases sufficiently, perhaps as early as mid-March as in other seasonally ice-covered seas. Beyond the influence of surface meltwater, thermal heating of the central regions that are mixed to the sea bottom in winter may not produce sufficient stability for a bloom to be initiated until much later in the summer—even as late as mid-June.

Coastal water near and above the continental shelf retains some stability even during winter and here also a spring bloom begins as soon as local irradiance is sufficiently strong. Though blooms in ice-melt surface water toward the boundary with BPLR rapidly form a strong deep chlorophyll maximum at close to 50 m depth, the bloom in coastal water often has uniform chlorophyll throughout the mixed layer (Mitchell *et al.*, 1991a).

The Barents Sea bloom is well described by Skjoldal and Rey (1989), who suggest that it may be initiated either by diatoms or by *P. pouchetti*, although small flagellates dominate postbloom conditions in the classical manner. Winter nitrate in the photic zone is lower than in other polar seas (12–14 $\mu$M) and may be reduced to low or even undetectable levels after the bloom. A deep chlorophyll maximum develops by midsummer within the upper 50 m, with great between-year variability, forced by changes in stability and pycnocline formation in the surface layer. In June 1982 there was no deep chlorophyll maximum and residual nitrate was about 1.5 $\mu$M, whereas in 1980 at the same place and in the same month, nitrate was undetectable and there was a strong DCM at 30 m. In 1981, the situation was intermediate.

Progressively during the summer, the deep chlorophyll maximum deepens to come to lie on the upper slope of the nitracline. Very high sedimentation rates of algal cells may occur during and just after the spring bloom, especially if this is dominated by *Phaeocystis*. Rates of up to 1 g C $m^{-2}$ $d^{-1}$ have been recorded, though the average rate in spring is about one-third of this value (Wassman *et al.*, 1991). Part of this pulse of sedimentation will be the result of zooplankton grazing and the release of rapidly sinking fecal pellets, and there is some evidence that interannual variability extends also to the percentage of total seasonal primary production consumed by primary herbivores, mainly *C. finmarchicus*. In the summer of 1980, the calanoid herbivores consumed about 70% of the total primary production, though they took a much smaller percentage in 1983 and 1984, when massive sedimentation occurred. Eilertsen *et al.* (1989b), on the other hand, suggest that the greater part (80–95%) of the phytoplankton biomass produced in spring settles unconsumed to the sea floor, though later in the season they find much closer coupling between production and consumption. This rain of organic material to the sediments, whether directly as aggregates or indirectly as fecal pellets, is surely the key to the rich benthic fauna and demersal fish stocks of the Barents Sea (see later discussion).

The species of herbivorous plankton in the SARC part of the Barents Sea and their life cycles are similar to those of the Norwegian Sea and at OWS I. The principal difference must be (though this seems not to have been investigated) that since their normal overwintering depths are not available in the Barents Sea, the population probably aggregates into the few deep basins, such as the Bear Island Channel. This is how similar species manage their affairs in the Gulf of Maine, where they encounter the same problem (see NWCS). Of course, copepods are only a part of the herbivore biomass and larger krill are also important: *Meganyctiphanes norvegicus* in the Atlantic water, *Thysanoessa inermis* in more coastal regions, and *T. longicaudatus* in Atlanto-Arctic water. All perform diel migration, but *Meganyctiphanes* spends much of the year in the benthic habitat, swarming near the surface only to reproduce. The filtering mechanism of these organisms restricts their intake to the larger autotrophic cells.

Studies of the balance between production of zooplankton and its consumption by fish, and the dynamic balance between different fish species biomasses, have been well developed in this province and will repay a brief review. Apart from herring and blue whiting, the significant species are cod (*Gadus morhua*), mackerel (*Scomber scombrus*), horse mackerel (*T. trachurus*), and capelin (*M. mallotus*) as well as a range of small mesopelagic species. During summer, the feeding migrations of oceanic herring (and of more coastal mackerel) involve a substantial intake of zooplankton food. It is especially the larger, mature herring that perform the most extensive feeding migrations. The relationship between *Calanus* and herring is close. Kaartvedt (2000) computes the balance between the total energy requirement of the Atlanto-Scandia herring and the total production of *Calanus*, one of its principal food items, in the Norwegian Sea. The 10–15 × $10^6$ t of herring require 4–7 times their own weight of food annually, and this consumption is not distant from the 60–75 × $10^6$ t annual production of *Calanus*, suggested from various sources.

Associated with changes in the primary production cycle, there is also strong between-year variability in the primary herbivore population and hence in herring food; in 1980–1982 there were 2–5 × $10^5$ ind $m^{-2}$ of *C. finmarchicus* in the central Barents Sea. In 1983 and 1984, there were just 1 × $10^4$ ind $m^{-2}$. The same is seen in the OWS I data: Irigoien's analysis of these shows that adult *Calanus* were late in surfacing in spring 1972 and were very sparse, but in 1973 were abundant and early. Conversely, the OWS I data for fish larvae show great differences in between-year abundance: in 1971 and 1972 they were greatly more abundant than in the two subsequent summers.

Herring, of course, are not the only predators of *Calanus*: during the winter, and during their rise in spring, *Calanus* encounter mesopelagic fish (e.g., *Benthosema glaciale*) together with blue whiting, and these predators probably account for much of the significant biomass loss suffered by the overwintering population. Kaartvedt (*op. cit.*) suggests that the seasonal vertical migration pattern of *Calanus* removes the vulnerable late copepodites into a deep sanctuary, but this is a difficult thesis to prove. It is at least as likely that the colder water in which the copepods overwinter is selected for its temperature regime, conducive to physiological stasis.

The balance between stock biomass of zooplankton and fish is dynamic and mutually varying. The stock biomass of capelin in the Barents Sea since the mid-1970s has varied from <0.5 to 7–8 million tons over quite short periods (Gjosaeter *et al.*, 2002); recruitment failure of capelin, as occurred in 1983 and 1991, is attributed to consumption of capelin larvae by herring stocks, themselves at peak abundance in these years. After the 1991 population crash of capelin, the biomass of zooplankton in the Barents Sea more than doubled, from about 5 to about 14 g $m^{-2}$. Biomass of euphausiids (*Thysanoessa* spp.) is also inversely related to capelin biomass in the Barents Sea. Similar trophic dynamics link cod biomass to the relative abundance of components of the pelagic ecosystem. After a juvenile period when the codling feed largely on benthic hard-ground invertebrates, growing cod have a very catholic diet, taken largely from the pelagos. In periods when capelin biomass is low, as in 1991–94, the level of cannibalism increases significantly while polar cod, herring, and amphipods are more dominant in cod diet than at other times. During these years, the cod growth rate declines and appears to be dependent on capelin abundance. Also in such years, nesting success is low in surface-feeding sea birds, such as fulmar and kittiwake. Nor should we forget that krill and clupeid fish (capelin, herring) are consumed in very large quantities by baleen whales, representing transfer to the highest trophic levels along notably short food chains.

It is, of course, herring (*Clupea harengus*) that are most associated in our minds with the SARC province, because this is the home of the highly variable Atlanto-Scandian stock that performs extensive migrations each year throughout the Norwegian Sea. This stock is a paradigm for the balance between longevity and recruitment variability in fish (e.g., Longhurst, 2002): these herring produce a significant year-class only every 10 years or so (Fig. 9.4). The 1950 year class formed >50% numerically of all fish after 1955 until at least 1963. In 1922, the great 1904 year class still formed 22% of all fish >2 years old. Clearly, it is only an exceptional conjunction of events that permits this large stock of herring to maintain its occupation of the Nordic seas. What is not so widely known is

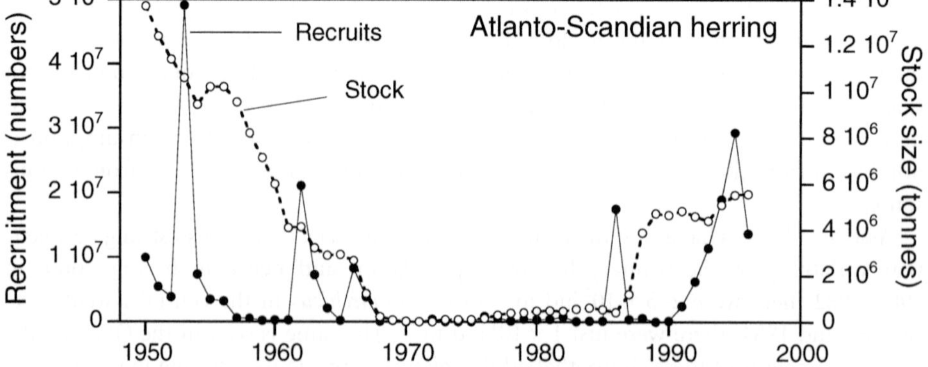


**Fig. 9.4** Stock size and recruitment pattern at decadal scale for the Norwegian Sea stock of herring (*Clupea harengus*); observe the lack of relationship between the size of each successive parental stock and the number of young-of-the-year recruits produced by that stock.

that a small pelagic gadoid, the blue whiting (*Micromesistius poutassou*) that also abounds in the Norwegian Sea, also has strikingly intermittent recruitment. In the final 30 years of the last century, recruitment was usually of order 5–10 $\times 10^9$ 1-group fish each year but peaked very much higher during four brief periods (1983–84 $= 20 \times 10^9$, 1990 $= 22 \times 10^9$, 1996 $= 48 \times 10^9$, and 2000 $= 58 \times 10^9$), as reported recently by Skjoldal and Saetre (2004).

So, it is perhaps in this very well studied province that we have as good a chance as anywhere of finally unraveling the enigma of how variability in recruitment to fish stocks is forced, and how it may be predicted. Even so, it is obvious that this is no simple matter. A recent study of the relative influence of environment and size of the parent stock on recruitment of cod, using a 60-year data set (1930–1990), suitable to support age-based modeling, simply found that the effect of parent stock size on subsequent recruitment is less clear the longer the time series that is examined. Unfortunately, as Matishov *et al.* (2003) point out, even in well-studied regions such as the Barents Sea it is very difficult to separate the natural dynamics from the consequences of many decades of heavy industrial fishing. They emphasize that the biomass of each component of what they call the "consortive" ecosystem (because many components depend on the performance of a central species or group of species) is currently very different from the pristine state, individual species and groups being reduced by 5–50% of pristine biomass.

## Regional Benthic and Demersal Ecology

It is, of course, only in the shallow Barents Sea that significant benthic habitat exists in this province, and here it must be significantly modified by decades of intensive industrial trawling for demersal fish with heavy equipment. We must assume that, as in the North Sea, the larger, long-lived lamellibranchs and echinoderms will have suffered heavy damage. The pristine benthic ecosystem resembled that of BPLR and ARCT (see earlier discussion) with a *Venus fluctuosa* community on sandy bottoms, and a *Macoma calcarea* community down to 130 m where silt content is higher. An unusual bivalve-polychaete association (the *Yoldia hyperborea* community, *sens.* Thorson) also occurs in the inner Barents Sea. There is some evidence of the effect of changing oceanographic conditions in the progressive extension of Atlantic benthic species northward along the shelf west of Spitzbergen between the 1930s and the 1950s.

Here, the benthic invertebrates support a fauna of demersal-feeding fish including plaice (*Pleuronectes platessa*) and halibut (*Reinhardtius hippoglossoides*) and—another human intervention—a burgeoning population of Kamchatka crabs (*Paralithodes*), introduced in 1932 and 1961 on Russian initiative; this population was estimated at 12.5 million crabs already by the year 2000 (Matishov *et al.*, 2003). It is not credible that a population size of 15 million individuals is "acceptable for ecosystem stability" as is claimed: such statistics cannot be computed, they must be obtained by experiments—irreversible, in this case.

## Synopsis

*Case 1—Polar irradiance-mediated production peak.* $Z_m$ undergoes deep winter mixing until shallow thermal stratification is reimposed in April; $Z_{eu}$ is always shoaler than $Z_m$ except for the 4 months when the pycnocline is illuminated (Fig. 9.5). Seasonal evolution of P is symmetrical about a strong midsummer maximum, the vernal increase responding to the light field more closely than to the initiation of shoaling of $Z_m$. A change in the rate of declining P occurs with autumnal deepening of $Z_m$. Biomass accumulation in spring is rapid (and varies strongly between years) and the subsequent decline after midsummer is brief, because a major second period of accumulation occurs in the second half of the year; this is highly variable between years. Renewal of accumulation in autumn and early

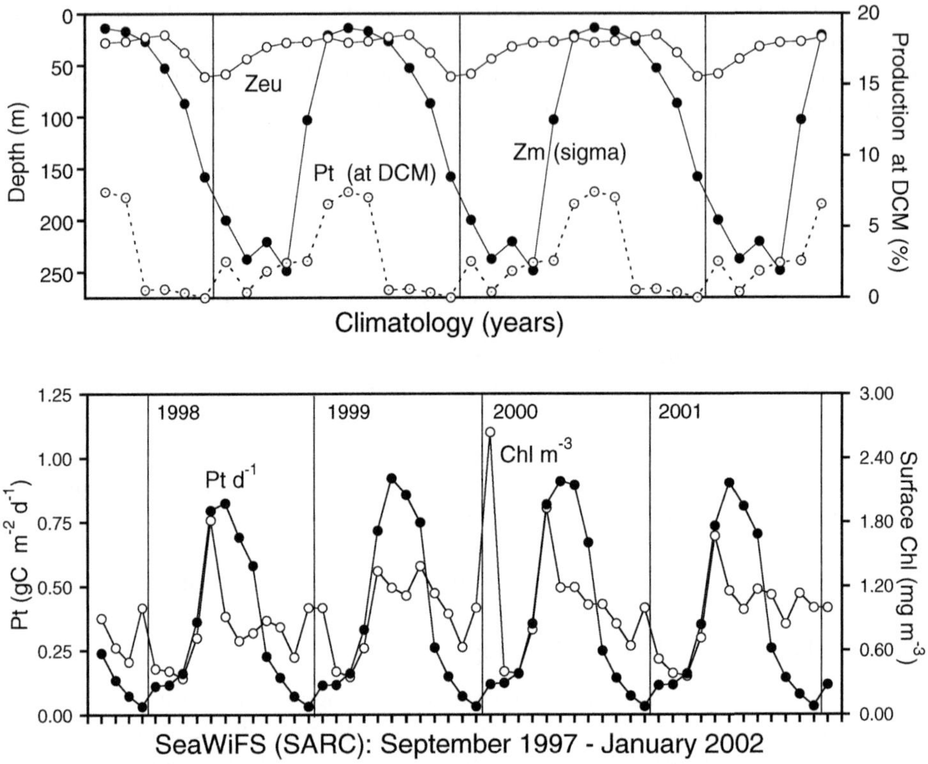


**Fig. 9.5** SARC: seasonal cycles of monthly surface chlorophyll and depth-integrated autotrophic production for the years 1997–2002 from SeaWiFS data together with characteristic seasonal cycles of mixed-layer depths from Levitus climatological data and photic depths computed from characteristic irradiance and the archive of chlorophyll profiles discussed in Chapter 1.

winter is interpreted as response to the descent of the new generation of herbivores out of the photic zone in July–August.

# Atlantic Westerly Winds Biome

The global pattern of regional wind stress exhibits two striking anomalies: the first, of course, are the westerly winds that howl around the Southern Ocean south of the Subtropical Front; these are the strongest and most sustained winds anywhere on the planet. The second is in the North Atlantic, where winter wind stress at 50–55°N is equivalent to that in the Southern Ocean: here, mean scalar winds are of order 10 m $sec^{-1}$.

Deep winter mixing by these, together with thermal convection, sets up conditions throughout the poleward portion of the anticyclonic gyral of the North Atlantic basin that result in a phytoplankton bloom sufficiently anomalous at the global dimension that it must be central to any synthesis of North Atlantic pelagic ecology.

The Subtropical Convergence, which lies across the North Atlantic subtropical gyre at 30–35°N under the conjunction between the westerly winds and the wind systems of lower latitudes, forms a surface frontal system and a rational southern limit to the winter-mixing regions of the westerly winds biome. It lies, *grosso modo*, from Florida to southern Spain. The equivalent front in the southern hemisphere lies across the South Atlantic from Argentina to South Africa at 35–40°S. The Subarctic Front, discussed earlier, and its austral counterpart in the Southern Ocean form the poleward limits of the biome.

# North Atlantic Drift Province (NADR)

## Extent of the Province

NADR comprises part of the west wind drift region of the North Atlantic in the sense of Dietrich (1964), including the jet current along the OPF and the broader, slower flow in the more eastern part of the ocean. Thus, the poleward boundary is the OPF itself along the south of the ARCT, that extends from 55°N at the Grand Banks of Newfoundland to 60°N, south of Iceland: from here, an arbitrary line is taken zonally to the edge of the Shetlands shelf, separating NADR and SARC provinces.

To the south, the boundary of NADR separates the northeasterly North Atlantic Current from the flow of Gulf Stream water into the southeasterly Azores Current (previously known as the "southern branch" of the Gulf Stream). This line of separation is poorly defined, but lies across the ocean at about 42–44°N, the latitude of NW Spain. In the east, the edge of the European continental shelf is taken as the boundary of the province.

## Defining Characteristics of Regional Oceanography

Bifurcation of the Gulf Stream extension occurs in the western part of the ocean; prior to reaching the Mid-Atlantic Ridge (30°W) the flow is clearly in two streams, the North Atlantic Current toward the northeast and the Azores Current toward the southeast around the northern limb of the subtropical gyre (Krause, 1986). NADR thus comprises the slow northeastward drift of surface water toward the Iceland-Faeroes Channel, together with the North Atlantic Current (NAC), the frontal jet of the OPF along the northern boundary of the province.

The province coincides very closely with that part of the North Atlantic where seasonal surface temperature differences are greatest (>5°C), exceeding those in the provinces to north and south (Louanchi and Najjar, 2001). It is also characterized by significant presence of mesoscale eddies originating in the very active vorticity of Gulf Stream and in the NAC that are even more important in the GFST province (see later discussion); however, the consequences in NADR are sufficient that, at any location, conditions tend to have significant variability at all scales. The strength, location, and longevity of mesoscale eddies are now readily monitored as SLA surfaces obtained from the TOPEX-POSEIDON data; along meridional sections obtained with undulating, towed instrument instrumentation (such as the "Sea Rover") it is easy to demonstrate that the axes of both the Azores Front (see NAST) and the OPF are each associated with strongly increased vorticity (Strass and Woods, 1988).

As noted previously, wind speed in autumn and winter is greater in this province than anywhere else except in the Southern Ocean, and the consequent stress at the sea surface, combined with rapid seasonal heat loss, induces anomalous deepening of the mixed layer, so that by midwinter the pycnocline slopes down from about 300 m in the central ocean to more than 500 m along the European continental edge. It should be noted that the ratio $Z_{cr}/Z_m$ in this province takes relatively low values of about 0.3–0.4 compared with 0.8–0.9 in the subtropical gyre to the south (Dutkiewicz *et al.*, 2001). The great depth of mixing in winter ensures a sufficiently deep and sufficiently permanent surface stratification, such that $Z_m < Z_{cr}$ shall be a necessary condition to initiate growth of the phytoplankton. That this condition is forced by complex, and hence highly variable, local meteorological conditions is clear from the analysis by Taylor and Stephens (1993) of the consequences of the difference between depths of diurnal and nocturnal mixing due to the effect of increased buoyancy by solar heating during daytime. The Sverdrup criterion can only be satisfied if and when sufficient buoyancy is induced during the daytime to overcome the loss of cells below $Z_{cr}$ by mixing at night. Thus, development of a vernal shoal pycnocline is associated with increasing sun angles and relaxation of wind stress.

Although the evidence is somewhat circumstantial, this appears to occur patchily and as a response to local meteorological conditions that change from year to year, rather than as the smooth poleward progression that might be inferred from mean conditions.

The general nutrient regime of the Atlantic has been discussed in Chapter 5 and, as noted there, comparison with that of the North Pacific is profitable; the younger deep water of the Atlantic and the very deep winter mixing cause nutrient concentrations at standard depths to be lower than in the North Pacific. Moreover, it does not appear necessary to invoke the smaller size of the North Atlantic basin, and hence a supposed greater supply of continental Fe-bearing dust (as has been done) to explain its much stronger spring bloom: the strikingly different stratification regimes of the two northern oceans is a sufficient explanation.

## Response of the Pelagic Ecosystems

The general seasonal pattern of phytoplankton growth responds to the succession of deep winter mixing followed by vernal stratification, while the details of the sea-surface chlorophyll field are determined by the distribution of eddy kinetic energy and local short-term changes in irradiance and wind stress. It has been assumed in the past that there is a simple northward progression of the spring bloom in the NADR, and that sufficient stratification is also induced progressively northward. However (as is so often the case), we find that reality is rather more complex, now that we have the tools for satisfactory observation of the evolution of the chlorophyll field.

Formal analysis of oceanwide SeaWiFS images has been used to evaluate the date when events in the chlorophyll record occur (Siegel *et al.*, 2002). The date of the initiation of the bloom at each pixel is taken to be when chlorophyll biomass increases 5% above median conditions; this occurs in early February south of about 40°N and propagates northward, so that at 50°N it occurs in early June. Further north still, initiation occurs almost simultaneously right across the northern part of NADR and in SARC Province. Computed critical depths ($Z_{cr}$) on the day of bloom initiation derived from this analysis shoal slightly northward from ~30 m at 40–45°N to ~25 m at 55–60°N.

In any case, the seasonal development of phytoplankton biomass can now readily be followed in SeaWiFS or MODIS 7-day and monthly sea surface chlorophyll images, especially with the reprocessing No. 4 set, in which cloud masking is significantly reduced. The seasonal evolution of MODIS surface chlorophyll in 2004 (see Color plate 5) is typical. All January images show that surface chlorophyll is lower in NADR than to the south near the Azores Front. February images suggest that chlorophyll enhancement in that month is initiated by the increase of surface irradiance right across the ocean in the eddy field associated with the subtropical frontal zone of the NAST province, at 35–40°N. In March, the entire region up to 50–52°N shows evidence of accumulating chlorophyll that is significantly enhanced—but not very much extended—during April. From May onward, the bloom in NADR becomes continuous with those in the adjacent provinces, SARC to the northeast and ARCT to the north. In late summer and autumn, the pattern continues to evolve so that high surface chlorophyll values are increasingly biased northward with southern areas becoming increasingly oligotrophic until October, when a minor increase in surface chlorophyll once more appears across the ocean at 40–50°N, progressively to weaken as winter approaches, wind-mixing deepens, and surface irradiance weakens.

The images confirm that patchiness does occur at all scales of chlorophyll enhancement, although comparison of all available images for each month does suggest some recurrent pattern. Two major linear regions of enhancement are seen in April 1999, one aligned with the eddy field of the subtropical font in the NAST province, the other apparently along the line of the meandering jet of the North Atlantic Current. This pattern of two regions of high chlorophyll, one zonal, the other diagonal across the North Atlantic, is repeated in many monthly images. Similarly, the "blue holes" in the chlorophyll field to the west of the British Isles seen in the May–July 1999 images are repeated in many summer months in other years; this appears to be a very "quiet" region with little mesoscale eddy activity. A region of high chlorophyll is repeatedly associated with Rockall Bank, lying across the boundary between NADR and SARC to the northwest of Scotland. Here, it is known that upwelling occurs onto the bank. It is clear, from examination of these images, that the assumption often made that major biological gradients in the North Atlantic do not exist in the zonal sense, but only in the meridional sense, is incorrect.

The between-year variability now revealed is significant so that the overall patterns remain, but their sequence differs strongly: the null hypothesis must be variability in meteorological forcing. Observations made at the BIOTRANS station (47°N 20°W) have revealed very clearly the instability of the early spring bloom (Koeve *et al.*, 2002). In April, 1992, when the doubling time of mixed-layer phytoplankton biomass was of order 7 days, the development of a stable, near-surface mixed layer was far from progressive, but rather was interrupted by episodes of deep mixing during the passage of storms. Heat flux into the mixed layer was reversed during these episodes, and the loss of cells into the interior of the ocean represented as much as 60% of total spring bloom production; thus, the sign of the heat flux across the sea surface was constantly changing during March and April, and became permanently positive only in May. Reevaluation of the NABE data for 1989 suggest that convective loss followed the same pattern and was equivalent to about 40% of total new production. Model studies of the same phenomenon suggest that where the $Z_{cr}/Z_m$ ratio is relatively low, because of deep winter mixing, the early bloom is very sensitive to irruptive mixing events and hence to meteorological modulation.

The University of Kiel "Sea Rover" meridional transects along 30°W that were obtained during spring and summer of 1984–1986 (Strass and Woods, 1991) provide us with synoptic sections of chlorophyll biomass to 150 m, and very usefully serve support the satellite data in the recognition of three phytoplankton seasons:

*The development of a spring bloom.* The first indications (April 18–24) of a bloom occurred where meandering flow was strongest, across 44–47°N, with a small southerly outlying patch at about 40°N to the south of this province, in NAST. Elsewhere, the bloom had not started. By April 24–29, consistent with rapid onset of density stratification, a near-surface bloom with values >1.25 mg chl $m^{-3}$ extended clear from 39° to 50°N, a distance of 1500 km. This seems to confirm, then, that initial stratification (and the first patches of surface bloom) is related to the effect of increasing irradiance on cells brought near the surface in mesoscale eddy dynamics, rather than on density stratification, which occurs almost simultaneously across a rather wide zonal swath of ocean.

*Transition to summer oligotrophy.* By the end of June, the transition to oligotrophic conditions had occurred, though patchily, poleward to 46°N. There was residual nitrate in the photic zone only in small areas having exceptionally shoal pycnoclines, suggesting an ephemeral event. Maximum chlorophyll values occurred near the OPF and from there south to 46°N the deep chlorophyll maximum lay at, or shallower than, the mixed-layer depth. Further to the south yet, nitrate was depleted and the chlorophyll maximum was significantly deeper than the bottom of the mixed layer. Oligotrophy propagates poleward during the summer at about 3° of latitude a month, whereas $Z_m$ deepens at about 10 m $month^{-1}$.

*Late summer regrowth.* By the end of August, everywhere south of the Oceanic Polar Front, which lay at 52–54°N, a chlorophyll maximum occurred within the seasonal pycnocline, deepening toward the south at about 3.5 m across each degree of latitude. Only to the north of the OPF was the chlorophyll maximum within the mixed layer, and here it took maximal values. Late summer blooms of coccolithophores occur consistently in the northern part of NADR and are especially frequent bordering the coastal boundary biome, such as over the Rockall Channel. Strass and Woods point out that the southward slope of the deep chlorophyll maximum is coincident with the southward deepening of the nitracline but is disjunct from $Z_m$, sloping downward and southward at a much slighter angle. New production continues at the deep chlorophyll maximum during the oligotrophic summer period after the termination of the spring bloom in the mixed layer. Some enhancement of surface chlorophyll extends into early autumn, producing a shoulder on the regional, seasonal graph of satellite-derived chlorophyll.

Because this province lies centrally in the area affected by the North Atlantic spring bloom, it has generally been believed to be the ideal case of the diatom-copepod food chain. However, we now know, thanks to the results of the 1989 North Atlantic Bloom Study (NABE; Ducklow and Harris, 1993) that nano- and picoplankton contribute significantly to total primary production in the Atlantic spring bloom and that diatoms are not consistently the dominant large cells. Silicate was reduced to limiting values before nitrate limitation occurred at the NABE time series stations at 18° and 40°W, representing the zonal extent of the province at about 45°N, so this unexpected result must be typical. Before silicate limitation, the dominant large cells were diatoms (mostly *Rhizosolenia, Fragillariopsis, Thalassema, Thalassiosira*, and *Nitzschia*) at 18°W, but dinoflagellates (perhaps facultative heterotrophs) dominated in optically counted samples at 45°W. Dominance shifted rapidly to an abundance ($10^4$cells $ml^{-1}$) of small (2–5 $\mu$m) flagellates as soon as silicate limitation occurred, without reduction of overall chlorophyll biomass. The unconsumed, silicon-depleted diatoms sank out, leaving behind an abundant mucopolysaccharide residue in the photic zone. In repeated zonal transects between April and October, Li and Harrison (2001) observed that the carbon biomass of the picoautotrophic fraction represented only 6% of the total autotrophs in NADR, heterotrophic bacteria representing 16% of autotrophs.

During the spring bloom studies at the western JGOFS-NABE sites (Harrison *et al.*, 1993) in NADR, microheterotrophs (ciliates and other protists) represented 11% of total living organic carbon compared with 4% for zooplankton. Protists were the principal consumers of primary production, taking 25% of the standing stock of plant cells per day and 90% of primary production based on phytoplankton growth rates of about 0.4–0.7 doublings $d^{-1}$, whereas grazing by zooplankton herbivores took less than 10% of daily production. Of the copepods, small forms (e.g., *Oithona*) took as much chlorophyll as the medium (*Metridia*) and large-size classes (*Calanus* and *Pleuromamma*) combined.

Biomass and grazing impact of three size classes of zooplankters (0.2–05, 0.5–1.0, and 1.0–2.0 mm) was investigated at the eastern NABE stations at 47°N 20°W during May 1989 (Dam *et al.*, 1993). Group biomass was inversely related to body size, the smallest class forming >50% of the total biomass that itself increased by a factor of 3 during the month, even as individual size increased sixfold. Accordingly, 66% of all consumption of autotrophic biomass was performed by the small fraction at the start of the period, and 44% by the medium fraction at the end. The active flux of DON out of the photic zone during diel migration was about 25% of the passive PON flux.

During the postbloom period, interaction between biota rather than physical forcing may—at least in some years—dominate the seasonal development of spatial and temporal variability of the pelagic ecosystem. Popova *et al.* (2002) discussed such a situation in April–May of 1997 some weeks after the peak of the seasonal bloom to the west of the British Isles (48°N 17°W). Here, deep wind-induced mixing during spring after deeper-than-usual winter mixing created a situation that approached the so-called HNLC condition, which—as the authors of the study remark—is unusual here. The application of a coupled 3D physical-biological model to this situation suggested that what was observed was a phytoplankton that was grazer-controlled rather than limited by nitrate availability—this did not drop below 1.0 $\mu$M $m^{-3}$ during the observations, and biomass of autotrophic cells remained relatively low. Moreover, it was apparent that the small-scale patchiness observed in the distribution of zooplankton resulted from top-down control through interaction between small and large size classes inducing predator-prey oscillations that originate, at least in part, in diel vertical migration patterns.

In fact, an unusually large range of zooplankton depth strategies obtains here because the NADR copepod fauna includes both northern and southern species (Williams and Conway, 1988), although both *C. finmarchicus* and *C. helgolandicus* perform seasonal ontogenetic migrations, wintering at 400–900 m as C5 and rising to the mixed layer in spring and summer, although the former remains a little deeper than the latter. *Neocalanus gracilis* remains within the upper 200 m throughout the year as does *C. tenuicornis*, which consistently avoids the near-surface layer and occurs at 20–200 m in all months. These two, and the other southern species, reproduce year-round.

The abundance of boreo-arctic *C. finmarchicus* is significantly lower here than further poleward and, indeed, NADR is the southern limit of this species, is the central range of *C. helgolandicus*, and is at the northern limit of the ranges of *C. tenuicornis, Neocalanus gracilis, Nannocalanus minor*, and *Calanoides carinatus*. Beaugrand *et al.* (2002a) are able to specify, on the basis of CPR data, those copepods that are characteristic of the NADR province, enabling it to be thus distinguished from ARCT and SARC, although unfortunately, these data do not permit of such distinction southward. Any statement about the distribution of zooplankton species must be tempered by recognition that these distributions are highly variable between years and between decades. It is clear from the CPR data (e.g., Colebrook, 1986; Planque *et al.*, 1997) that both long-term trends and year-to-year variability are responses to changes in physical conditions, even if the relationship is not always clearly understood. The North Atlantic Oscillation (NAO) has proved to be an excellent predictor in the NADR of *Calanus* abundance over a 38-year time series of CPR data, reflecting the response of copepod populations to ocean physics either directly or through food chain effects (Planque and Reid, 1998). As Beaugrand *et al.* (2002b) have shown, the CPR data for 1960–2000 strongly suggest a long-term shift in the distribution of key copepod species: biota with warm water affinities shifting further north on the eastern side of the North Atlantic, and cold water species shifting south in the west. What is the ultimate cause of the shift in the physical forcing that is thus expressed remains unclear, although the state of the NAO is certainly implicated.

The vertical ecology of mesozooplankton in NADR was examined by means of an 18-station, meridional LHPR section obtained in August 1975 from 42°N (west of Galicia) to 62°N, near OWS I (Longhurst and Williams, 1979). The profiles to 1000 m are dominated, day and night and all taxa compounded, by an abundant epiplankton within the mixed layer, which is separated by a biomass discontinuity from sparser plankton below. The bottom of the epiplankton layer always occurred just shallower than $Z_m$ so that, following this feature, the depth of the epiplankton deepened northward. Epiplankton biomass always represented >50% of total biomass to 1000 m that showed a general progression toward higher values poleward: off western Iberia, biomass was 200–800 mg $m^{-3}$ dry weight, compared with 900–2800 mg $m^{-3}$ at 55–60°N. The depth of maximum abundance was usually clearly observable, and shallower than the deep chlorophyll maximum that itself usually lay within the upper pycnocline. This represents a looser correspondence between herbivores and the phytoplankton profiles than occurs in tropical, permanently stratified water columns (see PNEC, for example).

Almost 95% of all biomass to 1000 m comprised only 12 sorted categories: the herbivores *C. finmarchicus, Pseudocalanus elongatus*, and *Limacina retroversa*, the omnivores *Metridia lucens, Pleuromamma robusta*, and *Acartia clausii*, and predators represented by total coelenterates, annelids and hyperiids, together with *Euchaeta norvegica*. Of these, *Euphausia, Pleuromamma*, and *Metridia* were interzonal migrants, with daytime depth layers exceeding 200 m everywhere along the section. These migrants lay, by day, in species- and stage-specific strata of only a few meters depth each, readily separable in the LHPR profiles. Diel vertical migration of these interzonal forms added 125–150% of biomass to the epiplankton at night.

Parin remarks that holoepipelagic fishes are relatively sparse in the boreal North Atlantic, and perhaps the most significant species in the open ocean is the saury (*Scombresox saurus*), which appears to fill the same ecological niche as the flying fishes of tropical seas. This species spawns in subtropical seas, especially the Caribbean, but makes extensive summer feeding migrations throughout the open North Atlantic. The very large bluefin (*Thunnus thynnus*) that spawn in the Sargasso Sea of the NAST province, adjacent to the south, are the only tuna species in NADR; like saury, bluefin perform seasonal feeding migrations along the general path of the oceanic Gulf Stream.

## Synopsis

*Case 2—Nutrient-limited spring production peak.* $Z_m$ undergoes extremely deep boreal winter excursion, with early spring near-surface thermal stratification so that the pycnocline lies within the euphotic zone from June to September. Rate increase of P begins in January–February as light increases, 60 days before establishment of shoal $Z_m$, reaching a (nutrient-limited?) maximum in May, subsequently declining progressively with irradiance to a winter low in December (Fig. 9.6). There is no response of P to the

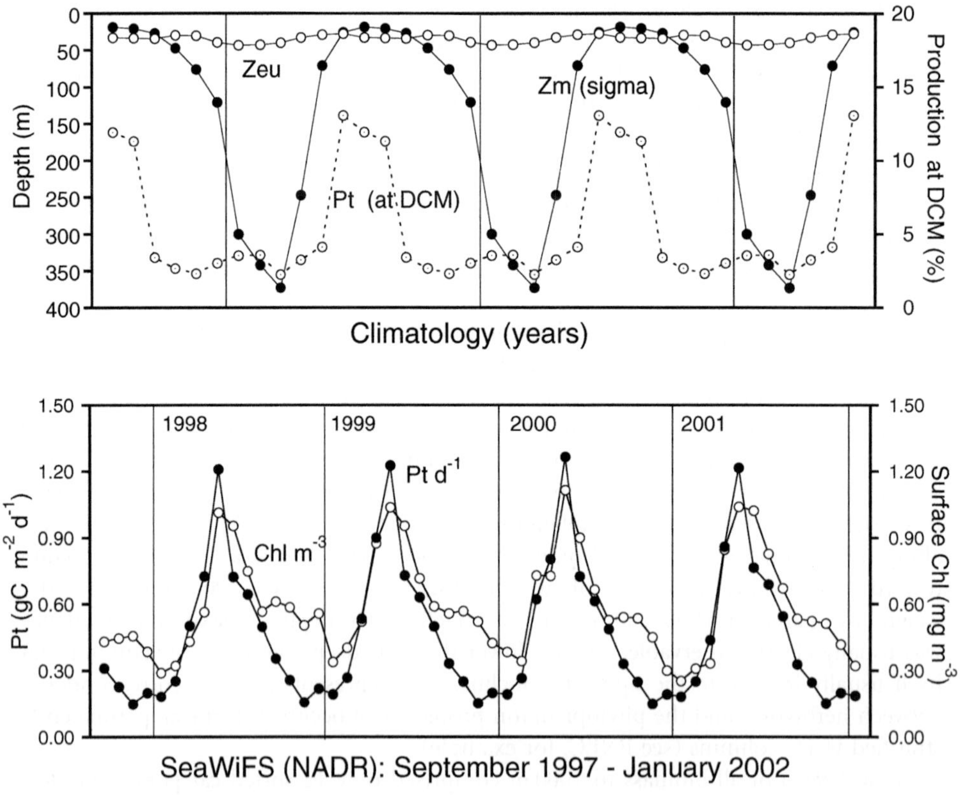


**Fig. 9.6** NADR: seasonal cycles of monthly surface chlorophyll and depth-integrated autotrophic production for the years 1997–2002 from SeaWiFS data together with characteristic seasonal cycles of mixed-layer depths from Levitus climatological data and photic depths computed from characteristic irradiance and the archive of chlorophyll profiles discussed in Chapter 1.

autumnal $Z_m$ deepening. Chlorophyll accumulates in the period March–May during the P increase, and in September–October the rate of seasonal decrease is reduced, consistent with a reduction in consumption as copepods migrate down out of the photic zone in July–September.

## Gulf Stream Province (GFST)

### Extent of the Province

The Gulf Stream Province (GFST) includes the offshore Florida Current and the Gulf Stream from Cape Hatteras to the Newfoundland Basin, where partition of flow occurs into southeast and northeastward streams (see NADR). The landward margin of the province is the North Wall of the Gulf Stream off the eastern United States and the shelf-edge front farther to the south. The line between the field of cold-core eddies and the warm water of the jet current defines the northern limit of the province after the Gulf Stream has detached from the shelf. The southeastern boundary is more difficult to define, but is best thought of as constrained by the distribution of energetic mesoscale eddies.

### Defining Characteristics of Regional Oceanography

Western boundary currents owe their relative strength to the zonal propagation of Rossby waves, which accumulate energy on the western margin of oceans, and its subsequent dissipation within the current. The Florida Current/Gulf Stream continuum is the western boundary current of the North Atlantic and is prominent as a group of closely spaced isotherms not only in hydrographic sections (e.g., Schroeder, 1965) but also in the surface thermal field.

Originating in flow of North Equatorial Current water through the Florida Straits, the velocity maximum of the Gulf Stream is topographically locked to the edge of the continental shelf as far north as Cape Hatteras, where the coastline veers westward, and it separates from the continental edge, to pass toward the northeast across deep water. The region of separation is dynamic, and episodes of surfacing of nutrient-rich North Atlantic central water occur in response to offshore meanders. After separation, the Gulf Stream behaves as a free inertial jet until it encounters the Newfoundland Rise, around which it continues (Fig. 9.7). A semipermanent loop passes around the Newfoundland Basin before continuing eastward into the NAC: this loop, or "ring meander," intermittently pinches off a mesoscale eddy that carries warm-water organisms far to the north into the Labrador Sea.

Although slope water forms a prominent, cold North Wall to landward of the stream, it is important to note that a shelf-slope front (marked by the 34% isohaline) within the slope water lies landward of the North Wall of the Gulf Stream so that confusion is possible between the two features. This region has what must be the strongest horizontal temperature gradient of SST anywhere in the oceans: in winter, 18°C Sargasso water and the pack ice of the Gulf of St. Lawrence are only 300–400 km apart. No wonder, then, that the GFST province is a region of rapid weather change and strong mesoscale variability. Summer stratification occurs right across Slope water, Gulf Stream, and Sargasso Sea and obscures the surface thermal gradient across the cold wall of the current.

The free inertial jet forms both cyclonic and anticyclonic meanders, and when these turn acutely enough to become pinched off, isolated mesoscale features are formed. Each is either a ring of isolated Gulf Stream water enclosing an area of cold slope water

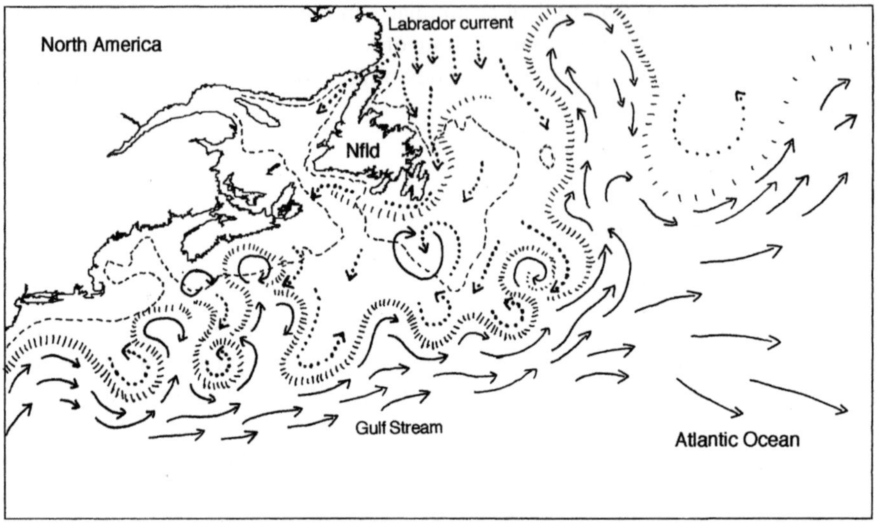


**Fig. 9.7** The conjunction between cold (dotted arrows) Labrador Current water with the warm (solid arrows) Gulf Stream. The dotted line is the 200-m isobath and the hatched lines represent fronts between the cold and warm water. Between the Labrador water, which dominates over the shelf, and the offshore Gulf Stream lies the mixed body of slope water. Irruption of a warm-core eddy onto the shelf paradoxically exports shelf organisms out to sea, because shelf water is entrained around the eddy. Note also the indication of bifurcation of the Gulf Stream in midocean.

(cold-core cyclonic eddy on the seaward side of the current) or a ring of warm Sargasso water (warm-core anticyclonic eddy on the landward side). These rings, both cyclonic and anticyclonic, are deep (1000–3000 m), large (100–300 km), and long-lived (6 months to 1 year), and they propagate contrary to the flow of the current. At any one time, two to 10 rings may be observed in satellite data (thermal, surface elevation, or chlorophyll) on each side of the current. The water contained in a ring is progressively modified to the characteristics of the surrounding water mass.

Though similar eddy fields occur in many other areas—notably in the Pacific homologue of the Gulf Stream—it is in the GFST province that these rings have been most closely investigated (e.g., Joyce and Wiebe, 1992). Some will quarrel with the logic, but this province is best defined as comprising not only the frontal jet of the Gulf Stream, from Cape Hatteras to the Newfoundland Basin, but also the field of cold-core eddies associated with its flow. Small numbers of eddies will travel beyond the boundaries of the province (Richardson, 1983) violating the logic, though not the convenience, of the definition. The few warm-core eddies that become inserted in slope and even shelf water are treated as a special characteristic of the coastal boundary province NWCS.

## Response of the Pelagic Ecosystems

The ecology of the warm water of the Gulf Stream resembles that of the adjacent Sargasso Sea (see North Atlantic Subtropical Gyral Province) from which it differs largely because of the consequences of the meandering of the frontal jet and by the associated population of mesoscale eddies. The ecology of these eddies is the outstanding interest of the GFST province for most biological oceanographers.

We shall therefore concentrate our attention on them and on some other special features associated with the meandering frontal jet current. In fact, a glance at SeaWiFS of

MODIS images will usually demonstrate some chlorophyll enhancement within the eddy field of the Gulf Stream as far to the east as the Mid-Atlantic Ridge almost continuously from spring to autumn, and this is especially noticeable when compared with the more transient enhancement in the eastern Atlantic at similar latitudes, where mesoscale eddies are less numerous and less energetic. It had commonly been thought that the passage of the Gulf Stream across the New England seamount chain induces the formation of particularly energetic meanders and eddies, but thermal imagery now shows that this is not the case (Cornillon, 1986).

The ecology of cold-core eddies was well studied in the late 1970s by the Ring Group (1981), mostly out of Woods Hole. Young rings contain water having the ecological characteristics and biota of slope water, and the central doming brings the 15°C isotherm close to the surface. The rate of primary production within a cold-core eddy is, at least initially, higher than the surrounding oceanic water by a factor of about 1.7; this is similar to the general ratio between slope water and the open Sargasso Sea. If it is shed at the end of winter, a spring bloom may occur within the young ring. The deep chlorophyll maximum layer thus formed weakens (by a factor of 5–10 in maximum chlorophyll values) and deepens to about 100 m as the summer progresses, coming to lie on the upper part of the nutricline within the ring. This evolution is due to both seasonal processes and, as the ring ages, progressive replacement of the slope water flora and fauna by species typical of the Sargasso Sea.

John Woods (1988) has explored and modeled the biological consequences of the dynamic changes in pycnocline topography that are associated with instability of mesoscale jets, such as the Gulf Stream. He points out that mesoscale jets are inherently unstable, with high isopycnal potential vorticity, and soon develop meanders with along-stream wavelengths of 10–100 km. Vortex contraction on the flanks of the anticyclonic warm-core meanders should cause upwelling to occur, with downwelling within the cyclonic meanders. This vertical motion is reflected in the sea-surface temperature field, and primary production rates are higher in the anticyclonic meanders where both pigment maximum and the nitracline rise along the upward-sloping isopycnals; a patchy distribution of mixed-layer chlorophyll develops at the same scale as the meanders (Lohrenz *et al.*, 1993). According to Rossby's theory of the Gulf Stream, a general cross-stream effect on production rate and standing stock was proposed by Yentsch (1974), by which water is drawn into the right side of the jet and discharged toward the left along the upward-sloping isopycnals. Together with cross-frontal mixing of relatively pigment-rich slope water, this should result in the cold north wall of the Gulf Stream being observable at the surface as a chlorophyll front.

However, more recently, Anderson and Robinson (2001) have assimilated data from two surveys in autumn 1988 (BIOSYNOP 21 and GULFCAST) into a 4D simulation of physical processes and biological effects, and some of the earlier suggestions are not confirmed. Primary production and chlorophyll concentration are not enhanced at the front over values in the adjacent Slope Water; high chlorophyll concentrations observed at sea on BIOSYNOP appear to be caused primarily by advection and convergence, rather than by *in situ* biological growth. The result from the Anderson/Robinson model appears to be confirmed by the biweekly high-resolution images for chlorophyll and sea surface temperature produced routinely at Bedford Institute of Oceanography.

Within the cold-core rings that become embedded in the Gulf Stream in winter or spring, the response of the entrained slope water mesozooplankton species, enclosed within the warming ring, is to descend progressively into cooler water as the summer advances. In this way, mature cold-core rings, late in the year, may come to have warm-water species above and shelf species below. Vertically integrated biomass of cold-core rings may thus exceed that of surrounding water because of the addition in such data of a fully developed immigrant Sargasso biota above and the remains of a refugee slope-water community below.

More generally, mesozooplankton species distributions echo the distribution of water types in this complex physical system. Using recurrent group and cluster analysis, Ashjian and Wishner (1993) examined how 22 categories of 18 species of copepods responded to the offshore downsloping of physical properties—pycnocline, nutricline, and downstream velocity profile—along a section across the Gulf Stream in May 1988. Four principal groups of taxa were identified, of which one (Group 3) comprised large, active diel migrants. Group 1 comprised epiplanktonic *Nannocalanus minor, Neocalanus gracilis*, and *Lucicutia ovalis* (and their growth stages) that layered in the upper 100 m, in water of about 25°C right across the stream. Group 2 (*Calanus finmarchicus, Rhincalanus nasutus*, and *Metridia lucens*) lay deep, in water <10°C, and always below the downsloping pycnocline. Group 4 (*Calanus tenuicornis* stages, and *Lucicutia gemina*) lay horizontally across the stream below Group 1, at 100–200 m, in water of 20–24°C.

A seasonal series of mesoplankton samples at 36°N reveals the general annual cycle of vertical migrations in the Gulf Stream system that will modify the foregoing simple pattern (Allison and Wishner, 1986). In winter, biomass in the upper 200 m is low both day and night in the slope water and in the adjacent north wall of the stream relative to biomass that is about one order of magnitude greater in spring and early summer. In the Gulf Stream itself and the adjacent Sargasso water, there is little seasonality in near-surface biomass. In May, biomass is concentrated at 0–100 m both by day and night, progressively deepening offshore, and little diel migration occurs. By September, diel migration is strongly established from the slope to the Sargasso water, with a subsurface maximum at night at about 80–90 m in all areas, near the DCM that develops during the summer months. Such strong diel migration to 400–600 m in a region with such active horizontal advection must cause very significant horizontal redistribution of biomass.

### Synopsis

*Case 2—Nutrient-limited spring production peak.* $Z_m$ undergoes moderate boreal winter excursion while $Z_{eu}$ remains fairly constant at 35–50 m, so that thermocline is illuminated from May to October. Rate increase of P begins in February, before a shoal (thermal) $Z_m$ is established in March–April, and reaches a (nutrient-limited?) maximum in April, subsequently declining progressively to an annual minimal rate in November (Fig. 9.8); a shoulder on this decline in August–October appears to be a response to the deepening of the mixed layer as it passes down out of the photic zone. Dynamic chlorophyll biomass range tracks the variations in P so that maximum accumulation occurs in April. Relations of biomass and P suggest that consumption and production are balanced, and that seasonal vertical migration of consumers is insignificant.

## NORTH ATLANTIC SUBTROPICAL GYRAL PROVINCE (NAST-E, NAST-W)

### Extent of the Province

There is no continental shelf in this province save for the island platforms of Bermuda and the Azores. NAST is bounded to the west and northwest by the eddy field of the Gulf Stream and to the northeast by the bifurcation of flow between the Azores Current and the North Atlantic Current; that is, at about 40–42°N. To the south, it is defined by the Subtropical Convergence (STC) that lies below the convergence between the trade winds and the westerlies, along about 25–30°N; this frontal zone is the equatorward limit of

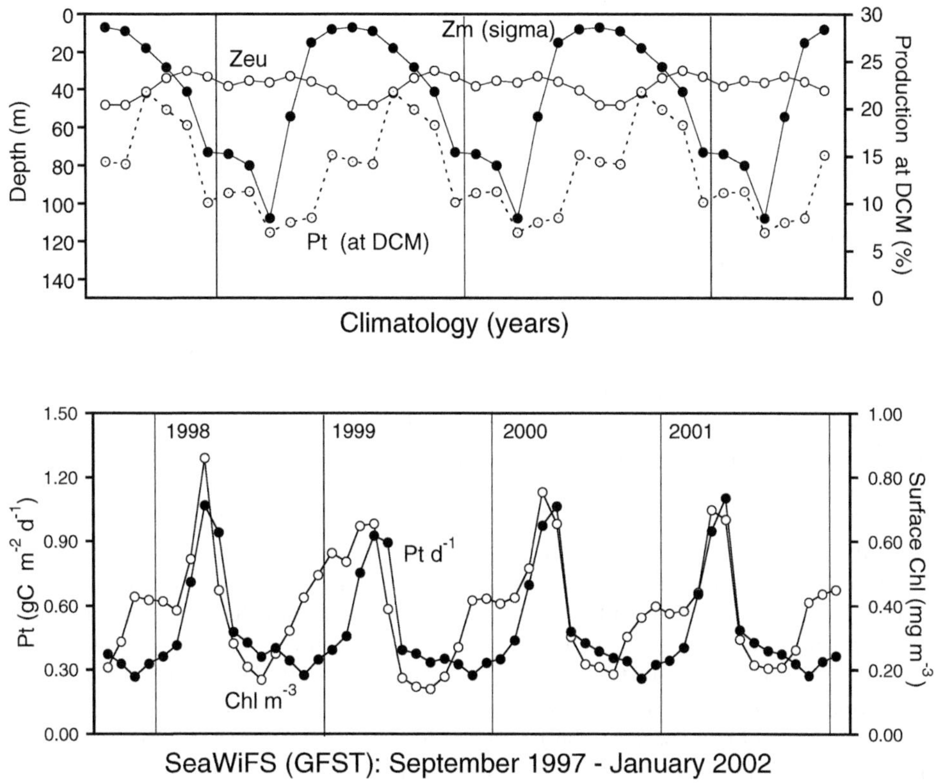


**Fig. 9.8** GFST: seasonal cycles of monthly surface chlorophyll and depth-integrated autotrophic production for the years 1997–2002 from SeaWiFS data together with characteristic seasonal cycles of mixed-layer depths from Levitus climatological data and photic depths computed from characteristic irradiance and the archive of chlorophyll profiles discussed in Chapter 1.

significant winter mixing, for which a useful marker would be the end-of-winter surface isotherm for 20°C.

## Defining Characteristics of Regional Oceanography

Errant cold-core eddies originating in Gulf Stream meanders may propagate into the NAST province beyond the average eddy field, which is considered to be part of GFST (see earlier discussion). Isolated seamounts support Taylor columns that may also spawn cyclonic, warm-core eddies that are observable in the SLA field. Such eddies occur, for instance, downstream from the Corner Rise seamounts (Richardson, 1980) and also from the Canaries, where we have good information on their biological effects (Aristegui *et al.*, 1997); here, eddies of both signs are generated downstream of the islands at intervals of several days to a few weeks at all seasons, often having elliptical or irregular form, suggesting that they are not yet in geostrophic balance.

This province represents that part of the anticyclonic subtropical gyre that lies below the influence of the westerly winds, although these are relatively weak at these latitudes. Therefore, although winter mixing does occur, it is weaker than further to the north, not only because of lower wind stress there but also, as noted in Chapter 3, because wind stress at the sea surface is preferentially transformed into momentum rather than mixing progressively equatorward.

Winter deepening of the surface mixed layer of this province is initiated by the passage of atmospheric cold fronts across the subtropical ocean in autumn, eroding the seasonal thermocline. The consequent convective mixing deepens $Z_m$ to 125–150 m, a process leading to the formation of the 18°C subtropical mode water (Worthington, 1986); in winter, this extends from 30°N almost to 40°N, then sinks and spreads, effectively separating the seasonal from the permanent thermocline. Toward the center of the Sargasso Sea, in the vicinity of Bermuda, the upper 200 m of the water column may become thermally uniform from January to April, after which stratification is reimposed not only by increasing irradiance and reduced wind stress at the surface, but also by rainfall that induces a shallow, freshened layer that is enriched in atmospheric nitrogen (Michaels *et al.*, 1993). As the summer progresses, the increasing frequency of tropical storms progressively modifies this simple shoal mixed layer; cooling occurs rapidly after early October.

Even though here we are far from the unpredictable higher latitudes, there remains significant between-year variability in seasonal mixing and stratification: maximum winter $Z_m$ varied over a 9-year period from >300 m in 1993–94 to only 170 m in several later winters, although in each year the maximum occurred predictably in February. Surface nitrate in most years briefly exceeds 1.0 μmol $kg^{-1}$ at the surface at this time, but this enrichment is not observed in those years when winter mixing is relatively shallow.

Many of the observations on which we depend for understanding the NAST province have originated from the Bermuda Atlantic Time Series (BATS) program, but it must be emphasized that BATS is placed right on the edge of the region of the North Atlantic where winter mixing penetrates sufficiently deeply to induce a weak spring bloom; as Nelson *et al.* (2004) point out, just south of the BATS site there is a strong meridional gradient in surface properties, beyond which the tropical conditions of the NATR province obtain. Because the BATS site is marginal to the winter-mixing regime of the North Atlantic, we should not be surprised that between-year variability here is strong, as has been observed.

Along the southern limit of the province, the transition between midlatitude westerlies and trade winds from the east drives convergence of surface water along the STC (Iselin, 1936; Voorhuis, 1969). The thermal fronts comprising the STC thus separate the anticyclonic gyre into a northern subtropical and a southern tropical portion at about 25–30°N. To the south of Bermuda, such a transition has long been known to occur, across which subtropical, winter-mixing conditions are replaced by tropical conditions in which surface water in winter never falls below 20°C. Schroeder (1965) traced the thermal front of the STC from 70°W eastward to 40°W, always between 20 and 31°N; at 30°W, to the south of the Azores, Gould (1985) mapped the same front carefully. The STC is found to be rich in eddy activity at all scales; shear instability and random superposition of internal waves produce small-scale structure along the front (Voorhuis and Bruce, 1982; Toole and Schmidt, 1987). Wavelike baroclinic eddies (of scale 800 km, 200 days) induce persistent westward-propagating sea surface temperature anomalies at the STC (Halliwell *et al.*, 1991), whose movement along the front is spaced 3–5 days apart (Leetma and Voorhis, 1978). The instantaneous location of the STC can readily be found in basin-scale TOPEX-Poseidon data.

The topography of the Mid-Atlantic Ridge constrains the main recirculation of western boundary current water within the western basin (e.g., Richardson, 1985), and the biological properties of the eastern and western basins differ significantly. Therefore, for some purposes, it may be useful to consider treating the western and eastern basins as distinct subprovinces: NAST-E and NAST-W. In this case, we should draw the boundary along the topography of the Mid-Atlantic Ridge, running southwest from the Azores.

## Response of the Pelagic Ecosystems

Because the geographical Sargasso Sea corresponds with the subprovince NAST-W, there is a very appropriate marker to demonstrate that the line between the eastern and the western subprovinces is not arbitrary, but biologically significant. Floating aggregations of gulf weed (*Sargassum* spp.) are abundant to the west but sparse to the east of this line, or approximately over the Mid-Atlantic Ridge. It is thought that this is consequent on the partially closed nature of circulation within the Sargasso Sea. There are also significant morphological and ecological differences between *Sargassum* in this province and in the NATR to the south of the STC (Niermann, 1986; Butler *et al.*, 1983, and references therein).

As mentioned earlier, there are >300 individual seamounts (not all associated with the Mid-Atlantic Ridge) of sufficient topography to sustain Taylor columns, so it may be expected that some areas of enhanced surface chlorophyll will be topographically controlled, even in midocean. Saltzman and Wishner (1997a) briefly review the ecological effects that may be anticipated where seamounts populate a deep ocean region. Uplifting of isotherms and upwelling of nutrients, interaction with diel vertical migrant zooplankton and consequent patch formation, and induction of relatively high biomass of pelagic fish are among the more frequently noted effects.

NAST(E) includes a sector of the offshore Canary Current, clear of the field of eddies and filaments generated within the coastal boundary zone. Here, mixed-layer depths are shoaler, and enhanced chlorophyll biomass is indicated by the satellite images, particularly south of the Canaries, though it will be important to separate the effects of eddies from the filaments of high chlorophyll induced by tidal mixing (and other coastal processes) at the islands. Vertical nutrient flux in cyclonic eddies occurs centrally by isopycnal transport, and therefore it is strongest where the pycnocline dome is shoalest and where some diapycnal mixing may occur across the shoaled pycnocline. As in many upwelling situations, chlorophyll enhancement occurs most strongly a little downstream of the surface nutrient maximum and so, in this case, chlorophyll is highest around the edge of the cyclonic eddies. Anticyclonic eddies frequently interact with filaments of water having enhanced chlorophyll, causing the chlorophyll signal to spiral inward toward the center of the eddy, as also occurs in Gulf Stream rings.

The STC front along the southern flank of NAST has been studied to the south of the Azores by Fernandez and Pingree (1996), who show that it supports local enhancement of primary production and biomass accumulation, as is generally the case for oceanic fronts. Winter productivity is in the range 0.8–0.9 g C $m^{-2}$ $d^{-1}$ or about twice the rate for those parts of this province not influenced by frontal or mesoscale eddy processes. These authors suggest that because of the great spatial extent of the frontal signature in the Azores Current system, primary production within the subtropical front is of major significance for regional carbon budgets. A DCM occurs everywhere across the NAST province at about 50–100 m across the province, following the depth of the nutricline rather than that of the pycnocline in the density gradient. It lies progressively deeper to the south and east across the province. As usual, the chlorophyll maximum is somewhat deeper than the depth of maximum photosynthetic rate and of the biomass maximum for photosynthetic cells. This must follow from the generalization that the C/chl ratio decreases with increasing depth of the DCM. Li (1995) shows that the DCM for cyanobacteria and prochlorophytes can be traced oceanwide across NAST (Fig. 9.9).

It is in the west of the province, in the Sargasso Sea, that the most detailed studies of ecosystem response have been performed, over a long period of years, because of the proximity of the "Oceanographic" at Woods Hole, and the existence of the Bermuda Station for Biological Research at St. George's. Gordon Riley (1957) obtained data for 2 years at OWS E at 35°N 48°W, to the northeast of Bermuda, and laid the groundwork for much of what has followed. The most comprehensive description of pelagic ecosystem here derives from the Bermuda Atlantic Time Series Study (BATS) studies at 31°N, just south of Bermuda over deep water, at a station that was occupied for several days on each of 111 occasions from 1989 to 1994. This extraordinary data set allows almost weekly

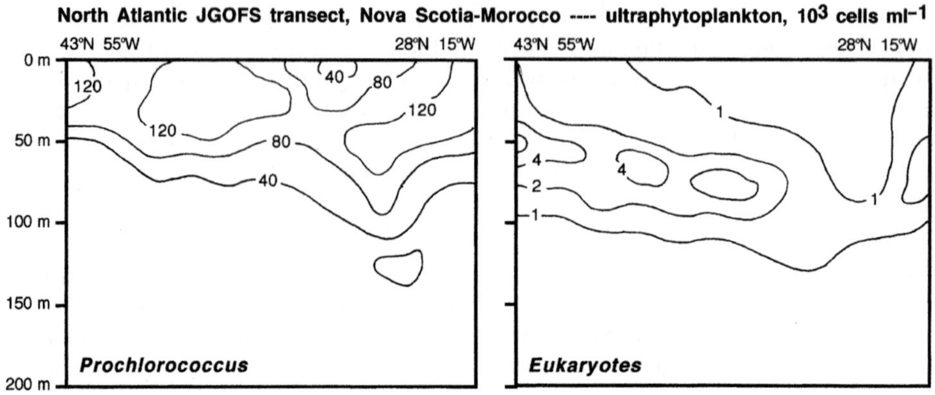


**Fig. 9.9** This zonal JGOFS section across the NAST province from Nova Scotia to Morocco shows how the depth of maximum abundance of cyanobacteria deepens eastward.

*Source: Redrawn from Li, 1995.*

precision over this 6-year period (Steinberg *et al.*, 2001): seasonal and between-year signals are unambiguous in the data. Because the BATS station is very close to OWS S, also studied since the 1950s, decadal variability in pelagic production is also recoverable for this region.

The seasonal evolution of the pelagic ecosystem of NAST responds to the seasonal evolution of mixed-layer depth and has been followed with precision in the multidepth profiles of bulk properties obtained at BATS (Steinberg *et al.*, 2001). As noted earlier, nitrate is available in the mixed layer briefly (usually 0.5–1.0 $\mu$M, and for <30 days) when mixing is deepest. The $NO_3/PO_4$ ratio is anomalous because it exceeds the Redfield ratio of 16, both generally in the Sargasso Sea and specifically at the BATS station. It is thought that balance must be obtained by nitrogen fixation, either by vertically migrating nitrogen-fixing *Trichodesmium* colonies that obtain phosphate at depth, or by vertical transport of deep nitrogen by vertically migrating *Rhizoselenia* mats, perhaps an important source of new nitrogen in other oceans.

A short winter bloom of varying duration occurs between February and April, and the period when chlorophyll is enhanced is usually shorter than the period of enhanced rate of primary production. During this brief seasonal bloom, high chlorophyll values extend from the surface to 150 m depth, this episode being imposed, as it were, on a permanent DCM at about 100 m. McGillicuddy and Robinson (1997) have suggested that the observed rate of new production in the mixed layer here in the Sargasso Sea is consistent with an independently derived net flux of 0.5 mol N $m^{-2}$ year into the mixed layer. Most of this flux is thought to occur during the formation and intensification of cyclonic eddies in a manner consistent with the observed eddy field; about 30% may result from the interaction between mesoscale flow and wind-driven surface currents. These authors point out that it is important to consider the interaction between the light field in the oligotrophic ocean and vertical motions, of both signs, induced within eddy fields. "Cold" cyclonic eddies import nutrients up into the euphotic zone, whereas "warm" anticyclonic eddies carry phytoplankton-rich, nutrient-depleted water down below the lighted layer.

At OWS S, also near Bermuda but less comprehensively monitored than the BATS station, depth of winter mixing varies (as at BATS) from 150 to >250 m and the duration of the bloom (as indicated by enhanced chlorophyll) varies from 40 to 120 days; mixing was especially deep and of long duration in 1991–1993, apparently associated with the ENSO events of those years. During the period 1958–1971, because it was as deep as

350–450 m, mixed-layer nitrates in the range of 2–5 μM $kg^{-1}$ have been hindcast for that period. HPLC analysis of phytoplankton accessory pigments at BATS enables characterization of the taxa that contribute seasonally to blooms and to the permanent deep maximum. It is, of course, the prokaryotic picoplankton that dominate the autotrophic biomass almost year-round, with some seasonal segregation between groups. Prochlorophytes are seasonally variable, being important from late spring, and through summer, while cyanobacteria are more regular in their biomass, never falling to such low values as prochlorophytes (DuRand *et al.*, 2001).

Coccolith blooms occur in late winter and spring (<100 cells $liter^{-1}$) at OWS S, with much lower concentrations during summer and with some species succession: *Emiliana huxleyi* dominates near-surface in spring and *Florisphaera profunda* at the DCM in late summer (Haidar and Thiersen, 2001). Prymnesiophytes and pelagophytes numerically dominate the eukaryote phytoplankton and dinoflagellates are relatively less important. Diatom blooms occur, often late in the summer but, even so, their cell numbers may be quite high. It has been suggested that the phytoplankton community, and especially the eukaryotes, are resistant to changes in oceanographic conditions as these change seasonally. Perhaps this is an observation of general relevance in warm open oceans?

The DCM is a site of especially intense trophic activity in this province as it is elsewhere, and it is a preferred depth zone for herbivorous mesozooplankton. Consumption follows a strong diel cycle not only for diel migrant micronekton and copepods but also for those resident in the photic zone. For the diel vertical migration of micronekton and large copepods, consumption is lowest in late afternoon and peaks around midnight. It is in this province that the role of small heterotrophic nanoflagellates in both "nurturing and grazing on planktonic bacteria," as Sieburth and Davis (1984) put it, was worked out. As elsewhere, the bulk profiles at the BATS show that bacteroplankton dominate (if cyanobacteria are included) the living biomass, and also that the heterotrophic bacterial biomass is maximal in the deep chlorophyll maximum layer, though it is somewhat more seasonal there than the autotrophic cells. Likewise, most of the biomass of heterotrophic organisms at BATS comprises the nanozooplankton (2–5 μm): acantharia, radiolaria, and foraminifera are, as usual, the most abundant forms. As elsewhere, this was a difficult community to assess at BATS, given the high level of production by autotrophic symbionts within the cells of these organisms.

The seasonal biological cycle at the 1989 JGOFS station (32°N 20°W) in the NAST-E subprovince reported by Jochem and Zeitzschel (1993) appears to be similar to the weak spring bloom found at the BATS mooring. However, we must be very careful when comparing data sets not obtained simultaneously, because the 1955–1994 data from OWS S show strong decadal and between-year variability, principally forced by the relative length of the winter mixing period and the depth of mixing (Michaels and Knap, 1996); these factors determine the nutrient levels in the mixed layer when stratification is reestablished.

One may be surprised that modern ecological studies, such as those derived from the BATS series of observations, do not appear to consider seriously the macroalgae. Here, the existence of floating *Sargassum* (Fucacae) should not pass without notice, for it is a unique feature of this province. Several species (*S. natans, S. fluitans, S. hystrix*, and *S. polyceratus*) are involved, having differential distributions (e.g., Stoner, 1983), while different morphological forms of at least one species are recognized. The occurrence of masses of gulf weed dense enough to stop sailing ships were probably only the stuff of old sailors' yarns, and *Sargassum* normally occurs in yellow, disc-shaped clumps 10–50 cm in diameter at the surface of the ocean. These may be randomly dispersed (as I have seen them) or arranged in windrows, accumulated by the Langmuir circulation, as much as 50 m wide, and aligned parallel to the direction of the wind. It has not been easy to derive satisfactory estimates of biomass from surface net tows, though many have been made at OWS S, for the individual data vary from 2.3 kg $m^{-2}$ wet weight down to a few milligrams per square meter. Associated with the floating gulf weed are epiphytic, phycoerythrin-rich cyanobacteria of which at least some are capable of N fixation. The mean C turnover time of *Sargassum* biomass is about 40 days but the relative productivity of both gulf-weed and its associated epiphytes contributes no more than 0.5% to total regional primary production in the western Sargasso Sea (Carpenter and Cox, 1974).

A specialized fauna of fish and epibionts is, of course, associated with *Sargassum* and this varies regionally: for instance, stronger encrustation of epibionts (hydroids, calcareous algae, spirorbid polychaetes, etc.) occurs in cooler water. Young-of-the-year amberjack (*Seriola dumerlii*) are associated with offshore *Sargassum*, and these individuals have faster growth rates than those that remain close to their coastal spawning region.

Seasonal mesozooplankton biomass at BATS in the period 1994–98 was maximal in late spring and (in some years) again in autumn and early winter while, between years, biomass is positively correlated with autotrophic biomass and productivity (Madin *et al.*, 2001). The diel migration factor is from 1.7 to 3.4 and, as is usually the case, migration is more prominent in the larger size fractions: mean biomass values (0–200 m) for this period were 418 and 659 mg dw $m^{-2}$ for day and night respectively: diel migration, then, penetrates rather deep in these clear waters. Migrants are dominated by copepods (*Pleuromamma, Euchirella*), euphausiids (*Thysanopoda, Nematobrachion*), amphipods (*Scina, Anchylomera*), and shrimps (*Sergestes, Sergia*).

The mesozooplankton from BATS have not yet, I think, been formally published taxonomically, but previous studies in the NAST province indicate the existence of a typical warm-water assemblage dominated numerically by small epiplanktonic copepods: *Clausocalanus, Oithona, Farranula, Ctenocalanus*, and *Lucicutia*. In the northern Sargasso Sea, Grice and Hart (1962) found 92 epiplanktonic copepod species, 9 chaetognaths, 20 euphausiids, 23 siphonophores, and so on—a typical warm-water zooplankton. Smaller species dominate the summer zooplankton, more efficient at using the small autotrophic cells and protists that then abound.

It is in this province that we first meet the large deep-water tunas in any abundance. Fontenau (1997) shows that there is a significant east-west difference in tuna habitat. In NAST-W, albacore (*Thunnus alalunga*) is present year-round, but maintains a larger population during the cooler months, whereas swordfish (*Xiphias gladiator*), bluefin (*Thunnus thynnus*) and bigeye (*Th. obesus*) together formed only 10–30% of the longline catches in the period 1984–1993. This is a rather unselective gear and we may assume the catches indicate rather well the actual species association of large predatory fish present. But, in NAST-E, albacore are a minority component and the most abundant species are bluefin and swordfish. Adult albacore, of course, occur preferentially around the subtropical sectors of the principal oceanic gyral systems and are ecologically distinct in their habitat requirements from the cooler-water adult bluefin. I shall reserve discussion of their ecology until we get to the discussion of the Pacific subtropical provinces.

## Synopsis

*Case 3—Winter-spring production with nutrient limitation.* $Z_m$ undergoes a moderate boreal winter excursion, slightly deeper in the west, while Zeu remains at 45–55 m, so that the thermocline is illuminated for 6–7 months. An increase in P is initiated in December, continuing until the maximum P rate is reached in April–May (Fig. 9.10) when thermal stratification is fully reestablished at 25–30 m. Dynamic range of chlorophyll biomass is small, reaching an annual maximum in March–April with higher values in the east (<40 mg $m^{-3}$) than in the west (<25 mg $m^{-3}$), after which the standing stock is rapidly depleted. The chlorophyll cycle is therefore consistent with a close match between consumption and production, and with the seasonality in zooplankton biomass.

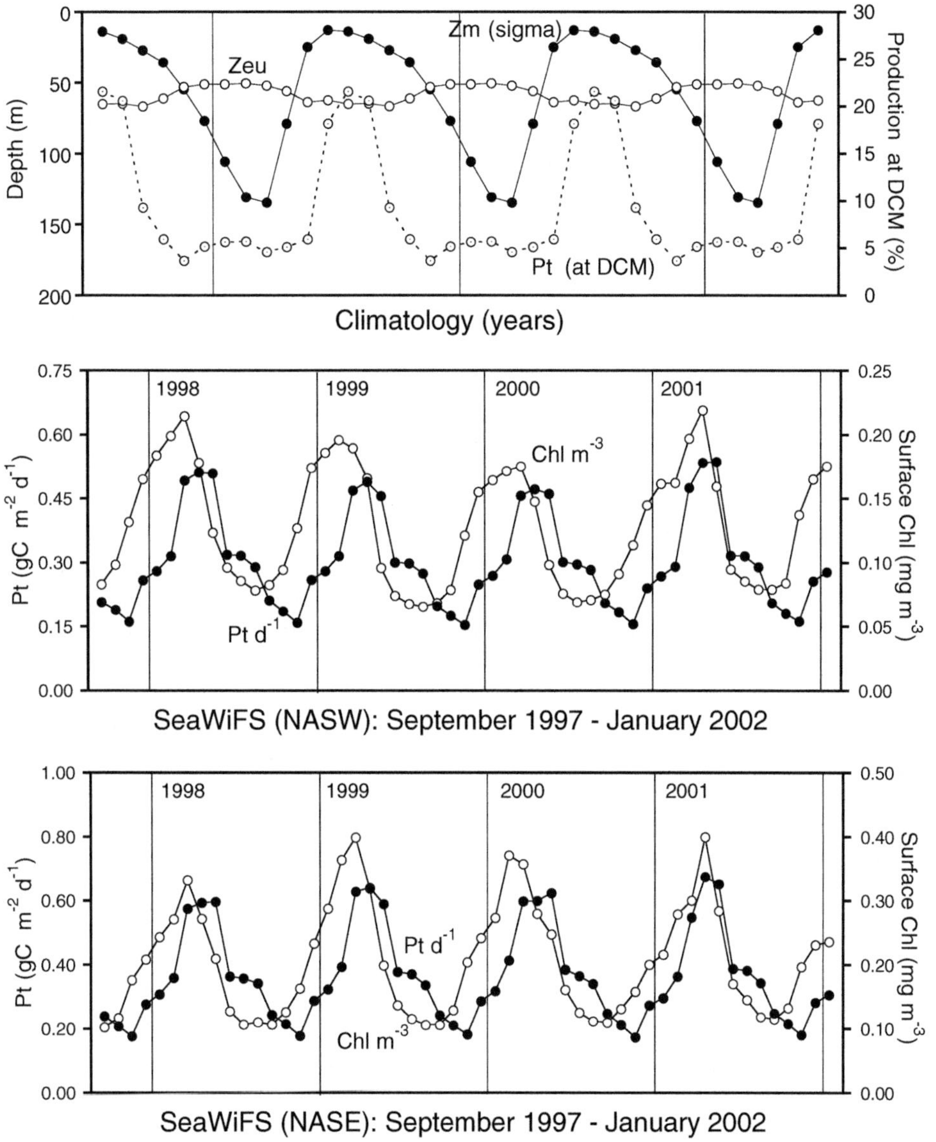


**Fig. 9.10** NAST: seasonal cycles of monthly surface chlorophyll and depth-integrated autotrophic production for the years 1997–2002 from SeaWiFS data together with characteristic seasonal cycles of mixed-layer depths from Levitus climatological data and photic depths computed from characteristic irradiance and the archive of chlorophyll profiles discussed in Chapter 1.

# Mediterranean Sea, Black Sea Province (MEDI)

## Extent of the Province

The Mediterranean Sea, Black Sea Province (MEDI) includes the whole of the Mediterranean basin, distinguished by the ancients from the "Ocean Stream" lying beyond the Pillars of Hercules at Gibraltar. Included are the marginal seas (Adriatic, Aegean, and Cretan) and also the Black Sea and its marginal Azov Sea. The landlocked seas of Asia (the Caspian and the fast-disappearing Aral) are treated as saline lakes, no longer part of the ocean.

## Continental Shelf Topography and Tidal and Shelf-Edge Fronts

The Mediterranean Sea comprises two deep basins, connected with the Atlantic Ocean across a sill depth of only 290 m at the Straits of Gibraltar. The Black Sea is landlocked except for its connection with the Mediterranean through the Bosphorus; this connection is slender, having a shore-to-shore width of only 725 m at the choke point, and a midchannel sill depth of only 40 m.

The Mediterranean is characterized by rather narrow continental terraces that accommodate the deltas of a few large rivers: Nile, Po, Rhone, and Ebro; extensive shelf areas occur only in the northern half of the Adriatic, and on the eastern coast of Tunisia where the shelf widens to about 300 km in the Gulf of Gabes. Most of the shelf is rocky or carries quartz sand deposits with a biogenic calcareous component, and mud is restricted to coastal embayments. Some mud occurs, of course, on either side of the Nile delta, and here the edge of the shelf conforms to the line of the out-built delta.

The coasts of the Black Sea are steep-to only in the southern part of the basin. To the north and west, there are extensive shelf areas that are, in general, much muddier than those of the Mediterranean. The peninsula of the Crimea stands on a wide shelf that reaches 200 km wide off Odessa; the edge of this shelf runs east-west, so almost directly across the northern part of the Black Sea. The effects of the great deltaic region, and of the effluents, of the Danube and Dniester on the western coast must be integrated into an understanding of this sea. The Sea of Azov is a very shallow continental shelf embayment, almost completely isolated from the main basin of the Black Sea, to the east of the Crimean peninsula.

In both Mediterranean and Black Seas the principal currents are marginal, adjacent to the coastline or shelf break, so that coastal and shelf-break fronts are a common feature that isolate this flow from the slower gyral and subgyral circulations. Tidal effects, mostly semidiurnal, are small in both the Mediterranean and Black Seas, because of the relatively small size of each basin.

## Defining Characteristics of Regional Oceanography

It will be convenient to discuss the Mediterranean and Black Seas separately because of their very different characteristics. The Mediterranean is an evaporative basin constrained by a shallow sill at the Straits of Gibraltar, whereas in the Black Sea the salt balance is approximately in equilibrium despite its shallow sill. The fast and continuous surface flow from the Black Sea into the Mediterranean through the narrow passages of the Bosphorus and Dardanelles is a unique feature in the circulation of the ocean. The Danube contributes two-thirds of the 304 $km^3$ of freshwater that enters the Black Sea annually. One of the most useful and comprehensive accounts of Black Sea ecology is still that of Caspers, to be found in Hedgepeth (1957); this is particularly useful now as a reference for the near-pristine state of a now highly modified ecosystem.

***Mediterranean Sea*** It is in winter, when the zonal band of the westerlies lies farthest equatorward, that the planetary wind field is most effective in the Mediterranean; wind stress at the surface is strongest and most uniformly distributed from December to March. For most of the year, although the wind-stress climatology is dominated by northwesterly winds, the dominant effects are local and orographic. In winter there are local irruptions of especially strong northerly winds, cold and dry, that in the western Mediterranean are the adiabatic "mistral" winds channeled down the Rhone valley, while the similar "bora" is not appreciated by the residents of Venice and Trieste.

In the absence of large-scale geostrophic circulation, the physical oceanography of the Mediterranean is strongly influenced by the local wind fields (Theocaris *et al.*, 1998). The circulation of surface water is characterized especially in the eastern basin by a number

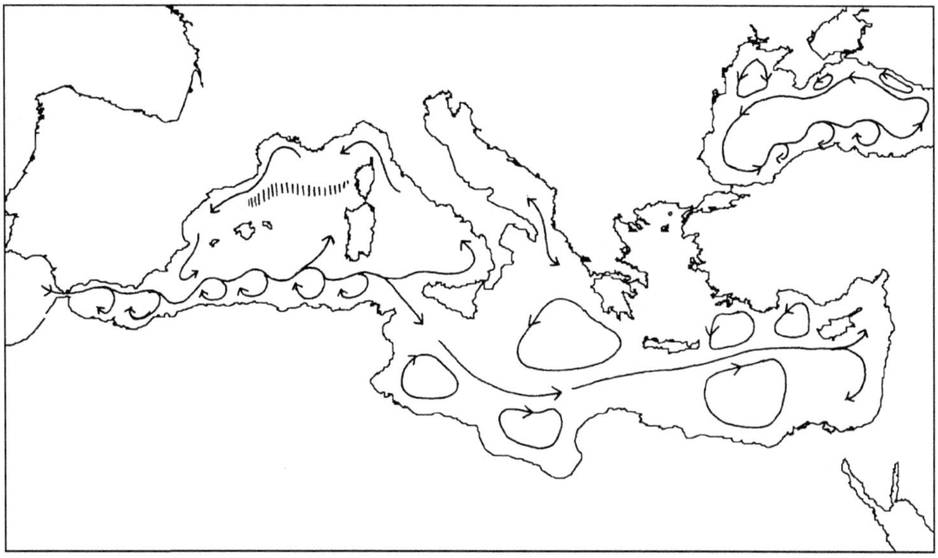

**Fig. 9.11** Cartoon of the general circulation of the Mediterranean and Black Seas to illustrate the dominance of coast form on the location of both persistent and intermittent eddies and fronts. Comparison with any suitable sea surface chlorophyll image will show the extent to which coastal eddies and dipoles are the locations of chlorophyll enhancement.

of semipermanent gyres resembling in scale the large mesoscale eddies of the open ocean. However, these are not errant eddies and their location, together with their associated fronts and jet currents, is generally predictable. The general circulation pattern is complex (Fig. 9.11) because of many factors: lateral thermohaline fluxes due to acceleration through narrow straits, flux of freshwater from river flows, topographic effects of the complex continental and insular coastlines, and a Rossby internal deformation radius of 10–15 km. Although we shall not be concerned here with the details of the deep overturning circulation, we should note that deep, cold, dense water-mass formation occurs in the Adriatic Sea and in the Gulf of Lions, forced by the effects of bora and mistral, respectively. The signature at the sea surface of this process is easily enough misinterpreted as the effects of divergent upwelling of cold water.

The Mediterranean Sea comprises two partially isolated basins within each of which a cyclonic surface circulation occurs (see Robinson and Malanotte-Rozzoli, 1993; Minas and Nival, 1988), and the details of coastline alignment impose many smaller, semipermanent gyres. The narrowness of the Sicilian Channel (140 km) partially isolates the gyral circulations of the eastern and western basins, which Millot (1992) regards as two separate Mediterranean seas. The Tyrrhenian Sea, partially enclosed by Sicily and Sardinia-Corsica, and the Adriatic Sea, behind the narrow (70 km) Strait of Otranto, each have a partially enclosed cyclonic gyral circulation.

The two gyral circulations of the western and eastern Mediterranean are only partially isolated, so there is a general cyclonic flow around the whole basin, with the surface water becoming progressively saline and the return flow progressively deeper. This process preconditions the water that enters the Ligurian Sea and the Gulf of Lions so that mistral wind episodes in winter (that strongly cool the surface water) readily induce deep convection and the formation of Mediterranean Deep Water. It is this mechanism that is responsible for the relative vertical uniformity of Mediterranean water masses. Wind-induced divergent upwelling occurs on the western coast of Sicily, at several places along the eastern coast of the Adriatic and Aegean Seas, and on the western coast of Crete. In these places, it is observable as a cold signature in sea surface temperature images but, interestingly, not always in the sea surface chlorophyll images.

Of more ecological interest, perhaps, than the general circulation of the semipermanent gyral system is the more active series of coastal rim currents that proceed cyclonically around the Mediterranean and are especially continuous around the western basin. This flow is forced by topographic steering and perhaps by deflection of motion to the right, or toward the coast, by the positive curl of the wind stress. Although this has long been the accepted explanation for rim currents in both Mediterranean and Black Seas, the actual forcing mechanism may be more complex, according to Saur *et al.* (1994).

Through the Straits of Gibraltar, Atlantic water enters the Mediterranean at the surface, and this perpetual inflow attracted the attention of the ancient philosophers and was not resolved until surprisingly late—toward the end of the 18th century: this was, with tidal streams, the first problem in oceanography to attract scientific attention. Local fisherfolk, of course, knew from the earliest times that a deep countercurrent was the answer—although perhaps they did not know that there was a problem to be solved. The Atlantic water passes eastward along the African coast as the density-driven, topographically locked Algerian Current, which continues into the eastern basin through the Sicilian Channel (Millot, 1992) as the Ionian-Atlantic stream. Instability in this coastal flow generates a field of anticyclonic eddies and meanders, perhaps especially rich along the Algerian coast. These are of order 50–100 km in diameter and drift eastward along the coast at a rate of several kilometers per day. After a lifetime of several months they tend to enlarge and depart progressively from the coast, often trailing a filament of upwelled water with them, originating around their SW flanks. The Western Alboran Sea is occupied by a persistent anticyclonic gyre forced by the orientation and topography of the Straits of Gibraltar. In the eastern part of the Alboran Sea, circulation is more variable (Millot, 1987).

In the eastern Mediterranean basin, the already meandering Ionian-Atlantic stream at the Sicilian Channel becomes a mid-Mediterranean jet that can be traced far into the Levantine Basin passing between flanking cyclonic eddies (Cretan, Rhodian, and West Cyprus) to the north and anticyclonic eddies to the south (Shikmona and Mersa Matruh). The quasipermanent position of these eddies is determined by topography (POEM Group, 1992). A coastal anticyclonic loop current around the Ligurian Sea between Sicily and Greece creates a series of mostly cyclonic eddies.

Although tidal streams in the Mediterranean are generally weak, the Venturi effect over the shallow sill of the Straits of Messina produces tidal currents that are unusually strong, and violent, local upwelling occurs. This is, of course, the whirlpool of Greek antiquity lying between six-headed, dog-barking Scylla, and Charybdis, who swallowed the sea and vomited it back again thrice daily.

***Black Sea*** With its anoxic interior, the Black Sea is the most extreme case of a meromictic basin in the present-day ocean (for modern reviews, see Murray, 1991, and Özsoy and Ünlüata, 1998). Salinity nowhere exceeds 17‰, and excess precipitation together with runoff from the rivers Danube, Dniester, and Don creates a surface low-salinity layer overlying a halocline at about 100 m that is sufficiently strong to prevent ventilation of the interior of the sea. Below the halocline, the basin lacks oxygen and has high concentrations (increasing with depth) of dissolved $H_2S$. Stratification includes a summer thermocline at 10–30 m, shoaler than the main discontinuity that combines nutricline, pycnocline, and oxycline.

The hydrology of the Black Sea depends strongly on its mass water budget because of its near enclosure. Since precipitation and river discharge exceed evaporation, a fast, permanent flow pours southward through the Bosphorus—making a ferry crossing at Istanbul a memorable affair. Some Mediterranean water (representing <50% of the outgoing flux) enters the Black Sea as a bottom flow along the narrow channel, though much of this undercurrent water is entrained back into the fast outflowing surface current (Caspers, 1957; Sorokin, 1983). On entering the Black Sea, the remainder spills over the shelf break into deeper water.

The basin-scale circulation is cyclonic and appears as a coastal current that is the analogue of the coastal flow of the Mediterranean. It is more continuous in the Black Sea, however, because of the less complex coastal topography there. From the mouth of the Bosphorus a strong current runs eastward along the Paphlagonian coast, forming the strongest flow of the Rim Current (40–80 km wide) that circles the whole Black Sea at the 200-m depth contour. The Rim Current is associated with two principal cyclonic gyres that occupy the eastern and western basins (divided to the south of the Crimea). These are constrained by the shelf edge that runs zonally across the basin at the latitude of southern Crimea, so that flow on the northern shallow shelves themselves is more variable. Smaller anticyclonic gyres lie between the main gyres and the coast, trapped by topographic features. During winter, the two-gyre circulation may break down, to be replaced with a single, more elongated cyclonic gyre, and, by the end of winter, very cold shelf water has been formed on the wide shelf regions to the northeast and to the west of the Crimea. This water is progressively advected around the western side of the Black Sea and its effect may be traced even along the southern coast. Especially along the southern and eastern coasts, there is strong mesoscale vorticity in the meandering and filamentous flow, whose features propagate eastward at 10–15 km a day. This field of vorticity widens at Cape Baba.

This circulation pattern explains the topography of the halocline and the oxic/anoxic interface that lies at about 150 m near the centers of circulation of the two main cyclonic gyres and deepens to >200 m around the coastal margins. Only below a small permanent anticyclonic gyre in the southeastern part of the sea does the oxic/anoxic interface deepen away from the coast. Because the chemistry of the oxic/anoxic interface is so intimately connected with biological processes, we shall defer discussion of it to the following section.

## Response of the Pelagic Ecosystems

The ecological characteristics of the two seas are sufficiently different that to place them in a single province is largely a matter of convenience. Both, however, were significantly modified during the 20th century, not only from land-based sources of contamination but also by reduced runoff from the major rivers entering the basins. Nitrate values in the mixed layer of the Black Sea have increased significantly in the past 25 years. Also very significant has been the loss of the annual Nile flood, held in recent decades behind the Aswan High Dam, resulting in a very significant modification of the ecology of the eastern Mediterranean. The artificial Lessepsian connection between the Mediterranean basin and the Red Sea is of great significance for taxonomic biogeography because of immigration of Indo-Pacific species through the Suez Canal. This transport, and the introduction of exotic species in ballast water of tankers, was discussed in Chapter 2.

The ecological response of the two seas to seasonal changes environmental forcing is, for all these reasons, quite different as is clearly demonstrated by the seasonal chlorophyll images from the SeaWiFS and MODIS sensors since 1997. The Mediterranean shows a clear seasonal winter–spring bloom, stronger in the western than the eastern basin, and from June until October almost the whole sea is deeply oligotrophic. The Black Sea, on the other hand, appears now to have an almost uniformly high level of algal biomass over deep water—a green field that is relatively invariant seasonally. Even higher biomass is consistently indicated over the northern and western shelf areas, with permanent "hot spots" in the Azov Sea and at the margin of the Danube-Dniester deltaic region.

***Mediterranean Sea*** The seasonal cycle of primary production and consumption resembles that of the subtropical Atlantic. Winter mixing causes nitrate to become available in the photic zone and a relatively weak late-winter bloom ensues, which is followed by a long period in which the profile includes a DCM. The western basin consistently supports a more active algal response to wind stress than the relatively oligotrophic eastern basin. The western winter–spring bloom is patchy, and differs in its distribution rather significantly between years. For instance, in March 1999 there was a strong accumulation of chlorophyll between southern France and Corsica that persisted through the following month. Although the winter–spring bloom in each year tends to accumulate more chlorophyll along the south coast of France than elsewhere, such a strong offshore event did not occur in 1998, 2000, or 2001 and only weakly, to the west of Corsica, in 2002.

It is only in winter months that positive net community production exceeds respiration so that a winter production pulse (mid-January to mid-February) occurs off Southern Spain (Rodriguez *et al.*, 1987), both near-surface and subsurface chlorophyll maxima being dominated by small autotrophic cells. At this season, 80–100% of the biomass passed a 10-mm mesh and 20–60% passed even a 1-μm Nuclepore filter. In the Adriatic in summer, Revelante and Gilmartin (1994) found a twofold higher biomass of larger cells in the DCM compared with the rest of the water column, even though picoplankton formed 50% of the total biomass. By late spring, a DCM is established in both the eastern and western Mediterranean basins; typical profiles show that this and the nutricline occurs at the base of the thermocline at 75–80 m, with the topography of the density surface following geostrophic flow. In anticyclonic features, the DCM generally coincides with the nitracline rather than with density surfaces, suggesting that primary production is limited primarily by nutrient supply and only secondarily by light and other factors. At, or close to, the DCM is the expected layer of abundant zooplankton. All these features deepen through midsummer, at a rate of about 15–20 m a month (Estrada *et al.*, 1993).

Because deep and intermediate water masses are formed within the Mediterranean basin by the modification of surface water, subpycnocline nutrient levels are significantly lower than those in the open ocean, with maximum values in mid-depths of 9.5 μM at 250 m in the western basin (Coste *et al.*, 1988). In fact, at the Straits of Gibraltar, the balance between nutrients transported in the incoming and outgoing water masses translates into a net gain of nutrients for the Mediterranean basin. The presence of a discrete Atlantic water body in the Alboran Sea, separated across a density gradient from the shoaler water mass, somewhat complicates observations. Rodriguez (*in litt.*) suggests that the variable thickness of the Atlantic-Mediterranean water interface controls the thickness of the DCM as well as its maximum chlorophyll concentration. In the eastern basin, Yilmaz (1994) observed the control of DCM formation and maintenance by variance in nutrient levels and irradiance. DCMs were shallower (50 m) and contained more chlorophyll in late winter, and were deeper (100 m) and weaker in summer.

An accessory mechanism for vertical transport of nutrients has been described in the western Mediterranean. Here, diel migrant herbivorous copepods, *Centropages typicus*, are observed to feed continuously within the DCM by day but to occur near the surface at night in food-poor conditions; this must force an active flux of organic nitrogen up from the DCM into surface water. It has also been shown that individual weather systems may now induce transient blooms by delivery of nitrate and other nutrients in rainfall: rain collected in the smoggy northern Adriatic contains $<80\,\mu M$ nitrate and $<36\,\mu M$ ammonium (Malej, 1997). The response of the phytoplankton to such anthropogenic inputs is a dynamic that has been unduly neglected for—given the widespread turbidity of the lower atmosphere in recent decades—it is surely not restricted to the Adriatic Sea alone?

The majority of the available chlorophyll images, at all seasons, show significant enhancement over the wide continental shelf on the eastern coast of Tunisia. This is strongest just inshore of the small archipelago off Sfax, but in the oligotrophic season it occupies the whole shelf area that is then significantly greener than the deep water beyond. I suggest that this permanent chlorophyll hot spot may represent benthic macroalgal meadows in the very shallow water surrounding the archipelago that lies 20–30 km offshore. Other coastal regions with higher-than-background chlorophyll accumulation lie along the Nile delta, in the NW Adriatic Sea off the Po delta, on the French coast off the Rhone delta, and in the inner Aegean Sea. The entire shelf off Israel lies within the photic zone and chlorophyll profiles show near-bottom maxima; in such areas, as has been shown in the northern Adriatic (Ott, 1992), the high biomass of suspension-feeding macrobenthos appear to control pelagic biomass. Such benthic organisms have a very high weight-specific metabolic rate. This process may be interrupted by summer stratification, in which case a DCM forms some distance above the bottom.

Major upwelling events occur in the Sicilian Channel that appear to be wind induced; they occur mostly with a lag of 3 days after a westerly gale (often originating in a Provençal mistral event) passes through the area. They are usually based on the southwestern tip of Sicily (near Cape Granitola), although they often involve the entire southern coast of the island. They may also extend across the entire Sicilian Channel to the African coast (Piccioni *et al.*, 1988). In the eastern Mediterranean, important surface enrichment was induced by the annual Nile discharge prior to the closure of the Aswan High Dam and the control of the annual flood. This feature is now very restricted, and in the Levant Basin, offshore water is now of clarity equal to that of the central oceanic gyres. The pelagic fisheries of the region are much diminished. On the Israeli shelf, there is a subsurface algal bloom in bottom water over the narrow shelf, and here, the chlorophyll profile consistently shows a maximum value just above the bottom deposits (Townsend *et al.*, 1988). Whether this is induced by the episodic upwelling of nutrients across the shelf break, or whether it is due to regeneration of nutrients within sediments, and their subsequent release, is not clear.

There are also persistent surface chlorophyll features in the Alboran Sea related to upwelling, both along the east coast of Spain and around the Alboran eddies. A jet of Atlantic water along the Spanish coast induces geostrophic upwelling and a very shallow chlorophyll maximum. Upwelling is forced around the anticyclonic gyre that occupies the western Alboran Sea (Minas *et al.*, 1991). Farther east, a divergent front between Spain and the Balearics between southward coastal and northward offshore flow is associated with upwelling and chlorophyll enhancement (Estrada and Margalef, 1988). Cyclonic eddies generated at the coast within the Algerian Current are not wind induced but nevertheless force upwelling over sufficient spatial and temporal scales to induce algal growth and surface enhancement of the chlorophyll field (Millot, 1987). But they are not prominent in the chlorophyll images.

The copepods of the Mediterranean basin resemble those of the subtropical Atlantic, and since the sill depth excludes all abyssal biota that do not perform vertical migrations, the vertical ranges of many species extend much deeper here. Diel vertical migrations are unusually extensive: *Pyrosoma atlanticum*, *Clio pyramidata*, and *Cymbulia peroni* rise from 600–800 m to around 45–50 m at night in the western basin, for example.

The copepods of mass occurrence in the photic zone are typically *Oithona* spp., *Clausocalanus* spp., *Neocalanus gracilis*, and *Centropages typicus*, whereas *C. helgolandicus* is restricted almost completely to much greater depths. Several studies of species distributions in the Adriatic show that the northern shallow region is dominated by coastal species (high numbers, low diversity), and the southern, deep-water region by oceanic species (low numbers, high diversity).

***Black Sea*** Any account of the ecology of the Black Sea must start with the extraordinary fact that below some depth between 80 and 200 m, shoaler in mid-gyre but deeper near the coast, oxygen is absent and hydrogen sulfide concentrations increase progressively downward. At this redox interface, there is a "null zone", tens of meters thick, in which both gases are present at very low concentrations. At this interface, photosynthetic oxidation of $H_2S$ by sulfur-oxidizing bacteria (*Thiobacillus*) occurs, resulting in high concentrations of elemental sulfur. Other phototrophic, green (bacteriochlorophyll-β) sulfur bacteria, mostly *Chlorobium phaeobacterioides*, are also active in the redox interface. The interface between oxygenated water and water that contains $H_2S$ is, of course, the lower limit of zooplankton in the water column and of benthos in the sediments.

The pristine Black Sea plankton was overall much less diverse than that of the Mediterranean, including, for example, only 15 species of copepods compared with 304. It also contained a significant element of freshwater plankton over the northern continental shelves and in the Sea of Azov, where salinity is very low. The mesozooplankton assemblage in the 1940s comprised only 7 hydromedusae, 2 scyphomedusae, 1 ctenophore, 12 cladocerans, 15 copepods, 1 isopod, 2 chaetognaths, and 6 appendicularians. Thus, coccoliths, radiolarians, siphonophores, salps, and pteropods were absent. The distribution of mesozooplankton in the Black Sea was also unique (Vinogradov *et al.*, 1985). By day, the dominant species formed a layer of high abundance (2.5–38.0 g $m^{-3}$ wet weight) over a depth interval of only 5–20 m, coinciding exactly with the isopleths for 0.4–0.5 ml $O_2$ $liter^{-1}$ that occurs in the null zone (or redox interface) discussed previously. Following this isopleth, the actual depth of the zooplankton layer varied from 50 to 150 m and the distribution of organisms within this narrow layer was predictable with precision. The upper part was occupied by the tentaculate ctenophore *Pleurobrachia pileus*, the middle zone by late copepodite and adult *C. helgolandicus*, and the lower zone was occupied by the chaetognath *Sagitta setosa*. The small *Pseudocalanus elongatus* might also be present. At night, diel migration altered the pattern. The copepods and chaetognaths rose into the near-surface layers (0–20 m), whereas the ctenophores adjusted their depth only slightly, rising by about 20 m. Thus, during each diel cycle, the copepods ran the minefield of ctenophores passively awaiting their passage while at the same time avoiding the hunting chaetognaths, which tracked their migration. Seasonal vertical migration also occurred, but rather anomalously. The cold-water forms, such as *C. helgolandicus, P. elongatus*, and *Oithona similis*, occurred throughout the oxygenated layer in winter but descended to depths around 50 m during summer. In general, their depth distribution was constrained between the oxycline and the isotherm of their maximal temperature, usually 10–14°C. Following the precautionary principle, and lacking current information, I have written this paragraph in the past tense.

In the oxygenated water above the redox layer, a thermocline develops in summer, with which is associated a DCM due to normal autotrophic phytoplankton. However, within the oxygenated surface layer, nitrate has a very unusual profile. In the redox zone the nitrogen cycle is extremely active, and this layer is a sink for nitrate, nitrite, and ammonium. Consequently, rather than a nitracline, we find a subsurface nitrate maximum in the Black Sea. Below this maximum, nitrate is utilized by the activity at the redox layer and, above, by autotrophic cells of the euphotic zone (Sorokin, 1983; Murray and Izdar, 1989; Murray, 1991). This nitrate maximum consistently had values of about 7.0 μM in 1988 compared to only 3 μM in 1970. A similar increase in the integrated rate of primary production (100 to 300 mg C $m^{-2}$ $d^{-1}$) and chlorophyll concentration (0.2 to 0.4 mg chl $m^{-3}$) has also been observed between 1970 and 1990 (see later discussion). In past decades, a winter bloom has occurred rather generally between November and March in the Black Sea, replacing the summer community of coccolithophores and dinoflagellates with a surge of diatoms (Vedernikov and Demidov, 1991; Krupatkina *et al.*, 1991).

No wonder, then, that chlorophyll images show that the range (both temporal and spatial) of chlorophyll values is now relatively limited. Over the deep basin of the Black Sea, indicated surface chlorophyll concentration varies only from 0.5 to

1.5 mg chl $m^{-3}$ in winter (January) and, during the summer period (June), only from 0.25 to 2.5 mg chl $m^{-3}$. Obviously, this situation differs significantly from the pristine condition of the sea, prior to the massive eutrophication of recent decades and the catastrophic decline of herbivores. As late as 1980, the eddy field along the Anatolian coast carried strong filaments of high chlorophyll, observable in the CZCS images and quite distinct from offshore blue water. In fact, during early part of the period 1978–1986, many CZCS images revealed a response of the chlorophyll field to the dynamic instabilities of meanders within the Rim Current (Saur *et al.*, 1994). I am unable to observe this response in the SeaWiFS or MODIS images for 1998–2005.

Consensus appears to have developed concerning the forcing of these changes and, once again, diverse human interventions are the cause. The Black Sea has been massively fertilized by nutrients from agricultural and domestic runoff in recent decades and concurrently phytoplankton biomass has increased by an order of magnitude over the western shelves and by a factor of 3–4 over deep water, and has become taxonomically less diverse. Exotic ballast-water species (*Noctiluca, Pleurobrachia, Aurelia*, and *Mnemiopsis*) now dominate the zooplankton biomass to more than 90% by weight, having largely eliminated the omnivorous and herbivorous copepods and cladocerans whose consumption formerly balanced algal growth. It is suggested that prior to 1990, and the massive upgrowth of *Mnemiopsis*, the intense winter blooms that now characterize the plankton calendar were unknown. The continuous algal blooms of today's Black Sea are very different from the healthy state that existed until the 1960s, when algal biomass stood at much lower levels and apparently varied more strongly with the seasons. The present situation is extremely complex and dynamic: the newly arrived ctenophore *Beroe ovata* now feeds largely on the longer-established zooplanktivore *Mnemiopsis leidyi*. This predation appears, at least in the northern Black Sea, to have reduced the impact of *Mnemiopsis* on the indigenous zooplankton herbivores (Finenko *et al.*, 2003). How the dynamics of this new situation will work out is anybody's guess.

Discharge of nutrients from rivers, especially of the Danube and Dniester, is associated with massive local blooms of the newly arrived coccolith *Emiliania huxleyi*. These blooms are observable in satellite imagery as a dominant feature of the western part of the Black Sea, especially between May and July. In this region, the contributing organisms and the magnitude of blooms have progressively changed in recent years toward blooms dominated by nanoplankton, including coccoliths (Mihnea, 1997). More generally, the northwest shelf is a major source of nutrients for the southern Anatolian coast and for the open western basin. When transport around the rim current is interrupted, chlorophyll biomass on the southern coast diminishes rapidly. So far as it can now be discerned, the pristine phytoplankton calendar began in early winter with algal growth near the base of the euphotic zone that by November encompassed the entire euphotic water column, to be terminated in January or February by nutrient depletion and consumption by mesozooplankton.

There have been other major perturbations in the ecological balance of the pelagial in the Black Sea in recent decades. The stocks of pelagic clupeid fish have almost completely collapsed, and declared landings of *Engraulis* dropped by an order of magnitude between 1979 and 1990. To what extent this was due to overfishing in the classical sense, or to the concomitant collapse in zooplankton food organisms, has not been established.

## Regional Benthic and Demersal Ecology

Because the coastal regions were well investigated by the early benthic naturalists from Monaco, Naples, Split, and other marine stations, we know that the pristine benthic macrofauna of the shelves of the Mediterranean was in no way singular. That of the Black Sea followed the usual pattern, though with a somewhat attenuated cast of characters.

Thorson reviewed the characteristic macrobenthic species associations that are specific to mud, muddy sand, and sand deposits in these seas. He recorded that these communities had been identified on the coasts of North Africa and Israel, in the Adriatic, on the west coast of Italy, and on the shelf of the western Black Sea. So, on sandy deposits down to 50 m, we would have expected to find a *Venus* community and in shoaler water, or where the sand was very compact, a *Tellina* community. In muddy sediments from 10 to 100 m, the boreo-mediterranean *Amphiura* community occurred. Closer inshore, or where organic content of muddy sediments was very high, a *Syndosmya* association was usually found. Off the Danube delta, extensive midshelf and highly organic deposits have been described as "*Mytilus*-mud" from the abundance there of that bivalve.

The northern shelves of the Black Sea have a very gentle slope so that water as shallow as 30 m extends out to midshelf. Meadows of the red alga *Phyllophora* reach this depth and, at least until the 1950's, covered much of the shelf between the Crimean peninsula and the Danube delta. Shoreward, *Zostera* meadows are very extensive. It is possible that these macrophytes may be the origin of some of the high chlorophyll patches to be observed in satellite images in the northwestern Black Sea and the Sea of Azov.

In the shallow northern Adriatic the suspension-feeding activity of the bivalve mollusks of the benthic communities is sufficiently active to control phytoplankton biomass during periods when mixing extends to the bottom. In spring, after shallow stratification is established within the winter mixed layer, this balance is interrupted and if algal blooms extend into this period of the year, massive sedimentation events of dead cells may occur. Such events provide organic material to the sediments at a faster rate than deposit-feeding organisms can process it. Consequently, local anoxia and mass mortality of the benthic fauna may occur. The relationship between production of phytoplankton in the water column and the nature of the subjacent benthos has been investigated in front of the Danube delta. An inner zone, characterized by large sinking rates of organic particulate material, is dominated by deposit-feeding benthic infauna. A zone lying to the north, below a permanent anticyclonic gyre, is characterized by low sedimentation rates and is inhabited by a macrobenthos dominated by suspension feeding organisms: here, benthic remineralization rates are high. Finally, to the south of the delta, sedimentation rates are also low, as are rates of benthic remineralization, and here also the macrobenthos is dominated by suspension-feeding bivalves. Off the Nile delta, the rim current carries organic particles in the river effluent principally eastward and, consequently, biomass and productivity of benthic communities is almost twice as high to the east of the delta than to the west.

In these two basins it is difficult to recover the ecology of the fish populations in their pristine state. Invasions, introductions, and largely uncontrolled fisheries that have devastated the benthic habitat all make such a task very difficult. Taxonomically, of course, this province is a meeting place for Atlantic subtropical fish that we shall meet again on the NW African coast, and some elements of the boreal Atlantic fauna. To these are now added a Red Sea element of Lessepsian migrants, of which some have become abundant in the eastern Mediterranean to the point of supporting new fisheries.

## Synopsis

*Case 3—Winter-spring production with nutrient limitation—(MEDI only).* Although $Z_m$ undergoes a large boreal winter excursion (10 m June–July, 100 m February–March), the thermocline is illuminated from April to November. Rate increase in P is initiated in December and is sustained until annual maximum occurs in March–April (Fig. 9.12); the rate begins to decrease in May–June when $Z_m$ is very shoal, and declining rates are sustained until November. Chlorophyll accumulation begins much earlier and reaches maximal values between December and March, from which a decline is initiated even though the P rate continues to increase. Annual minimum values of integrated chlorophyll

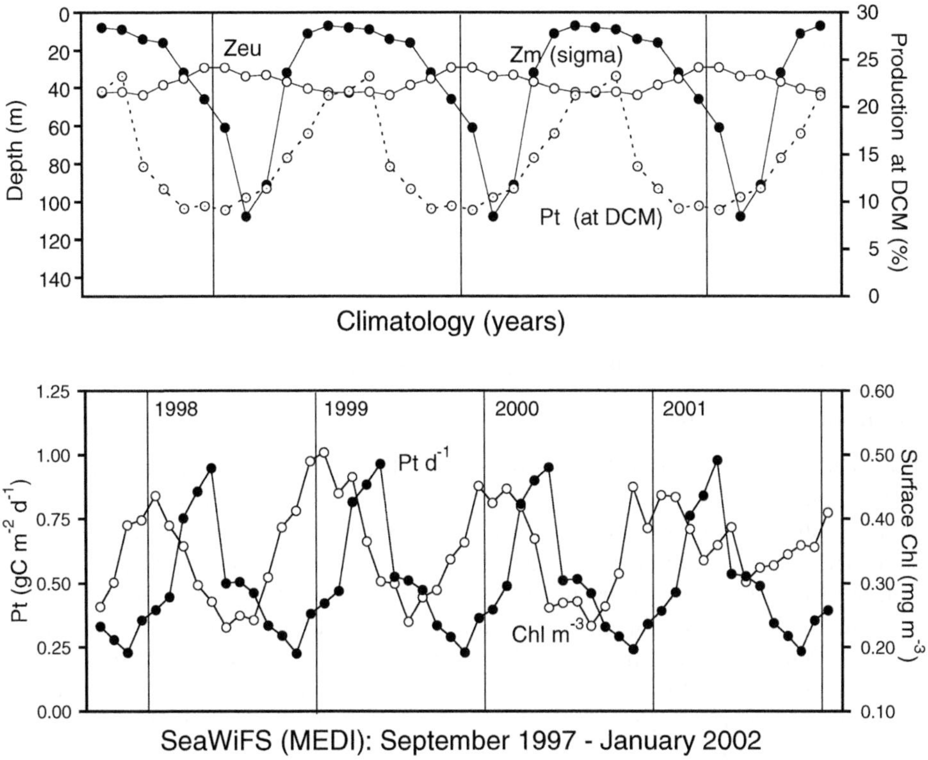


**Fig. 9.12** MEDI: seasonal cycles of monthly surface chlorophyll and depth-integrated autotrophic production for the years 1997–2002 from SeaWiFS data together with characteristic seasonal cycles of mixed-layer depths from Levitus climatological data and photic depths computed from characteristic irradiance and the archive of chlorophyll profiles discussed in Chapter 1.

biomass occur during boreal summer. A seasonal increase in herbivore biomass by factor of 3 between the December–January minimum and April–May maximum may be the typical seasonal cycle here, in which case chlorophyll accumulation is generally consistent with observed production and inferred consumption. In the Black Sea, of course, stratification is anomalous and while the P rate follows solar irradiance seasonality, there is no strong seasonality in chlorophyll accumulation.

# Atlantic Trade Wind Biome

Here we encounter for the first time in our descriptions of biogeochemical provinces the effect of distant physical forcing on algal dynamics that characterizes the tropical ocean, as discussed in Chapter 4. The intensification of the western jet current along the northern coast of Brazil and the Guianas by the trade winds over the western Atlantic during boreal summer requires that geostrophic balance be maintained by the tilting of the equatorial pycnocline about a meridional hinge line at 20–25°W. Mixed-layer depths increase in the west and decrease in the east, where a seasonal algal bloom may be induced by the local effect of moderate seasonal intensification of meridional winds. Thus, local biological responses to changes in mixed-layer depth are distantly forced, not locally forced as they are in the Westerlies and Polar biomes. The seasonal tilting of the density stratification of the equatorial Atlantic westward, to deepen the mixed layer in the west while shoaling it in the east, is a function of the dimension of the Atlantic basin and cannot occur in the Pacific. The equilibrium time of tropical oceans to low-frequency wind forcing is a function of how long the baroclinic waves require to pass along the equatorial wave guide. The Pacific basin (15,000 km) is too wide for this to occur seasonally as it does in the Atlantic (5000 km) although the seasonal changes in wind forcing are similar. In the Pacific, therefore, westward tilting of the pycnocline occurs—and can only occur—at the scale of ENSO events, with their long-sustained modification of regional wind stress over the ocean (Longhurst, 1993). These facts have fundamental consequences for the seasonality of phytoplankton growth—and all that stems therefrom—in the tropical oceanic biome.

## North Atlantic Tropical Gyral Province (NATR)

### Extent of the Province

The North Atlantic Tropical Gyral Province (NATR) comprises that part of the North Atlantic gyre that lies south of the subtropical convergence that runs zonally across the ocean at about 30°N (see North Atlantic Subtropical Gyral Province). The southern boundary of NATR is the limit of westerly flow along the thermocline ridge associated with the North Equatorial Current (NEC) at about 12–14°N. The western boundary is taken to be the edge of the coastal boundary biome seaward of the Antilles and the Bahamas. NATR thus includes the offshore segment of the Canary Current south of the Canaries, the flow of this into the NEC, and then the continuation of this flow into the western limb of the gyre. The only shoal water in this province is the platform of the Cape Verde Islands.

### Defining Characteristics of Regional Oceanography

Here, we lack the wealth of information that is available for the provinces to the north and for the equatorial ocean to the south. Perhaps because NATR consistently has the lowest surface chlorophyll of the North Atlantic—well seen as a big blue hole in seasonal chlorophyll images—it has attracted little attention from biological oceanographers. It was, for instance, ignored by McClain *et al.* (1990) in their CZCS-based models of phytoplankton growth in 11 boxes intended to characterize the regional ecology of the North Atlantic. Like many other provinces that follow in this and later chapters, we shall just have to do the best we can with what we have.

It is in this province that we first encounter the open ocean regions where phytoplankton biomass and productivity are minimal at the global scale. Although the usage is incorrect oceanographic terminology, these regions are often known as the "subtropical gyres," as by McClain *et al.* (2004), who specify the seasonally changing size of each "gyre" by reference to chlorophyll concentrations of $>0.07$ mg m$^{-3}$. This isoline defines a region that is very much smaller than the subtropical gyres with which we are familiar in descriptive physical oceanography; this, of course, in the Atlantic includes flow around both boundary currents together with flow in the Azores and the North Equatorial Current (NEC). I believe that this has led to some confusion in the past in how biological oceanographers think about subtropical gyres. Although NATR is, ideally, one of the regions studied so helpfully by MacClain *et al.* (see later discussion), it will not be useful to restrict our attention to the much smaller box used by those authors.

The NATR lies below the trade-wind belt of the North Atlantic, separated from the westerlies by the Azores High, which is often expressed as a series of high-pressure cells lying southwest-northeast across the ocean at about 30°N and is best developed in boreal summer. Thus, NATR comprises the ageostrophic, 200-m-deep, wind-driven flow (Fiekas *et al.*, 1992) around the southern half of the anticyclonic gyre of the North

Atlantic; analysis of the surface wind field of the North Atlantic demonstrates that this province corresponds to a zone of negative Ekman upwelling flux, of order $-10$ to $-20 \times 10^{-7}$ m $sec^{-1}$. Both to the north and south of NATR, Ekman vertical flux takes positive sign (McClain and Firestone, 1993) and there also the seasonal variability of surface dynamic topography is greater. Finally, changes of sea surface temperature between Niño and Niña conditions are significantly lower in the quiet NATR than in the more active regions to north and south.

The critical characteristics of the province of interest to us are simple: in the western part of the province, both sea-surface elevation and pycnocline depth take their maximum values and, from here, the pycnocline slopes upward toward the edges of the province; to the north, a relatively steep slope occurs at the STC, discussed for the NAST province, whereas from May through December, a thermocline ridge lies along 10°N, trending somewhat NE-SW and deepening to the west (Hastenrath and Merle, 1987). During boreal winter, when the NECC become discontinuous west of 20°W, separation is difficult between westward flow of the NEC and of the equator-crossing South Equatorial Current (SEC), though the conjunction still occurs at about 10°N.

In NATR, mixed-layer depth varies seasonally, but not as a result of wind mixing as in higher latitudes beyond the subtropical convergence; it is rather a response to changes in the regional wind field, and hence to baroclinic adjustment and changes in Ekman pumping. Essentially, here, pycnocline topography reflects transport so that a deep thermocline trough lies southwest-northeast across the province at about 20–25°N in most months, with greatest mixed-layer depths at 70–80 m in winter (December–May) and rather shoaler (25–40 m) during the rest of the year.

The vertical displacement of the pycnocline carries with it the nutricline and, as McClain *et al.* emphasize, there is a concurrent horizontal transport of nutrients peripherally around the subtropical gyre. It is to be noted that the location of the greatest pycnocline depth, and hence the deepest nutricline, and the highest sea-level anomaly differ by a significant distance, although both occur in the western part of the province. The province is distinguished from the WTRA to the south by a zonal band of maximum salinity (>37 ppt) that lies across the entire ocean, clearly differentiated across a salinity gradient at 10–15°S from the less saline (<36.25 ppt) equatorial water of the WTRA that carries a seasonal Amazon signal (Dessier and Donguy, 1994).

The Cape Verde Islands (17–19°N) lie within the eastern limb of the anticyclonic gyre. Their downstream eddies therefore modify the thermocline topography of the offshore Canary Current as it passes to the southeast, away from the African continent (see NAST Province for a discussion of this process at the Canaries).

## Response of the Pelagic Ecosystems

There is little comprehensive information on the mixed-layer nitrate field, but values appear to be uniformly low (0.2 μM) at all seasons (Wroblewski, 1989; Glover and Brewer, 1988, ICITA Atlas). Higher values in wintertime (<2.0 μM) occur along the southern boundary of the province at about 10°N, probably attributable to distant-field effects of the zone of Ekman suction in the NECC of the adjacent WTRA (see later discussion). Unfortunately, the repeated Atlantic Meridional Transect passes through the province at its extreme eastern end and the "seasonal" signal for chlorophyll observed in AMT data, and specifically attributed to this province, is probably induced by intermittent transit of the survey ships across parcels of eutrophic surface water upwelled in the Canary Current.

The NATR has a consistently low and rather featureless surface chlorophyll field with a seasonal cycle of very small magnitude and much irregularity, despite the significant seasonal change in solar irradiance (500–800 W $m^{-2}$ in February and >1000 W $m^{-2}$ in June). In fact, satellite chlorophyll, integrated over the entire province, shows greater annual variability than seasonality: an unusually strong regional maximum occurred in late 1998, whereas in subsequent years the seasonal maxima were much weaker. It is only during the second half of the year, when the mixed layer is shoaler than the depth of 1% surface irradiance by about 20 m, that higher vertically integrated production rates are computed. The satellite data support this result because there is some evidence of a progressive accumulation of chlorophyll over most of the area of the province during this period. This conclusion is not invalidated by the distant-field effect of the high chlorophyll in the NECC that occurs in winter along the southern edge of NATR, and dominates the integrated NATR chlorophyll field.

Perhaps clearer is the seasonal analysis of McClain *et al.*, who determined that the areal extent of chlorophyll values >0.07 mg $m^{-3}$ changes seasonally, being minimal in December–January and maximal in June–September, following the irradiance cycle. McClain *et al.* enquire how nutrients can be supplied to the euphotic zone to support the observed productivity in such regions as this, where downwelling is characteristic and nutriclines are deep. They suggest that the probable mechanism is surface Ekman lateral flux and they compute for this region (and the four other similar subtropical gyral regions) that nitrate and phosphate horizontal fluxes in the upper 50 m are net positive into the gyre interiors. Here, in the North Atlantic, 85% of this lateral flux originates across the southern boundary from the NECC. Despite this supply, the regional pattern of phytoplankton biomass appears to be dominated by the local effects of nutricline depth; as McClain *et al.* point out, the region of minimal surface chlorophyll corresponds not with the focus of the gyre at the point of maximum sea level elevation but >1000 km distant, where pycnocline depth is maximal.

Bulk analysis of chlorophyll pigments at the 1991–92 French "Eumeli" mesotrophic station in NAST and the oligotrophic station in NATR show clearly the distinct characteristics of tropical phytoplankton. In NATR, low bulk chlorophyll values had a relatively large zeaxanthin (cyanobacterial) component and a relatively small fucoxanthin (diatom) component. But, the peridinin (dinoflagellate) component had a similar range of values in the two provinces. Microzooplankton biomass and abundance (~1.0 mg C $m^{-3}$) are as small as anywhere in the oceans and 5–10 times smaller than in the neighboring subtropical provinces.

In case we should become too confident that the NATR (and similar trade-wind provinces in other oceans) are highly predictable, relatively invariant oceanic environments, the now-routine repetitive observations of the sea surface chlorophyll field by satellite imagery may have some surprises for us. Consider the case of surface chlorophyll images of the NATR in the fall of 2001 (see color plate 6). Here, we are surprised to see a very large dendritic region of chlorophyll enhancement (<0.5 mg chl $m^{-3}$) about the size of Spain, lying far offshore in the oligotrophic region, centered about 32°W, and at the latitude of Cap Blanc. The like occurs in no other monthly image during the period 1997–2003. One might suppose it to have been a large parcel of upwelled water that had taken an unusual offshore trajectory except for the fact that in the previous month one can just discern the first indications of chlorophyll enhancement at the same location. The following month, it is entirely absent. This feature superficially resembles the midocean bloom that I find in the South Indian Ocean (see ISSG province), which exhibits the same mesoscale swirls and rings, translates consistently eastward, and can be observed over a period of several months. But the October 2001 feature was a far briefer phenomenon than the other, which has a largely predictable return period each year, in greater or lesser intensity. Further, as can be seen from the images, any correlation between surface chlorophyll swirls and underlying mesoscale eddy features in the NATR bloom is much weaker than that of the Indian Ocean.

During boreal summer, the initiation of retroflection of the NECC from the coast of northern Brazil (see GUIA, later) occupies along a greater length of that coast than when the feature is fully formed; at that time, retroflection off Demerara (at about 8°N) may induce a major chlorophyll enhancement in the extreme southwestern corner of NTRA, and it is such episodes that are probably responsible for the anomalous peaks in regional chlorophyll biomass obtained within the "statutory" boundaries of this province.

It was in this region that the "Typical Tropical Structure" (TTS) for profiles of phytoplankton and associated measures was explored and described by Herbland and Voituriez (1979); the key characteristic of the TTS is, of course, the manner in which the depth of the deep chlorophyll maximum is determined by pycnocline, nutricline, and irradiance, and how the profiles of heterotrophic bacteria and of herbivores match that of autotrophs in a manner suggestive of consumption and recycling *in situ* of organic material. These relationships have been best worked out in the eastern Pacific and are discussed later.

Apparently, we know almost nothing of the dynamics of herbivore consumption and its seasonality in this province and can only speculate about the mechanism that leads to the weak accumulation of chlorophyll in winter when the primary production rate appears to be at a minimum. The NATR is the southern part of a region of very sparse zooplankton that occupies the central part of the North Atlantic gyre; integrated biomass from 0 to 100 m, indicated by profiles obtained with bioluminescence probes, is significantly lower (always <100 mg $m^{-3}$) than in the WTRA to the south where a zone of high values stretches across the ocean along the NEC/NECC ridge (Piontkowski *et al.*, 1997). Incidentally, these 2000-odd Soviet observations over the North Atlantic confirm the basin-scale model of North Atlantic plankton dynamics offered by Wroblewski, Sarmiento, and Flierl (1988); this model solution, based on mixing rates, nutrient profiles, and mixed-layer depths, indicates a zone of very small phytoplankton and zooplankton biomass in NAST and NATR, clearly differentiated from higher values lying across the ocean to the south of NATR.

The French "Eumeli" oligotrophic station showed clearly that macrozooplankton and nekton numbers in NATR were 5–10 times lower (263 ind $m^{-2}$, 0–965 m) than at the mesotrophic and eutrophic stations in the Canary Current (Andersen *et al.*, 1997). The size-selected assemblage used by these authors comprised siphonophores (*Chelophyes*), pteropods (*Clio*), chaetognaths, euphausiids (*Stylocheiron, Euphausia*), and fish (*Cyclothone*). In NATR, compared with some other trade-wind provinces from which we have zooplankton profiles, the pycnocline is relatively weak and, I think in consequence, the vertical distribution of individual organisms is rather dispersed. Daytime residence depths of diel migrants were found to be more various than the depths entered at night, which are consistently within the upper 50–75 m. As usual, the bathypelagic fish *Cyclothone* remains at depth day and night and, also as usual, their species partition space among themselves: *C. braueri* lies 200 m shoaler than *C. acclidens*. Euphausiids are active diel migrants, with the exception of the genus *Stylocheiron*, which resides day and night at species-specific depth ranges within the upper 200 m.

This province, together with the Gulf of Mexico and the Gulf Stream, is within the ambit of the summer migrations of tropical yellowfin (*Thunnus albacares*) and skipjack (*Katsuwonus pelamis*) tuna. That is to say, these large predators were abundant here in the summer months until at least the 1950s and 1960s, but are now reduced to relatively very small populations.

## Synopsis

*Case 4—Small-amplitude response to trade wind seasonality*—$Z_m$ based on the density criterion shows the effects of winter mixing in the northern part of province (regionally, 25 m June–October, 50 m in January–February). $Z_{eu}$ is consistently at between 50 and 60 m, so thermocline is permanently illuminated. Productivity has very weak seasonal

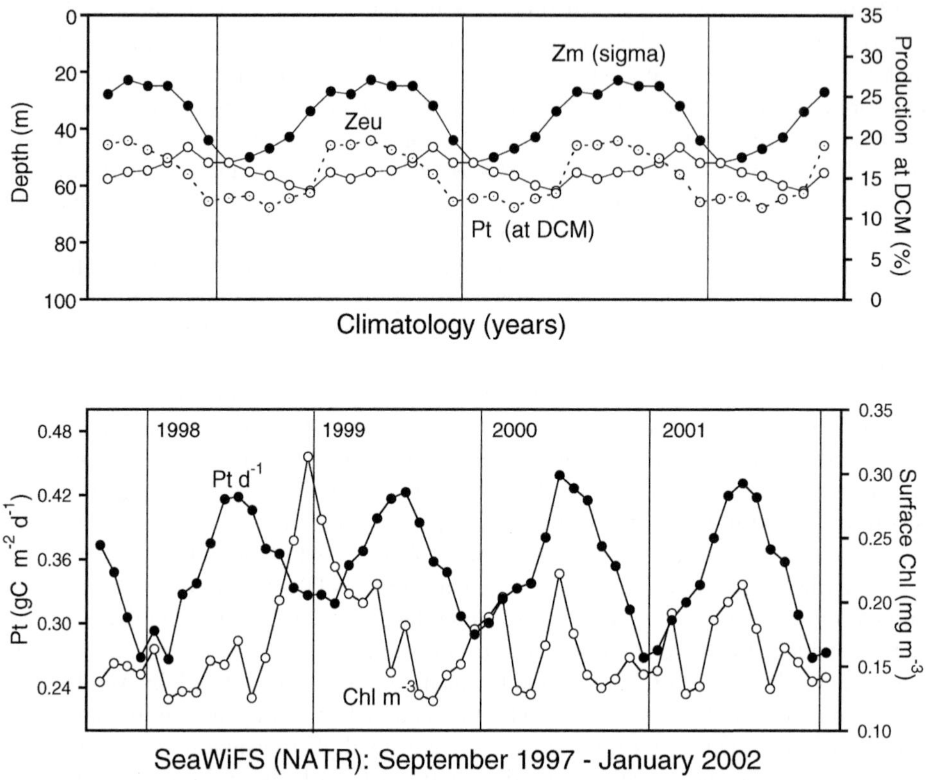


**Fig. 9.13** NATR: seasonal cycles of monthly surface chlorophyll and depth-integrated autotrophic production for the years 1997–2002 from SeaWiFS data together with characteristic seasonal cycles of mixed-layer depths from Levitus climatological data and photic depths computed from characteristic irradiance and the archive of chlorophyll profiles discussed in Chapter 1.

change, being somewhat enhanced in boreal summer, corresponding to the period of deepest mixed layers and strongest irradiance at the surface. Seasonality in the dynamic range of chlorophyll is unclear, biomass varying from 0.12 to 0.32 mg C $m^{-3}$ with annual maxima occurring at all seasons (Fig. 9.13).

# CARIBBEAN PROVINCE (CARB)

## Extent of the Province

This province comprises the Gulf of Mexico and the northern Caribbean Sea, bordered by the American coastline from Florida to Nicaragua and by the Antilles arc from the Bahamas to Barbados (15°N) where an arbitrary line is drawn across the Caribbean to isolate the southern Caribbean shelf, resembling the Guianean shelf ecologically, from CARB. The province is divided by the constriction of the Yucatan Channel into two basins, Gulf of Mexico and northern Caribbean Sea. It might be more logical to consider these to be two different provinces, but their circulation is so integrated that it is better to recognize their similarities rather than their differences. The name used for this region is, I acknowledge, not logical: it would have been better in the first place to have used "Intra-Americas Seas Province," following current oceanographic usage; but it would be counterproductive to change it now, since many readers are familiar with the existing name.

## Continental Shelf Topography and Tidal and Shelf-Edge Fronts

The entire coastline has a significant shelf width, from the West Florida Shelf (250 km) to the equally wide Campeche Bank off Yucatan. The northern coast is dominated by the bird's-foot delta of the Mississippi River, so that riverborne silt nourishes the sedimentary regime of the northern and western coasts of the Gulf of Mexico. The northern coast is also characterized by the existence of an almost continuous series of coastal lagoons situated behind bar beaches; the existence of these extensive features has consequences for this entire shelf. The DeSoto canyon separates the sedimentary regime of the northern coast from the coralline sediments that dominate the West Florida Shelf; these sediments may serve as a model for depth zonation on similar coastlines elsewhere. Pure quartz sand lies outside the beaches with foraminiferan silt prominent in shallow water, and quartz-shell sand mixtures dominate a little deeper. At midshelf depths, there is a band of pure shell sand while, deeper yet, a zone of sand formed of particles of calcareous algae gives way at about midshelf to sand formed almost entirely of oolite particles. At the break of shelf, almost everywhere in this province, there is a "mud line" at the limit of a highly organic, sedimentary regime, the origin of extensive slumping of sediments into deep water.

The low-lying coastline from western Florida to southern Yucatan is largely free of coral formations because of limiting winter temperatures and excessive runoff of freshwater and deposition of riverine material. Fringing reefs are abundant only in the Caribbean, whereas an offshore barrier reef occurs along the eastern coast of Yucatan and extends south onto the Mesquite Bank. Several carbonate banks of continental shelf dimension and depth lie in deep water: the Campeche and Bahamas Banks and the Mesquite Bank on the eastern coast of Honduras. The terraced outer platforms of these bear calcareous and pellet muds, with oolites and coral-algal fragments in very shallow water. These white, calcareous sediments are forced into waves and ripples by tidal streams. The Antilles islands are typically endowed with leeward reefs on their western coasts.

The northern coasts are close to the critical latitude for the formation of diurnal tides so that tidal streams and fronts are relatively weak through the province. Storm surges driven by the hurricanes typical of the West Indies region are, on the other hand relatively common and devastating to coastal ecosystems.

## Defining Characteristics of Regional Oceanography

The major source of energy for water motion in this province is wind; the seasonal migration of the ITCZ over South America is (relative to that over Africa) rather small, so easterly Trades blow almost constantly over the Caribbean, although more strongly during boreal spring and summer. At this season, coastal upwelling is induced along the northeast coasts of the Yucatan peninsula by the conjunction of high current speed (of order 2.5 m $sec^{-1}$) and strong winds from the southeast.

A very important component of wind forcing in this province are the "weekly easterly waves" that may develop into full-blown hurricanes, propagating generally northward across the province; in such events, surface wind stress is twice that associated with normal trade winds. This stress creates transient circulation, open ocean upwelling, and storm surges and may, in exceptional circumstances, reduce surface temperature by as much as 3–4°C and cause a 30- to 40-m deepening of the mixed layer (Mooers and Maul, 1998). The wind regime of the Gulf of Mexico is characterized by winter westerlies along the northern shelf, although their effects are less than those of the quasiweekly "northers" that cool shelf waters seasonally: in January, the 30-m temperature over the shelf here may be taken down to about 20°C by these winds. Nevertheless, mixed-layer depths are not strongly variable across the province although the circulation is sufficiently complex as to make the derivation of mean mixed depths unsatisfactory: be that as it may, the regional mean mixed layer appears to be of order 20 m in July, 40 m in January.

Circulation within this region is dominated by the throughflow of water from the equatorial currents of both North and South Atlantic (NEC, SEC) through the passages of the Lesser Antilles, and the delivery of this water to the North Atlantic in the Florida Current. A useful account of flow-through circulation is that of Johns *et al.* (2002).

The water entering the Caribbean through the Lesser Antilles arc is largely of NEC origin, of high clarity, and with little vertical structure in the upper 100 m (Borstad, 1982), and this forms about half of the total influx. Nevertheless, the satellite images show that in many months when active retroflection of the NBC is occurring, the water mass immediately outside the Lesser Antilles is highly eutrophic and has been retroflected from the coast. The images suggest that, in this situation, water containing significant algal biomass enters the northern Caribbean far seaward from the mainland shelf. Water from the Guiana Current, which originates in the SEC, is turbid and carries much DOM and passes along the southern coast of the Caribbean (see GUIA) and thence, after losing much of this material, flows northward along the Nicaraguan coast as the Yucatan Current and so into the CARB province. These multiple sources for flows of surface water through this province result in a rather complex pattern of circulation that is detectable in sea surface topography and in surface chlorophyll. Flow through the various passages in the island arc induces eddies and vortices, initially of about 100 km diameter but increasing in dimension downstream (Kinder, 1983); indeed, the ubiquity of mesoscale eddies throughout this province is one of its principal characteristics.

Flow through the Yucatan Channel, and then along the northeast edge of the Campeche Bank briefly forms a true western boundary current that detaches from topography and passes northward across the Gulf of Mexico over deep water as the Loop Current. Once detached, it becomes unstable and forms an anticyclonic loop around the eastern Gulf, passing from the tip of the Campeche Bank to the western edge of the Florida shelf. Flow then follows the shelf edge south to the Florida Straits, whence it emerges as the Florida Current and the Gulf Stream. The Loop Current penetrates furthest into the western Gulf when it separates from the Campeche Bank at a relatively western position (Molinari and Morrison, 1988). When the Loop Current penetrates sufficiently far to the north, the effluent of the Mississippi may become entrained in the flow.

The dynamic shifts of the Loop Current induce the pinching-off of very large meanders that become major features of the circulation and nutrient budget of the Gulf of Mexico. These warm-core, anticyclonic rings are generated at about 6- to 18-month intervals from the Loop Current, mostly in winter (Vukovich, 1988), and move into the western Gulf, subsequently to pass around the northern coast as they age. They are 100–300 km in diameter, have a lifetime of about 1 year, and may interact dynamically with the continental shelf edge (Vukovich and Waddell, 1991). After such a feature is pinched off, the flow of the Yucatan Current briefly passes directly toward the Straits of Florida. Upwelling of subpycnocline water in the Gulf of Mexico, due to instabilities in the Loop Current, occurs at the shelf edge and along the outer shelf in many places throughout the province, especially off northern Yucatan, but also around Campeche Bank, and southwest of Florida (Mooers and Maul, 1998).

Cold-dome cyclonic eddies (80–120 km) occur on the eastern/southern side of the Loop Current, evolving into tongues or quasipermanent meanders of cool water (Vukovich and Maul, 1985) that are a characteristic feature of the eastern Gulf. Over the shelves of the northern Gulf of Mexico, cyclonic cells are expected to occur for most of the year, but will be interrupted by extreme wind events; some may be detached from the isolated warm-core rings.

## Response of the Pelagic Ecosystems

The chlorophyll images show several regions of permanently high pigment concentration, probably due to benthic processes, that must be distinguished from features in the chlorophyll field due to phytoplankton biomass. The largest of these permanent anomalies represents the underwater platform of the Bahamas Bank, overlain with extremely clear oceanic water, which appears as a high-pigment anomaly that exactly matches its outline. This is surely due to symbiotic and benthic algal chlorophyll showing through a few meters of the clearest ocean water. Similar patches of apparently benthic chlorophyll are associated with the Florida Keys and several regions along the southwest coast of Cuba, especially around the Isle of Pines.

The regional, seasonal cycle of productivity indicated by satellite data is of small magnitude, but is not simple: there appear to be chlorophyll maxima in both winter and summer: in each, the indicated biomass is between 0.4 and 0.6 mg chl $m^{-3}$. Relative productivity, integrated by various models, follows a simpler pattern following the cycle of irradiance, ranging from 300 to 500 mgC $m^{-2}$ $mo^{-1}$. The serial images enable a distinction to be made between the two seasons of maximal surface chlorophyll biomass: in summer, there is a very clear difference between oligotrophic regions having very clear surface water and regions of chlorophyll enhancement, whereas in winter this distinction is more obscure, because the entire CARB province then experiences uniform, but rather slight, chlorophyll enhancement.

A constant feature of the chlorophyll field in the Gulf of Mexico reflects the offshore entrainment of shelf water and the upwelling of nutrients induced by the anticyclonicity of the Loop Current in the eastern Gulf (Paluskiewicz *et al.*, 1983). Anticyclonic eddies significantly modify the nutrient budget as they propagate into the western Gulf because they have, of course, a bowl-shaped pycnocline topography so that nitrate isopleths are brought up into the lighted zone around their perimeters (Walsh *et al.*, 1989). Nitracline doming in cyclonic eddies shed from the Loop Current into the eastern Gulf also produces features in the chlorophyll field. Thus, both families of rings exhibit appropriate chlorophyll enrichment, observable in satellite images (Yentsch, 1982; Salas de Leon and Monreal-Gomez, 1986; Trees and El-Sayed, 1986).

Hobson and Lorenzen (1972) explored the deep oligotrophic chlorophyll profiles that are characteristic of the Gulf of Mexico. They noted the significant nonuniformity of the depth of the pycnocline, even at the same season, due to complex flow structure at the mesoscale. This was one of the earliest observations of how maxima of chlorophyll, phytoplankton carbon, and microzooplankton carbon coincide within the pycnocline and above a nitracline. As has subsequently been found to be the general case, when the pycnocline was deeper than about 100 m, the DCM was extremely weak.

Associated with the oligotrophic profiles, the vertical distribution of mesozooplankton follows the usual pattern. The >150-μm fraction is dominated by copepods (90% numerically and >50% of biomass) that are concentrated in the mixed layer; these are a typical suite of warm-water genera (*Clausocalanus, Euchaeta, Scolecithrix*, and *Nannocalanus*). Diel migrants (especially *Pleuromamma* and *Sergestes*) shuttle between daytime depths of 300–400 m and the photic zone at night (Hopkins, 1982). Though this study is not explicit in this regard, we may assume that the usual relationship exists between residence depths of mesoplankton and the features of the phytoplankton profile.

At upwelling sites along the coast of northern Yucatan, the mesozooplankton respond in a predictable manner to upwelling episodes, as they do at larger scale in eastern boundary current upwelling zones (Suárez-Morales, 1995). Bray-Curtis diversity is relatively low and biomass relatively high in regions of strong upwelling and high chlorophyll, where the dominant copepod (*Temora stylifera*) may reach abundances of >20,000 ind $m^{-3}$. Beyond each upwelling cell, diversity increases and biomass decreases, and both neritic and oceanic species mingle in this dynamic coastal region.

Several recent investigations have demonstrated the effects of cyclonic and anticyclonic eddy features on pelagic ecology in this province. The distribution of zooplankton and micronekton biomass reflect nutrient enhancement, so that relatively strong algal growth occurs in cyclonic eddies and in the confluences of flow within eddy pairs. This phenomenon translates to the highest trophic levels, so that whales and other cetaceans are not only concentrated along the shelf break (where we would otherwise expect them to be most abundant) but also associated with cyclones and flow confluence between eddy pairs. Such a preferential distribution has also been related to an observed relationship between the distribution of zooplankton biomass and of cephalopod paralarval numbers. Cephalopods are, of course, a mainstay of the diet of the sperm whales, pygmy sperm whales, pilot whales, and other cetaceans.

## Regional Benthic and Demersal Ecology

Although coral reefs will not directly concern us, it must be noted in passing that the coasts of the Bahamas and of the Antilles islands are largely characterized by this benthic habitat. Throughout the Gulf and Caribbean, wherever the organic content of deposits is low, patches and banks of deeper-water corals occur down almost to the shelf break, as well as in the more familiar shallow-water reefs. This is quite different from the situation in the eastern tropical Atlantic, as we shall see. The Atlantic reef fauna, which has its most generous expression here in CARB, is still relatively depauperate: there are only 35 coral species in 26 genera in the Atlantic, whereas in the Indo-Pacific there are 700 species in 80 genera. There is significant endemism at the generic level in the Atlantic and significant regional differences between the reef fauna of the Caribbean and of isolated reefs along the coast of Brazil. In passing, however, it should be noted that the Caribbean reef systems are under severe pressure from exotic marine organisms, including pathogens, perhaps introduced in ballast water of ships passing the Panama Canal. The now-classical case is that of the die-off of the previously abundant sea-urchin *Diadema* spp., associated with the identification of a lethal pathogenic bacterium (*Clostridium*). These sea urchins were the dominant algal grazers on reef systems and, in their absence, macroalgae have increased enormously in coverage and size, smothering corals to depths of 10–15 m (Richards and Bohnsack, 1990). It is perhaps not accidental that this unusual mortality of urchins was first noticed in the vicinity of the canal itself.

Walsh (1988) reviewed the role of the nutrient-laden effluent of the Mississippi River in the productivity of the northern Gulf of Mexico. The Mississippi water has a nitrate content of $<150\,\mu$g-at. $NO_3$ liter$^{-1}$ during the spring floods. He concluded that only about 21% of the 250–350 gC m$^{-2}$ y$^{-1}$ phytoplankton biomass produced over the Texas-Louisiana continental shelf is taken up into the pelagic food chain. Of the remaining 79% of phytoplankton biomass that sinks into the benthic environment, more than half goes directly to burial or to export into deep water by slumping at the shelf edge, and only about 80 gC m$^{-2}$ y$^{-1}$ enters the benthic food chain. We shall find that a similar conclusion was reached, on a much smaller scale, for the fate of the phytoplankton biomass produced in the discharge of a West African river. Despite this, there is a very direct relationship between river discharge and biological productivity on the adjacent continental shelf, and it is in the Gulf of Mexico and Caribbean that this relationship has been as well studied as anywhere else. Not only does a spatial correlation exist, but there is also a positive correlation between annual discharge rate and the productivity of the associated neritic fish stocks whose biomass is dominated by species of rapid growth (Deegan *et al.*, 1986).

Two features appear to be important in forcing the ecology of benthic communities and those of demersal fish: the depth of the thermocline that underlies the tropical surface water mass, which here lies near the break of slope, and also the consequence of river effluents. The community structure of the benthic infauna has been little investigated in this region, despite the great commercial importance of the macrobenthic epifauna, dominated by *Penaeus setiferus* and *P. aztecus* and the swimming crabs *Callinectes* spp. As noted earlier, the De Soto canyon isolates the muddy "shrimp ground" of the Mississippi sedimentary regime from the west shelf of Florida, populated by a variety of hard-ground, warm-water epibenthic species.

In tropical seas, the nature of the fish fauna is dominated both by deposit type and by water mass. Because the tropical surface water extends so deep on the shelf here, there is no equivalent in the Caribbean to the passage of a subtropical sparid fauna through the tropical region, deep on the shelf and in cooler water, as occurs in the eastern Atlantic. The turbid waters and muddy deposits of the coastal regions are populated especially by catfish, threadfins, and small sciaenids and everywhere it is the nature of the deposits that chiefly determines the characteristic species associations of fish: softer deposits are inhabited by a fauna of large sciaenids (*Micropogon, Cynoscion*), grunts (*Haemulon*), and threadfins (*Polynemus*), much more diverse than that of the eastern Atlantic. On sandy deposits and stony bottoms species associations occur in which lutjanids, sparids, and pomadaysids are important. On the Campeche Bank, for instance, three species associations having different core species of these families have been found to occur in different microhabitats that are distinguished by deposit type and relative depth.

Here, we first meet the mass occurrence of the phytophagous, filter-feeding clupeid fishes that we shall find are typical of warm seas where strong blooms of large diatoms occur. In the case of the CARB province, this niche is occupied by the Gulf menhaden (*Brevoortia patronus*), especially typical of the northern coast of the Gulf, where local diatom blooms associated with river discharges are rather widespread. Menhaden have a very fine filter-basket formed by their gill rakers, and diatoms such as *Coscinodiscus* dominate their diet. Rather extensive migrations are undertaken to maintain the fish in regions of appropriate blooms and when these are not accessible, they resort to a maintenance diet of microzooplankton. The stocks of *B. tyrannus* of the eastern coast of North America enter the Gulf of Mexico to spawn. Unlike the ecologically similar *Ethmalosa* of the eastern Atlantic, *Brevoortia* does not occupy the lowest latitudes. A very large biomass is maintained and has supported major industrial fisheries.

### Synopsis

*Case 4—Small-amplitude response to trade wind seasonality*—The density-dependent mixed layer undergoes weak boreal winter excursion from <20 m in June–September to 35 m in February while $Z_{eu}$ remains 45–50 m in all months; the thermocline is therefore permanently illuminated. Productivity follows the irradiance cycle with a shoulder (as so frequently is observed elsewhere) on the descending limb of the curve. Seasonality in chlorophyll accumulation is complex and somewhat variable between years, with two regional maxima in boreal summer, the second corresponding to the shoulder on the productivity curve. However, the accumulation of chlorophyll in data aggregated for the entire province (Fig. 9.14) is largely influenced by the late summer events in the south-eastern Caribbean, when the highly pigmented water of the Orinoco discharge plume passes into this province, and this may account for the second "chlorophyll" peak.

## Western Tropical Atlantic Province (WTRA)

### Extent of the Province

The WTRA comprises the tropical Atlantic west of 15°W, lying between NATR to the north and SATL to the south. The northern boundary is the southern flank of the NEC

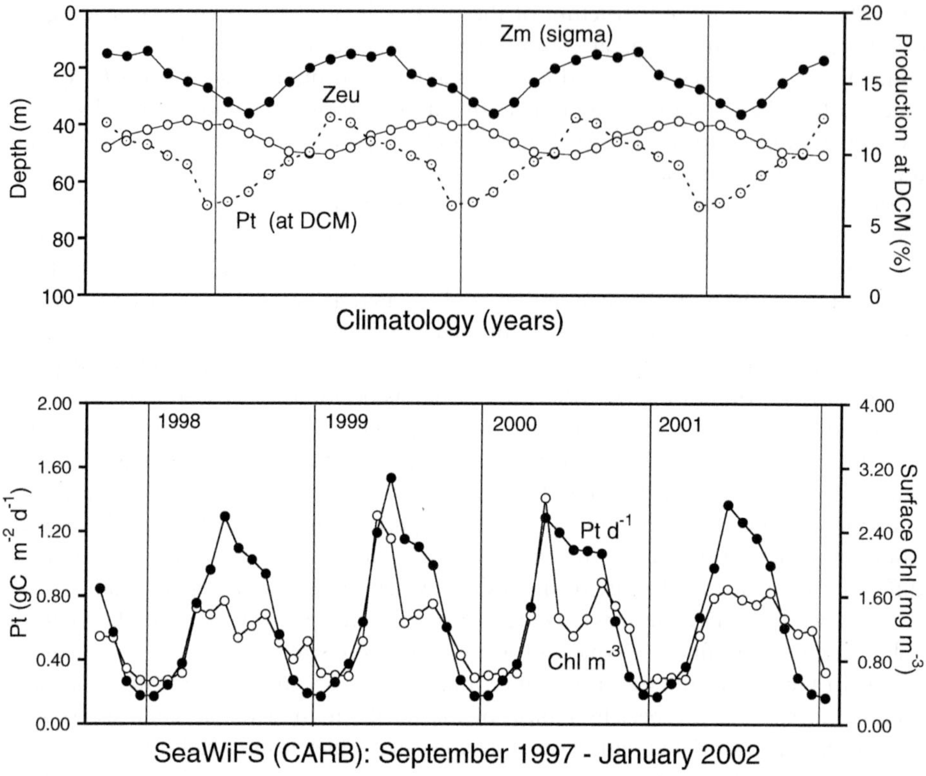


**Fig. 9.14** CARB: seasonal cycles of monthly surface chlorophyll and depth-integrated autotrophic production for the years 1997–2002 from SeaWiFS data together with characteristic seasonal cycles of mixed-layer depths from Levitus climatological data and photic depths computed from characteristic irradiance and the archive of chlorophyll profiles discussed in Chapter 1.

across the Atlantic at about 12–14°N. In the south, the boundary is set at 5°S at 15°W, diagonally northwest across the ocean to the equator at 40°W.

## Defining Characteristics of Regional Oceanography

Our knowledge concerning physical processes in the WTRA is rich, far richer than for ecology, because the tropical Atlantic has been the site of several recent international studies (GATE, SEQUAL, and FOCAL) of ocean-atmosphere dynamics. The boundaries of the province are set so as to include the dominant physical processes of this part of the tropical Atlantic that are critical to understanding its pelagic ecology: (i) the divergence at the equator that penetrates seasonally into the western half of the ocean, (ii) the zonal field of tropical instability waves in the western ocean that propagate into the North Brazil Current, and (iii) the tropical cyclonic gyre, or Guinea Dome in the eastern part of the province.

This is a region of complex and dynamic surface circulation forced by the seasonally varying strength and location of the trade winds, and by the consequence of the disappearance of the Coriolis acceleration at the equator. It is here that we meet the complex equatorial current system for the first time (see, for example, Stramma and Schott, 1999) that exists in more ideal form in the Pacific Ocean. Recall that in the Atlantic, unlike in the ideal Pacific, much of the flow of the South Equatorial Current (SEC) passes across the equator into the North Brazil Current (NBC) and on into the northern hemisphere, rather than returning poleward around the South Atlantic gyre. Finally, the eastward flow of the North Equatorial Countercurrent (NECC) from northern Brazil to Senegal creates a zone of shear with the westward flow of the SEC to the south.

The seasonal ecology of the WTRA clearly reflects how these circulation features respond to the seasonal strength of the trade winds, and the meridional shift of the intertropical convergence zone (ITCZ) that lies between them. There are two major responses to this seasonality with which we shall be concerned: (i) the thermocline of the equatorial Atlantic tilts seasonally about its zonal and meridional axes, and (ii) the eastward flow in the NECC is seasonally variable. Although, as noted earlier, there is a seasonal signal in regional surface salinity attributable to the Amazon effluent, this is a trivial effect compared with the low-salinity plume from the Congo that spreads westward across the eastern tropical Atlantic (see ETRA later).

Seasonal changes in thermocline depth in WTRA are related to the migration of the trade-wind belts in a complex manner. In boreal summer, June to August, when the ITCZ is in its most northerly location, westward wind stress over the equatorial Atlantic is at its seasonal maximum and induces a double seasonal tilt of the equatorial thermocline: one toward the west, the other northward (Merle, 1983; Houghton, 1983; Hastenrath and Merle, 1987). These motions induce a major deepening of the thermocline from August to October to the north of Brazil and force the mixed-layer climatology of the entire WTRA province. Because this phenomenon has greater ecological importance in the eastern part of the ocean, I shall return to it in the section devoted to the ETRA.

In boreal summer, the contribution of the equator-crossing SEC to the coastal North Brazil Current (NBC) is maximal (Boisvert, 1967; Müller-Karger *et al.*, 1988), so that the strengthened NBC retroflects eastward at about 5°N, topographically locked to the Demerara Rise. This process is associated with the propagation along the South American shelf of mesoscale, anticyclonic eddies with a periodicity of about 50 days (see GUIA Province); the similarity of these eddies to the "Great Whorl" of the Somali Current (see ARAB Province) with respect to latitude, season, and formation is striking (Bruce *et al.*, 1985). The consequences of the retroflection for the ecology of the western oceanic regions of WTRA are important and are discussed below.

Important seasonal change in ocean physics occur in a zone that lies transversely right across the ocean at 10–12°N, where the NECC flows eastward along the southern flank of a geostrophic ridge (Garzoli and Katz, 1983, Hastenrath and Merle, 1987). Sverdrup (1947) showed that it was the curl of wind stress rather than wind stress itself that drives generic ECCs (Philander, 1985), a fact that has important ecological consequences. Ekman suction (Isemer and Hasse, 1987) creates divergence along the axis of the NECC, and in the Atlantic this may attain a vertical velocity at the base of the Ekman layer of $\sim 20 \times 10^{-5}$ cm sec$^{-1}$.

The rather complex changes in the NECC that are occasioned by seasonal changes in wind forcing have recently been modeled by Elmoussaoui *et al.* (2005) and may be summarized briefly as follows. In August and September, when the ITCZ is at its most northerly position, NECC flow is continuous, eastward across the ocean at about 10°N; when it nears the African coast, part continues to the east as the oceanic Guinea Current, while part flows around the tropical cyclonic gyre of the North Atlantic, the Guinea Dome. In boreal winter, when the ITCZ is at its most southerly position, the NECC lies under the northeast trades so that in October, a discontinuity in eastward flow appears at about 35°W, and the two now-separated areas of eastward flow regress respectively toward Africa and America. By March, the origin of eastward flow in the NECC is at about 18°W not far from the African coast, just outside the coastal Guinea Current; from here it continues right into the Bight of Biafra (Boisvert, 1967; Bruce *et al.*, 1985; Garzoli and Richardson, 1989). Surface flow in the entire WTRA is now westward, is indistinguishable from the SEC, and is in approximately the same direction as the trade winds. A small portion of this flow passes around the cyclonic gyre just west of

Senegambia. This, the Guinea Dome, is a permanent feature that has been compared to "a mountain sitting on the end of the thermocline ridge between the NEC and NECC." Its relative development appears not to be related to variance in the local Ekman vertical velocity fields responding to local wind, and it is one of the principal distantly forced features of the regional oceanography of the tropical Atlantic (Siedler *et al.*, 1992).

Vertical flux and upwelling along the equator are forced by divergence in the easterly wind field caused by the change in sign of the Coriolis force at the equator. Unrelated to upwelling, current shear between the fast eastward flow of the equatorial undercurrent (EUC), and the slower, broader westward surface flow of the SEC produces exchange between deep, cool EUC water and the warmer SEC. These two dynamic processes contribute to significant vertical nutrient flux at, and close to, the equator. Chlorophyll images show that, just as in the Pacific, the consequences of equatorial divergence take the form of waves associated with Rossby or tropical instability waves (TIWs) that are more perfectly developed in the Pacific; nevertheless, they also occur in the Atlantic both north and south of the equator—and not only in summer, when they are most readily observed in surface fields of SST and chlorophyll (see Color plate 7). They are generated by barotropic instability in the shear between the equatorial undercurrent and the SEC and are particularly energetic in the central part of the WTRA province where, at 15–35°W, annual mean eddy kinetic energy is $>200\,cm^2\,sec^{-1}$ in the 20- to 50-day band (Jochem *et al.*, 2004). The region of 1°N 15°W concentrates the largest variability in TIWs when a long time series (1998–2001) is averaged over years (Catabiano *et al.*, 2005).

TIWs are observed in both SST and chlorophyll images as cusp-shaped, or lunate, features of wavelength 1000–2000 km and period of 20–40 days, that reveal both their anticylonicity and their westward progression at velocities of $\sim 0.5\,m\,sec^{-1}$. Thus, they propagate across the WTRA province and into the NBC in periods of only several months. The distribution of convergence and divergence within a TIW is discussed later, in relation to their effects on biota.

## Response of the Pelagic Ecosystems

Generally, the chlorophyll field of the equatorial Atlantic responds to the westward tilt of the equatorial thermocline in boreal summer by showing a demarcation near the 15–20°W pivot line: to the west, near-surface chlorophyll values are consistently lower than those to the east, a fact recently confirmed at sea by Pérez *et al.* (2005). This line is close to the boundary between WTRA and ETRA.

The pelagic ecology of the WTRA responds to this and to the other physical processes discussed in the previous section; indeed, the readily available sea surface chlorophyll images, both 7-day and monthly, are the most effective tool for characterizing instantaneous circulation patterns simply and effectively. In these images, two seasonally changing patterns of surface chlorophyll capture our attention: (i) along and to the north of the equator, especially east of the BCC retroflection area where the TIWs, discussed earlier, are most strongly expressed (ii) along the axis of the NECC and over the Guinea Dome. I shall discuss these features individually next, based on examination of 64 consecutive monthly SeaWiFS images from September 1997 to December 2002.

### *The Equatorial Divergence, Tropical Instability Waves, and the NBC Retroflection*

During most of the year, the consequences of equatorial divergence are evident in the chlorophyll field as a band of chlorophyll enhancement symmetrically aligned with the equator. This feature is absent only briefly each year, for 2–4 months, usually from January to April. For some months before and after this, it may be weak and diffuse. In later boreal summer, usually in the period June–September when westward zonal wind stress is maximal in the western ocean, chlorophyll enhancement is also maximal. Of course, this feature strengthens eastward following the general eastward shoaling of the thermocline.

During periods when it is fully developed, the western termination of the divergence leaves the equator to curve toward the northwest, parallel to the coast; this represents, I believe, the effect of horizontal current shear along the northeast flank of the flow of the equator-crossing SEC. This occurs in the same period when part of the flow of the NBC is retroflected into the open ocean. The fate of the Amazon and Orinoco plumes in the NBC and the retroflection into the NECC of some of this water are discussed in more detail in the section devoted to the BRAZ province; here, it will be appropriate to discuss only the subsequent fate of water that is retroflected, usually from August to October. Between-year differences in the intensity of the retroflection are due to changes in trade-wind intensity (Müller-Karger *et al.*, 1988).

It is frequently suggested that the retroflection of the NBC sheds large anticyclonic eddies, apparently generated within the retroflection. In reality, eddies that are shed into the NBC are actually TIWs that have been entrained into the retroflection after their long passage across the tropical Atlantic. This encounter creates the now-familiar cusp of high-pigment water around the northern arc of the retroflection feature and is readily observed in serial images.

Motion, and hence transport, around the northern arcs of these anticyclonic features is eastward; thus, on encountering the offshore pigment field of the NBC, water having an Amazon signature in salinity and dissolved organic matter (DOM) is entrained offshore into the open ocean (e.g., Hu *et al.*, 2004). Chlorophyll images in late summer may show two to four contiguous lunate arcs, each representing an individual TIW, between the NBC retroflection and the Guinea Dome. The entrainment, or retroflection, sequence is initiated in June or July with the appearance of a curved tongue of high pigment that passes northeast across the Demerara abyssal plain around an anticyclonic eddy that has recently encountered the topography of South America. Toward the end of each year, the lunate arcs become increasingly diffuse until they are indistinguishable from the general chlorophyll enhancement characteristic of the NECC. Observe that these mesoscale features move progressively westward across the ocean contrary, that is, to the apparent flux of Amazon discharge water toward the east.

There has been some debate over how to interpret these pigment features. It has been suggested that high-nutrient, high-turbidity water discharged from the Amazon accounts for the observed pigment enhancement in the retroflected eddies (e.g., Müller-Karger *et al.*, 1988; Johns *et al.*, 1990; Signorini *et al.*, 1999). However, because it seemed unclear how unutilized nutrients could be conserved so far from the river mouth over oceanic depths, it was suggested (Longhurst, 1995) that what was observed was the result of eddy upwelling (e.g., Woods, 1988). The matter is now more satisfactorily resolved by the use of algorithms for SeaWiFS data suitable for both Case 1 and Case 2 water (Hu *et al.*, 2004); this analysis enables the separation in the data of colored DOM (CDOM or "gelbstoffe") from chlorophyll. In low-pigment, oligotrophic water in the center of a TIW, which was being entrained into the retroflection, chlorophyll dominated absorption at 443 nm, whereas in the high-pigment retroflection plume, CDOM dominated and here there was no apparent correlation between the two pigment types. Observation (by PALACE floats) of the eastward transport of low-salinity water in the NECC appears to clinch the argument: the multiple lunate features observed by SeaWiFS across the tropical western Atlantic are at least in part the consequence of CDOM from the Amazon forests.

But to some unknown extent they must also represent enhanced phytoplankton growth. Fortunately, we have an excellent description of physical and ecological processes within a tropical instability wave, that suggested that these features should be characterized as "whirling ecosystems" (Menkes, 2002). An impeccable study at sea of a TIW was undertaken near 3°N 19°W in June 1997; this location is in the central Atlantic, so that what was observed had nothing to do with the eventual encounter of this feature with the NBC retroflection far to the west and what was observed may be therefore taken as representative of processes within the TIWs discussed earlier. Water from the equatorial divergence (cold, nutrient-rich, and of high chlorophyll content) was found to be advected northward and downward around the western side of the feature; as the circulation is completed back to the south, the pycnocline, nutricline and DCM were found to move progressively surfaceward to enter again the equatorial water mass. So, as the authors emphasize, "a fully three-dimensional circulation ... dominates the distribution of physical and biological tracers in the presence of tropical instabilities and maintains the cusp-like shapes of temperature and chlorophyll observed from space." The relative vertical and horizontal distribution of SST and velocity, nitrate, primary production, and *in situ* chlorophyll, together with zooplankton and micronekton biomass, were all consistent; upwelling at depth within the vortex may not imply cross-isopycnal flux that would supply new nutrients that may rather be supplied in the poleward surface flow from the equatorial upwelling.

These observations suggest what must be the fundamental mechanism within TIWs that produces their characteristic signature in the surface chlorophyll field: the fact that, in the western part of the ocean, after retroflection of the NBC is initiated, water rich in CDOM is entrained around the northern cusps does not require that this mechanism should not function. We await examination of such a TIW with the double algorithm technique used by Hu *et al.* in the NBC retroflection itself.

***The Guinea Dome and the NECC*** The entire region occupied by the NECC exhibits enhanced chlorophyll, both patchy and diffuse, that is dissociated from the lunate features and the retroflection of the NBC discussed earlier, although enhanced chlorophyll within the NECC reaches its maximum westward extension in the same months that the retroflection eddies are strongest, and remains so even as they diffuse toward the end of the year. It is only in February or March that the expression of the NECC in the surface chlorophyll field retreats to midocean. In exceptional years, as in 2001, this may not occur until May.

Perhaps a sufficient explanation for this zone of high chlorophyll values lies in strong vertical Ekman flux that occurs from 20°W to 40°W, particularly from June to October, causing divergence along the crest of the thermal ridge between NECC and NEC (Isemer and Hasse, 1987). Furthermore, as pointed out by Yentsch (1990), underlying the NECC is a baroclinic ridge in the subsurface nitrate field taking values of 16.0 μM at 150 m, similar to concentrations at the same depth south of Greenland. Such high concentrations will render any physical mechanism that tends to draw subsurface water toward the surface unusually effective in supplying nutrients to the photic zone. In addition, the strongly meandering flow of the NECC itself induces vertical motion within cyclonic eddies and due to eddy/eddy interactions, and this motion is thought to be a nonnegligible source of nutrient flux to surface waters (Dadou *et al.*, 1996).

Seasonal changes in vertical Ekman velocity support this model: the greatest vertical flux occurs when chlorophyll values are highest. In January, vertical Ekman velocity along the NECC is weak but variable in sign, and by April, a broad band of zero vertical transport corresponds with the area to be occupied later by the NECC. When the ITCZ is in its northernmost position (July–September) and the southeast trades dominate the wind field, a zone of exceptionally strong positive curl of the wind stress lies along 10–12°N from 20°W to 40°W (Isemer and Hasse, 1987; Hastenrath and Lamb, 1977). This translates into Ekman suction (upwelling) rates of roughly 2 m $d^{-1}$ along the whole length of the NECC during this season. Toward the end of this trimester, the locus of highest values comes to lie nearer the African coast. Averaged over the year, the band of positive Ekman suction remains a prominent feature from northern Brazil to Senegal and is the most prominent feature of Ekman vertical velocity in the North Atlantic Ocean at any latitude (Isemer and Hasse, 1987).

Although subthermocline, high-nitrate water does not surface in the Guinea Dome, as it does in the Costa Rica Dome (Voituriez and Herbland, 1982), this feature is observable in the chlorophyll imagery, although not consistently. At the eastern termination of the NECC, large irregular patches of very high values that appear not to be plumes of coastally upwelled water may occur to the southwest of Mauritania at 20–25°W, and this appears to represent the region in which the Guinea Dome is most generally expressed at the surface. Such features are most prominent during the boreal winter for a period of 4–7 months, and there is significant between-year variability in both duration and strength.

The ecological profile observed in WTRA is usually characteristic of an oligotrophic situation, resembling the "typical tropical situation" in the sense of Herbland (1983); only within the equatorial divergence is there significant divergence from this pattern. Depending on the depth of the mixed layer—from almost 100 m in the west to 20–30 m in the Guinea Dome—the chlorophyll and productivity maxima and the subjacent nitracline lie either at the top of the pycnocline or deep within it. As Voituriez and Herbland (1982) note, the Guinea Dome exhibits a remarkable stratification of physical, chemical, and biological parameters. Here, the situation is sufficiently stable that the nitrate-deficient surface layer is significantly deeper than the wind-mixed surface layer. The nutricline, then, is located at some depth within the pycnocline and is associated with very sharp maxima of algal growth rate, of chlorophyll biomass, and of numbers of herbivorous zooplankton. It is at this depth that optimal conditions of irradiance, nutrient supply, and rate of vertical mixing occur. The general relationships between depths of maximum grazing rate, of maximal algal growth rate, and of maximal algal biomass are complex but the tropical ocean offers the best site for their investigation. It is found that depths of highest plant biomass, of chlorophyll, and of growth rate become progressively more separated in the vertical the deeper they lie in the tropical ocean. Thus, maximal growth rate becomes progressively shallower than the chlorophyll maximum the deeper they lie. Depths of maximal mesozooplankton are usually close to the depth of maximal algal growth rate, implying a trophic link involving the regeneration of plant nutrients by excretion of zooplankton (Longhurst and Harrison, 1989).

We have little information on the vertical distribution of individual biota here, but there is no reason to believe that the eastern tropical Pacific profiles (see Chapter 11, North Pacific Equatorial Countercurrent and North Pacific Tropical Gyre Provinces) will not serve as relevant models for the tropical Atlantic. The progressive shoaling of horizons brings oxygen-deficient water ($< 2.5\,\text{ml liter}^{-1}$) to within 75 m of the surface in the Guinea Dome, and this has important effects on the distribution of mesopelagic fish species. Myctophids dominate the biomass of this component and are represented by species having northern, subtropical, and tropical distributions, whereas other families of mesopelagic fish (Gonostomatidae, Sternoptychidae, and Scopeloarchidae), and some of the deeper-living myctophids are curiously absent. It has been suggested that these anomalies may be accounted for by the unusual vertical distribution of temperature, and of other critical properties, in the doming region and also by larval retention within the semiclosed circulation south of the Cape Verde Frontal Zone.

For the horizontal distribution of biota, apart from bulk chlorophyll data, we have recourse to rather few data: first, from the EQUALANT surveys of the 1960s, performed in support of the U.S. tuna-fishing industry, and, second, from the accumulated Soviet survey data (Piontkowski *et al.*, 1997). Both these data sets emphasize the existence of enhanced zooplankton biomass associated with the NECC and Guinea Dome, while the

EQUALANT data demonstrate a band of high zooplankton biomass from 5°N to 5°S, with patches of higher values right at the equator. The basin-scale model of Wroblewski *et al.* (1988) predicts relatively high zooplankton biomass in the NECC/Guinea Dome region.

It is no surprise, therefore to find in the Fontenau tuna atlas that the NECC is a favored location, especially in the third quarter of the year, for the tuna long-line fishery. The absence of fishing effort along the equatorial divergence is perhaps due to the difficulty of working long-line gear in an area of such strong vertical current shear.

## Synopsis

*Case 4—Small-amplitude response to trade wind seasonality*—$Z_m$ forced by density shows weak geostrophically forced seasonality when meaned over a whole area of province (20 m in June, 30 m in October-December). $Z_{eu}$ is consistently deeper (45–50 m) so the thermocline is illuminated in all months, while the P rate also has a very weak dynamic range (Fig. 9.15), around 0.35–0.50 gC $m^{-2}$ $d^{-1}$ year-round, with two seasonal maxima in boreal spring and autumn. This province is probably too complex for the mean values clearly to reflect the Ekman suction that occurs along NECC in boreal summer. Seasonal changes in chlorophyll biomass are rather slight (from 0.15 to 0.25 mgC $m^{-3}$), and with highest abundance at the time of the autumn peak in productivity. This situation is consistent with a rather close balance between consumption and production.

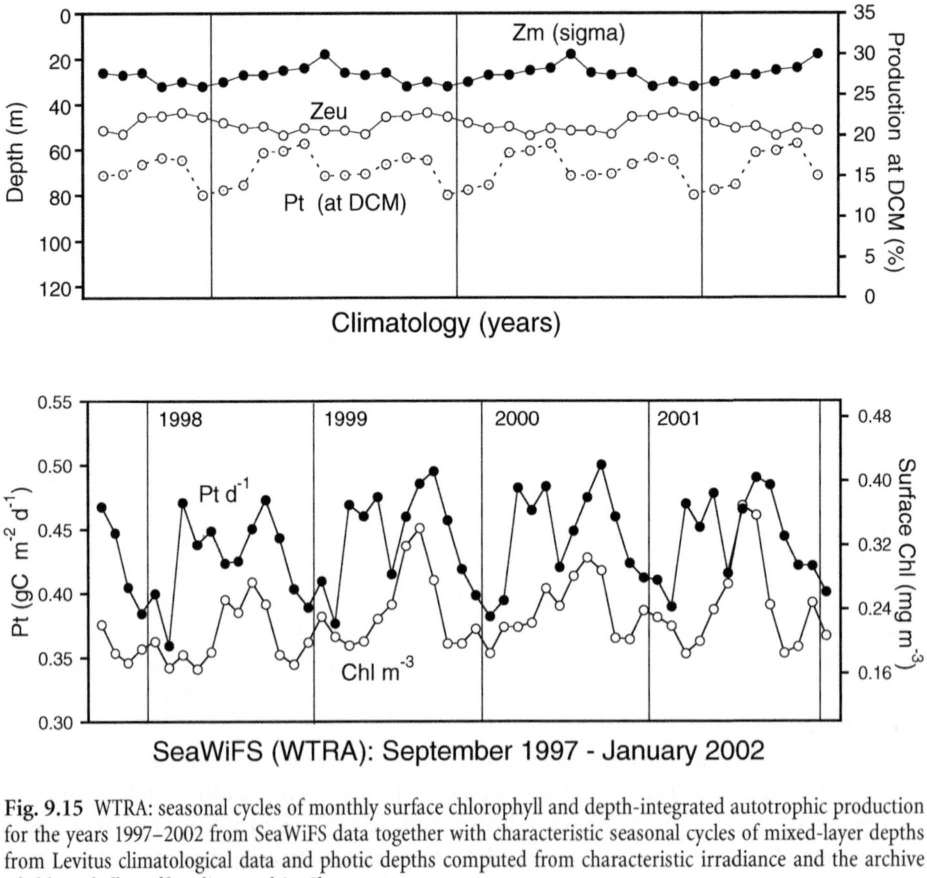


**Fig. 9.15** WTRA: seasonal cycles of monthly surface chlorophyll and depth-integrated autotrophic production for the years 1997–2002 from SeaWiFS data together with characteristic seasonal cycles of mixed-layer depths from Levitus climatological data and photic depths computed from characteristic irradiance and the archive of chlorophyll profiles discussed in Chapter 1.

# Eastern Tropical Atlantic Province (ETRA)

## Extent of the Province

This province represents the oceanic Gulf of Guinea (*sens. lat.*) to the east of the boundary with WTRA that is placed at 15°W. Unlike the WTRA, this province is almost symmetrical about the equator, lying between the shelf-edge front of the east-west coast of West Africa (Guinea to Nigeria) along about 5–10°N and a line drawn diagonally (NW-SE) across the ocean between 5°S and 10°S. The eastern boundary is, of course the shelf-break front along the north-south coast from Cameroon to Congo. As discussed later, this arrangement is intended to include in one province the tropical gyral circulation (SECC and Angola Dome) of the southern hemisphere to match the same elements of the northern hemisphere included in WTRA. It is entirely oceanic, save for the chain of small islands (Annobon to Fernando Po) in the Bight of Biafra.

## Defining Characteristics of Regional Oceanography

The circulation of this province, within equatorial currents and the tropical gyre, is largely a mirror image of that in the WTRA Province. Geostrophic flow, and hence the topography of the thermocline, is rendered the more complex by the change of sign of the Coriolis force at the equator, which the province straddles. As in other equatorial provinces, seasonal variation in the circulation and topography of the thermocline is forced by the overriding influence of the variable trade winds.

The ITCZ remains almost year-round over WTRA, lying above the northwest corner of ETRA only in winter when in its most southerly position; southeasterly trade-wind stress is thus more consistent in its direction than in WTRA, although wind stress at the equator in ETRA at 4°W varies from about 0.2 dyn $cm^{-2}$ in February-March to 0.8 dyn $cm^{-2}$ in July–September when the ITCZ moves rapidly northward so that by August reaches 15°N. This causes the rapid extension of the strong (5–8 m $sec^{-1}$) southeasterly trade winds across the equator that surge into the western basin of the tropical Atlantic, strongly intensifying the zonal wind component there. Meanwhile, wind stress in the eastern Atlantic strengthens principally in its meridional component (Hastenrath and Lamb, 1977). Almost the entire province lies under negative wind curl during this season, with the southern limit of the SEC approximating to the change of sign in vertical motion (Gordon and Bosley, 1991). It is to these seasonal changes in wind-stress and to the effect of divergence in the wind field at the equator that we look for explanations of seasonal changes in circulation and regional ecology.

From May to October, as a direct consequence of the impulsive intensification of zonal wind stress, the equatorial thermocline shoals in the ETRA and deepens in the western Atlantic, as already noted in the account of the WTRA province (Fig. 9.16). To the east of a pivot line at 15–20°W, the base of the thermocline (defined as the depth of the 15°C isotherm) shoals from 145 to 90 m, whereas the upper thermocline (the 20°C isotherm) shoals by about 20 m. Thus, both the thermocline and the mixed layer above it are vertically compressed in the eastern ocean.

This process is accompanied by a cooling of the mixed layer in the ETRA province from 29°C down to about 23–24°C (e.g., Pérez *et al.*, 2005). The seasonal cooling, now that we can inspect its progress in 9-km, 8-day MODIS SST images, lies less strictly along the equator than in the Pacific. Rather, it is a broad zone that narrows westward and lies diagonally across the eastern Atlantic from the northern Angola Bight to the equator at about 2–3°W, and thence along the equator; initial appearance occurs in May or June, close to the coast at about 2–4°S. Within this somewhat diffuse cool zone may be discerned some indication of a separate cool feature associated with the equator itself. The

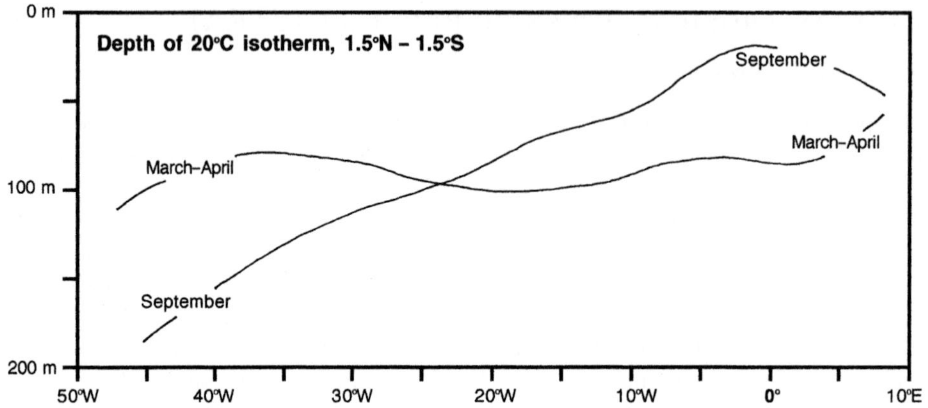


**Fig. 9.16** Cartoon to illustrate the extent of the seasonal zonal tilt of the equatorial Atlantic thermocline in response to changing trade wind stress at the sea surface.

*Source: Redrawn from Houghton, 1983.*

general pattern is retained as it evolves over the following weeks, although the divergence at the equator itself becomes progressively evident as the cooling extends further west.

That the shoaling of the thermocline is distantly forced is confirmed by the cooling of the mixed layer by 2–5°C even while net heat gain across the sea surface remains positive (40–80 W $m^{-2}$) in the months from August to October from the African coast down to >5°S (Hastenrath and Lamb, 1977): elsewhere in the tropical Atlantic, sea surface temperature changes are phase-linked to net heat gain across the sea surface (Houghton, 1991). Note that the aggregated data for $Z_m$ (defined by the shallow thermal stratification) for the ETRA province represent not only the near-equatorial regions where the boreal summer-autumn shoaling is strong, but also regions to the south of 5°S where the climatology differs.

Response of the mixed layer to anomalous, rapid collapse of the trade winds is additional confirmation that shoaling of the thermocline in the Gulf of Guinea is indeed distantly forced and is not a response to local wind stress (Verstraete, 1992): between 1960 and 1990 nine such episodes occurred. At irregular intervals, SST in the eastern tropical Atlantic undergoes anomalous warming episodes during boreal summer, each usually exceeding 0.5°C. These are associated with a relaxation of the westward component of wind stress in the next summer after a year in which such wind stress was unusually strong; this leads to an unusual accumulation of tropical surface water in the western ocean that later surges back into the east along the equatorial wave guide. The consequences of this are discussed later.

Here in the ETRA, the ideal equatorial circulation pattern is much influenced by the shape of the western coast of the African continent so that the Gulf of Guinea is bounded to the north by 2000 km of zonally oriented coastline. The regional expression of the South Equatorial Counter Current (SECC) is by no means a simple mirror image of its counterpart in the northern hemisphere, the NECC discussed earlier. This lies permanently across the northern part of ETRA, where it is the dominant eastward flow at all times; as noted previously, west of the ETRA/WTRA boundary, the NECC is seasonal, whereas here in the ETRA it forms the permanent Guinea Current (NECC-GC).

The northern limb of the anticyclonic gyre of the South Atlantic Ocean (counter-clockwise in the southern hemisphere) passes across this province as the SEC, flowing westward to split at Cabo de Sao Roque; here, most of the flow passes north into the NBC while a smaller part turns south along the Brazilian coast and so around the gyre. Stramma and Schott (1999) discuss the partition of the SEC into several streams separated by variable eastward flow, mainly subsurface. Symmetrically below the equator is the eastward flow of the Atlantic Equatorial Undercurrent (EUC), originating north of Brazil; this is 250–300 km wide, has a vertical thickness of 150–250 m, and slopes progressively upward to the east. Depth of the EUC core is 50–75 m, and inversion with westward flow of the SEC lying above it occurs at only 15–35 m (Neumann, 1969; McCreary *et al.*, 1984; Voituriez, 1981). Note that the water carried in the EUC has high salinity and is of nutrient-depleted origin.

Within the flow of the SEC is embedded the eastward flow of an SECC at about 5°S. This has been described as weak, narrow, and rather variable. It is in no way comparable to the NECC in the WTRA and is captured neither in 90-d surface velocity analyses, nor in drifter-derived climatology, nor yet in ship-drift archives: the NECC is, of course, prominent in each of these. The topography of the thermocline below the SECC flow should, theoretically, carry a ridge parallel with the equator and close to it, but this is hardly observable, perhaps because of divergence at the equator that shoals the isotherms for >20°C.

Thus, except for the eastward NECC–Guinea Current, surface flow over the entire eastern Atlantic is predominantly, and year-round, toward the northwest and west: this applies even across the subsurface Angola Dome, although some meandering is evident there. The dominance of this vector in the regional surface flow has important consequences for the motion of shallow features, such as the Congo plume of light, low-salinity water.

This situation is modified in anomalous years of "Atlantic El Niño," when weakened trade winds may come to have an eastward component; in such episodes, the SEC is weakened while NECC-GC, the EUC, and (in exceptional events, termed "Benguela Niños") the SECC strengthen. This anomalous flow results in the accumulation of warm, near-surface water in the eastern part of ETRA that "discharges" (to use the term of Binet *et al.*, 2001) southward along the Congo-Angola coast. Such events occur during positive anomalies of the Southern Oscillation Index and hence the Pacific cold phase, as in 1934, 1963, 1984, and 1995; during the 1984 event, eastward flow dominated from 0° to 5°S. Such anomalous SECC flow passes south, over water upwelled in the Benguela Current, as well as equatorward along the Congo coast (Shannon *et al.*, 1986).

As in the other tropical oceans, the poleward turn of eastward flow on encountering the continent creates a cyclonic dome in the thermocline topography: here, this is the subsurface Angola Dome, which lies west of the Angola Bight, centered at about 13°S 5°E. Although the details of circulation in this undersampled region are unclear, this large cyclonic gyre is forced by the poleward turn of both EUC (as the Angola Current on the eastward side of the gyre) as well as by the SECC that feeds directly into its northern slope (Shannon, 1985; Mercier *et al.*, 2003). At the axis of the tropical gyre, at the Angola Dome, the pycnocline reaches closer to the surface than it does in the Guinea Dome, perhaps with significance for algal growth processes (Voituriez, 1981). This feature is associated with Ekman upwelling, but here this is significantly weaker than in the NECC and Guinea Dome of the WTRA province: seasonal climatological upwelling velocity in the SECC/Angola gyre field is only 25% of that in the comparable North Atlantic situation. Below this tropical gyre lies the most strongly developed subsurface oxygen minimum in the entire Atlantic Ocean (Bubnov, 1972; Chapman and Shannon, 1985). It may be useful to point out that the gyre is not a circular feature lying in the Angola Bight, as represented in the usual diagram of this feature, but the area over which Ekman upwelling is important (and that is what is interesting in the present context) is a much larger, somewhat triangular region based in the Angola Bight but extending NW beyond the Greenwich meridian (McClain and Firestone, 1993).

A further complication in the interpretation of this province is the fact that the water of the Congo River, with a rate of discharge second only to the Amazon, enters the

Atlantic at about 6°S and thus very close to the Angola Gyre, on its northeast side. The plume of the Congo is of strongly colored water and discharge is maximal in June and December as water from different parts of the drainage basin reaches the sea. The plume normally passes west or northwest across the northern part of the tropical gyre, but during Benguela Niños, as in 1995, it may pass to the south along the shelf into the BENG province (Gammelsrd *et al.*, 1998). This plume is a prominent feature in the surface pigment field of ETRA, whereas the smaller discharge from the Niger River into the Bight of Biafra is much less consequential. The surface salinity signal of this discharge is much stronger than that of the Amazon: the EQUALANT surveys revealed a plume at the surface near the equator in which values as low as 33 ppt stretched well past the Greenwich meridian to the longitude of Ghana (see Fig. 9.17, later); from its location and shape, this plume appears not only to carry Congo water, but also to transport water from the Bight of Biafra (Niger and Cameroon river effluents) toward the west. The core of the plume carries very fresh water, down to 31‰; in comparison, the surface water of the western Atlantic in WTRA is seldom lower than 35 ppt despite the apparent offshore retroflection of the Amazon plume.

## Response of the Pelagic Ecosystems

Both the EQUALANT (1963–65) and CIPREA (1978–79) surveys suggested that there was only a very minor seasonal signal in the chlorophyll biomass or of primary production in the open ocean (Voituriez, 1981), a conclusion that was refuted by the first serial satellite images of sea surface chlorophyll that became available. Data from the

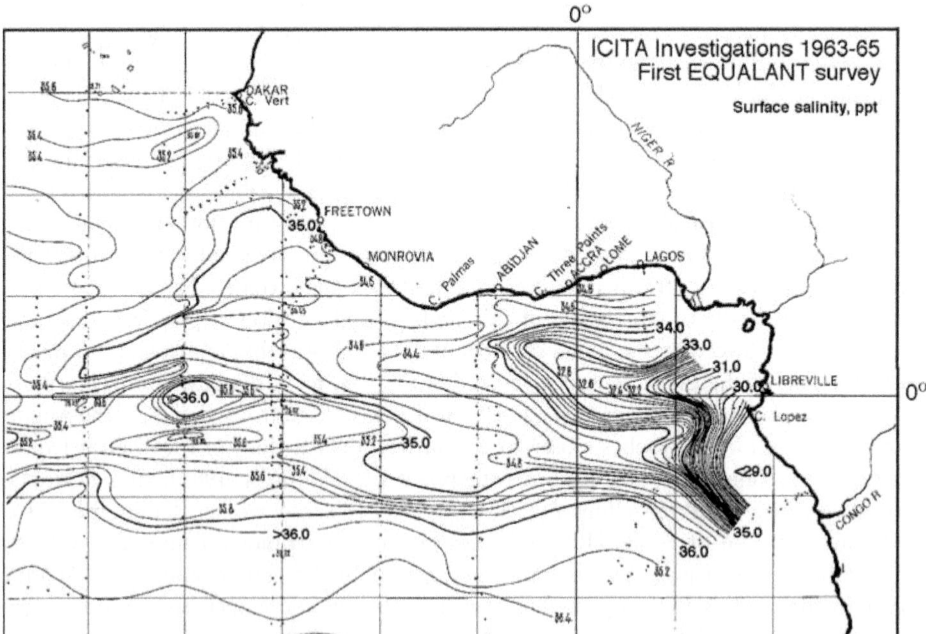


**Fig. 9.17** The regional distribution of surface salinity during the first EQUALANT survey (including the first oceanographic voyage by a Nigerian research ship) in boreal winter. Note the apparently unremarked integrity of the low-salinity plume that originates near the Congo mouth and is traceable clear across the Gulf of Guinea.

*Source: Modified from the ICITA Atlas.*

CZCS instrument demonstrated that significant seasonality does occur, a fact abundantly confirmed by more recent satellite data. I can suggest no explanation for the contrary indications from the CIPREA data, but EQUALANT data were obtained during a year of a Benguela Niño. The CZCS images were read as suggesting that relatively high chlorophyll ($0.3$–$0.5\,mg\,m^{-3}$) "occurs almost uniformly over the eastern tropical Atlantic in June–August, east of a meridional boundary at about 20–25°W" (Longhurst, 1993). However, the superior quality and data flow of the images from later sensors shows that the real situation is more complex than suggested by my initial, and quite erroneous, interpretation; the first satisfactory interpretation of the main features of phytoplankton response is that of Signorini *et al.* (1999). This pattern has recently been confirmed from both satellite observations and measurements made at sea by Pérez *et al.* (2005); these authors show empirical relationships between AVHRR SST and SeaWiFS chlorophyll that differ in ETRA (10°W) and WTRA (25°W) such that, for ETRA, chl = 3.50–0.12 SST, $r^2 = 0.67$, $n = 48$, $P = <0.0001$, and for WTRA, chl = 2.87–0.10 SST, $r^2 = 0.51$, $n = 48$, $P = <0.0001$. This inverse correlation, the authors noted, confirmed interpretations made earlier (e.g., Longhurst, 1993).

This knowledge, and examination of many images, shows that there are several different processes whose effects must be disentangled before each can be understood, and the form of the basin of the eastern equatorial Atlantic makes this no simple task. There is, indeed, a rather diffuse seasonal enhancement of surface chlorophyll almost everywhere in ETRA, caused by a variety of responses to:

(i) *Wind-driven linear equatorial divergence*, with which is associated a chlorophyll enhancement field that is clearly centered upon linear, equatorial upwelling that is, compared with the Pacific, rather episodic

(ii) *Ekman upwelling above and around the Angola dome*, with which is associated a second, broader zone of chlorophyll enhancement that narrows westward and may be continuous in the early months of the cool season with enhancement at the equatorial divergence

(iii) *The highly pigmented, shallow plume of the Congo River* that lies diagonally across the tropical gyre, rooted at about 5°S, and is readily confused with both of the above

(iv) *Offshore effects of upwelling processes within the NECC-GC* that occur south of the east-west coast of tropical West Africa (see GUIN) and thus along the zonal flow of the Guinea Current

It will be necessary to deal with these individually, while recognizing that the effects of (i), (ii), and (iii) are superimposed spatially and hence difficult to understand individually.

Chlorophyll enhancement associated with equatorial divergence is weak or even absent in boreal winter and appears as a threadlike feature, usually in June but sometimes, as in 2001, in May. It usually takes maximum zonal dimension in July and August when mixed-layer depths are minimal, and weakens and diffuses in the following months.

The evolution of the canonical seasonal equatorial bloom, and some of its between-year variance, could be traced only incompletely in the CZCS images, 1978–1986, but, in the continuous imagery available from SeaWiFS and MODIS sensors since 1997, the seasonal and between-year changes can be followed more clearly. In each year, the surface chlorophyll field is weakest in March and progressively strengthens from April to June, the latter being the first month when it is fully developed. The onset of the regional bloom was a little earlier and stronger in 1998 and 2000 than in the other years. In each year, the maximum extent of the bloom is reached during July and August, with the maxima of 2000 and 2001 being somewhat stronger and more dispersed meridionally than other years. Finally, in September of each year the northern border of the equatorial bloom retreats southward, leaving a zone of low-pigment biomass between it and the enhancement associated with the NECC-GC. This zone is continuous with a "blue hole" in the outer Bight of Biafra (see earlier discussion) and lies above the trough in pycnocline topography between the NECC-GC and the northern part of the SEC. This pattern persists and the blue hole increases in size until the equatorial bloom disperses in January and February. From the evidence of surface chlorophyll during the SeaWiFS period, the 1984 warm event in the Gulf of Guinea (see earlier discussion) was exceptional as suggested by the 1964–1993 monthly SST archives (Delacluse *et al.*, 1994). This cycle was followed and illustrated by Pérez *et al.* (2004), who were able to obtain a satisfactory empirical relationship between SeaWiFS chlorophyll and *in situ* observations in the ETRA.

As already noted, the equatorial chlorophyll enhancement associated is—at first glance—asymmetric and extends much farther south than would be anticipated. It now seems probable that this represents, at least in part, the southern hemisphere equivalent of the chlorophyll enrichment that lies along the NECC and Guinea Dome in the WTRA province. Not that it is a simple mirror image of that feature, because of the asymmetry of the African continent in the two hemispheres, and because rather than the broad SW-NE feature of the NECC, here we find a variety of features. Most typically, it is a broadly triangular region that narrows to the northwest and is based within the Angola Bight; there are many permutations of this feature, the most extreme being a rather narrow, zonal band of enhanced chlorophyll parallel to the equator, but at about 7–8°S. Its eastern termination betrays it for what it is, for here it merges imperceptibly around the northern edge of the Angola Dome, subsequently becoming increasingly wide and increasingly strong. It may, then, represent the result of the same processes in the SECC that were discussed earlier in relation to the zonal bloom in the NECC. It is not easy to use the chlorophyll images to support the earlier assumptions concerning the ecology of this region, based on investigations near the axis of the Angola Dome (Voituriez and Herbland, 1982).

Rather, a relatively simple model has been invoked to explain the observations (Monger *et al.*, 1997). Phytoplankton growth is enhanced in May by distant uplift of the regional thermocline, forced by winds in the western Atlantic, and the consequent entrainment of nutrients into the photic zone, together with some mixing by wind stress although this is minimal at very low latitudes. Residual nitrate along the equatorial divergence during boreal summer represents the normal situation in which the rate of supply temporarily exceeds the utilization rate. By August, the thermocline has returned to depth but, because the EUC is now deeper than prior to uplift, such vertical entrainment as occurs delivers water richer in nutrients than previously. Thus the active growth of phytoplankton continues for some months until, in December, the EUC returns to its boreal winter location above the thermocline. Any divergence after this time brings only nutrient-depleted water to the surface.

The model proposed by Monger *et al.* was intended to apply only between 3°N and 3°S, being tested against survey and satellite data from that zone; thus, it was intended as a model for the equatorial divergence cell, rather than for the regions to south and north discussed earlier. Nevertheless, it requires that the nutricline should bear a shoal ridge at some distance (say 5° latitude) to north and south of the equatorial divergence, and that some deeper water should be drawn into the photic zone there. None of which is inconsistent with the suggestion made above that the pigment enhancement apparently associated with the elongated tropical gyre is delivered by a process distinct from equatorial divergence.

Now, the third component of pigment enhancement to take our attention is attributable to the Congo or Zaire plume. Wherever the indicated pigment biomass in relevant satellite images exceeds, say, 2 mg chl $m^{-3}$ at the surface, is linear, and is based at the Congo mouth, then we may say with some confidence that it is indeed the Congo plume we are looking at. Such a feature is present in a high proportion of both 8- and 30-day images but has a seasonality that is not easy to generalize; thus, in the 3 years 2000–2003, it was especially strong and extended far across the tropical gyre in three periods: January–March 2000, July–September 2001, and January–March 2002, the last being shown in Color plate 8. On the contrary, the images for November 2001 show general pigment enhancement of the tropical gyre with an imposed meandering plume of blue water that corresponds in form and location to what we would expect of the Congo plume; but the March 2003 image shows a well-developed plume extending west across blue water in the complete absence of any general enhancement of the Angola Gyre. This suggests that the plume remains more discrete than diffuse, and that is very largely Ekman upwelling that causes the diffused chlorophyll enhancement observed in the gyre south of the equatorial divergence. Figure 9.17 shows that regional surface salinity patterns confirm these interpretations of the chlorophyll field: both the Congo plume and the freshwater from Niger and Cameroun rivers that discharges toward the west from the Bight of Biafra are identifiable.

Finally, we must notice the fourth mechanism that may lead to pigment accumulation, along the southern flank of the NECC-GC; we may attribute this simply to the same processes that force a bloom in the transoceanic section of the NECC. As already noted, in many of the serial surface chlorophyll images a very clear zone of oligotrophic water is seen between the effects of the equatorial divergence and what we can attribute to processes associated with the NECC-GC; this may occur in any month, but is especially evident as the divergence bloom recedes during boreal winter months.

How far can the inferences made from satellite images be supported by *in situ* observations? During the FOCAL voyages, meridional chlorophyll sections were worked across the equator at 4°, 23°, and 35°W (Herbland *et al.*, 1987; Oudot and Morin, 1987). At 4°W the DCM shoals from 50 m to about 25 m at the equator while chlorophyll values within it double. The same occurs, but less clearly, at 23°W. At 35°W, in the WTRA, the DCM remained at about 60–70 m in both seasons with essentially unchanged values. The nutricline and the DCM were found to co-occur in the upper few meters of the thermocline in both seasons, with nitrate at limiting concentrations in the mixed layer, except in the equatorial divergence zone itself during the cool season. The cool water that lies along and just south of the equator is indeed associated with enhanced ($<5\,\mu M$) nitrate levels at the surface because the shoaling of the thermocline above the EUC coincides with the occurrence of stronger turbulence there than to north or south (Hebert *et al.*, 1991).

Meridional transects worked recently between 10°N – 10°S in the western part of ETRA (Pérez *et al.*, 2005) showed that although production rates were higher by a factor of 6 near the equator, total cell biomass was remarkably similar along the entire length of the transects (19–22 mg chl $m^{-2}$); chlorophyll enhancement near the equator was strongly correlated with increased mixed-layer $NO_3$($>1.0\,\mu M$ compared with $<0.1\,\mu M$ to north and south). Cell size fractions were also remarkably uniform along each transect, picophytoplankton contributing 80% of biomass in the oligotrophic zones and 60% at the equatorial divergence, suggesting strong top-down control of primary production by herbivores. Cell numbers of all fractions were higher in the divergence zone, peaking at around 2°S, thus confirming the general observation noted earlier of a general bias toward the south of higher chlorophyll biomass at the surface in the ETRA. Of course, it has been known since the early 1980s that the seasonal blooms here are dominated by small cells, though the data reported then did not yet reflect the revelation, later derived from sea-going flow spectrometry, that cyanobacteria and prochlorophytes dominate the chlorophyll biomass at the bottom of the photic zone. In 1985, Herbland reported that in the nitrate-depleted mixed layer, cells $<1\,\mu m$ dominate the chlorophyll biomass (mean = 71%), whereas at the nitracline such cells comprise only 50% of autotrophic biomass.

Finally, an SST climatology has been specified by Pérez *et al.* (2005) from the AMT and other studies at sea that closely confirms the computations made from satellite data as shown, for example, in Fig. 9.18: Pérez *et al.* show a seasonal cycle for both WTRA

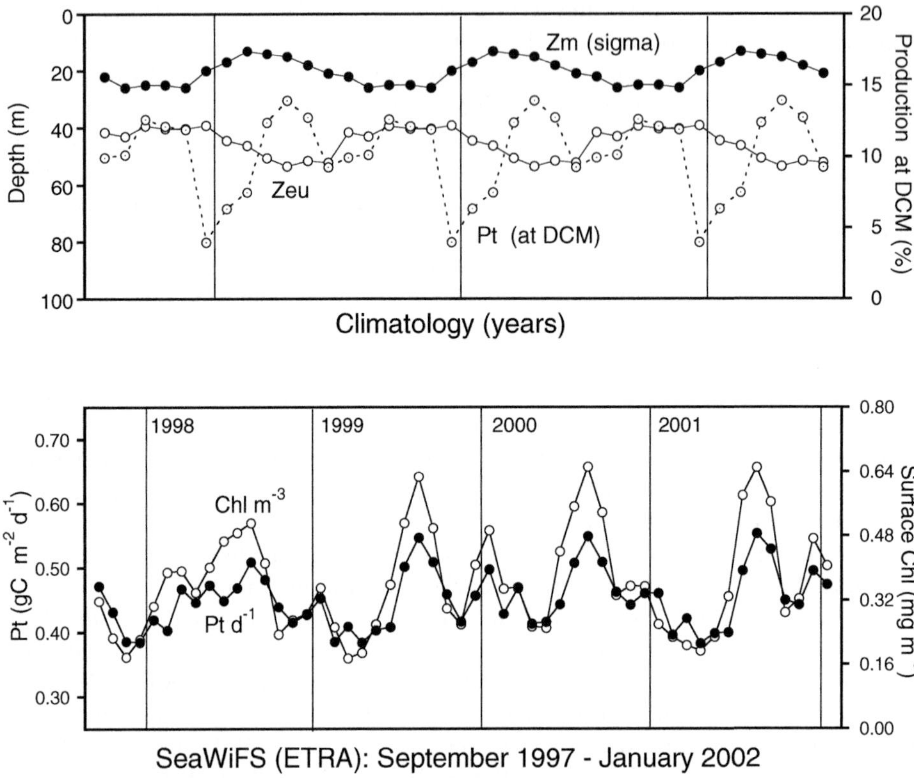


**Fig. 9.18** ETRA: seasonal cycles of monthly surface chlorophyll and depth-integrated autotrophic production for the years 1997–2002 from SeaWiFS data together with characteristic seasonal cycles of mixed-layer depths from Levitus climatological data and photic depths computed from characteristic irradiance and the archive of chlorophyll profiles discussed in Chapter 1.

and ETRA, as defined here, in each of which maximum values occur in August and minimal values in March–April. The ranges are, however, very different: WTRA from 0.10 to 0.23 mg $m^{-3}$, and ETRA from 0.10 to 0.50 mg $m^{-3}$.

There are few modern studies of higher trophic levels in the ETRA, but Le Borgne (1981) was able to correlate standing stocks (expressed as integrated mixed-layer dry weight) of mesozooplankton with chlorophyll and found a positive correlation ($DW_{zoo} = 74.5\,chl^{.05}$) with a slope not different from unity. The zooplankton biomass distribution reported by the EQUALANT investigations support this conclusion and, more widely, the analysis of accumulated Soviet data of phytoplankton and zooplankton biomass in the tropical Atlantic by Finenko *et al.* (2003). Based on several thousand measurements the relationship of chlorophyll to zooplankton biomass was 2.2 in ETRA and 1.4 in WTRA: this is a higher ratio than found in subtropical provinces (e.g., NAST, SATL). Le Borgne found that mesozooplankton biomass was aggregated at the DCM and when detailed mesozooplankton profiles are obtained in the eastern tropical Atlantic, it is predictable that they will resemble the many that are already available for the eastern tropical Pacific. At upper trophic levels, there is some evidence in catch-rate maps for tuna that ETRA produces more of the two shallower-swimming tropical species (yellowfin and skipjack) than does the deeper and more oligotrophic mixed layer of the WTRA. Conversely, the deeper-swimming bigeye (*Thunnus obesus*) shows less east-west difference in abundance in the tropical Atlantic. As with all such fishery-dependent evidence, we must note that perhaps this observation simply reflects the fact that purse seiners and bait boats will obviously find their work easier above the relatively shoal thermocline of the ETRA.

### Synopsis

*Case 5—Large-amplitude response to monsoon-like reversal of trade winds*—integrated over the entire province, $Z_m$ shows very weak distantly forced seasonality from 17 m in March–May to 25 m in September–November, while $Z_{eu}$ is consistently deeper (40–50 m), so that the thermocline is illuminated in all months. Seasonal changes in P rate are complex but generally are inverse to changes in $Z_m$, being highest from boreal summer to early winter. Chlorophyll biomass closely tracks P rate change (Fig. 9.18). Herbivore biomass is known to respond in a predictable manner to chlorophyll accumulation, confirming the assumption that close coupling occurs between production and consumption.

## South Atlantic Gyral Province (SATL)

### Extent of the Province

The SATL comprises the anticyclonic circulation of the South Atlantic, excluding the coastal boundary currents that are treated as separate provinces (see BRAZ, BENG). The east and west boundaries of the province lie at the edge of the eddy fields associated with these two currents: to the north, along the southern boundaries of ETRA and WTRA and to the south, along the limit of the biological enhancement of the meandering Subtropical Convergence Front that defines the SSTC province.

### Defining Characteristics of Regional Oceanography

As will become clear, this is one of the least well-researched regions, far from the major oceanographic research institutes and the troubles of the world. To the long-haul work of the South American institutes have been added in recent decades some international initiatives, such as the CONFLUENCE program of Argentina, France, and the United States of 1988–1990 (Anonymous, 1990), although these have mostly looked at the margins of this province. We may also expect some assistance from the Atlantic Meridional Section (AMT) project, worked by Antarctic supply vessels sailing between the UK and Antarctica (e.g., Aiken and Bale, 2000) that pass down the western part of SATL. WOCE, of course, contributed significantly to our general knowledge of the physics of the entire South Atlantic, and these data are now readily available to all through Java OceanAtlas. Integrated accounts of the circulation within this ocean basin go back to the 1960s, although I have leaned heavily on the very accessible review of upper-level circulation processes of Petersen and Stramma (1991). Of course, the characteristics of the region that will be of interest to ecologists are dominated by the fact this is a major anticyclonic subtropical gyre; consequently, the bowl-shaped isopleths for nutrients and of other properties of interest slope upward toward the margins of the ocean.

The South Atlantic gyre is not entirely the homologue of that in the North Atlantic, because of the geographical differences in the shapes of the two ocean basins. The consequences of this are not trivial: the heat equator, and hence the ITCZ between the northern and southern trade-wind systems, lies north of the true equator. Nor is the South Atlantic fully enclosed to the east since Africa extends south only to 35°S. Perhaps most importantly, because the Andes extend further both poleward and equatorward than do the western mountains of North America, they form a more complete barrier to the flow of the planetary westerlies: this extends almost 20° latitude further poleward than does the tip of Africa.

For these reasons, the confluence between the westerlies and the trade winds across the South Atlantic is not simple; the westerlies sweep up into the South Atlantic around Cape Horn, to pass to the east across the ocean in the "roaring forties." Thus, at all seasons, westerlies dominate across the southern margin of the ocean from just north of the Falklands plateau to just south of Cape Town near 35–45°S. Although the subtropical gyre of the South Atlantic is more persistently under the influence of the trade winds than that of the North Atlantic a significant seasonal variation in its mixed-layer depth is nevertheless forced by seasonal variation in wind stress and surface heat flux. This effect is zonally asymmetric, reaching deeper in the western part of the ocean than in the east. Winter heat loss at the sea surface is remarkably uniform (ranging between only −100 and −125 W $m^{-2}$) in July over the whole region from 10°S in the SEC right down to 45°S. It is not surprising, then, that even at relatively low latitudes a deepening of the mixed layer should occur during the austral winter (Levitus, 1982), when the ITCZ is at its most northerly position and highest wind speeds and wave heights occur from the equator to 25°S. The greatest mixed-layer depths (<100 m) occur along a zonal trough at about 20°S in July, across the west-central part of the gyre.

Flow of the Benguela Current along the African coast begins to turn to the northwest (thus leaving the region designated as the BENG Province) and passes across the equatorward side of the gyre as a broad stream at about 15–30°S. The axis of this flow lies across the ocean in a southeast/northwest direction to join the intensified western boundary current that then returns the flow southward as the Brazil Current. As occurs in the northern hemisphere, prominent tropical instability waves, associated with the equatorial divergence, may be observed in the surface chlorophyll field along the northern border of SATL, these being most prominent when trade wind stress across the equatorial zone is maximal in late austral winter (August–October).

The confluence between the Falkland and Brazil Currents at about 38–40°S, which forces each to turn eastward across the ocean within the Subtropical Convergence zone, is a highly energetic region that dominates the circulation pattern of the southern Atlantic Ocean and is rich in mesoscale features that are prominent in the SLA field (Fig. 9.19). This distracts our attention from the fact that the South Atlantic subtropical gyre actually exists as a double-cell circulation (Tsuchiya, 1985), the foci of the two "pinched" (to use Tsuchiya's word) gyres lying close to the western coasts. A zone of eddying, also visible in the SLA field, lies southwest across the ocean: is this a surface indication of a subtropical countercurrent that, as Qiu (1999) suggests (see SPSG), is to be anticipated at this latitude in each subtropical gyre? Although Tsuchiya does not mention it, we should note that his two subgyres lie alongside the Rio Grande Ridge that rises to within 2000 m of the sea surface at around 29°S. Is this topography involved in locating the circulation features?

Despite this possibility, it is the Confluence region itself and the region of heavy meandering and SST anomalies at the boundary of the Subtropical Convergence zone of the South Atlantic Current (SAC) across the southern part of this province that takes our attention, even though the area of meanders itself is attributed here to the SSTC Province. Significant changes in the latitude at which the Confluence occurs, and flow separates from the continent, have been observed almost throughout the last century (Olson *et al.*, 1988); this variability appears to be forced by variability in the Brazil Current and so by the relative strength of the seasonal trade winds associated with the Southern Oscillation. SST anomalies here are weaker than those associated with eastern Pacific ENSO variability, even if associated with drought conditions over South America.

Because the retroflection loop of the Agulhas Current southwest of the Cape of Good Hope (see EAFR) carries more eddy kinetic energy than anywhere else in the southern hemisphere, large warm-core eddies are shed when an Agulhas intrusion into the Atlantic occurs, usually 5–10 times each year. These unusually large (300-km) eddies have very long lifetimes, and some survive to reach the Brazilian coastal boundary where they have been resolved in altimeter data. They are therefore of potential significance in the structure of the whole subtropical gyre (Peterson and Stramma, 1991; Shannon, 1985).

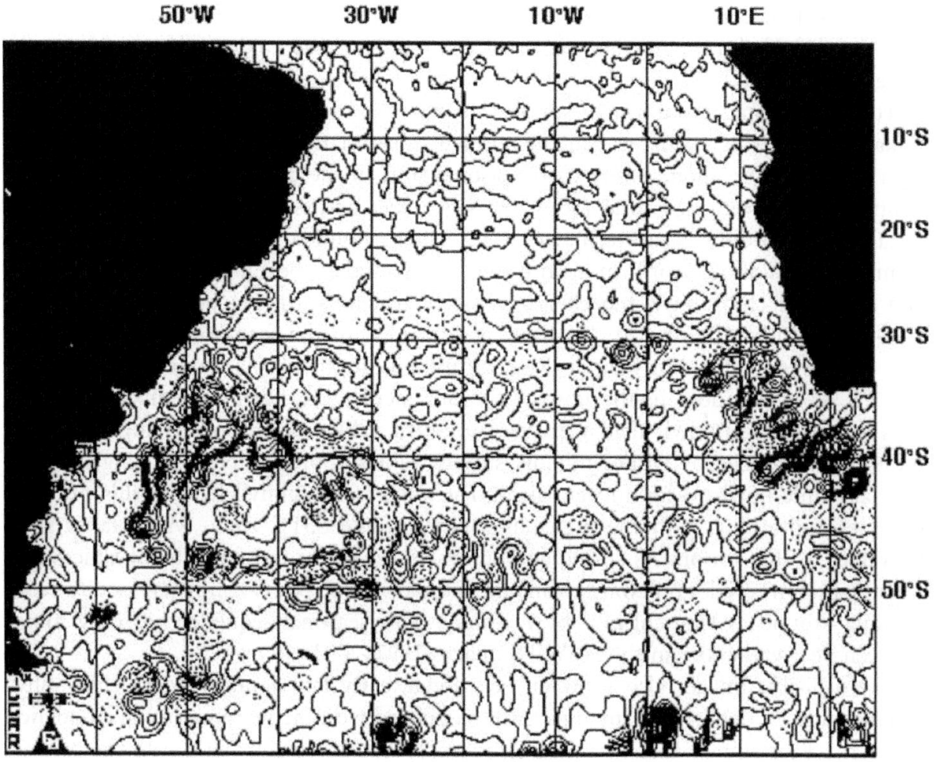


**Fig. 9.19** Sea surface elevation over the South Atlantic to emphasize the extreme southwest position of the focal axis of the gyre that is not, as is usually assumed, located centrally. Ephemeral, dendritic blooms are observed in sea surface satellite imagery near the northern edge of the focal eddy field; it is suggested that this is a similar, but less intense, phenomenon to that of the dendritic bloom of the southwest Indian Ocean, whose gyral focus is also in the extreme southwest. (TOPEX-POSEIDON data for 1 December 2002, courtesy of the University of Colorado.)

## Response of the Pelagic Ecosystems

Modern information on the ecology of the open South Atlantic is singularly lacking, although Piontkowski *et al.* (2003) have reviewed results of the exploration of the ocean by Russian biogeographers during the Soviet period. Some information is also accumulating from the routine voyages of Antarctic scientific supply ships, the AMT project discussed earlier. These data confirm that phytoplankton biomass does accumulate during austral summer, with the expected greater importance of a DCM equatorward. It is also clear from the AMT observations that the general pattern of phytoplankton growth is consistent with the baroclinicity of the nitrate field, with higher chlorophyll values consistently around the periphery and minimal values in the central regions (Marañon *et al.*, 2000).

Unfortunately, the AMT tracks pass only along the western margin of the South Atlantic gyre, so it is to satellite data that we must turn for a conceptual model of productivity in this province as a whole. This is simply stated: basin-wide surface chlorophyll and SST cycles correspond very well, so that the date of greatest spread of oligotrophic conditions matches very closely the date of the SST maximum in late February. The surface chlorophyll data indicate a basin-wide maximum in late August, coinciding with minimum basin-wide SSTs. An overall pattern for the relative distribution of SST is, again, simply stated: this is always higher in the western part of the province, cooler in the eastern part. Such a pattern reflects the sources of the water entering the western and eastern limbs of the subtropical gyre: the warm SEC and cold SAC, respectively. The summer SST field reflects the spread of warm conditions eastward across the province, the winter field the progressive westward cooling from the cold water adjacent to the Benguela Current.

For the seasonal evolution of surface chlorophyll, the satellite images are very illuminating, and from them we can infer much more than we can observe. There is, in each year of observation, a very clear seasonal shift in the chlorophyll field. The South Subtropical Convergence zone (SSTC Province) lies across the ocean as a permanent zonal high-chlorophyll feature, having its northern boundary at about 40°S in austral summer. From about April, in austral fall, this boundary becomes increasingly diffuse and enhanced chlorophyll moves progressively equatorward, reaching almost 25°S in mid-winter (July–August). At this time, associated with the focus of the gyral circulation southwest of the Rio Grande Rise, diffuse chlorophyll enhancement reaches even to 15°S, as it also does in the flow of the SEC across the northern limb of the gyre. This seasonal chlorophyll field with very uniform values of around 0.3 mg chl $m^{-3}$ is presumably induced by winter mixing, surface values appropriately matching depth of the baroclinic bowl in the nutrient field. The winter chlorophyll field of this province becomes continuous with the high chlorophyll of the SSTC zone, which takes similar values at this season. As spring and summer conditions develop, the oceanic South Atlantic bloom recedes poleward, while the biomass in the SSTC zone increases, to reach maxima around 1–2 mg chl $m^{-3}$ by February. This seasonal sequence has only minor between-year variability in the SeaWiFS images 1997–2003.

The available images also show an occasional, early fall (January–May) chlorophyll enhancement of dendritic form apparently associated with incoherence in the circulation above the Rio Grande Rise at about 45°W 45°S that is seen in Fig. 9.19. This was an exceptional feature in April 2000, dominating the chlorophyll field of the western ocean. The location of this feature closely matches the ridge between the double-cell circulation of Tsuchiya (and of cold anomalies in the SST field) sufficiently well that we may be confident of the reasons for its formation. In late summer, early fall of 1998, 1999, 2001, and 2002 much fainter and more obscure linear enhancement is observed in the same region. This feature would bear investigation for it matches, both in months of occurrence and location in the ocean basin, a similar dendritic chlorophyll feature in the western Indian Ocean (see INDW Province). That feature is not associated with any topography, but it is intriguing that the months of greatest development of the Rio Grande Rise bloom in 2000 correspond with the formation of a very similar dendritic bloom in the SW Indian Ocean. It seems very probable that at similar latitudes in both oceans, and near their western margins, equivalent south subtropical countercurrents take their origins, being observable at the surface as highly eddying features in the chlorophyll field.

The detailed meridional section run by Agusti and Duarte (1999) down the western side of the gyral circulation in October–November 1995 confirmed that the chlorophyll profile has the anticipated relationship with physical structure. The shallow (60–80 m) DCM of the tropical region persisted until 12°S, although here it was underlain by a much deeper chlorophyll feature at 100–150 m, just below the depth of maximal Brunt-Väisälä frequency; this wider DCM deepens progressively to about 20°S and then shoals toward the south. At 35°S, at the southern end of the section, approaching or within the SSTC, a near-surface (40-m) DCM is interposed above the remnant deeper feature. The relationship between depths of DCM and 1% surface light is very close along the entire section, as is that between surface and integrated chlorophyll concentration. A meridional section reported by Dufour and Stretta (1973) close to the Greenwich meridian shows a very similar situation, with the phosphatocline coinciding very closely to the southward-deepening pycnocline.

The AMT sections for October (austral spring) and May (austral fall) confirm these patterns down the western side of the SATL province. However, as Marañon *et al.* (2000) point out, the DCM is neither a biomass nor a productivity maximum, as is often assumed, but rather results from a local increase in the chl/C ratio and represents only a small contribution to total integrated primary production in the water column. Marañon *et al.* found that the phytoplankton turnover rate in the oligotrophic SATL gyre (at 50 mg C $m^{-2}$ $d^{-1}$), measured as the microalgal growth rate, was unexpectedly low (0.21 $d^{-1}$ at the surface) representing <20% of maximal expected growth. They suggest that this finding argues against the general assumption of high turnover rates in oligotrophic gyral situations. They did, however, observe an unexpectedly high variability in phytoplankton dynamics in the oligotrophic gyre so that productivity and growth rates varied by a factor of 8, while microbial biomass remained relatively constant. This is consistent with the suggestion of Mahaffey *et al.* (2004), also associated with AMT data, that nitrogen supply in oligotrophic gyres is predominantly from the subthermocline source, probably mostly due to fine-scale upwelling in mesoscale eddies and frontal systems. General diapycnal transfer and convection, they suggest, is a minor source in such locations.

The October 1996 AMT section (Zubkov *et al.*, 1998) provided unequivocal information on the relative distribution of microautotrophs across the subtropical gyre (Fig. 9.20); the differential distribution of *Prochlorococcus* in the subtropical gyre and of *Synechococcus* and picoeukaryotes (together with heterotrophic bacteria) in the more eutrophic regions to north and south is very striking in these data, as is the relationship of each with the

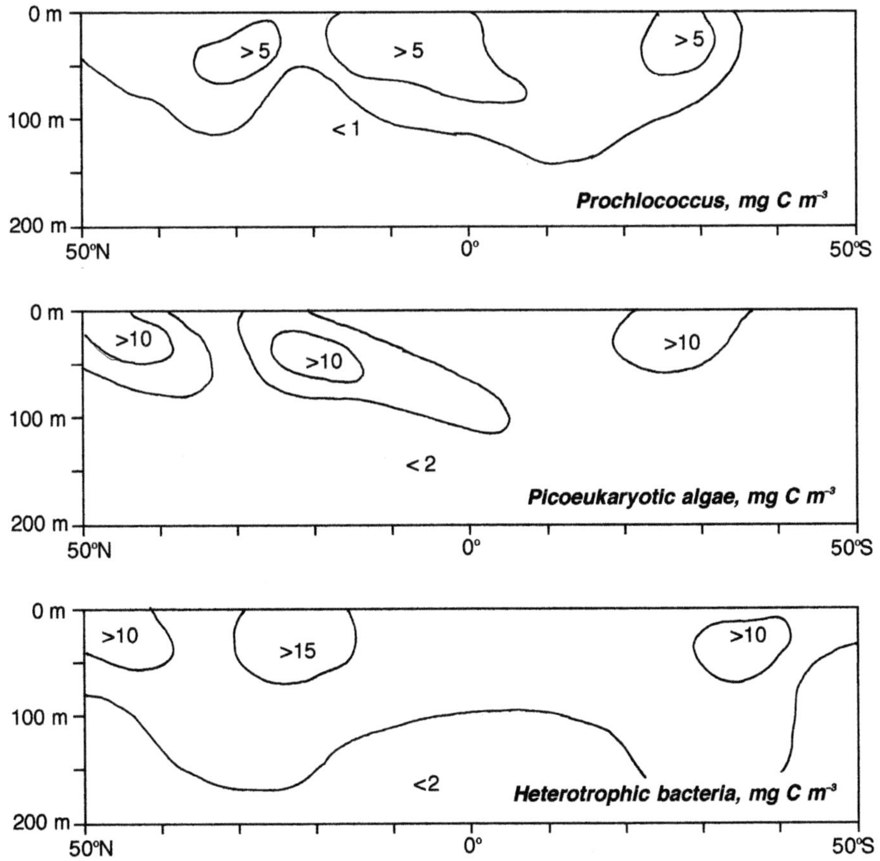


**Fig. 9.20** An Atlantic meridional section, to show the differential distribution of heterotrophic and autotrophic bacterial flora and of picoeukaryotes in the upper water column.

*Source: Redrawn from Zubkov, 1998.*

depth of the mixed layer. This is an excellent illustration of the numerical dominance of *Prochlorococcus* in low-biomass oligotrophic situations and of the general dominance of biomass—because they are larger cells—of the picoeukaryotes.

The Russian biogeographic studies do not help us a great deal: an extremely detailed listing of the distributions of 574 species of phytoplankton obtained from visual cell counts and of mesozooplankton species counts from 1300 stations (including 416 species of copepods) was reduced by Piontkowski *et al.* (2003) to spatial distributions of indices of relative diversity, which is greatest, as we might expect, on the western, more dynamic side of the gyre. It is noted that although many species of both groups occur everywhere in the gyre, others occur only in specific regions of flow. The calculated ratio of group abundance is such that mesozooplankton carbon biomass may be expressed as an exponential function of phytoplankton carbon.

Thus, the emphasis in studies of zooplankton in the South Atlantic has so far been distributional in nature and most have been concerned specifically with the SW Atlantic. Distribution envelopes derived from presence-absence data have frequently been plotted, as by Boltovsky (1986) for many taxonomic groups, from which he inferred more than 20 different characteristic patterns of "biogeographic zonations, transition zones and communities" between equatorial and polar latitudes in the western South Atlantic. He compared these with the large-scale distribution of euphausiids in the Pacific according to Brinton, and emphasized that in both cases, many species cross what he describes as "boundaries between water masses." By far the most critical and useful study that I have seen is that of Gibbons (1997), who compared distribution of all euphausiids species on a 5° grid, as already discussed in Chapter 1. This emphasizes that the major changes in distribution occur across the subtropical convergence zone and that change is weaker between the gyral populations and those within the equatorial zones to the north. Gibbons emphasizes the differentiation of pelagic ecosystems by the characteristic relative abundances of species that occur in both entities: such differences are not exposed in the habitual presence/absence data of biogeographers.

### Synopsis

*Case 3—Winter-spring production with nutrient limitation*—All seasonal changes, integrated over the province, have very weak amplitude but are reasonably consistent. $Z_m$ undergoes a moderate austral winter excursion from 25 m in February–March to 60 m in May–September; because $Z_{eu}$ remains between 50 and 70 m, the thermocline is almost constantly illuminated. P rate increases consistently from austral spring August–September, reaching and briefly sustaining higher values in October–November; the production rate then declines, prior to austral midsummer, and the decrease is linear to April–May, just before austral midwinter (Fig. 9.21). Chlorophyll accumulation begins earlier and, in some years, maximum chlorophyll biomass occurs 60 days before maximum P rate is achieved.

# Atlantic Coastal Biome

## Northeast Atlantic Shelves Province (NECS)

### Extent of the Province

The Northeast Atlantic Shelves Province (NECS) comprises the continental shelf of western Europe, from Cape Finisterre in NW Spain to the Skagerrak north of Denmark, and thence into, and including, the Baltic Sea. The edge of the deep Faeroe-Shetland

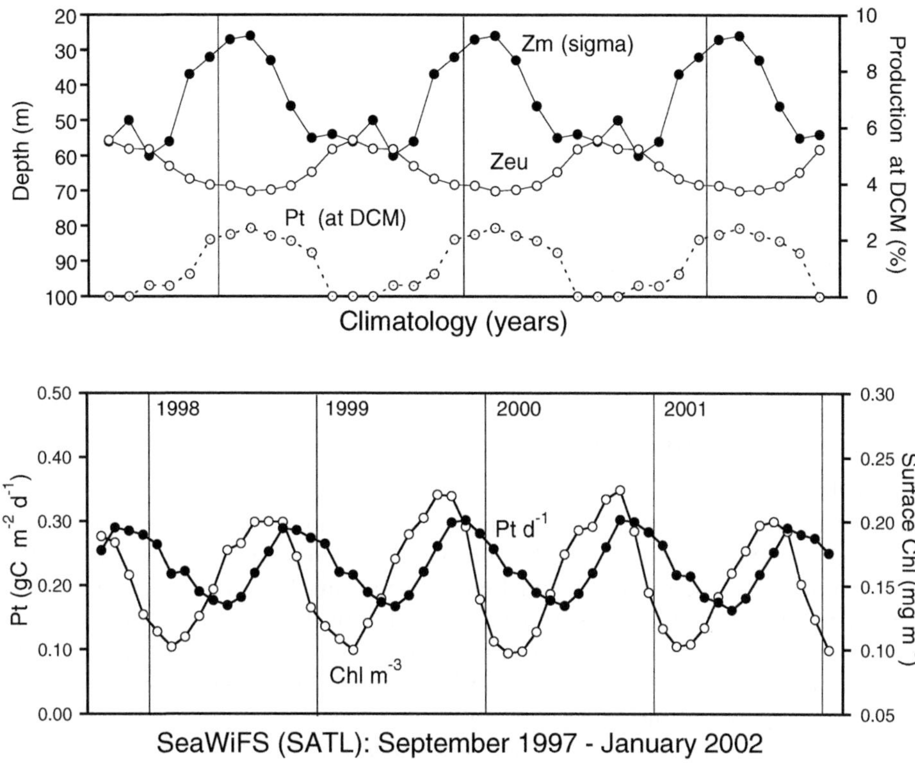


**Fig. 9.21** SATL: seasonal cycles of monthly surface chlorophyll and depth-integrated autotrophic production for the years 1997–2002 from SeaWiFS data together with characteristic seasonal cycles of mixed-layer depths from Levitus climatological data and photic depths computed from characteristic irradiance and the archive of chlorophyll profiles discussed in Chapter 1.

Channel and the Norwegian Trench forms the separation between this and the SARC province to the north.

## Continental Shelf Topography

The topography of this shelf region is dominated by the consequences of glacial activity that has cut many deep canyons into the shelf edge. Deep troughs occur in many areas far from the shelf edge as in the English Channel, and both relic and modern sand waves occur commonly, as in the Celtic Sea. The North Sea has a shallow central region of glacial deposits, the Dogger Bank, only about 20 m deep, as well as several relic deep basins. This province was, in fact, mostly dry land during the last Pleistocene glaciation.

Of the entire continental shelf region ($1.63 \times 10^6\,km^2$), the Baltic Sea represents about a quarter and the North Sea about one-third. The width of the shelf from the Danish coast to the shelf west of Ireland is ~1000 km and it is only in the extreme south, along the northern coast of Spain and beyond the Cap Ferret Canyon at 45°N, and also at the entrance to the Baltic, that the shelf is very narrow.

Like other coastal regions, this province could rationally be subdivided almost infinitely, and a first-level subdivision may be useful for some purposes. If so, the following would probably be the candidates for the primary divisions: (i) The *Celtic Seas* from Cape Finisterre to the Shetlands, (ii) the *North Sea* from Dover to the Skagerrak, and (iii) the *Baltic Sea*, including the Gulfs of Bothnia and Finland.

The relative distribution of sediment types is controlled here, as on other shelves, largely by the regular cycle of resuspension and subsequent deposition of fines that, as

I discussed in Chapter 1, will give us difficulties in interpreting chlorophyll images, in addition to the problem presented by CDOM. Where tidal flows are rapid, fines are not deposited, and gravel and coarse sand dominate. Where wind stress at the surface is strong, at least seasonally, fines are resuspended so that bottom deposits tend also to be sandy. Marine geologists have long recognized a "mud line" at 70–80 m in this province, below which one may expect to encounter bottom types dominated by fine, muddy material. Thus, on the Aquitanian shelf, much of the solid material originating in the effluents of the Loire and Gironde rivers is segregated in a muddy belt at midshelf depths, the "Grande Vasière" of French benthic ecologists. Beyond 100 m in the northern basin of the North Sea and in the Norwegian trough, mud also dominates. In shoaler water, soft muddy sediments occur only in some restricted depressions off the Gironde, in the central North Sea, and also in estuaries and some coastal embayments.

## Defining Characteristics of Regional Oceanography

The hydrography of this province is dominated by tidal forces in the west, and by freshwater buoyancy in the east; for this reason, I have chosen not to discuss tidal fronts in a special section, as for some other provinces. As Simpson comments in his 1998 review of the Celtic Seas, the large tidal input there from the Atlantic represents about 12% of all tidal energy that is dissipated globally. The tidal wave, coming from the deep ocean, interacts individually and differently with the topography of each of the many basins comprising this province. Thus, tidal amplitude varies from almost zero at the amphidromic points of the $M_2$ tide to >10 m in cone-shaped embayments such as the Bristol Channel or the Gulf of St. Malo. Simpson also noted, "Particularly strong tidal responses occur in the English Channel, the Irish Sea and the Bristol Channel with the largest ranges in each case occurring on the right side of the basin when viewed from the ocean." In the Irish Sea, for example, 80% of the incident energy is dissipated, and the frictional stress at the seabed may exceed 4 Pa and is largely in excess of 0.25 Pa, which is equivalent to the surface stress of a wind of 13 m $sec^{-1}$.

For the reasons briefly outlined in Chapter 3, those regions where tidal streams are sufficiently strong that the mixing parameter $\log_{10} h\, u^{-3}$ remains in excess of a critical value (around 2), summer stratification induced by surface heating does not occur. Mixed and stratified regions are separated by convergent "tidal-mixing fronts" that are variable in position during the lunar cycle (see Fig. 3.4). These fronts usually separate a cool water mass from one with warm water above and cold below. Flow along such fronts is associated with a two-cell transverse circulation, often marked by convergent slicks and a linear cold temperature anomaly, and may break down into a series of mostly cyclonic eddies. Convergence in tidal fronts has, as we have seen, important consequences for the physical accumulation of biota.

Of course, during winter, when the open North Atlantic is mixed to a depth of several hundred meters, the water column over this shelf is everywhere mixed to the bottom. It is only after increasing irradiance in the spring has imposed stratification on the deeper parts of the shelf that we may expect tidal fronts to occur. The progressive seasonal development of stratification causes such fronts to move onshore in the spring and offshore in the autumn, over a distance of >200 km in the western Channel (Pingree, 1975; Pingree and Griffiths, 1978; Pingree *et al.*, 1982). In summer, the principal tidally mixed regions are the eastern Channel, the southern North Sea and the Irish Sea. Tidal fronts, therefore, occur predictably in an arc across the southern North Sea from Yorkshire to the Friesian coast, across the western entrance to the English Channel, and at northern and southern mouths of the Irish Sea. The location of tidal fronts is sufficiently predictable that they could be used for a regional partition of this province that would be sensitive to the ecological consequences of stratification and mixing regimes respectively. In tidally mixed regions, here and elsewhere, I shall assume that the "chlorophyll" signal in SeaWiFS images largely represents the relative distribution of suspended fines and CDOM rather than phytoplankton biomass. This is, of course, an approximation: the mapping of total suspended material (TSM) for coastal water quality control is a delicate matter and techniques are not yet resolved. Consult the Dutch POWERS-II Report if this problem is of concern to you.

In addition to tidal fronts, a variety of other shallow-water, coastal, and estuarine fronts must be considered here; estuarine fronts, characterized by turbidity and salinity gradients, occur in many places along all coastlines, although not many have been studied in detail. Because the hydrography of the Baltic Sea is dominated by large freshwater inputs (see later discussion) and has a narrow entrance to the North Sea through the Kattegat, a very strong salinity gradient occurs, having the characteristics of a frontal zone, within this channel to the east of Denmark.

The spring-summer shelf-edge front appears as a meandering belt of relatively cool surface water (1–2°C anomaly) just beyond the shelf edge and overlying the upper slope. Satellite AVHRR shows this to be a quasipermanent feature, at least 800 km in length and wider at the end of summer than in spring. For this retrograde shelf edge front, explanations have evolved more slowly than for tidal fronts although, following Pingree *et al.* (1986), it is now understood that an internal tide is generated at the 200-m isobath and propagates from there both offshore and onto the shelf. As Joint *et al.* (2001) suggest, the resultant progressive waves force the vertical transport of cool, nutrient-rich water into the euphotic zone. An understanding of these processes were among the objective of the OMEX I studies (1993–1995) of the ocean margin of Western Europe: both appear to contribute to the linear zone of enhanced productivity above the shelf break of the Celtic Sea and Bay of Biscay.

The continental shelf of Western Europe lies beside a region of weak poleward drift so, as noted earlier, transport over the shelf is dominated by local winds: strong geostrophic flow along the continental slope is lacking. The outer shelf is largely occupied by homohaline water that is thermally stratified in summer, except from northern Denmark to the west of Scandinavia, where low-salinity Baltic outflow induces haline stratification. Influx of Atlantic water occurs principally as southerly flow in the Norwegian Trench, and on either side of the Shetlands; flow through the Straits of Dover is trivial. Interannual variability in the supply of Atlantic water is thought to be a major factor in forcing regime shifts in the North Sea ecosystem.

On the middle shelf regions to the west of Brittany, where values of the stratification parameter are extremely high (so that stratification occurs very early in the year), nutrient dynamics in summer are strongly influenced by several zones of remnant "winter water." This is colder and denser than the surrounding water and remains trapped in the areas where summer stratification is most strongly developed; these "*bourrelets froids*" (Le Fèvre, 1986) are dome-shaped and have the potential to act both as a nutrient reserve and as an accumulator of nutrients during the summer months. Underlying the Armorican bourrelet lies the region of the "*Grand Vasière*", or big mudhole, noted earlier.

Especially in the Bay of Biscay, and along the Friesian coast, the effluents from the Gironde, Loire, and Rhine form buoyant plumes that extend both seaward and coastwise from the river mouths: the Bay of Biscay has a very generalized near-coastal region of lowered salinity, whereas the Loire plume extends northwest into the mouth of the Channel under conditions of exceptional discharge. As will be discussed later, these buoyant plumes have important consequences for the development of planktonic biota.

In the northeastern part of the province the circulation pattern is buoyancy-dominated, although even here, tidal energy dissipation may far exceed the energy input by surface wind stress (Rodhe, 1998). Beyond the region of tidal fronts in the southwestern North Sea, climatological salinity is progressively reduced from 34.5‰ to 10–15‰ in the

Kattegat and thence down to 2–3‰ in the Gulfs of Bothnia and Finland. Stratification is thus imposed progressively by buoyancy as well as by surface heating. Outflow of low-salinity water released from the Kattegat forms the narrow (10 km wide), baroclinic Baltic Current that conforms to the Swedish coast and, after modification within the Skagerrak, subsequently follows the topography of Norway as a wider coastal current.

The Baltic itself is a relatively small mediterranean sea of mean depth 57 m, having a shallow sill depth (<20 m) at the Kattegat. Water balance (and an estuary-like circulation) depends on river discharge and episodic ventilation of the deeper water mass by irruptions of saline water over the sill. Permanent salinity stratification occurs through the entire Baltic, with a halocline at 30–40 m, the actual depth depending on the sill depths between the different basins. The inner Baltic has a very low surface salinity (1–3‰), permitting ice cover to develop over the entire northern region in winter. Surface irradiance in summer reinforces stratification. The Gottland Deep (about 200 m depth) is anoxic, because of the shallow sill and weak ventilation of the central Baltic Sea: the extent of the anoxic area is rather variable, depending on the relative amount of ventilation that occurs each year through the Kattegat. General surface circulation in the Baltic is cyclonic, weak, and replete with mesoscale eddy features; the dominant southwesterly winds generate both coastal jets, and more or less permanent upwelling at some coastal topographic features.

### Response of the Pelagic Ecosystems

It should be borne in mind, when thinking about the ecology of this province, that here the marine environment has been strongly modified by farming ashore and by mining, oil drilling, and fishing at sea. Massive extraction of sand and gravel in the southern North Sea has recycled significant amounts of previously buried inorganic nutrients into the water column. The intensive agriculture of the European fields releases very large quantities of nutrients, pesticides, and herbicides: phosphate input to the North Sea from such sources increased by a factor of about 7 from 1950 to 1980. Industrial fishing has strongly modified not only fish stocks, but also disturbed the surficial geology and the benthic invertebrates of the sea floor.

In discussing the physical environment, I have emphasized the distinction between mixed and stratified regions, and the existence of tidal fronts, because these are fundamental elements in the functioning of continental shelf ecosystems. Because of the displacement of these fronts with the lunar tidal cycle their exact distribution is not observable with precision in the 7- and 30-day composite satellite images. Nevertheless, one of the most prominent features in almost every image of this region is the Friesian front between the shallow region where bottom stress from tidal currents is very strong and the deeper southern North Sea, where stratification develops in summer. This seemingly "high chlorophyll" feature is, in fact, a shallow region of relatively very turbid water. Several exemplary TSM images of this region are available, each representing an individual day during 1999, and can be compared with the SeaWiFS images for the same 7-day period to support this assumption. These images show that not only is tidal bed stress responsible for resuspension, but so also is the passage of a storm system. Anyone who has seen the brown seas off the East Anglian coast in windy weather knows what the satellite is seeing under such conditions.

Nevertheless, the chlorophyll images serve very well to illustrate the difference between instantaneous reality and the canonical model for the phytoplankton seasonal cycle in this province; the complex spatial pattern of apparent chlorophyll enhancement changes strikingly from month to month, and between years. A review of the 7-day composites for 1999 shall serve to illustrate the whole. The shelf-break front and slope current often bears a linear band of high chlorophyll, of much higher concentrations than over the adjacent shelf. This is seen well in early June, when the slope current bloom extended from

Norway to the west of Ireland. Early in the year, until the end of February, chlorophyll biomass was uniformly higher over the shelf than over the open ocean, so that almost the entire shelf break was delimited by the abrupt difference in biomass. During March 1999, a discrete spring bloom occurred on the deep shelf to the southwest of Ireland, and in April the northern North Sea sustained a spring bloom. By the middle of July 1999, the entire central North Sea had evolved very low surface chlorophyll while the Norwegian Coastal Current remained a prominent high-chlorophyll feature, separated from higher chlorophyll over much of the North Sea by blue water that follows the line of the Norwegian Trench. Finally, in the last months of 1999, chlorophyll biomass in the Central North Sea increased from the very low values of July and remained in this state through the following winter.

In April 1998, a strong bloom occurred over the ocean west and north of the British Isles that was restricted very precisely by the shelf edge, landward of which chlorophyll took much lower values. This was the opposite condition from that obtaining a few weeks earlier, when values over shelf depths exceeded those over the deep ocean. But the shelf edge is by no means always a line of demarcation between different conditions. In June 1999, an incursion of clear blue water Atlantic water flooded landward over the wide shelf regions to the west of Cornwall. And on many occasions, large areas of high chlorophyll concentration cover both shelf and contiguous deep ocean regions.

Such a diverse range of processes represents the interaction between an ideal seasonal production cycle, modeled as a response to climatological conditions, and between-year differences in physical forcing. In the ideal model, we can recognize four ecological seasons: (i) autotrophic growth constrained by light limitation in winter, (ii) a nutrient-limited spring bloom, (iii) stratified conditions during summer with localized dynamic zones of high chlorophyll, and (iv) renewed autotrophic growth in autumn if stratification breaks down while surface irradiance is still relatively strong. This is, of course, a shorthand version of the classical plankton calendar for the continental shelf that was worked out many decades ago at the old European marine biological stations.

We should not expect to observe this ideal sequence in many regions: for instance, in areas of permanently mixed water where tidal streams are too strong and water is too shallow for summer stratification to develop. Here, because tidal friction is constantly supplying nitrogen to the water column from benthic regeneration processes, and because the limiting tidal fronts are constantly transporting nitrogen into the mixed areas, we should not expect nitrogen to limit a bloom. Nor, under such circumstances, should we expect the bloom to begin as early in the spring as it does offshore, because of greater light limitation due to suspensoid: rather, we may expect a midsummer bloom in which the rate of primary production is a simple function of irradiance. In the mixed area of the western English Channel, in March the rate of primary production starts to increase from low values in winter and this increase is maintained steadily until a maximum rate is achieved in July. An isotonic decrease is subsequently observed until the winter minimum is reached again in November. During the whole period, diatoms dominate the large cells and dinoflagellates are negligible. The same seasonal cycle occurs in the permanently mixed, highly turbid Severn estuary and the macrotidal Bristol Channel, with a progressive seasonal bias toward a spring bloom in the outer, less turbid region.

Another situation that diverges from the ideal sequence is offered by the anomalous blooms that may be observed wherever and whenever stratification is locally imposed on a previously mixed water column; this occurs most frequently as a result of freshwater buoyancy, imparted by river effluents, provided that surface irradiance is sufficient at the time to support plant growth. This may occur very early in the year, as discussed later.

A very useful compendium of location-specific time series (1960–1984) of nitrate, productivity, and zooplankton biomass is offered by Bot *et al.* (1996). The ideal phytoplankton cycle seems to occur most frequently in the central North Sea. In the southern

North Sea the spring and autumn peaks may be of about the same magnitude, except in the mixed region off the Dutch coast where a single summer bloom is the rule. In the northern North Sea and the western English Channel, the spring bloom is stronger, relative to the autumn bloom, than elsewhere. The Baltic Sea has a double-bloom cycle, and the autumn bloom may reach even higher chlorophyll values than those during spring (Kullenberg, 1983). In the Irish Sea, and probably elsewhere, the spring bloom starts in shallow embayments (as in a recent study of Dundalk Bay) and occurs progressively later offshore and farther to the north.

Seasonal succession has been followed particularly closely in the western Channel, where Holligan and Harbour (1977) distinguished a near-surface spring bloom ($<4$ mg chl $m^{-3}$, 0–15 m, April) from a summer subsurface bloom in the thermocline (2–4 mg chl $m^{-3}$, 20–25 m, May–September, fueled by regenerated $NH_4$). An autumnal near-surface bloom ($<2$ mg chl $m^{-3}$, 0–15 m, late September to October) followed. Diatoms initiated the spring bloom and were abundant until May, when they were progressively replaced by dinoflagellates and flagellates; this process was completed by midsummer. In the autumn bloom, diatoms again became important. The spring bloom of diatoms develops faster than herbivores can increase their consumption rate by population buildup; consequently, much of the plant biomass sediments to the sea floor to provide at least a part of the regenerated nitrogen utilized by the microalgae of the summer phytoplankton. Such an imbalance of copepods and diatoms has been observed in several locations in NECS.

The summer subsurface chlorophyll maximum may be concentrated within a depth range of only a few meters, as in the Skagerrak, where it has been observed to dome centrally, following the density contours very precisely around the gyral circulation. In the Belt Sea, a study by Smetacek *et al.* (1984) revealed what is probably a typical seasonal cycle in shoal water. The spring bloom utilized nitrate that had accumulated in winter, during which large-scale sinking of plant cells occurs. This bloom was followed by an early summer population maximum of herbivorous zooplankton and consequently very little sedimentation of plant cells. By midsummer a complex food web had developed, based largely on regenerated ammonium. Finally, during autumn and after the seasonal increase in wind strength, a bloom developed, based on nitrate that had accumulated in subthermocline water during the summer.

Given the diverse range of characteristic situations in large shelf regions such as this, it is probably not entirely satisfactory to categorize the relative significance of autotrophic cell fractions. Nevertheless, the relatively new generalization that autotrophic pico- and nanoplankton are a vital and important component is also valid in these shelf waters. Joint and Williams (1985) computed that 36% of primary production over the western shelf is the work of the 0.2 to 1.0 $\mu$m cell fraction and that 77% is produced by the 0.2 to 5.0 $\mu$m fraction. Even more recently it was suggested that production by autotrophic picoplankton accounts for 50% of production prior to the spring bloom, but thereafter the absolute production rate of the fraction changes little, subsequent seasonal increases in productivity being due almost entirely to cells $>2\,\mu$m.

Much attention has been given in recent years to progressive evolution of ecosystem structure and functions here, as in the North Pacific: a recent review by Alongi (1999) is a good introduction to these issues. Here, it is enough to note that it is suggested that there has been a shift of dominance from diatoms toward dinoflagellates and also to earlier spring blooms with very persistent summer blooms of dinoflagellates.

The summer profile on the shelf may be typified by that of the Celtic Sea, a two-layered system in which the upper water is at summer temperatures and the lower is at about 8 or 9°C, or close to winter values. At the interface, in the thermocline, a DCM is associated with maxima of microflagellates and ciliates. Bacterial biomass is uniformly high in the upper layer and uniformly low in the lower layer. Mesozooplankton partition this vertical space: the Atlantic-boreal *C. finmarchicus* is restricted to the lower layer, the temperate *C. helgolandicus* is restricted to the upper layer, and the euphausiids (*Nyctiphanes couchi* and *Meganyctiphanes norvegica*) make diel migrations between the two layers. In the western English Channel two chaetognaths, like the two species of *Calanus*, partition the water column: *Sagitta elegans* occupies the cold, lower layer and *S. setosa* the upper layer. None of these findings are at variance with observations of vertical profiles in summer in the central North Sea or in the FLEX box in the northern North Sea, where *C. finmarchicus* is the dominant calanoid at all depths.

The OMEX I investigations of the shelf-break ecosystem west of Brittany (e.g., Joint *et al.*, 2001a,b) revealed daily productivity of ~150 gC $m^{-2}$, with an f-ratio of <0.25 during summer months, but higher during the spring bloom. During winter, heterotrophic demand (bacteria to mesozooplankton) was significantly lower than autotrophic production, although during summer the heterotrophic demand exceeded production of plant cells. The annual budget suggests that 40–60% of total autotrophic production is not utilized by the heterotrophs and is available for export to midwater organisms and the benthos. These direct measurements during OMEX I were compared by Joint *et al.* (*op. cit.*) with computations that they made from SeaWiFS data: the two methods produced very similar results. Joint and Williams had already estimated in 1985 that the demands of herbivores in the Celtic Sea could only be met on the assumption that they can take particles in the 1- to 5-μm range; microplankton appear to consume 10–40% of the standing stock of autotrophs daily.

The pelagic ecosystem of the Baltic is spatially complex (Segerstrale, 1957). In this small sea, there is a gradient from estuarine organisms in the west (Kattegat: salinity, 10–15‰) to lacustrine organisms in the east (Gulf of Bothnia: 2–3‰). The glacial relic copepod *Limnocalanus grimaldii* occurs in >97% of plankton tows in the Gulf of Bothnia and in <10% in the Belt Sea east of Denmark where the salinity gradient is sharpest. The marine medusa *Aurelia aurita* is abundant in the western Baltic but occurs rarely and does not reproduce in the Gulf of Finland. Another glacial relic mysid (*Mysis relicta*) occurs in deep water. Brackish water copepods (*Eurytemora hirundoides* and *Acartia bifilosa*) and cladocera (*Bosmina maritima*) are an important component of the fauna, with the last organism at times being the most abundant planktonic crustacean in summer and an important food for herring. Other cladocera (*Podon* and *Evadne*) and rotifers are especially important in the gulfs in summer. Large cells of the phytoplankton are dominated by diatoms and Cyanophycae, especially *Nodularia, Anabaena*, and *Aphanizomena*.

In the Landsort Deep, in the main basin of the Baltic Sea, the mesozooplankton exhibit some resource partitioning in a water column having a pycnocline at 70–80 m, associated with a nutricline and (presumably) a DCM (Ackefors, 1966). Small abundant copepods are specialized to depth horizons: *Acartia* spp., 15 m; *Temora longicornis*, 20–25 m; *P. elongatus*, 50–100 m.

Fisheries science in this region has been much concerned with the consequences of between-year changes in the physical environment for the recruitment to fish populations, mediated through the variability of the pelagic ecosystem that is physically forced. This concern goes back to Johann Hjort's hypothesis, expressed in 1914, of the probability of massive mortality of larval fish in the event that their food was—in some way—insufficient or unavailable. I mention these problems here only because it was in the NECS province that the issue of the determination of year-class strength first attracted serious attention. It is, of course, a global problem at the roots of the current fishery crisis.

It is increasingly evident that uncertainty in the survival of larval fish populations, and hence in subsequent recruitment to the adult stock, is closely tied to uncertainty in the seasonality and strength of the plankton cycle. This, in turn, is a consequence of variability in atmospheric forcing of circulation, stratification, and mixed-layer temperatures over the continental shelf here as elsewhere.

Although it must be an oversimplification, the timing and abundance of planktonic copepods has become the paradigm for the factor most directly involved in the match-mismatch model of the survival of larval fish. This is thought to occur only when the date of occlusion of fish eggs matches the date of sufficient copepod abundance in their immediate surroundings (e.g., Cushing, 1990). By extension, it is supposed that if copepod ecology is forced by a single environmental factor, perhaps the relative westerly wind stress in winter indicated by the value of the NAO, then this factor should serve as an indicator of future survival of young fish. Many people have now attempted to hindcast the recruitment to fish stocks by reference to a simple environmental index.

Unfortunately for such predictions, the factors determining the timing of copepod abundance are very complex and bear no simple and direct relationship with the timing and strength of phytoplankton growth, itself rather directly forced by wind and sun. But the timing of the appearance of the new generation of *Calanus* is dependent not only on nutrition, but also on the fate each year of the deep, overwintering generation. Because the winter environment of these copepods varies strongly, the reappearance of *Calanus* in the surface waters is unpredictable and irregular.

If this is the general case, and if survival of fish larvae does depend importantly on the availability of sufficient copepod food without delay after occlusion, then it would be surprising that subsequent recruitment should be a simple function of potential egg production of the adult stock. In fact, it was for many decades a mantra of stock assessment methodology that recruitment was maximal at some intermediate stock size, as in the conventional but mythical relationship for Arcto-Norwegian cod that you will find in text books. The reality, usually concealed behind log-normal plots, is seen in the subsequent year-class strengths of each stock: in the case of North Sea cod, 1965–1985, this varied from 7 to 493 fish at unit age expressed as catch per unit effort (Garrod and Schumacher, 1994). During the same period the total catch, a weak indicator of total stock biomass, varied by little more than a factor of 2. Even more striking is the variability of North Sea haddock, which in the period 1944–1971 had year-class strength at unit age varying from <50 to 28,152 fish also expressed as catch per unit effort. A longer view of recruitment to the same stock (1900–1971) shows how recruitment in most years is modest, or effectively a failure. Interspersed among these, and at intervals ranging from 3 to 6 years, are the 10 or so "super-year-classes" that must have formed the basis of the stock during this period. By far the three strongest year-classes were recruited in the years of the "gadoid outburst" of the 1960's.

This event itself illustrates very well the sequential effects of environmental forcing of physical oceanography, through the planktonic ecosystem and on to higher trophic levels. During the period 1960–1990 the seasonal cycle of abundance of *Calanus finmarchicus* and *C. helgolandicus* in the northern North Sea underwent a progressive evolution: *C. finmarchicus* abundance progressively decreased, and its period of maximum abundance occurred later, and extended later, from 1960 to 1980 (Beare and McKenzie, 1999b). Using preliminary data on copepod abundance, Cushing (1990) noted that a stepwise multiple regression of cod recruitment against *Calanus* abundance and timing ("delay") and March water temperatures showed that recruitment was correlated with the index of delay, and modified by *Calanus* abundance and March water temperatures.

This finding is probably typical, but not directly exportable to other species. Note that each fish species has a characteristic relationship with its ecological environment during the critical period of planktonic larval life. Thus, each has an individual period of maximum occlusion rate from plankton eggs, ranging from February for plaice (*Platessa*) to late April for whiting (*Merlangius*). We may presume that each is matched to the period of maximal probability that its most suitable larval food will be available in satisfactory abundance.

A very recent demonstration of the same issue in the Bay of Biscay may be noted; here, the anchovy (*Engraulis anchoa*) population comprises only three year-classes, and population size is strongly variable between years. However, there is no significant relationship between adult spawning stock size and subsequent recruitment success: this is now known to be forced by the effects of strong or weak seasonal coastal upwelling on the success of each year class (De Oliveira *et al.*, 2005). This finding requires that "classical" techniques for setting catch quotas that incorporate some level of assumption of the elusive "stock-recruitment" relationship be replaced with novel techniques for prediction of future stock sizes. Of course, this generality is not yet accepted by the fishery science community.

## Regional Benthic and Demersal Ecology

Here in NECS, exploration of benthic ecology has had a long history—and has perhaps been carried further—than in most other coastal provinces. I shall concern myself mostly with the macrobenthos, though recognizing the significance everywhere of the benthic meio- and microfauna that utilize the particulate organic material—small crustaceans, annelids, and nematodes, together with protists and bacteria. It is, of course, these organisms that are responsible for much of the benthic nutrient regeneration that occurs in shallow seas.

Many studies in recent decades have extended our knowledge of the distribution of the benthic communities, both offshore in the North Sea and in coastal regions, especially off England, where detailed ecological studies of *Amphiura, Venus, Abra, Macoma*, and *Modiolus* reef communities were undertaken in relation to tidal stress and other factors (e.g., Warwick and Uncles, 1980). These studies emphasized the trophic links between macro- and meiofauna and the relative importance of deposit and suspension feeding in relation to bottom type.

The Peterson-Thorson system for the description of benthic species groupings, though often challenged (see Chapter 3), remains a useful basis for thinking about how these organisms are arranged on continental shelves. It is subjectively simple to reach concordance between this and the more holistic system of benthic "*étages*" of J.-M. Pérès (see Glémarec, 1973). Just as for the pelagic ecosystem, we find that the regional ecological response of the benthos is a function of the regional oceanographic regime, in this case mediated through surficial geology. It is the grain size and organic content of a sediment that principally determine what benthic organisms will occur within it, and this sedimentary regime itself is a function of physical mixing and resuspension by wind and tides, as I have discussed in previous sections.

It is relatively simple to integrate the results of modern studies, such as the ICES North Sea benthos survey (1986), with earlier summations by Thorson, Glémarec, and others (e.g., Duineveld *et al.*, 1991) into an informal recognition of this relationship, *grosso modo*, for the NECS Province. Wherever sandy ground occurs, the benthic fauna will be some version of a *Venus* association, or, if the median size of fine particles is $<200\,\mu m$ it will be a *Tellina* association. The transition from coarse sand to gravel will be accompanied by progressive changes in component species; where median size of small particles is around 1 mm, then we may expect *Amphioxus* to abound with *Venus fasciata*. The biomass of sandy-ground benthos is dominated by suspension-feeding organisms—*Venus gallina, Mactra corallina, Tellina fabula, and Pharus legumen* in shallow water. The deeper sands below the mud line are usually not so clean as in shoaler water, and here we should expect organisms such as *Venus casina, Astarte sulcata, Spatangus purpurea, Dentalium entalis*, and, specifically, *Hyalonoecia tubicola*. The annual production/biomass ratio for *Venus* species associations is relatively low, in one case around 0.5.

Where organic material becomes an important component of the deposits, and mud (particles <60 μm) comprises >25% of the whole sediment, we shall expect some version of an *Amphiura* association of sediment-feeding organisms dominated by ophiuroids feeding on the sediment surface, and polychaetes feeding within it. Once again, increasing mud content is associated with changing composition of the fauna. In softer sediments (30–70% fines) we shall find *Maldane* and *Clymene*, and in very soft mud (>80% fines) there will also be the polychaetes *Pectinaria, Sternaspis* and *Maldane*, and the bivalve mollusks *Abra nitride*. These *Amphiura* associations seem to have a higher turnover rate than the *Venus* community, in one case the annual P/B ratio being reported as very close to unity.

These are dynamic species associations, and proportional representation of component species is neither stable, nor uniform from place to place. It is this that renders the common-sense classification of species associations so vulnerable to statistical criticism. A typical *Amphiura* community (*sens.* Thorson) was studied by Buchanan and Warwick (1974) for many years in the North Sea. In 1958, at the start of this period, the nominate species, the ophiuroid *Amphiura filiformis*, was very abundant (>12–15 $m^{-2}$) and dominated the contents of a grab-haul. But by 1971, there were only 2 ind $m^{-2}$, because no recruitment had occurred for many years. Nevertheless, the authors of the study found no contradiction later on in using the shorthand of an "*Amphiura* community" to describe their object of study.

### Synopsis

*Case 2—Nutrient-limited spring production but modified by shallow-water processes.* The mixed layer undergoes boreal winter excursion from 10–20 m (May–August) to 50–60 m (December–March), while $Z_{eu}$ is consistently at 20–25 m. The thermocline is thus illuminated only during boreal summer. The winter mixing carries the pycnocline to the bottom from midshelf shoreward and to moderate depths seaward of this line. P rises rapidly in boreal spring to a May maximum, the subsequent decline showing only weakly the habitual autumn change in rate. Significant accumulation of chlorophyll in boreal winter October–March is perhaps largely an effect of river discharge and winter resuspension on satellite imagery, and seasonal maximum chlorophyll accumulation is usually 30 days prior to the peak P rate (Fig. 9.22).

## CANARY CURRENT COASTAL PROVINCE (CNRY)

### Extent of the Province

The Canary Current Coastal Province comprises the southerly coastal flow of the eastern boundary current of the North Atlantic from Cape Finisterre in northern Portugal to Cape Verde in Senegal, its seaward boundary being the convergent front at the outer limit of the zone of anticyclonic curl of wind stress approximately 200–400 km offshore. The province includes the field of eddies associated with this convergence. It also includes the Canary Islands that lie about 100 km offshore from Cap Juby (28°N).

### Continental Shelf Topography and Tidal and Shelf-Edge Fronts

Although the continental shelf here has nowhere the same width as is typical of NECS, especially along the western coast of Iberia, it is somewhat wider along the northwestern coast of Africa than in any other eastern boundary current province; here, there are exceptionally wide areas in the bights between Capes Ghir and Jubi and between Capes Bojador and Blanc where the 200-m contour lies 100–120 km offshore, whereas off the capes themselves it narrows to 20–30 km. The Arguin Bank just south of Cape Blanc

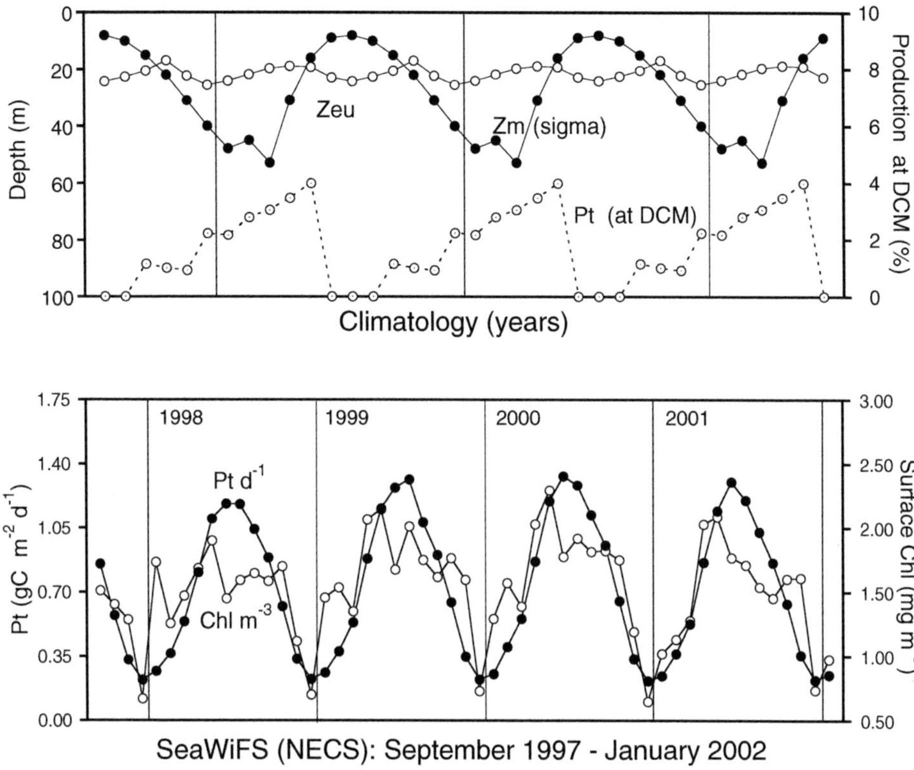


**Fig. 9.22** NECS: seasonal cycles of monthly surface chlorophyll and depth-integrated autotrophic production for the years 1997–2002 from SeaWiFS data together with characteristic seasonal cycles of mixed-layer depths from Levitus climatological data and photic depths computed from characteristic irradiance and the archive of chlorophyll profiles discussed in Chapter 1.

at 20°N requires special attention. Sandy deposits dominate this entire region, the sand having both aeolian quartz and biogenic calcareous components, and there are some outcrops of uncovered rock. Following the usual generalization, organic-rich deposits occur mostly in shallow water.

The dimensions of the continental shelf define the inshore part of the province, whereas the seaward edge of the upwelled water mass takes a complex form in the surface temperature field that defines the outer boundary. Tidal currents, and tidal mixing on the shelf, are relatively weak throughout the region, so continental shelf tidal fronts are not to be expected.

## Defining Characteristics of Regional Oceanography

A useful modern account of the oceanography of the Canary Current System is that of Barton *et al.* (1998), who describes how the broad equatorward flow around the eastern limb of the North Atlantic subtropical gyre passes equatorward along the coasts of Portugal and northwest Africa as several individual streams that originate in the northern limb of the subtropical gyre. Flow is seasonally continuous from Cape Finisterre to the major separation of the Canary Current from the coast near Cape Blanc, except that some of the flow along the Iberian coast passes eastward into the Gulf of Cadiz and thence into the Mediterranean. Beyond Cape Blanc lies a region of more complex circulation that extends south to Cape Verde at 15°N. Flow is significantly reversed only off Portugal and only in winter, when poleward flow is induced along the inner shelf. Some poleward flow also occurs at the coast in semipermanent eddies in the bight south of Cape Ghir (31°N), at Cape Jubi (27°N), and wherever mesoscale cyclonic eddies form on the landward side of the main flow of the offshore current (Fedoseev, 1970). The Canary Islands may induce a counterclockwise meander of the offshore part of the coastal boundary zone around the archipelago at about the 2000-m isobath, a feature that then influences the surface temperature field nearer to the coast.

These coasts lie below the seasonally migrating trade wind belt that reaches the Iberian Peninsula during boreal summer and retreats southward in winter. The wind field over the Canary Current System is typical of those over eastern boundary currents (Bakun and Nelson, 1991); zero wind-stress curl coincides with the offshore zone of maximum equatorward wind velocity and with the main flow of the offshore equatorward current. Landward, cyclonic curl strengthens upwelling, whereas anticyclonic curl forces a convergent front offshore. The annual mean alongshore component of wind stress off northwest Africa is $-1.5\,\text{dyn cm}^{-2}$, compared to only $-0.5\,\text{dyn cm}^{-2}$ off Oregon, or $-0.8\,\text{dyn cm}^{-2}$ off Peru (Barber and Smith, 1981). In addition, the winds are upwelling-favorable for almost the entire year off Africa, but for only half the year off Oregon. Off northwest Africa, wind mixing frequently reaches to the bottom and this has consequences for the accumulation of the organic material formed in the euphotic zone during upwelling blooms.

The seasonal meridional migration of the trade wind system, and hence of the velocity and curl maxima in the coastal regime, drives the basic seasonal cycle of upwelling, so that this is most intense in boreal summer (June–October) off Portugal and in boreal winter (January–May) off Senegal; in the central region, from 20°N to 25°N, upwelling is more continuous (Wooster *et al.*, 1976; Mittelstaedt, 1991). Event-scale changes of wind strength and direction may, for a few days, induce or suppress upwelling (van Camp *et al.*, 1991; Joint *et al.*, 2001b) and thus modify the surface thermal field within a short period. Cyclonic curl maxima are associated at the coast with the major upwelling centers discussed later, whereas some anticyclonic curl in the coastal winds may occur as a local response to coastal orientation off Portugal. A persistent lobe of anticyclonic curl also interrupts the coastal field of cyclonic curl between Cape Bojador and Cape Sim in autumn and winter, although strong upwelling occurs here during other seasons (Bakun and Nelson, 1991).

The climatic seasonal mean fields of temperature and density (e.g., Mittelstaedt, 1991) indicate that even where upwelling occurs continuously, there are small topographically locked cells of greater intensity. Cool filaments, exhibiting strong vorticity, develop from the persistent upwelling cells (Wooster *et al.*, 1976) and may extend far beyond the shelf edge, especially at 30–35°N and near Cape Blanc at 20–22°N (van Camp *et al.*, 1991). These cool filaments, although narrow and shallow, may extend for several hundred kilometers and have attracted much attention recently in this region during the European CANIGO voyages (Barton *et al.*, 2004), because of their potential for export of organic material and nutrients to the oligotrophic ocean. An unusually large and persistent filament occurs south of Cape Blanc; this feature is independent of local winds and is forced by convergence between the Canary Current and the poleward flow of the NECC, which loops to the southwest on encountering the Canary Current, thus forming a quasipermanent cyclonic eddy. This poleward flow is permanent but, during boreal winter, when the trade-wind belt is at its most southerly, a narrow equatorward current passes inside it, close along the coast, that continues the flow of the Canary Current from the north down to Cape Verde and even a little beyond. Upwelling at the coast between Cape Blanc and Cape Verde occurs only at this season.

The principal foci for upwelling in this province are the following:

*Cape Ortegal/Cape Finisterra (42–43°N)*: Though upwelling occurs in summer along the whole Galician coast, and within the mouths of the rias, it is most intense where the coast turns southerly between these two capes (Fraga, 1981). Here, intensification occurs because of the conjunction of different subsurface water masses. Some confusion is possible in color images between algal production due to upwelling south of Cape Finisterre and production within the adjacent coastal rias due to river discharge of nutrients.

*Cape St. Vincent (37°N)*: Upwelling behind this southwest-jutting cape is topographically reinforced, strong, and extends along the coast of the Algarve, where red tide blooms are frequent accessories to upwelling events. Off Huelva, a front often extends in a southwesterly direction and marks the limit of the upwelling process.

*Capes Spartel (35°N), Cantir (32.5°N), Ghir (31.5°N), and Juby (28°N)*: Upwelling occurs the length of this coast essentially year-round, although peaking in frequency and intensity during summer. It is strongest behind these capes because of favorable angles between the coastline and the direction of the trade winds (Mittelstaedt, 1991). Discharge from rivers draining from the Atlas Mountains may cause some difficulty of interpretation along this sector of the coast, where upwelling is not such an intense phenomenon as further south. The Canary Islands lie just beyond the continental edge at Cape Jubi and their presence must be noted when examining satellite images of this region.

The *Canary Islands (27–29°N)* archipelago spans the southward flow of the Canary Current; a recurrent cyclonic eddy south of the islands entrains mesoscale eddies of both signs formed behind individual islands as well as cold filaments arising at the coast. Recirculation of these features within the persistent cyclone constrains the export capacity of individual filaments (Barton *et al.*, 2004). Water may be recirculated within this system for some weeks, surface layer drifters completing several circuits before being released into the complex flows further south.

*Cape Bojador (26°N)* in Western Sahara is an upwelling focus in both spring and summer and, from here south to Cape Blanc, upwelling occurs more consistently year-round than elsewhere in this province. Perhaps onshore flow may restrict the spread of the upwelled water to a rather narrow zone along the coast (Mittelstaedt, 1991). Especially between Cape Bojador and Cape Barbas, upwelled NACW water forms cells of bottom water on the shelf, within which nutrient levels may be unusually high (Hughes and Fiuza, 1982; Minas *et al.*, 1982).

*Cape Blanc, Cape Timiris (19–21°N)* is a major upwelling focus where the greatest intensity and frequency of upwelling occurs during the first months of the year. It is also a special case because a shelf-crossing submarine canyon may serve to focus the upwelling flow (Roy, 1990) so that an immensely productive, plumelike ecosystem develops, as noted earlier, that is similar to that off Peru at 15°S (see Chapter 11, HUMB).

*Arguin Bank (20°N)* occupies the bight to the south of Cape Blanc and is a special case. There is at times a strong excess of evaporation over precipitation in the shallow water around this biologically rich area of shallows and offshore banks. When this occurs, episodic cascades of the dense water that is formed on the bank down the continental slope are replaced almost immediately with cool, recently upwelled water, bringing nutrients to the shallows (van Camp *et al.*, 1991), this being, perhaps, a unique model of the general process of upwelling.

*Cape Verde, Cape Roxo (12–15°N)*: In this most southerly focus of upwelling, activity is restricted to late months of boreal winter when the trade-wind belt is at its most southerly position, and Harmattan winds reach the sea. In boreal summer and fall, the region is entirely under the influence of the NECC and of tropical water. At Cape Roxo, the coast bends eastward into the Gulf of Guinea and a stable, tropical coastal regime is met (see the GUIN Coastal Province).

Because this province lies between the Atlantic and the Sahara desert, a special characteristic of the wind regime is the dust-laden nature of the northeast trades, strongest in boreal winter, wherever their back trajectory lies over the desert (Hastenrath, 1985). These Harmattan winds bear a heavy burden of mineral dust over the ocean from Morocco to the Gulf of Guinea and may deposit as much as 25 g $m^{-2}$ $y^{-1}$ at the sea surface off Senegal. This dust deposition is among the heaviest anywhere in the oceans and is similar to deposition of loess clays in the northwestern Pacific; such aeolian deposits may be a major source of turbidity in inshore waters and have, of course, been invoked in releasing autotrophic cells from Fe limitation, as was discussed in Chapter 5.

## Response of the Pelagic Ecosystems

Although the North Atlantic bloom of late winter or spring importantly affects the shelf waters of Iberia south to the Gulf of Cadiz and is forced by stratification, growth of phytoplankton in the CNRY province is overwhelmingly controlled by wind-driven variations in vertical transport of nutrients into the euphotic zone. An intuitive and simple negative relationship between chlorophyll and temperature was apparent in the earliest satellite images for which both fields were available (van Camp *et al.*, 1991), similar to that established off South Africa (Shannon, 1985; Lutjeharms *et al.*, 1985), and is abundantly confirmed by relevant images available today. The dimension of upwelling cells is smaller than the area over which appropriate wind stress is applied, because water depth restricts upwelling to a band only 10–20 km wide (Barber and Smith, 1981) except in the case of shelf-edge upwelling.

The continuous series of chlorophyll images available since 1997 demonstrate unequivocally how the consequences of upwelling dominate phytoplankton ecology in this province. The 30-day SeaWiFS and MODIS composites cannot reveal the details of the evolution of individual upwelling events and conceal the form of individual filaments, but they unequivocally show how upwelling differs seasonally and in relative strength along the coast between Iberia and Senegal. This series of images confirms quite clearly the influence of the meridionally migrating belt of trade winds but also shows that even in seasons when NE wind stress is expected to be minimal, upwelling may not be entirely absent. The images strikingly confirm the far wider extent of the effect of upwelling to the south of Cape Blanc and how, along the Iberian, Moroccan, and Saharan coast, surface chlorophyll in excess of 2.5 mg $m^{-3}$ is restricted to a narrow coastal belt, of order 30–40 km wide. The filaments that extend seaward from the upwelling centers along these northern coasts usually include much lower chlorophyll values (around 0.2 mg $m^{-3}$) than to the south of Cape Blanc where the biological consequences of upwelling are far more extensive. In winter months, as in November 1999, it is not unusual for a discrete high-chlorophyll feature (5–7 mg $m^{-3}$) to extend as much as 400 km offshore from the Mauretanian bight. In September–November 1998, this feature was continuous with a high-chlorophyll, eddylike enhancement in the NECC, as already discussed. In October and November 2000, the divergence of the NECC between poleward flow and eastward flow into the Guinea Current appears to have been unusually far to the south. In those months, upwelling seems to have occurred as far south as Cape Roxo, and enhanced surface chlorophyll extended over much of the Bissagos shelf off Guinea. This was an unusual but not unique event during the 8-year series of images now available.

Perhaps one explanation of these observations lies in the different nutrient content of water that is upwelled on either side of Cape Blanc. To the south, upwelled South Atlantic Central water (SACW) is relatively nutrient rich ($NO_3 = 14$–$20\,\mu M\ kg^{-1}$), whereas, from Cape Blanc to the Iberian Peninsula, relatively nutrient-poor North Atlantic Central water is upwelled and may contain as little as 2.6 μM $kg^{-1}$ (Alvarez-Salgado *et al.*, 2001). A complex front between the two regimes occurs off Cape Blanc so that SACW may at times be upwelled as far north as Cape Barbas (Minas *et al.*, 1982). Calculations by Alvarez-Salgado *et al.* show that the low nutrient levels off Iberia have a direct effect on rates of new production there. But Barber and Smith (1981) remind us that we should think not only about new production fueled by upwelled nitrate in eastern boundary currents but also about nitrogen regenerated there as $NH_4$. Because of the episodic nature of upwelling pulses, and hence of blooms, much of the pelagic organic matter that is produced sinks unconsumed to the sediments. Here, in the Canary Current, the relatively high wind stress and strong equatorward and cross-shelf currents prevent accumulation and ensure that benthic utilization and remineralization rates are high. A further consequence of the relatively high wind stress and deep wind mixing is that only weak near-surface stratification develops during the immediate post-upwelling production phase of each episode. In the other eastern boundary currents, accumulating phytoplankton cells are more frequently mixed down, even below the euphotic zone, and their concentration diluted. One of the findings of the CUEA project, in which the Oregon, Peruvian, and northwestern African regions were compared, was that because of high wind stress, primary production rates are relatively lower in this province than off the western coasts of America (Barber and Smith, 1981).

The giant filament south of Cape Blanc, reaching as much as 400 km offshore, may appear as a persistent chlorophyll feature or it may be seen principally in the temperature field. To explain the appearance of a chlorophyll feature so far offshore, Gabric *et al.* (1993) proposed a mechanism for the conservation of nutrients in the upwelled water as it advected to the west. For this to occur, the algal cells seeding the upwelling must be entrained from below the photic layer when upwelling occurs over the shelf break: in this event, a delay occurs before a bloom can be initiated, as the cells undergo physiological conditioning to a photic environment. Probably more typical, however, is the sequence observed by Joint *et al.* (2001b) during two 5-day Lagrangian drift studies in a filament at the shelf break off northern Portugal; during the first 5-day period, within the body of the filament as it drifted along the shelf edge, nutrients were progressively reduced; this occurred first at the surface, then progressively deeper as a DCM was established and as a shift occurred from small flagellates and dinoflagellates to a diatom-dominated community; during this period the f-ratio declined from 0.7 to 0.5. The second 5-day drift followed the head of the filament as it moved out over the deep ocean, when it comprised extremely oligotrophic water, in which $NH_4$ concentration exceeded $NO_3$, and which was dominated by autotrophic picoplankton that were responsible for 65% of total production, diatoms being almost absent. Bacterial biomass remained little changed during both drifts, but growth rates were significantly higher in the first than in the second experiment.

Because the Canary Current, to the north of Cape Blanc, is not compensated by a countercurrent over the shelf break, offshore advection in the resulting single upwelling cell is sufficiently rapid (a residence time of about 10 days) that ecological succession is significantly lagged. Consequently, the processes of production and consumption are not tightly linked and herbivore development lags the growth of phytoplankton biomass by several weeks. Thus, the rate of remineralization lags the rate of new production of algal biomass, and diatom biomass is therefore continually lost by sedimentation. Further, the rate of excretion of N and P by herbivores is insufficiently matched by excretion rate of Si, so that diatom growth is limited by the upwelling supply of this latter element. South of Cape Blanc, during the winter upwelling season, the poleward flow of the NECC performs as a shelf-break countercurrent beyond the southward, upwelling flow of Canary Current water. Two upwelling cells are thus formed of which the inner, over the shelf, is partially closed so that it retains the elements of the pelagic ecosystem and reduces offshore advection. In this way, primary and secondary production within the shelf break are coupled more effectively than in any regions to the north of Cape Blanc.

In a classical series of papers, Margalef suggested that in upwelling regions the phytoplankton species associations are arranged sequentially (and successionally) around each upwelling core, to expand or contract spatially with it. To verify this thesis, Blasco *et al.* (1981) extracted PCA components from the occurrence of the most abundant 104 species of diatoms, dinoflagellates, and coccoliths along serial cross-shelf sections in this region. Three components accounted for almost 40% of the total phytoplankton variability. Two of these were related to the upwelled water mass as it gradually moves offshore, although PC2 has highest scores inshore and close to the bottom. The third component, PC3, has highest scores during periods of relaxation of upwelling and is associated with the coastal warm water current in the bight south of Cape Blanc. This arrangement conforms with earlier informal suggestions that the phytoplankton species associations are arranged in linear, coastwise bands.

Most of the current information about transfers within the pelagic ecosystem for the Canary Current is at the diatom-copepod level, although a recent study off Cape Ghir at 31°N found very high bacterial consumption of phytoplankton in the upper water column. About 50% of the total production of 1.0–2.5 gC $m^{-2}$ $d^{-1}$ was consumed in this way, and 0.2–0.4 gC $m^{-2}$ $d^{-1}$ could be accounted for by the activity of copepods (Head *et al.*, 2002). Another study in this region found that the dominant two large herbivorous copepods (*Calanoides carinatus* and *Centropages chierchae*) fed preferentially on the small centric diatoms, together with some pennate species, which are characteristic of such situations. However, in offshore situations where upwelling blooms were not occurring, the same species of copepods mostly had empty guts, a fact clearly related to the dominant phytoplankton assemblage of microflagellates and very large species of diatoms.

We should especially note here the occurrence of *Calanoides*, a specialist of low-latitude upwelling systems. We may expect, by analogy with what occurs elsewhere, that during nonupwelling periods the population of *C. carinatus*, dominated by stage C5 copepodites, will descend into deep water. Otherwise, over deep water, the mesozooplankton profile closely resembles those for the eastern tropical Pacific: biomass is highest in the upper 100 m (not well specified in profiles from this region) and a secondary maximum at 200–400 m represents diel migrants at their daytime residence depths. Over shallow water, in the same region, a sharp chlorophyll maximum at about 10–15 m containing many diatoms was associated with maximum abundance of a wide range of herbivore genera: *Corycaeus, Oithona, Oncaea, Euterpina, Muggaeia, Oikopleura*, and others.

In the southern region off Senegal, two species groups of macrozooplankton dominate the biomass. One (copepod-rich) is associated with cool upwelled water masses as their contained phytoplankton evolves, the other (thaliacean-rich) with warm tropical surface water. During the boreal winter upwelling period, the former comes to dominate the biomass, whereas, during the summer quiet period, the latter dominates. At transitional seasons between the two states, biomass is relatively low. In fact, the contrary transport of the coastal current from the north in winter, and the offshore "countercurrent" (as it is sometimes termed) of NECC water from the south leads to a very complex distribution of plankton species off Senegal and Mauretania (e.g., Weikert, 1984). This is further complicated by the intercalation of eddies originating in different water masses. From the south, in NECC water, come *Calanoides carinatus, Nannocalanus minor*, and *Penilia avirostris*, whereas *Calanus helgolandicus* is a clear indicator species, associated with upwelling, of water of northern origin. With *Calanus*, we may associate *Mecynocera clausii, Candacia armata*, and *Pleuromamma piseki* as typical NW African coastal species.

The fish component of the pelagic ecosystem is rather more diverse than in some other eastern boundary currents, because of the unusual meridional extent of the Canary Current system. The principal species, excluding large predators, are anchovy (*Engraulis enchrasicholus*), sardine (*Sardina pilchardus*), sardinellas (*Sardinella aurita* and *S. maderensis*), horse mackerel (*Trachurus trachurus* and *T. trechae*), and Spanish mackerel (*Scomber japonicus*). A peculiarity of this province is the relatively recent invasion of the African coastline by the "European" sardine (*S. pilchardus*). Until the 1920s, this fish was restricted to regions north of the Gulf of Cadiz and was first noticed as abundant off Morocco only in the 1950s. Twenty years later, these sardines had appeared abundantly off Senegal and were heavily fished there, only to retreat northward again in the early 1980s.

Apparently, the abundance and distribution of European sardines along the African coastline responds directly to the relative strength and location of upwelling between years, and hence to changes in the strength of the trade winds. In recent decades, there have been two periods of high sardine abundance (1976–77, 1989–90), and each of these periods followed several years of intensified northeast trades, and hence of upwelling strength (Binet *et al.*, 2001). The actual relationship remains somewhat uncertain, although *S. pilchardus*, unlike the zooplanktivore *S. aurita*, is capable of feeding directly on the phytoplankton that will have been unusually available during these years. However, recruitment tends not to succeed when strong offshore advection occurs during the first 3 months of larval life, so it seems probable that the spawning potential of the existing stock is enhanced during years of strong upwelling, and that recruitment is consequently enhanced in subsequent years when offshore transport of larvae is relatively weak. Off the Iberian segment of the coastline, *S. pilchardus* also exhibits periodic changes in population abundance but the cause of these is not yet very well understood. But, however it is determined, the variance in Iberian sardine year-class strength is very significant—for a long period, the fishery was sustained solely by the strong 1983 and 1987 year-classes. Subsequently, apart from the moderate 1991 year-class, there was very little new recruitment to the stock and the fishery once again entered crisis.

How clupeid recruitment responds to variance in environmental forcing in eastern boundary currents has been a principal focus of fisheries ecology since the 1970's. I shall discuss it as a special topic when considering the CALC province, where much of the critical work has been done. It will be necessary to consider both the consequences of water column stability in allowing sufficient concentration of food particles for first-feeding larvae, and subsequently their retention in upwelling regions where sufficient food for later, more mobile larvae is produced. It is the latter mechanism that has been most often invoked for the control of recruitment in the CALC province, but I suspect that that fact merely reflects the interests of individual oceanographers who have worked there.

## Regional Benthic and Demersal Ecology

The distribution and ecology of benthic macroorganisms on the continental shelf has been little studied in this province, most attention having been given to the consumption of sedimenting organic material from upwelling cells in deep water beyond the shelf. One survey of benthic macrofaunal biomass on the inner shelf suggested that this decreased with the seasonal arrival in shoal water of demersal fish.

I can find no general ecological studies of the benthos other than my own brief survey, performed with quantitative grab samplers at 19 stations over the continental shelf south of Cape Blanc at 13–14°N in December 1956. Considering the unusually high rate of primary production in the overlying water column, I was surprised to find that the shelf was almost entirely covered with clean shell-sand that extended from 10 m off the Gambia River right out to the break of slope at 180–200 m. The color of the contained organic material was olive-brown, indicating the dominance of the fecal pellets of macrobenthos, but this material formed (as settled volume) only around 18% of the deposit samples over most of the shelf down to 75 m, increasing to 44% at 180–200 m. In very shallow water in the mouth of the Gambia, organic material was still unusually low for such situations—only around 35%. These values are much lower than measured with the same technique on the Guinea shelf in comparable situations (Longhurst, 1958).

The macrobenthic species associations were essentially identical with those on the Guinean shelf (see GUIN), so that—because of the nature of the deposits—a typical *Venus* community dominated down to midshelf depths. At only four stations close to the mouth of the Gambia River, in slightly muddier sediments, was an *Amphioplus* community found. A very reduced deep shelf community, rich in pennatularians and sharing many other species with the same association in the Gulf of Guinea, extended from 75–80 m down to the shelf break.

Not only is the species composition of the macrobenthos of this province a function of depth and sediment type, but so also is the species composition of the demersal fish community. The deep sparid community, which will be introduced in the account of the GUIN province, typically occupies the shelly and sandy grounds of the continental shelf the length of the African coast from north of Cape Verde right up to the Mediterranean. Dominant species are *Dentex canariensis* and *D. filosus, Pagrus ehrenbergi*, and species of *Upenaeus, Smaris, Lepidotrigla*, and *Brotula*.

## Synopsis

*Case 6—Intermittent production at coastal divergences*—The synopsis of archived data offered in Fig. 9.23 does not represent any single upwelling cell. Rather, all upwelling cells have shorter upwelling seasons than is suggested by the integrated data, which best fit the central region from 20°N to 25°N. General seasonality of $Z_m$ is consistent with both boreal winter mixing (50 m January–February) and summer stratification and upwelling (20 m August–September). Because $Z_{eu}$ remains between 50 and 70 m in all months, the thermocline is illuminated except for 5 months in boreal winter, and presumably also in

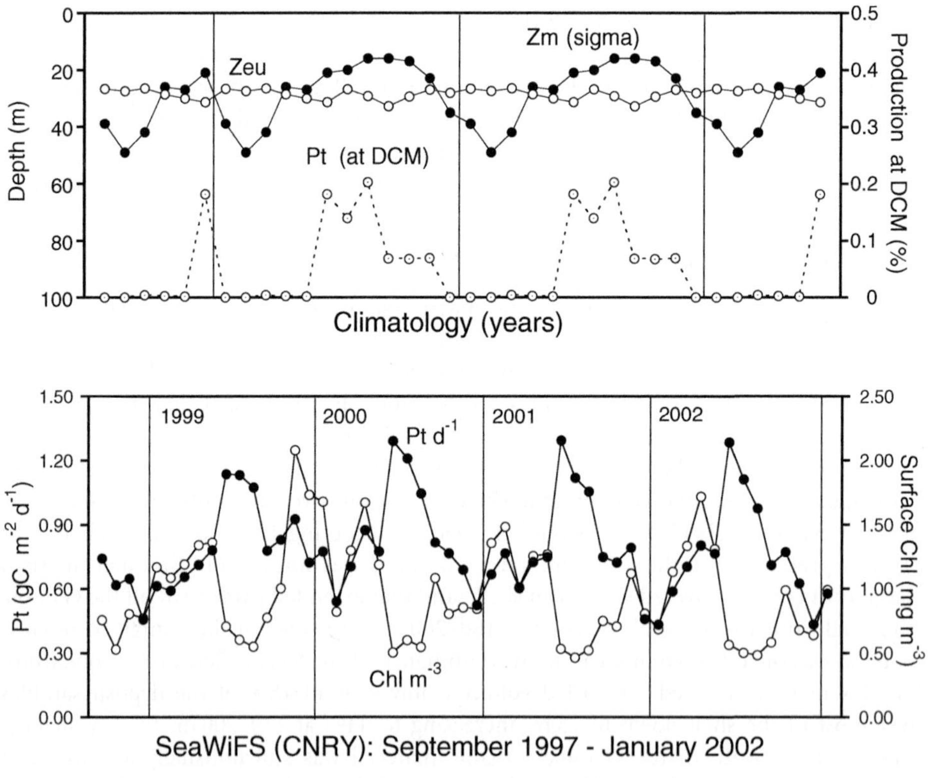


**Fig. 9.23** CNRY: seasonal cycles of monthly surface chlorophyll and depth-integrated autotrophic production for the years 1997–2002 from SeaWiFS data together with characteristic seasonal cycles of mixed-layer depths from Levitus climatological data and photic depths computed from characteristic irradiance and the archive of chlorophyll profiles discussed in Chapter 1.

strong upwelling cells, the effects of which are concealed in these mean values. The P rate increases continuously from the December minimum to an annual maximum in March–April with an unusually prominent secondary maximum in September–November due to upwelling processes. Accumulation of cells approximately matches changing P rates year-round. Consumer biomass (or sinking rate) appears to increase rapidly after midsummer when chlorophyll biomass takes annual minimum values despite high summer upwelling P rates. This would be consistent with rise of seasonal vertical migrant consumers.

## Guinea Current Coastal Province (GUIN)

### Extent of the Province

The GUIN province extends from Cape Roxo (12°N) eastward along the coastline of tropical West Africa (the Guinea Coast hereafter) to the Bight of Biafra, and thence south along the Congolese and Angolan coast to Cape Frio (18°S), the Central African coast hereafter. Along the east-west Guinea Coast the seaward boundary is the shelf-edge front within the NECC-GC (at 2–3°N 4°W). Along the coast from Cameroon to the Angola Bight, *faute de mieux* (see below), the line of the Angola/Benguela Front is used, though this (enclosing the large offshore cyclonic gyre) really carries us too far seaward from the coastal boundary zone.

### Continental Shelf Topography and Tidal and Shelf-Edge Fronts

The continental shelf of the Guinea Coast is narrow (about 25 km) except from Cape Roxo to Sherbro Island in the extreme west, where a width of about 250 km is reached. Lesser extensions of the shelf occur off Ghana and between the Niger delta and the large island of Fernando Po. Deep canyons cut across the shelf off the Ivory Coast (the "Trou sans Fond") and at the mouth of the Congo River.

Especially on the western and central shelves of the Guinea Coast, linear relic coral banks occur deep on the shelf, usually close to the shelf break. Otherwise, the shelf is generally flat and the coastline straight, but the nature of the sediments varies significantly according to coastal geology and the distribution of river effluents. Two major (and many minor) rivers open onto the shelf: the Niger, through its wide deltaic region on the eastern Guinea Coast, and the Congo, through a narrow mouth over the Congo canyon. The wide western shelf has the form of a dissected plateau and carries the relic submerged delta of the Bissau River, whose flow is much reduced in the present era; emergent parts of this feature form the offshore Bissagos archipelago. On this segment of the shelf, sediments are presently accumulating only inshore, so that muddy deposits dominate at depths shoaler than the permanent thermocline, below which the bottom is sandy out to the linear fossil coral banks of Pleistocene age on the outer shelf.

East of Sherbro Island and as far as Cape Palmas, the narrow shelf is largely mud-covered with a narrow coastal strip of sand in water <20 m deep. In the central upwelling region, the extent of sandy substrate is much greater and may extend out to the shelf edge as it does to the east of Sassandra. To the east of Lagos, and increasingly into the Bight of Biafra, the sedimentary flux of the Niger River loads the shelf with extensive deposits of soft mud so that here sandy ground is rare, even close inshore. Off the eastern Niger delta, from Bonny to the Rio del Rey, black mud is widely distributed and is associated with anoxic bottom water.

On the Central African coast, the Congo River canyon, and the Walvis Ridge just south of Cape Frio are the dominant topographic features; the 200-m contour lies a little farther offshore than on the Guinea Coast with very little change in shelf width the entire length of this segment of the coastline.

Reef corals exist only on Annobon, the most seaward of the Biafran islands; otherwise, the coastline is fringed with sandy beaches, and with mangrove vegetation wherever there is protection from surf.

## Defining Characteristics of Regional Oceanography

The regional oceanography of this province is dominated by two factors: (i) the relatively shallow mixed layer above the sharp and permanent thermocline that is characteristic of the eastern Atlantic and (ii) the strongly seasonal reversal of the wind systems associated with the migration of the ITCZ so that along the Guinea Coast dry NE trades dominate in boreal winter, and wet SE trades in boreal summer. The Central African coastline is dominated by SE winds the entire year.

Most reviews of its oceanography have partitioned the Gulf of Guinea, between the subtropical cold fronts at Cape Roxo and Cape Lopez, into a central upwelling area off Ghana and Ivory Coast and two nonupwelling regions to the east and west. This is an oversimplification of the real environment of the Gulf of Guinea, and for reasons that will become apparent, I have rather followed Ajao and Houghton (1998) in partitioning this region into only two sections outlined earlier: (i) the Guinea Coast, from Cape Roxo east into the corner of the Bight of Biafra, and (ii) the Central African coast, down to Cape Frio at the Namibian border. We have a great deal more information about the Guinea coast than the Central African coast, much of it from the colonial period and immediately afterward.

***Guinea Coast*** The continental shelf of the 2000 km east-west coastline is occupied by the eastward flow of the coastal component of the NECC-GC (see NATR), here usually termed simply the Guinea Current. This receives a minor contribution in boreal winter from the Canary Current, close inshore around Cape Roxo, and, after intensifying its flow relative to the NECC between 8°W and the Greenwich meridian, it then weakens progressively toward the east. The Guinea Current carries warm, light water and occupies a shallow (30 m), warm (27°C) mixed layer above a very sharp thermocline ($<1°C\ m^{-1}$). Below this eastward flow there is a contrary, westward flow of cold, highly saline water originating at the termination of the EUC; this undercurrent, which I shall call the Forcados Current following the usage of canoe fishermen on the Nigerian coast, weakens toward the west. Recall that during boreal summer the thermocline of the whole eastern tropical Atlantic shoals (see ETRA) and slopes upward toward the coast; it is at this season that the Forcados Current flow is likely to surface intermittently, so that the Guinea Current apparently moves offshore (Verstraete, 1992). I have monitored this apparent reversal of flow at the coast off Nigeria, finding that each episode lasts for periods from one day to several weeks.

Although a shelf-break front is often identifiable, the Guinea Current above the shelf is continuous with the eastward flow of the NECC-GC during most of the year; however, satellite-tracked RAFOS floats now suggest that reversal of flow at the shelf edge in the western part of the Guinea Coast occurs in October with the onset of the NE Trades; at this time, several eastbound, surface-drogued, satellite-tracked floats reversed direction and moved closely along the shelf-break from Liberia to Guinea over a period of several months (Dave Hebert, pers. comm., 2005). This is a region of very complex circulation, and it is no surprise that long-term changes do occur although they are only rather dimly observed in the presently available data: it appears that coastal SST was anomalously high during the 1980s, for example, both at the surface and deep on the shelf after an unusually cold period during the 1970s.

Surface water masses are under the influence of the alternation of dry and wet seasons along the Guinea Coast. Seasonal cloud cover modifies the cycle of solar irradiance, from the 400 g cal $cm^{-1}$ $d^{-1}$ of boreal winter to about 250 g cal $cm^{-1}$ $d^{-1}$ in boreal summer. Surface temperatures consequently fall to 24–25°C and the isotherms for 19–23°C become more widely spaced. Lutjeharms and Meeuwis (1987) described as an "upwelling cell" the region of cool surface water seen in satellite thermal imagery between Fernando Po and the coast, mostly in March to August, during the period of greatest rainfall on the Cameroons. This interpretation must be incorrect because these supposed "cells" occur over an area of unusually shallow and muddy continental shelf. At this season, river discharges are often visible in ocean color images as pigmented plumes, especially in the Guinea–Sierra Leone and Nigeria-Cameroon sections of the coast. Thus, during the June–September wet season, surface salinity is strongly reduced by the extremely heavy rainfall both in the Bight of Biafra and off Sierra Leone, in both places associated with hilly country behind the coast. During the dry, Harmattan season of NE trades (January–March), significant falls of aeolian dust occur at sea in the western part of the region and a secondary, evaporative cooling of sea surface temperature occurs over the whole region. During the dry season, a sharp pycnocline lies below a shallow (30 m) mixed layer at ~27°C that in the wet season is almost eroded by seasonal wind stress, so that surface temperatures fall to 24–25°C and the isotherms for 19–23°C become widely spaced.

From eastern Liberia to western Nigeria, episodic shoaling of the thermocline is sufficiently strong that it constitutes coastal upwelling. This process is highly variable between years and is strongest along the coastlines of Ghana and the Ivory Coast, contiguous with the "Dahomey Gap" in the forest cover ashore. Upwelling occurs especially frequently off Cape Palmas (7°W) and Cape Three Points (2°W). Though a general cooling associated with the shoaling and eroding of the thermocline occurs each year, significant localized upwelling occurs only sporadically except off this central region, and even here there are exceptional years, such as 1968, when it does not occur. Upwelling is not a continuous process and the general season of cool water at the coast from June to September is punctuated by a series of isolated cold events, each lasting for a week or two (Longhurst, 1964; Philander, 1979; Houghton and Colin, 1986).

The exact mechanism that causes cold water to upwell along the central part of the coast is still quite obscure, though very much studied, and has been reviewed recently by Ajao and Houghton (2002). I had previously suggested, along with others, that this was a classical case of offshore Ekman flux driven by coastwise winds, but apparently we were quite wrong: there is not, as Ajao and Houghton demonstrate, even a qualitative relationship between wind speed or direction and the onset of upwelling on the central coast. Several other mechanisms have been proposed, and it is now thought that a mechanism involving coastally trapped internal waves must be involved. A fortnightly coastal wave propagates westward at around 0.5 m $sec^{-1}$, having a period exactly that of the lunisolar tide, and this appears to have its origin on the shelf of the Niger delta, where $M_2$ and $S_2$ tides interact in a nonlinear fashion through bottom friction. These waves travel westward along the Guinea coast, doming the thermocline and nutricline as they pass (especially where steep or protruding topography is encountered), forcing local upwelling events that may cause the westward undercurrent to surface, so that the events can be tracked westward along the coast using individual coastal station data. Though local wind stress is usually of the correct sign to generate coastal Ekman divergence, it is of quite insufficient strength to induce the observed vertical movement that occurs down to 500 m at the start of the cooling season. Whether these shelf waves, or a free equatorial wave propagating poleward along the coast, or even another form of trapped coastal wave, are responsible for the observed upwelling remains moot at this time.

There is some evidence from satellite IRT images of sea surface temperature of shelf-break fronts, and of shelf-break trapping of the flow from the Canary Current south of Cape Roxo and along the edge of the Guinea shelf, especially in boreal summer. Biological observations suggest that off both the Guinea and Nigerian shelves an amplification of the height of internal waves occurs by the shoaling of the continental slope and may be a persistent feature, as it is elsewhere.

Finally, we should remember that Cape Palmas, in western Ivory Coast, lies at only about 4°N, or just about 250 km north of the equator. We may therefore expect some interaction between upwelling dynamics here and the strong meandering of the equatorial divergence related to Kelvin waves passing east along the equatorial wave guide (see ETRA). These meanders are of an appropriate scale to interact with coastal processes, and I shall present some evidence later to suggest that such interaction can, in fact, be observed in the ecological response to vertical fluxes.

***Central African Coast*** Here, the coastal oceanography has not been so fully investigated, and the current streams are rather variable, especially from Fernando Po down to Cape Lopez. Here, the now very weak Guinea Current meets variable northward wind-forced surface flow from equatorial regions.

Between Cape Lopez and Cape Frio (18°S) the circulation is theoretically equivalent to that along the Guinea Coast, except that here the coastline is oriented north-south. A weak equatorward surface current lies above a strong, poleward subsurface countercurrent originating, as does the Forcados Current, in the terminal bifurcation of the EUC. This is the Angola Current that, on meeting the strong equatorward flow of the Benguela Current at Cape Frio, turns offshore and passes around the southern flank of the Angola Dome (see ETRA) to form the Angola-Benguela Front.

The response of the Atlantic Ocean to the ENSO cycle was discussed earlier, but it is in this region that the consequences of Atlantic Niño-like episodes are perhaps most important. The strengthening of the SECC during "Benguela Niños," as in 1995, drives an influx of warm water to the coast of Angola, with ecological consequences to be discussed below. When it is the NECC that is strengthened, during a "Guinea Niño," it is Bight of Biafra water that floods southward along the central African coastline, again with important ecological consequences (Binet *et al.*, 2001).

Surface wind drift over the whole of the Congo-Angola Bight is toward the northwest, especially during the boreal summer and between Cape Lopez and Luanda. At this season, the low-salinity plume of the Congo River, originating at 6°S, turns seaward and may be traced as far out into the South Atlantic as 5°E. In boreal winter, this plume passes northward along the coast toward Cape Lopez. South of this cape, coastal upwelling cells occur down nearly to 11°S. These upwelling events are not as well documented as those along the northern coast, but they occur during the same season and have been presumed to be similarly forced, perhaps by Rossby waves propagating southward along the shelf edge, although, in fact, the mechanism that forces this upwelling remains—as Ajao and Houghton remark—"baffling," especially because the timing of coastal upwelling events precedes events observed along the equator; the explanations concerning an origin in equatorial waves that have been offered for the Guinea upwelling lack conviction here.

The Congo effluent dominates the features that influence the ecology of the Central African coast; this is the largest discharge of any river after the Amazon, from which it differs in significant ways. The Congo carries a relatively small load of suspended material ($<120\,t\ 10^{-6}\ y^{-1}$ or more than an order of magnitude less than the Amazon) because of the relatively low gradients of the drainage basin, so that chemical rather than physical erosion dominates upriver; organic material comprises 32% of the suspensoids, and phosphate content is relatively high compared with other tropical rivers. The estuary of the Congo is deep compared with that of the Amazon, which has extensive shoals, so that the nutrient-rich Congo plume is less dynamically mixed with nutrient-poor Atlantic water and retains its high nutrient status much farther seaward than the Amazon (Cadee, 1979). Like the Amazon, the discharge rate of the Congo is weakly and negatively correlated with tropical Pacific SST anomalies and the ENSO cycle; tropical rainfall is exceptionally low during the warm ENSO phase.

## Response of the Pelagic Ecosystems

This province provides us with an unusually simple model of how the characteristics of a water column may structure a pelagic ecosystem and, assisted by its control of the distribution of sediment types, may also determine the characteristic ecology of the benthos and demersal fish. The simplicity of this model depends on the fact that the shoal, but very abrupt, pycnocline typical of this province lies at midshelf depths over the shelf, or even shoaler.

This province is characterized by a vertical arrangement of properties typical of oligotrophic situations, the depth of 1% irradiance being usually between 20 and 45 m over the shelf. During boreal summer, studies at coastal stations from Sierra Leone to Nigeria have recorded increases in mixed-layer chlorophyll and in phytoplankton cell numbers and a decrease in water clarity consistent with phytoplankton blooms during this season. The pattern is consistent with the various coastal upwelling regions discussed earlier (Bainbridge, 1960a; Longhurst, 1964; Sevrin-Reyssac, 1993). The FOCAL meridional section at 4°W (Oudot, 1987) also associated the upsloping of the deep chlorophyll maximum toward the coast in July with a shoreward increase in chlorophyll values. Off the Central African coast, the nitracline shoals into the upper 20 m in July, inducing there a seasonal shallow chlorophyll maximum (Binet, 1983).

The interpretation of sea surface chlorophyll images is relatively simple along the northern Guinea coast, and these clearly confirm the existence and ecological consequences of upwelling cells on the central Guinea coast, east of Capes Palmas and Three Points, and also the intermittent intrusion of chlorophyll-rich Canary Current water onto the Bissau shelf. These images show that the seaward extent of the ecological consequence of upwelling is much greater than had been anticipated. The most extreme case that I have found is that of the July 1998 upwelling event east of Cape Palmas that induced a 300-km filament ($<2.5\,\text{mg chl m}^{-3}$) that extended toward the southwest around the NE quadrant of an equatorial meander centered at 2°N, just east of the Greenwich meridian. A scan of the MODIS and SeaWiFS 30-day images will yield other examples of chlorophyll filaments far out over the deep ocean that are rooted in the coastal upwelling cells situated near the two prominent capes. It will also reveal that upwelling cells are, as we expect them to be from coastal observations, especially active from June to October each year.

On the Central African coastline, surface chlorophyll is far more difficult to interpret, principally because of the presence of the Congo effluent and of the Angola Dome not far offshore: the shelf environment along this coast is strongly influenced by the region of high chlorophyll values that is present in the Angola bight from Cape Lopez to Mossamedes; coastal processes are with great difficulty isolated from processes on the eastern, African side of the Angola Dome. The available chlorophyll images, not all of which can be shown here because of licensing policies, will repay close examination, although they cannot entirely resolve the difficulties; for instance, to what extent does the image of the buoyant plume of the Congo (see Fig. 9.16) represent CDOM like that of the Amazon? I can find no survey data representing transects across the plume to resolve this issue. In April 2001, we see what appears to be a simple river plume turned offshore by the then-prevailing winds from a source region between Cabinda and Louanda, apparently at the Congo mouth. February 1998 images show a much larger plume, rooted in approximately the same part of the Angola bight, which appears to be a response to meanders in the equatorial current system. Its active nature appears to be confirmed by the existence of cloud cover specifically associated with it, presumably in response to lowered sea surface temperatures. More enigmatic is the coastwise plume seen in images for September 2000, when both Congo effluent and terminal equatorial enhancement appear to be separable. To add to the difficulties induced by the several dynamic processes leading to chlorophyll enhancement in the adjacent open ocean, and despite seasonal cloud cover during the appropriate seasons, the images do suggest that upwelling cells occur along the coast both north and south of Cape Lopez.

Of course, over most of the shelf in this province, most of the time, upwelling is exceptional and the water column is stratified, with phytoplankton biomass concentrated in a DCM at the depth of the pycnocline, encountering the continental shelf at 15–45 m. Most attention has been given to the production cycle in locations where upwelling events occur, so that for the larger part of the open coast where upwelling does not occur, it is much more difficult to locate a satisfactory account of the production calendar. From data on water clarity and color at Lagos, obtained on a standard monthly transect (Longhurst, 1964), results at midshelf beyond any influence of creek effluents, suggested that a seasonal bloom occurs during the wet season, when strong onshore winds occur, during boreal summer months; this is supported by zooplankton biomass data obtained on the same transect. Copepod, cladoceran, and appendicularian numbers were one to several orders of magnitude greater in August and September than during the remainder of the year. This seasonal cycle suggests that the bloom on the open nonupwelling shelf of Nigeria, or off Sierra Leone and Guinea, is initiated by nutrient flux and later terminated by nutrient limitation.

In some of the larger embayments, such as the wide Sierra Leone estuary where tidal streams are strong, the seasonal production cycle is the reverse of that of the open sea, and algal cells are probably limited by light rather than by nutrients. Phytoplankton growth (about 850 mg C $m^{-2}$ $d^{-1}$) is maximal in the boreal winter dry season, when estuarine water is relatively clear, and it is greatly reduced (<75 mg C $m^{-2}$ $d^{-1}$) in the rainy season of boreal summer when inshore water carries a heavy load of silt. Furthermore, during the dry season, phytoplankton biomass is greater at neap than at spring tides, and thus when the coastal water is clearest (Bainbridge, 1960b). Herbivore consumption of this coastal bloom is small relative to autotrophic production. In the wet season, the apparent mesozooplankton demand is computed at about 91% of the relatively small daily primary production, but in the dry season only 5% of the larger bloom is required to sustain the zooplankton. Because in this case the demand of the benthic community is even less (about 1%), much of the inshore dry season diatom bloom must be exported from the system or buried in the coastal mud banks as organic matter (Longhurst, 1983). However, the persistence of large diatoms in inshore blooms province supports a rich population of phytoplankton-feeding clupeids (*Ethmalosa dorsalis*) in the inshore waters. These are the ecological equivalent of the Indian oil sardine (see INDW).

A peculiarity of the zooplankton of the Gulf of Guinea are seasonal "blooms" of cladocera (mainly *Penilia avirostris*, but also *Evadne tergestina*) that sporadically dominate the biomass in nonupwelling situations, this being observed at all seasons. Thus, off Sierra Leone from October to February (Bainbridge, 1960b) and off western Nigeria in September (Longhurst, 1964) cladocera may briefly dominate the zooplankton population, especially inshore. The zooplankton profile is typical of tropical seas; at night, biomass in the 0–25 m zone is 5–10 times greater than below 25 m, whereas by day the biomass is equally distributed from 0 to 50 m. Genera that perform these minor diel migrations are chiefly *Nanocalanus, Paracalanus*, and *Neocalanus*. Except over water >500 m deep, the major diel migrants *Metridia* and *Pleuromamma* are absent.

It is remarkable that even the very minor upwelling cells along the Guinea coast should support populations of the paradigmatic large copepod of the major upwellings in eastern boundary currents: *Calanoides carinatus*. This species has been recorded extensively over the outer shelf during boreal summer, and preferentially during upwelling episodes, from Ghana to Angola (Petit and Courties, 1976). In fact, as Bainbridge (1960a) demonstrated, the global biomass of zooplankton along the shelf from Fernando Po to Ghana is structured by the relative abundance of *Calanoides* rather than by the numerous warm-water species that dominate the zooplankton taxonomically. The fidelity of *Calanoides* to upwelling cells is shown by the fact that I observed it abundantly on the shelf off Lagos, Nigeria, on only one occasion—during a brief upwelling episode in a region where upwelling episodes are very rare.

Off Ghana, Mensah (1974) showed that only the C5s survived the end of the upwelling season, during which three or four generations appeared, with the development from egg to adult taking only 14–18 days. During the nonupwelling season, the population is maintained as C5s that remain dormant beyond the shelf, at deeper than 500 m. Because this period of 7–8 months could not be sustained on the oil sac accumulated during diatom feeding during the upwelling period, Mensah investigated feeding at depth and found this species could also capture bacterioplankton. In the C5 instar, maxillary setules have an interval of 1–4 $\mu$m. The same cycle was observed on the Central African shelf by Petit and Courties (1976), who followed six to eight generations during the upwelling period but found only C5s, and only at 800 m off the shelf, during the remainder of the year. The flexibility of life history schedules with the particular seasonal sequence of phytoplankton production each year off the Central African coast, and the fidelity between *Calanoides* numbers and patchiness of phytoplankton biomass, is remarkable.

These large calanoid copepods are a principal food of the West African sardine (*Sardinella aurita*), whose annual appearance off Ghana and Ivory Coast (Cury and Roy, 1987) coincides with the rise of *Calanoides* to the surface waters. In fact, here as elsewhere in some minor low-latitude upwelling regions, these two species (or their ecological equivalents) are closely linked. There is a simple relationship between SST, zooplankton abundance (dominated by *Calanoides*), *Sardinella* gonad maturation, and egg production off Ghana (Quaatey and Maravelas, 1999). From June, as the seasonal SST minimum approaches, sardine gonads mature. Sardine ova take maximal abundance in the water column in July–August, corresponding closely with the peak zooplankton abundance of August–September. In 1986, the SST minimum occurred in June, and in 1988–91 in July. Sardine ova extrusion followed the same pattern. In 1987, upwelling failed, zooplankton abundance was minimal, but sardine ova were produced in normal numbers. One wonders what became of that year-class of sardines?

On both the Guinea and Central African coasts, the changing relative abundance of sardines and other pelagic fish has attracted much attention (Gammelsrd *et al.*, 1998). During "Benguela Niños," the advent of warm water off the Congo and Angola is associated with mortality of *Sardinops sagax* and *T. trachurus* and a southward displacement of sardine stocks. Off the Congo, the cold-water *Sardinella maderensis* dominated the fishery from 1964 to 1983, when substantial warming was associated with the replacement of this species by *S. aurita*. The effect of a "Guinea Niño," associated with the transport of warm Bight of Biafra water south along Gabon, may be to force *S. aurita* offshore and *S. maderensis* even further south to the Congo fishery. Along the Guinea coast, it is postulated that the more extensive Ekman upwelling events of the 1980s may have favored an expansion of the Ghana stock of *S. aurita* eastward to the Ivory Coast.

## Regional Benthic and Demersal Ecology

The ecology of the benthic communities and of the demersal fish associated with them responds primarily to the presence of the permanent, shallow thermocline along almost the entire continental shelf; there are basically two habitats—one of warm, sunlit water at shallow depths and the other of cool water, deeper on the shelf, and dark.

In this province, we are fortunate in having the results of what is certainly the most complete survey of an almost pristine demersal fish fauna that has been performed anywhere in the oceans: the Guinean Trawling Survey (GTS) comprised 480 one-hour standard tows made by two matched French trawlers at eight standard depths on 63 transects, placed 60 km apart along the 4000-km shelf from Guinea to Angola, and it was performed in both seasons. The data, submitted to recurrent group analysis, revealed 20 species groups with clear affinities for specific deposit and depth habitats (Fager and Longhurst, 1968). These were objectively assembled into a warm-water fauna above the thermocline on level bottoms usually of mud or sandy-mud, a fauna on sands below the thermocline down to 100 m, and shallow (30–40 m) and deep (70–200 m) hard-ground species groups. There was a remarkable homogeneity in these associations the entire length of the coast and also in their relative abundance, with the single exception of the northernmost section immediately south of Cape Roxo, which was influenced by the Canary Current, and where biomass was significantly higher than elsewhere. It was also clear that the objective grouping supported almost entirely the well-known subjective groups of (i) a fauna of brown or silvery Sciaenidae, Drepanidae, Polynemidae, and Cynoglossidae above the thermocline on muds and muddy sand, and (ii) a fauna of red or pink Sparidae, Triglidae, Scorpaenidae, and Platycephalidae below the thermocline on sandy deposits. At and beyond the shelf edge, a fauna of *Chlorophthalmus, Scomber, Bembrops, Peristedion*, and *Antigonia* lay above the deeper-yet *Merluccius* and *Halosaurus*. On hard ground above and below the thermocline, large red and brown snappers (*Lutjanus, Lethrinus*) were characteristic of a reef fauna. The objective grouping identified very few eurybathic species that were not associated with one of the major species groups.

The distributions of invertebrate benthic communities similarly follow both the nature of the deposits and also the distribution of water types. The primary divisions are (i) between the organisms that occur only on soft muddy bottoms or those on sand, shell sand, and shell gravel, and (ii) between those that occur only in tropical surface water above the thermocline or those of the cooler, deeper water (Longhurst, 1958, 1963). In the tropical surface water, above the thermocline, a *Macoma* community, dominated by small bivalves and polychaetes occupies the soft muddy deposits in shallow water. An *Amphioplus* community occurs on slightly deeper muddy deposits, comprising many ophiuroids, polychaetes, and burrowing crustacea, while a *Venus* community occurs on shelly sands, largely of their own making, and includes forms such as *Branchiostoma* and echinoids. A *Tellina* community occurs on mineral sands inshore and includes many specialized echinoderms and polychaetes. Deep benthic communities below the thermocline on shelly sands and muds differ strikingly from the shallow forms, being dominated by genera such as *Hyalonoecia, Leptometra, Ophiothrix*, and *Pennatula*. At the shelf edge, both of these benthic communities are replaced by cold-water forms that are largely conspecific with those at similar depths off Western Europe.

Preliminary budgeting of autotrophic production and benthic and pelagic consumption on the Sierra Leone shelf suggests that benthic turnover, as measured from the biomass of macroinvertebrates comprising the communities discussed earlier, is sufficiently slow that the autotrophic biomass unconsumed by pelagic herbivores at most seasons is grossly in excess of the demand of macrobenthic organisms. This supports the now-familiar concept that in the benthos, as in the pelagic ecosystem, it is the microbial organisms that perform the bulk of community respiration.

This concept was reinforced by data from a survey performed by "Meteor" in 1988 of meiobenthos biomass distribution on 13 box-corer transects from the mid-shelf depths to the deep ocean, along the coast from Guinea to Angola (Soltwedel, 1997). This survey demonstrated a clear forcing of meiobenthic biomass by both regional phytoplankton production and depth: on upwelling coasts, biomass was higher but, everywhere, decreased with increasing depth. So, foraminiferan, nematode, harpacticoid, and polychaete abundance is clearly forced by the relative supply of POC, of all sorts, raining from the pelagial onto the benthic ecosystem. This is not exactly news, but has not often been observed in tropical seas.

## Synopsis

*Case 5—Large-amplitude response to monsoon-like reversal of trade winds, but modified by coastal processes*—Generalizations concerning the calendar of this province are unsatisfactory, because of the major differences between the Guinean and Central African coasts, and the difficulty of separating processes in the latter from the effects of the Angola Dome, offshore. In hindsight it would have been preferable to recognize two separate provinces for the east-west and north-south coasts. The integrated data for the province presented in Fig. 9.24 show two seasonal peaks of productivity and chlorophyll abundance that appear to represent the events on the two coastal segments: a lesser Guinea coast maximum in boreal spring, and the greater Central African coast maximum in boreal fall. Here, $Z_p$ is consistently deeper (25–45 m) than the mixed layer, so the thermocline is illuminated in all months. Production rate is seasonally light limited, being low from April to September, during the coastal rainy season, whereas a significant and sustained P rate increase occurs in September–October during the transition from wet season solar radiation to dry season, when values are higher by a factor of 4. Chlorophyll biomass very closely traces P rate changes; this is consistent with a close match between consumption and production rates.

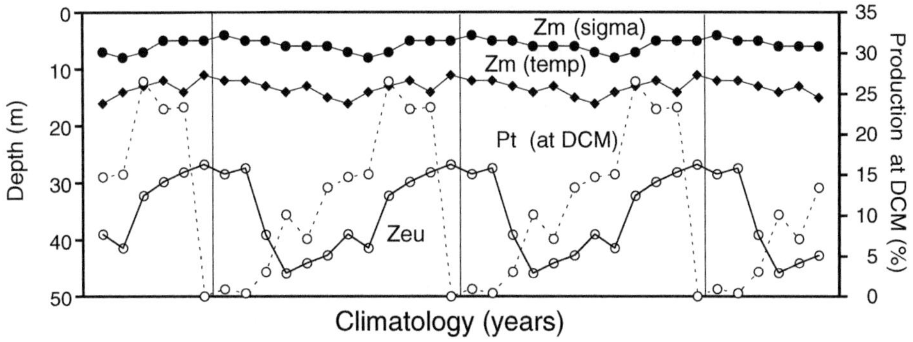


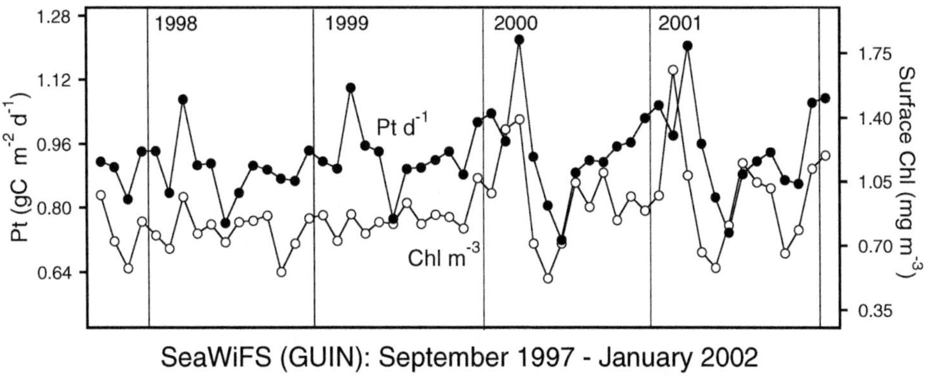


**Fig. 9.24** GUIN: seasonal cycles of monthly surface chlorophyll and depth-integrated autotrophic production for the years 1997–2002 from SeaWiFS data together with characteristic seasonal cycles of mixed-layer depths from Levitus climatological data and photic depths computed from characteristic irradiance and the archive of chlorophyll profiles discussed in Chapter 1.

# Benguela Current Coastal Province (BENG)

## Extent of the Province

BENG is the shelf of southwestern Africa from the Cape of Good Hope north to the region of Cape Frio (18°S). Excluded is the retroflection region of the Agulhas Current south and southwest of the Cape of Good Hope itself that is included in the EAFR coastal province. Included is the upwelling eddy field of the Benguela Current.

## Continental Shelf Topography and Tidal and Shelf-Edge Fronts

This continental margin is a trailing edge (recollect the mid-Atlantic ridge sea-floor spreading center) so we may expect wider shelf areas than off the western coast of South America: and so they are. The shelf of southwestern Africa is about 150–180 km wide, and deep at that, with the break of slope at about 400 m. It is only at salient capes that the shelf becomes narrower, although at Cape Frio, at the northern end of the province, the isobaths deeper than 500 m become incorporated into the topography of the Walvis Ridge that extends deep into the Atlantic in a southwesterly direction. Muddy deposits lie inshore and the outer shelf carries relic quartz sands and rock. The shelf also includes anomalously deep zones; double shelf breaks are common at about two depths: 150–200 m and 300–500 m. Shelf topography plays a significant role in determining the locations of anomalies in the shallow circulation. Maximum range of spring tides is about 2 m, but I have found no references to tidal fronts between mixed and stratified areas of shelf; a shelf-edge frontal feature is discussed later as part of the cross-shelf circulation that results from upwelling at the coast.

## Defining Characteristics of Regional Oceanography

The Benguela Current forms the eastern limb of the anticyclonic gyre of the South Atlantic and originates in the flow of the South Atlantic Current around the southern limb of the gyre, but also receives intermittent intrusions of Indian Ocean water from retroflection of the Agulhas Current.

This system is complex, but is well explored and has been the subject of several recent reviews (Chapman and Shannon, 1985; Shillington, 1998). Because the continent of Africa does not extend so far southward as the American continent, the Benguela is unique among the four eastern boundary currents because it interacts, poleward, with the western boundary current of the adjacent ocean. Thus, the Agulhas Current passes partly around South Africa prior to retroflection back into the Indian Ocean. Because this mechanism is unstable, some Agulhas retroflection rings are lost into the southwest Atlantic, later to interact with the outer part of the Benguela Current off Namibia; the frequency of such intrusions tends to be variable at the decadal scale.

The wind stress field over the Benguela Current is typical of those overlying other eastern boundary currents and represents the interaction between the atmospheric anticyclone of the southeast Atlantic and the deserts behind the coast of southwestern Africa. Winds are perennially favorable to upwelling south of about 15°S, and cyclonic wind stress curl lies inshore of the wind velocity maximum. A coastal zone of cyclonic wind stress originates between Cape Point and Cape Columbine (33–35°S), where it is narrow, but widens to the north until at about 20°S the cyclonicity becomes very patchy, especially in austral winter and spring. Cyclonic curl maxima and hence exceptionally strong upwelling-favorable winds occur near Luderitz and Cape Frio; in the latter region, these conditions are persistent both seasonally, and also at shorter time scales. This persistence distinguishes the northern from the southern Benguela region, where upwelling is modulated at periods of about 1 week by relaxation or reversal of coastal winds, in response to the passage of cyclones poleward of the Cape peninsula.

Equatorward flow in the Benguela Current is carried in a series of jet currents that are strongest in summer, and in eddying flow that responds to coastal and continental shelf topography and incorporates a series of wind-forced upwelling cells (Shannon, 1985; Lutjeharms and Meeuwis, 1987). The broad equatorward flow has two components that diverge off Cape Columbine at 33°S: (i)—aligned with the shelf edge is a meandering, offshore core of maximum velocity, that represents a shelf edge jet, and (ii)—closer to the coast, the weaker Columbine jet is transformed progressively northward into variable coastal flow. Flow of the shelf edge jet is topographically steered so that it lies consistently over the continental slope from the Cape up to Luderitz (27°S), where some separation from the slope first occurs.

Separation into the northern limb of the southern subtropical gyral circulation is completed in the vicinity of Cabo Frio at 18°S, after the offshore flow has passed through a gap in the deep topography of the Walvis Ridge at 20°S; here, warm water in the seasonally shifting Angola Front is encountered (Shannon *et al.*, 1987). This front represents the convergence zone between the poleward surface flow of tropical surface water in the Angola Current and the equatorward flow of cool Benguela water and is marked at the surface by a very strong temperature gradient. Tonguelike mesoscale intrusions of warm water, especially in austral summer, pass across the front that also marks the transition between strong stratification in the tropical water mass to the north and weak stratification to the south. Especially in late austral winter, the chlorophyll signature of upwelled water that passes toward the west along the Benguela/Angola Front forms a diffuse feature across the South Atlantic to at least 10°W. This region of high chlorophyll biomass is separated from the chlorophyll signature of the equatorial current system in satellite images by a zone of blue water.

As also occurs in the Canary Current, there is significant deposition at the sea surface of aerosol particulates; off southwestern Africa, aeolian dust events are associated with offshore katabatic "berg" winds from the deserts of southwestern Africa, which blow strongly for periods of several days, often simultaneously over as much as 1500 km of the coastline between 20 and 30°S. Such offshore wind events result in a succession of small upwelling cells, restricted to <10 km from the coast.

Day-night alternation of the land-sea breeze system is strongly developed (as it is off all coasts that are backed by major deserts) and the sea breeze has a fetch of 100–150 km. As Shannon (1985) comments, this variability can be expected to modulate diel processes in the upwelling system. At the secular scale, equatorward wind stress, and hence cool conditions in the Benguela, was strong in the 1920s and 1930s but a warm period intervened in the 1940s; upwelling wind stress then progressively increased again until the 1990s (Shannon and O'Toole, 2003). Such large-scale changes in wind field intensity and direction significantly modify the intensity and distribution of upwelling, stratification and surface temperature. The consequences of "Benguela Niños" in the equatorial Atlantic have been discussed in Chapter 8: like the intrusions of Agulhas influence from the south, these events transport warm, nutrient-poor tropical water across the borders of the Benguela province.

Coastal upwelling in the Benguela Current, as in other eastern boundary currents (Mooers *et al.*, 1978), results in cross-shelf circulation and a set of fronts related to rotary motion within the upwelling cells. Near the shelf break, there is frequently a prominent front where upwelled water sinks along the shoreward side, to create a coastal cell, while Atlantic water diverges on the seaward side. Where upwelling is especially vigorous, the upwelled water may be circumscribed by a surface front closer inshore than the shelf-break convergent/divergent front (20–30 km wide) that can be traced north to Cape Frio, becoming more diffuse toward the north. The outer part of the divergence zone is associated with surface slicks above internal waves (Shannon, 1985). Particularly pronounced divergence zones, and jet currents, occur off Cape Columbine and the Cape peninsula and cold-core eddies appear as isolated areas of chlorophyll enhancement along this frontal zone; beyond the shelf-break front, an eddy field marks the edge of the clear waters of the subtropical gyre of the South Atlantic.

Shillington (1998) points out that strong interaction occurs between coastal upwelling cells and Agulhas rings that have defied retroflection, and subsequently pass northward well beyond the shelf break; these rings may entrain cold water so as to extend a filament of upwelled water anomalously far out into the South Atlantic. These, ending in a pair of contrary-rotating gyres, may extend even as far west as the dateline from central regions of the Benguela upwelling area (see Color plate 9).

The upwelling centers identified by Shannon (1985) and Lutjeharms and Meeuwis (1987) are not always observable as entities in the satellite images, although coincidentally, one of the earliest SeaWiFS images was of an upwelling center below cloudless skies that represented an atmospheric high-pressure cell. For practical purposes, the upwelling centers can be grouped as follows:

*Cape Peninsula (34° S) and Cape Columbine (32° S)*: Upwelling is largely restricted to the austral summer trimester (December–February), and the average temperature of upwelled water in both cells is about 17°C. The upwelling cells off Cape Columbine may generate cold filaments that extend 500–700 km offshore and have a longshore spacing of 200–300 km, whereas those off the Cape peninsula are about half that dimension. Each of these upwelling cells is frequently bounded seaward by a jet current, and off Cape Columbine this jet is isolated from the shelf edge jet by a clear divergence zone (Shannon, 1985). In summer, upwelling may extend around the Cape peninsula, as far to the east as Cape Agulhas (35°S 20°E).

*Hondeklip Bay, Namaqualand (29° S)*: A regional maximum of cyclonic wind stress curl and upwelling occurs persistently in the bight south of the Orange River; the shelf here is narrow and deep, and the source of water brought to the surface in coastal upwelling is also deep. A persistent mesoscale cold filament or tongue originates at this upwelling center. Upwelling response to wind stress is slower than that in the regions to the south; maximum upwelling occurs in October–December, and minimum upwelling occurs in May–July.

*Luderitz to Walvis Bay (24–22° S)*: This is the most intense upwelling area, especially off Luderitz itself, and forms a major environmental barrier in the Benguela Current system. Upwelling occurs throughout the year, being seen in >80% of images scanned by Lutjeharms and Meeuwis (1987) over a 3-year period; nevertheless, there is a clear seasonal cycle with minimal upwelling in December–January and maximal in July–September. SST at 200 m accordingly varies between 15 and 19°C. Upwelled water is quite cool (mean temperature is 16.5°C at Luderitz and 17°C in the Namaqua cell) because the shelf break is exceptionally deep and the source of upwelled water lies at 200–300 m. The seaward extent of upwelling is greatest here, reaching almost 300 km off Luderitz, and the upwelling event scale is long compared with that of regions to the south. Shannon (1985) comments that here the event scale resembles Peru and northwest Africa rather than Oregon or the southern Benguela.

*Central (21–23°) and northern Namibia, especially Cape Frio (19–17° S)*: Upwelling is continuous throughout the year but again is strongest in the austral winter trimester (June–August), with some extension into spring (September–November). During summer the water column is sufficiently stratified so as somewhat to constrain upwelling. Upwelled water is >16°C (mean of upwelling cells is 19.5°C at Cunene and 18.5°C in the Namibia cell). The continental shelf off central Namibia is narrower and shoaler than farther south, having a shelf break at 140 m about 100 km offshore. For this reason, very cool water is unavailable as the source of upwelling. The upwelling cells in this sector extend about 150 km offshore. There may be a persistent convergence between these upwelling cells and those off Luderitz to the south.

Although the upwelling centers just described are the foci of enhanced algal growth and surface chlorophyll, the Benguela Current is sufficiently dynamic that upwelling occurs seasonally essentially the whole length of the coast.

## Response of the Pelagic Ecosystems

Attention here, as in the other eastern boundary provinces, is necessarily directed at the response of the pelagic ecosystem to physical processes associated with upwelling: particularly how autotrophic cells modify their nutrient demand and growth characteristics when brought to the surface, thereby experiencing changes in irradiance and temperature. Three sequential zones can be recognized in a Benguela upwelling cell, each corresponding to a stage in the evolution of the response of the plankton, both producers and consumers, to the upwelling process: (i) an inshore zone where nitrate and silicate are supplied and utilized by an algal bloom; (ii) a middle zone where one of these nutrients becomes limiting; and (iii) an outer zone, associated with the meandering equatorward jet current, where autotrophic cells become seriously nutrient limited and changes occur in floristic composition. Sinking of aggregates and particles occurs principally in the middle zone, and if this is over shelf depths, remineralization will be important. In general, the chlorophyll-enhanced water follows the outline of the shelf edge except where it narrows to the south of Luderitz. During upwelling episodes there is often a near-coastal band of relatively clear, recently upwelled water, with the maximum chlorophyll concentration of the inshore zone in midshelf about 15–25 km offshore. Chlorophyll biomass, after upwelling, forms a near-surface maximum (Shannon and Pillar, 1986) and accumulates at the upwelling front, where $<9$ mg chl-*a* $m^{-3}$ is observed, associated with high bacterial abundance. The relative abundance of autotrophic and heterotrophic cells changes as upwelled water ages and heterotrophic activity comes progressively to dominate.

The production and subsequent remineralization of organic material from phytoplankton and fecal pellets formed after the advection of nutrients into the photic layer in upwelling cells is associated with the formation of oxygen-deficient water over the shelf (Chapman and Shannon, 1985) so that, in the quiet periods between upwelling events, red tides are common all along the Benguela. The main southeast Atlantic subsurface oxygen minimum layer lies north of the Benguela-Angola front, between Cape Frio and Luanda, that is associated with the cyclonic gyre of the Angola Dome. However, off Namibia, especially Walvis Bay, and off Luderitz, as well as to the north of Cape Columbine, oxygen deficiency may be induced *in situ* in shelf waters (Chapman and Shannon, 1985). Both oxygen-deficient ($<2.0$ ml $liter^{-1}$) and anoxic water that may even contain $H_2S$ may be brought to the surface by coastal upwelling, and this has significant consequences for shelf biota, including fish kills, especially at Walvis Bay.

The filaments and eddies (chlorophyll values of 3–10 mg chl $m^{-3}$) of upwelled water may merge with the next upwelling center so that it is often visually impossible to separate the production from each individual center in satellite color images. Under these circumstances, the whole of the Benguela Current appears as a coastal band of high chlorophyll. Investigation of a mature filament off Namibia revealed that it was composed of cool, aged upwelled water in which utilization of $NH_4$ by nanoplankton and picoplankton dominated nitrogen uptake, and in which salps were very abundant; the filament lay between warm, oligotrophic water of a detached Agulhas ring and oligotrophic Atlantic water (Shillington *et al.*, 1990).

The biological response to upwelling differs significantly between the northern and southern Benguelan regions. Comparing Namibia with the Cape Province, Chapman and Shannon (1985) suggest that upwelling in the south occurs mostly as a series of short pulses, whereas in the north it is more continuous during the upwelling season. That the southern upwelling should be more dominated by diatom growth than the north is probably a consequence of the short time scale of the upwelling pulses. In the north, high diatom concentrations (2–5 μg chl liter$^{-1}$) occur within a shallow euphotic zone of 20–35 m in which the colonial *Phaeocystis* lies preferentially close to the surface while *Coscinodiscus* is deeper: this partition typically creates a bimodal chlorophyll profile (Vordelwuebecke *et al.*, 2000).

Especially in the north, coupling between herbivore biomass and plant growth is close and this results in rapid silica depletion, as diatom frustules are incorporated in copepod fecal pellets, then to contribute to the large deposits of biogenic silica that occur in the shelf deposits off Namibia. In the south, more silica is recycled in the water column. Consequently, it is probable that nitrate is the limiting nutrient in the south, off the Cape peninsula, whereas silica limits plant growth in the north, off Namibia, as is reported to be the case in the Peru-Chile Current (see HUMB).

Mesozooplankton of the Benguela system has low diversity relative to that of the Agulhas Current to the east and there is little endemism among the individual biota,; these characteristics were attributed by Gibbons and Hutchings (1996) to the episodic and intermittent nature of the upwelling process; such pattern is inimical to stability and diversity. They note also that a large fraction of total biomass is concentrated in a small fraction of the total area of this province. They also remark that the identity of many of the herbivore species could be predicted from studies elsewhere: thus, *Calanoides carinatus* is the dominant copepod, associated with species of *Paracalanus, Clausocalanus, Centropages*, and the rest. As there is a different emphasis in the dominant large algal cells between neritic (diatoms) and offshore (dinoflagellates) and between northern (dinoflagellates) and southern (diatoms) regions, similarly, for the mesozooplankton, there are different regional patterns; offshore and to the north, there is high diversity and low biomass with many gelatinous organisms, whereas inshore and to the south, crustaceans dominate a zooplankton of low diversity and high biomass.

Despite such differences in relative abundance of different organisms, there is very little replacement of species between northern and southern regions; the dominant foraminifera, siphonophora, hydromedusae, scyphomedusae, decapoda, copepoda, amphipoda, and chaetognatha are similar in both. Only among the euphausiids is a species-replacement important: *Nyctiphanes capensis* dominates in the north, *Euphausia lucens* in the south. Within these species lists, Gibbons and Hutchings distinguish neritic, frontal, and oceanic assemblages of differing diversity: neritic low, oceanic high, while frontal plankton is a mixture of the other two groups and attains higher densities than either. Within each assemblage, there is equally a vertical segregation that provides niche partitioning among rather similar species, as between *Euphausia hanseni* and *Nematoscelis megalops*, the former at the shelf break and the latter inshore; even at fronts, vertical segregation is maintained. Pioneer forms, such as *Calanoides carinatus*, display ontogenetic migrations that place their resting stages in locations that ensure that upwelling water will be sufficiently seeded with individuals capable of very rapid growth and reproduction as to ensure, in turn, a sufficiency of resting individuals. This pattern has been proposed for other pioneer species in the Benguela, such as *Sagitta friderici* and *Euphausia lucens*.

The Benguela current is the habitat—as are all the other eastern boundary currents—of important pelagic clupeid stocks that consume calanoid copepods and diatoms. Clupeids often exist as species pairs (one large and one small) in eastern boundary currents and here *Sardinops ocellatus* and *Engraulis capensis* share this habitat with horse mackerel (*T. trachurus*) and red-eye (*Etrumeus whiteheadi*). These shoaling fish occur preferentially along the upwelling front over the outer continental shelf, and much of the patchiness of observed mesozooplankton abundance may be attributed to intense and localized consumption by their dense shoals; models of the relationship between *Engraulis* biomass, sea surface chlorophyll and temperature, and *Calanoides* abundance show that this relationship is highly sensitive to parameters difficult to determine empirically, such as predator-specific zooplankton mortality rates. The other side of this coin is that the recruitment of these species is, in turn, controlled at least in large part by the intensity of upwelling and the consequent strength of the plankton blooms.

The anomalously large year-class of Cape anchovy resulted from the late 1999 spawning in the southern Benguela of an unexceptional adult population at a time when the southeast wind anomaly (and hence upwelling intensity and negative SST anomaly) was the strongest of the previous 40 years for which records exist (Roy *et al.*, 2001). Although this event raised as many questions as answers, the association between physical forcing and subsequent recruitment is not in question; what is in question is why, in this case, the previously established inverse relationship should fail. It had previously been determined that strong upwelling was detrimental to anchovy year-class strength, because it was thought that increased offshore advection would lead to increased loss of eggs and larvae to the open ocean. However this may be, there has been a significant long-term shift in relative abundance; since 1965–1966 the biomass of *Engraulis* has been very significantly greater than of *Sardinops*—indeed, in the late 1980s by an order of magnitude—although during the 1950s the biomass of the two species was essentially equivalent.

## Regional Benthic and Demersal Ecology

All linear shelves having intermittent but strong coastal upwelling risk benthic anoxia as the sediments are supplied with DOC produced in the pelagial at rates faster than this material can be oxidized by heterotrophic microbes. The Benguela system is no exception, and widespread depletion of oxygen occurs, especially in the northern part of the province; in exceptional events, such as in 1993–1994, benthic hypoxia may extend the entire length of the shelf and may result in mass mortality of benthic invertebrates and demersal fish, although species such as hake may be able to respond by actively changing their distribution.

Demersal fish surveys, such as those of MacPherson and Gordoa (1992, 1996) and Mas-Riera *et al.* (1999), reveal partitions of the benthic habitat primarily into shelf and slope components and secondarily with relation to bottom type and bottom dissolved oxygen and temperature. Once again, species replacement to ensure habitat partitioning is very evident: *Merluccius paradoxus* and *M. capensis* inhabit southern and northern shelf environments respectively: *M. polli* is a species of lower latitudes. The entire shelf/slope habitat may be partitioned into six regions, of which the shelves north and south of 28°S will be of greatest interest to us here. The species replacement is striking between these regions: in 1997–98, *M. paradoxus* formed 55–80% of the entire biomass in the south, depending on season, and *M. capensis* about 75–80% of all biomass in the north. What it is today, I am not prepared to guess. The biomass runners-up were, respectively, *Nematogobius barbatus* in the north and *Helicolenus dactylopterus* in the south. Diversity of each was found to decrease in regions of poorly oxygenated bottom water. The earlier surveys of MacPherson and Gordoa performed in 1983–1990 suggested a somewhat different partition and revealed significant between-year changes in biomass. *M. paradoxus* was assigned by this analysis to a southern slope association, and of the "southern shelf" species only *Lepidopus caudatus* was sufficiently abundant to bring significant biomass to the total for the entire province. A general decrease of demersal biomass was recorded during this period.

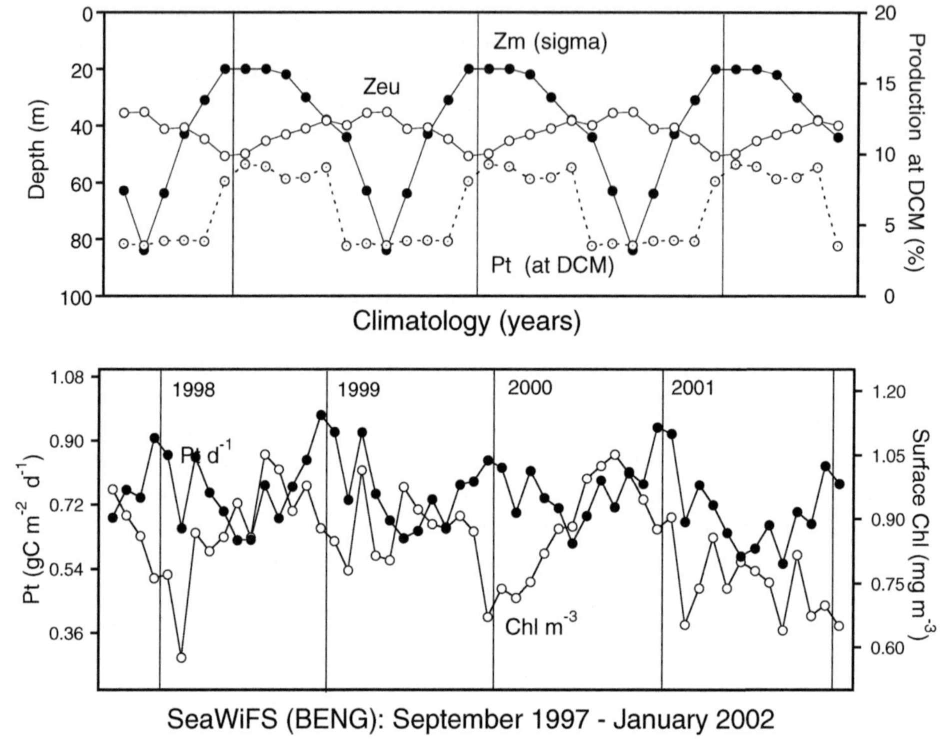


**Fig. 9.25** BENG: seasonal cycles of monthly surface chlorophyll and depth-integrated autotrophic production for the years 1997–2002 from SeaWiFS data together with characteristic seasonal cycles of mixed-layer depths from Levitus climatological data and photic depths computed from characteristic irradiance and the archive of chlorophyll profiles discussed in Chapter 1.

## Synopsis

*Case 6—Intermittent production at coastal divergences*—The climatology offered by archived data does not correspond with any of the individual upwelling cells, but shows the generally deeper austral winter pycnocline and the effects of austral summer upwelling on primary production rate (Fig. 9.25). Seasonality of $Z_m$ responds both to austral winter mixing and to coastal upwelling in austral summer. $Z_{eu}$ lies consistently at moderate depths so thermocline is illuminated in all months except July–October (austral winter-spring) and presumably also in strong upwelling cells, the effects of which are concealed in mean values. Rate of P increases from winter minimum to annual austral summer maximum in May during the maxima of trade wind stress and coastal divergence. Seasonal dynamic range is strong (0.5–1.75 mg chl $m^{-3}$) but does not closely track seasonality in rate of P.

# Northwest Atlantic Shelves Province (NWCS)

## Extent of the Province

The NWCS province comprises the continental shelf and slope water from the Florida Keys to the Strait of Belle Isle between Newfoundland and Labrador. The offshore boundary is either, as convenient, the north, inshore wall of the Gulf Stream from Florida to Newfoundland or the shelf-edge front; arbitrary lines are drawn around the east and north sides of the Grand Banks. It is traditional to partition the geography of this region into Acadian, Virginian, and Carolinian provinces, but this does not fully respond to the variety of habitat included in NWCS and a more useful arrangement would be the following:

(i) Newfoundland shelf, including the Grand Banks and Flemish Cap
(ii) Gulf of St. Lawrence
(iii) Northeast shelf and Gulf of Maine, from Cabot Strait to Cape Hatteras
(iv) South Atlantic Bight, from Cape Hatteras to the Florida Keys

It will be useful in this account to recognize this partition for some of the issues to be discussed; the fourth compartment is required on account of its transitional character with the CARB province to the south.

## Continental Shelf Topography and Tidal and Shelf-Edge Fronts

The continental shelf of eastern North America is a typical trailing-edge shelf and was glaciated as far south as Cape Hatteras during the Pleistocene. The shelf shoals progressively southward from the anomalously deep regions off Labrador until, beyond Cape Hatteras, it merges with the coastal plain ashore. In the northern regions, two entrants greatly extend the area of shelf-depth water: (i) the Gulf of St. Lawrence that lies behind Newfoundland, opening eastward above the glaciated deep trough of the Cabot Strait and (ii) the more open Gulf of Maine, lying behind Georges Bank, and including the Bay of Fundy.

The continental shelf is of great width (~500 km) at the Grand Banks, which are isolated from coastal processes by the Avalon Channel along the eastern coast of Newfoundland. The sediments and topography of the Grand Banks reflect the effects of ice formation, and extensive gravel beds represent the terminations of the continental glaciers of North America. Lying further to the east is the isolated shallow water of the Flemish Cap. Northeast of Newfoundland and north of the Grand Banks, a large flat area with depths exceeding 200 m is often miscalled the Newfoundland Shelf.

Below Saguenay, the St. Lawrence River follows the deeply glaciated Laurentian Trough that is the largest deep re-entrant on the American coastline. This trough continues across the Gulf and passes out across the shelf through the Cabot Strait (45–47°N) as the Laurentian Channel, which is both deep (<400 m) and wide (<100 km); this has consequences for circulation and also for shelf-edge upwelling.

The topography on the Scotian Shelf is irregular, resembling that of the Grand Banks, and there are several unusually deep (<200 m) basins at midshelf, whereas shallow sandbanks are widespread on the outer shelf, of which one forms Sable Island. The Gulf of Maine continues the glaciated topography and deposits, including the ice-cut Northeast Channel between Georges Bank and the Scotian Shelf. Georges Bank, seaward of the Gulf, is a large, shallow feature surrounded by deeper water and bears parallel sand ridges that are sufficiently shallow as to force linear wave trains at the sea surface.

The Middle Atlantic Bight, between Cape Cod and Cape Hatteras, at all depths shallower than the shelf break, bears terrigenous sediments of rather homogenous grade. The flat, linear shelf of the South Atlantic Bight is dominated by the input of sediments from many rivers and is quite unusually shallow, the break of slope occurring at only 80 m. The existence of a coastal frontal zone inhibits cross-shelf transport, so riverine sediment accumulates inshore, while coarse-grained sediments dominate the outer shelf.

Tidal streams may be sufficiently rapid in some places, and water sufficiently shoal, that stratification is rarely or never achieved; this may occur both inshore and also offshore, as on the top of Georges Bank. The well-mixed shallow region of central Georges Bank is separated from the surrounding stratified water in summer by an annular tidal front that passes around the bank and has important consequences for the general productivity of the bank ecosystem (Loder and Greenberg, 1986). Other tidally mixed areas occur in the Gulf of Maine, across the mouth of the macrotidal Bay of Fundy, around Grand Manan Island, and also on the Nantucket Shoals south of Cape Cod (Townsend, 1992). Within the Gulf of St. Lawrence, tidal mixing and tidal fronts occur mainly in the channels between Anticosti, Prince Edward Island, and the mainland and also around the Magdalen Islands (Pingree and Griffiths, 1980). Further south, in the Middle Atlantic Bight, tidal fronts parallel the coast and separate an inner neritic from an outer open shelf zone. Beyond Cape Hatteras, the water column over the shelf in the South Atlantic Bight is permanently stratified except as noted below.

A shelf/slope front, marked by the 34% isohaline within the slope water, reflects the instability between the shelf and slope water masses and is a consistent feature of this province although, unfortunately, confusion is possible between this front and the North Wall itself (Smith and Petrie, 1982). The shelf/slope front is periodically modified when warm-core eddies, which have been shed from the meandering Gulf Stream, flood in across the outer shelf and create instability and jet currents in the shelf water. At such times, bursts of topographic Rossby waves propagate across the shelf as internal bores or solitary waves (Smith and Sandstrom, 1986); this mechanism is responsible for significant transport of nutrient-replete slope water up onto the shelf. Where the Gulf Stream lies over the slope in the South Atlantic Bight, energetic frontal eddies frequently intrude into shallow water.

## Defining Characteristics of Regional Oceanography

Circulation is extremely dynamic and variable in this province, a not surprising observation in the light of the extraordinary conjunction here of polar and subtropical conditions; in winter, the pack ice of the Gulf of St. Lawrence is only about 300 km from the north wall of the Gulf Stream and hence of subtropical surface water. I believe that this is a unique condition.

The Labrador Current transports low-salinity, Arctic water and separates into an inshore and offshore component on encountering the topography of the Grand Banks. The inshore component passes along the Avalon Channel close to the east coast of Newfoundland and contributes to shelf-water formation on the Grand Banks; some also enters the Gulf of St. Lawrence through the Straits of Belle Isle. The offshore component passes around the continental slope of the outer Grand Banks and contributes to the slope water that lies between the Gulf Stream and the shelf water.

Despite the complex geography of its continental shelf, and its important meridional extent, the NWCS province is nevertheless unified by equatorward flow of cool water originating in the Labrador Current, which may be followed as far as south as Cape Hatteras at 35°N, always above the slope and with maximum velocity just beyond the shelf break: at Newfoundland, southward flow over shelf, slope, and deep water is respectively 0.8, 5.7, and 30 Sv. The freshwater component of the flow over the shelf is relatively minor, being measured in millisverdrups (Loder *et al.*, 1998), despite the effluent from the St. Lawrence River.

Thus, the entire province north of Cape Hatteras is characterized by meandering southward coastal flow throughout the year, though this responds to variability in sustained wind direction. During winter, southerly winds frequently confine the coastal water zone to a narrow, vertically homogenous feature along the inner shelf. Conversely, northerly winds spread the inner shelf water seaward and thus induce stratification. Indeed, during periods of sustained longshore wind from the southwest, local upwelling cells may form at many points along the coast. Coastal upwelling occurs off southwest Nova Scotia persistently, and also intermittently along the southeast-facing coast when wind forcing is from a favorable direction.

Circulation in the Gulf of St. Lawrence passes around a complex eddy field, which is induced by the several large islands and the complex coastline and is influenced by the discharge of the St. Lawrence River, which reaches $30 \times 10^{-3}\,m^{-3}\,sec^{-1}$ at times of peak flow in summer. This flow, together with Labrador Current water that has entered directly from the north through the Straits of Belle Isle, drives an outward stream at the surface through the Cabot Strait at all times. Ice cover forms during winter in the Labrador Current and over much of the Gulf of St. Lawrence, and during the period of spring breakup, fractured pack ice is transported south out of the Gulf and along the eastern coast of Nova Scotia. Ice regimes on the northern parts of the province differ strongly depending on the origin of the flow in which freezing occurs. Unlike the Gulf of St. Lawrence where pack ice is formed *in situ*, the winter ice cover off the Labrador coast, which extends south to the northern Grand Banks, is characterized by the frequent presence of icebergs drifted from the glacier fronts of the Canadian arctic. These may survive individually much farther to the southeast than does the seasonal ice cover, as *Titanic* discovered to its cost.

Since the entire Gulf of St. Lawrence has an estuarine circulation, so that the discharge of brackish water at the surface toward the open ocean induces a landward flow of deep water through the Cabot Straits and, consequently, the upwelling of nutrient-rich water at the head of the Laurentian Channel north of the Gaspé Peninsula (Dickie and Trites, 1983). The water that passes out onto the Scotia shelf from the Gulf of St. Lawrence moves southward along the coast with the shelf water mass, passes cyclonically around the Gulf of Maine, rounds Georges Bank, and so moves on southward past Cape Cod to form the shelf water of the Middle Atlantic Bight.

The southern end of the province beyond Cape Hatteras is transitional to the CARB province to the south; here, the northward flow of the Gulf Stream water above the slope is contrary to the generally equatorward drift of slope water in this province. Shelf water in the South Atlantic Bight, although it still passes southward along the coast, is therefore heavily influenced by admixture of warm, tropical water with major consequences for nutrient flux onto the shelf (Boicourt *et al.*, 1998). The slope of the South Atlantic Bight, at about 31°N, bears a NE-trending ridge—the "Charleston Bump"—that deflects the deep flow of the Gulf Stream, inducing meanders that may form cyclonic frontal eddies that turn back over the shelf and flood it with oceanic water. Because the Gulf Stream at these latitudes is bistable, lying along one or the other of two persistent tracks, the frontal eddies take two modes: smaller and larger. This bistability determines the consequences for shelf ecology because, in the large eddy mode, warm oceanic water passes around a cold dome and may then propagate right in to the coastline.

The development of a summer stratified regime from the winter conditions of near-uniform properties to 50–100 m depth that obtain everywhere except in the South Atlantic Bight is anything but simple. On the northern Grand Banks, buoyancy is enhanced in spring by the presence of superficial water of low density, the result of melting pack ice, and also by increasing surface irradiance and consequent heat gain. The same occurs in the Gulf of St. Lawrence, although here meltwater is relatively less important (Doyon *et al.*, 2000). On the Scotian Shelf, warm slope water episodically floods across the shelf; because it is of significantly higher salinity than the shelf water, it is denser and so remains near-bottom. Such fluxes are an important nutrient input to shelf ecosystems, and are akin to the transport of nutrient-rich slope water (2–5°C) into the head of the Gulf of St. Lawrence below the cold intermediate water at −1° to 2°C.

At the southern end of the province, in the South Atlantic Bight, the timing of stratification in spring is dependent on the direction of wind forcing (Flagg *et al.*, 2002); downwelling-favorable winds maintain the shallow plume of light water from Chesapeake Bay close to the coast as it passes southward in the general shelf circulation, but upwelling-favorable winds cause it to spread out across the shelf. This provides the extra stability needed to enable the ambient surface heat flux to stratify the water column so that mixing occurs again only at the end of summer.

In this province, the effect of changes in the North Atlantic Oscillation (NAO, see Chapter 8) is as important as anywhere. The NAO Index characterizes the strength and location of the westerly winds over the North Atlantic in winter and, consequently, the meridional displacement of the Gulf Stream: during periods of low NAO Index, the flow of the Labrador Current is increased and the area occupied by Labrador Slope Water is enlarged as the North Wall of the Gulf Stream shifts southward and comes to lie farther offshore. This is shown in Color plate 10, illustrating the differences between early spring of 2000 when the NAO Index was high (+2.8), and spring of 2001 when it was low (−1.89). Zwanenberg, *et al.* (2003) notes that annual changes in bottom water temperature on the Scotian Shelf are among the most variable in the North Atlantic; a 50-year record shows periods of 5–10 years of sustained anomalies of both signs with very rapid switching between cold and warm conditions. An influx of cold water into Emerald Basin in 1998 dropped temperatures there by 3°C. To follow such events, it is very helpful that a library of matched 2-km, 15-day AVHRR SST and SeaWiFS chlorophyll images for the area from Cape Cod to northern Labrador is now readily accessible on-line (http://www.mar.dfo-mpo.gc.ca/science/ocean/ias/seawifs).

## Response of the Pelagic Ecosystems

In terms of primary production and (at the other end of the trophic chain) the potential production of demersal and pelagic fish, this is one of the richest regions of the ocean; unfortunately, it is also here that the failure of fishery science to provide adequate methods for the management of stocks has been most acutely demonstrated.

Because of complexity of the geography of the NWCS it is difficult to generalize the pelagic ecosystem and, without the satellite data now available, much would remain obscure. At the regional scale, archived chlorophyll images show that the Grand Banks and the Scotian Shelf, together with the Gulf of St. Lawrence and the Gulf of Maine, are the principal centers of productivity, and also of seasonal change. Although there are major differences between these regions, the seasonal productivity cycle in each involves the occurrence of a spring bloom, which is observed in the images as a major accumulation of chlorophyll biomass. In at least the northern regions, a second period of higher chlorophyll biomass occurs in late autumn or early winter; Harrison *et al.* (1999) detected bimodal seasonal cycles of primary production, modeled from surface chlorophyll, on the southern Labrador shelf, the Newfoundland shelf, the Grand Banks, the Scotian Shelf, and Georges Bank. The timing and magnitude of maxima at these locations varied significantly: spring bloom is stronger than that of autumn off southern Labrador, spring and autumn are similar off Newfoundland, and autumn is stronger than spring at the more southerly locations.

Of course, chlorophyll enhancement does not occur uniformly across any of the subregions listed earlier and, between years, the blooms change very significantly both as to pattern and timing, although the perpetually high pigment signal observed in places such as the Bay of Fundy, the inner Gulf of St. Lawrence, and Chesapeake Bay is assumed to be at least in part the consequence of high loads of suspended sediments and of CDOM.

It will now be appropriate to review these regions individually and in greater detail. A comparison of the pattern of SST and surface chlorophyll on the **Grand Banks** and on the **Scotian Shelf** is very instructive. In the former region, the patches of strong chlorophyll enhancement tend to be linear and change their location in response to the distribution of Gulf Stream eddies and of cold Labrador Current water, so that there is little commonality between years; on the Scotian shelf and in the Gulf of Maine, the shallow and tidally mixed banks such as Georges Bank are routinely observed as areas of enhanced chlorophyll where seasonality is weaker and chlorophyll remains relatively high during winter months. Chlorophyll enhancement in the Gulf of St. Lawrence is also clearly shown by between-year comparisons of satellite images to respond to the mesoscale pattern in the cyclonic circulation around the Gulf. The response of phytoplankton growth rate to dynamic physical events at the mesoscale is self-evident: a quick comparison of a few pairs of matched SST and surface chlorophyll images is worth a thousand words.

The regional and between-year variability that we can now observe makes it all the more satisfactory that the early serial data obtained across this region by the Continuous Plankton Recorder (CPR) does yield apparently correct seasonal patterns. The CPR database (Colebrook, 1979) differentiates seasonal cycles for (i) the Grand Banks (strong spring blooms, weak autumn blooms), (ii) the Scotian Shelf (weaker and more similar spring and autumn blooms), and (iii) the Cape Cod region (algal abundance barely higher in summer than winter). The CPR algal data are based only on large phytoplankton cells but, even so, they suggest that the physical processes discussed previously do determine the regional pattern of algal growth dynamics in the province and that the pattern of chlorophyll enhancement observed in the SeaWiFS images can mostly be related to frontal dynamics.

A spring bloom occurs over the whole western **Gulf of St. Lawrence** as soon as ice cover clears in spring, and algal growth is no longer light-limited; it has particularly high values ($<200$ mg C $m^{-2}$ $hr^{-1}$) in the region of upwelling and also further down the Gulf where it is reinforced by strong tidal mixing over the Magdalen shallows, and also in the region of the tidal front between Anticosti Island and the Gaspé peninsula. Knowledge of the pelagic ecology of this region was advanced by the Canadian JGOFS process study performed there in 1992–94 (Roy *et al.*, 2000a). This revealed yet another example of the now-familiar pattern of seasonality in the large and small autotrophic cell fractions; the small fraction forms a "background" of autotrophy that has a relatively small seasonal cycle of relative abundance, because the growth of protistan herbivores is so closely matched to autotrophic cell-division rate. Against this background, episodic blooms of larger ($>5\,\mu$m) cells are responsible for the much more variable total chlorophyll biomass. At least in part, the strength of the chlorophyll enhancement here is a consequence of the climatological strength of wind mixing. Strong mixing induces strong nutrient flux from below, but limits the access of autotrophic cells to surface light, whereas weak mixing restricts the cells to a shallow, well-lit mixed layer with weak nutrient flux: moderate mixing, says Doyon, et al (2000), is responsible for strong phytoplankton growth here.

The pelagic ecosystem of the Gulf of St. Lawrence is dominated during winter and spring by large phytoplankton cells that are heavily grazed by mesozooplankton, whereas the heterotrophic loop dominates in summer when nanophytoplankton and heterotrophic dinoflagellates and ciliates replace the diatoms and copepods of winter. Despite this, the flux of sinking organic material is approximately similar summer and winter (Savenkoff *et al.*, 2000). Of the total POC flux, ~50% comprised mesozooplankton fecal pellets—with some seasonality—while <10% comprised intact phytoplankton cells. The composition of this flux responds to the taxonomic composition of the superjacent zooplankton population, so that where the smaller copepods (e.g., *Temora*) were more relatively numerous, the fecal pellet fraction was relatively larger (Roy *et al.*, 2000b).

In the **Gulf of Maine** (see Townsend *et al.*, 1992) upwelling occurs at many small estuarine fronts and is thought to be a major factor in overall productivity, because the water in the deep basins is constantly renewed by episodic pumping from deep slope-water sources. However, the effects of tidal mixing dominate the distribution and strength of algal blooms, and the principles discussed already for the northeast Atlantic continental shelf (see Northeast Atlantic Shelves Province) apply here. Blooms are initially light limited in tidally mixed areas, then persist longer because of the constant renewal of nutrients due to benthic regeneration. In stratified areas, an oligotrophic profile rapidly forms when the initial nutrient charge is utilized. In winter, very dense aggregations of *C. finmarchicus* and the euphausiid *M. norvegica* occur in the deep troughs and basins, both in the Gulf of Maine and on the eastern Scotian shelf. As in the northeastern Atlantic, these are essentially deep-water expatriates—though they may still be the dominant organisms of their trophic group—and the shelf basins are their deep-water refuges. In winter, when seasonal migration would have carried them to depths of 1000 m in the open ocean, extremely dense aggregations of oceanic copepods and euphausiids occur close to the bottom in these basins, whose depths exceed 200 m.

The spring bloom over the deep water of the Gulf of Maine can precede water column stratification (Townsend *et al.*, 1992), and this may be a general but overlooked process wherever spring blooms occur over deeper water. The principle involved is that deep penetration of light in a clear, winter-mixed water column may support cell growth rates sufficiently high as to overcome the vertical excursion rates induced by wind stress at the surface, especially in calm weather. This process is assisted by the nature of the spring-bloom cells in some cases, including gelatinous colonies and diatom chains with very low sinking rates. In such situations, it is also possible—as has been demonstrated for the Arabian Sea—that the presence of a layer of light- and heat-absorbing chlorophyll may be a contributing factor in the eventual establishment of stratification. This phenomenon should be watched for in the open North Atlantic.

The winter-spring transition on **Georges Bank** was followed closely in the critical period of 1997 during a JGOFS study of that habitat (Townsend and Thomas, 2001). Initiation of the bloom occurs very early: in January, chlorophyll biomass was at a seasonal low but already in February had risen to 2–3 μg chl liter$^{-1}$ inside the 60-m isobath, where nitrate and nitrite were together already about 4 μM. The bloom survived a sustained incursion of Scotian Shelf water and continued during the month of March and spread over deeper parts of the Bank. Silicate appears to have been the overall limiting molecule, although there was some recycling. In May and June, high near-surface concentrations (>5 μg chl liter$^{-1}$) of chlorophyll occurred around the deeper margin of the Bank, although these were patchily distributed. As already noted, cross-frontal transfer of nutrients plays a major role in supplying nitrate to support the summer bloom in the tidally mixed parts of Georges Bank (Loder and Platt, 1985; Horne *et al.*, 1989). A leading candidate for cross-frontal transport is the residual circulation associated with tidal current interactions. The computed rate of this transfer is sufficient to support observed new production at the front itself (67% of total production is nitrate based) and within the enclosed mixed area on the bank, where only 27% of production is fueled by nitrate in summer.

On Georges Bank, and further south, *C. finmarchicus* is scarce and the dominant copepods, both numerically and by biomass, are *Oithona* spp. that are distributed throughout the water column, even in stratified water. A conceptual model was developed to answer the question, "Can Georges Bank larval cod survive on a calanoid diet?" (Lynch *et al.*, 2001), and this concluded that the smallest (4–6 mm) classes of larvae cannot do so, although larger classes (>10–12 mm) should grow satisfactorily at indicated copepod densities. On the other hand, the apparent abundance of *Pseudocalanus* on the Bank should provide sufficient food for all larval size classes, but is appropriately distributed in neither space nor time. Since it is known that *Calanus* undergo episodic starvation on Georges Bank, it is not surprising that cod larval survival, and hence subsequent year-class strength, should be as uncertain as it is known to be. At a larger scale, *Calanus* abundance in the entire Gulf of Maine is also known to be variable over the long term, responding to changes in the value of the NOA index and hence to changes in temperature within the Gulf.

Most areas south of Cape Cod have a rather variable seasonal production cycle, with highest rates tending to occur toward the end of summer, and only the slope water has a classical seasonal cycle of productivity with a single spring peak. There are seasonal changes in the vertical distribution of mesoplankton herbivores relative to the DCM:

during summer, there is a coincidence in depth between chlorophyll and larger herbivore abundance, whereas at the end of summer the copepods begin to descend into the lower layer below the pycnocline.

In the **Mid-Atlantic Bight** the winter-mixed region is linear and neritic so that chlorophyll increases regularly seaward: at the same time, the percentage of total production contributed by nanoplankton also increases seaward. Consistently higher rates of primary production occur in the New York Bight, associated with discharge of urban sewage into an area of strong tidal fronts and mixing, and on the northern part of Georges Bank. Productivity, and primary production in these two areas is 25% higher than in any other part of the Mid-Atlantic Bight (O'Reilly and Busch, 1984). Below the warm, clear surface water over the shelf in summer there may be a cold, well-lit, chlorophyll-rich layer of bottom water; this originates in the Gulf of Maine and on the Nova Scotia shelf when shallow summer stratification isolates the cooler bottom water. Advected off-shelf at Cape Hatteras and vertically mixed in the Gulf Stream flow, this water body contributes significantly to the nutrient budget of the Gulf Stream (Wood *et al.*, 1996).

Off Cape Hatteras, recent studies (e.g., Lohrenz *et al.*, 2002) have confirmed that maximum chlorophyll biomass, measured *in situ*, occurs in March and is associated with a strong DCM in which highest values (>10 mg chl $m^{-3}$) occur near the shelf edge front, associated with nutrient flux there; this feature is dominated by large cells (>8 μm) and, once again, by July the lower biomass is dominated by small cells. Anomalously high chlorophyll values in midsummer are associated with influxes across the shelf of high-nutrient slope water and are located close to mesoscale meander features associated with the Gulf Stream flow in the same way as this occurs in the South Atlantic Bight (see later discussion). Herbivores have been found to remove ~40% and ~60% of large- and small-fraction autotrophic cells, respectively, during the spring shelf edge bloom, so that small autotrophs (which contribute ~50% of total primary production) contribute very significantly to overall production and are significantly incorporated into the microbial food web (Verity *et al.*, 2002).

In the **South Atlantic Bight** (Carolina Capes region) the western, cyclonic edge of the Gulf Stream (as noted earlier) generates wavelike perturbations that propagate northward along the continental edge and so induce upwelling of cool nutrient-rich Atlantic Central water at topographic features, especially during summer (Blanton *et al.*, 1981). Because the upwelled water mass has great clarity, subsurface algal blooms may occur across the shelf in the bottom water, and this process supports dense concentrations of small copepods. Thus, abundance of most mesoplankton species on the Florida shelf increases progressively toward deeper water, which has high particulate load (Paffenhofer, 1983).

As Schollaert *et al.* (2004) discuss, the actual track taken by the bistable Gulf Stream (see earlier discussion) determines the relative strength of seasonal chlorophyll bloom in the slope water area beyond the very narrow shelf of the South Atlantic Bight. Over a 4-year period, the SeaWiFS images showed a consistent strengthening of this bloom, with which was associated a progressive northward shift of the axis of the offshore Gulf Stream. Analysis of this observation shows that the nutrients on the South Atlantic Bight shelf are not delivered from a distant source by the extension of Labrador Current water southward, as had previously been assumed, but rather that they are delivered into slope water from subsurface Gulf Stream water. This mechanism is more effective when the Gulf Stream takes a more northerly track.

## Regional Benthic and Demersal Ecology

The community structure of the invertebrate benthos here is not as well described as on comparable European continental shelves, and on trawlable grounds we must assume that it has been highly modified; however, experimental trawling on grounds on the Grand

Banks that had been left untrawled for 12 years recorded minimal disturbance, and it was found that structural changes in the benthic community mimicked changes due to natural causes: no distinctive trawling signature could be detected (Kenchington *et al.*, 2001).

It is assumed that the pristine benthic fauna exhibited a parallel series of species associations to those better explored off Europe in the early 20th century and before to the start of heavy industrial trawling; indeed, some examples of isocommunities from this region were reported by Thorson (1957). The recent JGOFS investigations in the Gulf of St. Lawrence included studies of the benthic communities (Desrosiers, 2000) that characterized trophic relations and microbial activity in relation to depth, slope, and substrate: deep settling basins are dominated by surface deposit–feeding invertebrates, whereas on sloping ground, subsurface deposit feeders either dominate or have equivalent biomass to surface feeders. These two trophic groups comprise >60% of total benthic biomass within the Gulf. Polychaetes (*Capitella, Orbinia*) and bivalves (*Nucula*) largely dominate the subsurface feeding biota, with some seasonal replacement. Surface-feeding polychaetes and bivalves are more diverse: *Laonice, Paraonis, Terebellides* and *Chaetoderma, Nuculana*, and *Antalis*, respectively. The presence of these species associations has a very strong impact on the temporal composition and granulometry of the deposits and of their characteristic microbial populations.

A strange form of benthic-pelagic coupling was uncovered during the JGOFS study of Georges Bank (Concelman, *et al*,. 2001). Abundant individuals of the benthic hydroid *Clytia* spp. were found in the mesozooplankton samples over the central parts of the Bank; examination of archived plankton samples showed that these are present in some years, absent in others. The source of these individuals is located on the NE peak of the Bank, whence they are detached from the epibenthos by anomalously strong wind mixing during the passage of major storm systems across the Bank, to be progressively retained in very high concentrations in the central low-advective regions. Here, they can form ~90% by numbers of all zooplankton, with copepods, chaetognaths, and others thus forming ~10%. Growth of pelagic hydranths, which require four to five *Artemia nauplii* daily *in vitro*, is enhanced by turbulence. When present, they form an important item in the diet of some fish, especially winter flounder, and are a major predator of other mesozooplankton.

It is difficult to know how to write about the demersal ecology of this province, especially of the northern regions that dominate the whole: the shock of the collapse of the cod stock of the Grand Banks and other areas of this province in 1992 is still with us. That this also had severe impacts of local human demography in the coastal fishing communities (Hamilton, 1998) is perhaps not so well understood except to those who live in the Atlantic Provinces of Canada.

There are, of course, lessons for ecologists to be learned from the event. For instance, one of the reasons that stock assessment data were somewhat misleading in the years prior to the crash is that we now know that species distributions may be changed not only by between-year environmental changes, but also by excessive fishing pressure: as the northern cod stocks (*Gadus morhua*) declined, so the range occupied by the remnant shrank progressively, and the area of highest concentration moved progressively southward. Like other fish with associative behavior, local densities varied significantly less than the overall reduction in biomass of the stock, which fell by about six orders of magnitude during this period (Atkinson *et al.*, 1997). Under such circumstances, it is not surprising that assessment surveys at sea did not return accurate estimates of population size.

However it happened, the remnants of the Grand Banks cod stock that had sustained a fishery for 500 years was declared commercially extinct in 1992 and the fishery was closed, as was that on the Scotian shelf shortly thereafter. Despite this moratorium, the biomass of the northern cod stocks continued to decline even further, from 11,700 t in

1994 to only 900 t in 1996; a corresponding decline in incoming year-class strength was attributed principally to the reduction in offshore spawners, but also to increasing natural mortality in the planktonic phase.

There was almost certainly an environmental component in the stock decline, associated with the extreme variations in temperature and salinity recorded in the early 1990s after a long period of generally declining temperatures over the Grand Banks. But this is not to say that political pressure to maintain catch levels, and widespread avoidance of regulation at sea, together with application of inappropriate models of stock dynamics were not the primary cause of the crash. From this experience, a fundamental reevaluation of fish stock management science is in order.

One lesson learned from this event is that it is not sufficient simply to stop fishing in order to reestablish the status ante quo, because not only did the stock continue to decline after the moratorium, but a decade later it still shows only very slight recovery: had a federal election not been called in 2005, it seems doubtful if the quotas introduced for some stocks would have been permitted. Associated with the lack of population regrowth, there is evidence that major shifts have been induced in the entire structure of the benthic ecosystem. As the large fish stocks declined, and the area they occupied shrank on the Grand Banks, so the local population of northern shrimp (*Pandalus borealis*) increased both its biomass and area of distribution; this was attributed initially to environmental change but is now acknowledged to have been an ecosystem response to altered predation pressure. Metanalysis demonstrates that the reciprocal abundance exhibited by cod and northern shrimp over long periods and large areas is a top-down, predation effect (Worm and Myers, 2003). That ecosystem structure is unstable seems an inescapable conclusion: consider the evidence for a major shift in trophic links in the Gulf of St. Lawrence in the period 1959–2000. During the first 30 of these years, in stomachs of 30- to 60-cm cod, 6–70% of prey mass was composed of euphausiids: but only trace amounts were found after 1990. During the same period, herring consumption by cod shifted from 1.3% of prey mass to almost 20%, matching changes in the relative abundance of herring over the period. It would be a brave ecologist who would confidently predict future ecosystem structure on these shelves, yet such is the stated objective of a new thrust of fishery science toward ecosystem-based stock management.

### Synopsis

*Case 2—Nutrient- limited spring production peak but modified by shallow-water processes.* The mixed-layer $Z_m$ undergoes significant boreal winter excursion from 10–20 m (June–September) to 50–100 m (February–March), while $Z_{eu}$ lies consistently at 20–30 m. Winter mixing carries the pycnocline to the bottom along the midshelf, and to moderate depths seaward of this line. The thermocline is therefore illuminated during boreal summer from May to September. The rate of P begins to increase in late winter, matched rather closely by chlorophyll accumulation, with some indication that in autumn chlorophyll accumulates by reduction of herbivore pressure (Fig. 9.26). Alternatively, we cannot exclude the possibility that the significant chlorophyll accumulation in some years during boreal winter (October–March) in satellite imagery may be partly the effects of river discharge and winter resuspension.

## Guianas Coastal Province (GUIA)

### Extent of the Province

The Guianas Coastal Province (GUIA) extends from Cape de Sao Roque (5°S) in Brazil to Trinidad (10°N) within the offshore limits of the North Brazil (or Guiana) Current

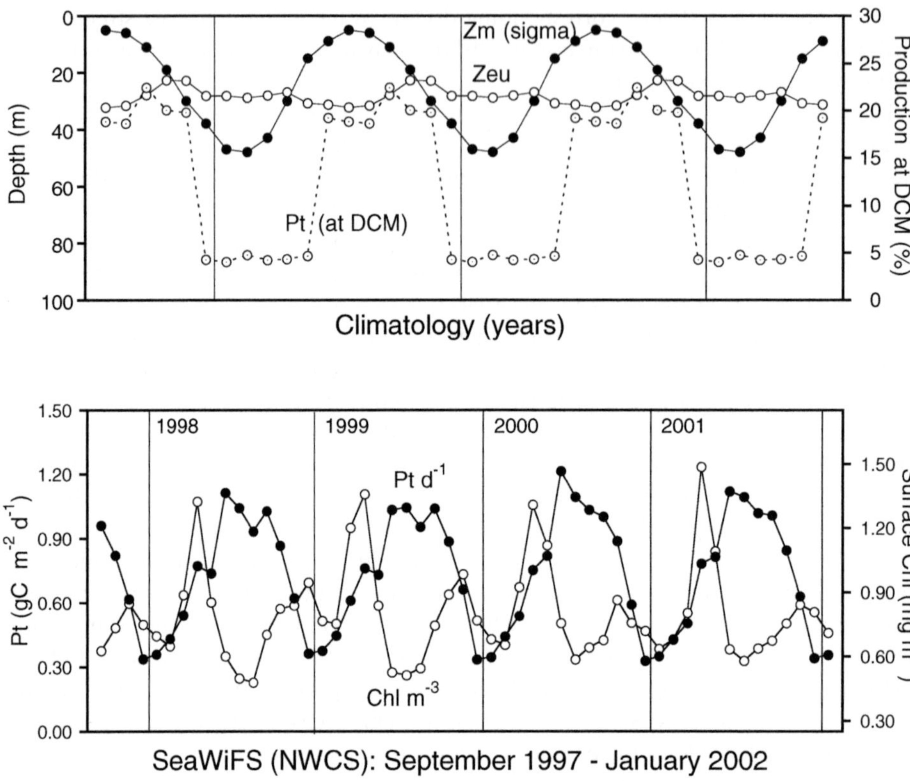


**Fig. 9.26** NWCS: seasonal cycles of monthly surface chlorophyll and depth-integrated autotrophic production for the years 1997–2002 from SeaWiFS data together with characteristic seasonal cycles of mixed-layer depths from Levitus climatological data and photic depths computed from characteristic irradiance and the archive of chlorophyll profiles discussed in Chapter 1.

that runs along the northern coast of Brazil and Guiana as far as Trinidad. Westward from here, the shelf region along the entire northern coasts of Venezuela and southern Panama in the southern Caribbean is included.

## Continental Shelf Topography and Tidal and Shelf-Edge Fronts

Almost the entire province has a significant width of continental shelf, this being widest in the region of the Amazon mouth where the prograde shelf break front lies about 200 km offshore. Many rivers open into this province, draining the rain-forest regions of continental South America: the Para, Amazon, Essequibo, and Orinoco rivers being the largest. The coastal regime is consequently highly turbid, especially north of the Amazon mouth, where migratory mud banks are established (Wells, 1983). South of the Amazon mouth, the sediments become increasingly calcareous and the shelf narrows to only 20 km at Cap San Roque. Off the mouth of the Orinoco, clay and mud give way at midshelf depth to sand and calcareous sediments and the calcareous sands of the outer shelf form linear banks (or "bioherms"). These topographic features, intimately connected with the seasonal discharge of sediments, will be discussed later.

## Defining Characteristics of Regional Oceanography

The North Brazil Current (NBC) crosses the equator northward along the coastline of South America, carrying SEC water that has originated in the South Atlantic subtropical gyre. The NBC is a warm, shallow (<200 m) stream and is one of the sources for the equatorial undercurrent (Metcalf and Stalcup, 1967); its flow is augmented to the north of the equator by confluence with the weaker NEC (Boisvert, 1967). The NBC changes character seasonally, forming a series of coastal eddies in boreal winter that transport only 10 Sv, but the surge of the southeast trades into the western Atlantic during boreal summer transforms it into a western jet that then transports as much as 35 Sv.

Although topographically locked to the continental edge, the NBC becomes unstable between the mouths of the Amazon and Orinoco so that a small part of the flow which overlies the continental slope is retroflected eastward into the open Atlantic in a series of very large, persistent eddies: the Amazon eddy is formed at 4°N, and the Demerara eddy at 8°N (Bruce *et al.*, 1985). It is convenient to consider these eddies, which form prominent features extending far seaward, as part of the western tropical Atlantic province (see WARM); they are, as Fratantoni and Glickson (2002) remark, the largest oceanic rings formed anywhere in the ocean, which is not surprising given the low value taken by the Coriolis parameter at only 2-3° from the equator.

It is the effect of the discharges from the Amazon on the ecology of the Atlantic shelf, and from the Orinoco on the Caribbean shelf, that will take most of our attention in discussing this province. The Orinoco is not a negligible river, having the fourth largest annual discharge of any, but the Amazon, of course, is a special case that is well described by Geyer *et al.* (1996) both because it discharges more water than any other river but also because it lies directly on the equator; moreover, unlike the Congo, it is not associated with a deep cross-shelf canyon. The freshwater discharged by the Amazon has a nitrate concentration of about 15 μM, which is small compared with the discharge of other great rivers: Yangtse, Yellow, and Mississippi rivers discharge freshwater containing 50–100 μM nitrate. Perhaps when the Amazon rain forest is entirely cleared, and the basin reduced largely to agriculture as in the other examples, its freshwater drainage will come to have similarly high nitrate values. Nevertheless, the Amazon discharge does have a very strong signature of CDOM, not distinguishable from the chlorophyll signature in satellite imagery unless the data are processed with a specific algorithm (Hu *et al.*, 2004).

At peak flow in May, the rate of discharge from the Amazon is about 0.2 Sv (or ~220,000 $m^{-3}$ $sec^{-1}$), about twice the minimum flow in November. Such strong discharges prevent seawater from entering the river mouth, so that a strong salinity front occurs about 100 km offshore, coincident with a turbidity front, and lying above a series of transverse shoals formed by deposition of sediments. Even at this distance offshore, water depth is only about 15–20 m above a submarine delta that is being built out across the shelf. Although tidal streams are sufficiently strong that cross-shelf velocities reach 200 cm $sec^{-1}$ in the frontal zone, the suspended load in the bottom water is so great as to dampen vertical mixing by tidally induced bottom stress.

The shallow, low-salinity plume of river water passes northwest along the coast, the form taken by its bounding salinity front being wind-dependent. Southeast winds induce a "fast" plume of brackish water to head along the coast, whereas northeast wind produces a "slow" plume that balloons out across the shelf, with very little northward transport from the river mouths. Here, another interesting example of the consequence of the low value of the Coriolis acceleration at low latitudes (see Chapter 4) is observed: in middle to high latitudes, as Nittrouer and DeMaster (1996) point out, the Coriolis force would induce the plume to turn anticyclonically across the shelf (clockwise, that is, in the northern hemisphere), but here the dynamics of the plume respond almost directly to the sustained landward wind stress. This is sufficient to maintain a general flow along the coast to the northwest, with anticyclonic retroflection occurring only in relation to topographic features on the slope (Fig. 9.27).

Drogued, satellite-tracked drifters have been deployed in the NBC just offshore of the Amazon shelf and near the river mouths. Most of these buoys subsequently track northwest along the shelf, with those nearest the shelf edge describing a series of tight anticyclonic eddies only as they approached and entered the Caribbean near Trinidad. Several passed at a shallow angle toward the edge of the shelf and entered the retroflection

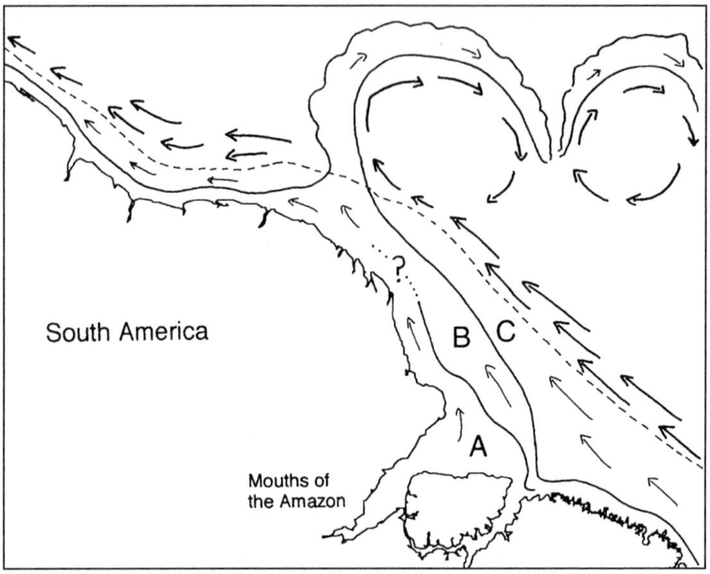


**Fig. 9.27** Idealized diagram of the processes involved in the retroflection of the Amazon plume during boreal summer when the coastal jet current is most strongly developed. Shown is the inshore zone (A) of high turbidity with suspended matter >100 mg liter$^{-1}$ and an outer zone (B) of high chlorophyll values of 2–4 mg m$^{-3}$. On the outer shelf, these two values become 1.0 and 0.1, respectively. The paired retroflecting arcs are persistent features at this season, carrying high chlorophyll only on their northern flanks, and are of controversial origin.

eddies discussed earlier before passing into the North Atlantic Equatorial Current and heading off toward the Caribbean or Gulf of Mexico. One, placed in the NBC over the slope off the Amazon, entered a retroflection eddy and passed from there into the NECC before reaching the coast of West Africa. This technique confirms the details of Amazon water flow along the Atlantic shelf in a series of lenses that may be partially retroflected seaward into the NECC. Guiana shelf water, containing a very large proportion of Amazon water, passes into the Caribbean as a band 150–500 m wide, with a flow maximum in June and July.

Coastal upwelling occurs from northern Brazil to Surinam, as indicated by the upslope of isopycnals toward the coast and the occurrence of relatively cool water in the upper 10 m close inshore. In fact, this is a predictable consequence of the development of a coastal wind-driven jet current along the western boundary of an ocean and of the prevailing vector of coastal wind stress during the boreal summer, when southeast trades are strongest (Gibbs, 1980). On the northern coast of Venezuela, the Carioca Basin exhibits strong coastal upwelling along the linear east-west coast of the Paria Peninsula; this responds very closely to the field of intensity of the westward stress of the trade winds. Climatologically, these vary by a factor of two between July–October (low stress) and November–May (high stress), and SST minima are positively correlated with wind speed, lagged by about 15 days (Müller-Karger *et al.*, 2004).

## Response of the Pelagic Ecosystems

Three ecological zones are generally recognized in studies of the ecology of the Amazon shelf (e.g., Smith and DeMaster, 1996):

(i) A coastal zone inshore of the transverse shoals, where turbidity is sufficiently high as to inhibit photosynthesis and where nitrate remains high (~7 μM) so that here CDOM dominates the "chlorophyll" that is observed by satellite sensors.

(ii) A linear zone along and just seaward of the turbidity front, where primary production ceases to be light limited and chlorophyll accumulates ($<25\,\mu g$ chl $m^{-3}$) in a near-surface ($<5$ m) bloom while CDOM remains a significant fraction of observable "pigments." It is possible that chlorophyll accumulation in the second zone is enhanced by the upwelling of shelf water along the salinity front, as in an estuarine circulation; in shelf water not directly influenced by the Amazon discharge, CDOM is not present in unusually high concentration.

(iii) Seaward of the chlorophyll plume, still only in about 40–50 m depth, nitrate becomes limiting and is always $<0.5\,\mu M$ so that chlorophyll values are reduced to 0.2–0.6 mg chl $m^{-3}$.

Currently, there appears to be very little evidence for the availability to algal cells of nitrogen regenerated by either benthic microbial activity or water column metabolism. Nevertheless, it has been suggested recently (Subramanian, pers. comm, 2005.) that during periods of high discharge of both Amazon and Orinoco, diatoms having endosymbiotic diazotrophic cyanobacteria dominate the autotrophs; these support high productivity and sedimentation rates ($>150$ mg $m^{-2}$ $d^{-1}$) within the river plumes.

The coastally trapped discharge from the Amazon and the subsequent input from the Orinoco modify the ecology of the eastern shelf of Venezuela and the Gulf of Paria (Bonilla *et al.*, 1993). Most of the nutrients discharged by the Orinoco remain in this area, where benthic regeneration is very active, so that their contribution to the eastern Caribbean is problematical, though the low-salinity signal of Amazon and Orinoco water can be detected as far away as Puerto Rico. Significant amounts of nitrogen probably do not survive the passage through the coastal ecosystems and the Gulf of Paria (unless recycled nitrogen is entrained in the regional flow) so that the pigment feature so prominent in the eastern Caribbean is probably not caused by riverborne nutrients, as suggested by Müller-Karger *et al.* (1989), but rather is now thought to be caused by upwelling in the southern Carioco Basin.

Because the varved sediments of the deep Carioco Basin offer such an exceptional archive of past climate changes, extending back for 600,000 years, revealing changes in North Atlantic Deep Water formation during the glacial periods, an unusual amount of scientific effort has been put into understanding the pelagic processes in the water column that are revealed in the varved sediments. Here, we have unrivaled sea truth for the available sea surface chlorophyll images and very complete monitoring of the variable wind regime and the water column processes thereby forced (Müller-Karger *et al.*, 2004). There is a very regular seasonal cycle in bloom area (and concomitant area of reduced SST) between $>10,000\,km^2$ in March or April and an order of magnitude smaller in October; both shipborne measurement and satellite images deliver a single period annually (January–May) of high productivity, the integrated annual productivity varying from 372 to 650 g C $m^{-2}$ $y^{-1}$ over the 6 years 1996–2001. These measurements have enabled a very accurate estimate to be made of the resultant flux into the varved sediments at depth. Satellite images show that the cold water and high chlorophyll plumes are directed toward the northwest from the coastal region from the east-west coastline that lies to the west of the island of Trinidad. The sea truth investigations also assisted in separating the effects of CDOM-laden Amazon shelf water passing between this island and the mainland from the chlorophyll associated with cold water at the coast a little to the west.

## Regional Benthic and Demersal Ecology

The benthic habitat on the Amazon shelf is anomalous because of the unusually high rate of deposition of riverborne organic particulates that occurs here seasonally; during the period of rising river flow in February–March, there are extreme episodes of settlement whose blanketing effect is sufficiently strong as to take down planktonic organisms from the water column into the sediments (Aller and Todorov, 1997). Calanoid copepods and other planktonic biota are rapidly buried, along with the *in situ* benthic fauna, to depths of ~25cm; in the AMASSEDS surveys, extraordinarily high abundances of buried copepods were observed at all stations, reaching >5000 copepods $m^{-2}$ at the river mouth station, in water ~20 m deep. The copepods were intact and had fresh phytoplankton in their guts, so they must have been carried down and been buried very rapidly. During the period of falling river discharge (August–October) this phenomenon occurred only close to the river mouth; at this period, the deposits below the inner river plume are at their most stable.

More generally, the seasonality of benthic communities within the ambit of the Amazon plume responds directly to discharge rate: this is observed in the abundance and size frequency distribution of individual species, as well as in their behaviors as expressed by burrowing and feeding depths, and also in the specific diversity of communities (Aller and Stupakoff, 1996). During the period of rising river flow, these indices all take minimal values, associated with very dynamic overturn and resuspension of at least the top 100 cm of the soft sediments; in the opposite season of low discharge rates and weak wind stress, the soft sediments become consolidated and are very rapidly invaded by burrowing polychaetes that reach high density (~100 ind $cm^{-2}$), while meiofauna (nematodes, harpacicoids and foraminifera) and bacterial numbers increase by about two orders of magnitude (from ~3 to ~$220 \times 10^9$ cells $cm^{-2}$) through the top 10 cm of the sediments. With increasing depth, both diversity and abundance of meiofaunal groups increase.

The macrofauna is dominated by polychaetes and crustacea, whereas mollusks are relatively unimportant. Early colonizing polychaetes (e.g., capitellids, spionids) are especially characteristic everywhere, with a wide range of other taxa, of which Lumbrinereidae appear to reach the highest densities. Crustacea are mostly burrowing forms: cumaceans (*Eudorella*), burrowing carideans (*Ogyrides*), thalassinid mud shrimps (*Callianassa* and *Upogebia*), and tube-dwelling amphipods (*Erichthonus* and *Ampelisca*). Sediment reworking rates of order 100–200 g $m^{-2}$ $d^{-1}$ were computed for these and related organisms. Relatively insignificant numbers of brachyuran crustaceans are also present. Mollusks include both surface (*Tellina*) and subsurface (*Nucula, Yoldia*) deposit feeders, together with a few predatory gastropods. These species associations recall those that inhabit very soft muddy deposits off estuary mouths in the Gulf of Guinea, where the same genera of burrowing crustacea are equally important; the resuspension-deposition of the cycle deposits recalls that of the mobile mud banks ("chakara") of the Malabar coast of western India where comparable mortality and recolonization of the benthic macrofauna is induced.

The demersal fish associations of the shelf regions from the Amazon to the Orinoco were well surveyed in their near-pristine state in 1957–59 (Lowe-McConnell, 1962), clearly establishing their general resemblance to those of the Gulf of Guinea shelf, with a few interesting differences. In fact, we do not lack for early survey data of this region by local and colonial governments, and also by fishery agencies of the United States and the USSR (see Longhurst and Pauly, 1987). The dominant coastal species association, on soft coastal mud, the length of these coasts comprises 11 species of sciaenids (*Cynoscion, Macrodon, Nebris, Stellifer*) as well as threadfin (*Polynemus*), spadefish (*Chaetodipterus*), flatfish (*Paralichthys*), sea catfish (*Bagrus, Arius*), and various elasmobranchs. Further offshore and on firmer muddy deposits down to at least 60 m, sciaenids (*Micropogon, Macropogon, Cynoscion*) dominate a demersal fish fauna that also includes many coastal genera characteristic of the Gulf of Guinea: *Polynemus, Larimus, Vomer*, and so on. Beyond 60 m, although the sandy deposits resemble those of the outer shelf of the Gulf of

Guinea, a "sparid fauna" is absent, and *Caranx*, *Sphyraena*, and some Lutjanids occupy the grounds. It has been suggested that the western Atlantic sparid community, typical of the eastern coast of the United States and the South American coast south of Cape Frio, lacks sufficiently illuminated, sufficiently cool subsurface water for these subtropical species to penetrate through the Caribbean and the Guianian region by tropical submergence—as they do in the eastern Atlantic.

## Synopsis

*Case 4—Small-amplitude response to trade wind seasonality, strongly biased by variability of the Amazon/Orinoco discharge rates.* Satellite data indicate a seasonal cycle with lower productivity in boreal winter; the accumulation of chlorophyll suggests close coupling between production and consumption (Fig. 9.28). The effect of freshwater discharges is seen in the permanently shallow pycnocline, whereas the deeper thermocline lies close to the photic depth, at between 35 and 45 m in all months, $Z_{eu}$ being deeper than thermocline from March to September. Chlorophyll biomass is unreliable because of CDOM effects of river effluents; consequently, the continental shelf north of the equator has apparently high pigment values, peaking in late boreal summer with maximal effects of Orinoco-Amazon effluents and retroflection eddies.

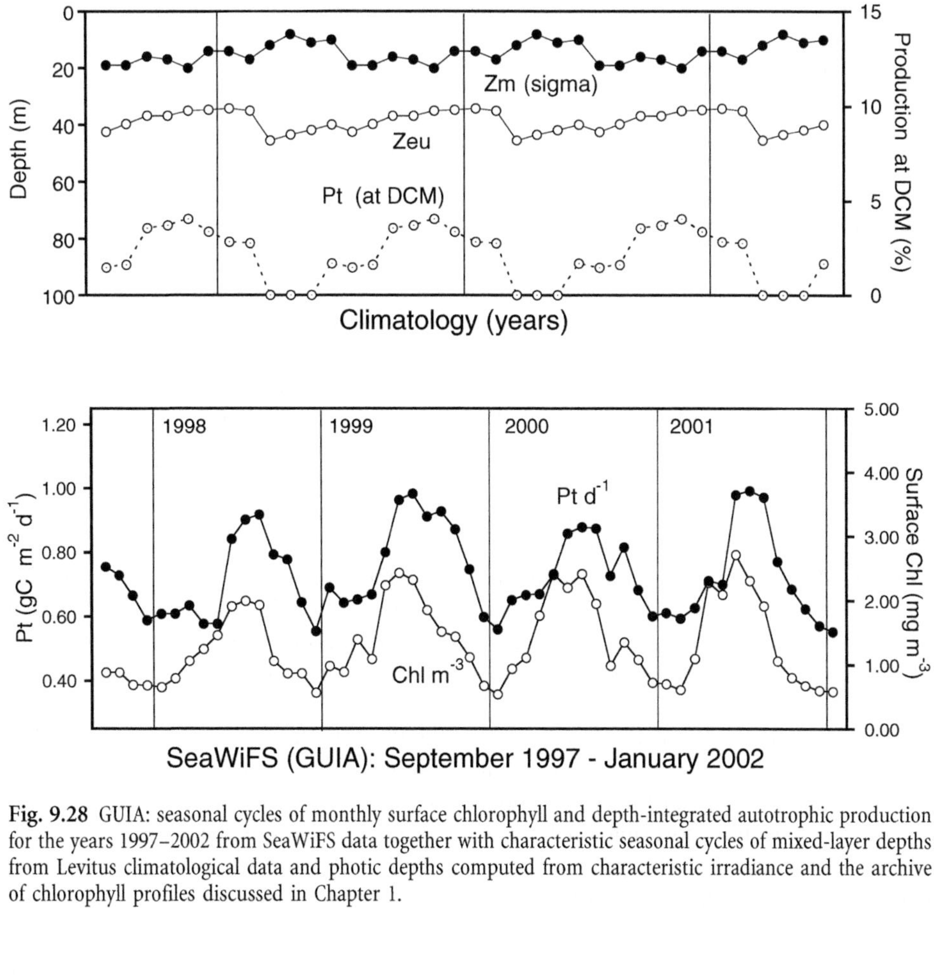


**Fig. 9.28** GUIA: seasonal cycles of monthly surface chlorophyll and depth-integrated autotrophic production for the years 1997–2002 from SeaWiFS data together with characteristic seasonal cycles of mixed-layer depths from Levitus climatological data and photic depths computed from characteristic irradiance and the archive of chlorophyll profiles discussed in Chapter 1.

# Brazil Current Coastal Province (BRAZ)

## Extent of the Province

The northern boundary of the BRAZ province is placed near Recife (8°S), where the coast of South America changes orientation, and close to the partition of the flow of the transoceanic SEC between this province and Amazon Shelf of the GUIA province. The BRAZ province extends down the east coast of South America to about 35°S, off Argentina, where the northward flow of the Falkland Current is encountered. This important confluence region is discussed in the next section, devoted to the FKLD province. The outer boundary of the rather narrow BRAZ province seldom occurs beyond the 2000-m contour.

## Continental Shelf Topography and Tidal and Shelf-Edge Fronts

This coast is remarkably straight and the shelf usually rather narrow: in the north it reduces to a width of only 10 km. The only topographic features of importance are near 25°S, where there are two prominent capes: Cabo Sao Thomé and Cabo Frio (both this and the similarly named cape in the Benguela Province owe their name to the same phenomenon—upwelling). At 16–17°S, the shelf has a complex topography of valleys and submarine canyons, and the wide Abrolhos and Royal Charlotte Banks emerge from great depths and bear hermatypic coral formations, almost the most southerly reefs in the South Atlantic. In the bight to the south of Rio de Janeiro, the shelf widens to 200 km, and the sediments lack the calcareous components characteristic of lower latitudes. There are no canyons here, but numerous topographic depressions, holding soft sediments, characterize this section of shelf. Along the Argentinian coast, deep water lies unusually close inshore.

## Defining Characteristics of Regional Oceanography

The Brazil Current (BC) is the weakest of the western boundary currents because only about one quarter of the westward flow across the equatorial southern Atlantic of the SEC enters the BC. As noted previously, the remainder flows northward into the incipient NBC, rather than remaining in the southern hemisphere subtropical gyre (Peterson and Stramma, 1991; Gordon and Greengrove, 1986). South of 30°S, the BC strengthens (at about 5% per 100 km from 20°S to 35°S) by entrainment from a recirculation cell lying between the continent and the Rio Grande Rise. This gyre originates at the confluence of the BC and the northward flow of the Falkland Current (Castro and de Miranda, 1998).

Wind stress over the BC responds to the seasonal migration of the ITCZ that lies between trade winds and westerlies; during austral summer prevailing winds are from the E or NE over the entire region, whereas during austral winter, to south of Cabo Frio, winds are coastwise and dominantly from the SW while, to the north of this cape, the dominant direction is increasingly from the SE, or directly onshore. Repeated westerly waves and cold fronts propagate equatorward along the coast at 5- to 10-day intervals, reach Recife in austral summer, and dominate the synoptic weather pattern at all seasons.

As the oligotrophic, nutrient-poor water of the Brazil Current proceeds southward, its maximum flow lies over the shelf edge but transport extends across the shelf, especially in austral summer. In austral winter, flow retreats to beyond the shelf break. Seasonal variation in mass transport (maximal in austral summer and minimal in winter) is directly related to the seasonal change in wind stress curl. The strongest (weakest) transport occurs when the confluence region is furthest south (north), according to Matano *et al.*

(1993). The latitude of the confluence, which determines where separation of the Brazil and Falklands Currents from the continent occurs, is farther to the north during austral winter and spring; this may be related to the general seasonal shift of wind systems and to a seasonal meridional shift of the subtropical gyre (Peterson and Stramma, 1991). Imposed on this seasonality, there is also some short-term variability in the southward extent of the Brazil Current. When a loop that has pushed unusually far south retreats to the north, it may be shed as a series of warm-core eddies that pass into the Antarctic Circumpolar Current (Partos and Piccolo, 1988).

Although intrusions of SACW, and also shelf-edge upwelling, occur off southeastern Brazil and on the shelf in the vicinity of the Abrolhos and Royal Charlotte Banks, these processes have been most intensively studied at Cabo Frio (23°S), in relation to the offshore meander that occurs there and the upwelling that regularly occurs to the northeast of the cape (Peterson and Stramma, 1991). This is a complex process, combining wind stress and the topography of a major canyon that cuts across the shelf (Mascarenas *et al.*, 1971; Moreira da Silva, 1971). Northeast winds force water to upwell across the plateau that lies east of the cape and to surface either at Cabo Frio itself, or somewhat to the north. To the south of Cape Frio, a coastal cyclonic eddy forms an eastward compensation current. Upwelling may be inhibited during periods when the Brazil Current remains for a period topographically locked to the coastline, so that the Cabo Frio meander does not form.

Upwelling-downwelling episodes occur very quickly in response to changes in wind direction, which themselves are very variable and episodic, responding to the cyclical development and collapse of the atmospheric anticyclone east of southern Brazil. A pulse of northeast winds will very rapidly draw SACW, of 18°C and 10 μM $NO_3$, to the surface from about 300 m. Such episodes occur more frequently in austral summer than in winter, during which the frequent passage of cold fronts and southwest winds maintains an almost continuous downwelling, oligotrophic situation along the outer shelf.

In the extreme south of this province, the single major source of freshwater is the Parana-Plate system at 35°S that discharges sediment at a rate of $\sim 90 \times 10^6$ t $y^{-1}$; this is a significant flux of solids and has major consequences for the ecology of the inner shelf of the southern part of the BRAZ province, because it transported equatorward and alongshore. The seasonal variability of the equatorward extension of the low-salinity plume is forced by changes in wind stress rather than by changes in the discharge rate of the River Plate and from the Patos Lagoon.

## Response of the Pelagic Ecosystems

Although the upwelling at Cape Frio has occupied much of the attention of regional studies here, the satellite chlorophyll data for this coast reveal that it is not a major feature and is only with some difficulty identified in the routine 8-day images; high chlorophyll values appears to occur only very close to the coast and offshore filaments are relatively very weak compared with those from many other upwelling sites. The austral spring-summer bloom that occurs over the shelf dominates seasonal changes observed in chlorophyll images that represent the entire BRAZ province; this bloom involves the entire region that lies poleward of the ITCZ, and takes higher values of chlorophyll biomass over the shelf than offshore. Consequently, during austral spring and summer, the entire shelf area is progressively occupied by water of relatively high chlorophyll biomass that is separable from offshore water that remains relatively oligotrophic.

An unusually prominent high-chlorophyll feature occurs on the inshore zone of the area of wide shelf south of 28°S, and this probably represents both production on the shelf and also the discharge from the River Plate; the plume appears to be constrained by a sharp turbidity front, visible in NOAA-AVHRR and in chlorophyll imagery. This migrates seasonally: upstream and westward in summer (minimum discharge) and downstream and eastward in winter (maximum discharge).

Perhaps because it is not one of the world's major upwelling sites, the physical mechanism that forces upwelling episodes at Cape Frio is better known than the ecological consequences, although the physiological parameters of phytoplankton involved in upwelling blooms here have been observed to respond appropriately to light, nutrients, and temperature (Gonzalez-Rodriguez *et al.*, 1992; Gonzalez-Rodriguez, 1994). The upwelling of SACW is episodic, and the biological response is rapid, though the upwelled water does require conditioning by superficial heating and stratification before primary production rate increases significantly.

The anticipated physiological conditioning of algal cells has been followed in this upwelling, showing that the light-saturation curve responds to conditions of upwelling and downwelling with changing values. A typical upwelling event lasts from 5 to 10 days, during which successive reductions in wind speed induce surface heating so that the rate of primary production starts to increase by about day 5 to introduce the "production phase" of the event. During this period, the mixed layer (typically 10 m deep, 15–20°C at the surface, and 15–20 μmol $NO_3$) accumulates only about 5–6 mg chl $m^{-3}$ from a production rate of about 10 mg C $m^{-3}$ $hr^{-1}$. Subsequently, as the system returns to a downwelling state, chlorophyll biomass rapidly falls by an order of magnitude, though productivity remains at about 2 or 3 mg C $m^{-3}$ $hr^{-1}$ for some time. This suggests that the upwelling bloom is rapidly and heavily grazed, a suggestion for which there is some experimental support. There is also some indication that the magnitude of the production cycle may respond to the initial inoculum of cells in the upwelling parcel of water rather than to its nutrient load. The newly upwelled water contains a sparse population of benthic diatoms that are rapidly replaced by an intense bloom of neritic and opportunistic diatoms.

Farther poleward, at 25–30°S (or from Santos to Cape St. Marta), recent studies during austral summer over and beyond the shelf have confirmed the oligotrophic nature of the tropical water (0–200 m) carried by the Brazil Current, and the ecological effects of intrusions of SACW across the shelf (Metzler *et al.*, 1997). These surveys of the Brazilian COROAS expeditions found highly variable rates of nitrate uptake by phytoplankton in this hydrographically dynamic region. Differences between inshore and oceanic profiles of production rate and chlorophyll biomass, and the concomitant physiological constants, were predictable; over the shelf, <32% of the nitrogen utilized by autotrophic cells was obtained from $NO_3$. At coastal stations and midshelf at 30°S, the DCM was found by Odebrecht and Djurfeldt (1989) to be dominated by larger autotrophic cells (>20 μm) while the surface layer was populated by the smaller pico and nano fractions; chlorophyll maxima in the DCM were due to the presence of large centric diatoms (*Coscinodiscus, Thalassiosira*) having high cellular chlorophyll content. Here, the DCM flora is sustained by nutrients supplied by bottom-driven turbulence during both upwelling and downwelling conditions.

There appears to be a clear distinction between zooplankton populations that are characteristic of SACW irruptions onto and across the shelf and the shelf plankton proper, which is dominated by neritic species; several accounts speak of *Calanoides carinatus* and *Ctenocalanus vanus* as indicators of SACW water as far south as Rio de Janeiro. At Cabo Frio, Valentin (1984) describes two characteristics species groups associated with various nearshore neritic conditions: (i) herbivores—*Ctenocalanus vanus, Calanoides carinatus, Penilia avirostris*, and (ii) omnivores, predators—Siphonophora, Chaetognatha, *Corycaeus amazonicus, Eucalanus pileatus, Clausocalanus acuicornis*. An intermediate offshore neritic zone is characterized by a great abundance of appendicularians, whereas the oceanic zone of the Brazil Current has a sparse but very diverse mesozooplankton of typical warm-water species—*Clausocalanus furcatus, C. pavo, Mecynocera clausii*, and the rest.

Over the Abrolhos Bank and the other wide shelf regions to the north of Cape Frio, a much more oligotrophic regime is represented by mesozooplankton of low biomass and high diversity with very minor intrusions of the upwelling fauna from the southwest.

The low-trophic level pelagic fish of the region are dominated by the presence of *Sardinella brasiliensis* and *Engraulis anchoa* that are both especially associated with the Cape Frio region; the sardine population off Brazil comprises a larger number of year-classes (six to seven) under normal circumstances than does *S. anchovia* that occurs farther north off the Venezuelan coast and comprises only two to three year-classes. In both species, spawning peaks correspond with upwelling seasons off Cape Frio and in the Carioco Basin, respectively. Anchovy schools appear to occur preferentially over the slope rather than the shelf.

## Regional Benthic and Demersal Ecology

Benthic studies in this region rather commonly emphasize the process of carbonate sedimentation on the substrates, principally due to the presence of foraminiferans; north of Cape Frio, the presence of *Halimeda*, of bryozoans, and of coralline algae is emphasized. South of Cape Frio, the carbonate-rich sediments are restricted to the outer shelf where coralline algae and foraminiferan masses contribute largely to carbonate formation. Inshore, a "mud line" is described, within which foraminifera are not abundant.

The very unusual abundance at such latitudes of rhynchnonelliform brachiopods (two groups: *Terbratulina* + *Argyrotheca* and *Bouchardia* + *Platidia*) occurs on the outer shelf of the South Brazilian Bight, always on carbonate sediments to which they themselves contribute. The former group especially favors sediments with high (70–95%) $CaCO_3$ content, and both groups occur over depth ranges of 100–200 m, at very high abundance (terebratulids alone may occur at densities of >100 ind $m^{-2}$), preferentially in areas that lie under the influence of shelf-edge upwelling. These assemblages appear to be unusual relics of ecological conditions that were dominant in Paleozoic seas.

The ecology of the demersal fish here, as elsewhere, responds directly to water mass and deposit characteristics; as in the eastern Atlantic, the superficial subtropical water mass is occupied by species associations in which Sciaenidae dominate: a tropical sciaenid community penetrates southward inshore of a subtropical sciaenid community and is characterized by species of *Ctenosciaena, Cynoscion*, and *Paralonchurus*. Offshore, on sandy and shelly deposits, and in cooler water, yet another version of the subtropical sparid fauna occurs, including *Pagrus pagrus* and *Cheilodactylus bergi*. There is, of course, a high diversity of demersal fish on the Abrolhos and Royal Charlotte Banks associated with the hermatypic coral formations and other rocky grounds; these will include a very much greater range of genera and a higher diversity than elsewhere in the province.

## Synopsis

*Case 2—Nutrient-limited spring production but modified by shallow-water processes.* This province has an overall production cycle closely resembling that of NECS, but shifted by 6 months with maximum accumulation of chlorophyll biomass in early austral summer and a shoulder on the descending limb of the biomass cycle in fall. The mixed layer defined by the pycnocline $Z_m$ has moderate seasonality, deepening to 40 m in austral winter from summer depths of about 10 m. $Z_{eu}$ lies at about 30–50 m so that the thermocline is illuminated in all months (Fig. 9.29).

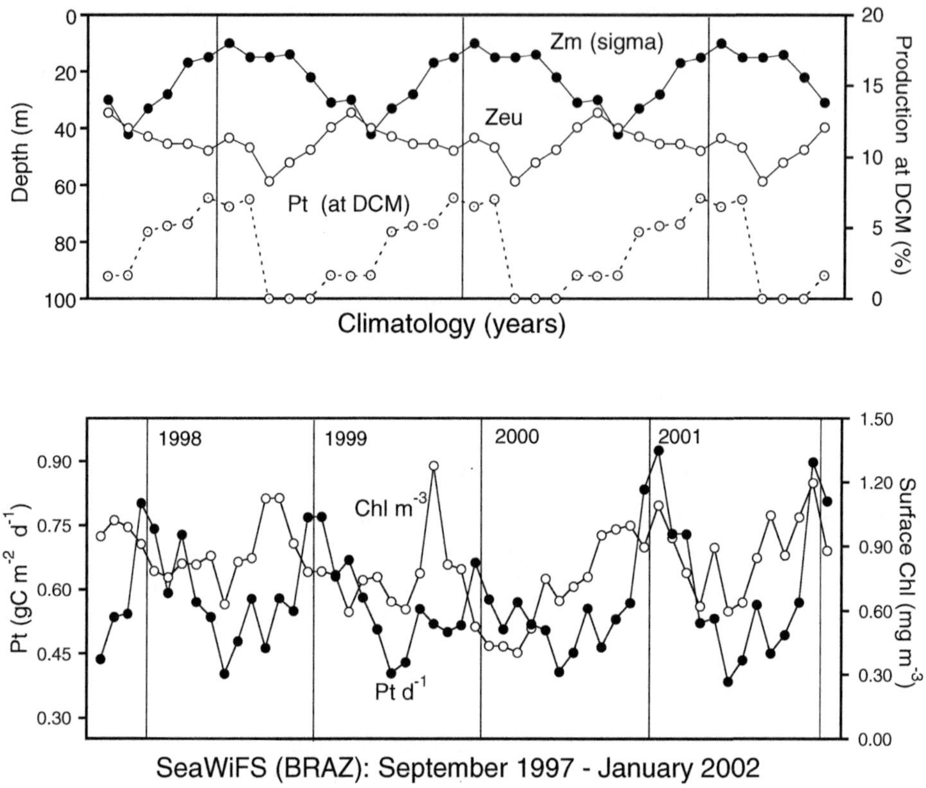


**Fig. 9.29** BRAZ: seasonal cycles of monthly surface chlorophyll and depth-integrated autotrophic production for the years 1997–2002 from SeaWiFS data together with characteristic seasonal cycles of mixed-layer depths from Levitus climatological data and photic depths computed from characteristic irradiance and the archive of chlorophyll profiles discussed in Chapter 1.

## Southwest Atlantic Shelves Province (FKLD)

### Extent of the Province

The Southwest Atlantic Shelves Province (FKLD) comprises the Argentine shelf and Falklands plateau from Mar del Plata (38°S) to Tierra del Fuego (55°S), including the oceanographic confluence region between the Falklands and Brazil currents. The seaward boundary is perhaps best approximated to the deep-water contours around the Falklands plateau and northward along the Argentinean continental shelf to about 35°S, or wherever in real time the southerly flow meets equatorward flow at the shelf edge.

### Continental Shelf Topography and Tidal and Shelf-Edge Fronts

From the River Plate south to Tierra del Fuego, this is one of the widest and flattest continental shelves anywhere in the oceans, and for this fact alone it is useful to separate it from the rest of the South Atlantic coastal zone. The Falkland Islands lie on the eastern edge of the shelf: they are rocky, glaciated, and contested.

The FKLD province comprises both the Argentine-Patagonia shelf and the adjoining Falklands plateau, where the continental shelf widens to about 800 km at 50°S; there are several major embayments on the coast (Bahia Blanca and the Gulfs of San Matias and San Jorge). South of the Falklands, a trough having depths of <700 m extends along the southern and western sides of the Falkland Plateau. The Straits of Magellan near the southern end of the province have a glaciated, fjordlike topography unlike the eastern coast of Patagonia and Tierra del Fuego that lack glaciated terrain: here, the Pleistocene ice flowed largely down the Pacific slope.

A shelf-break front occurs at least along the whole north-south section of the Patagonian shelf edge from 37°S to 47°S (Glorioso, 1987), separating the cool subantarctic water offshore from the warmer shelf water. Tidal dissipation rates are high and an inshore tidally mixed zone lies shoreward of the 40-m isobath, where a slight escarpment on the shelf topography may occur, and is especially well developed off headlands, as at the Valdes peninsula at 42°S. Seaward of the mixed area, and separated from it across typical tidal fronts, summer stratification occurs. Macrotidal conditions occur south of the Gulf of San Jorge (50°S) down to the southern end of the province; here, observations of tidal fronts have been related successfully to the stratification parameter (Glorioso and Simpson, 1994). Elsewhere, except in some embayments, mesotidal conditions prevail.

## Defining Characteristics of Regional Oceanography

This province has a very complex circulation that accommodates two singularities (Fig. 9.30): (i) the Subantarctic Front, associated with the Antarctic Circumpolar Current, is entrained into lower latitudes on rounding Cape Horn from the southeastern Pacific and flows equatorward as the Falkland Current; (ii) the confluence of the poleward Brazil Current and the equatorward Falkland Current at about the latitude of Bahia Blanca requires a combined offshore flux whose location is seasonally variable between 38.5°S (July–December) and 36.5°S (January–March). The instantaneous location of the confluence front is determined principally by local wind forcing. Unfortunately for present purposes, the most intensively studied part of this shelf is the extreme northern parts, on the borders of the BRAZ province, and it is these studies that dominate modern reviews of regional circulation, such as that of Piccolo (1998).

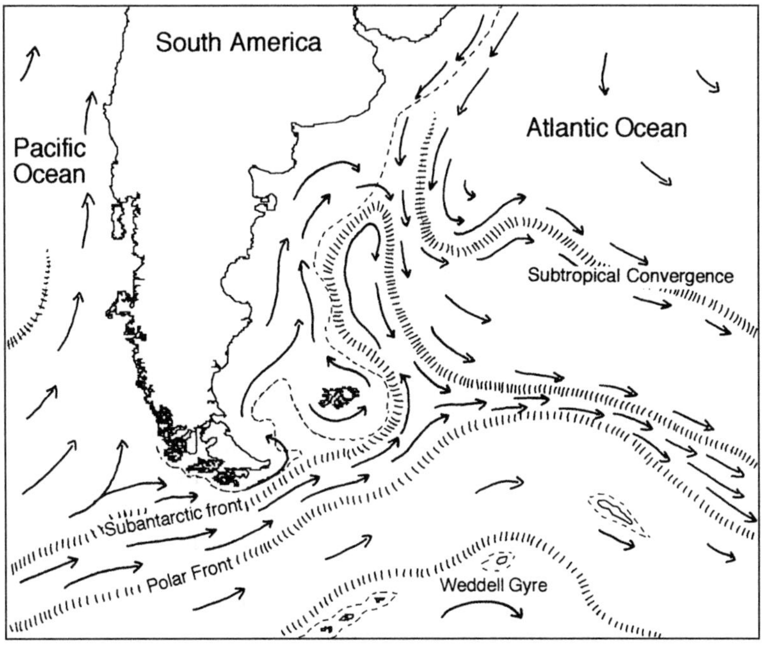


**Fig. 9.30** This cartoon of circulation around the Falkland Plateau and through the Drake Passage emphasizes the retroflection of the Falkland Current and of the Subantarctic Front. The 200-m isobath is represented by a dashed line, and fronts are represented as hatched lines.

The Falkland Current carries subantarctic water northward along the continental slope from the latitude of Cape Horn, first around Birdwood Bank and then the Falkland Plateau along the 1000-m isobath. Flow continues north along the continental slope to the confluence with the Brazil Current at about 40°S (Peterson and Whitworth, 1989). Cold-core eddies are shed, particularly across the Falkland plateau, and the more saline shelf water originating in the Brazil Current is thereby modified. Eddy generation is also especially active at the confluence zone, in the general latitude of Bahia Blanca, because here the Falkland Current makes an abrupt cyclonic loop and returns toward the southeast, alongside the seaward flow from the Brazil Current. This confluence then continues eastward across the ocean as the Subtropical Convergence and stands out in satellite imagery as a region rich in eddies.

The same interaction occurs here between tidal mixing and the seasonal cycle of mixing and stratification that has been investigated principally in the NECS province. The neritic area is mixed year-round by tidal stress, whereas deeper water over the shelf is mixed in winter and stratified rapidly in spring as buoyancy is accumulated within the upper 30 m of the water column. Because the shelf has so little relief, the tidal front separating the mixed from stratified water in summer runs parallel to the coast along the 40-m isobath and clearly separates coastal from outer shelf water. The water column is almost everywhere mixed down to the sediments by seasonal wind stress during winter when the warmest and least dense water occurs near the coast, and a moderately steep shelf-break front forms between the 80-m and 100-m isobaths, separating shelf from Falkland Current water. Stratification is induced in spring and summer in the classical manner, by increased insolation and decreased wind stress together with some advective transport into the region (Piccolo, 1998); only inshore, in the macrotidal regions, does stratification not occur. On the southern part of the shelf, just north of the Falklands, a summer thermocline at 50 m induces density stratification in a water column that is essentially isohaline: salinity ranges across the shelf from 32‰ near the Tierra del Fuego coast to about 34‰ at the shelf edge north of the Falklands. In the extreme south, beyond 52°S, summer stratification is rarely observed.

The presence of the subantarctic water of the Falkland Current along the shelf edge is very significant for the distribution of productivity in this province. Shelf-edge upwelling brings nutrient-rich water either to the surface or, alternatively, causes it to pass westward below the thermocline so that, with the breakdown of stratification in the autumn, nutrients the newly mixed shelf water are replenished with nutrients at deep ocean concentration.

## Response of the Pelagic Ecosystems

Though we have little direct information from regional studies, there is no reason to believe that the description of processes associated with midlatitude continental shelves having active tidal frontal regions (e.g., NECS and NWCS provinces) is not also be generally relevant to here.

It is to the serial chlorophyll images that we must turn to deduce a characteristic climatology of productivity in this province and, once again, they do not fail us. Clearly revealed are the effects of the mixing and stratification regime described earlier and, just as clearly, the consequences of the shelf-edge flow of subantarctic water in the Falkland Current. In midwinter, the entire shelf region has <1 mg chl $m^{-3}$, weakest in the south off Tierra del Fuego, but this is, nevertheless, higher than the offshore oceanic regions; this value begins to increase generally, and overall, as early in the sun-year as the middle of July. However, the first signs of major blooms occur in the north, off Bahia Blanca (38–40°S) along the shelf break in mid-September. This linear bloom rapidly extends southward until by early October it has reached the edge of the Falkland Plateau and has extended to the shelf break at the head of the deep water SW of the Falklands (see Color plate 20). The development of major blooms is subsequently induced in early November across the width of the southern shelf regions down to at least 50°S. The final progression in midsummer, in January, takes the strong midshelf blooms down to the shelf east of Tierra del Fuego.

Already by the end of February, regression toward the north is well underway, and by early April no strong blooms exceeding 2–3 mg chl $m^{-3}$ remain and the effects of the shelf-break frontal upwelling are minimal. The winter period of low chlorophyll overall is thus relatively brief and, in any case, difficult to observe because of heavy seasonal cloud cover.

The tidally mixed area along the inner half of the shelf also appears as a consistent high-chlorophyll feature (sometimes, off the Gulf of San Jorge, with clearer water inshore), as does the shoal water around the Falkland Islands, where tidal mixing presumably also occurs. It has been demonstrated that SeaWiFS-derived surface chlorophyll concentrations are too high relative to *in situ* observations made at sea, and once again, the CDOM problem is invoked. I have been unable to locate any detailed studies of phytoplankton processes on the shelf to the south of the transition zone.

Mesozooplankton variability was studied in a four-season, multistation survey of the entire shelf south of 44°S and was found to respond closely to the seasonal and spatial pattern of productivity discussed earlier (Sabatini and Colombo, 2001). The changes in numerical abundance of copepods followed a very significant pattern: in the winter months, biomass at all stations was <100 mg $m^{-3}$; in the spring trimester, biomass exceeded 1000 mg $m^{-3}$ in the north-central shelf regions and in coastal embayments; in the summer trimester, similarly high biomass was concentrated in a linear zone above the shelf break from 45°S to 50°S, with remnant high biomass in southern neritic regions. Finally, in autumn, very high biomass occurred in neritic regions, but unfortunately the middle and outer shelf regions were not surveyed. For neither euphausiid nor amphipod biomass were high concentrations observed over the shelf break.

The dominant large copepod in the shelf regions that are most liable to strong summer stratification is *Calanus australis* (Sabatini *et al.*, 2000). This is a circumpolar austral species that is often associated with upwelling fronts that here appears to produce a single generation annually, so that by March—toward the end of austral summer—the population is heavily biased to C5s that are then entering diapause. Off Patagonia, it occurs neither in coastal, tidally mixed water, nor offshore beyond midshelf where it is replaced by *C. simillimus* and *Neocalanus tonsus.* The former contributes very heavily to a linear zone of high copepod biomass along the shelf break in spring and summer months.

## Regional Benthic and Demersal Ecology

This is an anomalous shelf region because, although it lies between latitudes that are equivalent to those of southern Labrador, it is nevertheless equatorward of the oceanic Subtropical Front that passes around almost the entire margin of the shelf topography. It is also by far the largest high-latitude shelf region in the austral oceans.

The studies of the benthic ecology of this shelf that I have been able to locate are not very coherent, so that it is difficult to present a unified account of the benthic and demersal ecosystem. What must be said at once is that its isolation, far from global centers of human population, has not protected it from the pressure of industrial fishing, so that what is observed today must differ very significantly from the pristine state. Catches increased to a maximum removal rate of about $2.0 \times 10^6$ t $y^{-1}$ during the latter half of the 20th century, after which a progressive decline set in; stratified trawl surveys suggested that longtail hake (*Macronurus magellanicus*) at one time represented ~60%

of total fish biomass, while common hake (*Merluccius hubbsi*) contributed the second largest biomass, although today it dominates the fish landings. These two are followed by southern blue whiting (*Micromesistius australis*), by pink cuskeel (*Genypterus blacodes*), and by various notothenids, such as the semipelagic toothfish (*Dissostichus eleginoides*). What seems unusual here is the great abundance of several species of squid; approximately 25% of the entire world catch of around $3 \times 10^6$ t $y^{-1}$ was taken from the shelf of this province in the late 1990s. One may infer either that the large-scale removal of demersal fish has created a regime shift, or that the pristine ecology of this shelf was rather unusual.

Appropriately for such a high-latitude shelf region, two large teleosts, each about 1 m in maximum length and each with a potential longevity of about 15 years, are dominant in the demersal biomass. As elsewhere, these fish (*Macronurus, Merluccius*) are high-trophic-level species, depending largely on amphipods, euphausiids, small fish, and cannibalism. Hake are critically dependent on the shelf-break, thermohaline, and tidal fronts discussed earlier. These prevent the larvae from winter spawning in the northern part of the species range from drifting into subtropical waters; summer spawning sites in more coastal waters to the south are associated with the thermohaline and tidal fronts off the Valdes peninsula and in the Gulfs of San Matias and San Jorge. Eggs and larvae are preferentially distributed in frontal and stratified areas where surface and bottom temperatures differ by about 4°C. Further, the seasonal migrations of the stock carries it between the northern and southern frontal regions where nektonic biomass is high relative to nonfrontal regions.

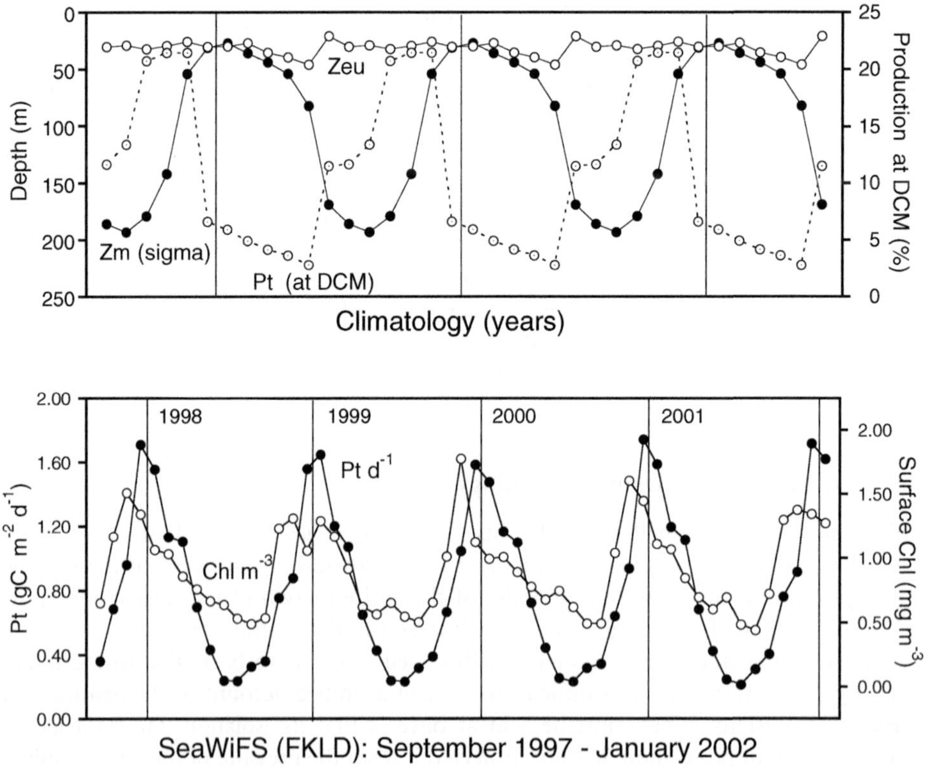


**Fig. 9.31** FKLD: seasonal cycles of monthly surface chlorophyll and depth-integrated autotrophic production for the years 1997–2002 from SeaWiFS data together with characteristic seasonal cycles of mixed-layer depths from Levitus climatological data and photic depths computed from characteristic irradiance and the archive of chlorophyll profiles discussed in Chapter 1.

## Synopsis

*Case 2—Nutrient-limited spring production peak, modified by shallow-water processes.* The seasonal production cycle is very well captured by the data used to represent the entire province (Fig. 9.31). The mixed layer as defined by the pycnocline $Z_m$ undergoes strong austral winter excursion (25 m January, 190 m August) from which it shoals very strongly in March. Since $Z_{eu}$ lies consistently at 20–40 m, the thermocline is illuminated only very briefly in austral summer (December–January). The P rate increases linearly from austral winter minimum to sustained summer maximum, after which the rate declines rapidly to winter values, this change matching the long winter deepening and brief summer shoaling (October–March) of $Z_m$ over a large depth range (30–225 m). Significant near-surface chlorophyll occurs at all seasons, and whereas the period of significant accumulation in austral winter (June–July) is largely coastal, that of summer (November–February) is also offshore, especially at the shelf break. These observations are consistent with only an approximate production-consumption balance year-round.

Chapter 10

# The Indian Ocean

The Indian Ocean, north of the Subtropical Convergence zone, is the smallest ocean basin (about $50 \times 10^6$ km$^2$) and has some special characteristics that must be accommodated when it is partitioned into biogeochemical provinces. It has become much better known in recent years, although the new research has been concentrated mostly in the Arabian Sea, adjacent to Madagascar and in the Agulhas Current. The subtropical gyre of the southern hemisphere—like that of the other oceans—has been relatively neglected. This uneven coverage is especially unfortunate because of the unusual complexity of the circulation of this ocean that is forced by the characteristic seasonal reversal of the monsoon wind systems.

Our organized knowledge of the Indian Ocean begins with results of the classical John Murray Expedition (1933–34) and of the International Indian Ocean Expedition (IIOE) of 1959–1965 that was an early version of the international multiship investigations that have since come to dominate the progress of oceanography. Unfortunately, though much biology was done during the IIOE, very little ecology emerged and, because the wide grid of stations was sampled quite irregularly, no uniform field of values for any property was obtained. In fact, this was expressly not the intention of the organizers; rather, their intention was to permit complete freedom of action to the scientists aboard the 11 participating ships. A centrally planned grid of stations and agreed procedures to wring answers from critical problems would have been "ideal but unrealistic," according to the principal coordinator. Nevertheless, the Indian Ocean Oceanographic Atlas of Wyrtki (1971) remains the best source of maps of the physical oceanography of this ocean. The results of the IIOE were discussed at a SCOR-IBP Symposium in 1971; the resultant volume (Zeitzschel, 1973) is an important document, both historically and as a source of information on subjects that have not been studied subsequently over the entire basin of this ocean. A glance will demonstrate the extent to which the southern subtropical gyre was relatively neglected, and the great difficulty of deriving any comprehensive information for that region.

Fortunately, we have come a long way since the days of the IIOE, because of the intense and disciplined exploration of the ecological consequences of monsoon reversal during the Arabian Sea component of Joint Global Ocean Flux Study (JGOFS, 1994–96), and the Netherlands Indian Ocean program (1992–93). The Arabian Sea JGOFS was one of the most successful investigations in recent decades (see Smith *et al.*, 1998a); five ships from Europe and the United States, one from India, and several from Pakistan undertook about 20 voyages along standard meridional transects across the northern Arabian Sea that were arranged to cover the critical periods of the monsoon cycle. Stations were specially chosen to investigate the seasonal upwellings that are a consequence of the Southwest Monsoon; of particular interest was the spatial extent of the biological consequences of this process and the distribution at the surface of oxygen-depleted, upwelled water. Special emphasis was given to the response of biological carbon flux to the reversing monsoon winds, and a highly instrumented mooring at 15°S was maintained to monitor this process. Shipborne studies of all components of the pelagic ecosystem—including nutrient dynamics, productivity, the organization of the microbial food web, and the dynamics of herbivore consumption—were given prominence in the scientific program aboard the participating ships.

The results of many of these investigations were published during the period 1998–2002 in a series of issues of *Deep-Sea Research Part II*, and in 2005 in a special edition of *Progress in Oceanography*; these present an excellent description of the dynamic response of the pelagic ecosystem of this region to the seasonal uplifting and advection of nutrient-rich water. The changing community composition of producers and consumers is now as well understood here, far from the great oceanographic institutions, as anywhere else in the oceans. Obviously, the organization and execution of the JGOFS Arabian Sea Expedition was a far cry from that of the IIOE, and this shows in the results obtained. However, nothing is perfect and an ideal synthesis remains difficult to achieve, principally because so many key studies and observations were not performed in each critical environment, in each season. Unfortunately for the possibility of taking another look at these problems, the world has changed since 1994–1996: after the U.S.S. *Cole* incident, the Mogadishu affair, and the invasion of Iraq, it is unlikely that further research can be undertaken in these waters aboard Western ships—or, at least, not for a very long time.

The anomalous circulation and ecology of the Indian Ocean are principally forced by two factors: its basin form, and the consequent wind regime. The Indian Ocean extends only marginally into the northern hemisphere, adjacent to the high terrain of the Asiatic landmass, and this is the key to the singularity of this ocean because its northern region is dominated by the most striking example of a seasonally reversing wind regime that occurs anywhere. The reversing wind regime here has given the Arabic name of *monsoon* to such wind systems elsewhere—as in western Africa and southern Argentina. The monsoons carried the dhows from Arabia to Zanzibar and then home again "laden with sandalwood, apes and ivory" and human merchandise that we would rather forget. Monsoon winds peak in February and July; the winter high-pressure belt over central Asia feeds the cool, dry Northeast Monsoon that sweeps down across the Arabian Sea and forces seasonal circulation patterns across the equator and into the northern part of the southern subtropical gyre. At this season, the atmospheric intertropical convergence zone (ITCZ) is located well to the south of the equator, a fact having important consequences for the development of the equatorial current system. The Southwest Monsoon is induced by the low atmospheric pressure that forms over Arabia and Pakistan during the heat of summer and forms a strong jetlike stream that is constrained to the west by the highlands of eastern Africa. Although the Southwest Monsoon has important effects along the coastline, wind speed is maximal in the Findlater Jet that passes SW-NE across the northern part of the Arabian Sea, generally aligned from Somalia to Bombay. Because the meridional pressure gradient is significantly greater across the relevant latitudes in July than it is in February, the Southwest Monsoon winds are stronger and steadier than those of the Northeast Monsoon. The effects on the pelagic ecosystem induced by the Southwest Monsoon were a principal interest of the 1994–96 JGOFS ecologists.

Any partition of the Indian Ocean must accommodate two seasonal circulation states, because the seasonal monsoon wind reversal rapidly induces reversal in both shallow and deep currents because (as noted in Chapter 3) wind stress in low latitudes induces momentum rather than mixing. Indeed, the reversal of the Somali Current and the rapid spin-up of a deep boundary jet in response to the onset of the Southwest Monsoon is the classical model for this phenomenon in low-latitude oceans.

The Northeast Monsoon forces weak westward flow across the whole ocean between India and the equator so that some southerly flow must occur at the coast of Africa, while the Arabian Sea and Bay of Bengal are occupied by cyclonic flow. The strong, equator-crossing Southwest Monsoon of boreal summer reverses this situation, spinning up anticyclonic gyres in the Arabian Sea and Bay of Bengal, and establishing eastward flow across the ocean north of the equator. Most importantly, it forces a jet current, deep and fast, northwest along the Somali and Arabian coasts. Though the response of the ocean is swift, often occurring within 1 month, there is some lag so that the extremes of current speed occur in February and August.

Because the equatorial near-zonal winds are relatively weak, compared with the Atlantic and Pacific Oceans, and are not symmetric about the equator, there is no significant upwelling at the equator in the Indian Ocean. Wind stress along the equator is greatest during the intermonsoon periods when westerly winds prevail and force a convergent eastward flow in the ocean—the fast Indian Equatorial Jet (600 km wide). Shallow overturning cells in the equatorial circulation from 5°S to 10°S are driven by wind curl in the zonal wind field and, although isopycnals rise toward the surface here, no actual upwelling appears to occur at the surface; however, some chlorophyll enhancement is observed and will be discussed later (Miyama *et al.*, 2003). Note in satellite images how the Maldive Island chain, which lies at right angles across the equator, perturbs the chlorophyll pattern there.

The present scheme of biogeochemical provinces of the Indian Ocean owes much to the analysis of Colborn (1975), who partitioned the Indian Ocean by reference to the monthly evolution of the mixed layer, and the gradient in the thermocline. He thus obtained 274 "sub areas" which were aggregated into 40 "primary areas having distinct thermal characteristics" and these again were gathered into a smaller number of "major geographic provinces." Banse (1987) and Brock *et al.* (1991) both interpreted their studies of thermal structure and algal blooms in terms of Colborn's classification of the northwest Arabian Sea. For convenience of comparison, the equivalence of the biogeochemical provinces proposed here with Colborn's areas is noted in each section.

A monthly thermocline topography for the Indian Ocean that was compiled by Rao *et al.* (1989) followed Colborn in defining mixed-layer depth solely by thermal criteria. Where heavy rainfall occurs at sea (as in some parts of the monsoon provinces) or where river runoff lowers the salinity of the surface layers, a purely thermal definition of mixed-layer depth is not satisfactory, as was pointed out by Banse (1987). Global mixed-layer topography based on density criteria has been computed by Levitus (1982), but these have only a limited information content for the Arabian Sea. Despite all these difficulties, changes in the topography of the thermocline and in sea surface elevation do reflect the major seasonal changes in circulation during northeast and Southwest Monsoons (Hastenrath, 1989).

When discussing the thermal structure of the Arabian Sea, as of other areas, we should note that the presence of phytoplankton modifies the heating-cooling cycle of the upper part of the water column. The extremely clear water, and strong DCM, of the Arabian Sea has enabled a calculation to be made of an inverse effect of biology on physics: the relative heating of seawater by differential absorption of short-wavelength solar radiation by pigmented phytoplankton cells in the DCM (Sathyendranath *et al.*, 1991). A maximum rate of 4°C per month (August) was calculated, which is not insignificant compared with cooling due to upwelling of about 2.5°C per month (July); thus, phytoplankton pigment enhances the rate of heating during the period of warming surface water and reduces the rate of cooling during upwelling periods. This effect should be watched for in other regions, perhaps especially where stratification is initiated in spring-bloom situations.

The northern Arabian Sea is more difficult to partition rationally than the remainder of the Indian Ocean. Several different schemes could be proposed following the principles outlined in Chapter 7, and, although none is entirely objective, the reasons for the choices made are as follows. The continental shelf and borderland along the west coast of the Indian continent has special characteristics that make it sensible and simple to recognize this as a separate province (Western India Coastal Province, INDW). From Somalia to the Gulf of Oman, the Southwest Monsoon forces a wide zone of coastal divergence; this feature is continuous with an offshore region of Ekman suction forced by the same wind system. Both are also areas of strong biological enhancement. It is unreasonable to separate these two related processes, so it is best to group part of the deep northwest Arabian Sea together with the Somalia-Arabian coastal boundary in a single province (Northwest Arabian Sea Upwelling Province, ARAB).

The two adjacent evaporative basins, the Red Sea and the Arabian Gulf, are taken together simply as a matter of convenience and because together they make an interesting and instructive comparison (as the Red Sea, Arabian Gulf Province, REDS).

# INDIAN OCEAN TRADE WIND BIOME

## INDIAN MONSOON GYRES PROVINCE (MONS)

### Extent of the Province

The MONS province extends from the hydrochemical front at 10°S (see later discussion) north to the offshore limits of the coastal provinces ARAB, INDE, and INDW. The province also includes the central Bay of Bengal and the southern part of the Arabian Sea. Thus, it is similar to Colborn areas 2 and 12–21.

### Continental Shelf Topography and Tidal and Shelf-Edge Fronts

Though continental shelf topography is not relevant to this province, the shallow water of the Maldive Islands aligned along 73°E, from the equator north to the Indian continental shelf, has consequences for both circulation and biological response.

### Defining Characteristics of Regional Oceanography

This province is synonymous with the reversing circulations of the two monsoon gyres of the northern Indian Ocean, with relatively low salinity and high surface temperatures, as befits an extension westward of the warm-water pool of the Pacific Ocean (see Chapter 11, WARM).

The monsoon gyral circulation occupies the whole northern Indian Ocean south to the hydrochemical front of Wyrtki (1973a) that lies above a zonal thermocline ridge at 10–15°S separating high-oxygen, high-salinity regions to the south from low-oxygen, low-salinity regions to the north. This is a convergent front and lies on the equatorward flank of the perpetual westward flow of the South Equatorial Current (SEC). In the eastern part of the province it is less well marked and the tropical-subtropical distinction (based on subsurface nitrate) occurs at 20–22°S along the 110°E IIOE section of Tranter (1973).

When the Northeast Monsoon becomes established in boreal winter, westward flow across the ocean as the Northeast Monsoon Drift or the North Equatorial Current (NEC) is established between about 2°N and the Indian subcontinent. This flow originates in the Bay of Bengal and in water that has been transported westward through the Indonesian archipelago; accordingly, surface salinities are low in the eastern part of the province during this season. Because of the different characteristic latitude here of the intertropical convergence zone than in Pacific and Atlantic Oceans, divergence of the surface water mass at the equator is very weak, or does not occur. Surface chlorophyll is a very good tracer of this process and a glance at the SeaWiFS serial chlorophyll images will convince you that the linear equatorial feature so prominent elsewhere is absent in the Indian

Ocean. The climatology of the chlorophyll field also demonstrates that strong between-year variability characterizes the surface circulation of this ocean.

The southward coastal current that develops along East Africa during the Northeast Monsoon (see later discussion) returns eastward across the ocean as the South Equatorial Countercurrent (SECC) along the northern slope of the thermal ridge at 5–10°S (Reverdin and Fieux, 1987). Convergence thus occurs at about 2–5°S between the NEC and the SECC (Wyrtki, 1973a).

In boreal summer, the establishment of the Southwest Monsoon collapses the winter circulation and, by April or May, forces the reversal of the NEC eastward as the Southwest Monsoon Current (SMC) from about 5°S up to 10°N. At the same season, an anticyclonic gyre is spun up in the Arabian Sea, and flow from the now-northward Somali Jet passes around the eastern limb of the gyre and then eastward as the SMC along the same latitudes that the NEC occupies in winter. The equatorial undercurrent of the Indian Ocean is singular in that it is not an oceanwide, permanent feature of the circulation: there is no persistent westward wind stress and thus no persistent westward wind drift to be compensated by subsurface return flow. We need concern ourselves only with the shallow Equatorial Jet (600 km wide) that passes eastward along the equator briefly at each transition between monsoon seasons: in April–May and in September–October. To judge from the evidence of the surface chlorophyll field, this is a highly variable feature. It appears to have been exceptionally strong in October–November 1998; at this time, there is evidence of the development in the eastern part of the ocean of Rossby waves, as occurs in equatorial features in the Pacific and Atlantic Oceans.

In fact, the circulation of the entire northern Indian Ocean is extremely variable between years, being closely linked to ENSO events. Sea-level anomalies (SLA) from TOPEX-POSEIDON altimetry have been used to demonstrate variability at all appropriate time scales in the Bay of Bengal (Somayajulu *et al.*, 2003). Circulation was found to respond to all phases of ENSO events, so that both SLA and SST anomalies are negatively correlated with western Pacific values. The ENSO signal in the oceanic, equatorial region was found to be transmitted to the Bay of Bengal by coastally trapped Kelvin and radiated Rossby waves along the Sumatran and adjacent coasts.

The entire province has a permanent thermocline, usually at 30–50 m, except from Sumatra to the south of Sri Lanka where it deepens in the Southwest Monsoon to 100 m (Colborn, 1975). In the area of the oceanwide zonal monsoon currents, the thermocline lies shallower (also thinner and steeper) in the west and deeper (also thicker and weaker) in the east (Hastenrath, 1989).

An important low-oxygen layer underlies the thermocline at intermediate depths through much of the eastern Arabian Sea, weakening progressively southward toward the hydrochemical front discussed previously. This deficiency is maintained by the slow passage through these depths of southern water, already oxygen deficient, within which there are very high rates of oxygen consumption resulting from progressive bacterial decomposition of large masses of organic material sinking from the superjacent algal blooms (Olson *et al.*, 1993; Kamykowski and Zentara, 1990; Vinogradov and Voronina, 1961). It is one of the three major denitrifying regions in the oceans, and the source of significant flux of $N_2$ to the atmosphere. There is some evidence, from apparently increasing abundance of *Pleuromamma indica*, a vertical migrant copepod characteristic of such conditions, that the oxygen minimum layer has increased in importance during the last 30 years or so (Smith and Madhupratap, 2005).

Because of their unique characteristics, the two northern embayments require individual attention.

***Arabian Sea Gyre*** This account should be read alongside the discussion of seasonal upwelling and other coastal processes forced by the reversing monsoon regime in the adjacent coastal provinces: ARAB and INDW. The reversing monsoons create a reversing gyral circulation that occupies the entire Arabian Sea: cyclonic in boreal winter, anticyclonic in summer. The topographically trapped gyres that form adjacent to the Somali coastline and the island of Socotra are here treated as being within the coastal provinces.

At the onset of the Southwest Monsoon winds ($< 15\,m\,sec^{-1}$) of boreal summer, the pycnocline in the central Arabian Sea is shallow and the density profile is typical of oligotrophic ocean regions. The mixed layer of the central part of the gyre deepens to 100–110 m as an anticyclonic gyre is spun up, with the center of the bowl lying rather centrally in the basin at about 10°N 63°E; the seasonal response of the pycnocline of the central Arabian Sea to physical forcing has been recently analyzed by Kumar and Narvekar (2005), based on serial observations along a meridional transect from the coast of Pakistan down to the Equator. The rate of deepening is greatest just after monsoon onset and is associated with anticyclonic wind-stress curl. The central Arabian Sea is cooled both by downward transfer of heat, and by evaporative heat flux despite the relatively high atmospheric humidity at this time. In the southeastern Arabian Sea, the mixed layer deepens by 25–35 m and cools by 2°C (Colborn, 1975; Rao, 1986; Hastenrath, 1989; Bauer *et al.*, 1991). Thus, even beyond the effects of coastal upwelling generated by the Southwest Monsoon, the entire Arabian Sea experiences some cooling and freshening.

During the cooler Northeast Monsoon winds of boreal winter, the wind maximum forms a weaker (c. $6\,m\,sec^{-1}$), broader jet that forces relatively insignificant Ekman dynamics, though some downwelling is predicted (Bauer *et al.*, 1991) within the general circulation that now proceeds in a cyclonic sense. The cool, dry winter monsoon forces negative heat flux, and deep penetrative convection reaches the pycnocline at around 125 m. This results in a significant flux of nutrients into the mixed layer so that nitrate concentrations >4 nM have been observed by the end of winter in the central Arabian Sea at depths shoaler than 50 m (Wiggert *et al.*, 2000). These observations have suggested high-*S* conditions to some people, and this possibility will be discussed later.

***Bay of Bengal Gyre*** Compared with the Arabian Sea, circulation in the Bay of Bengal is weaker and less predictable, and its response to monsoon reversal more complex, because it is partially open on its eastern margin where there is connection with the western Pacific through the Indonesian archipelago. Circulation and mixed-layer depth in the north of the Bay of Bengal is determined by winter cooling and by the very large fluvial input of freshwater from the mouths of the Ganges and Irrawaddy rivers, as well as by the reversing monsoon winds and the influence of the zonal currents lying across the south of the embayment. The interaction between these diverse forcings is complex and results in highly variable circulation. The northern area has very low surface salinities year-round, so that the 33.5‰ isohaline lies zonally, and almost permanently, at the sea surface across the Bay of Bengal at 10–15°N.

During the Northeast Monsoon, a cyclonic gyre forms in the Bay of Bengal (Wyrtki, 1973b; Rao and Sastry, 1981). With this is associated the East Indian Winter Jet of Tomczak and Godfrey, a shallow current passing equatorward along the eastern coast of India, rounding Sri Lanka and continuing into the Arabian Sea from October to March; this jet can be traced in the surface thermal field as a warm feature. The Southwest Monsoon spins up an anticyclonic gyre, though weaker than that of the Arabian Sea, which transports warm water of southern origin (Shetye, 1993), observable in satellite IR imagery, northward along the coasts of both India and Sri Lanka. Late in the season, there is some flux from the west coast of India and around the east coast of Sri Lanka that extends, while maintaining anticyclonic curl, into the cooler water of the northeast sector of the Bay of Bengal (Legeckis, 1987).

The topography of the mixed layer of the central Bay of Bengal is less predictable than in the Arabian Sea, perhaps because of undersampling, but also because this region is characterized by abundant, strong, and variable mesoscale eddies of both signs. Thermal stratification is strongest in April and May because of the northern excursion of the sun and cloudless skies, though because salinity distribution largely determines mixed-layer depth, the pycnocline remains rather shallow and abrupt. With the onset of the stronger winds of the southwest summer monsoon, this density structure is broken down by mixing, overturn, and cooling that are due to heavy seasonal cloud cover. During the same period, the anticyclonic gyral circulation breaks down, and the Bay of Bengal then contains two minor cyclonic eddies, which must include pycnocline uplift at their centers. One is in the northern part, in July–September, centered at 20°N 90°E. The other appears in October off Madras at 85°E, moves south, and dissipates off Sri Lanka in December (Banse, 1990a).

The consequence of these processes is that mixed-layer depths in the north of the Bay of Bengal (15–17°N) deepen by 50–75 m as the Northeast Monsoon season progresses (Colborn, 1975) and a thermal ridge develops, lying southwest-northeast from Madras to Burma, separating the Bay of Bengal circulation from an area of deeper mixing to the northwest of Sumatra. Minimum mixed-layer depths along this ridge are 30–40 m and south of the ridge, mixed-layer depths reach 90 m during the Northeast Monsoon. Beyond the shelf regions, the mixed layer is permanently devoid of nitrate and silicate, the main nutricline lying within with the principal thermocline at 50–100 m and responding to shoaling of the density gradient in cold-core eddies (Madhupratap *et al.*, 2003). The thermocline is also coincident with a strong oxycline, leading down into a deeper water mass having around 0.5 ml liter$^{-1}$ dissolved $O_2$. This extends down to around 500 m.

## Regional Response of the Pelagic Ecosystem

The serial chlorophyll images now available, more or less continuously since 1978, leave no doubt that the phytoplankton responds strongly to the forcing of the reversing monsoon wind systems. Although there are significant between-year differences, the main features of this response are readily described in simple terms, as below; for a more detailed account, based on 8 years of CZCS data, and partitioned within three open-ocean boxes representing the entire Arabian Sea, see the masterly synthesis by Banse and English (2000). It is to be noted that in that account, as in what follows, the division of the early cycle into four periods is not always very clear: it is often stated that the two intermonsoon periods both represent oligotrophic seasons, although it is clear that they are far from equally so. Seasonality of chlorophyll biomass is dominated by the period of clear, blue seawater that succeeds the winter monsoon during the period April–June: this is the period of maximal solar irradiance at the sea surface, the minimum occurring at the height of the Southwest Monsoon under very cloudy skies. The study by Kumar and Narvekar (see earlier discussion) of the central Arabian Sea to physical forcing concluded that seasonal changes in the productivity of the pelagic ecosystem were a direct response to seasonal changes in incoming irradiance, wind stress at the sea surface, and the role of external forcing in the seasonal variation of nutrient supply to the euphotic zone. They also reported that westward propagation of Rossby waves, independent of the seasonal reversal of monsoon winds, appears also to be involved in seasonal changes in mixed-layer depth.

Simple examination of serial chlorophyll images shows that the Arabian Sea almost invariably has a higher surface chlorophyll biomass than the Bay of Bengal and that the strongest response of the phytoplankton is aligned approximately below the Findlater Jet—that is to say, rather in the northwest quadrant of the Arabian Sea than toward its center. During the Southwest Monsoon, the region of higher chlorophyll (~1.0 mg chl m$^{-3}$) moves progressively toward the southeast across the Arabian Sea to reach its maximal extent in the intermonsoon period; then, some reduction of biomass occurs to levels sustained during the period of the Northeast Monsoon. At the onset of the spring intermonsoon period, chlorophyll biomass is rapidly decreased, so that from April to June the entire offshore Arabian Sea is extremely oligotrophic ($<0.1$ mg chl $m^{-3}$). Only a meridional line of higher chlorophyll at the Maldive Islands along 73°E forms a recurrent feature in surface chlorophyll images at this season: westward flow across this line creates chlorophyll enhancement downstream to the west, a situation that is reversed in boreal summer, when flow is in the opposite direction. At the same season, the regional deep chlorophyll maximum of the anticyclonic Bay of Bengal gyre deepens progressively toward the center of the gyre.

The onset of the winter monsoon coincides with a minor buildup of phytoplankton biomass around the periphery of the Bay of Bengal, including around the southern limb of the cyclonic gyre that spins up at that season; this region of relatively high chlorophyll is rooted at Sri Lanka. The equatorial region generally shows as a low-chlorophyll zone, adjacent to higher chlorophyll lying around the north of the south subtropical gyre, beyond the 10°S hydrochemical front (see earlier discussion). The eastward equatorial jet of intermonsoon periods does, in some years, appear to be associated with high chlorophyll biomass that is aligned with the equator, but this is more diffuse than the typical narrow, meandering equatorial divergence signal seen in the other oceans. Thus, this feature was especially strongly developed in November–December 1997 from midocean across to Sumatra exhibiting, particularly in the latter month, evidence of meandering at the Rossby scale.

You may have wondered about the observation that I discussed briefly earlier, concerning a rather high concentration of $NO_3$ in the mixed layer of the central Arabian Sea during the Northeast Monsoon; this might seem, at first sight, anomalous in view of the continued relatively high chlorophyll biomass there—high, that is, for a tropical open-ocean situation—at this season. Both SeaSoar observations and models of this process give the same result: during the Northeast Monsoon, diurnal shallow stratification within the mixed layer permits accumulation of phytoplankton biomass that is mixed down each night, thus reducing the number of cells in the near-surface zone. Ammonium is regenerated each night within the "microbial loop" as cells are consumed and digested, and is available each morning within the mixed layer in sufficient concentration as to inhibit the uptake of $NO_3$ (Wiggert *et al.*, 2000).

Thus, phytoplankton response (productivity of order 1.0–1.5 $gC\,m^{-2}\,d^{-1}$) to deep convective mixing is instantaneous while, at the same time, the $NO_3$ that is entrained into the deepening mixed layer across the eroding nutricline is not fully utilized and accumulates there. Prior to the bloom, during the oligotrophic period, $NO_3$ utilization (at about 0.15 $\mu M\,d^{-1}$) is balanced with the entrainment of this molecule into the euphotic zone. This mechanism supports the finding of Watts and Owens (1999) of the suppression of nitrate assimilation here and more generally the observation of Harrison *et al.* (1996) concerning nitrate and ammonium interaction at nanomolar concentrations in the oligotrophic ocean.

Because, in occasional years, chlorophyll biomass remains below 1.0 $mg\,m^{-3}$, some have suggested that this is yet another high-*S* region (see Chapter 5). There is no reason, however, to postulate either Fe or $SiO_3$ limitation here because, for one thing, if Fe is delivered by dust in nonlimiting quantities anywhere at the oceans surface, it must be here: Sharon Smith has described this region as "Mother Nature's Iron Experiment." Nor is it necessary to suggest that it is grazing control that restrains phytoplankton accumulation. The daily stratification cycle appears to be a sufficient explanation because, once shallow stratification is established permanently after monsoon wind stress at the surface is reduced, then phytoplankton biomass accumulates, the $NH_4$ pool is consumed, and plant growth begins to take up the previously entrained $NO_3$ and to reduce its concentration.

Sections taken across the northern part of the Arabian Sea in September and November show that the dominance of large cells (>18 μm), mostly diatoms, does not extend beyond the coastal upwelling regions. Across the open ocean, during monsoon and intermonsoon periods, even when chlorophyll biomass was approximately equally distributed between phytoplankton size fractions there was, in general, a maximum of production in the smallest (0.2–2.0 μm) class (Savidge and Gilpin, 1999). Nevertheless, phytoplankton biomass dominated by the prokaryote fraction, among which *Synechococcus* was particularly abundant, reaching $>10^8$ cells liter$^{-1}$. During the intermonsoon period, prochlorophytes increased in relative abundance, and in the center of the gyre, where conditions became extremely oligotrophic, these cells dominated the autotrophic community. Prochlorophytes appear to be preferentially distributed within a secondary fluorescence maximum at the oxic-anoxic interface at the bottom of the euphotic zone, and well below the depth of the primary chlorophyll maximum (Johnson *et al.*, 1999). Here, the population was composed almost exclusively of *Prochlorococccus* spp. This population is strongly light adapted, having instantaneous growth rates of order $\mu = 0.001\,d^{-1}$. Even shallower-living populations of *Prochlorococccus* have doubling times of less than once daily, whereas *Synechococcus* have maximum growth rates of $>2\,d^{-1}$ that respond rapidly to nutrient availability.

During this intermonsoon season, toward the center of the gyre, coccolithophores also become more relatively abundant than elsewhere (Tarran *et al.*, 1999). At this season, also, the chlorophyll profile more frequently exhibits a pigment maximum at or close to the nutricline at around 75 m, and deeper than the top of the pycnocline, rather than within the mixed layer above. In the central gyre, producer and consumer biomass covaries seasonally within each of the pico, nano, and micro fractions.

During the Southwest and Northeast Monsoons, photosynthetic prokaryotes were responsible for ~25% and ~50% of productivity, respectively, and whereas the growth of *Prochlorococccus* is in balance with its consumption by protests, that of *Synechococcus* largely escapes consumption, >50% compounding daily during the Southwest Monsoon (Brown *et al.*, 2002).

Among the consumer groups, including herbivores, the nano- and microzooplankton fractions form a relatively constant biomass while mesozooplankton biomass fluctuates more strongly with the seasons and responds to the onset of upwelling (Stelfox *et al.*, 1999). Along two sonar transects across the entire Arabian Sea, Luo *et al.* (2000) found a fivefold increase in sonic scattering from zooplankton and mesopelagic fish during the Southwest Monsoon compared with the spring intermonsoon. Heterotrophic nanoflagellate biomass ($<855$ cells ml$^{-1}$) exceeded that of the microheterotroph biomass, dominated by aloricate ciliates and dinoflagellates ($<333$ cells ml$^{-1}$). These organisms were an order of magnitude more abundant than sarcodines and crustacean nauplii. During and after the Southwest Monsoon of 1994, in the open gyre, consumption rates of microzooplankton grazing represented up to 20–30% of daily primary production that ranged from 1.0 to 1.5 gC m$^{-2}$ d$^{-1}$.

During the intermonsoons, daily consumption represents 30–50% of a rather smaller productivity (0.3–0.8 gC m$^{-2}$ d$^{-1}$). Mesozooplankton herbivory was judged by Stelfox *et al.* to represent around 25–30% of that of microzooplankton. The grazing dynamics of mesoplankton were addressed independently by Roman *et al.* (2000), who found, overall, that the copepod mesozooplankton (dominated by ~50 species of copepods) represented a biomass in the range of 10–50 mgC m$^{-2}$ that ingested 10–40 mgC m$^{-2}$ of food daily, representing 10–60% of daily depth-integrated primary production. Lowest ratios of ingestion/primary production were found during the Northeast Monsoon. The seasonal response of copepod biomass to autotrophic productivity was rapid, so that in the central gyre during the spring intermonsoon this was of order 20–50 mgC m$^{-2}$, whereas during the Southwest Monsoon it increased to 30–200 mgC $m^{-2}$. Copepod diversity responds appropriately: of a total of about 185 copepod species that occur frequently in the pelagos, 57 species were encountered near-surface under the Findlater Jet in the Southwest Monsoon, but only 43 during the Northeast Monsoon. In the central gyre, smaller species dominate the reproducing copepod populations: *Paracalanus aculeatus, Undinula vulgaris, Cosmocalanus darwinii, Undinula vulgaris, Calanus minor*, and others. The large *Calanoides carinatus* is restricted to the near-shore upwelling regions of the ARAB province. According to Madhupratap and Haridas (1986) the Indian Ocean copepod fauna has closer affinities to the Pacific than to the Atlantic fauna: this what one might expect, given the relative permeability of the Indo-Pacific boundary.

The deep oxygen minimum (at about 200–500 m) of the Arabian Sea extends over much of the northern Indian Ocean, though it apparently has little impact on the vertical distribution of mesoplankton beyond the central Arabian Sea itself (Madhupratap and Haridas, 1990). Elsewhere in this province the vertical distribution of typical tropical genera closely follows the pattern we shall encounter in the eastern Pacific and as was described for the region north of the Findlater Jet by Luo *et al.* (2000). However, further south, in the presence of the subsurface strongly anoxic layer ($O_2$ concentration is near zero below 160 m and $H_2S$ may be present) has major consequences for the distribution of plankton in all size classes; the classical investigations of this phenomenon were those of the Soviet academicians Vinogradov and Voronina (1961). The oxygen-deficient layer suppresses the diel vertical migration that would otherwise be expected to occur here, the diel migrants remaining in the surface layers. On the other side of the anoxic barrier, the deep-living copepod *Lucicutia* sp. is restricted below the lower layer of the zone; it occurs, therefore, only deeper than 400 m. Here, it consumes sinking particles—detritus, fecal pellets, clumped bacteria, phytoplankton and microheterotroph cells, and so on. Whole cells form a smaller proportion of its diet during the spring intermonsoon period, when more intense reworking of organic material within the surface microbial web must occur (Gowing and Wishner, 1998).

However, the ocean always reserves surprises for us: during the JGOFS investigations at the central mooring site, where very strong anoxic conditions obtained at all seasons below 200 m, Mincks *et al.* (2000) reported that abundant crustacea occur in this zone and perform limited diel migrations into the surface layers above. Thus, pelagic decapod shrimps (*Gennades, Sergia*, and *Eupasiphae* spp.) and portunid crabs (*Charybdis smithii*) live largely within the anoxic layer and must have major respiratory modification for that habitat. Feeding on mesoplankton occurs principally within the anoxic layer, and is continuous day and night. The population of pelagic crabs is at times very dense (< 13 g $m^{-2}$, 0–500 m, has been observed) and the species appears to have an annual life cycle in which the pelagic phase is returned passively to the shelf regions for reproduction. Should this transport fail, very dense sedimentation of immature crabs may ensue in the open ocean; in one such fallout, up to 1 crab $m^{-2}$ was observed on the deep ocean floor. Such falls may represent 20–30% of the total regional POC sinking flux and create regions of intense microbial activity in the sediments as the carcasses are integrated into the deep-sea food web (Christiansen and Boetius, 2000).

Locations of high abundance of higher trophic level organisms, exemplified by tuna that are capable of long-distance migrations, respond to the relative productivity of different parts of this province. Yellowfin, skipjack, and bigeye all avoid the Bay of Bengal, except the area of the coastal current along the eastern coast of India and Sri Lanka, but are abundant in the eastern equatorial part of this province. The central Arabian Sea is largely avoided, perhaps because of the oxygen minimum layer, and only yellowfin penetrate the Arabian Sea itself in large numbers, concentrating mostly in the Omani upwelling region.

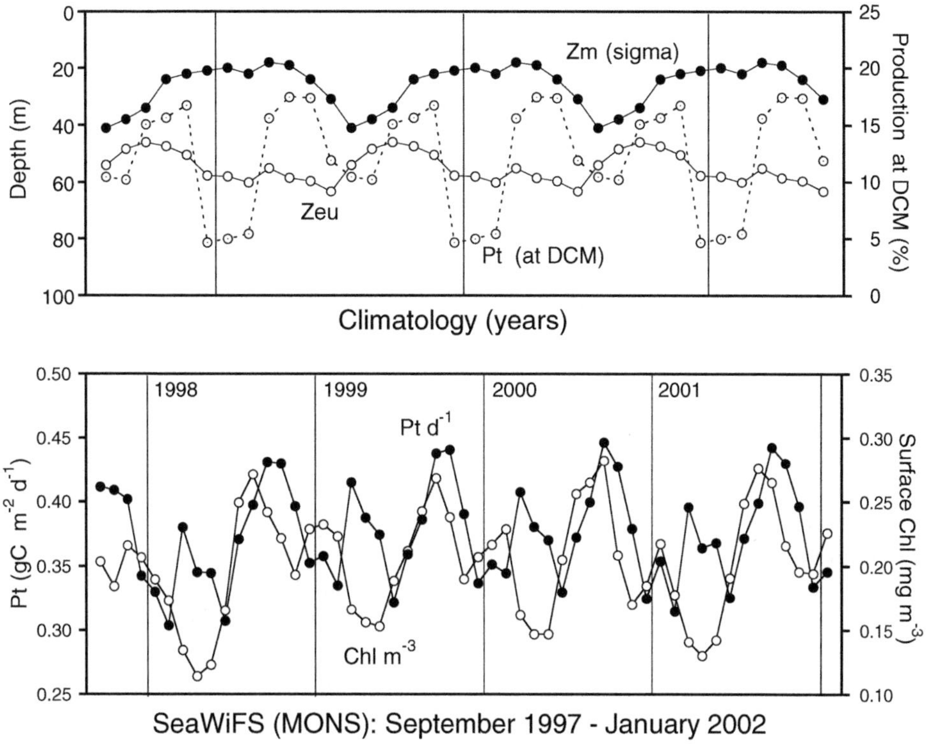


**Fig. 10.1** MONS: seasonal cycles of monthly surface chlorophyll and depth-integrated autotrophic production for the years 1997–2002 from SeaWiFS data together with characteristic seasonal cycles of mixed-layer depths from Levitus climatological data and photic depths computed from characteristic irradiance and the archive of chlorophyll profiles discussed in Chapter 1.

## Synopsis

*Case 5—Large-amplitude response to monsoonlike reversal of trade winds*—The climatology (Fig. 10.1) offered by archived satellite and other data show that the mixed layer responds to flow in Somali Current, deepening from 20 m to 50 m as eastern flow develops, and shoaling again during NE monsoon as westward flow of the North Equatorial Current intervenes. $Z_p$ is consistently deeper (50–60 m) so the thermocline is illuminated year-round. The double seasonal increases in rate of P for relatively brief periods are forced by this seasonality in wind stress; the major response is from June to August (from 200 to 400 mgC $m^{-2}$ $d^{-1}$) corresponding with onset of SW monsoon conditions. Chlorophyll accumulation tracks the change in productivity at this season, but fails to do so during the increase in productivity with the onset of the NE monsoon.

# Indian South Subtropical Gyre Province (ISSG)

## Extent of the Province

The Indian South Subtropical Gyre Province (ISSG) extends from the hydrochemical front at ~10°S to the Subtropical Convergence at ~40°S. The eastern margin is the Australian coastal boundary at the outer edge of the Leeuwin Current, and to the west at the outer edge of the Agulhas and East Madagascar Currents and, thus, comprising Colborn areas 22, 23, 25–29, and 30–32 (in part).

## Continental Shelf Topography and Tidal and Shelf-Edge Fronts

Major areas of flat shallow topography, <200 m deep, lie along the Mauritius-Seychelles Ridge; these are the Nazareth, Saya do Malha, and Seychelles Banks.

## Defining Characteristics of Regional Oceanography

We have very little organized knowledge about the subtropical gyre of the southern Indian Ocean; some of the IIOE investigations did survey this province, especially by means of a few long meridional sections by the then-Soviet ships, which generated scattered data on primary and secondary production. In the southeastern part, the well-planned Australian IIOE sections produced a serious body of organized ecological data, seasonally, although this is really more relevant to the AUSW province. South African interest in the SW Indian Ocean has illuminated some dark corners of ISSG in recent years, and the recent review of Schott and McCreary (2000) is a most useful tool.

The zonal thermocline ridge at about 10°S is the limit of westward flow of the low-salinity SEC on its southern slope and of the eastward flow of the SECC along its northern slope. Below the southern slope of the ridge, and thus below the salinity minimum, Wyrtki's hydrochemical front slopes down southward and separates the high-nutrient/low-oxygen water of the monsoon gyre from the low-nutrient/high-oxygen water of the subtropical gyre. The location of the ridge changes seasonally; as the SEC strengthens during the Southwest Monsoon, the thermal ridge moves progressively northward (to 5°S) and shoals ($Z_m = < 30$ m) in the Southwest Monsoon, whereas during the Northeast Monsoon it is at 10°S and deeper ($Z_m = < 40$ m).

During boreal summer, the southeast trades cross the equator northward together with the ITCZ, which comes to lie along the southern flank of the Himalaya in August; the wind-driven anticyclonic circulation south of the equator is then fully established. Anticyclonic wind stress curl extends from about 10–15°S to the southern limb of the gyre, associated with deepening of the mixed layer centrally in the gyre; this effect is strongest in austral summer and autumn, reaching maximum effect at 30°S. Increasing wind stress and decreasing insolation during austral winter impose negative heat flux (–100 to $-140\,W\,m^{-2}$ in August) that deepens the SW Indian Ocean mixed layer to about 75–100 m. In austral summer, this shoals to 50–75 m. The ISSG province is yet another case of a subtropical anticyclonic gyre that has nitrate isopleths deepest in the central regions and shallowest around the periphery. As elsewhere (see SATL and SPSG provinces), this peripheral shoaling of nitrate is clearly reflected in the surface chlorophyll field.

Circulation within this subtropical gyre differs from those of the Atlantic and Pacific Oceans because the eastward flow of the South Indian Ocean Current, along the South Subtropical Convergence zone, does not generate a boundary current on encountering the Australian continent (Tomczak and Godfrey, 1994). Instead, much of the flow around the southern limb of the gyre progressively recirculates northward into the center of the gyre as two main streams (Stramma and Lutjeharms, 1997). Even prior to reaching 70°E, as much as 40 Sv of the original 60 Sv is recirculated into the SW Indian Ocean subgyre that is both smaller and more dynamic than the equivalent feature in other oceans. The remainder of the flow is drawn into the main subtropical gyre in a stream centered on about 90°E, perhaps associated with the meridional ridge that occurs there: this is prior to reaching the Australian continent, in any case. Between this recirculation stream and the poleward Leeuwin Current at the Australian coast there forms an eddy field within which Rossby waves are formed and from which they are strongly propagated westward across the ocean (Andrews, 1977). Another eddy field is generated to the east of Madagascar, comprising anticyclonic eddies with a periodicity of about 50 days. TOPEX-POSEIDON images reveal that a complex band of mesoscale eddying arcs across the ocean from western Australia to Madagascar: eddies are of order 100 km diameter, have SLA signatures of around 15–25 cm, and—because cyclonic features lie equatorward of anticyclonic features—the whole forms a ridge-trough system in the SLA field.

The low salinity of the water mass transported in the SEC originates partly from the southerly, eastern limb of the Bay of Bengal gyre, partly from the low-salinity water of the Southeast Asian archipelago, and partly from the effects of a belt of very heavy rainfall ($>200\,cm\ y^{-1}$) from 3°N to 10°S, east of 60°E. Finally, there is input to SEC water from rainfall and rivers as it passes north of Madagascar. Mixed-layer depth in the north of the province may be influenced more by salinity distribution than by temperature.

## Regional Response of the Pelagic Ecosystems

Persistently high water clarity extends over much of this province, for much of the year, and surface chlorophyll of $<0.05\,mg\ chl\ m^{-3}$ is widespread. Chlorophyll images suggest four regions of higher chlorophyll: (i) a zonal region south of the equator across the entire ocean corresponding to the SEC, (ii) in the eddy field to the west of Australia, (iii) to the east of southern Madagascar, and (iv) at midocean locations of shoal water and banks.

***South Equatorial Current*** Apparently coincident with the SEC is the most persistent region of high chlorophyll in this province, which dominates many of the serial monthly images. It lies almost permanently across the ocean south of the equator, being initiated in January or February and strengthening progressively as the austral winter progresses; it is most strongly developed during the period of the Southwest Monsoon in July–August, when it becomes continuous with the high chlorophyll in the ARAB province in the western, coastal regions of the Arabian Sea. At its initiation, it lies zonally across the ocean at ~10°S, but as it strengthens it shifts progressively southward and comes to lie somewhat diagonally from about 20°S to the west of Australia to 10°S, or even equator-crossing in the west of the ocean. Except at this season in the extreme west, an equatorial band of clear water, lacking chlorophyll biomass, is maintained. This bloom must represent the consequences of persistent eddying within the permanent westward flow of the SEC, because the entire region having chlorophyll higher than background is rich in Rossby-scale eddies with low-chlorophyll cores and streamers of high chlorophyll around their periphery. Examined in movie mode, the serial images show how individual eddies in the chlorophyll field may be followed across the western part of the ocean, subsequently crossing the equator to proceed up into the Somali Current. Although the eddy field is very well defined in the sea surface chlorophyll images, these do not suggest that very high concentrations are reached: around $0.2–0.5\,mg\ chl\ m^{-3}$ appears to be normal, although $1.0\,mg\ chl\ m^{-3}$ may be reached in the western part of the ocean.

These simple observations are compatible with the divergent flow and potential upwelling that occurs on the 10°S thermocline ridge and should be reflected in biological enhancement at the surface. Recall that here the geostrophic ridge along the equatorward side of the subtropical gyre brings the nutricline relatively close to the surface, as is confirmed by 100-m nutrient distributions. Production/loss balance computed from the surface satellite chlorophyll suggests that coupling between herbivores and phytoplankton is extremely close: accumulation tracks the rise in primary production rate only very briefly, and then declines well before productivity slackens.

The IIOE samples of higher trophic levels such as mesozooplankton were not sufficiently closely spaced to identify the local consequences of this zonal band of high productivity. However, Tranter (1973) discusses the seasonal reproductive migrations of southern bluefin tuna that carry them into the region of 20°S to the west of Australia, where they must encounter sufficiently high abundance of squid and other large nekton to maintain the adult stock and sufficiently abundant mesozooplankton and nekton to support the growth of their larvae and young-of-the-year. Fontenau's maps of tuna fishing effort support this observation, although this fishery apparently failed after about 1970, perhaps because of stock failure. Fontenau also shows that fishing effort for yellowfin and bigeye tuna continues higher than background in a linear zone across the southern Indian Ocean that corresponds well with the location of the SEC eddy field. The distribution of these top pelagic predators may be sufficient evidence for us to be reasonably confident that the bloom in the SEC supports—as we would expect it to do—a pelagic ecosystem with biomass significantly higher than background.

Suda (1973) examined the records of tuna catches in the Indian Ocean in the relatively early years of the fishery (1966–68), when we may assume that the relative distribution of species had become little modified by the fishery. He shows a distribution of the longline hooking rate of albacore (*Thunnus alalunga*) that includes a very striking concentration of fish across the entire southern Indian Ocean at about 30–32°S, apparently corresponding to the eddying flow around the southern limb of the gyral circulation. Concentration of fish is highest at about 90–100°E, or at about the longitude of the major reflux into the interior of the gyre (see earlier discussion). The linear concentration of fish across the ocean is quite distinct from a weaker linear zone at about 45–50°S that corresponds very well with the eddying flow in the SSTC province.

***Eastern Eddy Field*** During the late austral winter, from July to September, the eddy field west of Australia exhibits higher chlorophyll biomass than during the first half of the year and becomes continuous with the eastern region of high chlorophyll in the SEC. During the early phase of this bloom development, one may observe in the images many long, undulating streamers of relatively high chlorophyll that are rooted in the western Australian eddy field but extending sufficiently far as to join the SEC region in midocean. It is here that the only serial observations of the major components of the pelagic ecosystem were completed during the IIOE (Tranter, 1973). The Australian meridional transect along 110°E, at the longitude of central Java and from 10–32°S, was worked so to give 16 day and 16 night stations every 2 months. By today's standards, the observations were not sophisticated but this is the only organized, seasonal data set we possess of nutrients, productivity, chlorophyll, zooplankton, and micronekton at comparable time and spatial scales.

The seasonal cycle obtained by the Australian IIOE matches very well what we now observe in the serial chlorophyll images: an austral winter bloom occurs from March–April until September–October, the period when $NO_3$ values, integrated to 100 m, are almost an order of magnitude higher than in the period December–February. Productivity, integrated over the same depths, increases from $<20$ to $>60$ mgC $m^{-2}$ $hr^{-1}$, and this permits $>25$ mg $m^{-2}$ of chlorophyll biomass to be accumulated. Seasonal variability of night zooplankton is relatively weak with an increase of $<20\%$ to $\sim2.0$ mg $m^{-3}$, and there appears to be a lag of several months in their response to the phytoplankton bloom and decline. Only a few species (e.g., *Eucalanus concinna, Rhincalanus nasutus*) appear even to double their biomass seasonally. Tranter (1973) used a correlative matrix to determine if the changes observed represented an ecosystem response or whether they were artifactual, caused by physical translocation of biota through each station position. He concluded that nitrate, primary production rate, and zooplankton biomass all vary seasonally by about 40–50% of the annual mean value, whereas phytoplankton biomass (on the evidence of chlorophyll) varies only by about 20%. This conclusion agrees with our general assumptions about pelagic ecology in warm seas. However, looking more closely at evidence for trophic succession, Tranter found the following pattern: in tropical water the timing of zooplankton biomass correlates well with productivity and nitrate, but in subtropical water chlorophyll correlates positively with productivity but negatively with zooplankton. He infers that consumption response is instantaneous in low latitudes but is lagged by one cruise interval (several weeks) farther poleward.

***The Madagascar Basin*** An enigmatic seasonal dendritic bloom in the southwest Indian Ocean escaped observation until revealed by satellite images. These show that chlorophyll biomass may reach 1–3 mg chl $m^{-3}$ against the oligotrophic oceanic background of order 0.05 mg chl $m^{-3}$ and that this bloom may, at its maximum extent, cover as much as 3000 km in zonal extent, from 45°E to 80°E—that is to say, halfway to Australia, as occurred in 1999 (Longhurst, 2001). It takes the form of a series of arcuate patches lying north and south on either side of a zonal axis along about 25°S (see Color plate 11).

The dendritic bloom originates in February near the retroflection of the East Madagascar Current and very rapidly extends eastward zonally across the ocean to reach its furthest extent in April, at which time the chlorophyll signal near the origin has attenuated significantly. The individual features appear faintly, and subsequently strengthen as the bloom progresses. The evolution of juvenile to mature features is characteristic: rapid spreading occurs of arcuate wisps of chlorophyll enhancement to form broader regions of high chlorophyll that eventually overlie the entire feature around which they were first developed. The rapid extension of the bloom toward the east led Srokosz *et al.* (2004) to propose that the entire feature represents a "plankton wave," caused by "the swirling motions of the eddy field" that "diffuse the plankton against both the mean flow and the eddy and Rossby wave propagation direction." Further analysis of this suggestion was not forthcoming.

The bloom is highly variable between years: it was strong in 1997 (Polder), 1999, 2000, 2002, and 2004 (SeaWiFS) and absent or very weak in 1998, 2001, 2003, and 2005 (SeaWiFS, MODIS). I can now find no relationship between this variability and any readily available climatic index; earlier, I suggested a relationship with the SOI, but this seems untenable now with better access to data.

There appear to be two probable explanations for the initiation of the bloom, although others have been suggested; both probable explanations depend on a change in the depth of the regional mixed layer. If this shoaled, then an existing subsurface bloom at the chlorophyll maximum might become visible to the satellite sensors rather rapidly. However, this is unlikely because a period of heat loss from the surface begins after austral midsummer so that a seasonal deepening of the mixed layer is initiated in February that takes the 40-m mixed layer of austral summer down to about 80 m by May and even deeper prior to the next austral spring. Such conditions, to the east of Australia, are known to initiate entrainment blooms there (Dandonneau and Gohin, 1984). I have noted in Chapter 9 the probable existence of a homologous—but much fainter and less frequent—bloom at exactly the homologous location in the South Atlantic. Further, we should note that the deepening of the mixed layer progresses eastward, as it must given the overall bowl form of the thermocline in the subtropical gyre. Srokosz *et al.* present their "plankton wave" concept largely to explain the eastward movement: it seems to me that this suggestion is unnecessary.

The deepening of the mixed layer provides us with a possible explanation for this bloom by the simple entrainment of nutrients into the euphotic zone caused by downward erosion of the nutricline. Nor should the rapidity of onset of this entrainment bloom surprise us if, as seems probable, it is dominated by picoautotrophs. In this highly eddying region, drifting buoys (and observations of the swirls) suggest a generally westward flow (Lutjeharms *et al.*, 1981); there is therefore no question that nutrient transport from the East Madagascar Current might be involved as has been suggested. Given the probable mechanism, one would anticipate a negative relationship between chlorophyll biomass and the SLA in cms and, indeed, this is what we observe. There is, as has already been suggested, the required coincidence between elements of the chlorophyll field (lenses, whirls) and the SLA field (anticyclonic elevation, cyclonic depression).

***Shallow Topography of the Central Regions*** Finally, the serial chlorophyll images clearly reveal discrete seasonal blooms that are associated with the shallow banks (Nazareth, Hawkins, and Saya de Malha banks on the Mauritius-Seychelles Ridge, 5–20°S) and down the Chagos archipelago, especially during austral winter. This appears to confirm previous suggestions that the Mauritius-Seychelles Ridge causes divergence in the SEC during this season, leading to nutrient enrichment of surface water on its western side; similar effects occur at the Seychelles themselves, which lie athwart the thermal ridge at 10°S (Ragoonaden *et al.*, 1987; Vethamony *et al.*, 1987).

However, these banks are sandy flats that have depths of only 15–60 m and lie atop the topography that rises from abyssal depths. The flats are dotted with coral heads but not occupied by massive reef-coral development as elsewhere, and in their shallower regions they support very extensive sea-grass meadows. Examination of 4-km MODIS images confirms that the regions of enhanced chlorophyll associated with these midocean banks have a permanent "core" that represents the shape of the banks themselves, surrounded by a variable "aura" of apparently lower chlorophyll biomass. This is especially evident on the Chagos banks where central, permanent patches of apparently rather high chlorophyll biomass are observed.

We have very little organized knowledge of the ecology of the Indian Ocean banks, beyond surveys done for fish resource assessment prior to exploitation: these are reviewed by Longhurst and Pauly (1987). Obviously, the fish fauna is dominated by hard-substrate genera: Serranidae, Lutjanidae, Lethrinidae, and Labridae.

### Synopsis

*Case 3—Winter-spring production with nutrient limitation*—$Z_m$ undergoes an austral winter excursion (20–25 m December–January, 75 m September) and since $Z_{eu}$ is always at 50–70 m, the thermocline is illuminated in all except 3 winter months, July–September. P responds to the period (June–August) of nutrient input due to winter deepening of the mixed layer with a sustained rate increase, though over a small dynamic range to an annual maximum in austral spring (August–October). The period of decreasing P rate (November–August) to the summer minimum coincides with period when $Z_{eu}$ lies within thermocline. Chlorophyll biomass takes a very low dynamic range, with only a brief (June–July) period of accumulation to slightly higher (0.20 mgC $m^{-3}$) biomass from summer low biomass of 0.05 mgC $m^{-3}$ (Fig. 10.2). Failure to sustain chlorophyll accumulation after the initial increase in productivity is consistent with vertical migration or other response of herbivores.

# Indian Ocean Coastal Biome

## Red Sea, Arabian Gulf Province (REDS)

### Extent of the Province

This province comprises the Red Sea north of the Straits of Bab-el-Mandeb and the Arabian Gulf within the Straits of Hormuz. Note that in older literature, what is now generally called the Arabian Gulf may be given as the Persian Gulf. The Gulfs of Aden and Oman are considered to be a part of the adjacent coastal boundary province (ARAB).

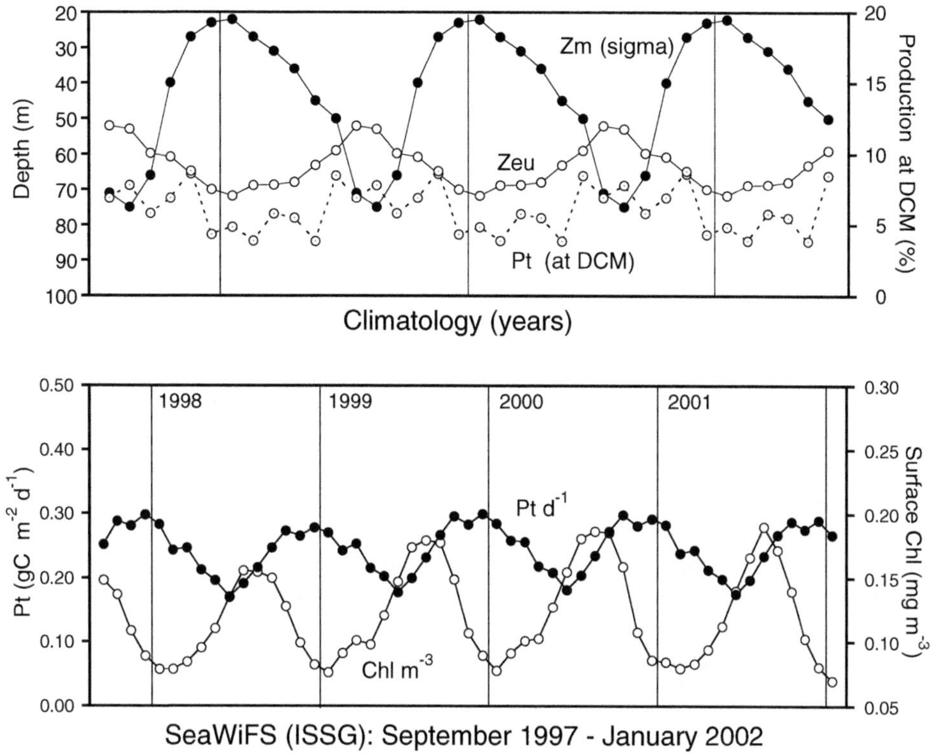


**Fig. 10.2** ISSG: seasonal cycles of monthly surface chlorophyll and depth-integrated autotrophic production for the years 1997–2002 from SeaWiFS data together with characteristic seasonal cycles of mixed-layer depths from Levitus climatological data and photic depths computed from characteristic irradiance and the archive of chlorophyll profiles discussed in Chapter 1.

## Continental Shelf Topography and Tidal and Shelf-Edge Fronts

The Red Sea comprises a shallow coastal shelf surrounding a very narrow deep basin geologically continuing the East African rift valley. Depths of < 100 m occupy 41% of the total area of the basin and the shallow area is wider south of about 20°N, reaching 120 km at Massawa (Head, 1987). Over much of the Red Sea, the shallow shelf is occupied by active coral reefs. It is these reefs, and the effect of the heavy brine filling the deep rift, which mostly have occupied marine ecologists who have had the chance to work in here.

The Arabian Gulf comprises a series of deeper basins, reaching to 80 m along the north coast, and a series of shallower banks (often < 20 m) along the eastern and southern coasts; mean depth overall is ~40 m. The south coast is an area of active coastal accretion by biotic carbonate sedimentation in shallow "sabkha" lagoons and offshore organic reefs (Shinn, 1987).

## Defining Characteristics of Regional Oceanography

These two evaporative basins, one deep and the other shallow, have little in common other than their proximity, so their circulations are discussed independently. As already noted, in an entirely logical system it would be well to separate these two entities as individual provinces; they are maintained together here purely as a matter of convenience.

***Red Sea*** The Red Sea is a long, narrow rift valley occupied by the sea, in which evaporative processes drive the deep circulation so that surface flow may be contrary to prevailing wind stress. The circulation pattern is strongly influenced by the existence of a shallow sill of only 110 m at the Straits of Bab-el-Mandeb. Excess of evaporation over precipitation in the basin of about 2 m $y^{-1}$ forces a constant influx of surface water through these straits from the Arabian Sea. Winter cooling to about 18°C at the northern end of the Red Sea of water already having a salinity of 42‰ creates strong deep convection cells in which the deep water (anomalously warm and saline at 21.5°C and 40.6‰) of the Red Sea basin is formed (Dietrich *et al.*, 1970). At the Straits of Bab-el-Mendab, deep water flows out below the surface influx at a rate of about half that of the Mediterranean outflow over the sill of the Straits of Gibraltar. Surface wind-driven streams are generally weak and eddying, but flow occurs toward the south along the whole length of the western coast of the Red Sea during the season of northerly, summer winds; at other times the density-driven flow will cause surface drift contrary to these winds, especially up the eastern coast. Thus, the general surface circulation is cyclonic.

The relative depth of the mixed layer is approximately the inverse of surface temperature, except where it is destroyed by winter mixing or tidal effects in shallow water. In boreal spring, thermocline depths decrease, until the summer situation is reached, with minimum depths of about 30 m on the eastern side of the rift. In the southern Red Sea, the summer thermocline is sustained above the uplift of the weak permanent thermocline forced by the very active flow of a cool-water core northward through the Straits of Bab-el-Mandeb under the influence of the Southwest Monsoon. This shallow permanent thermocline persists to about 22°N in the Red Sea. In the northern Red Sea, the summer thermocline, at about 25 m, lies above the nearly isothermal deep water mass.

An important layer of oxygen deficiency (0.9–1.3 ml $O_2$ $liter^{-1}$) occurs at approximately 300–650 m throughout the Red Sea and is exceptional among such areas because both temperature and salinity are relatively very high: ~22°C and 40.5‰, respectively (Weikert, 1984). This strange habitat will require (later) some discussion of the reaction of zooplankton to its characteristics.

The monsoon reversal has only minor influence over the Red Sea, where the prevailing winds are along the axis of the rift valley and from the north; during winter the wind is reversed over the southern part so that a wind convergence occurs at 18–22°N (Edwards, 1987). Surface inflow at Bab-el-Mandeb is therefore stronger in winter than in summer. At times of stronger than usual northerlies, the water column at these straits may comprise three layers: a thin (40 m) surface, wind-driven outflow; a deep density-driven outflow; and an intermediate low-salinity inflow. Interannual variability in inflow through Bab-el-Mandeb is also strong and can be detected by differential spreading of Arabian Sea water (Ganssen and Kroon, 1991).

***Arabian Gulf*** The shallow Arabian Gulf is oceanographically an extension of the surface water of the Arabian Sea, and there is no shallow sill between it and the Gulf of Oman as there is between the Red Sea and the Gulf of Aden. Evaporation greatly exceeds both precipitation and the input of river water from the Euphrates and Tigris, and salinity reaches >50% in shallow water on the Arabian coast. A slow cyclonic circulation is maintained having low-energy regions toward the head of the Gulf; evaporation (<1.5 m $y^{-1}$) occurs in the southern bight lying to the east of the Qatar peninsula; the dense water formed in this way sinks to the deeper regions of the Gulf.

In winter, a weak thermocline is briefly established in the outer part of the Arabian Gulf at 30–40 m, but during the intense heating of the boreal summer an isothermal layer effectively ceases to exist, with surface temperatures reaching 32°C above a thermal gradient to bottom water of 22–24°C. Salinity reaches 40‰ along the Iranian coast and off Arabia, where the outflow of dense bottom water occurs, especially in the deeper channels.

Surface drift occurs in response to the strong wind events that sweep the Gulf, especially northeast Shamal winds along the axis of the gulf that break down the density stratification to at least 60 m and temporarily destroy the mean cyclonic circulation.

## Regional Response of the Pelagic Ecosystem

These two basins are both, in their own ways, extreme habitats for the organisms of the pelagic ecosystem. Nevertheless, they remain quite productive and, as Sheppard and Dixon (1998) suggest, for this reason they are very useful laboratories for the study of environmental stress on the marine organism.

Salinity >49‰ is lethal to planktonic organisms, and because the concentrations just quoted are beginning to approach that level, it is not surprising that periods when there was apparently no planktonic ecosystem should be recorded in the sediments of the Gulf of Aqaba (Fenton *et al.*, 2000). Further, given the extreme water clarity at some seasons, it is not surprising that UVBR-induced DNA damage should occur in bacterio- and phytoplankton; in the upper 15 m of the water column, DNA damage to such cells accumulates during the daytime is not entirely repaired during the hours of darkness. This suggests that photomortality is a potential loss parameter among the picofraction of autotrophs (Boelen *et al.*, 2002).

The single significant source of new dissolved nutrients in the Red Sea is the surface water flowing in from the Gulf of Aden, though tidal mixing may transport regenerated nutrients into the photic zone. In the two northern gulfs (Suez and Eilat-Aqaba) local inputs of industrial effluents have induced local blooms. Over most of the Red Sea, the strong pycnocline is an effective and permanent barrier to the vertical mixing of nutrients from below. Consequently, the overall productivity of the mixed layer decreases progressively northward.

Offshore Red Sea water has great clarity and primary production is generally extremely low (about 100 mgC $m^{-2}$ $day^{-1}$), though locally, in coastal situations, rates may be higher. It is only in the southern region that there appear to be significant offshore algal blooms; the chlorophyll images from CZCS to MODIS all agree that the entire southern area of wide shallow shelf (that is, south of about Massawa) has relatively high surface chlorophyll during the Southwest Monsoon. Highest chlorophyll concentration occurs over the shelves on both coastlines, but during the boreal winter over the southern Red Sea surface chlorophyll is characteristically >1.0 mg $m^{-3}$, although over the adjacent shelves it is somewhat higher. In the summer, rather elevated chlorophyll extends farther north and shows evidence of enhancement around major eddy features. Generally, the coastal regions always show slightly higher chlorophyll compared with the central region down the axis of the Red Sea, and a curious feature has been described at the coast (Niemann *et al.*, 2004): in the Gulf of Aqaba, the intense cooling that is induced by offshore breeze in the winter cools surface water sufficiently that it sinks down submarine gullies at the shelf edge, carrying with it the surface phytoplankton to depths of >300 m. Such gravity current cascades may provide a previously overlooked pathway of nutrients to the deeper zooplankton. It should also be noted, however, that the extensive coastal coral reefs, especially on the southern shelves, support major consumers of phytoplankton cells, the herbivorous soft corals; surface water masses that have passed across such reefs are phytoplankton-deficient.

In the extreme north, in the Gulf of Aqaba, a strong spring bloom has been described with a weaker autumn bloom (Labiosa *et al.*, 2003). Here, a gradient of chlorophyll concentration across the Gulf is attributable to upwelling on the eastern side that appears to be a stronger source of nutrients than convective entrainment.

Though the Red Sea has an anomalous density profile, the profiles of algal biomass and nutrient distribution closely resemble the typical tropical situation with a significant DCM (Weikert, 1987). Picoplankton (0.2–2.0 $\mu$m) are the dominant size fraction of phytoplankton and contribute about 75% of both autotroph biomass and primary production (Gradinger *et al.*, 1992). At the other end of the algal size spectrum, prominent blooms of *Oscillatoria erythraeum* are frequent in the open parts of the Red Sea and are perhaps the reason for its name; however, the significance of these blooms for basin-scale carbon fixation and for the general pelagic ecosystem has not apparently been assessed.

The Red Sea is a net importer of zooplankton by the density-driven influx in the south, though many species of expatriated Indian Ocean species do not long survive the extreme conditions. It has been suggested that this nitrogen flux may partly balance the negative nitrate balance associated with the two-layered pattern of exchange at the entrance to the Red Sea. Obviously, zooplankton diversity is attenuated northward, and in the extreme northern Red Sea relatively few oceanic species survive.

The effects on vertical profiles of plankton of the strong subsurface oxygen minimum ($O_2 < 1.3$ ml liter$^{-1}$) in the Red Sea between 100 and 700 m have been investigated. Whereas epiplankton avoid the layer so that biomass rapidly declines from 50 to 100 m above the sharp pycnocline, the subsurface zooplankton biomass maximum lies within the $O_2$ minimum layer. Here, diel migrants (e.g., *Pleuromamma indica*) have their daytime residence depths, as this genus does elsewhere, along with apparently resting populations of *Rhincalanus nasutus* (Weikert, 1984).

The extremely deep mixed layer, which may exceed 500 m in the Red Sea during the winter mixing period, also has consequences for the vertical distribution of zooplankters (Farstey *et al.*, 2002). The epiplankton at this time is distributed homogenously throughout the layer, but not as a passive tracer: only the smaller organisms behave thus, while the larger organisms (e.g., adult *Pleuromamma* sp.) retain their diel migration pattern, although their nocturnal depth range is extended under such conditions.

Producer-consumer dynamics in the Red Sea are probably typical of extremely oligotrophic conditions. Protist consumption of small cells appears to exceed the consumption rate of mesozooplankton by about two orders of magnitude, taking heterotrophic bacteria and small algae at turnover rates of 0.7–1.0 and 0.7–1.3 d$^{-1}$, respectively. Curiously, removal of *Synechococcus* seems to suffer only very much lower levels of daily mortality (Sommer *et al.*, 2002). This observation may be related to the relatively weak top-down control of larger nano-plankton by herbivores that characterizes the northern Red Sea, compared with a strong bottom-up control of their population levels by nutrient concentrations. Other observation in the southern Red Sea suggest removal by protists of ~50% of the standing stock and ~100% of the primary production daily.

Satellite images suggest that the Arabian Gulf is relatively turbid throughout the year with maximum clarity in the winter quarter (January–March) and we may expect this to be due, in part, to suspended carbonate particulates. But it is not easy to be confident, for there is a lack of studies of this region compared with the Red Sea and much of the attention of researchers here has been on the consequences of coastal development, urbanization, and oil discharges. Comparatively, the studies of regional ecosystem dynamics are immature.

Nitrate concentration varies strongly seasonally and is undetectable at the end of summer, whereas surface chlorophyll values in the range of 0.75–1.4 mg chl m$^{-3}$ occur throughout the central part of the Gulf (Al-Saadi, 2001), with even higher local maxima along the Trucial Coast (Dorgham and Moftah, 1989). Note that during the summer complete mixing of the water column occurs over much of the Arabian Gulf, as noted earlier. These chlorophyll values are almost an order of magnitude higher than those in the adjacent Gulf of Oman at the same season. Once again, as in the Red Sea, it is reported that in coastal regions the chlorophyll biomass accumulates to significantly higher levels than in the central Gulf, and this observation is confirmed by the available satellite imagery. The southern bight, into which the Qatari peninsula protrudes, sustains very much higher concentrations of chlorophyll than elsewhere in many images.

Of the larger algae, *Trichodesmium* and *Anabaena* form a major fraction of total biomass, especially in areas where surface nitrate is undetectable, and it is reported that a gradient exists in the diversity of phytoplankton taxa. Low diversity and high biomass characterize the estuarized northern region of the Shatt-al-Arab, whereas in the southern regions off Kuwait, diversity is much higher and biomass lower. It is also reported that, over the long term, the diversity of dinoflagellates has increased—from 34 species observed in 1934 to 211 in 1990. In these polluted waters, the probability of toxic blooms is thought by regional biologists to be high and the valuable penaeid shrimp fisheries to be at risk.

Characteristic of the Arabian Gulf is the combination of high temperatures and strong evaporation that leads to the precipitation, as gypsum, of calcium sulfate, and this may have consequences for our interpretation of chlorophyll images, as noted earlier. Gypsum precipitation is, of course, the mechanism for the formation of the characteristic coastal form here, represented by the "sabkha" concretions that form on the inner shelf and gradually coalesce to enclose extensive coastal lagoons. The extensive shallow shelf regions are dominated by sandy deposits within which have been recognized local representatives of the globally distributed *Amphiura* and *Venus* isocommunities of benthic invertebrates. The shallow water of the Gulf makes coupling especially strong between pelagic production and benthic consumption and there is, in consequence, a diverse and abundant population of detrital-feeding penaeid shrimps throughout the Gulf; seven species occur, of which *Penaeus semisulcatus* is the dominant. Demersal fish are dominated by sandy-ground forms, of which the lutjanids, lethrinids, and serranids formed 50% of the pristine standing stock—exactly as occurs on other very low-latitude sandy shelf regions (Longhurst and Pauly, 1987). The diatom-filtering oil sardine of the coastal regions of the Arabian Sea appears not to enter these two basins in significant numbers.

### Synopsis

*Case 5—Large-amplitude response to monsoonlike reversal of trade winds*—Mixed-layer seasonality is dominated by boreal winter mixing to just below photic depth from 10–20 m April–October, to 35 m in February. Since $Z_{eu}$ is consistently between 30 and 40 m, the thermocline is illuminated in all months except perhaps briefly in January–February. Seasonal productivity resembles seasonal changes in ARAB, but the SW Monsoon P peak rate occurs later (October, 0.9–1.0 gC $m^{-2}$ $d^{-1}$) and is sustained longer (Fig. 10.3). Accumulation of chlorophyll occurs only during the onset of the SW Monsoon, but accumulated biomass is sustained longer than in ARAB (June–November). The relationship between chlorophyll biomass and P rate is consistent with an overall close match between consumption and production, and also with some increase in consumer biomass after the onset of the SW Monsoon.

## Northwest Arabian Sea Upwelling Province (ARAB)

### Extent of the Province

The ARAB province includes the coastal areas from central Kenya (12°S) to Pakistan, and also the northwest Arabian Sea. This arrangement permits the processes on either flank of the Southwest Monsoon Jet to be discussed as a single system, especially off the coasts of Somalia and Oman.

### Continental Shelf Topography and Tidal and Shelf-Edge Fronts

The African and Arabian coasts of this province have very narrow (< 5–10 km) shelves above steep continental slopes, relieved only by a small area of shallow banks southwest

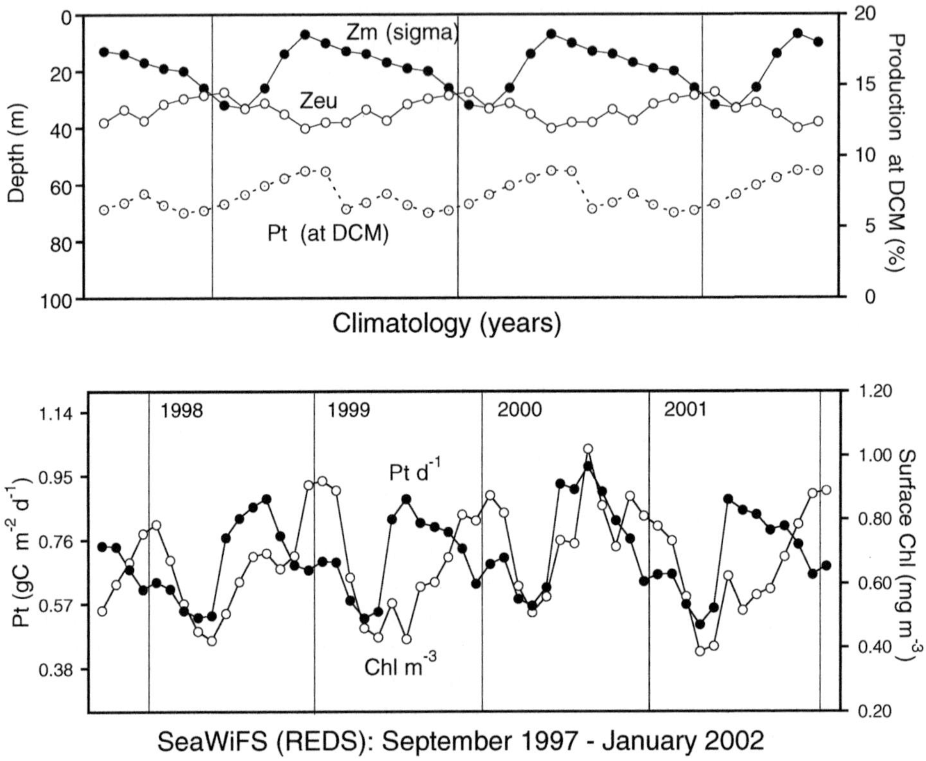


**Fig. 10.3** REDS: seasonal cycles of monthly surface chlorophyll and depth-integrated autotrophic production for the years 1997–2002 from SeaWiFS data together with characteristic seasonal cycles of mixed-layer depths from Levitus climatological data and photic depths computed from characteristic irradiance and the archive of chlorophyll profiles discussed in Chapter 1.

of Socotra Island, and some shallow bights on the coast of Oman, within which the shelf widens to about 75 km. The shelf widens significantly only at about 66°E, where it turns southeast into the coastal boundary province of western India (INDW). Across the mouth of the Gulf of Oman lies the Murray Ridge, which is sufficiently shallow (< 1000 m) as to modify circulation and the structure of the upper water column.

## Defining Characteristics of Regional Oceanography

Circulation and mixed-layer dynamics, which result in upwelling and algal blooms during boreal summer, are a complex response to reversing monsoon wind stress. This regime has already been discussed, but here we are concerned principally with the near-coastal consequences of seasonal changes in wind direction and strength. Entry into the extensive recent studies on this region could well begin with Banse (1984), Burkill *et al.* (1993a), Brock *et al.* (1992), or Schott (1983). Understanding the ecology of this province was a major objective of the JGOFS Arabian Sea expeditions.

The wind field of boreal summer over the Somali Current is a mirror image of winds over the eastern boundary currents of the Atlantic and Pacific. Here, the coast lies to the west of the wind maximum, which is the poleward, low-level Findlater Jet (see MONS). Nevertheless, the arrangement of wind-stress curl here is similar to that over eastern boundary currents—negative wind-stress curl lies seaward, and positive curl landward of the axis of maximal wind velocity (see Charts 46 and 48 in Hastenrath and Lamb, 1979). The surfacing of deep water along the Somali and Omani coasts during the Southwest Monsoon is a principal interest of this region for oceanographers.

Although the Somali Current has volume transport almost equivalent to the Gulf Stream, it changes direction by 180° twice annually under the influence of the seasonally reversing monsoon winds (Fig. 10.4). At the onset of the Northeast Monsoon of boreal winter, southward flow develops along the western margin of the Arabian Sea, especially from northern Somalia (10–11°N, near Cape Guardafui) where onshore flow feeds into the coastal current. This flow continues to 2–3°S, near the change of orientation of the Kenyan coastline, where northward flow in the East African Coastal Current (carrying low-salinity water of SEC origin) converges with it. Here, after merging, both flows, which

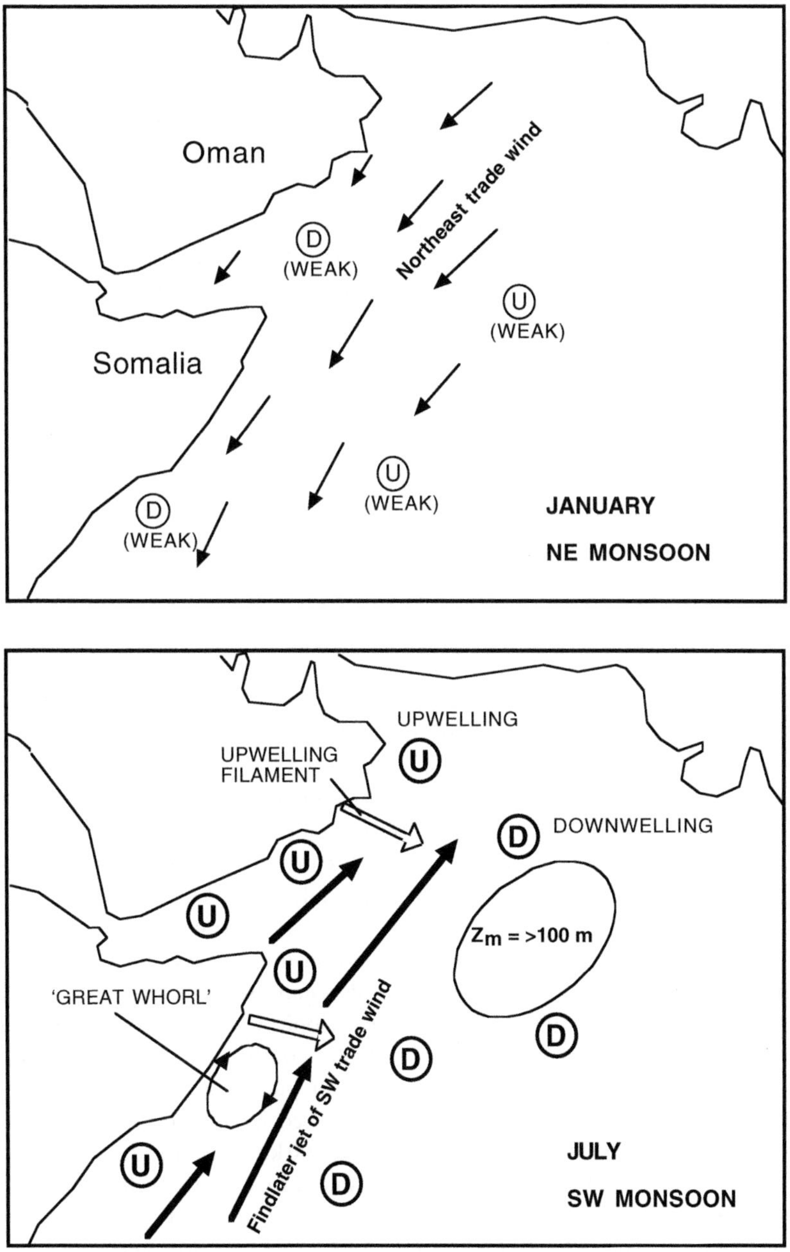


**Fig. 10.4** Idealized diagrams of the ecological consequences of the seasonally reversing monsoon winds over the Somali Current; redrawn from several sources.

reach to about 500–600 m (Schott, 1983), turn offshore at the origin of the SECC. North of Cape Guardafui, along the coast of Oman, flow at this season is weak and variable, although there is significant transport into the Gulf of Aden and the Red Sea (Halim, 1984). On the evidence of satellite chlorophyll images, there is significant upwelling along the eastern Omani coastline during winter, although this appears to be somewhat intermittent: I can find no record of this having been investigated. It is perhaps a simple case of offshore transport of surface water and its replacement from below.

With the onset of the Southwest Monsoon winds of boreal summer, response and spin-up of northeast flow in the Somali Current occurs within a few weeks. The current responds to progressive step increases in wind strength as the monsoon evolves in a series of surges, and upwelling is rapidly induced at several locations on the Somali and Omani coastlines by processes specific to each (Swallow, 1984). The first indication of the reversal of the monsoon is the onset in **March** of onshore winds off southern Somalia that cause the onshore flow to shift to about 5°N instead of at Cape Guardafui (the prominent tip of the Horn of Africa) and to diverge into northward and southward coastal currents. In **April**, southwest winds begin to blow along the coast south of the equator and within a few days a northward coastal current develops from there to at least 5°N, where offshore divergence occurs (Schott, 1983) and a wedge of cold upwelled water develops along the offshore flow around the north side of a large gyre. In **May**, the main onset of the Southwest Monsoon occurs quite suddenly: the response of the current is to turn offshore a little further to the south (3–4°N), creating a cold wedge at this latitude. In **June** the alongshore winds extend to northern Somalia with strong anticyclonic curl offshore of the Findlater Jet, and within about 14 days this situation induces an anticyclonic gyre, the "Great Whirl," south of Socotra and Cape Guardafui (Swallow and Fieux, 1982).

Finally, in **August**, this gyre intensifies, even as the offshore motion and cold wedge of surface water located at 3–4°N moves farther north and comes to lie on the southern side of the "Great Whirl." A second clockwise gyre is spun up under the influence of the Findlater Jet to the east of Socotra. The Great Whirl is associated with strong upwelling at the coast so that offshore transport around the northern side of the gyre forms a second cold wedge at the surface; the cool water is subsequently entrained into the Socotra gyre. The actual locations of the centers of these two gyres have considerable interannual variability that is apparently due not entirely to variability in the wind field, but also to an internal source in the chaotic nature of the dynamic circulation of the Arabian Sea (Wirth *et al.*, 2002).

Further northeast still, flow of the Somali Current (though weaker than off Somalia) continues along the coast around the head of the Arabian Sea to the Indian subcontinent (Swallow, 1984; Schott, 1983). Thus, the Somali Current itself, which has attracted so much attention for its rapid reversal, is actually only part of a much larger boundary current system along which coastal divergence occurs in several places.

The Findlater Jet of the Southwest Monsoon winds leaves the coast over Cape Guardafui and continues northeasterly toward Pakistan, well offshore from the Arabian coast (Brock and McLain, 1992). The area of cyclonic wind curl to the north of this jet creates significant Ekman suction off the Omani continental shelf, whereas the anticyclonic curl field to the southeast creates Ekman pumping and deepens the mixed layer (Lee, 2000). The total upwelling region induced by Ekman wind-stress curl off Arabia is very large: 1000 km along the coast and 400 km offshore through which pass $8 \times 10^6\,m^{-3}\,sec^{-1}$ at the 50-m level which is equivalent to a vertical velocity of $1\text{–}2\,m\,d^{-1}$ (Swallow, 1984). The innermost 30–40 km of upwelling occurs as a direct response to coastal wind effects, especially off Ras Madraka and at the Kuria Muria islands, and brings cooler water (24.2°C) to the surface than farther offshore. The most prominent upwelling feature is the offshore jet that develops at Ras al Hadd, the prominent cape at the eastern end of the Omani coastline, which is visible as a cold plume in the surface temperature field throughout the Southwest Monsoon season from May to October (Böhm, 1999; Manghnani *et al.*, 1998). The thermal contrast between this and the adjacent Gulf of Oman waters increases as the monsoon season develops. Off Somalia, upwelled water may be as cool as 13°C, indicating source depths of at least 200 m, although upwelling occurs only in a few relatively restricted locations; along the Omani coast, upwelling has a relatively shallower origin but is more widespread.

On the Omani coast, upwelling plumes are not located primarily in relation to features of coastal topography, as they are off Somalia, but rather in relation to the form of the offshore sea-level anomaly field, being advected preferentially through areas of low sea level. Generally, the offshore extent of upwelled water along this coastline increases during the Southwest Monsoon, so that the cold front may come to lie as much as 120 km offshore. It is the Murray Ridge, lying across the mouth of the Gulf of Oman, that induces the persistent offshore field of mesoscale eddies which intensifies during the Southwest Monsoon. Filaments of cool upwelled water from the Oman coast pass along the flanks of the ridge, and the eddy field is more prominent to the north, where the surface chlorophyll field suggests that a significant anticyclonic eddy persists in the central Gulf of Oman (Weaks, 1984).

The thermocline of the northwest Arabian Sea is shallower at the coast than offshore but shoals again above the Murray Ridge. In the mouth of the Gulf of Oman the thermocline is extremely strong and shallow (a 4°C change occurs over about a 5 m depth range below a 10 m deep surface mixed layer). Farther into the Gulf of Oman the mixed layer deepens progressively, responding to the outflow of highly saline surface water from the Arabian Gulf (Owens *et al.*, 1993). The regional mixed layer deepens in response to both monsoons so that depth maxima occur in August and February, respectively, in response to the southwest and northeast monsoons. The JGOFS studies of the 1990s found that the effects of convective mixing induced in the Arabian Sea by the onset of the Northeast Monsoon were more significant than had previously been thought (Marra and Barber, 2005).

Seasonal nutrient availability in this province is dominated, of course, by the upwelling cycle and by changes in the depth of the offshore mixed layer, as discussed recently by Woodward *et al.* (1999). Nitrate concentration of water upwelled close to the coast is of order 18 $\mu$mol liter$^{-1}$, considerably higher than in water upwelled offshore and beyond the shelf (< 12.5 $\mu$mol liter$^{-1}$); these values may be compared with average ambient concentrations at, say, 1500 km offshore of 0.035 $\mu$mol liter$^{-1}$. N/P ratios suggest nitrate limitation almost everywhere. A linear relationship ($r^2 = 0.89$) exists between surface temperature and nitrate content for surface temperatures of 14–27°C (Smith, 1984); average daily warming rates for surface water and the indicated loss rate for nitrate suggest algal primary production of about 98 mgC m$^{-3}$ d$^{-1}$, which is approximately the same as that observed in experimental data. Rates of 2.5 gC m$^{-2}$ d$^{-1}$ are usual in the upwelling area off the coast of Oman compared with < 0.3 gC m$^{-2}$ d$^{-1}$ in the oligotrophic central gyre of the Arabian Sea (Mantoura *et al.*, 1993; Owens *et al.*, 1993).

The monsoon regime also induces major seasonal shifts in surface irradiance levels; the cloud cover during the Southwest Monsoon of boreal summer causes a shoaling of the euphotic layer from about 90 m to about 50 m between April and August; it is not until February that irradiance from clearing skies again begins to penetrate significantly deeper.

## Regional Response of the Pelagic Ecosystem

The seasonal reversal of monsoon winds clearly has major biological consequences over the entire northwest Arabian Sea; during the Southwest Monsoon a large area of high surface chlorophyll becomes prominent in satellite images and is clearly associated with upwelling centers. Such images suggest that the region occupied by this algal bloom is equivalent to the total area covered by the upwelling in all four eastern boundary currents. Even before CZCS images were available, it was suggested that Southwest Monsoon bloom of the Arabian Sea might be equivalent in productivity to all eastern boundary currents combined (Smith, 1984). But, even if the Somali upwelling does have such potential, it is not realised. This deficit has been attributed to the relatively short residence time of upwelled water in the Somali system, before being dispersed out over the Arabian Sea; surface velocities in the Great Whirl are of order 250 cm $sec^{-1}$ and chlorophyll concentrations reduce rapidly from 2–3 $\mu g$ $liter^{-1}$ to 0.5–1.5 $\mu g$ $liter^{-1}$ (Hitchcock *et al.*, 2000).

The reversal of the monsoon winds dominates biological response to mixed-layer dynamics in the upwelling province, and this coupling is so tight that between-year variability in timing and strength of monsoon winds is directly reflected in the seasonal ecology. The weak monsoon of 1982 generated a coastal phytoplankton bloom that had maximum pigment values only about one-fourth of those in a strong monsoon. Surface chlorophyll fields, both from ship-based data (e.g., Halim, 1984) and from satellite images (e.g., Banse and McClain, 1986), match the upwelling pattern computed from wind stress. During the Southwest Monsoon, many of the anticipated mesoscale circulation features can usually be identified in the surface chlorophyll field: the persistent anticyclonic eddy off northern Somalia, the broad field of upwelling off Arabia, and the lower chlorophyll over the Murray Ridge and in the central Gulf of Oman (see Color plate 12).

An understanding of algal dynamics in this complex province requires integration of the vertical relations of density, nutrients, and light. Chlorophyll profiles respond predictably to deepening of the pycnocline and the level of ambient radiation remaining at the nutricline, which usually lies in the upper pycnocline. Meridional chlorophyll sections through the Arabian Sea in boreal summer show that to the right of the atmospheric jet the pycnocline and nutricline are deep (approx 80–100 m) and are associated with the chlorophyll maximum. To the left of the jet, maximum chlorophyll occurs in the mixed layer, and the nutricline approaches the surface.

The distribution of subsurface light in relation to mixed-layer depth can be used to model the seasonality of primary production in the Arabian Sea. During the relatively windless but brief intermonsoon periods, the Arabian Sea becomes oligotrophic with a deep chlorophyll maximum just above the nutricline at $<50$ m. The 1% isolume lies deeper than this because the mixed-layer water has low chlorophyll and is very transparent (Brock *et al.*, 1993). At the onset of both monsoon seasons, the 1% isolume rises into the mixed layer, as algal growth reduces transparency. In the oligotrophic seasons, production at the DCM exceeds production in the mixed layer.

The general pattern of the response of autotrophic organisms to monsoon forcing is logical and well-known: larger cells, mostly diatoms, are more important during periods of upwelling, while the pico and nano fractions dominate in the oligotrophic, intermonsoon periods and comprise 40–80% of the standing stock of POC (Burkill *et al.*, 1993b; Veldhuis *et al.*, 1997). Production and consumption are unbalanced during upwelling periods, so that strong flux of cells to the sea floor occurs, but during the intermonsoon periods, balance is struck; nevertheless, it is argued (e.g., Marra and Barber, 2005) that physiological rate parameters and productivity measurements suggest that column-integrated phytoplankton growth is strongly limited by neither irradiance nor nutrient supply. This argument suggests that variance in biomass depends importantly on mixing at all scales, including upwelling and the presence of mesoscale eddies.

During oligotrophic periods, even close to the coast, diel variation in mixed-layer depth is small and nitrate is almost totally depleted; analysis of critical depths suggests that active growth continues at the deep chlorophyll maximum, which lies ~50 m and hence below the effective depth of satellite remote sensing. At this period, *Prochlorococcus* is abundant in the pico fraction, although during upwelling periods this organism occurs only at offshore, oligotrophic situations (Campbell, 1998). The pico fraction forms >90% of phytoplankton biomass at oceanic stations, but only 35% in the diatom-dominated coastal zone during monsoon periods, at which times *Prochlorococcus* is effectively absent, and *Synechococcus* dominates the picoplankton (Brown, 1999).

Larger cells (1–2 μm), on the whole, lie deeper than smaller cells (0.7 μm) so that in the deep chlorophyll maxima, small diatoms have relatively greater importance, as they do in the near-surface chlorophyll maxima of upwelling cells close to the coast. In such situations, production rates of around 1–2 gC $m^{-2}$ $day^{-1}$ are commonly observed, of which about 50% is generated by the pico fraction, about 10% by the nano fraction, and about 40% by larger cells. Offshore, in oligotrophic situations this is reversed: from 60 to 75% of the daily production of 0.4 to 0.5 gC $m^{-2}$ $d^{-1}$ is generated by picoplankton, principally phycoerythrin-rich chlorococcoid cyanobacteria and prochlorophytes that occur at concentrations of ~110 cells $liter^{-1}$.

Dilution experiments (Landry *et al.*, 1998) showed that phytoplankton growth in the upwelling areas during the Southwest Monsoon is based on a division rate 0.9–2.2 $d^{-1}$ compared with < 0.5 $d^{-1}$ in the oligotrophic region offshore; enhanced growth rates inshore during the Northeast Monsoon were significantly lower, at 0.3–1.3 $d^{-1}$. Microzooplankton grazing rate was of order 45–50% of phytoplankton growth rate inshore during both monsoons, compared with parity in the oligotrophic condition. The relationship between phytoplankton growth rate and $NO_3$ concentration in these same experiments show a discontinuity at ~0.2 μM $NO_3$, below which growth rates fall very rapidly to very small numbers, and above which high growth rates are sustained. The highest growth rate observed (2.7 $d^{-1}$), at the inshore station off Ras al Hadd in August-September represents, as Landry *et al.* point out, almost four doublings per day. This is the highest rate observed anywhere by the dilution technique and approaches the maximum potentially observed in cultures: it is ascribable to "the unique combination of high temperature (>22°C), high nutrients (>19 μM $NO_3$) and the almost complete dominance of a single fast-growing diatom species, *Chaetoceros curvisetus*." The imbalance in such situations between herbivore consumption and production is such that the equivalent of one doubling per day is left unconsumed. Smith *et al.* (1998b) suggest that no more than 70% of daily primary production is likely to be consumed by mesozooplankton in the upwelling regions.

This analysis reflects the established view of the response of the autotrophic cells to upwelling in this province, that of a "classical" upwelling process in which nutrient limitation during the oligotrophic, nonupwelling season is relieved by the vertical entrainment of a sufficient supply of micro- and macronutrients to induce a change of state in the phytoplankton cells whose rate of increase, to some extent, then escapes from the balance previously struck with herbivore consumption. However, the JGOFS studies suggested a novel mechanism to Marra and Barber (2005), who participated in the work at sea and who suggest (to use their own words) that "Vertical mixing dilutes both phytoplankton and micrograzers alike, but since phytoplankton are not mixed to greater than their critical depth, they continue to grow but experience less grazing pressure"; this, Marra and Barber suggest, is a sufficient explanation of the observed changes in phytoplankton biomass associated with regional changes in mixed-layer depth. Differential growth rates of autotrophs and consumers had been evoked earlier by Goericke (2002) to support a suggestion that top-down effects were responsible for the seasonal changes in phytoplankton biomass in the Arabian Sea. This mechanism must form part of the complex processes that control the dynamics of production and consumption after a deepening of the mixed layer here as elsewhere, but I believe that the relative importance of consumption rates and nutrient limitation in determining biomass of autotrophic cells is not yet resolved.

Mesozooplankton (>200 μm) biomass is greatest in the coastal regions and during the monsoons, especially the summer Southwest Monsoon when biomass reaches 400 mg C $m^{-2}$ requiring a consumption of <150 mg C $m^{-2} d^{-1}$ for maintenance (Roman *et al.*, 2000). Such ingestion rates are relatively high compared to measured rates in eutrophic regimes elsewhere. This upwelling province is characterized by the abundant presence of the ontogenetic migrant *Calanoides carinatus*, which has very little diel migration activity but has a seasonal ecological cycle compatible with observations made on the same species off West Africa (see Chapter 9, GUIN) and elsewhere. *C. carinatus*, together with two other large copepods (*Eucalanus monachus* and *E. crassus*), dominates the epipelagic mesoplankton in water <25°C, during the southwest monsoon; at this season the abundance of these species is inversely related to water temperature (Smith, 1982; Smith *et al.*, 1998b). In the Somali upwelling centers at 4–6° and 10–12°N, *C. carinatus* has been observed at concentrations of 6000–9000 ind $m^{-2}$, but between these centers <1000 ind $m^{-2}$ have been found. During the intermonsoons and the Northeast Monsoon, these species are sparse at depths shoaler than 200 m.

Both females and progressively older copepodites of *Calanoides* occur together in upwelled water in a spatial sequence in ageing upwelled water, which indicates that reproduction occurs there. Generation time is probably of the order of 25 days, and this suggests that no more than a single generation will occur before the population is forced below the surface by rising temperature as each parcel of upwelled water is advected horizontally. Smith suggests that subsurface circulation is apt to retain resting, lipid-replete stage 5 copepodites (C5s) at >500 m in a trajectory that will position them to return to the surface in a succeeding upwelling cell with reproductive products ready to mature. *Calanoides* consumes small diatoms in the range 25–75 μm (*Rhizoselenia, Nitzschia, Eucampia*, etc.), and Smith (1984) computes the diatom consumption of recently upwelled *Calanoides* as 25–45% of the lower primary production nearshore but only 1–15% of the higher daily rate offshore. Such consumption represents an ingestion rate of about 50% of body carbon daily, necessary to support the observed growth rate: ~25% of body weight daily during the juvenile instars that leads to generation times of ~15 days from egg to adult. Egg production rate is also very high, and as many as 70 eggs are produced daily by each female. Lipid content is as high as 45% in stage 5 copepodites that are already several months into their period of diapause: this copepod is magnificently adapted to life in seasonal coastal upwelling blooms. Some species of *Eucalanus* may likewise be very well adapted to such situations, being able to withstand oxygen stress for long periods: off Somalia (and presumably also off Oman) these covary with *C. carinatus*. Maximal concentrations of *Eucalanus monachus* (90 ind $m^{-3}$) and *E. crassus* (75 ind $m^{-3}$) may indeed rival the numerical dominance of *C. carinatus* (170 ind $m^{-3}$). Nevertheless, *Calanoides* is more sensitive to increasing water temperature than *Eucalanus*, for which reason it descends earlier into diapause.

The existence of blooms during both the Southwest and the Northeast Monsoons suggest that *Calanoides* will have a more complex life history here than in simpler upwelling situations. It remains to be seen if this involves a complex pattern of subpopulations specializing in each regional, seasonal bloom or whether (more likely) the deep subthermocline water of the Arabian Sea everywhere carries a sufficient population of resting C5s to seed any upwelling parcel of water. The fact that blooms occur in both seasons perhaps also explains a paradox of the Arabian Sea mesozooplankton: that there is relatively very low variability in overall plankton biomass between seasons (Madhupratap *et al.*, 1996).

Although I have found no reference to modern investigations of the ecology of euphausiids in this province, they must be a major component of the pelagic ecosystem, as indeed is suggested by Brinton's IIOE maps of euphausiid distribution. The JGOFS investigations did make some preliminary assessment of the consumption of copepods by myctophid fish, for which they form the principal item of diet; it has been suggested that this province harbors the largest single stock of mesopelagic fish anywhere in the oceans. It is likely that the dominant species (*Diaphus arabicus* and *Benthosema pterotum*) are major predators of *Calanoides* and the other large copepods. Their biomass is relatively invariant through the year, but with some increase in summer months (<400 mg C $m^{-2}$), and they have an annual life cycle. These fish feed principally at night and so probably fill their guts once each 24 hours. Their computed daily consumption is quite close to the potential daily growth of zooplankton, assuming a growth rate of 24% $d^{-1}$ for the latter. They are themselves probably a major component of the diet of the yellowfin tuna that migrate to the coastal region of Oman, while being essentially absent in the oligotrophic regions of the Arabian Sea (Fontenau, 1997).

It is also curious that the accumulated literature appears to be quite silent on the place of shoaling clupeid fish in this upwelling province, a silence all the stranger because of the abundant studies and documentation of the role of sardine and anchovies in other upwelling provinces: California, Benguela and so on. Raja (1969) records the presence of the oil sardine (*Sardinella longiceps*) from central Somalia eastward into the Bay of Bengal. I can personally vouch for its presence at Aden, because I watched them feeding from the deck of a troop ship in 1946. This species is unusual in its habit of swimming with its mouth agape, filter-feeding on large diatoms, resembling an animated plankton net. It will be discussed in the account of the INDW province because it is very abundant there, but FAO fisheries statistics suggest only very small catches west of Pakistan. It may well be, as has been suggested, that a lack of markets and of fishing traditions along the Omani and Somali coasts conceals its true abundance. It would certainly be very surprising if, as would be deduced from evidence presently available, this upwelling region were to prove to be unique in not being exploited by sardines and anchovies *en masse*.

### Synopsis

*Case 5—Large-amplitude response to monsoonlike reversal of trade winds* (Fig. 10.5)—From an intermonsoon depth of 20 m, the mixed layer undergoes two weak excursions, to 50 m by the seasonal cooling of the surface water mass in the NE Monsoon (January–February) and to 35 m by spin-up of Arabian Sea gyre in the SW Monsoon (July–August). $Z_{eu}$ is related to chlorophyll seasonality (50 m April–June, 30 m September), so thermocline illumination occurs principally during the intermonsoon period of March–July. The seasonal cycle of P is dominated by the massive response to coastal upwelling during the SW Monsoon with a sustained rate increase from the intermonsoon minimum (May, 800 mgC $m^{-2}$ $d^{-1}$) to the brief annual maximum (September, 1500 mgC $m^{-2}$ $d^{-1}$). There is a second rate increase (to 1200 mgC $m^{-2}$ $d^{-1}$) during the NE Monsoon and when winter cooling occurs in the north of the province. Chlorophyll biomass tracks P rate, accumulating 1.0 mgC $m^{-3}$ (NE Monsoon) to 1.9 mgC $m^{-3}$ (SW Monsoon). This cycle is consistent with a close match between consumer biomass and production rate, and also with the ascent of seasonal vertical migrant herbivores at the onset of the SW Monsoon.

## Western India Coastal Province (INDW)

### Extent of the Province

The INDW province extends from the mouth of the Indus at 25°N to the Gulf of Manar at about 7°N, comprising the continental shelf of western India and southern Pakistan

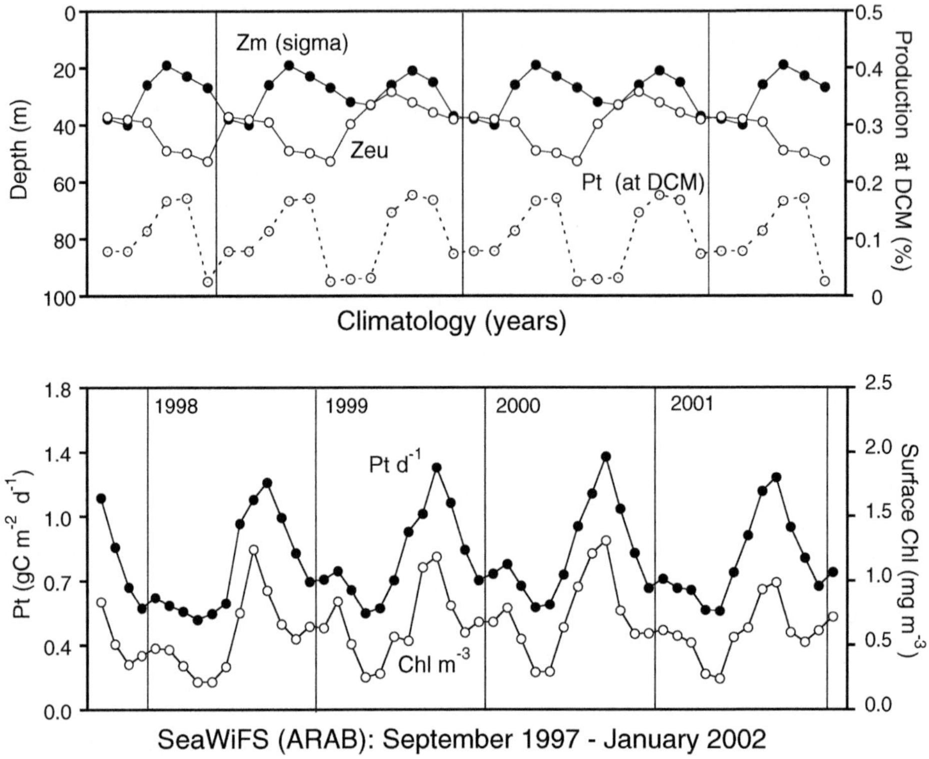


**Fig. 10.5** ARAB: seasonal cycles of monthly surface chlorophyll and depth-integrated autotrophic production for the years 1997–2002 from SeaWiFS data together with characteristic seasonal cycles of mixed-layer depths from Levitus climatological data and photic depths computed from characteristic irradiance and the archive of chlorophyll profiles discussed in Chapter 1.

(Colborn: parts of 4 and 11). It is also convenient to include the special conditions of the Laccadive Sea, lying between the continental shelf and the island chain that stands on the northern end of the Laccadive-Chagos Ridge.

## Continental Shelf Topography and Tidal and Shelf-Edge Fronts

The continental shelf along the western coast of the Indian subcontinent widens northward, reaching about 300 km at the Gulf of Cambay, after which it narrows again to about half that width. Where the coast turns west into the Gulf of Oman beyond Karachi it again narrows very significantly into the ARAB province. In the extreme south, the Palk Straits between Cape Comorin and Sri Lanka are uniquely shallow, having a sill depth of only about 10 m. In the Gulfs of Cambay and Kutch at 17–20°N to the south and north of the Saurashtra peninsula, the tidal range is very high—reaching 5 m at springs in the former. It has been suggested that circulation on the shelf around the Saurashtra peninsula is strongly influenced by tidal streams and decoupled from oceanic processes because the shelf width is greater than the local radius of deformation.

## Defining Characteristics of Regional Oceanography

Regional circulation responds both to local forcing and to remotely forced effects of the reversal of the monsoons. From November to April, light, dry northeasterly winds induce some cooling and mixing of shelf water in the northern part of the province off Pakistan.

During this season, the cyclonic circulation of the Arabian Sea causes downwelling of isopleths near the coast. Coastal currents respond to this local wind forcing but there is some question about the actual forcing of the southward flow that occurs during the Northeast Monsoon: thermohaline driving, river runoff, alongshore wind stress, and a continuation of the equatorward EICC have all been invoked.

The Southwest Monsoon is initiated in the south in May and June, then spreads northward and continues until October or even November. Accordingly, mean flow is reversed during this period and the longshore components of the coastal current suggest that it is mainly barotropic and wind-driven (Antony and Shenoi, 1991). It is also clear that what causes the reversal of flow at the end of the year is the reversal of the monsoon winds. The flow proper to each monsoon forms, of course, the outer edge of the eastern limb of the Arabian Sea gyre; during the Southwest Monsoon this is accompanied by upsloping of the thermocline toward the coast and intrusion of subpycnocline, hypoxic water onto the shelf which may persist during the entire Southwest Monsoon (Wooster *et al.*, 1967).

Local processes, such as wind-forced coastal upwelling, are complicated in this way by the distantly forced effects of the spin-up of the whole Arabian Sea gyre that occurs before the effects of local wind stress are evident. We should expect, therefore, that a simple model of coastal upwelling forced by Ekman divergence (e.g., Mathew, 1982) should not entirely fit observations and the relative significance of geostrophic and wind-driven upwelling on this coast should be examined holistically. Geostrophic upsloping of density contours starts in April, several months before local wind stress could force local upwelling, and occurs because offshore isopleths begin to slope upward toward the shore as the Southwest Monsoon spins up an anticyclonic gyre (Longhurst and Wooster, 1990; McCreary *et al.*, 1993). Neritic processes then erode the shoaled pycnocline. Later in the season, continuing equatorward wind stress transports the now-shallower surface layer offshore, leading to Ekman upwelling along the coastline. At such low latitudes as these, the required wind stress is relatively small compared with the situation at midlatitudes: off Cochin at 10°N, the same upwelling strength as that off California at 40°N is forced by a much weaker coastwise wind stress. Thus, upwelling is a consequence of both remotely forced baroclinic adjustment (Wyrtki, 1973a) and an equatorward component of wind stress along the coast during the Southwest Monsoon.

The Laccadive Islands provide a western barrier to circulation between the southern part of this area and the open Arabian Sea. Colborn (1975) characterizes this province as having a shallow permanent thermocline, which breaks the surface during the Southwest Monsoon—having a mixed-layer depth varying from 25 m (Southwest Monsoon) to 75 m (Northeast Monsoon).

## Regional Response of the Pelagic Ecosystem

This is yet another coast with a diatom bloom that responds to local seasonal upwelling, although satellite images show that the entire coastline is characterized by a strong and permanent nearshore band of relatively high chlorophyll concentration, even during periods when upwelling is not anticipated. This signal is generally attenuated toward the south and is widest at the Gulf of Cambay (20°S), where it occupies the entire gulf. It is likely that at least part of this signal is caused by DCOM, known to interfere with the interpretation of satellite images of the Amazon plume (q.v.). Nevertheless, the seasonal upwelling that is associated with the Southwest Monsoon does unequivocally force a bloom that is strongest on the southern segment of the shelf.

During the preupwelling period, when the pycnocline starts to slope upward toward the coast, equatorward flow along the coast has a low offshore component, and nutrient levels are low. The mixed layer is thin and subsurface water with low oxygen content has just started to appear on the shelf. When Ekman upwelling begins, surface nutrient levels increase, the mixed layer becomes even thinner, and very low oxygen concentrations (<0.5 ml $O_2$ liter$^{-1}$) characterize shelf waters below the thermocline.

The upwelling period of the Southwest Monsoon (August–October) is associated with algal blooms reaching biomass levels of as much as 4.0 mg chl m$^{-3}$, perhaps fueled not only to the advection of nutrients in the upwelled water, but also by remineralization of nutrients released from soft sediments. High chlorophyll concentrations, in the range 1–3 mg chl m$^{-3}$, extend out beyond the continental shelf across the full width of the Laccadive Sea, forming jets and spirals readily observed in satellite imagery (Lierheimer and Banse, 2002).

During these near-surface algal blooms, water clarity is very low (chlorophyll maximum at <5 m; water column chlorophyll <200 mg m$^{-2}$; Secchi disc, 2–3 m) and diatom cells take high concentrations of $<3 \times 10^5$ cells liter$^{-1}$ (Shah, 1973); these may be accompanied by blooms of *Trichodesmium*, especially in boreal spring. Although I have located no information on seasonal cycles of herbivore consumption in this region, Shah does suggest that the relatively high ratio of phaeopigments to chlorophyll during upwelling periods indicates strong coupling between herbivores and algal cells.

However, Madhupratap *et al.* (1994) have made another and very interesting suggestion concerning the support of planktonic herbivores. Perhaps, they suggest, the massive remobilization of green mud, heavy in POC/DOC from the inshore linear banks of this substance, the well-known 'chakara' of this coast, together with the extremely strong discharge of river water that occurs essentially the length of the western coast of India (see later discussion), carrying heavy loads of decaying organic material, and hence of bacteria, may directly fuel the zooplankton biomass via the microbial loop. In this model, it is a land-based microbial loop that provides larval sardines with copepod food, so that their recruitment success (see later discussion) would not be as closely correlated with upwelling and diatom production as is indicated by other evidence. This is a novel suggestion of great interest that is not contrary to the clear relationship that exists between diatom blooms and their consumption by herbivorous fish. The concept of a *land-based* coastal food chain that comprises DOC-bacteria-flagellates-microzooplankton-copepod-fish may perhaps be of wider application?

Nevertheless, Menon and George (1977) show that the mesozooplankton biomass over the Cochin shelf off southwest India (7–17°N) responds positively and rapidly to the upwelling cycle that was, itself, indicated by the depth to the 1.0 ml $O_2$ liter$^{-1}$ isoline at standard stations across the shelf. Over a period of five annual upwelling cycles (1971–1975), frequent observations over a grid of stations showed that there was very little lag in zooplankton population growth in response to the start of upwelling. Zooplankton biomass on this coast is routinely minimal (0.1 ml m$^{-3}$) from January to April and maximal (0.7–0.8 ml m$^{-3}$) from July to September.

The composition of mesozooplankton here is entirely typical of tropical seas but we should note how the distribution of species over the deeper shelf regions is constrained by the presence of the zone of very low oxygen tension. Even in the Laccadive Sea, the normal vertical distribution of the diel migrant copepods *Pleuromamma* spp. is strongly modified. Excluded from their normal daytime residence depths, these species are anomalously abundant throughout the 24 hours in the near-surface layers (Haq *et al.*, 1973). This occurs despite the fact that *P. indica* can tolerate oxygen tensions of <0.1 ml liter$^{-1}$, compared with the lower limit of around 0.5 ml liter$^{-1}$ tolerated by most other large oceanic copepods. Although most studies of mesoplankton in the Indian coastal seas have been taxonomic, enough is now known of regional ecology as to suggest that the onshelf penetration of oceanic species is relatively limited, perhaps because of the rather unusual conditions in the coastal zone. As on other tropical continental shelves, the relatively high abundance of cladocerans (*Evadne, Penilia*) would surprise a cold-water planktologist.

As along other upwelling coastlines, the seasonal diatom blooms support a pelagic food chain that includes shoaling, planktivorous fish—here, they are the oil sardine (*Sardinella longiceps*), the Indian mackerel (*Rastrelliger kanagurta*), and anchovies (*Stolephorus* spp.). But because of the reversing monsoon winds, and some special characteristics of terrestrial runoff, the pelagic ecosystem off western India is unique. Here, also, the direct link between diatom production and adult fish growth is perhaps better demonstrated than anywhere else so that—at short time-scales—the growth rates of both sardine and mackerel are strongly related to the availability of their planktonic food. Like some other coasts in warm seas where seasonal diatom blooms are exceptionally strong, this province supports a pelagic clupeid fish that depends almost entirely on diatoms for food, obtained by active filtration on gill-rakers: this is *Sardinella longiceps*, the oil sardine of the Indian Ocean. Species having comparable ecology occur in the western Atlantic (*Brevoortia*) and in the Gulf of Guinea (*Ethmalosa*). Note that here, as in the Gulf of Guinea, high concentrations of diatoms are associated with relatively higher numbers of large cells than occurs at lower concentrations, as was discussed by Banse *et al.* (1996). The larger cells presumably escape ingestion by small herbivores (e.g., small copepods) and it is hypothesized that population size of the larger copepods requires a few weeks to respond; oil sardine populations may respond more rapidly, being available to consume diatoms whenever upwelling occurs—and the larger the diatoms are, the better!

The oil sardine is extremely abundant in this province and supports a fishery that has been sustained since the early 1900s, and whose long-term variability was discussed in Chapter 8. The arrival of sardines at the coast coincides with the seasonal bloom of the diatom *Fragillaria* (= *Nitzschia*) *oceanica* and also of *Coscinodiscus, Pleurosigma,* and *Biddulphia.* The first indications of a bloom, at the start of the monsoon season, coincides with the arrival of prespawning adults, and the peak of the bloom, in September or October, coincides with the main fishery for juvenile sardines, these being the fish-of-the-year. Dinoflagellates and copepods, together with soft organic-rich material resuspended by tidal streams from the coastal banks of almost liquid mud, appear to sustain the sardines from October to January. The abundance of this rapidly growing fish is therefore closely linked to the productivity of the ecosystem; having long records of sardine abundance, we can hindcast the relative performance of the plankton over the same period. This was discussed in Chapter 8, in which statistical series of physical time series representing an upwelling index were compared with sardine landings.

The long-term data on landings of mackerel (*Rastrelliger*) and anchovies (*Stylephorus*) suggest that the stock biomass of these species is not so intimately dependent on upwelling—and hence on diatom blooms—as is that of the oil sardine. I tentatively venture to suggest that this may be some support for the significance of a land-based microbial web in fueling the mesozooplankton of this coastline.

## Regional Benthic and Demersal Ecology

The inner shelf off the alluvial coast of Kerala (8–12°S) is a mobile mud regime, the inshore mud banks forming during the boreal winter and remobilizing during the summer, when longshore currents and wind mixing are maximal. The ecology of the benthic fauna here can be presumed to resemble that in the deposition-resuspension mud regime off the mouth of the Amazon (see GUIA).

The continental shelf of western India is an example of a region that currently exhibits extreme hypoxia in the bottom waters and eutrophication in surface waters due, in large part, to increasing runoff of nutrients, especially nitrate, from terrestrial sources (Naqui *et al.*, 2000). A zone of anoxia develops in late summer and autumn: here, adjacent to the shelf, the highest accumulations of $H_2S$ and $N_2O$ yet recorded have been observed in recent years. The water upwelled onto the shelf (see earlier discussion) is already hypoxic and is capped by a shallow layer of low-salinity water formed by admixture with river effluents. Strong stratification ensues, imposing very reduced ventilation of upwelled water that is in contact with shelf sediments. This rapidly becomes anoxic, by oxidation of phytoplankton and other organic material sinking from the strong seasonal bloom (again, see the earlier discussion). This shallow anoxic layer was determined by Naqui *et al.* to be distinct from the deeper, perennial hypoxic zone of the Arabian Sea with which it joins when the latter is entrained onto the shelf as described previously. The formation of excessive $N_2O$ is perhaps the consequence here, as in lakes, of nitrification occurring in the near absence of oxygen.

The consequence of these various processes is that the benthic fauna and demersal fish populations along the entire continental shelf of western India and Pakistan may be essentially absent deeper than about 50 m. I have been able to locate several benthic surveys of this shelf, and each reports a preponderance of polychaetes—these comprised almost 60% of macrobenthic biomass on the inner shelf off Goa (Harkantra and Parukelar, 1981). Off Bombay, another survey showed that 61% of the meiofauna was composed of polychaetes, and that the meiofaunal biomass was dominated (>90%) by foraminiferans. Of the remainder of the meiofauna, almost 20% were mollusks, amongst which bivalves of the infauna were dominant: *Mactra, Arca, Tellina,* and so on. Epifauna comprised natant crustacea (e.g., *Metapenaeus, Mysis*) and echinoderms (e.g., *Ophiothrix*). These characteristics, and genera, are very similar to the *Amphioplus* isocommunity of very soft, muddy deposits in the Gulf of Guinea (Longhurst, 1958).

The remobilization of the predominantly soft deposits of the shallower parts of the shelf not only interrupts the continuity of the inshore chakara but also requires that the benthic fauna become re-established annually (Seshappa, 1953). Of course, infaunal polychaetes dominate the recruiting organisms.

All along the west coast of India, the demersal fish fauna is typical of tropical soft deposits and resembles, even if it is more diverse, the fauna of the muddy parts of the Gulf of Guinea. Sciaenids form 40–60% of the trawlable biomass and are dominated by species of *Pseudosciaena* and *Otolithoides*; these occur together with polynemids, ariids, and rays. This fauna, as in the eastern Atlantic, extends down to about 60–75 m across the shelf. Where deposits are sandier, toward the southern end of the coastline and into the Gulf of Oman, sciaenids are scarce and various breams and groupers take their place.

## Synopsis

*Case 5—Large-amplitude response to monsoonlike reversal of trade winds*—Mixed-layer seasonality is dominated by boreal winter mixing to 40–45 m in January–February; SW Monsoon upwelling slightly deepens the summer mixed layer. $Z_{eu}$ shoals rapidly to 30 m from 50 m at the onset of the SW Monsoon, but the thermocline is permanently illuminated except briefly in boreal winter. The seasonal maximum rate of P occurs in September (1.2–1.4 gC $m^{-2}$ $d^{-1}$) with a rapid decline to a postmonsoon minimum in December (Fig. 10.6). Chlorophyll accumulation occurs only during the onset of the SW Monsoon, declining from an August peak biomass of around 1.5 mg $m^{-3}$ to an intermonsoon value of about 0.25 mgC $m^{-3}$, after buildup of herbivore population causes a change of sign. This cycle is consistent with a close match between consumption and production rates, and perhaps some increase in consumer biomass after the onset of the SW Monsoon.

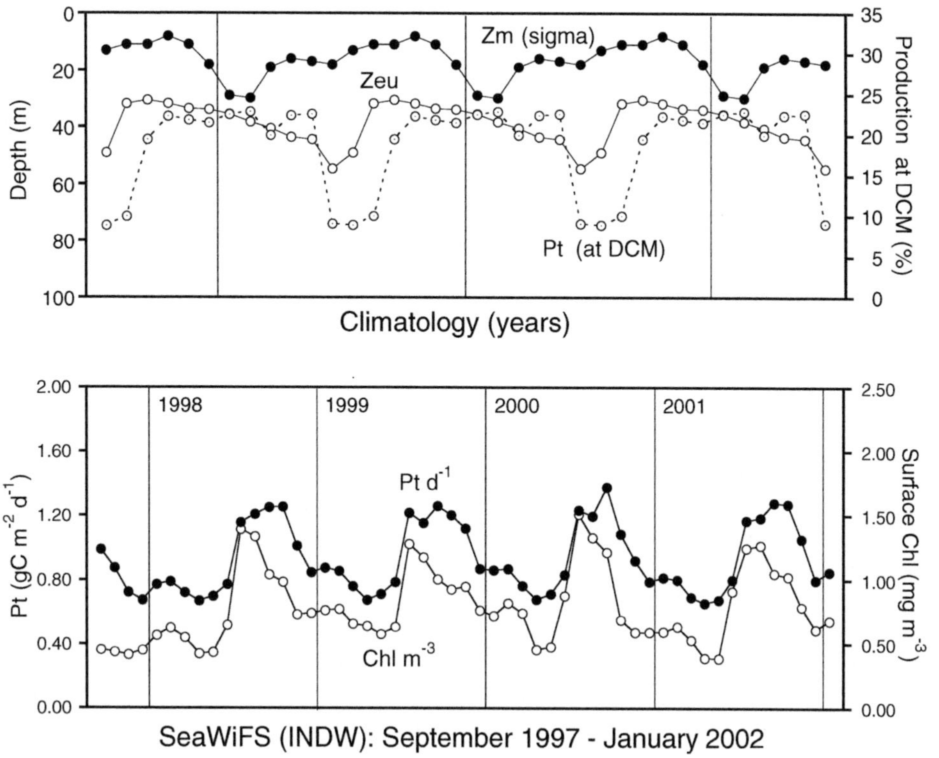


**Fig. 10.6** INDW: seasonal cycles of monthly surface chlorophyll and depth-integrated autotrophic production for the years 1997–2002 from SeaWiFS data together with characteristic seasonal cycles of mixed layer depths from Levitus climatological data and photic depths computed from characteristic irradiance and the archive of chlorophyll profiles discussed in Chapter 1.

## Eastern India Coastal Province (INDE)

### Extent of the Province

The Eastern India Coastal Province (INDE) comprises the coastal regions over and adjacent to the continental shelf of the Bay of Bengal from the east coast of Sri Lanka in the west to the Andaman Islands in the east (comprising Colborn 15 and 19).

### Continental Shelf Topography and Tidal and Shelf-Edge Fronts

Only off the mouths of the Brahmaputra-Ganges delta in the northwest and off the Irrawaddy in the east is there a significant width of shelf in this province. Both these areas have low-lying coastlines with mangroves, and the 200-m contour at the shelf break lies as much as 250 km offshore. A major canyon cuts in to the center of the Ganges delta and effectively bisects the shelf. In both these regions, we may expect that tidally forced fronts will occur because the shelf off these river mouths is subject to relatively high $M_2$ tidal constituents, of order 80–100 cm. The shelf of the remainder of the east coast of India, from the southern tip of Sri Lanka (Cape Comorin) up to about 19°N is much narrower than at the head of the Bay of Bengal. Here, because the shelf width is smaller than the radius of deformation, we should expect that coastal processes will be dominated by large-scale ocean dynamics. Another, but much less extensive, narrow shelf region occurs between the Brahmaputra-Ganges and Irrawaddy deltas.

## Defining Characteristics of Regional Oceanography

Winter cooling occurs in the northern part of the Bay of Bengal, and because of its semienclosed nature, coastal regions are strongly influenced both by the seasonal circulation of the whole bight (see MONS), and by regional freshwater inputs. The Brahmaputra and Ganges release freshwater at a rate of about $10^{12}\,m^3\,y^{-1}$, making this the fourth largest freshwater input, after the Amazon, Congo, and Orinoco. Even so, this is only a small fraction of the total freshwater inputs, that include heavy rainfall at the sea surface. The surface salinity of the entire northern Bay of Bengal strongly reflects the input of freshwater.

This entire region experiences wind-forcing seasonality similar to that of the ARAB and INDE provinces; perhaps the best modern guide to the coastal circulation of this region is that of Shetye and Guveia (1998), on which I have relied heavily in what follows.

Coastal flows follow the gyral circulation of the entire Bay of Bengal, although the mechanisms involved remain obscure. The best-observed flow is that of the seasonally reversing East India Coastal Current (EICC) that lies above the narrow shelf regions from Sri Lanka to Bangladesh. Forced by the Northeast Monsoon, a western boundary current is initiated on the northern part of the coast in October so that southward flow occurs along the east coast of India, observable as a meandering warm-water body in satellite imagery; warm-core eddies are detached at the edge of the current, resembling those of the Gulf Stream (Legeckis, 1987). As the Northeast Monsoon begins to fail in December, the current remains strongest on the southern part of the coastline, and at this season downwelling occurs the length of the coast and an inshore plume of low-salinity water hugs the coastline, originating on the northern shelf.

In February, during the winter intermonsoon period, the EICC starts to flow northward and by March–April, the period of strongest flow, the EICC is established the entire length of the coastline of Sri Lanka and eastern India. This flow decays by the end of June and is associated with an anticyclonic subtropical gyre that occupies the Bay of Bengal and is unique in its seasonality. Along almost the coast south of 20°N, the contours for density, temperature, and salinity slope upward toward the coast, and in a 40-km-wide band the isolines appropriate to 70 m offshore break the surface. Thus, the southern regime is upwelling, while the northern is downwelling. Modeling studies suggest that upwelling-favorable winds during the Southwest Monsoon may force Ekman divergence and upwelling south of Visakhatapam (17.5°N), but farther north, freshwater from the Ganges mouths (especially of the Hoogli) is sufficiently strong as to inhibit upwelling (Johns *et al.*, 1992). However, observations suggest that deep water may be brought to the surface south of Visakhatapam not only by upwelling but also perhaps by vertical mixing in March and April and again in June–August off Waltair (Banse, 1990b; Shetye, 1993). Whatever the mechanism, the inshore flank of this warm current is marked by cooler, high-salinity water brought to the surface by the eddying dynamics of the flow. Eddies of scale 150-km diameter are observed in satellite SST images to become detached from the flow of this western boundary current.

During this same season, on the eastern side of the Bay of Bengal the water column may be rendered isothermal over the continental shelf by tidal mixing and thermal convection (Sesama, 1989). Unfortunately, coastal dynamics have been much less well observed in the eastern part of the province, and about all that can be said is that perpetual northward flow occurs through the Straits of Macassar from the Indo-Pacific regions and that a small reversing gyre occupies the shelf to the west of Burma.

## Regional Response of the Pelagic Ecosystem

I have located only one recent regional-scale study using current techniques (JGOFS protocols) of the ecology of the Bay of Bengal shelves; this is a single coastwise line of stations worked recently during the Southwest Monsoon (Madhupratap *et al.*, 2003). I can only agree with the authors of this study when they lament that reliable information on ecological characteristics of this region is, as they write, "elusive." If we did not have recourse to satellite images, it would be very difficult to say anything useful about the ecology of this province.

Some early coastal shelf studies, such as that of Subba Rao (1976) off Waltair, suggested that there was a bimodal phytoplankton bloom, increased growth occurring during both monsoons, although all concur that the maximal accumulation of biomass occurs during the Southwest Monsoon. There is also general concurrence concerning the effects of estuarization of the northern part of the eastern India coast and of upwelling along the central part of the coast. What I have not found is any recognition of the fact that the entire shelf region of the Gulf of Manar and the Palk Straits supports relatively high chlorophyll biomass throughout the year, filaments of which are carried offshore and to the northeast. That this occurs is evident from satellite images, which show it to be especially strong during the Southwest Monsoon. The seasonal images also make it clear that the two shelf regions facing the Brahmaputra-Ganges and the Irrawaddy deltas each support apparently high chlorophyll biomass at all seasons; we must, of course, assume that at least a part of this signal is due to presently unquantifiable amounts of CDOM, as off the Amazon. At all seasons, on the other hand, the eastern coast of Burma between these regions lacks any significant bloom.

Although the east coast of India is aligned parallel with that of Somalia and Arabia and shares a generally similar seasonal wind regime, the ecological consequences of upwelling during the Southwest Monsoon off India are in no way comparable in the two regions. This is evident from a scan of the relevant seasonal and monthly chlorophyll images that indicate that this is a far weaker bloom than in the western Arabian Sea, although they confirm that boreal summer upwelling does result in enhanced near-surface chlorophyll along the shelf from Madras to about 20°N. Even during this season, the water column is sufficiently stabilized that a DCM occurs in all sections of this coast, at about 60–80 m at the edge of the western boundary current and sloping up (and weakening) toward the coast. This DCM has maximum chlorophyll values of about 0.75–1.0 mg chl $m^{-3}$ and is aligned with the nitracline and the ammonium maximum, suggesting that nutrient depletion occurs in the mixed layer above—as, indeed, we would expect (Sarma and Aswanikumar, 1991). The single modern coastwise transect revealed a mixed layer that decreased in depth progressively toward the north (30 m at 12°N, 5 m at 19°N); the chlorophyll maximum was everywhere deeper than this (40–80 m) with increasing biomass toward the north (0.15–0.4 mg $m^{-3}$). Productivity maxima (< 15 mgC $m^{-3}$ $d^{-1}$) were either coincident with the DCM or close under the surface. Phytoplankton of cell size >5 μm was dominated by diatoms (*Biddulphia, Chaetoceros, Thallasiothrix, Thallasionema*, and others) with some dinoflagellates (*Peridinium, Ceratium*), whereas autotrophic picoplankton counts were much lower than in the central Bay of Bengal except only at 19°N, where a maximum of $9.4 \times 10^{-7}$ $liter^{-1}$ was reached in an essentially unstratified water column below the freshwater plume. Phytoplankton diversity here is higher than in comparable situations in the Arabian Sea.

All this, as Madhupratap *et al.* (2003) point out, is consistent with the observed physical regime—relatively low ambient macronutrients and a very shallow (~5 to 10 m) euphotic zone on the shelf because of reduced irradiance and heavy sediment discharges during the Southwest Monsoon period. Other recent studies (Gomez *et al.*, 2000) have also found that below the heavy cloud cover of the Southwest Monsoon, even where high chlorophyll biomass has built up (~90 mg $m^{-3}$), production rates are very low (0.3 mgC $m^{-2}$ $d^{-1}$) and do not significantly increase until near the end of the monsoon period. Earlier in the season and before the heavy cloud cover is established, as well as during the Northeast Monsoon of January–February, primary production rates are significantly higher, especially where blooms of *Thallasiosira subtilis* occur.

The mesozooplankton of the shelf from 13°N to 20°N (Madras-Calcutta) was surveyed during the Southwest Monsoon with IIOE protocols in 1978 by Nair *et al.* (1981), who found significantly higher dry weight biomass (2.4 g $100\,m^{-3}$) south of 17°N than to the north of that latitude (0.77 g $100\,m^{-3}$). As was found also by the more recent investigations, biomass over the shelf was significantly higher than in the oceanic regions. Indeed, comment Nair *et al.*, the biomass in the first of these two zones was comparable with biomass taken off Arabia during other IIOE investigations. Later in the season, other contemporary studies showed that the high-biomass region extends to lower latitudes than 13°N, and it has been suggested that this is associated with upwelling on the coast between Madras and the Palk Strait. Other studies have suggested a significant decoupling between zooplankton biomass, the latter peaking significantly later than the local maximal algal biomass.

The IIOE samples (0.3 mm mesh) were dominated by copepods (~80% numerically) of 40 species in 31 genera, and the JGOFS transect sampled a mesozooplankton dominated, as expected, by copepods (*Para-* and *Clausoclanus, Eucalanus*) and cyclopoids. Some of the JGOFS samples were dominated by massive blooms of the mucous/filter feeding *Pyrosoma* that is otherwise more relatively abundant in the oceanic region of the Bay of Bengal. The numbers observed suggested that this species was perhaps capable of significantly lowering the ambient chlorophyll biomass. Other observations on the northern shelves off the deltaic region suggest that mesozooplankton biomass is maximal during, and just after, the Northeast Monsoon with relatively very low biomass during the Southeast Monsoon.

## Regional Benthic and Demersal Ecology

There is some information that suggests a general coupling between benthic biomass on the eastern Indian shelves and the productivity of the water column above. Parekular *et al.* (1982) suggest that macrobenthos biomass maxima (averaging >50 g $m^{-2}$) occur in three regions and that here the biomass of this fraction exceeds that of meiobenthos, otherwise relatively more important on the slope. These regions are adjacent to Palk Strait, in the central upwelling region (15–17°N) and on the shelf to the west of the Brahmaputra-Ganges delta. On loose, sandy soil, maximum biomass reached 150 g $m^{-2}$ but is much lower in the highly organic estuarine muds that are transported so generously onto the shelf in this province. Yet once again here, most of the ubiquitous genera of the infauna would be familiar to a temperate-zone benthic ecologist: *Eunice, Nephthys, Glycera, Neries, Cardium, Mactra, Ampelisca, Ophiothrix*—and so on (Harkantra *et al.*, 1982). It is among the vagile epifauna (e.g., *Penaeus*) and the demersal fish (e.g., Sciaenidae, Polynemidae) that specifically tropical genera dominate.

The demersal fish faunas of the shelves of this province are typical of those on other similar shelves. On muddy, estuarized shelves, especially inshore, we should expect sciaenids, polynemids, ariids, and various pleuronectiformes; on sandy deposits and usually deeper, sparids, lutjanids, serranids, and lethrinids are characteristic. This arrangement has been discussed in some detail for all tropical shelves by myself and Pauly (1987); the division between shallow and deep fish faunas occurs unusually deep at the head of the Bay of Bengal – at around 100 m. This is probably the consequence of the widespread distribution of soft deposits off the very large delta of the Brahmaputra-Ganges river system (Garces et al., 2006) A significant study of the effects of several decades of trawl-fishing on these faunas suggests serious modification to the pristine food web on the eastern coast of India (Vivekanandan *et al.*, 2005). On the southeastern sector, the dominant groups (both demersal and pelagic combined) are sardines, Indian mackerel, carangids, sciaenids, ribbonfish, rays, and penaeid shrimps. On the southern section of this coast, there is a clear negative relationship between commercial landings during the entire second half of the 20th century and the mean trophic level of the fish landed. This implies that fishing has induced a significant change in the ecosystem structure of this shelf over a period of 50 years.

## Synopsis

*Case 4—Small-amplitude response to trade-wind seasonality, but modified by major river discharges.* The climatology derived from satellite and other data confirms that INDE has a wide separation of pycnocline and thermocline induced by very shoal brackish surface layer. Pycnocline depth, induced by low-salinity river discharges, is very shallow (Fig. 10.7), whereas the thermocline has seasonal excursion from 20–30 m in April–June to 50 m in January–February and is illuminated most strongly in boreal summer, March–July. The P rate increase (500–800 mgC $m^{-2}$ $d^{-1}$) from June to November probably cannot be separated from effects of the SW Monsoon discharge from the Ganges and Irrawaddy. Chlorophyll biomass appears to track the PP rate relatively closely, consistent with biological effects, but coastal chlorophyll values are probably unreliable.

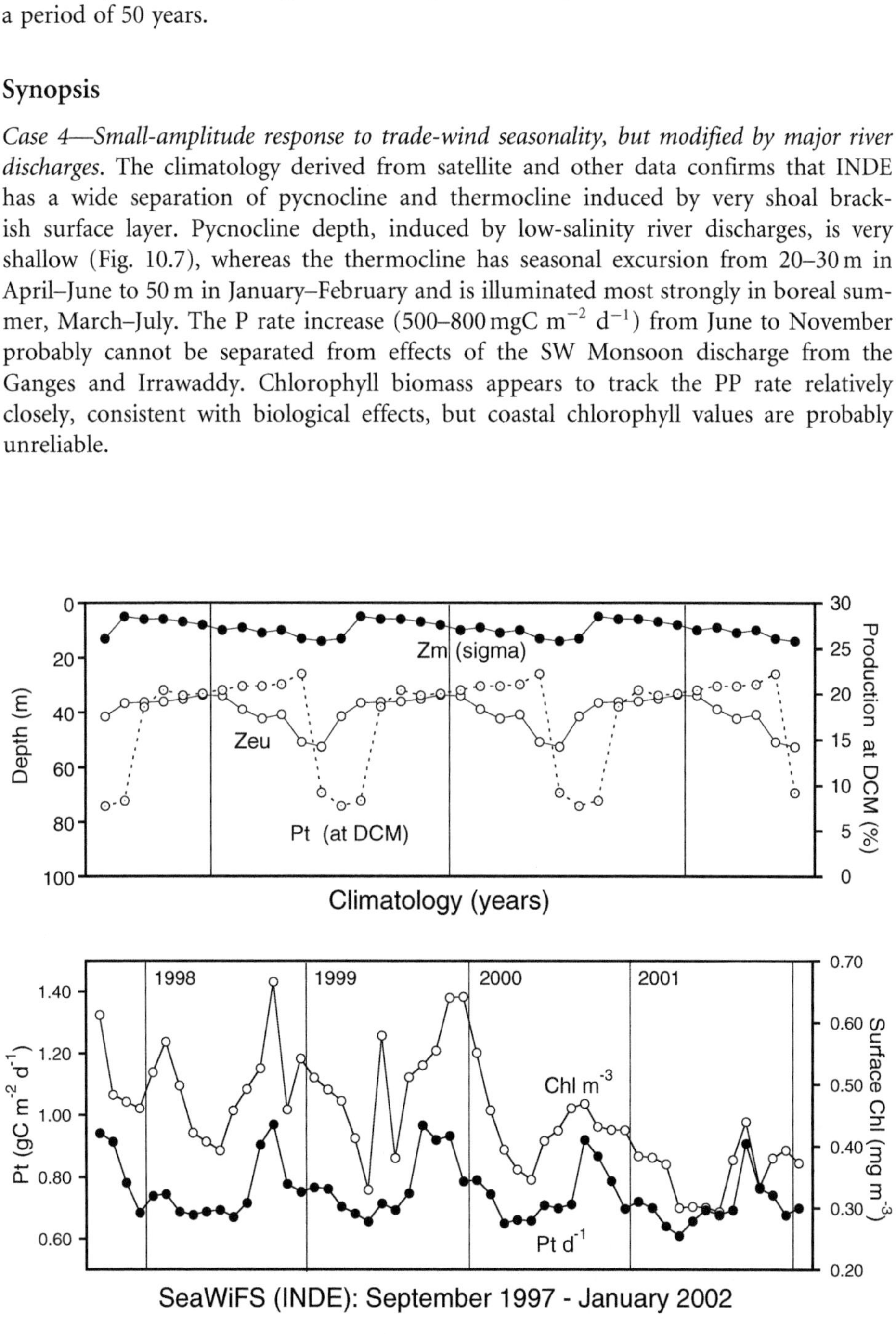


**Fig. 10.7** INDE: seasonal cycles of monthly surface chlorophyll and depth-integrated autotrophic production for the years 1997–2002 from SeaWiFS data together with characteristic seasonal cycles of mixed-layer depths from Levitus climatological data and photic depths computed from characteristic irradiance and the archive of chlorophyll profiles discussed in Chapter 1.

# Eastern Africa Coastal Province (EAFR)

## Extent of the Province

EAFR comprises the coastal boundary of the Indian Ocean from Zanzibar (5°S) to the Cape of Good Hope, including the Mozambique Channel and the east coast of Madagascar. The seaward, eastern limit of the province is the edge of significant flow in the western boundary currents (Colborn: 33 and parts of 27 and 30–32). For convenience, I include the Agulhas Retroflection, south of Africa. Were this province to be subdivided for other purposes into more natural regions, the following would be a suitable primary classification: (i) the equatorial East African coastline, (ii) the subtropical Madagascan region, (iii) the temperate Agulhas region, and (iv) the somewhat offshore Agulhas Retroflection. Sufficient differences could be found in their ecologies to support such an arrangement.

## Continental Shelf Topography and Tidal and Shelf-Edge Fronts

The continental shelves along most of the eastern coast of Africa are rather narrow and the slope is steep. Where there are fringing reefs, as in the bight between Tanga and Dar-es-Salaam (4–7°S) and the adjacent islands of Pemba and Zanzibar, and also along parts of the NE and SE coasts of Madagascar, the coral fronts stand above deep water and rather close to the shoreline.

However, in the wide bight (15–24°S) along the coast of Mozambique, the shelf is wider so that at Beira the break of slope is almost 150 km offshore. On the east coast of Madagascar, it lies about 100 km offshore. But the greatest area of shelf is the flat, triangular Agulhas Bank which lies between the Cape Peninsula (19°E) and Port Elizabeth (26°E) and reaches 180-km width south of Cape Agulhas. Several major rivers drain onto the shelf of this province, including the Zambezi just north of Beira at 18°S, the fourth river of Africa, and Kipling's "great, green, greasy Limpopo" at the southern end of the Mozambique Channel near Lorenço Marques at 25°S.

The tidal regime is dominated by the fact that the amphidromal point for the $M_2$ tide lies south of Cape Town; tidal amplitudes and the velocity of tidal streams increase, therefore, northward along the coastline. At Beira, the tidal amplitude may attain 6 m, and here the small-scale topography of the sediments suggests that tidal currents are more important than currents forced by processes over the deep ocean. As elsewhere, coastal processes determine the sedimentary regime in each locality. Major river discharges of organic material lead to a substrate with a high silt content, the presence of reef corals leads to white, highly calcareous sediments and, otherwise, sands together with shell fragments will dominate.

## Defining Characteristics of Regional Oceanography

Processes over continental shelves are everywhere directly influenced by local patterns of weather and wind stress at the sea surface. In this province, these factors respond to the seasonal, latitudinal progression of the atmospheric intertropical convergence zone (ITCZ) of generally easterly flow that lies across the northern Madagascar Channel in January, whereas during austral winter the ITCZ shifts into the northern hemisphere. Thus, during austral summer the Northeast Monsoon winds of the Arabian Sea penetrate down to at least 15°S over Mozambique and northern Madagascar. South of this latitude during austral summer, and over the entire province in winter, winds coming off the southern Indian Ocean predominate, because of a persistent high-pressure cell that lies southeast of Madagascar. In austral winter/boreal summer, this wind is continuous with the Southwest Monsoon of the Arabian Sea. Rainfall is associated with this situation and is maximal in the central and northern regions of the province from October to March. Beyond 30°S, the climate and wind-stress pattern is dominated by the westerlies of the southern oceans, and the passage of cold fronts. Compared to many coastal regions, seasonal changes in SST are here relatively small, because of the presence close to the coast from Madagascar to the Cape peninsula of subtropical surface water in the Agulhas Current.

It will be convenient to discuss the coastal circulation in relation to the four subregions of this province discussed earlier. The recent review of the coastal oceanography of southeast Africa of Schumann (1998) has been helpful, but excludes the first and last of these subregions.

***East African Current*** This is the partially reversing flow along the coast from northern Kenya to the Madagascar Channel fed by the South Equatorial Current, which meets the African continent at about 10°S and then diverges north and south throughout the year. During the boreal summer, the northward flow is continuous with, and enters, the deep flow of the Somali Current, as this is spun up by the strong Southwest Monsoon winds (see Arabian Sea Upwelling Region). In boreal winter, the northward coastal flow of water from the SEC encounters the southward flow of the reversed, slower Somali Current; the combined flow diverges from the coast at 2–3°S off Malindi and enters the South Equatorial Countercurrent to pass eastward across the ocean. Some upwelling may occur at the coast during this process.

***Mozambique Channel and Madagascar*** The flow in the Mozambique Channel is complex because southward transport along the African coast along the shelf edge is influenced by a persistent field of mesoscale gyres that lies further east (Saetre and Jorge da Silva, 1984): much of the flow through the channel is recirculated through these eddies. Along the remarkably linear east coast of Madagascar, southward flow occurs during the Northeast Monsoon and feeds into the Agulhas Current through a persistent field of mesoscale eddies (Lutjeharms, 1988). Part of this may pass directly into the subtropical gyre by retroflection south of Madagascar. At the southeast tip of Madagascar, upwelling is induced by the presence of a persistent cyclonic eddy, locked to shelf topography, and by favorable wind conditions (Lutjeharms *et al.*, 1981). This feature is intermittent and apparently associated with anomalously high wind stress, as occurred in February–March 2000, when a surface cold anomaly of 3–5°C was observed by DiMarco *et al.* (2000).

***Agulhas Current*** There is no seasonal variation in the strength of the Agulhas Current for the same reasons as for its larger analogue, the Gulf Stream of the North Atlantic (Pearce and Gruendlingh, 1982): at the latitudes occupied by these two western boundary currents, wind stress is transformed into mixing rather than momentum (see Chapter 4). Agulhas flow originates in the variable circulation at the southern end of the Mozambique Channel but is an essentially continuous jet to the southern extremity of Africa. About four eddies each year propagate southward through the Mozambique Channel and pass into the upstream region of the Agulhas Current (Schouten *et al.*, 2003). As in other boundary currents, three zones can be distinguished: a coastal zone of cyclonic shear, a jet of maximum velocities ($>2.5\,\mathrm{m\ sec^{-1}}$) at the shelf edge, and an offshore zone of anticyclonic shear. Gradients of velocity and temperature resemble those of the Florida Current and the Kuroshio. An inshore countercurrent, or recirculation flow, may be induced in the larger coastal bights (Schumann, 1982). Such flows may be associated with pulses of vortex shedding (Lutjeharms and Connell, 1989).

The Agulhas Current becomes progressively wider, and the dimensions of associated plumes and eddies progressively increase, southward (Lutjeharms *et al.*, 1989). Some meanders are persistent in location, as off Cape St. Francis at 34°S, where an episodic meander may form and attain a wavelength 100–150 km. This then propagates downstream at ~15 km per day (Swart and Gonzalves, 1983). Some mesoscale eddies may be shed by meanders at the seaward boundary of the current at any latitude, but they are most frequent in the terminal area south of South Africa.

***Agulhas Bank and Retroflection*** Retroflection occurs where the western boundary current runs out of topography at the southern extremity of Africa and encounters the eastward flow of the circumpolar zonal currents. This region is difficult to place in our system of provinces and is located here simply because its ecological effects are continuous with those on the Agulhas Bank. Otherwise, it might be better considered as part of the South Subtropical Convergence (SSTC) Province because the flow from the retroflection (see Color plate 13) feeds directly into the easterly jet current flowing along 40°S within the core of the SSTC as the South Indian Ocean Current into the zonal circumpolar current system (Stramma, 1992).

Eddies, originating in the "Natal pulse" at about 30°S, are shed as they pass around the retroflection loop and normally follow the general flow and pass eastward to the south of the Agulhas Return Current. However, some of the eddies shed from meanders on the western flank of the Agulhas pass into the South Atlantic at the origin of the Benguela Current system flow (Lutjeharms and van Ballegooyen, 1988). These are warm-core and anticyclonic in form, and their frequency is about four to five cycles per year, the location of shedding varying between years.

Although local forcing is different in each region of this province, the seasonal changes in the depth of the mixed layer follow two patterns: (i) in the northern part of the province, during the monsoon period of June–September (note, here this is a Southeast Monsoon), the pycnocline of the Indian Ocean tilts down westward, a motion that results in deepening from about 40–60 m to about 80–100 m along to the coast north of 25°S. Of course, this is also the austral winter so that, (ii) further to the south, wind stress and heat loss at the surface combine to deepen the mixed layer to about 100 m. Consequently, over the Agulhas Bank in winter, mixing penetrates to 65–75 m above a weak thermocline/nitracline, whereas in summer, stratification is established at around 30–35 m coincident with a strong nitracline (McMurray *et al.*, 1993). There is evidence (see later discussion) of episodic occurrences of coastal upwelling along the coast between Port Elizabeth and Mossel Bay, perhaps especially during summer.

## Regional Response of the Pelagic Ecosystem

A few ecological observations, made at 5–6°S, which should be typical of the whole of the East Africa Bight (0–10°S) were reviewed by McClanahan (1988). During the period of boreal summer monsoon winds (here largely from the southeast) of May–September, irradiance at the sea surface falls from 520 to ~380 langleys, surface temperature falls from 29.5 to 26°C, and wave height and rainfall are much increased. Salinity is minimal (34‰) in the same period, corresponding to the discharge cycle of the Tana and Sabaki rivers. The nutrient regime over the shelf is largely controlled by wind mixing, which delivers nutrients to the surface layers (Newell, 1959). Sediments and deep shelf waters are relatively nutrient-rich, although this is a new phenomenon caused by current farming practices in adjacent watersheds and consequent massive erosion.

In this region, water transparency is highest and chlorophyll biomass minimal just prior to the Northeast Monsoon, when primary production is maximal and large-celled algae are most diverse. Stability is strongest in the water column at this time, although the seasonal changes are not large. A reduction in chlorophyll and numbers of larger algal cells, and generally lower abundance of zooplankton, is associated with the deepening of the mixed layer in boreal summer. Mixed-layer chlorophyll is 0.3–0.6 mg chl $m^{-3}$ during the Southwest Monsoon and 0.5–1.0 mg chl $m^{-3}$ during the Northeast Monsoon. A nitrate-nitrogen maximum in the premonsoon period (February–March) may be associated with nitrogen fixation by heavy blooms at that season of *Oscillatoria erythraea*. Where the converging flows of the East African Current and the reversed Somali Current of boreal winter leave the coast, upwelling occurs in the retroflection area at about 2°S (Kabanova, 1968); an algal bloom associated with this retroflection is seen in some satellite images, at the base of an eastward filament aligned along the equator (Longhurst, 1993). The mesozooplankton population achieves maximum biomass toward the end of the Northeast Monsoon and—not coincidentally—it has been noted that fish and large invertebrates tend to maximize their reproductive effort at this season.

On the Madagascar shelves, at all seasons except the boreal summer monsoon, chlorophyll accumulation is stronger on the west coast, facing the Madagascar Channel, than on the east coast that faces the open ocean. The seasonal cycle of accumulation on the east coast is but a part of the strong bloom that occurs within the SEC in boreal summer (see ISSG), in which large eddies are seen to propagate toward the west, enveloping the entire eastern seaboard of Madagascar from May until October. Off the western coast, the chlorophyll biomass associated with the eddy field is almost permanent, and oligotrophic conditions obtain only very briefly during the boreal autumn intermonsoon period; by mid-December, accumulation recommences (Saetre and Jorge Da Silva, 1984). However, coastal blooms observed in level 3 images may simply reflect the effluent from the rivers draining the coastal lowlands on the western side of the island. Certainly, on the northeast coast, near Nosy-Bé, it has been observed that estuarine circulation on the shelf dominates nutrient dynamics, and the duration of the local bloom. The picoplankton fraction in the offshore blooms in the NE Madagascar Channel has been assessed at ~48% of biomass and as providing ~54% of total primary production (Magazzu *et al.*, 1985). The upwelling cell that occurs intermittently off the southeast cape of Madagascar (see earlier discussion) is quite frequently observed in satellite images to induce enhanced chlorophyll that is subsequently entrained toward the east.

In the south, over the Agulhas Bank, there is good information on the seasonal succession of primary production and how it is partitioned between nanophytoplankton and $>15-\mu m$ cells of the "net" phytoplankton. This is a typical temperate shelf production cycle even if, as Huggett and Richardson (2000) note, inshore coastal upwelling and dynamic upwelling along the eastern margin do support relatively high chlorophyll biomass. In winter and spring, the mixed layer is deeper than the euphotic zone; nitrate and chlorophyll are thoroughly mixed throughout the euphotic zone while productivity is low (3–4 mgC $m^{-3}$ $hr^{-1}$), and restricted to the upper 5–15 m of the mixed layer (McMurray *et al.*, 1993). In austral summer (December–January), the euphotic zone is thermally stratified and a deep chlorophyll maximum ($<6.0$ mg chl $m^{-3}$) forms at the depth of maximum nitrate gradient, at about half the euphotic depth of 70–80 m. The DCM lies at the nitracline and is approximately coincident with the depth of maximum production rate (5–6 mgC $m^{-3}$ $hr^{-1}$). Between these two seasons, a spring diatom bloom occurs when irradiance at the sea surface increases, and when turbulence in the mixed layer decreases before the summer, shallow thermocline is established. Thus, production is light-limited in winter, nutrient-limited in summer.

Imposed on this classical midlatitude cycle on the central Agulhas Bank, episodic uplift of the thermocline to $<30$ m occurs during shelf-edge upwelling events and the intrusion of cold bottom water up over the bank. These events result in episodic blooms of diatoms during the relatively oligotrophic summer period. During most of the year, small cells ($<15\,\mu m$) contribute 60–95% of all production, but during the onset of stabilization in spring and the summer bloom events, 60–85% of total production is by larger diatoms (*Chaetoceros, Bacteriastrum*, and *Nitzschia*). Examination of the serial chlorophyll images suggests that these events produce very prominent chlorophyll features at the retroflection of the Agulhas Current and that they are quite variable between years. Each event appears to be initiated by the appearance of a narrow zone of high chlorophyll along the coast from East London (32°) almost down to Cape Agulhas at 35°S, evidently forced by inshore upwelling. This feature very quickly begins to stream offshore along the eastern side of the Agulhas Bank. Such observations from the satellite images are consistent with other reports that a quasipermanent ridge of cool, upwelled water (12–18°C) lies toward the southwest from Mossel Bay and extends back along the coast past Port Elizabeth. This is on the coastal side of the narrow stream of warm subtropical water that is very clearly seen in serial satellite SST images of this region.

When it reaches the tip of the Bank, there is a tendency for the stream of high-chlorophyll water to continue offshore toward the southwest but, in most cases, the nascent eddy is captured in the flow of the Subtropical Convergence Zone and then passes off to the east as a mesoscale eddy. Algal growth within these warm-core rings is limited by convective instability, whereas algal growth outside them is light limited. Maximum chlorophyll concentration around the edges is consistent with ring-induced stability (Dower and Lucas, 1993). On the western edge of the Agulhas Bank, uplifting of the thermocline is episodic, associated with wind events that, during summer, may be associated with a linear bloom along the shelf edge. This is contiguous with the upwelling events in the Benguela Current and is perhaps best considered as part of that system.

Despite this very active production system, the pelagic ecosystem is not that of an upwelling coast. The dominant copepod, *Calanus aghulensis*, comprises 50–80% of all mesozooplankton and is a member of the *C. helgolandicus* group (Huggett and Richardson, 2000). The principal copepod along the upwelling coast of the Benguela region is, instead, *Calanoides carinatus*.

*Calanus aghulensis* is a diel migrant whose relatively rapid generation time (20 d at 15.5°C) is more sensitive to food availability than to temperature; diel vertical migration is linked to food abundance, and ontogenetic layering occurs at daytime depths. Relative population abundance is greatest around the periphery of the tongue of cold water discussed earlier, and the area occupied by this species extends from 32°S on the east coast to about 29°S, at the Orange River, to the west. This copepod is the main resource on which the Cape anchovy (*Engraulis capensis*) builds its population (Richardson *et al.*, 1997); the spawning and subsequent recruitment of anchovies is strongly dependent on the size of the population of *Calanus aghulensis*. Although it has been postulated that population fluctuations of *Engraulis* and *Calanus* should be mutually dependent, such balances are very difficult to analyze. Here, as off California, a negative functional relationship between population sizes of *Sardinella occelatus* and *Engraulis* has also been proposed—further complicated by the effects on each of varying fishing pressures. It has also been suggested that on the Agulhas Bank the transfer of biomass from primary and secondary production to fish biomass is, relative to regions such as Georges Bank, rather inefficient.

## Regional Benthic and Demersal Ecology

The unusually strong southward flow of the Agulhas Current induces bed-load motion on the shelf and the formation of linear shoals at around 50-m depth, running parallel to the coastline; these evolve geologically into elevated masses of sandstone that bear some of the most southerly coral reefs in this ocean, off KwaZulu-Natal at 26–27°S. The mass of these reefs is due, therefore, not to biogenic accretion of carbonate but to the sandstone hillocks that originate in the submerged sand dunes on the inner shelf. Only a few of these reefs break the surface. The warmth of the subtropical water (22–26°C, seasonally) of the Agulhas Current permits their growth so far to the south.

The coral reefs in the equatorial section of this province, off Kenya and Tanzania, suffered extreme heat stress and bleaching during the second half of 1998, 60–80% of live corals being lost in shallow areas; coral recovery is unexpectedly slow, due to removal of herbivorous fish by artisanal fishing and hence the proliferation of fleshy algae over the reefs. The fishing pressure on these reefs, near centers of population, is intense especially for spiny lobsters (*Panulirus* spp.). These, and several warm-water shrimps (Penaeidae), form an important part of the mobile benthic fauna of this province, at least as far as ~5°S; otherwise, this warm-water region supports a demersal fish fauna typical of tropical shelves—dominated by sciaenids (*Otolithes rubra* and others), together with *Thryssa, Pomadasys, Arius, Pellona, Trichiurus*—and so on.

## Synopsis

*Case 4—Small-amplitude response to trade-wind seasonality*—Here, the integration of the whole area in satellite chlorophyll data is not very satisfactory because of the great latitudinal extent of the province; as Fig. 10.8 shows, this suggests a rather simple cycle with maximal productivity in December–February (austral summer) and maximal biomass in June–August (austral winter). Mixed-layer $Z_m$ undergoes an austral winter excursion from 15 m (summer) to 60 m in June–July, forced by wind mixing in the south, and westward thermocline tilt in the tropical region. The pycnocline is illuminated only in boreal summer, June–September. There are very striking opposite seasonal trends in

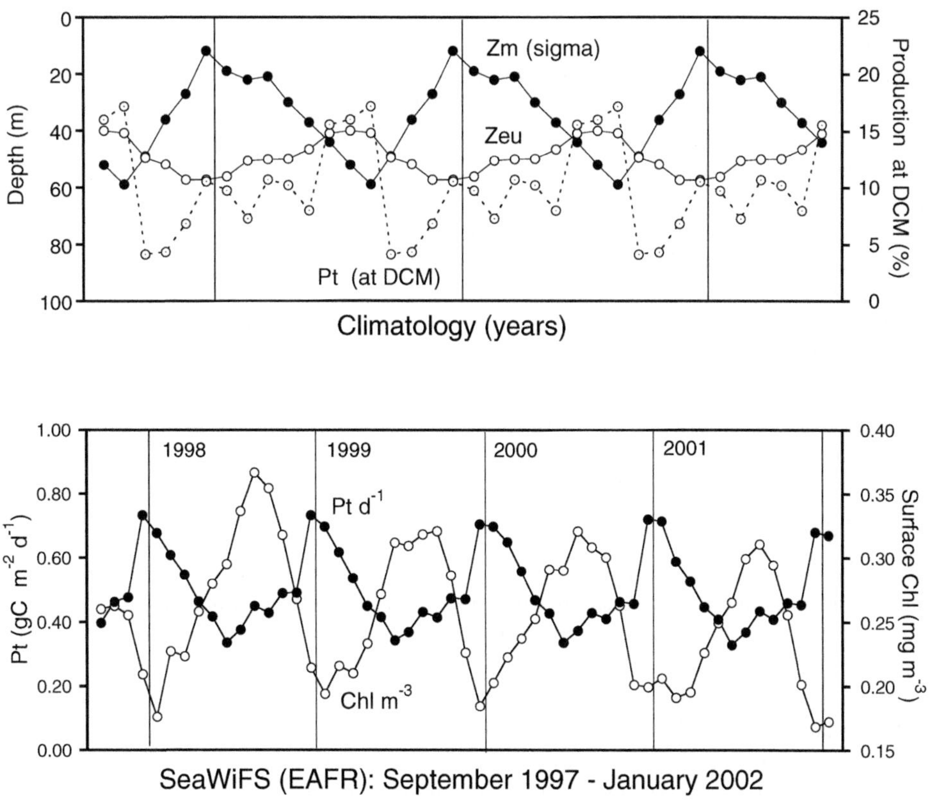


**Fig. 10.8** EAFR: seasonal cycles of monthly surface chlorophyll and depth-integrated autotrophic production for the years 1997–2002 from SeaWiFS data together with characteristic seasonal cycles of mixed layer depths from Levitus climatological data and photic depths computed from characteristic irradiance and the archive of chlorophyll profiles discussed in Chapter 1.

productivity and chlorophyll accumulation: the former responds to the seasonal irradiance cycle, whereas phytoplankton biomass is apparently maximal in austral winter. This apparently anomalous situation results from the dynamic upwelling events that were described earlier.

## Australia-Indonesia Coastal Province (AUSW)

### Extent of the Province

Like EAFR, the extent of this coastal boundary province is largely a matter of convenience to avoid excessive subdivision in this work. As defined here, it extends from northern Sumatra to southern Australia. Like EAFR, it could readily and logically be subdivided into several primary entities. A logical first cut would be the following: (i) the tropical southern coasts of Sumatra, Java, and the Lesser Sunda Islands; (ii) the subtropical western coast of Australia from Cape Bougainville (14°N) in the southeastern Timor Sea down to Cape Leeuwin at 34°S; and (iii) from Cape Leeuwin to the Bass Strait along the temperate coast of the Great Australian Bight.

The offshore boundary is identified in the usual manner, where limits of coastal flow can be identified (Colborn: coastal parts of 22, 23, 25, 26, and 34).

### Continental Shelf Topography and Tidal and Shelf-Edge Fronts

The southern islands of the Indonesian archipelago have a very narrow continental shelf, except off central to northern Sumatra where the chain of the Mentawai Islands 100–200 km offshore encloses a significant area of shallow water. The shallow shelf of the Timor Sea, not part of this province, is continuous with the wide Sahul shelf off northwestern Australia. The Sahul shelf extends from Cape Talbot (13°S) southwest to Northwest Cape (22°S), the shelf edge lying parallel to the coast and about 150–250 km offshore. Further south, from Northwest Cape to Cape Leeuwin (34°S), the shelf is narrower (50–100 km), especially off Perth, and the coastline lies almost directly toward the south. In the Great Australian Bight, on the south coast of Australia, there is a further region of significant shelf, reaching 150 km at about the center of the bight, south of the flatlands of the Nullabor Plains.

The tidal range on the northwest coast of Australia increases progressively northward so that at Broome (17°S) the spring tidal range is $> 6$ m; there are no reports, however, of thermal fronts between stratified and tidally mixed water on this wide section of the shelf. Instead, we should note that tropical cyclones are important in forcing storm surges that propagate southward as coastally trapped Kelvin waves.

### Defining Characteristics of Regional Oceanography

It will be convenient to deal with these three subregions individually, emphasizing that there is a fundamental difference in atmospheric forcing between northern and southern parts of this coastal province, corresponding to oceanic regions north and south of the tropical convergence at about 20°S. From the coast of Indonesia south to ~25°S, alternating seasonal monsoons dominate the wind regime, whereas further to the south the high-latitude westerly winds dominate. From ~25°S, therefore, a summer thermocline regime is appropriate, this feature forming each year in spring at ~40 m and subsequently deepening to ~100 m in autumn, above a permanent thermocline that lies at ~250 m.

***South Coast of Sumatra and Java*** The Java Coastal Current is reversed by the seasonal changes in orientation of boreal summer monsoon winds, and salinity is modified by rainfall on the highlands of the southern Indonesian archipelago at the same period, from May to September (Quadfasel and Cresswell, 1992). Precipitation exceeds evaporation (by 15 cm per month) during the wet season, whereas evaporation dominates (by 5 cm per month) during the dry season of boreal winter. The low salinity of coastal waters during the wet season induces a cross-shore pressure gradient and so forces a westward jet current at the coast, reinforced by the surface wind stress. This coastal jet forms part of the origin of the South Equatorial Current that is at its most northerly position in boreal summer. During boreal winter, when wind stress is toward the east, flow in the same direction is induced along the whole coast from the equator (off central Sumatra) into the western Timor Sea. The current reversal between seasons is influenced not only by local winds but also by distant forcing of strong westerly winds in the central Indian Ocean that induce eastward-propagating equatorial Kelvin waves.

Associated with the season of strongest westward flow of the Java Coastal Current under the forcing of the Southeast Monsoon, a period of significant coastal upwelling occurs from May to September along the coasts of Java and Sumbawa (Wyrtki, 1962, 1973b). This is a deep upwelling so that 200-m temperatures in the center of the feature fall to 12°C compared with 18–20°C at the same depth in the open ocean between Java and western Australia. The region of enhanced nutrient availability is larger than that off Arabia.

***Western Australia*** The Leeuwin Current is anomalous among all other eastern boundary currents in that it flows poleward, despite the equatorward wind stress over the eastern Indian Ocean and the general equatorward flow in the eastern limb of the subtropical gyre further offshore. It is a surface flow of warm, low-salinity tropical water above an equatorward undercurrent, with the level of no motion between the two flows lying at 200–300 m. It has significant seasonality, flowing south along the entire coastline more consistently during austral autumn and winter. The anomalous poleward flow is forced by the strength of the onshore geopotential anomaly caused by the poleward pressure gradient in the eastern Indian Ocean—this is sufficient to overcome equatorward wind stress (Church *et al.*, 1989). In the coastal zone, the eastward component of flow arising from the alongshore pressure gradient creates a downwelling situation in which isopycnals slope down toward the coast (Thompson, 1987).

The source waters for the Leeuwin Current are off the northwest Australian coast over the Sahul shelf, which is within the influence of the monsoon systems of Java and Sumatra; thus, a rainy monsoon period in austral summer (January–March) intervenes in an otherwise very dry area having exceptionally strong irradiance at the sea surface. Flow is maximal during the monsoon period and the velocity core of the current lies over the shelf edge; in austral winter it moves farther out to sea, forced by changes in local wind stress rather than changes in the longshore pressure gradient (Church *et al.*, 1989; Smith *et al.*, 1991). Satellite thermal imagery shows that the highly meandering flow extends at least 250–350 km offshore and frequently includes prominent warm filaments with strong vorticity. South of about 32°S, eddying extends even farther into the equatorward flow of cool, high-salinity water of the subtropical gyre, sometimes referred to as the West Australian Current. The temperature gradient between Leeuwin flow and the offshore gyral water becomes stronger progressively southward, despite the entrainment of subtropical water.

***Great Australian Bight*** The coastal boundary of the southern coast of Australia, between Cape Leeuwin in the west and Tasmania in the east, is under the influence of Leeuwin Current water, which is topographically locked to the continental edge as it rounds the Cape Leeuwin and proceeds eastward along the southern Australian shelf.

Nevertheless, maximal flow follows the shelf break of the Great Australian Bight as far as 130°E (Rochford, 1986) and is maximal in austral winter (May–October). After July, the eddying flow recedes from the eastern part of this coast. However, the warm and saline surface water mass that occupies the central and eastern part of the bight during much of the year is probably water from the Leeuwin Current modified by the arid and evaporative nature of the coastal climatic regime. Cold, high-salinity water of the west-wind drift of the Southern Ocean current system lies above the slope year-round, and when the Leeuwin Current flow is relaxed in austral summer this water mass intrudes over the shelf break and may flood onto the continental shelf. Off the eastern shelf, dense high-salinity water may cascade from the shelf and produce temperature inversions (Godfrey *et al.*, 1986).

The shelf edge here is thus a region of very active frontogenesis. The shelf edge current is fast ($<1.5\,\mathrm{m\ sec^{-1}}$) and strongly baroclinic in the west, especially after it rounds Cape Leeuwin, where its dynamics become nonlinear and current speed increases due to a Bernoulli effect. The fronts between warm Leeuwin Current water and the offshore cold water of the west wind drift are sharp and extend to $<200\,\mathrm{m}$. Large (200- to 300-km) cyclonic eddies and sickle-shaped vortex filaments curl back toward the west beyond the shelf edge and are especially prominent between Cape Leeuwin and Cape Arid. Several times each month there is a major irruption of filaments of warm water (often terminating in eddy pairs) into the west wind drift, especially when flow of the Leeuwin Current is strong (Griffiths and Pearce, 1985).

## Biological Response and Regional Ecology

By far the most comprehensive ecological coverage of this part of the ocean was the Australian contribution to the IIOE, which was reviewed in the description of the ISSG province that lies offshore of AUSW. To the south of Java, at 8–12°S, primary production rates are maximal during the southeast monsoon period (May–November) and, during the same period, zooplankton biomass was higher in the Java upwelling area than anywhere to the south. Though the observations are few, rates of primary production in a coastal cell extending from Java to about 10 or 11°S results in elevated chlorophyll ($<1.1\,\mathrm{mg\ chl\ m^{-3}}$) and productivity ($0.7\,\mathrm{gC\ m^{-2}\ day^{-1}}$) in austral winter (Humphrey and Kerr, 1969). In January and February, values are about 20% of those of the upwelling season May–October. Cushing (1973) evaluated this region as among the most productive of the entire Indian Ocean during the southerly monsoon period of boreal summer.

At the time of the IIOE, it was known that high chlorophyll biomass occurs seasonally over the Sahul shelf of northwestern Australia, and it had been thought that this was part of the same process as the simultaneous upwelling enrichment of the Southeast Monsoon on the coast of Java. However, a more recent investigation revealed a quite different mechanism (Tranter and Leech, 1987). The seasonal pulse of higher chlorophyll values was confirmed, especially subsurface and at the shelf break, but these were found to be unusual because the source of nutrients on the Sahul shelf is neither upwelling (as had been previously thought) nor riverborne nutrients. The source is rather episodic intrusions of cool, nitrate-rich slope water over the shelf edge below the warm, low-salinity surface layer. These intrusions result in near-bottom chlorophyll maxima ($>0.5\,\mathrm{mg\ liter^{-1}}$) over the shelf at 50–100 m that are continuous with the offshore DCM that lies permanently at ~75 m: such a situation also occurs below the Florida Current and perhaps also in other boundary currents. Chlorophyll enhancement is coincident with the base of the pycnocline and responds to vertical motion of this feature. Plant growth is possible in this situation because incident solar radiation on this coast is so high that the subpycnocline shelf water is sufficiently illuminated for algal growth to occur at all seasons. In austral winter, stratification is reduced and a "phytoplankton-dispersed" season (April–July) replaces the summer "phytoplankton-stratified" situation (August–March).

Another consequence of the extremely high solar irradiance is to impart sufficient stability to the upper water column to resist tidal mixing over much of the shelf; so, although tidal stream velocities have a fivefold amplitude range between spring and neap tides, the water column is uniform closer to the coast than the 40-m depth contour. Shelf sea fronts (and local eutrophication) are therefore not features of this otherwise apparently suitable wide and flat shelf (Tranter and Leech, 1987) and despite the fact that this is a region of strong potential internal tidal energy.

Further to the south, within the flow of the Leeuwin Current (both before and after its turn to the east around Cape Leeuwin) the thermal signals from the mesoscale offshore filaments and eddies are of similar magnitude to those observed seaward of eastern boundary currents. Here, of course, it is not cool, upwelled water that it transported away from the coast but rather warm, nutrient-poor water carrying low biomass of pelagic biota. Thus, it is only in the vorticity of the offshore eddy field that nutrients will be advected to the photic zone and where we can expect to observe algal blooms.

Though there is very little information on ecological seasons south of the Sahul shelf, we have to assume that a temperate cycle is appropriate here. This may also be inferred from the section at 110°E at relevant latitudes that, as noted previously, resembles the seasonal cycle in the Sargasso Sea.

Very low productivity in summer (< 18.2 mgC $m^{-2}$ $hr^{-1}$ compared with almost 10 times that value in the Gulf of Carpentaria) was observed by Motoda (1978). They believed this could be related to a regional lack of nutrient inputs—neither upwelling nor river effluents appear to have any importance in this region.

At the shelf break of the Great Australian Bight, there is a subsurface front that is capped by warm Leeuwin Current water in austral summer so that perhaps, at other times, shelf-edge upwelling may occur here (Young *et al.*, 1999). There is some evidence, too, of the aggregation of nektonic organisms here, which may support observed aggregations of southern bluefin tuna, but zooplankton biomass appears to be normally most abundant in the neritic zone.

Based on a temperature criterion, the mixed-layer depth of the Great Australian Bight deepens from < 50 m in summer to 100–150 m in winter. Such a seasonal range must surely be associated with spring bloom and summer oligotrophic conditions perhaps more like those of the North Atlantic model than those observed at the southern end of the 110°E section. However, this remains only speculation. Indeed, the serial chlorophyll images make it absolutely clear that there is a major seasonal production cycle here; from November until March, chlorophyll concentrations are very low except in a narrow neritic zone, where the satellite sensors indicate of order 0.2–0.3 mg chl $m^{-3}$ compared with totally oligotrophic conditions offshore. However, during austral winter (April–September) the entire region supports significantly higher concentrations (1.0–1.5 mg chl $m^{-3}$) from the coast south to the SSTC zone. At this season, the Great Australian Bight, including the shelf regions, is indistinguishable from the circumpolar bloom that lies just to the south of the continents.

Data collected as background for the Australian IIOE transect show that there are two centers where mesozooplankton biomass is significantly higher than background, and that these have a marked seasonal evolution. The highest observed biomass lies along the southern coast of Java, within the upwelling region from 105°E to 120°E. This is most strongly developed during the Southeast Monsoon (July–August) when biomass >100 mg $m^{-3}$ is widely distributed south of Java and on the Sahul shelf at 18–20°S. The Java zooplankton maximum (still >100 mg $m^{-3}$) is reduced to a linear, coastwise zone in October–November while over the Sahul shelf biomass has fallen to 25% of that figure. For the remainder of the year, only the Java region retains substantial mesozooplankton biomass (>50 mg $m^{-3}$).

The pelagic community off Java is strongly differentiated from that along the Australian shelf and also has greater endemism. The dominant copepod genera off southern Java include the upwelling coast specialist *Calanoides carinatus* but they also comprise many neritic forms that are recruited to the region of the Java Dome by being drawn through the interisland passages by the South Equatorial Current that flows strongly westward during the Southeast Monsoon (see earlier discussion). The typically neritic genera that are thus transported include *Acartia, Eucalanus, Undinula,* and *Euchaeta.* On the Australian coast, in the southward transport of the coastal segment of the Leeuwin Current, the tropical-subtropical element of the zooplankton community is progressively eliminated.

## Regional Benthic and Demersal Ecology

The Sahul shelf was surveyed in great detail by Okera (1982), who reported on the species associations of demersal fish but also on the surficial geology. This shelf has complex topography with sandy ridges at mid-depth, with silty basins between them, and near the shelf break there are numerous steep-sided banks; shoaler than 40 m, sandy deposits dominate. Using recurrent group analysis, he identified six assemblages of demersal species, associating each with its typical depth stratum and deposit type. Because these statistical groups are common to both the Sahul Shelf and the Gulf of Carpentaria on the north coast of Australia, I propose to discuss these only in relation to that location (see SUND province).

Tranter and Leech (1987) suggest that the unusual conditions of the Sahul Shelf discussed earlier would be expected to lead to an unusually direct utilization of the primary production by benthic organisms and, hence, by demersal fish. Filter-feeding benthos of the outer shelf deeper than 40 m, they point out, must have direct and

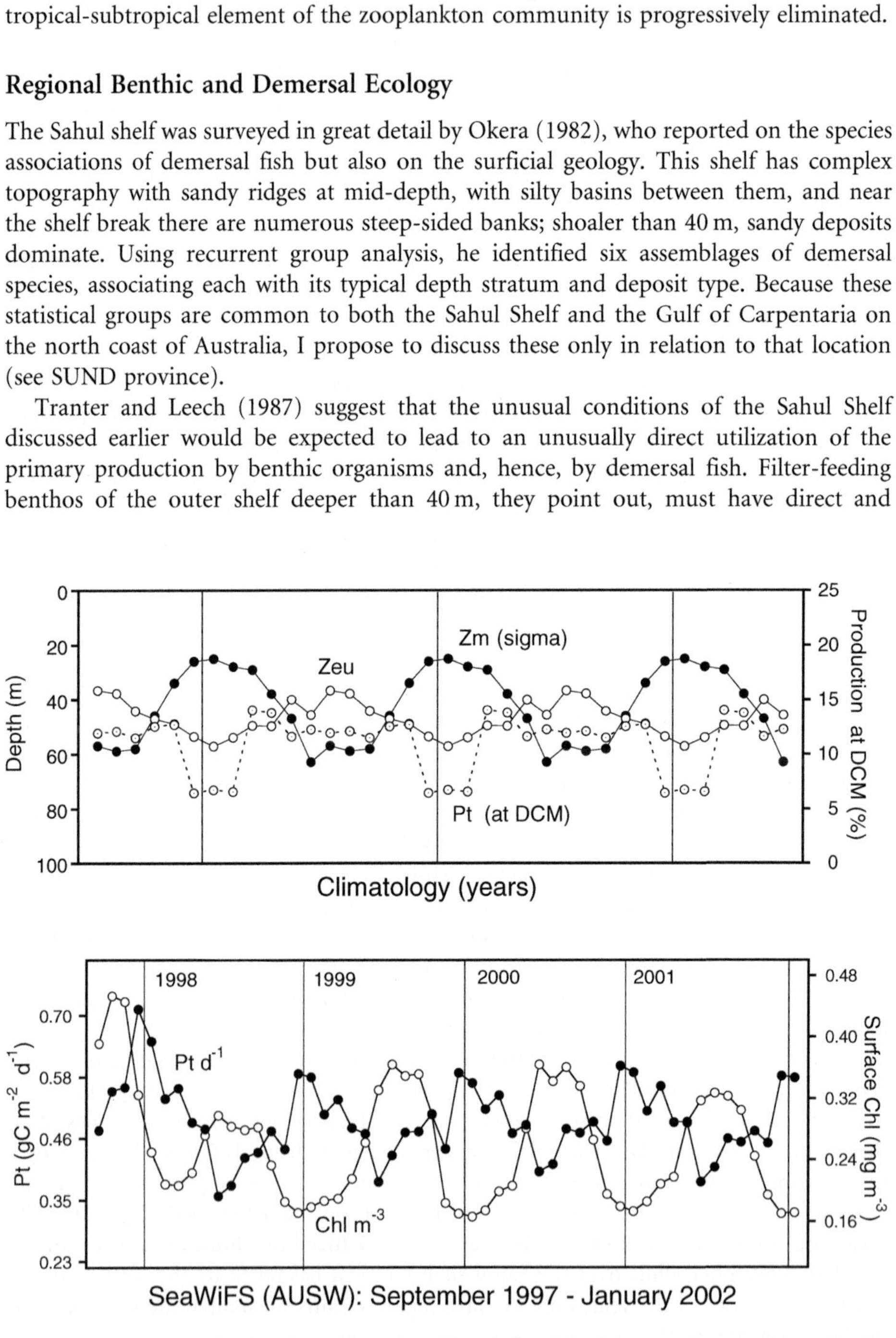


**Fig. 10.9** AUSW: seasonal cycles of monthly surface chlorophyll and depth-integrated autotrophic production for the years 1997–2002 from SeaWiFS data together with characteristic seasonal cycles of mixed-layer depths from Levitus climatological data and photic depths computed from characteristic irradiance and the archive of chlorophyll profiles discussed in Chapter 1.

immediate access to the subpycnocline phytoplankton production. Those living in deeper water may be expected to have access only to sedimentation of detrital material.

## Synopsis

The generalization offered by climatological data representing the entire province is not very helpful. It suggests merely that the shoal, permanent pycnocline is moderately deepened during austral winter from 30 m (summer) to 70 m June–July so that it lies shoaler than photic depth only in summer (Fig. 10.9). There is some increase of primary production rate with the entrainment of nutrients during the deepening phase of mixed layer, although chlorophyll accumulation occurs only during the period of declining productivity.

# Chapter 11

# THE PACIFIC OCEAN

The Pacific Ocean is clearly a special case, because it has an areal extent ($165 \times 10^3 \text{km}^{-2}$) that is almost half the total area of all oceans combined. It has a zonal dimension across 210° of longitude, or about 60% of the circumference of the earth. In the not-so-long-gone days of passenger travel by sea, 20 days steaming from Panama to New Zealand and always out of sight of land except only for Pitcairn Island, gave me a firsthand understanding of Pacific dimensions. The immense size of the Pacific is reflected in strong zonal differences in mixed-layer depth and other physical circulation features that are reflected in regional pelagic ecology.

A very important direct consequence of the great dimension and largely closed boundaries of this ocean is that here the major ocean circulation patterns can approach an ideal form in response to symmetric wind patterns in the two hemispheres and to the Coriolis effect. Here two almost-ideal subtropical gyres and their boundary currents are developed symmetrically on either side of a quasisymmetric development of an ideal equatorial current system having appropriate convergences, divergences, and countercurrents. If this ocean did not exist, our oceanographic textbooks would have been forced to imagine it. In understanding the other oceans, it is often convenient to make a comparison against the Pacific ideal: this is true both for physics and for biogeography, in the widest sense.

The Pacific Ocean is also rich in marginal seas, ranging from relatively small, evaporative basins (Gulf of California) to large epicontinental seas (Bering Sea). The Pacific also accommodates the vast Indo-Pacific archipelago that stretches from northern Australia to Luzon and Formosa; flow between two oceans (Pacific and Indian) is exchanged across this complex region of shallow shelves and small, semienclosed deep basins.

In the subarctic zone of the North Pacific, unlike its North Atlantic analogue, precipitation exceeds evaporation so that a surface layer of low salinity lies above a permanent halocline at 100–150 m. This feature therefore constrains the depth of winter mixing, and thus the renewal of mixed-layer nutrients. The North Pacific halocline therefore has unique biological consequences that distinguish the subarctic regimes of the two major oceans.

Unlike the Atlantic, the Pacific has no connection to the arctic regions in its northeastern quadrant, so the flow of the Kuroshio—the Pacific analogue of the Gulf Stream—does not, like the Gulf Stream, lose part of its flow into high latitudes. Instead, it flows more directly across the ocean once it has departed from the Japanese mainland, and this has important consequences for the comparative oceanography and ecology of the two oceans.

As in the Atlantic, the transition between the biomes of westerly and trade-wind zones is not sharp. Though the Subtropical Convergence zone or, in a more generalized sense, the zone of subtropical convergence fronts between the westerlies and the easterly trades, is useful to locate this transition, it is an imprecise and moving partition. Some winter mixing does occur equatorward of the Subtropical Convergence of the Northern

Hemisphere in both oceans. This is, again in a general sense, the zone of winter rather than spring blooms (McGowan and Williams, 1973; Banse and English, 1993), if indeed the term *bloom* is relevant to such very weak chlorophyll enhancement as occurs.

Perhaps most important, we must accommodate the fact that, more strongly than in the other oceans, Pacific circulation exists in two modes depending on the interannual variability of trade-wind stress (see Chapter 6). During El Niño-Southern Oscillation (ENSO) events, when trades weaken and are replaced by westerlies at low latitudes in the western Pacific, the heat balance of the entire ocean at low latitudes is perturbed for periods of a few months to 1–2 years. Here, for convenience, we regard this as the anomalous condition and consider the normal condition to be that obtaining when trade winds are well developed. The between-year, ENSO-scale alternation between the two states resembles in many important ways the seasonal response of the tropical Atlantic to between-season changes in trade-wind stress. This difference between the two oceans is a consequence of their different zonal dimensions (Philander and Chao, 1991; Philander and Pacanowski, 1981; Philander, 1985), a fact that has important consequences for their dynamic biogeography (Longhurst, 1993). We shall encounter reports of long-term alternations of state in circulation and ecosystem structure more commonly here than in the other oceans.

The provinces into which it is useful to partition the tropical eastern Pacific during normal conditions—*viz.*, the Pacific Equatorial Countercurrent (PEEC), Pacific Equatorial Divergence (PEQD), and North Pacific Tropical Gyre (NPTG) provinces—lose their defining characteristics partially or completely during an ENSO event. During such periods, rather uniform conditions with a deeper, warmer mixed layer obtain across the whole region and I shall refer to this partition, when convenient, as the WARM-ENSO Province, while recognizing its ephemeral nature and intermittent manifestation.

Because of the dimensions and hence global importance of the Pacific Ocean for many issues that have evolved over the years, from its fisheries' potential to ocean-atmosphere interactions, climate control, and global carbon fluxes, we are fortunate that it has attracted at least its fair share of oceanographic effort. We are also fortunate that much of this work has been centrally planned (and peer-reviewed in the planning) so that the issues of the day and—more important from our point of view—the key scientific questions have had a chance of being answered.

Much of this work has been multinational and multiship, an increasingly important aspect in recent years because the very extensive explorations by Soviet ships, which were characteristic of Pacific exploration in earlier decades, have unfortunately now almost ceased. Quite large regions of the ocean have been studied in a systematic way, along preplanned grids of stations, repeated seasonally where required. Because of the planned nature of much of the work that has been undertaken, the most important data sets are already generally available through national data networks.

One of the minor consequences of all this effort is the fact that it is in the Pacific Ocean that we can best appreciate the gross distribution patterns in the biogeography of pelagic biota. Here, one has the impression that they have room to breathe and are much more readily associated with regional oceanography than in the other oceans where the current systems are less rectilinear, or ideal. Perhaps the maps of John McGowan (1971) show this best. Here we see superimposed envelopes of distributions of groups of species of pelagic organisms—plankton to tuna—having subarctic, transitional, central, and equatorial distributions that, as we shall find, are in general agreement with the conclusions we reach here on quite different grounds.

A recent multiauthored and on-line publication of PICES reviews the ecology of all the major regional ecosystems of the North Pacific, but particularly from the point of view of their between-year and decadal changes of state. The authors assume a knowledge of the internal structure of each regional ecosystem, and document the changes that have occurred in the last half-century or so. Because of the availability of this work, I have not emphasized the aspect of changes of regional ecology in what follows, but have rather concentrated my text on the structural aspects of each region. "Marine Ecosystems of the North Pacific" may be downloaded from www.pices.int/publications.

# Pacific Polar Biome

## North Pacific Epicontinental Sea Province (BERS)

### Extent of the Province

This province comprises the Bering Sea and the Sea of Okhotsk, respectively enclosed by the arcs of the offshore Kuril and Aleutian Islands. There is considerable commonality between these two seas, but in a more dispersed classification of coastal provinces it would be logical to separate them. BERS is placed in the Polar, rather than in the Coastal biome, as a matter of convenience; marginal seas, such as the Mediterranean, with significant areas of both shelf and deep water, are perhaps better discussed holistically than partitioned.

### Continental Shelf Topography and Tidal and Shelf-Edge Fronts

In each of these marginal seas, the continental shelf occupies about half the total area. The shelf of the Bering Sea is a flat, featureless plain (as was noted in 1778 by Captain James Cook) and occupies almost one half of the area of the basin, the continental edge lying NW-SE across the basin (Coachman, 1986; Schumacher and Stabeno, 1998). A strong break of slope occurs at the shelf edge, with depths of 1000 m only tens of kilometers beyond the 200-m contour. The Okhotsk Sea shelf encircles the entire basin and is approximately the same width everywhere. It is narrower than the Bering Sea shelf, and the continental slope is less steep. These facts have significance for the occurrence of a relatively strong slope current in the Bering Sea and an unusually complex system of shelf break and shelf fronts there; tidal fronts are known to occur also on the continental shelf of the Okhotsk Sea (Zhabin *et al.*, 1990).

As Schumacher and Stabeno (1998) point out, tidal currents provide the dominant source of kinetic energy over very extensive regions of the inner shelf and may generate the observed residual flows. Although there is much variability in the exact tidal mechanism that dominates local forcing in different shelf regions: shoaler than 40–50 m, there is only very weak stratification so that tidally mixed conditions dominate.

Three convergent fronts, defining three shelf domains, occur in the open-water season over the continental shelf of the Bering Sea (Coachman *et al.*, 1980; Coachman, 1986). The depths of wind and tidal mixing overlap along an inner, coastal front that lies along the 50-m isobath for a distance of 1000 km from Bristol Bay to about 60°N, where it passes east of St. Lawrence Island and so north to the Bering Straits. This front encloses unstratified, low-salinity, high-nutrient water of the coastal domain that originates at the Umiak Pass. Flow in this domain is strongest adjacent to the front, where pumping of nutrients into the euphotic zone may occur (Kachel *et al.*, 2002).

A middle shelf front lies along the 100-m isobath and defines the central shelf domain, where mean flow is weak and variable, strongly influenced by subtidal fluctuations of several days duration. Here, in the absence of ice cover, winter wind stress mixes the water column to 100 m although, in spring, ice melt may provide sufficient stability for shallow stratification, before summer heating at the surface establishes stratification more securely. Stratification may break down in the middle shelf domain in summer after severe summer storms (Schumacher and Reed, 1983) even though there is a very strong thermal gradient across the pycnocline. The "cold belt," or bottom layer, may take temperatures <2°C even in summer; nutrient levels are high in the bottom layer, and rapidly exhausted in the surface layer.

A more complex (50-km wide) shelf-break front with a strong salinity gradient occurs along the 200-m isobath and can be traced into the western Bering Sea marking the transition between the isohaline oceanic water of the deep basin and the shelf waters that have important horizontal salinity gradients. Seaward of the shelf-break front, flow is consistently to the northwest around the central gyre, as part of the Bering Slope Current, whereas, inshore of this front, the flow is dominated by tidal forcing and is primarily cross-shelf. In the outer shelf domain, between middle and shelf-break fronts, strong flow generally follows the 100-m isobath, generating meanders that force eddies up over the shelf; subsurface water below the winter mixed layer is relatively warm.

## Defining Characteristics of Regional Oceanography

All modern accounts of this region emphasize the variability of conditions here and describe major interannual fluctuations in the physical—and hence ecological—regime; these fluctuations respond both to the Southern Oscillation and to the longer-term changes, decadal or secular, of mean global temperature.

Each sea lies behind a long island arc: the Bering Sea is separated from the Pacific by the Aleutian Islands, the Okhotsk Sea by the Kurils. Flow from the Alaska Stream into the central gyre of the Bering Sea occurs through the western passages; through Near Strait, the flow is almost directly into the East Kamchatka Current, whereas the Bering Slope Current, the deep eastern limb of the cyclonic gyre, is fed through Amchitka Pass (Favorite, 1974; Stabeno and Reed, 1994). Flow from the Bering Sea northward, forced by a 0.5-m difference in elevation between Pacific and Arctic Oceans, originates in continental shelf water passing around the northern limb of the central gyre and into the Gulf of Anadyr, then northward along the Asian coastline (Fig. 11.1). In 1990, the axis of the Alaska Stream shifted southward so that flow through the Near Strait into

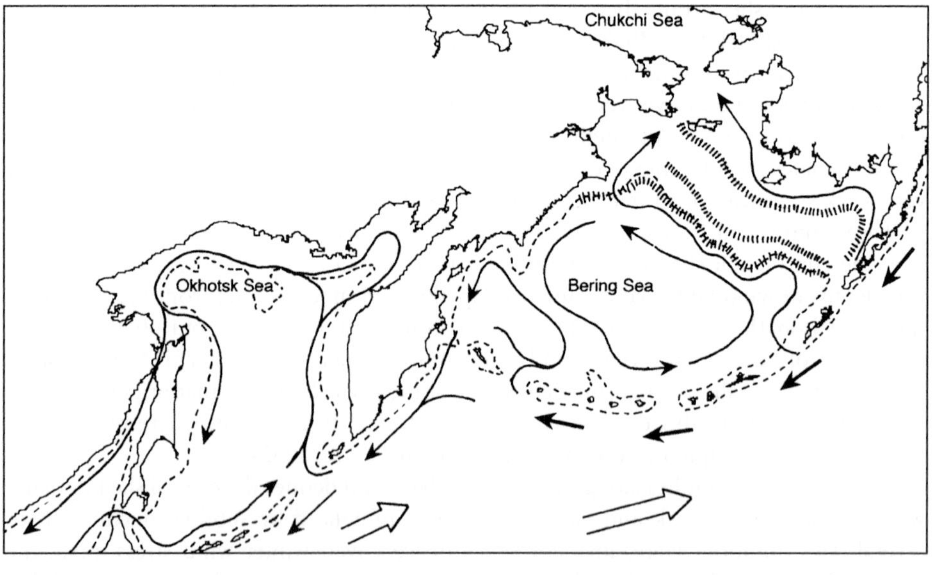


**Fig. 11.1** Idealized circulation of Bering and Okhotsk Seas, the 200-m isobath being the dashed line. Inner, middle, and shelf-break fronts associated with the very wide shelf are shown, as is also the flow of the Anadyr Current that feeds Bering Sea productivity north into the Chukchi Sea. The warm Soya current is shown passing from the Sea of Japan into the rim current of the Chukchi Sea.

the western Bering Sea ceased almost entirely, and the East Kamchatka Current was seriously weakened. This entirely unexpected change in the oceanographic regime must have modified significantly the ecology of the whole Bering Sea basin because subsurface water masses must have cooled strongly.

Although the Alaska Stream loses its characteristic highly stratified structure in passing between the Aleutian Islands, dispersion of low-salinity shelf water from the eastern quadrant results in strong stratification of the central gyre. Thus, the surface water mass of the Bering Sea, as in the Sea of Okhotsk, has low salinity and lies above a stable halocline at 100–300 m (Dodimead *et al.*, 1967; Takenouti and Ohtani, 1974). In both seas, there is a significant input of freshwater over the inner continental shelf both from ice melt (see later discussion) and from coastal runoff and rainfall. Over the shelf regions of the eastern Bering Sea, during summer stratification, the subthermocline water is subject to tidal mixing—strong or weak according to the velocity of tidal streams. The water above the thermocline remains wind-mixed.

Water from the western boundary current of the NW Pacific enters the Okhotsk Sea between islands in the Kuril chain and is circulated cyclonically around the basin. Additionally, the Okhotsk Sea receives water from the Japan Sea in the extreme south from the Tsushima Current. Over the Bering Sea shelf, flow is stronger inshore and also toward the shelf break than in the more slowly moving water of the midshelf domain. The general circulation of both seas is highly eddying and involves the formation of many quasipermanent major eddies. The surface circulation of each sea is dominated by a central cyclonic gyre situated over the deeper parts of the basin, the gyres of the two seas being connected by southward flow of the East Kamchatka Current along the east coast of the peninsula (Dodimead *et al.*, 1967). This subsequently passes around the central gyre of the Okhotsk Sea to augment the southward flow of the cold Oyashio. In both seas, cold water leaves to the southwest of the basin along the Asiatic coastline.

The Alaskan shelf receives significant inputs of freshwater from the Yukon and other rivers, whereas the Okhotsk Sea receives water only from the Anadyr, and that only into the effluent current from the basin, as this passes behind the island of Sakhalin.

These seas are subject to polar atmospheric conditions, with some influence of the Asiatic monsoon in the southwestern Okhotsk Sea in summer. Even though they lie at about the same latitude as the seas off Western Europe, they are partially ice-covered in winter and each, therefore, supports a seasonally migrating marginal ice zone (MIZ) with all that that implies for their ecology. Ice formation begins off the river mouths of the Bering Sea and, by the end of winter during cold years, ice extends over 75% of the entire sea. Ice coverage of the Sea of Okhotsk is relatively less complete and broken pack ice is more typical. During the formation of ice, the near-surface salinity increases, and this induces local density-driven flows that are sufficiently strong as to modify mean currents.

Ice cover duration is highly variable between years in the Bering Sea, varying from 2 to 28 weeks (mean 20 weeks) in the midshelf domain and the date when retreat is initiated in the south varies from mid-March to June (Schumacher and Stabeno, 1998); total coverage varies by as much as 40% about the mean. Ice coverage of the Okhotsk Sea also varies strongly between years, from complete to the line of the Kuril Islands to $<25\%$ coverage. These between-year variations appear to be forced by changes in storm tracks across the region that themselves represent the regional atmospheric response to changes in the Pacific Decadal Oscillation (PDO), positively correlated with the relative strength of the atmospheric Aleutian Low. The PDO was consistently negative from the early 1940s until 1976, when a change occurred in the Bering Sea that initiated a 12-year period of positive PDO values (e.g., Schumacher *et al.*, 2003). During this period, ice coverage was significantly reduced, and mixed-layer temperatures on the inner shelf were higher than previously. In 1989, the PDO returned to negative values and regional ice cover again increased, although—at least until the end of the century—it remained significantly smaller than before 1976 (Schumacher *et al.*, 2003). These changes are attributed to interaction between steplike changes in regional processes and the longer-term effects of the secular global warming trend.

## Regional Response of the Pelagic Ecosystem

In these two marginal seas, algal growth is both light- and nutrient-limited. Since the winter water column is well mixed, when irradiance reaches levels sufficient for growth of cells, nitrate concentration at the surface is high (10–25 $\mu$M $NO_3$ $liter^{-1}$), but is rapidly reduced to limiting concentrations after early summer algal growth (McRoy and Goering, 1974).

The satellite images are unequivocal, and record a regional maximum of chlorophyll accumulation (3–5 mg chl $m^{-3}$) at the surface, and as integrated over the euphotic zone (70–100 mg chl $m^{-2}$) in April of each year with great regularity. A secondary peak of chlorophyll accumulation (~1.5 mg chl $m^{-3}$) occurs in August–September. The highest primary production rate coincides with the spring phytoplankton bloom, though it does not decrease so rapidly as biomass; there is no increase at the time of the late summer biomass increase, which is therefore attributable to reduced herbivory. Although, in this analysis, the shelf and oceanic regions cannot be separated formally, examination of serial images suggests that the date of greatest chlorophyll accumulation in the central regions of each sea occurs 30–60 days later than on the shelves.

***Shelf Regions*** Epontic algal growth occurs below sea ice as early as reduction in snow cover and increase in sun angle permit. Although high biomass is achieved locally (~7 mg chl $m^{-3}$), integrated biomass in the water column is very small (<0.5 mg chl $m^{-2}$) compared with open-water blooms later in the season. Nevertheless, the local effect is important and the epontic algae support a wide range of other organisms.

Ice-edge blooms occur as soon as meltwater from the receding ice and increasing irradiance induces stability in the water column, although these blooms may be brief and are terminated by wind-mixing events. Such processes vary in location and timing from year to year. In coastal water, enclosed by the inner front, a single spring bloom occurs early in the season, probably fueled by nutrients remineralized during the preceding winter. In the pre-1976 cold period, this bloom was initiated during May and continued for about 1 month while its depth of maximum chlorophyll values deepened by about 1 m $d^{-1}$; it moved progressively seaward as stability was induced in deeper water across the middle and outer shelf domains. Ephemeral blooms occur throughout the summer, especially in the middle shelf, whenever wind events disrupt the two-layered density structure and allow nutrient-rich bottom water to be mixed into the euphotic zone; even during the bloom the rate of primary production is lower (< 20 g C $m^{-2}$ $day^{-1}$) than over the slope.

Nevertheless, the contemporary satellite images show very clearly how, as the season advances through May–June, the middle and outer shelf fronts of the southeastern Bering Sea each supports a major linear zone of high chlorophyll biomass, apparently >10 mg chl $m^{-3}$ (see Color plate 14). This phenomenon has been called, appropriately enough, the "Bering Sea Green Belt." The northern part of the shelf carries a rather uniform bloom, initiated earlier in the year, while, in the west, only minor blooms occur along the steep-to coast of Kamchatka. Nutrient pumping in the bottom water onto the shelf is induced by the passage of low-pressure systems, which also enhance the cross-shelf advection rates; despite such events, nitrate-fueled production is minimal over the shelf and maximal along the slope.

In May–June, the shelves surrounding the Okhotsk Sea support high chlorophyll more uniformly, although in some images, there is evidence of a shelf edge frontal bloom along the eastern shelf of this sea with lower chlorophyll in midshelf. On the southeastern shelf, high rates of primary production persist during the summer in the lee of the Aleutian Islands rather than in the highly turbulent flow through, for instance, Unimak Pass.

Although diatoms are thought to have dominated phytoplankton biomass in the Bering Sea for many years, this may have changed recently. During the very warm years 1997–98, there was an unprecedented incursion of the coccolithophore *Emiliana huxleyi* into the Bering Sea (Olson and Strom, 2002) and this organism has since become a major component of the autotrophic biomass there. Images obtained from the MODIS sensors confirm that the Bering Sea is now one of the very few places in the oceans where coccolithophore blooms reach biomasses >10 mg C $m^{-3}$. The Okhotsk Sea remains relatively free of such blooms.

In the inner and middle shelf domain the zooplankton fauna is characterized by the copepods *Calanus pacificus* (= *C. glacialis* of Heinrich, 1962a; Motoda and Minoda, 1974), together with several species of *Pseudocalanus* and *Acartia longiremis* and other biota. The euphausiid *Thysanoessa raschii* dominates the early spring zooplankton, followed by *C. pacificus* during early summer. In this area, *C. pacificus* passes the winter as stage 5 copepodites (C5s), presumably in deep water beyond the shelf edge. Springer *et al.* (1989) computed that the demand of these herbivorous copepods, and their associated biota, was such that they were capable of modifying standing stocks of algae and at times required the total daily production to satisfy their needs. Of course, classic food-web models of the Bering Sea assume that the dominant phytoplankton biomass is diatoms and that these are transferred to higher trophic levels largely through copepods: however, Olson and Strom (2002) propose another model that must be acknowledged. It has been shown that some diatoms are suboptimal food for some copepods in one part of the ocean, because of their aldehyde content, so Olson and Strom ask: "If diatoms are a sub-optimal diet for copepods, what supports the high crustacean biomass in the SE Bering Sea?" The assumption is based on a logical non sequitur and is perhaps another instance of William Dickinson's "mythic thinking" (see flyleaf); at the very least, it ignores the affirmation of Irigoien *et al.* (2002) that a diatom diet has no negative effect on the reproduction of pelagic copepods in a dozen other open ocean regions. Nevertheless, in the context of the Bering Sea, the enquiry of Olson and Strom must be considered. They invoke grazing rates of protistan plankton, obtained by dilution experiments, in late summer, that appear to balance population growth of both large and small phytoplankton cells, although bacterial cells were not included in their computations. I believe that the most that can be said at present is that all food-web diagrams and models are to be regarded as works in progress—and no more than that. Unfortunately, many resource ecologists and fishery biologists have yet to understand this depressing truth.

***Slope Regions*** Although the highest chlorophyll biomass associated with the outer front is along the southeast shelf edge, a linear zone of relatively high chlorophyll can be traced north to the Bering Straits. It is also characteristic that very large meanders should be induced out over the slope and even over deep water, chlorophyll biomass remaining very high in these excursions toward the central Bering Sea. The chlorophyll profile over the slope follows a typical seasonal evolution: after the near-surface spring bloom, algal growth during the summer (a 120-day growing period) continues in a deep chlorophyll maximum (DCM) near the pycnocline, fueled by nutrients in the deeper water of the Bering Slope Current.

The DCM of the Slope Current bifurcates around St. Lawrence Island. The eastern part subsequently spreads over the northeastern continental slope, inducing rich benthic fauna, and exists as a near-bottom chlorophyll maximum. The western part is significantly enriched in the western jet current (Anadyr Current), which flows through the Bering Straits and passes north into the Chukchi Sea. The shoaling of this flow as it enters the shallow water of the Anadyr Strait injects Slope Current nutrients into the photic zone as a cross-shelf flow (Springer and McRoy, 1993). This may induce a plume of intense growth of chain-forming diatoms (reaching 16 g C $m^{-2}$ $day^{-1}$ and <600 mg chl $m^{-2}$ in the center of the straits) that are transported into the Chukchi Sea through the Bering Straits. Springer and McRoy describe the Anadyr Current as a "north-flowing river of oceanic water . . . maintains a portion of these shelf waters in a eutrophic bloom summer-long," and Sambrotto *et al.* (1984) stated, "This phytoplankton production system from June through September is analogous to a laboratory continuous culture." In short, nitrate is continually supplied as slope water passes up onto the shelf, while cells are continuously lost by sinking, later to fuel high benthic production.

This domain of the Bering Sea supports an expatriate subarctic zooplankton assemblage, advected through the Aleutian chain and dominated by North Pacific biota: *Neocalanus plumchrus, Neocalanus cristatus, Eucalanus bungii,* and *Metridia pacifica* which comprise 70–90% of the copepod biomass, together with small copepods such as *Pseudocalanus* spp., the euphausiids *Thysanoessa* spp., the chaetognath *Sagitta elegans*, and others. These zooplankters support the immense flocks of diving planktivorous seabirds (mostly auklets, *Aethia* spp.) most of which nest in the northern part of the Bering Sea and whose numbers are evocative of a rich food supply. The larger copepods pass through one (*N. cristatus, E. bungii,* and *M. pacifica*) or two (*N. plumchrus*) generations a year (Heinrich, 1962b) and their abundance peaks during the summer.

Unlike the *C. pacificus* association of the eastern shelves, these oceanic zooplankton, which pass through the Slope Current and into the Anadyr Strait—that is, they are advected through the highly productive shelf-edge front and Anadyr Straits algal blooms—have a food demand that is far lower than the algal production rate, so the blooms are uncontrolled by herbivore grazing pressure. In fact, the advection of expatriate North Pacific planktonic biota into the Anadyr Straits is itself a significant carbon flux, headed for the deposits of the Chukchi Sea: Springer *et al.* (1989) compute this flux as $1.8 \times 10^{12}$ g C during a single summer.

***Deep Basins*** The deep regions of the Okhotsk Sea are less productive than shelf regions, although both are characterized by spring blooms in which production rates >5 g C $m^{-2}$ $d^{-1}$ are typical, and highest productivity occurs along the edges of the Sakhalin and Kashevarov Banks (Sorokin and Sorokin, 1999, 2002). During the course of the summer, episodic blooms are typical of the same regions and production remains quite high (~0.7 g C $m^{-2}$ $d^{-1}$) through July–August, associated with a typical DCM structure; the euphotic depth over deep water is 30–50 m, compared with only 12–25 m over the shelf. Diatoms form 50–80% of regional phytoplankton biomass, and bacterial and ciliate numbers are about 2–3 times higher over the shelf than in the open ocean. Although the coccolith bloom over the Bering Sea shelf appears to be a relatively novel phenomenon (see earlier discussion), blooms of *Coccolithus pelagicus* appear to be a feature of early spring blooms in the open Okhotsk Sea that occur very early in the season, as soon as the ice cover breaks up (Broerse *et al.*, 2000). These blooms support a typical cold-water zooplankton and euphausiids, comprising essentially the same genera and species as in the Bering Sea.

I can locate very little direct information on the production ecology of the oceanic part of the Bering Sea, perhaps because the shelf fisheries have almost exclusively occupied the attention of funding agencies. Thus, little modern research seems to have addressed the central cyclonic gyre since the review of Motoda and Minoda (1974), who gathered together scattered, earlier observations. It is clear from their study that (as could be expected) the populations of large oceanic copepods in the open gyral region resemble those advected into the Slope Current (see earlier discussion) with biomass (20–40 g wet weight $m^{-2}$) not different from values over the shelf; only in the Slope Current itself are values significantly higher. A zooplankton profile in the eastern part of the gyre, but seaward of the shelf-edge front, shows a biomass maximum at the pycnocline at about 45 m which suggests that the DCM described previously extends clear across the gyre in summer. The standing stocks of diatoms near the surface in the central gyre in early summer are lower by about one order of magnitude than at the shelf break ($10^6$ compared with $10^7$ cells $m^{-3}$).

## Regional Benthic and Demersal Ecology

It has long been assumed that the large demersal fish stocks of this province are supported by a productive shelf ecosystem, and this relationship has been the focus of two relatively recent fisheries-oriented programs: PROBES and ISHTAR (McRoy *et al.*, 1985; Hansell *et al.*, 1989). The classical supposition has been confirmed, as has the fact that the levels of phytoplankton production apparently exceed the demands of planktonic herbivores, and so support a very rich benthic ecosystem that is said to be the food base for the fisheries resources, but is probably not—as I shall discuss later.

For the Bering Sea, we do have access to holistic quantitative surveys of benthic organisms from Soviet investigations from the 1930s to the 1960s that covered the entire shelf region. These investigations showed that the benthic infauna is dominated by low-Arctic species associations, such as the circumpolar *Macoma calcarea* community (see Thorson, 1957) that occupies the midshelf region (40–80 m) from the Gulf of Anadyr right down to Bristol Bay. With the nominate species, there occurs also *Mya truncata, Cardium ciliatum, Ophiocten, Sericeum, Pectinaria granulata*, and *Astarte* spp. With increasing sand component in the substrate we would expect this community to be replaced by the arctic *Venus* association, whereas increasing silt would lead to a *Portlandia* community. The distribution of benthic biomass is very unequal, but is generally higher (220–950 g $m^{-2}$) on the northern shelves than on the eastern (55–169 g $m^{-2}$) shelf. Maximal densities are at midshelf depth (75–100 m) on all shelves. The higher biomass on the northern shelves is dominated by large epibenthic species (sand dollars, anemones, sponges) so that the percentage of the benthos there that is accessible to demersal fish is relatively low: 40–50% on the wide eastern shelf and the cold Gulf of Anadyr, 15–20% in the northern Chirikov basin and along the northwest coast.

The relations between the benthic infauna and the population biomass of nekton and fish are complex. The highest biomass occurs in the cold Anadyr region where the demersal fish fauna is depauperate, and where some populations of infaunal-feeding species (*Limanda, Hippoglossoides*) are small and perhaps not self-sustaining, as in the Chukchi Sea. On the main Bering Sea shelf, these and another flatfish, *Lepidopsetta aspersa*, are the only forms that are directly dependent on benthic infauna: many attempts have been made (some reviewed by Alton, 1974) to estimate benthic production from observed biomass and to balance this against demand by flatfish and epibenthic nekton, but the balances obtained remain unconvincing. What is clear, however, is the lack of direct trophic connection between the infauna of the benthos and the demersal fish that form the bulk of fish biomass here: pollack (*Theragra chalcogramma*), cod (*Gadus macrocephalus*), turbot (*Atheresthes stomias*), and others. These feed on smaller fish, which themselves eat natant invertebrates (shrimps, amphipods, euphausiids). These, in turn, are dependent on superficial meiobenthos and the microbial content of organic detritus.

However, the demersal ecosystem of this marginal region is in constant flux in response to changes in environmental forcing, just as is the pelagic ecosystem. But for the demersal system, there is an additional complication in interpreting observations: we have also to account for the consequences of removal by commercial fisheries of a major fraction ($\sim 1.5 \times 10^6$ t $y^{-1}$) of the higher trophic-level biomass, whether of fish (principally pollack and halibut) or crustaceans (principally the crabs *Paralithodes* and *Chionectes*). Long-term changes in overall biomass have been observed but not fully understood. An excellent example of the level of uncertainty is given by the analysis by Livingston *et al.* (1999) of trends since the 1970s in three trophic guilds: consumers (i) of pelagic fish, (ii) of benthic infauna, and (iii) of crabs and small fish. A number of case histories are examined to determine causal relationships forced by changes in environmental factors and by the fisheries: each analysis renders no better than further uncertainty. Is the long-term decline of Pacific halibut due to a contemporary rise in temperature, or is it due to increase in the pollack stocks and hence poor recruitment? Is the long-term decline in red king crabs due to the contemporary increase of yellowfin sole, or to high discard mortalities in the crab fishery? Those who believe that "ecosystem-based" fish stock management is just around the corner would do well to examine the literature on such problems on the east Bering Sea shelf, where analysis has been carried further than in most other major fishery areas.

## Synopsis

*Case 1—Polar irradiance-mediated production peak, strongly modified by shallow-water processes*—The pycnocline may be very shoal (< 10 m) in boreal summer after a rapid near-surface stabilization in April and May, which is not fully captured by archived data, and then deepens progressively to 90 m in January. $Z_{eu}$ is consistently between 20 and 40 m, so thermocline is illuminated only in summer and autumn, and thus during the declining phase of algal biomass (Fig. 11.2). The seasonal productivity cycle

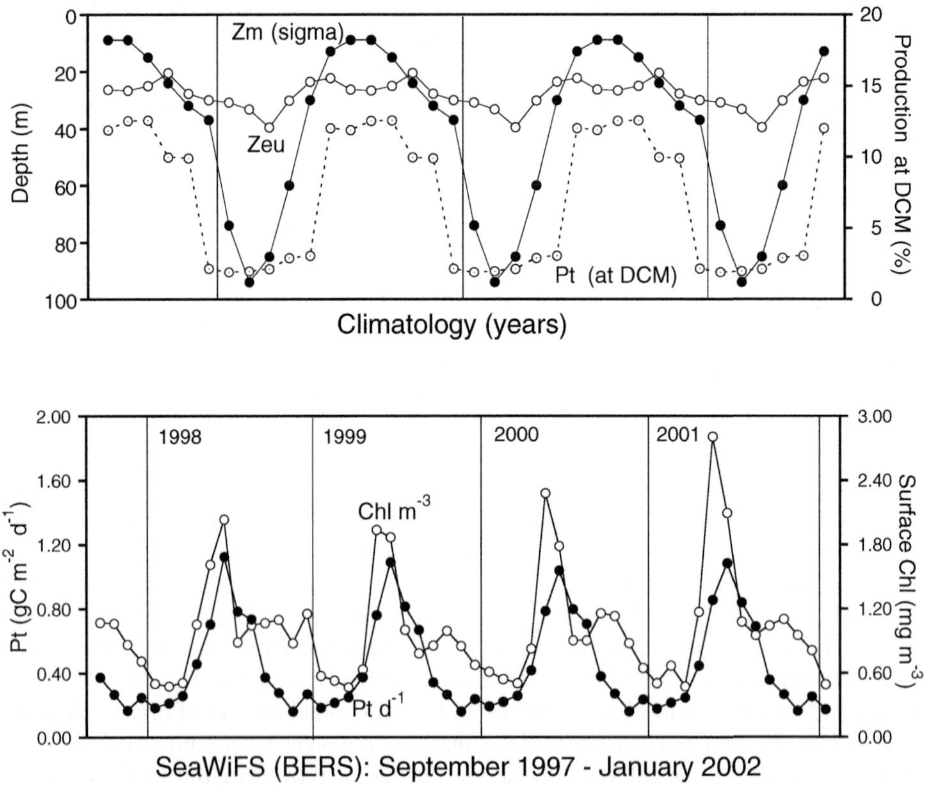


**Fig. 11.2** BERS: seasonal cycles of monthly surface chlorophyll and depth-integrated autotrophic production for the years 1997–2002 from SeaWiFS data together with characteristic seasonal cycles of mixed-layer depths from Levitus climatological data and photic depths computed from characteristic irradiance and the archive of chlorophyll profiles discussed in Chapter 1.

approximately tracks the light field, but the maximum rate occurs prior to midsummer, perhaps due to nutrient limitation, and the vernal rate increase is delayed until $Z_m$ begins to shoal in March. Accumulation is coincident with the rate increase in P, and decline is coincident with the decline in P through June–August. In September–October, accumulation is renewed and quickly reaches its annual maximum, values remaining high throughout winter in open water, perhaps representing sediments suspended during the period of deepening of mixed-layer depth. Renewed accumulation in September is consistent with the decrease in near-surface copepod biomass as the migrant copepods descend to overwintering depths.

# Pacific Westerly Winds Biome

## Pacific Subarctic Gyres Province, East and West (PSAG)

### Extent of the Province

This is a rather large province that should perhaps be partitioned zonally; nevertheless, in the first edition of this work it was thought better to treat it holistically because of a dearth of sufficient information to support a partition. Since then, however, we have access to the Canadian JGOFS studies in the NE subarctic gyre and also to the results of the JGOFS North Pacific Process Study at Station KNOT in the western gyre. These new studies will enable us to discuss the undoubted differences in the ecology of the two parts of this province. Despite this, I still prefer not to partition the gyre formally, so as to emphasize similarities rather than differences.

To the east and north, PSAG is enclosed by the offshore boundaries of the ALSK coastal province, and to the north and west by the boundary of BERS along the line of the Aleutian and Commander Islands, then south along the edge of the East Kamchatka Current, and finally along the Kuril Islands shelf to the eastern cape of Hokkaido. To the south, I have taken the divergence of surface flow along about 45°N (Uda, 1963; Ware and McFarlane, 1989), where North Pacific Current and West Wind Drift waters diverge. The boundary between the larger eastern and the smaller western subarctic gyres lies to the south of the westernmost group of Aleutian Islands.

### Defining Characteristics of Regional Oceanography

The subarctic (or subpolar) gyral circulation of the Pacific Ocean—Subarctic Current, Alaska Stream, and East Kamchatka Current—includes the partially isolated Alaskan and Western Subarctic gyres. This circulation loses some water to the Bering Sea, but the principal loss is into the California Current, because influx from the West Wind Drift cannot be balanced against the loss term without consideration of the entrainment of subhalocline water into the surface layer. This has significance for the ecology of the province through the continual supply of deep nutrients to the photic zone. These topics and others were reviewed in three now-classical regional studies: Dodimead *et al.* (1967), Uda (1963), and Favorite *et al.* (1976).

The eastern, Alaskan gyre has long been the site of intensive studies at Ocean Weather Station P (OWS P) at 50°N 145°W, whereas the western gyre is less well known. The Alaskan gyre forms an elongated (SW-NE) cyclonic dome in the halocline, shoaling to 75 m at the center. The dome, which is axial to circulation, can be identified by a surface salinity maximum in the range 32.8–33.0%. Seasonal changes in the depth of the surface layer of low-salinity water are slight, and it is effectively isolated from deep water by the permanent halocline at 100–150 m. Changes in SST lag heat input at the surface by about 2 months so that the maximum (13–14°C at OWS P) is reached in August and a minimum (5–6°C) in March. Within the isothermal surface mixed layer, a thermocline is established at 30–60 m in early summer, after which time the transient stratification, which occurs in the surface layer with increasing frequency during the summer, mixes down within a few days to join the main seasonal thermocline. The summer thermocline remains at about the same depth until it progressively deepens and is eroded by increasing wind stress (and decreasing heat input) during fall and early winter. Between this feature and the permanent halocline, salinity increases progressively during summer, because of exchange across the top of the halocline. Miller *et al.* (1991) suggest that Ekman suction of 1.5–3.0 m $month^{-1}$ is also the mechanism for a continual supply of new nutrients across the nutricline.

The western gyre is smaller, triangular in outline (look at the triangular space between the Commander and Kuril Islands), and it, too, has a central halocline dome although this is deeper than in the Alaskan gyre, reaching 200–300 m at 180°W. The western, equatorward limb of the western gyre forms a coastal jet, the Kuril Current, that heads southwest and is topographically locked to the topography of the Kuril islands.

The southern edge of the PSAG province (along 45°N for practical purposes) is intended to represent the division between meandering flow in the Subarctic Current and in the warmer, saltier water of the West Wind Drift or Transition Zone (see NPTG Province). Dodimead *et al.* suggest that the southern edge of subarctic conditions is indicated by the 7°C isoline at the depth of the halocline or maximal winter mixing. North of this line, temperature profiles in the subarctic water are consistently dicothermal, having a thermal minimum at about 100 m depth.

Because of the early interest in this region, and the serial observations at OWS P from the mid-1950s to 1981, we have almost unparalleled information on decadal and longer-term trends in the physical environment. As Whitney and Freeland (1999) point out, near-surface warming and salinity progressively declined during this period, and this will tend to decrease the density contrast in the upper part of the water column. Hence, we shall expect (for equivalent mixing) a reduction in annual nutrient flux, as indeed has been observed, and also a higher mean irradiance in spring within the now-shallower surface mixed layer.

Though nutrients will be discussed again later, it will be useful here to note that the southern boundary of PSAG is marked by a strong surface gradient in mixed-layer nitrate during winter. A contour of 0.4 μM passes across the whole North Pacific basin, just beyond the southern boundary of the PSAG province, which is almost entirely occupied by values in the range 5–20 μM. Maximum mixed-layer winter nitrates (15–20 μM) occur south of the western Aleutians, and to the south of Alaska (Anderson *et al.*, 1969; Glover *et al.*, 1994).

## Regional Response of the Pelagic Ecosystem

In the Pacific Subarctic, we get some confirmation of the ecological reality of the borders of a province from biogeographic data. Between the Polar Frontal Zone and the northern coasts, some subarctic biota have distributions that very closely match the boundaries of PSAG (McGowan, 1971). The most important species included in this assemblage are some copepod species that will be discussed later (*N. plumchrus*, *N. cristatus*, and *E. bungii*) together with the chaetognath *Sagitta elegans*, the euphausiids *Thysanoessa longipes* and *Euphausia pacifica*, and the mollusks *Lima helicina*, *Clio polita*, and *Clione limacina*. The distribution envelopes of these species match the borders of PSAG very well, with the important exception that some species also occur in the adjacent coastal provinces ALSK and CALC, and the polar province BERS. The oceanic region of the Pacific Subarctic is their center of distribution from which some individuals are lost in surface flow and by subduction. Of these, *E. bungii*, occurs just below the thermocline in the equatorial Pacific (see Chapter 2), whereas *N. cristatus* is transported equatorward at 600–800 m with the submergence of Oyashio water below the Kuroshio off Japan: but who knows how far some expatriate individuals of these and other species may not be transported?

PSAG(E) is the site of two of the great enigmas that puzzled biological oceanographers for many decades: (i) why should a spring bloom (in the sense of an accumulation of chlorophyll) not occur here as it does in the North Atlantic? and (ii) why does primary production during summer not utilize all the available mixed-layer nitrate? I shall return to these questions later, but readers will recall that the explanation that for many years seemed satisfactory was that the phytoplankton at OWS P was dominated by small ($< 20\,\mu m$) cells, consumed by copepods whose populations are sufficiently large at the end of winter as to suppress the accumulation of chlorophyll when productivity increases in spring. This model, of course, has been replaced by others that all involve some aspect of Fe limitation—it will be recalled that this region was the first high-*S* region to be recognized and the site of the first Fe-limitation experiments—but I suggest that the real system is far more complex than suggested by simple Fe-limitation models, such as that of Banse and English (1999). This was based on seasonal CZCS images in which neither spring nor summer blooms could be observed satisfactorily anywhere in the Subarctic Pacific, although some episodic autumn blooms were seen. These, it was suggested, were responses to iron in episodic falls of volcanic dust. Unfortunately for this model, the inferred lack of a chlorophyll accumulation cycle was artifactual, due to the very poor coverage of the region by the CZCS sensors.

That a seasonal cycle of production and chlorophyll accumulation does occur here is now reasonably well known, although our accounts of it are heavily weighted by observations at OWS P: almost all accounts of the Pacific Subarctic regime continue to emphasize the "lack of a spring bloom" or rather the lack of seasonal accumulation of chlorophyll. But, as the SeaWiFS and MODIS images make clear, there are significant differences between seasonal cycles in the two subgyres: in PSAG(E), as expected from OWS P data, chlorophyll accumulation and production rate both remain relatively high in summer, with a seasonal range of 0.25–0.6 chl $m^{-3}$. This corresponds well with observations at OWS P of seasonal changes in chlorophyll biomass of 0.2–0.4 mg chl $m^{-3}$. Peripherally around the gyre, however, this range increases and a spring maximum is observed in the SeaWiFS data; at 200 km beyond the shelf, this occurs in May, and the annual range is 0.5–1.4 chl $m^{-3}$ (Brickley and Thomas, 2004). In the data representing the entire Western gyre, a sharper peak in both chlorophyll biomass and production rate may occurs during May, as it did in four years (1998–2001) in the SeaWiFS observations and again in 2004 in MODIS data; this spring bloom appears to be a typical seasonal response. Further, the seasonal range of chlorophyll biomass in PSAG(W) is greater than in PSAG(E), increasing from winter values of 0.25 chl $m^{-3}$ to 1.3 chl $m^{-3}$ in spring.

The JGOFS section along the P line, and at OWS P, in 1992–1997 showed that small ($< 5\,\mu m$) cells dominated the phytoplankton everywhere, with some evolution offshore along the line toward a relative increase in the importance of very small cells. Further, even in the presence of a DCM, there was very little change in size composition down the profiles (Boyd and Harrison, 1999). These cells are predominantly autotrophic flagellates. Episodic contributions of larger cells ($>20\,\mu m$) to total biomass and productivity were noted, as at the inshore end of P line in March 1993, and in March 1996 and September 1995 at the outermost stations. In such episodes, small diatoms are dominant, and these cells are more important every year during the period of higher chlorophyll biomass.

In recent years, we have come to assume that protistan microzooplankton are the principal consumers of these small autotrophs and thus control their population biomass rather than, as was previously assumed, the large copepods. However, as discussed by Rivkin *et al.* (1999), there is other evidence to suggest that, on an annual basis, this consumption could account for no more than 50% of potential autotrophic production; dilution experiments demonstrated that protists really consume more heterotrophic bacteria than either *Synechococcus* or organisms containing chlorophyll-*a*. Grazing losses to protists of eukaryotic phytoplankton was less than 50% of the growth, leaving these cells with the potential to double their biomass every 3–6 days.

As Rivkin *et al.* comment, their results "do not support the paradigm" concerning the effects of protists grazing. Nevertheless, this remains the current textbook explanation of the OWS P phytoplankton cycle (e.g., Miller, 2004), as taught today in schools of oceanography. It is also the basis of recent models of processes at OWS P (e.g., Denman and Peña, 1999), which simulates $NO_3$ flux and utilization, the physiological effects of Fe limitation, and zooplankton consumption. I find more convincing the construct of Vézina and Savenkoff (1999) that follows the increasing food-web diversity that occurs at OWS P in late summer when chlorophyll is maximal and when the microbial loop food web is most strongly developed. It emphasizes the role of ammonium regeneration and hence the suppression of nitrate uptake by autotrophs, as well as the strong coupling between microzooplankton (dinoflagellates and ciliates) and mesozooplankton (copepods). Much of the autotrophic and bacterial production in late summer is not consumed and enters the detrital pool. This model exposes serious gaps in our knowledge, especially of the larger microzooplankton and the smaller mesozooplankton, including naupliar stages. Wheeler and Kokkinakis (1990) had earlier asked if ammonium recycling might not inhibit nitrate uptake. By careful budgeting, and investigating temporal variability in $NH_4$ and $NO_3$ concentrations, they concluded that this might be a significant factor that would impose an upper limit on $NO_3$ uptake in nutrient-rich, grazing-balanced ecosystems.

Results such as these, it seems to me, illustrate the dangers of jumping to conclusions on the basis of limited observations, as we all did in the early years of studies at OWS P: what are we to make, for instance, of the observations of Wong and Matear (1999) of episodic silicate limitation in long time series (1970–1980) of proxies for OWS P productivity? In the summers of 1972, 1976, and 1979, $SiO_3$ at OWS P was depleted to $< 1.0\,\mu M$ liter$^{-1}$. There was also unusual depletion of nitrate in some years, and the ratio of $\Delta NO_3/\Delta SiO_3$ for 1979 showed that diatoms dominated the nutrient draw-down, whereas in 1972 this was not the case, and the phytoplankton community was probably similar to those years when anomalous draw-down does not occur. Sediment trap data during the 1980s showed that in years of abnormally large flux, the opal (and hence diatom) content was not unusually high. This is indeed a cautionary tale.

None of this negates the possibility that Fe limitation may establish a phytoplankton community deficient in large cells, as suggested by Martin *et al.* (1989), who outlined a correlation between regional aeolian Fe inputs, nitrate depletion, and community cell size in the Gulf of Alaska. It may be, as suggested by Banse (1990a), that Fe limitation applies only to large cells (small surface area/volume ratio in the presence of low concentrations of Fe), which normally can outcompete nanophytoplankton for nitrate. If aeolian Fe limitation is eventually proved to be the critical factor in determining the composition of the dominant autotroph community in this province, it will no doubt have a more significant effect in the Alaska gyre than further west because aeolian dust from the exposed loess deposits of eastern Asia must be attenuated eastward (Duce and Tindale, 1991; Duce *et al.*, 1991).

Despite close coupling between autotrophic and heterotrophic microbiota, the subarctic ecosystem also supports an abundant population of several species of large herbivorous copepods, though their life histories do not resemble those of *Calanus finmarchicus* of the

North Atlantic. *Neocalanus plumchrus, N. cristatus*, and *E. bungii* are the most important of these and comprise 80–95% of total mesozooplankton biomass. *E. bungii* has a 2-year life cycle, but the others undergo maturation from C5s to adults below the surface layer without a long overwintering period. Consequently, as the recent JGOFS investigations confirmed, the annual biomass maximum of *Neocalanus* copepods in the surface layers occurs in spring and copepodites are present in surface waters at all stations along Line P during the winter (Goldblatt *et al.*, 1999).

*Neocalanus plumchrus* adults are present throughout the year at 400 m and some reproduction occurs in all months, peaking in winter. Female *N. cristatus* lie deeper (800–900 m) and reproduce in all months, though peak egg production occurs in November. *Eucalanus bungii* has a 2- to 3-year life cycle, the complex details of which place the whole population in the upper 100 m in summer and at 250–500 m in winter. Development from eggs to C4s occurs in the first year, with the C3s and C4s overwintering at depth and developing as far as C5s during their second summer. They pass their second winter as C5s and, maturing at depth, the new adults produce a new generation of eggs near the surface. Some of the females may enter diapause for a third winter and produce another batch of eggs near-surface the following spring.

The seasonal changes in mesozooplankton biomass in surface layers are largely forced by the annual life cycles of *Neocalanus flemingeri* and *N. plumchrus* that reproduce at depth. Their nauplii and early copepodites are present at the surface in winter, and the copepod biomass maximum occurs when 50% of the individuals of these species exist as the C4 stage and declines when the C5s migrate back to deep water in summer, there to mature and spawn. This migration reduces the near-surface mesozooplankton biomass by about 65%.

The important result of all these complexities is that some individuals of each species are present in the surface layers at all seasons. It is this fact that prompted the classical explanation of the OSW P observations that sufficient grazing pressure was already available to suppress the accumulation of phytoplankton biomass during the spring increase in production rate. As Goldblatt *et al.* (1999) show, mesozooplankton biomass in spring ($\sim$20 mg C $m^{-3}$) is high relative to that of autotrophic cells (10–20 mg C $m^{-3}$). Further, the biomass of microzooplankton in spring is about equivalent to that of the larger animals.

However, when the JGOFS investigators examined more carefully than had been done before the relative vertical distribution of copepods in relation to biomass of autotrophs and microherbivores that is concentrated in the upper 50 m, they found strong vertical niche separation among copepods. Thus, *N. cristatus* and *E. bungii*—both important contributors to mesozooplankton biomass—occur only deeper than 50 m and feed on sinking aggregates, whereas *N. flemingeri* and *N. plumchrus* occur shallower than this depth. Similar vertical separation at 50 m was observed among species of *Oithona*, whereas other copepods (e.g., *Microcalanus, Scolecithricella*) occurred mostly below 75 m. Consequently, at least in summer, much of the mesozooplankton can have little impact on phytoplankton and microbial biomass. Thus the JGOFS investigations of 1996–97 confirmed earlier studies that suggested that the large copepods probably have only a minor grazing impact on phytoplankton. However, small species of copepod ($<$ 1 mm) are not well sampled in standard nets and assumptions concerning their abundance may be too low by more than an order of magnitude. Their specific activity is high, so they may be a neglected source of mortality for small cells as they are in other comparable situations: Goldblatt *et al.* remind us that in the subantarctic, *Oithona* spp. are responsible for 50% of the mortality of small cells, though forming a much smaller biomass than the large copepods. *Oithona*, of course, is the most abundant mesozooplankter along Line P. Uncertainty such as this is our common lot when we come to examine a problem like that of the phytoplankton at OWS P closely, and with truly critical intent.

Turning now to the western gyre, PSAG(W), we have less information than for the Alaskan gyre although, as discussed earlier, the SeaWiFS and MODIS images show clearly that a spring bloom does occur here with all that that implies for ecosystem functioning. As already noted, we also now have information from the KNOT station investigations, although this station is somewhat anomalous because the influence of the Oyashio influx into the gyral circulation. There is also useful serial nutrient data, routinely obtained by container ships on the Vancouver-Japan route that passes directly across both subgyres of the Subarctic Province.

These data (Wong *et al.*, 2002) show that $NO_3$ is drawn down to lower levels in summer in the western gyre than at OWS P, especially in the southern part near Station KNOT (Kyoto North Pacific Ocean Time Series, 44°N 155°E), where June values of 2 $\mu$M liter$^{-1}$ are not unusual. Moreover, the $SiO_3/NO_3$ ratios in these data indicate that much of the nutrient utilization supports a diatom bloom, as captured in the serial observations at KNOT (Liu, 2002a; Mochizuki *et al.*, 2002). Here, centric diatom (*Thalassiosira, Coscinodiscus*) numbers dominate the large cell fraction year-round with seasonal species succession, whereas pennate diatoms (*Fragilariopsis, Neodenticulata*) are important only in spring. Biomass is dominated by the smaller cell fraction year-round against a seasonally varying abundance of diatoms that form >30% of autotrophic biomass in May, but <10% at other times.

Bacterial abundance at KNOT, like the picophytoplankton, shows greater seasonal change ($0.14–1.3 \times 10^{-6}$ cells liter$^{-1}$) than in the Alaskan gyre, the seasonal maximum being lagged several months behind the seasonal chlorophyll maximum. Seasonal changes occur in the composition of the pico fraction of the autotrophic cells, so that the spring bloom diatoms are followed first by picoeukaryotes, that are then succeeded by *Synechococcus* in late summer.

Here, it has been proposed that the spring diatom bloom is induced by Fe that is delivered in dust from the deserts of China; would it be facetious to suggest that fashion currently demands that wherever you need it, Fe is found to be right at hand? It is not hard to think of other reasons why diatoms here, as elsewhere, should build their population size in the spring. For one, the influence of Oyashio conditions in the region of KNOT would appear to be a sufficient explanation: here, after winter recharge of surface nutrient levels, silicate limitation apparently terminates a diatom bloom that is initiated by stratification and enhanced irradiance (e.g., Limsakul *et al.*, 2002). Here, unlike at OWS P, mixed-layers depths exceed 100 m in winter and are <20 m in summer. Current thinking among KNOT researchers also favors significant constraints on diatom biomass by herbivorous mesozooplankton (e.g., Fujii *et al.*, 2002).

The generally more productive ecosystem of the western gyre supports larger populations of higher trophic levels than the Alaskan gyre. The migrations of Southern Ocean shearwaters (*Puffinus* spp.) take them into the Pacific subarctic during boreal summer and they are (as reported by Springer *et al.*, 1999) an order of magnitude more abundant in the western gyre than in the east. Several species of seabirds that nest in the eastern region resort later to the western gyre to feed. These observations all suggest that the prey of these birds (euphausiids, saury, and Atka mackerel) is more relatively abundant in the western gyre. Baleen whales and plankton-feeding seabirds do not follow the same pattern but are more evenly distributed. Interannual variability in the abundance of the components of the oceanic ecosystem is critical for the breeding success of many seabirds around the coasts of the region.

## Synopsis

*Case 2—Nutrient-limited spring production peak*—Pycnocline undergoes a boreal winter excursion that is deeper in PASG (W) than in the east, but shoaling does not begin

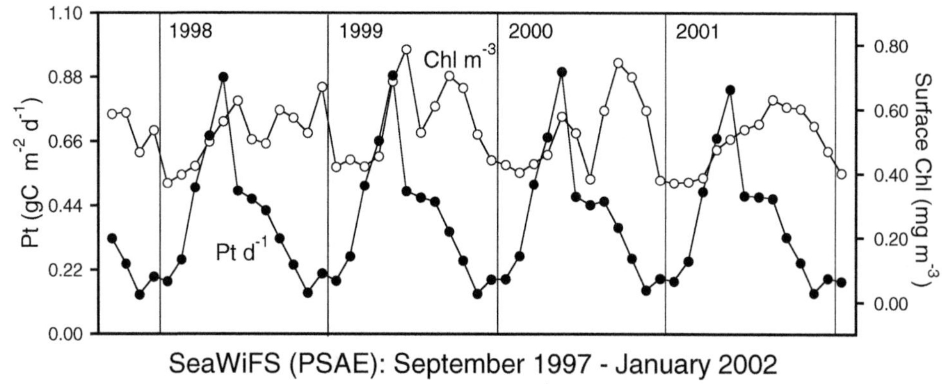


**Fig. 11.3** PSAG(E): seasonal cycles of monthly surface chlorophyll and depth-integrated primary production for the years 1997–2002 from SeaWiFS data, from Levitus climatological data, photic depths computed from characteristic irradiance, and the archive of chlorophyll profiles discussed in Chapter 1.

until April or May, so that thermocline is illuminated only from June until the October descent of $Z_m$ through $Z_{eu}$ (Fig. 11.3). Productivity is minimal in December–February, after $Z_m$ has deepened below the photic zone; productivity increases from February, prior to significant shoaling of $Z_m$, and a pulse of primary production occurs in March–May, thereafter decreasing rapidly to a summer-autumn low as $Z_m$ deepens. Chlorophyll biomass tracks the spring productivity increase, with a 30-day lag as accumulation is depressed by herbivores arriving at the surface.

A secondary accumulation occurs in September–October, weaker in PSAG(E) than in PSAG(W). After October, productivity takes minimal, winter values and chlorophyll is progressively lost, perhaps mainly by sinking. Cycles are similar in both eastern and western parts of province. Chlorophyll accumulation pattern is consistent with coincidence of spring rise of large biomass of seasonal migrants and their descent again in late boreal summer.

## Kuroshio Current Province (KURO)

### Extent of the Province

The Kuroshio Current Province (KURO) comprises the flow of the "black stream" from its origin as the Philippines Current at ~15°N to its passage eastward over the Shaskiy Rise at about 40°N 160°E, this being a convenient marker for the root of the North Pacific Current. Also included is the deep basin of the Japan Sea, excluding the shelf areas along the Chinese coastline of the China Sea Coastal Province (CHIN). The landward, warm-core eddy field along the shelf edge of the East China Sea and the flows from the Kuroshio over the shelf are considered part of the coastal boundary province of the East China Sea, whereas the eddy field of cold-core rings seaward of the axis of the Kuroshio is included in this province. Also included is the field of both warm-core and cold-core eddies from 35°N to 40°N, to the east of Honshu and southern Hokkaido, where the flow of both Oyashio and Kuroshio turns to the east. Other arrangements could be considered: perhaps because this eddy field is the root of the eastward flow in the NPTZ Province, it might have been just as logical to assign it there. Other arrangements were also considered for the Sea of Japan; the obvious alternative, which some may prefer, is to include it in its entirety as part of the CHIN coastal boundary province. The logic for retaining the eastern part in KURO is simply that this region is occupied by flow of unmodified Kuroshio water that passes west around the island of Honshu. Because this

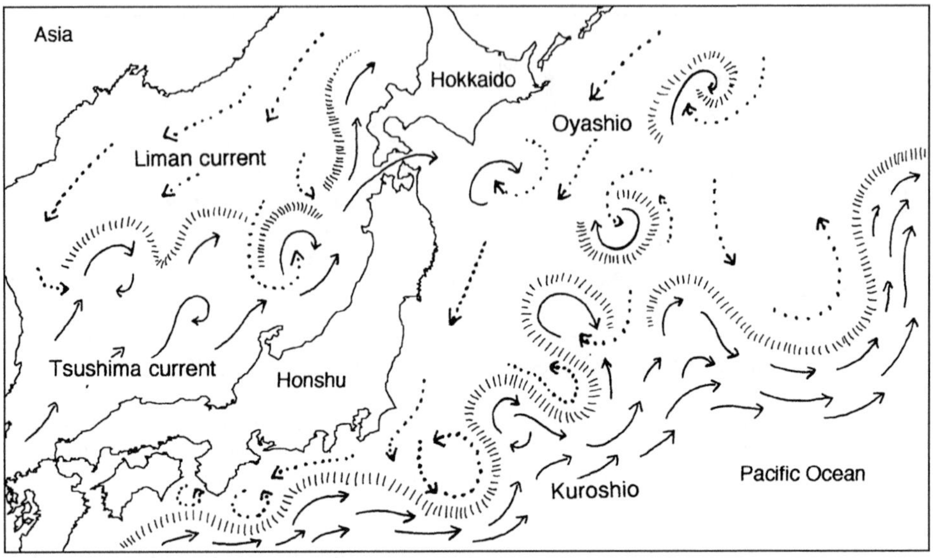


**Fig. 11.4** Conjunction between the flows of the cold Oyashio and warm Kuroshio. Hatched lines show the ideal location of fronts between cold (dotted) and warm (solid arrows) streams. The complexity of the system here (as at the confluence of Labrador and Gulf Stream flows) is the source of rich fisheries.

province spans a wide range of latitudes (10–45°N) it might, for some purposes, also be desirable to subdivide it meridionally: in this case, division could be made at the Bashi Channel (about 25°N) north of Taiwan. Figure 11.4 serves as a reminder of some of the geographical complexity of this province.

## Continental Shelf Topography and Tidal and Shelf-Edge Fronts

The continental shelf of the Japanese islands and the Sea of Japan is very narrow, and it is not useful to designate a separate coastal boundary province since the whole of the Kuroshio-Oyashio region is dominated by interaction with the topography of the islands and adjacent continent. Though the flow of the Kuroshio is, for much its length, locked to topography, it is only in the sector of the East China Sea that this topography is the edge of a shelf having significant width. The Sea of Japan has a small tidal range (0.2 m) and even on the Pacific coast of Japan tidal range is not large.

## Defining Characteristics of Regional Oceanography

Much of what is reviewed here derives from the work of Japanese oceanographers and fishery scientists, for whom the Kuroshio has special significance. I have relied heavily on the many reports of the Cooperative Studies of the Kuroshio (CSK) and especially on the review of Su *et al.* (1990). This region is, of course, monsoonal with southwest winds in boreal summer and northeast in boreal winter.

The Kuroshio is the western boundary current of the North Pacific subtropical gyre and invites comparison with its Atlantic homologue, the Gulf Stream. The beginning of the Kuroshio can be traced to the divergence of the North Equatorial Current (NEC) as it encounters topography east of Mindanao at about 10°N (Nitani, 1972). The divergent flow to the north then forms an intensified western boundary current that passes northwest along the continental edge coasts of the Philippines. This flow passes around a persistent anticyclonic eddy to the east at about the latitude (17–18°N) of Luzon, from whose shelf edge the velocity core separates before passing north to Taiwan. This separation is the site of active upwelling from October to January that is apparently unconnected to local wind forcing.

The Kuroshio is progressively strengthened by lateral entrainment of water from the subtropical gyre and, within it, the pycnocline slopes upward consistently toward the west. The main thermocline lies at 75–200 m near the shelf edge at 25°N and at 200–350 m out over deep water, the surface water mass of the Sea of Japan cooling more strongly in winter than the main core of the Kuroshio. The main flow is locked to the topography of the continental edge of the East China Sea, passing landward of the long Ryuku chain of islands and returning to the open ocean through the Tokara Strait south of Japan (Taft, 1972; Nitani, 1972). Although the main flow continues along the east coast of Japan, some passes as the Tshushima Current into the Sea of Japan; this flow responds to the alternating monsoon winds, becoming maximal in summer (Isobe *et al.*, 1994). Okhotsk Sea water enters the Sea of Japan from the Sakhalin Channel, so that the entire basin flows cyclonically: warm poleward flow along the Japanese coast and equatorward cold flow down the Sibero-Korean coast.

Velocity of the Kuroshio flow is topographically intensified north of Taiwan; the upwelling that is forced by this effect, and is reinforced by the advent of the Northeast monsoon, will be discussed in the account of the adjacent coastal boundary province (see CHIN).

Important and well-studied variation occurs in the main stream that flows south of the Japanese islands. Below southern Honshu (at about 135–140°E) a variety of paths are taken by the core, principally in two modes: (i) a simple alongshore path flowing along the shelf edge or (ii) more complex flow, passing around a large cyclonic meander that extends as much as 500 km offshore. In the second mode, which occurs during periods of high transport, the return limb of the meander lies along the western side of the topography of the Izu-Ogasawara Ridge, and the meander itself encloses a cold-core cyclonic eddy adjacent to the coast (Taft, 1972). Alternation between the two modes occurs on the decadal scale and has important biological consequences. Since 1960, the large meander has been established during five periods: 1960–63, 1975–79, 1982–84, 1988, and 1990. Since 1990, a strong warming trend in autumn-winter SSTs has occurred in the northern Kuroshio.

Retroflection out across the Pacific finally occurs at about 35°N (Su *et al.*, 1990), where strongly meandering flow passes northeastward adjacent to cold Oyashio water, so having a cold wall—like that of the Gulf Stream—on its landward side: this is the Kuroshio Front (Nagata *et al.*, 1986). The meanders and flow of Kuroshio water can then be traced eastward into the North Pacific Current as the frontal jet of the Transition Zone, or the NPTZ Province; the boundary between the Kuroshio and the NPTZ may conveniently be set at 155°E, above the topography of the Shatskiy Rise.

The eastward retroflection of the Kuroshio, and its encounter with the cool Oyashio flow, is a region of great complexity, well described by Kawai (1972). Part of the warm core (about 20°C) turns offshore near Tokyo and enters a region of meandering interactions. Thermal fronts between the cool Oyashio and the warm Kuroshio surface water are well marked and form the origin of the two fronts associated with the oceanic Polar Frontal Zone as it passes east across the North Pacific.

The Kuroshio, like the Gulf Stream, is a region of anomalously high values for the kinetic energy contained in the mesoscale eddy field (Dickson, 1983), and this is undoubtedly the reason for the high variability noted in shipboard measurements of the chlorophyll field (Taniguchi and Kawamura, 1972). The eddy field of the Kuroshio to the east of Japan resembles that of the Gulf Stream: cyclonic cold-core eddies are shed seaward and warm-core anticyclonic eddies landward (Kawai, 1979; Kitano, 1979). Eddies of both signs then propagate contrary to the direction of the current, for periods of up to several months. Warm rings form mainly in the transition area between the mean positions of the Oyashio and Kuroshio Fronts and about 120 km to the east of Hokkaido. The perturbed area between the cold and warm currents is rich in mesoscale features, including not only eddies but also warm streamers and filaments carried around cold eddies and secondary fronts (Kawai and Saitoh, 1986) and cold filaments carried around warm-core eddies (Sugimoto and Tameishi, 1992). Ephemeral episodes of warm water and strong currents in inshore waters and even bays along the south coast of Japan reflect the incidence of warm-core eddies encountering the coastline. These Kyucho events are therefore the equivalent of the irruptions of warm water at the Nova Scotia coast when a Gulf Stream eddy founders on the shelf.

Vertical structure of the water column as a response to heat exchange and wind mixing differs significantly along the course of the Kuroshio as atmospheric conditions and irradiance respond to latitude. Stratification is more permanent at the root of the Kuroshio and winter mixing to ~100 m is more sustained in the northern regions, near the Japanese islands. As in other coastal boundary currents, the summer pycnocline—and, hence, the DCM—slopes upward toward the coast.

## Regional Response of the Pelagic Ecosystem

Because of the great latitudinal extent of this province, it really comprises two regimes, here making a marriage of convenience: (i) one that is tropical, from 10°N to 25°N, having conditions very similar to the WARM Province, and (ii) another, from Taiwan northward, having a temperate, winter-mixing regime. Compared with available oceanographic analyses and with what is known of pelagic fisheries ecology (tuna, sardines, saury, etc.) there appear to be few modern studies of the primary production and consumption regimes. Most of what is available refers to the region north of Taiwan; therefore, for the southern region, the "beginning of the Kuroshio" of Japanese oceanographers, I can do no more than refer the reader to my discussion of production in the adjacent province, seaward of the Kuroshio itself (see NPTG, later). However, it is perhaps noteworthy that the serial satellite images indicate that productivity in the Kuroshio south of Taiwan exhibits no significant seasonality—a scan of the MODIS 2001 images, for instance, shows persistent oligotrophic conditions in deep water beyond the shelf break.

In the extra-tropical Kuroshio, there is a clear seasonal cycle in the mixed-layer depth (summer, 15–25 m; winter, 80–120 m) and because there is some information on the presence and differentiation of a DCM during summer, we expect a spring bloom and summer oligotrophy sequence to be widespread. The northern Kuroshio-Oyashio interaction region, east of Japan, is known to have strong seasonal algal blooms, though these are more restricted spatially than those of the North Atlantic, with regions of higher production values lying adjacent to the coastlines (Saijo *et al.*, 1970). Spring and autumn blooms were described in the central Sea of Japan by Nagata (1998).

The evidence of the serial chlorophyll images from SeaWiFS and MODIS is quite clear. The seasonal chlorophyll cycle is bimodal: a spring bloom does occur in KURO, both in the Sea of Japan and in the Pacific, everywhere north of Taiwan so that peak biomass is achieved in April, followed by a summer oligotrophic period from early June to September. Chlorophyll accumulates progressively during the last 3 months of the year only to decline again to an annual minimum in February. The spring bloom is initially strongest in the eastern half of the Sea of Japan and shifts progressively toward the north; maximum chlorophyll accumulation (< 8–10 mg chl $m^{-3}$) occurs in the "perturbed region" lying between the Kuroshio retroflection at 35–36°N and the Oyashio flow east of Hokkaido. Chlorophyll enhancement here is strongly influenced by eddy vorticity, upwelling along the shelf edge topography, mixing with nutrient-rich coastal water, and the encounter with Oyashio water (for further discussion of this region, see Shiomoto, 2000). The edges of warm-core rings are consistently associated with enhanced chlorophyll and higher-than-background abundance of other pelagic biota (Yamamoto and Nishizawa, 1986). This often appears also to result from the entrainment of slope-water biota in cold filaments drawn around the individual warm-core eddies; a typical example of the surface features of this region is that of the MODIS sea surface chlorophyll and temperature image for April 2004 (see Color plate 15). Depending on the precise definition of province boundaries, the perturbed region might be considered as the southwestern extremity of the Western Subarctic Gyre of PSAG, or as part of the KURO province. Not wishing to be dogmatic, I place it here for convenience.

There is, of course, some interannual variability in the seasonal cycle and, although we have rather few years to compare, the integrated KURO province data for both 2001 and 2002 show somewhat stronger chlorophyll maxima than the preceding 3 years. The strength of the summer DCM varies regionally, with the strongest features between Taiwan and the south coast of Honshu, where they lie deep (100–150 m), near the bottom of the euphotic zone, and contain chlorophyll maxima in the range 0.5–0.6 mg chl $m^{-3}$. In this region, daily primary production reaches 0.4 g C $m^{-2}$ $day^{-1}$. Close to the Tokara Straits themselves, the DCM is generally shallower and may extend to the bottom over the continental shelf of the East China Sea. In both spring and summer, *in situ* observations show that the Kuroshio main flow corresponds with a linear chlorophyll feature, and this is supported by satellite imagery. Chlorophyll sections across the stream southeast of Honshu in June show that the DCM has higher values within the axis of flow, and that it lies at a remarkably uniform depth (about 75 m and at the 1% isolume) from the coast to 400–500 km offshore. Predictably, the DCM lies in water with undetectable nitrate, but immediately above a strong nutricline (Takahashi *et al.*, 1985).

I have been unable to locate state-of-the-art discussions of the autotrophic cells, and of their growth and loss functions that produce the effects seen in satellite images: even recent accounts of the pelagic ecosystem, such as that of Terazaki (1999), recognize only the diatom-copepod-fish paradigm. This particular study discusses the differential distribution of diatom species in the cold- and warm-water sectors of the Sea of Japan, but omits both their seasonal succession and also the relative importance of the pico and nano fractions compared with the diatom fraction. An earlier study suggested that about 50% of autotrophic biomass in KURO is represented by $<8$-μm cells (Furuya and Marumo, 1983).

Terazaki indicates that major species of mesozooplankton resemble those of the open North Pacific, though the biomass is about one order of magnitude higher than in adjacent oceanic PSAG province. The copepod fauna is dominated by *N. cristatus, N. plumchrus, E. bungii, M. pacifica*, and species of *Pseudocalanus, Oithona*, and *Euchaeta/Pareuchaeta*. These perform seasonal and ontogenetic migrations similar to those described for the open-ocean provinces of the North Pacific, whereas during the summer they aggregate at 10–50 m below the surface. Knowledge of their daily production/biomass ratios (0.05 for six large species) is not very helpful. In both coastal and offshore regions of the Sea of Japan, mesozooplankton biomass (to $<200$ m) is maximal in April–May and is progressively reduced to a midwinter minimum.

The PICES volume mentioned earlier shows unpublished data describing large copepod winter biomass for each year in the period 1970–2000. One might take this as a proxy for the general status of the pelagic ecosystem and conclude that a general decline from high biomass during the period 1976–1994 and the strong recovery to high biomass thereafter was internally driven. However, this sequence may equally well have been imposed by mortality due to consumption by sardines: *Sardinops melanostictus* was involved in the Pacific-wide population explosion of these fish that occurred between 1970 and 1998, when catches rose from very small numbers to 1.5 million tons. This was the latest in a series of major shifts in population strength of this species. I note that the period of peak sardine catches, 1985–1990, exactly matches that of greatest decline in large copepod biomass in the Kuroshio. This species spawns preferentially in the path of the Tsushima Current off southern Honshu and in the eastern Sea of Japan.

The annual and decadal shifts in the core flow of the Kuroshio off the east coast of Honshu have long been known to have important biological consequences that can be traced through their effects on the recruitment of the small clupeid fish that collectively comprise the Iwashi fishery: principally *Sardinops melanostictus*, *Engraulis japonicus*, and *Etrumeus micropus*. Populations of these three planktivorous fish have undergone major fluctuations since the historical fishery was started in the 1500s and, as elsewhere, their relative abundances do not change simultaneously, or in the same sense. The post-1970 upswing in sardine population size represented a return to the large population that produced catches of about $1.1 \times 10^6$ tons in the 1930s, reduced to very small population size subsequently. *Sardinops* is a typical sardine in its food requirements, being able to eat diatoms in the larval and postlarval stages but not to subsist on them later: progressively larger zooplankton is utilized during its growth. *Engraulis* can utilize much smaller plankton organisms by active gill-raker filtration. It has long been known that these two biota tend to be mutually exclusive: many sardine and few anchovies, or vice versa. It has also long been known that a major population of sardines would only occur when warm oceanographic conditions obtained on the coast of Honshu.

The path taken by the Kuroshio loop off Honshu appears to control the relative abundance of species of pelagic fish. During the 1964–1971 period of rapidly declining *Sardinops* abundance, the Kuroshio meander to the southeast of Honshu formed a strong arc, enclosing a cold cyclonic eddy at the coast where conditions for the survival of sardine larvae were poor during almost a decade. Related to these observations is the fact that the conjunction between warm Kuroshio and cold subarctic water is also a boundary zone for the distributions of many pelagic species; the same must occur at the north wall of the Gulf Stream off eastern Canada. However, off Japan, the regional cultural interest in pelagic fisheries has led to a much greater knowledge of the distribution of pelagic invertebrates than in the Atlantic. For instance, maps of the distribution of the eggs and larvae of the Pacific saury *Cololabris saura* show that this species is restricted strictly to the warm Kuroshio water. Similarly, two species of chaetognath show specialization to cold and warm water: *Pterosagitta draco* in Kuroshio water and *Sagitta nagae* in coastal water.

## Synopsis

*Case 2—Nutrient-limited spring production peak*—The pycnocline undergoes boreal winter excursion (10–15 m June–September, 120 m February–March) and so the thermocline is illuminated for a relatively long period in boreal summer from May to October. Productivity increases in February and reaches its annual maximum in May, followed by a steep decline (nutrient-limited?) in June as $Z_m$ shoals above $Z_{eu}$, leading to an early seasonal minimum sustained from September to January (Fig. 11.5). Chlorophyll biomass consistently fails to track P rate in the second half of the year, even though most accumulation does occur in the period of maximum P-rate increase (February–April). Relations between P and chlorophyll are inconsistent with close matching of production and consumption. There is a strong apparent effect of seasonal vertical migrant biomass descending to overwintering depths.

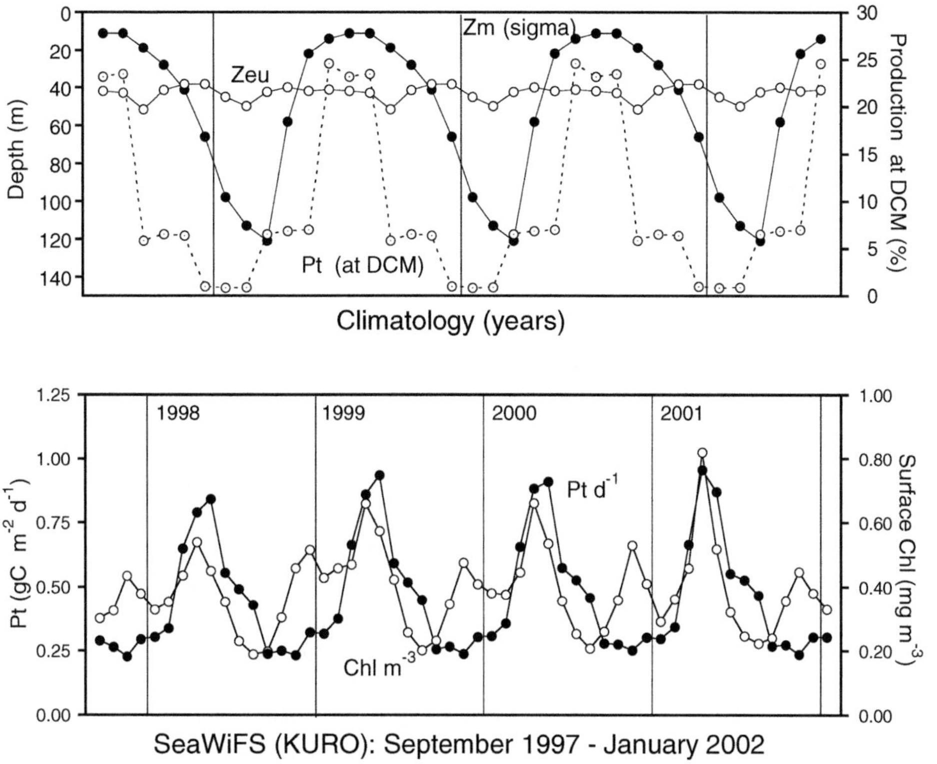


**Fig. 11.5** KURO: seasonal cycles of monthly surface chlorophyll and depth-integrated autotrophic production for the years 1997–2002 from SeaWiFS data together with characteristic seasonal cycles of mixed-layer depths from Levitus climatological data and photic depths computed from characteristic irradiance and the archive of chlorophyll profiles discussed in Chapter 1.

## North Pacific Subtropical and Polar Front Provinces (NPST and NPPF)

### Extent of the Province

The partition proposed here represents the only major change from the arrangement in the first edition of this book and is a return to my earlier proposals for a partition of the North Pacific into biogeochemical provinces (Longhurst, 1995). The definition of the North Pacific gyral province (NPTG) that was used in the first edition unfortunately merged the northern part of the central gyre, which is strongly mixed in winter, with the southern part where mixing is weaker. So I have now returned to the concept of a division of the great central gyre into a subtropical gyral province (NPST) and a tropical gyral province (NPTG). Because strong zonal differences occur in the subtropical portion of the gyre, NPST is formally subdivided—where this is useful—into eastern and western compartments.

Further, because the NPST and NPPF provinces have much in common, and should perhaps be considered as a single unit, I have included discussion of both in this section. These two provinces together lie diagonally across the ocean, and closely approximate what has come to be called the Transition Zone of the North Pacific. Because this term has been so loosely used, it is important to clarify what is *not* intended here. This is neither the transition zone of Roden (1970), nor that of McGowan (1971), who used the term for the broad flow of cool water from the North Pacific Current as it turns south into the California Current system. Nor is it the 50- to 150-m-deep halocline layer in the Alaska Gyre of Uda (1963), and, finally, it is not the Transition Zone of Dodimead *et al.* (1967), who apply the term to a zone to the north of the Subarctic Front. Here, it is used in the same sense as by the PICES oceanographers, who locate it climatically between 42°N and 32°N, or between the Subarctic and Subtropical Frontal Zones (SAFZ and STFZ, respectively): I shall use these same terms for the poleward limit of the NPPF province and the equatorward limit of the NPST province.

## Defining Characteristics of Regional Oceanography

The Transition Zone includes the area of meandering flow across the ocean that originates at the confluence of Kuroshio and Oyashio currents. It extends south to the Subtropical Convergence of Roden (1975) and north to the Subarctic Boundary that are, respectively, the SAFZ and STFZ. The warm and cold flows maintain their integrity, despite much meandering in the frontal zones, until they diverge on approaching the American continent to enter the eastern boundary current and the subpolar gyre, respectively. Both NPPF and NPST lie below sufficient wind stress of the winter westerlies that significant mixing occurs at this season, but more especially in the NPPF province. The NPST is exposed to weaker wind stress and so winter mixing is more moderate.

The extreme southern limit of the NPST province, at the subtropical front or STFZ, lies below the convergence between the westerly winds of the temperate zone and the subtropical easterlies or trades. Roden (1970) reminds us, however, that although climatic maps of wind stress suggest that the westerlies blow persistently toward the east, instantaneous maps reveal a series of depressions propagating toward the east at the crests of long planetary waves, of which two or three usually occupy the air mass over the North Pacific. Now that satellite images are commonplace, we are constantly reminded that instantaneous images reveal the extent to which climatic means are misleading. Nor is the STFZ itself as simple as portrayed in textbook diagrams; it is bounded by two meandering fronts between which surface water is characteristically 18.0–18.5°C, and between which there is a strong salinity gradient to water having salinity >35.1%, characteristic of the subtropical gyre. Statistical analysis of the locations of these thermal fronts, as observed by satellite sensors, suggests that they may be stronger and more numerous in El Niño years than at other times.

Near-surface Ekman transport in the ocean beneath the zones of westerlies and easterlies exhibits convergent flow at the STFZ (Roden, 1975). Here, low-salinity northern water descends below warmer and saltier subtropical water, so that individual fronts within the STFZ tend to exhibit strong salinity gradients at the surface. Consequently the STFZ is, in winter, both a temperature and salinity frontal zone, although, in summer, it includes only haline fronts: for this reason, the surfacing of the 35.1% isohaline is a useful indication of its location at all seasons.

The Transition Zone is energetic, and although density gradients across it are small, baroclinic shear is large, associated with the formation of eddies on the scale of 100–1000 km that are especially frequent in the western half of the ocean. The Transition Zone also differs from the subarctic and subtropical water bodies to north and south in the relatively low stability of the water column. To the north, the subarctic domain has a permanent shallow halocline (see PSAG) that is extremely resistant to winter mixing, whereas to the south in the tropical domain (see NPTG) the combined effects of the thermocline and halocline also confer relatively high stability (Roden, 1970, 1975). We can expect that these factors will have biological consequences distinguishing this province from those adjacent to it, poleward and equatorward.

## Regional Response of the Pelagic Ecosystem

The NPPF and NPST provinces have not attracted direct attention from oceanographers in recent decades so, lacking direct information on the characteristic seasonal evolution of the pelagic ecosystem, the obvious first step is to consult satellite imagery. This was first done by Glover *et al.* (1994), who assembled data from the CZCS sensors in which a steep transition zone was observed, between high chlorophyll to the north and low to the south, across the ocean at 30–40°N. This feature was also simulated by their simple model in which the critical process was the winter nitrate recharge of the upper layers, responding to mixed-layer seasonal geography according to Levitus' climatology. This shows a transition between mixing depths of >75 m and <50 m at about 30–35°S, corresponding to an end-of-winter surface nitrate transition from around 5.0 $\mu$M $m^{-3}$ to the north and <0.4 $\mu$M $m^{-3}$ to the south of the transition. Polovina *et al.* (2001) have expanded the study of this transition, showing that it can be traced 8000 km across the ocean and that it seasonally migrates 1000 km from 30–35°N to 40–45°N—or, essentially, across the Transition Zone as defined above.

All this may be so, but inspection of the available seasonal and monthly images reveals a more complex situation than could be described simply as a front. During the first half of the year, and most strongly in March–May, a narrow zone of high chlorophyll stretches across the ocean, continuous with the enhanced chlorophyll in the zone of interaction between Oyashio and Kuroshio east of Japan; its location corresponds very well with that of the NPPF province. The chlorophyll images may be matched with co-registered TOPEX-POSEIDON sea surface elevation images to confirm that the individual high chlorophyll features are indeed associated with meanders and eddies (see Color plate 16). The seasonal intensity of chlorophyll in this feature, as observed in the SeaWiFS 30-day images, matches climatological data aggregated for the NPPF province, which maintains somewhat higher chlorophyll values than in the NPST province. Although the CZCS images suggested significant differences between eastern and western halves of the NPST province, these are no longer observed in SeaWiFS or MODIS data. What is sustained, however, is the earlier chlorophyll accumulation in NPST(W) (peaks in February–March) than in NPST(E) (peaks in April–May).

Some previously unpublished sections for April and May along 158° and 172°W are offered in the PICES publication discussed earlier; these clearly show the anticipated DCM at between 50 and 100 m depth, deepening equatorward and, perhaps, shoaling progressively as the season advances. The meridional, late-summer URSA MAJOR chlorophyll section near 150°W shows two features of interest to this discussion: a rapid change to lower values in the DCM at 34–35°N (near the southern boundary as anticipated) and a discontinuity at 42–43°N in the slope of the DCM and of the chlorophyll values.

The zonal INDOPAC sections (July 1977) for temperature, chlorophyll, and nitrite ($NO_2$) provide useful information on the zonal structure of NPST, and they suggest important differences between NPST(W) and NPST(E), on either side of 160°W. Meandering is stronger in the west, as sea surface height anomalies from TOPEX-POSEIDON also now confirm, so that the western half of the section shows several large (diameter, 5° latitude) bowls in the mixed layer with depth anomalies ~250 m, whereas the eastern part has a more or less uniform mixed-layer depth at 75–100 m. Not surprisingly, the western part of the section has several deep excursions of high values of mixed-layer chlorophyll down to 200 m, whereas in the east the DCM lies consistently at 75–150 m, conforming to the mixed-layer depth. Finally, the nitrite section shows high values associated with the high chlorophyll values in the west, indicating rapid remineralization of DOC, whereas in the oligotrophic east values are very low. Passage over deep-ocean ridges (e.g., the Emperor Seamount Chain) modifies the characteristics of the western ocean.

Although these provinces have not been investigated comprehensively, we may take some comfort from biogeographical data because they do support a small but very characteristic assemblage of species for which this appears to be the unique distribution area. Some of the clearest demonstrations of this are the ZETES winter 1966 and the URSA MAJOR summer 1964 sections along 155°W (Venrick, 1974). Group analysis of 66 diatom taxa identified several recurrent groups of species whose distributions matched subarctic and subtropical gyres, and also a group of three species that are endemic to a "transitional domain" (our NPST Province at about 35–42°N) in winter. Two other groups of species comprise 99% of the diatom biomass in this zone in summer. Because the two meandering boundary fronts must be leaky, it is no surprise to find that the species groups endemic to the adjacent domains to the north and south overlap into this province. However, it is the relative fidelity of the "transition species" to their zone that is significant.

We have even better evidence from zooplankton. An early review of the biogeography of the Pacific Ocean (Reid, 1962) revealed a remarkable zonal strip of high biomass of mesozooplankton across the ocean at about 38–45°N that was very clearly separated by low biomass occupying most of the subarctic gyre from another zone of high values in the Bering Sea. Even more striking are the envelopes for the distribution of 12 "transition zone" species assembled by McGowan (1971); these lie in a tight group across the ocean with average meridional extents from 35°N to 45°N. They represent a wide range of taxa (euphausiids, copepods, mollusks, and foraminifera) and some, such as *Nematoscelis difficilis*, have a most remarkable fidelity to this province. These species go with the flow into the coastal currents off western North America (see CALC). How these species maintain their populations in this zonal river in the ocean is an unanswered question.

To all this early work, we can now add recent evidence, some obtained by satellite-monitored tags attached to large animals (Polovina *et al.*, 2001). Thus, loggerhead (*Caretta*) and Ridley (*Lepidochelys*) turtles are found to be associated preferentially with the Transition Zone, migrating seasonally with the temperature front, but displaying specific preferences within the zone. Ridleys are more southerly, in warmer water, and loggerheads more northerly. Distribution of the turtles responds to between-year differences in the location of the Transition Zone. It is also now known that albacore tuna travel along the frontal zones in the seasonal trans-Pacific migrations. A new but perhaps already overfished jig fishery by both U.S. and Japanese boats for the flying squid (*Ommastrephes*) is concentrated in the Transition Zone.

A remarkable study by Hyrenbach (2002) of satellite-tracked albatrosses (*Phoebastria* spp.) that nest on the central tropical Pacific islands show that both species fly fast and directly between there and their feeding grounds, where flight becomes slower and constantly changes direction. Feeding grounds for both Black-footed and Laysan albatross are in the Transition Domain (*sic*) and Subarctic Frontal Zone during the brooding period. Precise definition of conditions (temperature and chlorophyll) selected by the birds as they foraged along the transition zone was achieved by reference to contemporary satellite imagery. When both species are present, Laysan albatrosses exploit the colder (10–15°C) waters of the SAFZ (and are associated with the so-called Transition Zone Chlorophyll Front) while Black-footed albatrosses feed south of the STFZ in warmer water (>20°C).

## Synopsis

*Case 3—Winter-spring production with nutrient limitation*—The interaction between the boreal winter excursion of the pycnocline and the depth of the irradiance-driven photic depth is such that the pycnocline is illuminated for 1–2 months longer in NPST than in NPPF. The seasonal cycle of productivity exhibits a boreal spring peak (April–May) in both NPPF and in NPST, illustrated in Fig. 11.6. Chlorophyll biomass in NPPF (not shown) regularly exhibits a secondary autumn maximum in September–November and a spring maximum that is coincident with P maximum rate; in NPST, chlorophyll accumulation begins in boreal autumn and reaches a spring maximum that is a little later in NPST(E) than NPST(W). In each, there is the habitual shoulder on the descending limp of the productivity cycle during boreal summer.

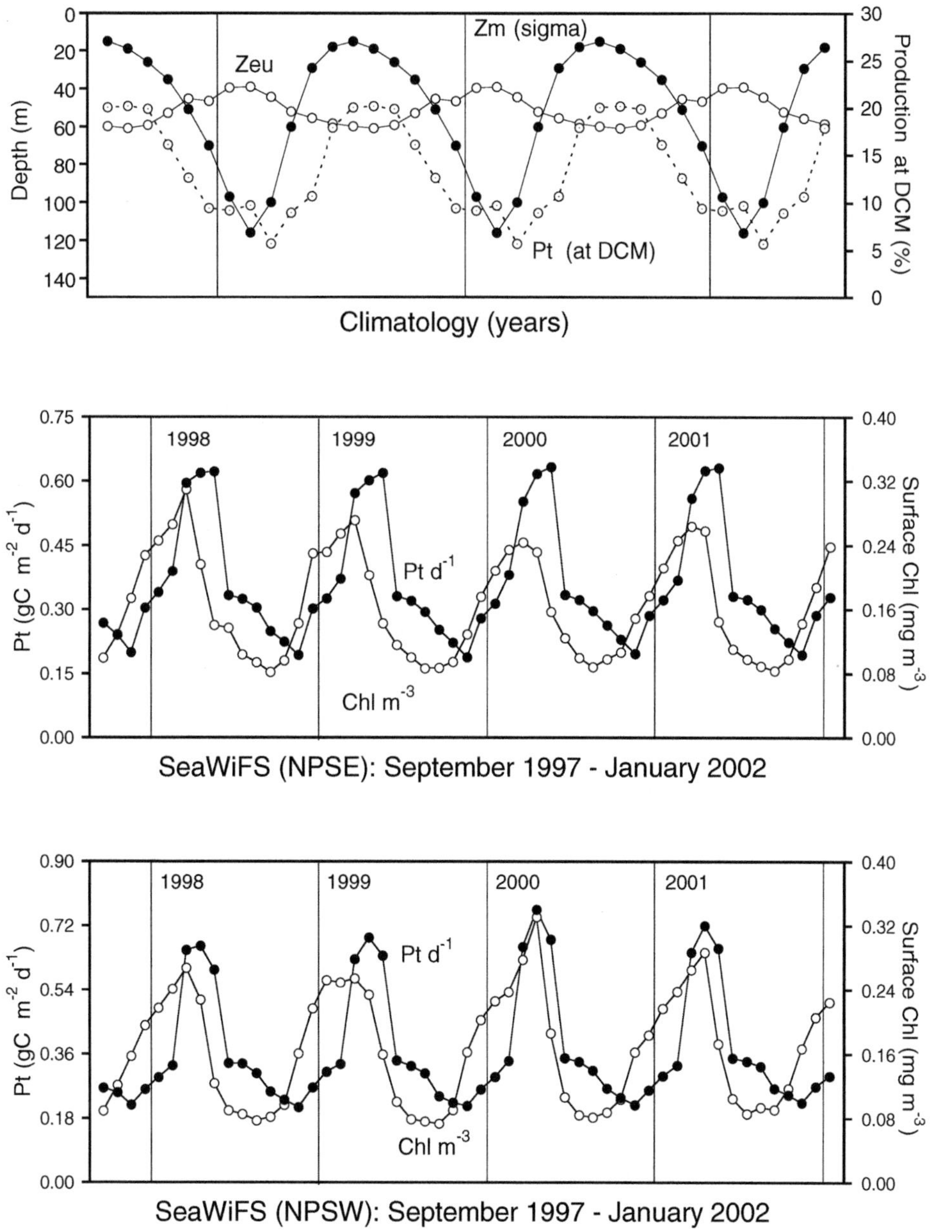


**Fig. 11.6** NPSW and NPSE seasonal cycles of monthly surface chlorophyll and depth-integrated autotrophic production for the years 1997–2002 from SeaWiFS data together with characteristic seasonal cycles of mixed-layer depths from Levitus climatological data and photic depths computed from characteristic irradiance and the archive of chlorophyll profiles discussed in Chapter 1.

## TASMAN SEA PROVINCE (TASM)

### Extent of the Province

The Tasman Sea Province (TASM) lies between the Subtropical Convergence Zone across about 45°S and the Tasman Front to the south of the Coral Sea at about 32°S and is enclosed by the coasts of Australia and New Zealand. This is a relatively small province, but because of its unique characteristics it is useful to recognize the Tasman Sea as an entity.

There is no shelf in this province: the eastern Tasman Sea coast of Australia and Tasmania is in AUSE and the western coastline of New Zealand is allocated to the New Zealand Coastal Province (NEWZ). Shelf-edge fronts, where they are found, represent the landward boundary of the province along these two coasts: they are especially prominent along the Westland upwelling strip of South Island.

### Defining Characteristics of Regional Oceanography

To place this province in the Westerlies Biome, when it lies entirely equatorward of the Subtropical Front, is somewhat at odds with the definition offered of this biome. I have chosen to place it here for pragmatic reasons, because—as a consequence of the arrangements of the coastline of Tasmania and Australia in relation to the circumpolar Southern Ocean—westerly winds do extend over most of the Tasman Sea. During austral winter, persistent high pressure over Australia induces strong westerlies, continuous with those over the Southern Ocean, across the entire province so that, in this sector alone, the STCZ does not have easterlies on its equatorward flank. In austral summer, the northern part of the province lies below northeasterly winds induced by the persistent low-pressure belt over northern Australia.

The Tasman is a windy sea, although it lies at only about 40° of latitude, so the seasonal cycle of winter wind mixing and summer stratification is strong. Wind stress and thermal convection drive the mixed layer down to 300 m in the southern half, though the regional mean depth of winter mixing is only about 125 m. This delivers end-of-winter mixed-layer nitrate in the range 2.0–4.0 $\mu$M. Incursions of subantarctic water in meanders isolated from the Subtropical frontal zone to the south have slightly higher (4.0–5.0 $\mu$M) values and those of subtropical water are nitrate-deficient.

Because this is a complex region of interaction between the general anticyclonic gyral flow and the instabilities induced in it along the topographic boundaries, the second principal characteristic of the Tasman Sea is a profusion of mesoscale eddy fields. The southern limit of the province is the edge of the eddy field associated with the South Subtropical Convergence that passes south of Tasmania and then around the Snares Bank, south of New Zealand. You may find old figures that show it passing diagonally across the Tasman Sea and around northern New Zealand: but modern data (Garner, 1967; Heath, 1981) show this to be incorrect. The eddy field associated with the northern edge of the convergence adds to the mesoscale activity of the central Tasman Sea, and the location of the convergence forces the general anticyclonic flow of the Tasman Sea to encounter the Westland coast of South Island where flow diverges north and south (see NEWZ).

The retroflection of the East Australia Current, the poleward western boundary current down the east coast of Australia, occurs near Sydney (32–34°S), whence it flows eastward across the Tasman Sea as a meandering frontal jet in which the individual eddies may be very large. Warm Coral Sea water lies to the north, and cool Tasman Sea water to the south of the Tasman Front that coincides with a zone of seasonal Ekman suction where upwelling reaches $2–4 \times 10^{-5}$ m sec$^{-1}$ during winter, though vertical motion in summer is weaker (Andrews *et al.*, 1980; Comiso *et al.*, 1993). Meanders within the

Tasman Front shed equatorward cyclonic meanders and poleward anticyclonic meanders, and these in turn shed warm and cold-core rings. The climatological origins of the eddies are topographically determined by the highest points of Lord Howe Rise and the West Norfolk Ridge (Hamilton, 1992), and hence preferred locations can be mapped for cold and warm features along the front. The actual location of the front as it passes across the Tasman Sea varies from 30°S to as much as 38°S (Mulhearn, 1987) before it rounds Cape North (35°S) to enter the poleward flow of the east New Zealand coastal currents.

The retroflection of the East Australian Current has major consequences for the Tasman Sea apart from bounding it to the north by the Tasman Front. The meandering flow along the front generates a field of Rossby waves that pass westward toward the Australian coast and induce the shedding of very large warm eddies: two or three are shed per year, and a standing population of six to eight in the Tasman Sea is normal. Within these eddies the seasonal cycles of mixing and stabilization are imposed on their initial structure. The surface thermal signature and shallow thermocline of Coral Sea water is lost during the first winter of the existence of an eddy, and in subsequent summers the surface mixed layer established over the whole Tasman Sea passes across the subsurface warm core of the eddy. A further source of eddies are the episodic irruptions of warm water of Leeuwin Current origin, transported along the south coast of Australia and occasionally penetrating the Bass Straits.

## Regional Response of the Pelagic Ecosystem

The southern Tasman Sea exhibits a typical spring bloom (Harris *et al.*, 1987) so that chlorophyll biomass in the south-central region is maximal in late September–early October, but regular seasonality does not extend across the whole region; the integrated chlorophyll data are, in this province, somewhat misleading because the effect of the southwest region dominates the whole (see Color plate 17). Monthly chlorophyll images indicate that higher surface chlorophyll occurs patchily across the whole southern Tasman Sea through austral summer and autumn associated with frontogenesis and other bloom-inducing processes; however, overall, this is the period when lowest surface chlorophyll biomass is observed in satellite data.

The spring bloom has been best described to the east of Tasmania, where it is a typical but rather uncertain process. Although strong westerlies may induce a regional mixed layer of almost 300 m, the actual depths achieved are dependent on the nature of the subsurface water mass, which has strong interannual variability caused by shifts in the location of the STCZ, and as mesoscale eddies come and go. There is a 40-day periodicity in wind stress in the westerly wind field and blooms occur intermittently whenever the westerlies slacken sufficiently to permit stabilization of the upper water column. The bloom and onset of nutrient depletion varies between years by as much as 4 months at a fixed station and is usually about 1 month earlier inshore than in the open Tasman Sea.

Austral winter chlorophyll values of $<0.5$ mg chl $m^{-3}$ rise to 2.0–2.5 mg chl $m^{-3}$ at the peak of the early summer bloom, usually between August and October, and then decline in late austral summer and early autumn from January to March. Nutrients become wholly depleted by about February, when the entire province is dominated by productivity and chlorophyll biomass that is sustained during austral summer by dynamic processes in the well-developed frontal region of the SSTC (see Color plate 17). A weak regional autumn bloom is evident in some images. All this, Murphy *et al.* (2001) suggest, is consistent with production being colimited by light and nitrate. Warm-core eddies associated with the retroflection of the East Australian Current may develop phytoplankton blooms later than in the surrounding water mass; one such, observed off New South Wales, developed highest biomass only in November–December (Tranter *et al.*, 1980); the DCM was near the summer thermocline at about 50 m. Investigations of other warm-core eddies confirmed that their productivity was high relative to the surrounding water mass and was fuelled by nutrients delivered by deep winter mixing within the core of each eddy.

Warm-core eddies containing water originating in the Coral Sea have a more complex horizontal distribution of surface chlorophyll, in which maximal values often occur near their western edges. One such eddy, tracked over a 19-month period, persistently had integrated chlorophyll of 60–80 mg chl $m^{-2}$, of which 60–70% was contributed by very small cells of the pico- and nanoplankton; this contribution reached 90% in some of the more oligotrophic stations worked within the eddy. Edge enrichment may occur also where warm-core eddies entrain crescents of cooler, nitrate-replete water (about 50 $\mu$M) from the coastal regions, especially where the edge of the eddy is comparatively straight. Here, populations of the upwelling copepod *Calanoides carinatus* may occur and the general biological enhancement ($NO_3$, 50–100 $\mu$M; chlorophyll, 1.5 mg $liter^{-1}$; copepods, 50–70 $m^{-3}$) may attract schools of the southern bluefin tuna (Tranter *et al.*, 1983). The eddies associated with the retroflection area, and those shed from the Tasman Front as it passes toward North Cape of New Zealand, may entrain expatriated subtropical organisms from the Coral Sea; this has been investigated for the case of several tropical hyperiid amphipods entrained in these features.

That this should be a very dynamic and changeable part of the ocean could be predicted from its boundary conditions, lying between the Southern Ocean and the Coral Sea and adjacent to the great landmass of Australia. This is confirmed by the observations of variability not only within seasons but also between seasons (Harris *et al.*, 1988, 1991). Though most relevant observations have been made on the east Tasmanian coast (see AUSE), they are generally applicable to the entire province. The observations compare the effects of different wind stress during two 2-year periods. The summers of 1986 and 1987 were cool and windy, whereas in 1988 and 1989 they were warmer and quieter. The cool years more closely resembled the subantarctic (deep winter mixing and high nitrate) and the warm years subtropical conditions (shallow winter mixing and low nitrate). Particle size of all planktonic components indicated a trend toward subtropical oligotrophic conditions in 1988 and 1989. Primary production was reduced, small copepods dominated, and all large zooplankters (especially salps and *Nyctiphanes australis*) were eliminated: these organisms appear to be dependent on large-cell new production rather than on small-cell regenerated production. In cool years, large populations of euphausiids occur, whose swarms support a rich *Trachurus* fishery, which collapsed during the two warm years. Harris *et al.* suggest that this entire chain of events was a regional manifestation of the Pacific-wide ENSO warm event of 1988.

## Synopsis

*Case 3—Winter-spring production with nutrient limitation*—The pycnocline undergoes a strong seasonal excursion from 25 m in austral summer to 150 m in August; $Z_{eu}$ varies only marginally, so that the pycnocline is illuminated only in austral summer (November–April). From a summer minimum (0.2 mg C $m^{-2}$ $d^{-1}$) in April–May, P shows a progressive fourfold rate increase through austral fall and winter (Fig. 11.7), to reach an annual maximum in October–November (early austral summer). Thereafter a progressive decline sets in, leading to the annual minimum in April. Chlorophyll accumulation follows the rate of P during the spring bloom but not the winter depression: at that season, a minor secondary accumulation occurs.

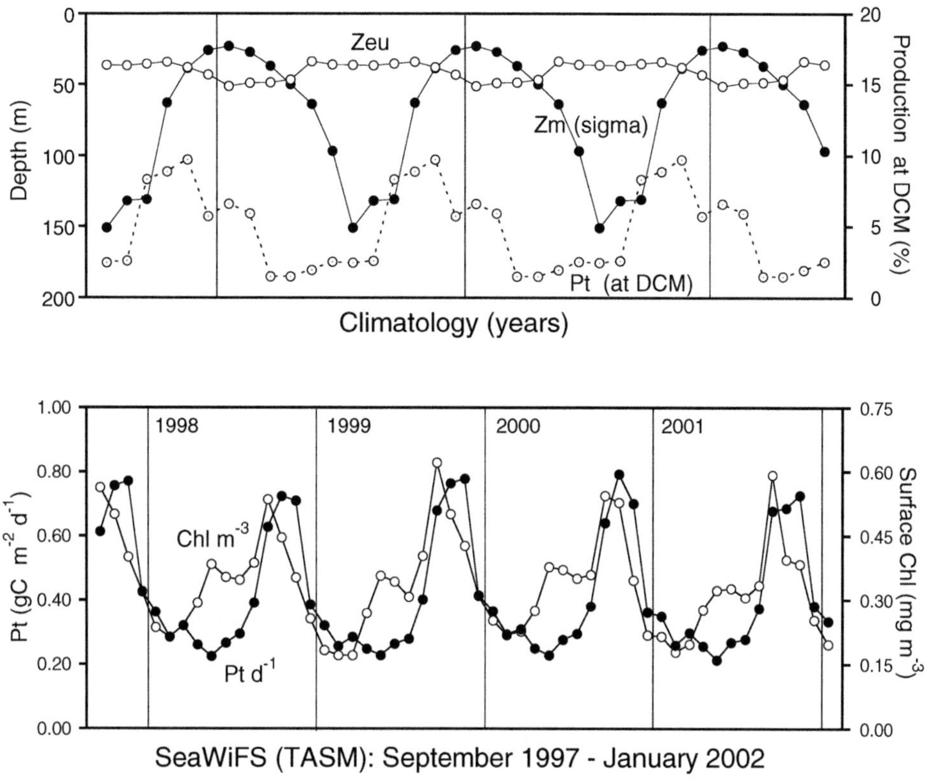


**Fig. 11.7** TASM: seasonal cycles of monthly surface chlorophyll and depth-integrated autotrophic production for the years 1997–2002 from SeaWiFS data together with characteristic seasonal cycles of mixed-layer depths from Levitus climatological data and photic depths were computed from characteristic irradiance and the archive of chlorophyll profiles discussed in Chapter 1.

# Pacific Trade Winds Biome

## North Pacific Tropical Gyre Province (NPTG)

### Extent of the Province

This large province is the equatorward part of the North Pacific central or subtropical gyre lying between the Subtropical Convergence, at about 30–32°N in midocean and 20–25°N in the west, and the northern Doldrum Front at about 10–11°N (Roden, 1975). The western and eastern boundaries of NPTG are taken to be the offshore edge of the boundary currents—of the Kuroshio flow along the eastern coasts of the Philippines and the offshore edge of the California Current system (for definition, see CALC). The Hawaiian Islands are placed almost centrally in the province.

### Defining Characteristics of Regional Oceanography

The great subtropical gyres are among the most difficult in which to sustain a logical partition based on the physics of circulation and stratification, and this province is not exceptional in that regard. Lying equatorward of the subtropical convergence, it is within the Trade Winds biome, with all that that implies, as was discussed in Chapters 4 and 6, but physical forcing is nonuniform with latitude. For instance, because of nonlinear change in stratification intensity, the effects of the equatorward increase of $R_i$, the internal radius of deformation, steepens at about 15–20° from the equator (Emery *et al.*, 1984).

This creates regional inequalities in the scaling, or horizontal dimensions, of mesoscale eddies and baroclinic Rossby waves.

The Subtropical Convergence along the northern edge of this province lies below the seasonally migrating conjunction of the westerly winds and the trades. This province, then, lies below the subtropical easterlies but nevertheless some winter mixing, forced by the frequent extratropical cyclones passing eastward (see NPST Province), extends as far south as the Hawaiian Islands at 20–25°N. As has been recently confirmed, an eddy-energetic Subtropical Counter Current induces zonal flow westward in this region at 18–25°N and is more consistent in boreal spring than at other seasons (Qiu, 1999).

At 22°N, the mixed layer deepens from 35–40 m in summer to 80–90 m in winter. But both climatology and ALOHA data (e.g., Karl and Lukas, 1996) suggest that such winter deepening may be a relatively brief affair, in some years involving no more than a single month, such as January 1993, that appears as an outlier on a more moderate seasonal cycle. Coincidentally, it is for this part of the province that we have the most abundant information from the HOT and ALOHA time-series stations. Elsewhere, and especially further to the southwest in the province, we should probably not encounter a significant winter excursion of mixed-layer depth. Of course, the permanent pycnocline lies still deeper and a permanent nitracline occurs across its density gradient. These features are continuous and at a nearly uniform depth, close to the 1% isolume, from the eastern edge of the province to at least the longitude of Hawaii. They are deepest in the northern part of the province, shoalest in the south, and slope up toward the American coast.

Both circulation and some characteristics of the water column in this province are somewhat variable, in response to the state of the ENSO index, although these effects are more important in the equatorial and warm pool provinces to the south. SST is significantly correlated with the SO index, and the westerly wind bursts that introduce an ENSO event modify mixed-layer depth and strength of the westward flow of the NEC. This lies between the Subtropical Convergence and the northern Doldrum Front of Roden (1975) at 10°N, where salinity decreases abruptly southward into equatorial water. Beyond 180°W, the southern part of this limb of the gyral circulation passes progressively under the influence of the heavy precipitation which characterizes the western Pacific "warm pool"; where these conditions are fully developed, to the south of 10°N and west of 160°E, it is proper to recognize a unique province (see WARM), some of whose characteristics will nevertheless be relevant to the western part of NPTG.

Flow perturbation past the Hawaiian island chain causes the formation of a von Karman vortex street, by the generation of alternate cyclonic and anticylonic eddies similar to those downstream of the Canaries (see NAST). However, in the case of the Hawaiian Islands, it is likely that eddies are also formed by the effect of curl of the wind stress, which induces upwelling and downwelling of surface water at the wind-shear lines by strong Ekman pumping. Off Hawaii it is suggested that this process induces sufficient vertical nutrient flux to induce features in the chlorophyll field (see the review of island eddy wakes by Aristegui *et al.*, 1997).

## Regional Response of the Pelagic Ecosystem

We are fortunate that for this province we have two sets of excellent time-series studies, together with some trans-Pacific meridional and zonal sections. Many expeditions, mostly from Scripps Institute of Oceanography in California, investigated the CLIMAX area (28–30°N) from 1968 to 1987 (Hayward, 1987). The recent HOT (Hawaii Ocean Time Series) investigations were established at the ALOHA station, at 22°N 158°W, in 1988 with the intention of obtaining a 20-year time series (Karl and Lukas, 1996). The following account relies very heavily on these two projects, which are among the very best data sets for the analysis of long-term changes in the structure of a pelagic ecosystem.

Large-scale gradients are lacking, as indicated by the depth and kind of the DCM, compared with the NPST province north of the Subtropical Convergence; zonal trans-Pacific sections of chlorophyll and primary production along 24°N (Venrick, 1989, 1991) have much greater uniformity than zonal sections along 47°N. The ridge-trough topography of the pycnocline associated with the zonal equatorial current system dominates the stability and vertical eddy diffusivity of the regional water column and, hence, the long-term supply of nutrients to the photic zone. The vertical ecological structure is modified by the development of a seasonal shallow thermocline (Hayward *et al.*, 1983), following the general model for a subtropical ocean with moderate winter mixing, as in the Sargasso Sea (see NATR).

The deep, nitrate-depleted euphotic zone of this province in summer, reaching even to 100 m, lies above a permanent nutricline. Year-round, and each year, and right across the subtropical gyre, a DCM occurs at 80–120 m, lying just above the nitracline and the permanent thermocline, and close to the 1% isolume (Venrick, 1991). In meridional sections (Venrick et al., 1973), the DCM is seen to conform to the general northward upsloping of the thermocline across the province. Some seasonal change occurs in the depth of this feature from < 100 m in winter to >100 m in summer. The DCM and nitracline remain deep when the weak summer thermocline is established at 35–45 m, well above them. The summer DCM contains maximum values of around 0.2–0.4 mg chl $m^{-3}$, whereas the winter DCM is a weaker feature, usually about 0.1 mg chl $m^{-3}$. Primary production rate is maximal rather shallower in the mixed layer, at 30–60 m. The profiles studied at ALOHA, and first reported by Letelier *et al.* (1996), integrate very well into Venrick's transpacific section and show that although primary production rates increase somewhat in summer, the interannual differences are at least as great as seasonal differences: the seasonal/interannual range is between 200 and 700 mg C $m^{-2}$ $d^{-1}$. The effect of seasonal changes in surface irradiance is very marked: there is no regular seasonality in production rate in the 0–100 m layer, but at 100–200 m the rate (about 20% of that in the shallower zone) shows very clear midsummer maxima in 4 of the 5 years reported by Karl *et al.*, (1996).

The vertical diffusive nutrient flux apparently required by these observations can only be partially satisfied by observed nitrate uptake, and the mechanism by which calculated new production is sustained above the summer thermocline remains unexplained. In the presence of a stable vertical ecological structure, as is observed, we would expect any upward physically driven nitrate flux across the nitracline to be taken up in the DCM (Hayward, 1987).

Karl *et al.* (2001) have integrated the observations at ALOHA with the data from CLIMAX. The very long series he obtained in this way runs from 1968 to 1997 and demonstrates that what these authors term a "domain shift" in the pelagic production system occurred during this period. The CLIMAX observations already showed that from 1968 to 1985 the rate of primary production by autotrophic cells had almost doubled during the summer (May–October). It remains uncertain if this was a continuous process or whether a step-function increase occurred between 1973 and 1980, partly because this was a period of great change in the methods of biological oceanography. Nevertheless, the data have been very carefully intercompared, and the results are convincing. Time-integrated profiles of chlorophyll and autotrophic production show that the significant increase in production occurred mostly within the 0–50 m layer, while the DCM did not respond to whatever forced the increase and remained close to 100 m throughout the series. Karl *et al.* (2001) suggested that these changes occurred in response to an unprecedented period of sustained values of the SOI (see Chapter 8) favorable to the development of El Niño conditions.

These changes were accompanied by a reduction in dissolved silicate and phosphate and a shift in the structure of the phytoplankton community that was convincingly indicated by the relative increase in chl*b* that occurred during this period. Such a shift toward an ecosystem dominated increasingly by prokaryotes carries major implications for altered nutrient flux pathways and for the trophic structure of the pelagic ecosystem, even up to higher levels. This has been characterized as involving selection for $N_2$-fixing cyanobacteria, including *Trichodesmium*, and a shift from large cells to *Prochlorococcus* and other very small autotrophs. Increased bacterial heterotrophic activity is also indicated. However, one should not infer from this discussion that the domain shift totally eliminated diatoms from the phytoplankton. As Scharek *et al.* (1999) showed, there remained in 1995 a highly structured diatom community at ALOHA, with distinct assemblages in the DCM and the mixed layer. Summer increases in numbers mainly involved lightly silicified species, such as *Hemiaulus hauckii* and *Masogloia woodiana.*

In this context, we should note the presence in this province of autotrophic cells such as the diatom *Hemiaulus hauckii* and also *Rhizoselenia cyclindricus* and *R. hebetata* that have nitrogen-fixing symbionts, principally the cyanophyte *Richelia intracellularis.* Venrick (1974) surveyed the occurrence and potential productivity of this symbiosis at the CLIMAX site and found it to be most significant in summer; Venrick computed a potential carbon fixation rate (37–77 mg C $m^{-2}$ day-) that was about 30–60% of the difference between total winter nonbloom and total summer bloom production at this station. Villareal *et al.* (1993) have investigated diatom "mats" composed of several species of *Rhizoselenia* that are very abundant at CLIMAX, containing >90% of all biogenic silica here; the mats appear to have the ability to modify their density from negative to positive buoyancy, and shuttle between the upper euphotic zone and the nitrate-replete deeper zone. It is suggested that these mats may be an important vector of new nitrogen inputs to the euphotic zone, representing as much as 50% of the required flux.

Profiles of microplankton biomass (mostly protists, monads, flagellates, and naked dinoflagellates) do not have a subsurface maximum corresponding to the DCM but rather a broad depth range of relatively high abundance (0–100 m, 5–10 mg C $liter^{-1}$) above a deeper zone of lower abundance (100–200 m, 1–5 mg C $liter^{-1}$). Separation of vertical habitat occurs also among mesozooplankton (Ambler and Miller, 1987), some species of which specialize in the DCM (presumably those that consume larger cells), whereas others preferentially occupy the upper mixed layer where primary production rates are maximal. In the western part of the province at 28°N 136°E, Tsuda *et al.* (1989) have observed that grazing by microplankton in the DCM (70–120 m) contributes 60–100% of total consumption and that this is balanced by production within a time scale of several days. It was reported that mesozooplankton contribute only <5% of all consumption; at shallower locations than the DCM, these organisms are largely secondary predators, consuming microplankton, though at the DCM they contributed to the consumption of algal cells. This statement must represent a great simplification of the taxon-by-taxon food selection of mesozooplankters concerning which there is a very large body of knowledge; it would be foolhardy to attempt to review this for each province because the common threads that runs through these studies concern (i) the diversity of the ranges of diet characteristic even of congeneric species and (ii) the flexibility of the response of each growth stage of each species to the ambient food environment in which it finds itself.

A strong seasonal signal in the abundance of all size fractions (0.2–20 mm) of the mesozooplankton can be detected despite the extreme internal incoherence of the data, with a seasonal doubling of both day and night abundance and biomass during summer months; interannual variability also occurs and a long-term doubling of biomass was observed in the CLIMAX-ALOHA time series (Landry *et al.*, 2001). This increase involves such forms as the harpacticoid *Macrosetella gracilis*, associated with mats of $N_2$-fixing *Trichodesmium* that attain maximal biomass in boreal summer.

The major consumers of the very small, dominant phytoplankton cells—at least after the domain shift—remain protists, with which the mesozooplankton cannot compete. Despite this, small copepods remain abundant and represent 80% of all zooplankton biomass by day and 77% at night (Landry *et al.*, 2001). Chaetognaths and larger crustacea (euphausiids, decapods, amphipods) dominate the larger size fractions. I refer you to the account of McGowan and Walker (1979) for an analysis of the taxonomic diversity of the 105 species of copepods that they observed in the central North Pacific, of which 89 taxa were recognized at CLIMAX-ALOHA. It is very difficult to obtain satisfactory analysis of this community, taxon by taxon, because each has a characteristic distribution that may not be resolved by standard sampling methods. For example, the most abundant copepod at CLIMAX is *Haloptilus longicornis*, a nonmigrant, distributed from the surface down to 400 m, and so not fully enumerated in standard, shallower plankton tows.

The envelopes of the distribution of 10 "central water mass" species plotted by McGowan (1971) provide an interesting point regarding the definition of this province: in the east of the ocean, and as far west as about 170°E, they match remarkably well the limits of this province as defined here. To the west of this longitude, the envelopes widen and eventually encompass the whole western Pacific from New Guinea to southern Japan. To the extent that the distribution of individual species necessarily indicates the distribution of common ecosystem-forcing characteristics, then perhaps further analysis of the biological oceanography of the western Pacific may suggest that the WARM province should be much more extensive than the limits used here.

## Synopsis

*Case 4—Small-amplitude response to trade-wind seasonality*—$Z_m$ has weak geostrophically forced seasonality when meaned over the whole area of province: 40–45 m in boreal summer (May–August) and 65 m in boreal winter (January–February). $Z_{eu}$ varies from 60 to 70 m so that thermocline is illuminated in all except two winter months. Productivity has very weak seasonality, with maximal rates in late summer when the pycnocline begins to deepen (Fig. 11.8). Seasonal changes of chlorophyll biomass are very small, with highest

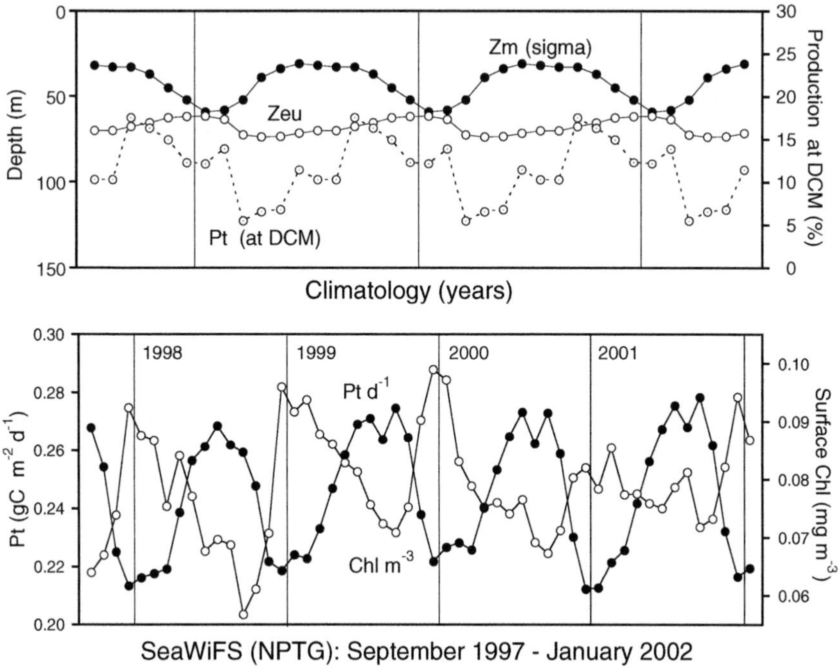


**Fig. 11.8** NPTG: seasonal cycles of monthly surface chlorophyll and depth-integrated autotrophic production for the years 1997–2002 from SeaWiFS data together with characteristic seasonal cycles of mixed-layer depths from Levitus climatological data and photic depths computed from characteristic irradiance and the archive of chlorophyll profiles discussed in Chapter 1.

values occurring during boreal winter when productivity is minimal. This is perhaps an effect of the seasonal migration to depth of herbivores in the northern part of the province.

## North Pacific Equatorial Countercurrent Province (PNEC)

### Extent of the Province

The North Pacific Equatorial Countercurrent Province (PNEC) is much smaller than the NPTG Province and lies between the northern and southern Doldrum salinity fronts, that is to say around 5–10°N in the central Pacific, widening toward the American continent. Thus, PNEC includes the triangular region of weak currents that lies between the influence of the equatorial divergence and the California Current extension that flows into the North Equatorial Current and the NPTG province. To the west, a meaningful limit to PNEC can be set around 180°E.

### Defining Characteristics of Regional Oceanography

The flow across this province of the NECC takes its origin in the Mindanao Dome at the termination of the western boundary current and terminates in the cyclonic flow around the Costa Rica Dome, off central America.

The western boundary of PNEC is set at the date line for the following reasons: despite the trans-Pacific nature of the North Equatorial Countercurrent (NECC), west of 180°E this flow lies below the western Pacific atmospheric convection cells, where excess of precipitation over evaporation induces a different ecological regime, that characteristic of the WARM province. Near the date line, a significant surface salinity front was clearly identified in the JGOFS Flupac sections (Le Borgne *et al.*, 2002a); here, salinity rises eastward rather abruptly by about 0.5‰, although surface temperature shows no change in its gradual eastward cooling trend. Here also, concentrations of $NO_3$ and chlorophyll change rather abruptly to higher values to the east. The boundary between WARM and PNEC is, therefore, significant for pelagic ecology.

The southern boundary lies at the oceanic Equatorial or Doldrum Front that crosses the equator near the Galapagos and is marked at the surface by a discontinuity in temperature, salinity, and in mixed-layer nitrate that is delivered by divergence at the equator.

The NECC flows eastward across the ocean, lying above the north slope of the equatorial thermocline ridge, all the way from 120°E to Central America where it turns northward. Flow of this countercurrent is weaker and more intermittent progressively toward the east; at 135°E it is 25 Sv, but near its termination in the eastern Pacific it is reduced to only about 10 Sv. Accordingly, mixed-layer depth must shallow toward the east. Generic NECCs are the product more of wind stress curl than of wind stress (Sverdrup, 1947; Philander, 1985), and when the Intertropical Convergence Zone (ITCZ) is in its northerly, boreal summer position, the NECC lies below a zone of exceptionally strong positive wind-stress curl that lies across the whole ocean, from the Indo-Pacific conjunction all the way to Central America above the thermal ridge that is associated with flow of the NECC (Lagler and O'Brien, 1980). From July to January, positive values of Ekman vertical velocity are higher in the eastern part of the ocean (from 140°W to the American continent), reaching local maxima of $60 \times 10^{-6}$ m $sec^{-1}$.

For these reasons, flow of the NECC is strongly seasonal in the eastern part of this province. During boreal winter, when the atmospheric ITCZ is at its most southerly, the countercurrent can still be detected in thermocline topography, but is sufficiently weak that flow at the surface does not occur east of about 110–120°W (Tomczak and Godfrey, 1994). In the eastern tropical Pacific, the ridge-trough system of the thermocline occupies a wider, triangular region between the coast and the convergent fronts of the North and

South Equatorial Currents where these turn westward away from the continent. Within this region of weak flow and shallow mixed layer, the eastward countercurrent bifurcates and the influence of the continent on regional wind stress and curvature results in several cyclonic domes of ecological significance: the Costa Rica Dome (CRD), offshore in the PNEC Province, and the Tehuantapec and Panama Bight domes in the coastal boundary province CAMR.

The doming of the thermocline off Costa Rica, with seasonal intensification, has regional ecological importance. Recent analysis of archived data and satellite imagery has now revealed the real nature and seasonal evolution of the CRD (Fiedler, 2002). Previously, it had been considered to be a quasipermanent feature located some 1000 km southwest of Honduras and Nicaragua, but we now know that it is formed *de novo* each winter. In February to April, it takes the form of a coastal shoaling of the thermocline in the Gulf of Papagaya, forced by Ekman pumping to the south of the wind jet through the sierra mountains behind the Gulf of Papagaya. In May–June it separates from the coast as the ITCZ moves to the north in boreal summer, and then, in July–November the CRD merges into the ridge associated with the NECC as this shoals beneath cyclonic wind stress on the northern side of the ITCZ. Finally, in December–January, the CRD deepens in response to strong trade winds as the ITCZ again moves to the south. Because of the ecological importance of the CRD, it is good finally to have a satisfactory account of its formation.

There is very little regular seasonality in the 20–35 m mixed layer of this province, and irradiance at the surface is sufficiently strong that the entire mixed layer and much of the thermocline lies perpetually within the euphotic zone. However, there is an eastward decrease in mixed-layer temperatures right along the axis of the NECC, which is strongest in boreal summer when very significant warming occurs in the western Pacific. A westward flow of cooler water, associated with the NEC, is encountered below about 200 m.

It is in this province that we meet the extensive seasonal survey data obtained by EASTROPAC in 1967–68 that, although not produced by the methods of today, still yield valuable insights. Consider Fig. 11.9, which shows the near-instantaneous location of the front between nitrate-depleted and nitrate-replete mixed-layer water on either side of the boundary between PNEC and PEQD provinces, or along about 2–5°N. The sharpness of this feature (which is, of course, the "Great Front" observed by Barber and his colleagues on R/V Thompson almost 30 years later) is quite obscured in data representations such as the well-known Ocean Atlas 2001, although it was mapped along 30° of longitude in both extreme seasons in the EASTROPAC data.

## Regional Response of the Pelagic Ecosystem

A seasonally and annually variable linear region of chlorophyll enhancement characterizes the NECC region, and may occasionally—in some satellite images—be traced clear across the Pacific Ocean to the origin of the NECC. Here, enhancement has been attributed to upwelling associated with current meandering (Christian *et al.*, 2004). But the curvature of wind stress that forces the flow of the NECC also induces enhanced algal growth in that flow (Longhurst, 1993), and Ekman suction is perhaps a sufficient explanation for the linear zone of chlorophyll enhancement associated with the oceanic NECC. Maximum curl values in winter produce a vertical velocity at the pycnocline of $0.75\,m\ d^{-1}$, which is a significant effect where the mixed-layer depth is only 25–50 m. The surface chlorophyll enhancement seen in satellite images in the eastern part of the NECC does not normally extend west of about 110°W, consistent with the progressive westward deepening of the mixed layer. We may expect enhancement of subsurface blooms along the NECC toward the west, but have no evidence for this.

Fiedler (1994) has analyzed monthly and between-year variability for the countercurrent from CZCS and wind-field data and has confirmed that wind stress, Ekman suction,

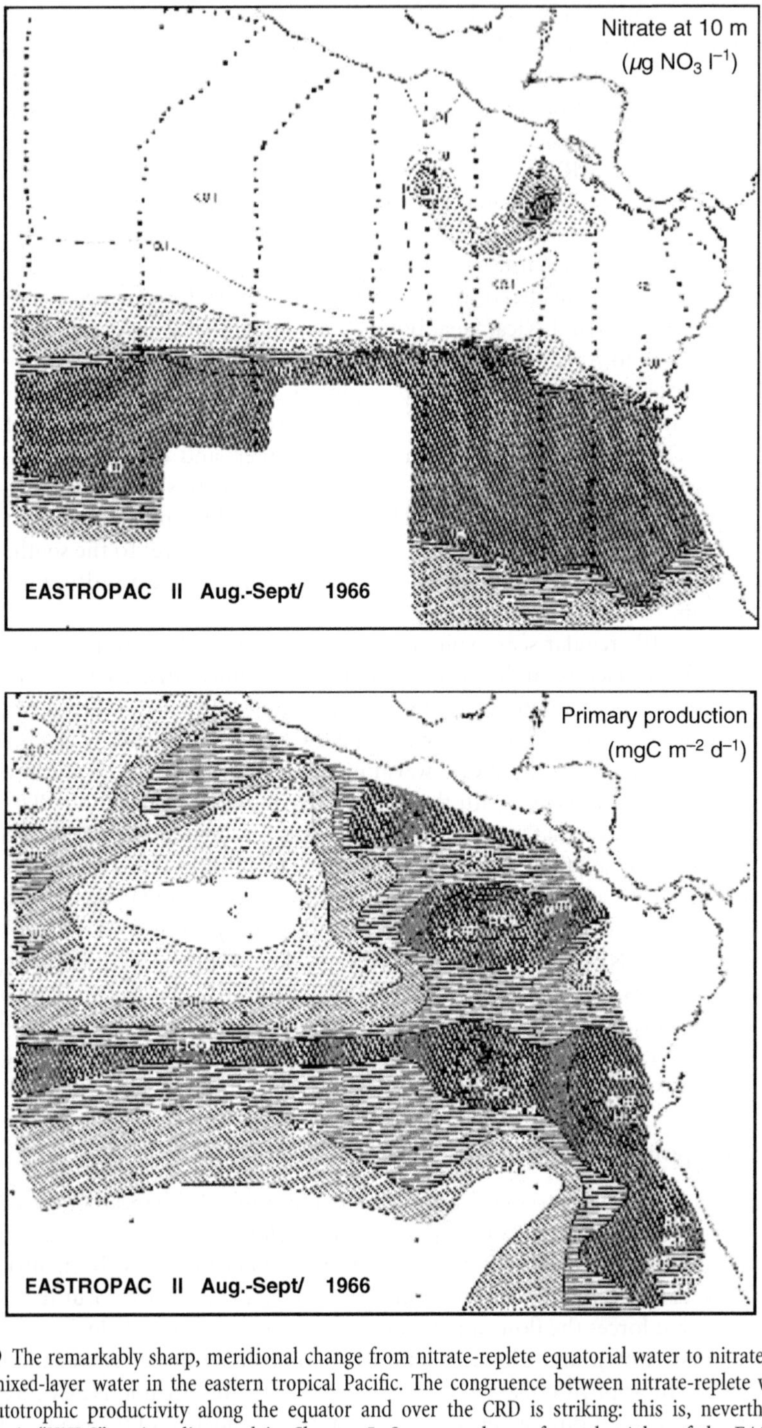


**Fig. 11.9** The remarkably sharp, meridional change from nitrate-replete equatorial water to nitrate-deficient surface mixed-layer water in the eastern tropical Pacific. The congruence between nitrate-replete water and higher autotrophic productivity along the equator and over the CRD is striking: this is, nevertheless, the paradigmatic "HNLC" region discussed in Chapter 5. Source: redrawn from the Atlas of the EASTROPAC surveys, 1967–1968.

and chlorophyll are maximal in winter (December–April). He also showed that there were significant between-year differences in the period 1979 to 1985. The 1983 El Niño warm event was marked by weak winds and low pigment from 90°W to 105°W, whereas the period 1984–1985 was marked by the return of winds and a very strong chlorophyll enhancement from the coast out to 98°W. The SeaWiFS sequence of monthly chlorophyll fields reveals the clear effects of the strong El Niño of 1997–98; surface chlorophyll was significantly lower than in subsequent years, and seasonality was strongly depressed. Recovery occurred at the end of 1998, and "normal" seasonality was re-established in winter 1999–2000 and continued for at least the following two winters. These two events remind us that investigations performed at sea in single seasons must be interpreted in the light of the existing condition of the Southern Oscillation Index.

Seasonal images also show that to the east of the region investigated by Fiedler, the NECC bloom exists in summer when it is absent farther to the west. In boreal summer the NECC bloom is truncated at 120°W but extends westward to about 145°W.

During the period of greatest chlorophyll biomass there is a clear continuity between enhancement in the PNEC and over the CRD, and there is clear separation between this linear zone of higher chlorophyll than background, and the enhancement that occurs within the PEQD province to the south. These observations are consistent with the observations at sea of high chlorophyll both at the surface and in the DCM at 10–12°N and 90–120°W during the EASTROPAC chlorophyll surveys in boreal winter. These seasonal grids in the NECC also confirm a boreal winter primary production maximum (300–400 mg C $m^{-2}$ $d^{-1}$) and a rather long period of high column-integrated chlorophyll (February or March to June or July) reaching values of 25 mg chl $m^{-2}$ (Blackburn *et al.*, 1970). The divergent upwelling effect may be so strong as to remove the entire mixed layer laterally and so expose the strong tropical thermocline (and its associated nutricline) at the surface with significant biological consequences. The location of the CRD is shown as values higher than background for nitrate, productivity, zooplankton volume in the EASTROPAC Atlas, as it is in data for blue whale sightings maintained by the Southwest Fishery Center, La Jolla, CA.

A typical expression of the ecology of the CRD was investigated in March and April 1981 (Herman, 1989; Longhurst, 1985a; Longhurst and Harrison, 1989), unfortunately before the importance of autotrophic picoplankton was fully understood. The main gradient of the thermocline mixed layer lay at 20–50 m and the DCM at 15–20 m with a maximum concentration of 3.0 mg $m^{-3}$, and primary production was maximal at 5–15 m with a maximal rate of about 8.0 mg C $m^{-3}$ $hr^{-1}$. Copepods were the principal mesozooplankton grazers, having their depth centroids closer to the depth of maximal primary production than to the DCM. There was, however, a broad mesozooplankton zone of high abundance from the surface down to about 50 m, below which layers of diel migrants by day occurred at 200–250 m, well above the top of the deep oxygen minimum zone. Repetitive profiles were much more variable within the CRD than beyond it. Below 250 m there was a remarkable association between sinking cells of large phytoplankton and of mesoplanktonic copepods; layers of high abundance of both occurred at 300–500 m, at 600–650 m, and again at 800–900 m. In the last case, up to 100,000 cells $m^{-3}$ were associated with around 60 copepods $m^{-3}$ (Subba Rao and Sameoto, 1988).

To the west of the CRD, the same arrangement in the upper 200 m was found at the BIOSTAT station, chosen to be representative of the unperturbed shallow-mixed layer of the eastern tropical ocean. Here the standing stocks and rates of all biological variables were lower, and features were deeper in the water column. The mixed layer was about 30 m deep, the DCM lay at about 40 m (< 1.0 mg $m^{-3}$), the primary production maximum was just slightly shoaler at 35 m (<2.0 mg C $m^{-2}$ $hr^{-1}$), and copepods occupied most of the mixed layer above, though their depth centroid lay close to the depth of maximum production.

Mesozooplankton at BIOSTAT could be partitioned among small herbivorous copepods whose highest abundance was just shoaler than the DCM, large *Eucalanus* spp. that lay mostly below the DCM yet within the thermocline, and omnivorous copepods and ostracods (*Conchoecia* spp.) that lay below the thermocline and well into the upper oxygen-depleted (< 1.0 ml liter$^{-1}$) zone. Individual predator species had characteristic distributions in each of these depth zones. The overall number of species was greatest in the pycnocline, where greatest stability (Brunt-Väisälä frequency < 30 cph) favored niche specialization among species groups both within genera and between related genera. Comparison of these data with 84 LHPR profiles obtained during the EASTROPAC surveys show that it is permissible to extrapolate from BIOSTAT over the entire eastern region, at least, of the PNEC province; the relations between depths of features of the mesozooplankton profile, of the MLD, and of the DCM are entirely predictable: significant correlation is obtained between the depth of the mesozooplankton maximum, mixed-layer depth, and the DCM across this region (Longhurst, 1976).

A curiosity of this province is the distribution of the cladoceran *Evadne* in the oceanic domain although this is usually considered to be a neritic genus. The EASTROPAC and other survey data show that both *E. tergestina* and *E. spinifera* are widely distributed in both NPTG and PNEC provinces and, while avoiding the equatorial divergence, are again found in the SPSG province to the south (Longhurst and Seibert, 1972). They occur mostly close to the surface (0–20 m) and are generally thought to consume bacterioplankton in preference to autotrophic cells.

We are not surprised to find that higher trophic level organisms respond to the distribution of mesozooplankton biomass—and hence to physical forcing—in an appropriate manner. Ichii *et al.* (2002) recently analyzed the distribution of "jumbo flying squid" (*Dosidocus gigas*) taken in this province by commercial fisheries. When the countercurrent is well developed, as it was in autumn 1997, active upwelling occurs along the

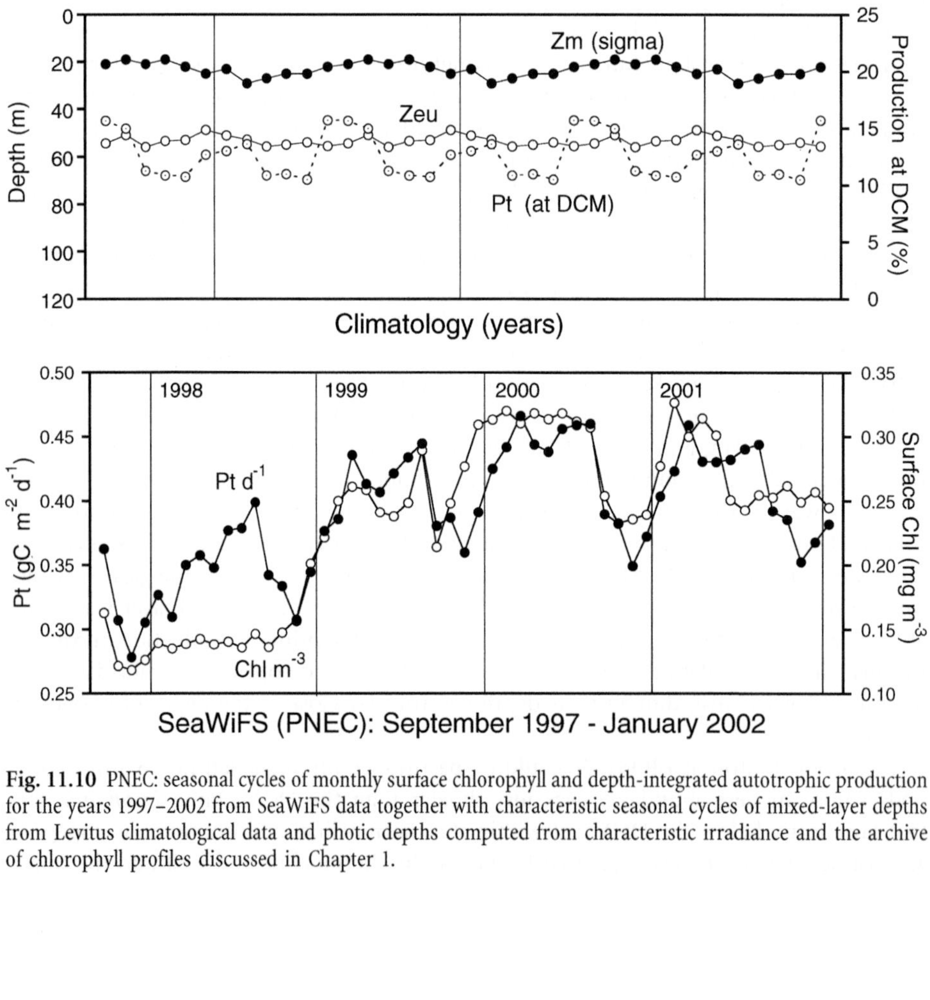


**Fig. 11.10** PNEC: seasonal cycles of monthly surface chlorophyll and depth-integrated autotrophic production for the years 1997–2002 from SeaWiFS data together with characteristic seasonal cycles of mixed-layer depths from Levitus climatological data and photic depths computed from characteristic irradiance and the archive of chlorophyll profiles discussed in Chapter 1.

ridge in the pycnocline west of the Costa Rica Dome, and accordingly the squid occur in high abundance and are spatially associated with a well-developed deep chlorophyll maximum. This did not occur in autumn 1999, when La Niña conditions prevailed. The strong salinity front (see earlier discussion) that characterizes the edge of the NECC retains the squid along the ridge, and also yellowfin tuna (*Thunnus albacares*) that—like *Dosidocus*—feed largely on micronekton.

### Synopsis

*Case 4—Small-amplitude response to trade-wind seasonality*—Mixed-layer depth has insignificant seasonality when meaned over the whole area of the province (20–30 m year-round) and because $Z_{eu}$ is consistently deeper (50–60 m), the thermocline is illuminated in all months. Productivity is seasonally almost invariant (0.30–0.45 mg C $m^{-2}$ $d^{-1}$) with a minor depression in June–July during strongest trade-wind season (Fig. 11.10); between-year differences are significant and respond to the value of the SOI. Chlorophyll biomass shows only slight enhancement in the first half of each year. The pattern is consistent with close balance in tropical seas between consumers and production.

## Pacific Equatorial Divergence Province (PEQD)

### Extent of the Province

PEQD lies between the Equatorial Convergent Front (ECF) to the north and the irregular flow of the South Equatorial Counter Current to the south. These were termed the Northern and Southern Doldrum fronts by Roden (1975). The northern front is, as we have seen when discussing the PNEC province, the southern boundary of the eastward-flowing NECC and lies above a thermocline trough at distances north of the equator that increase eastward: it lies at about 5°N at the date line, and 12°N in the eastern Pacific. The less distinct front of the SECC defines the southern limit of PEQD at about 5°S in the west and almost 20°S in the east. This front—or series of weak fronts—lies above a region where the thermocline deepens rapidly to the south; in the eastern Pacific it separates the warmer equatorial water from the cooler water of the Peru Current. PEQD is thus symmetrical about the equator, only from 115°W to 180°W, where it is narrowest (5°N to 5°S). Eastward from here it extends increasingly toward the south until it encounters the eastern coastal boundary current (CHIL province). The westward extent of PEQD is time-dependent, and the limit at 180°W is set simply as the climatological average condition, and for the same reasons as discussed earlier for the PNEC province.

### Defining Characteristics of Regional Oceanography

The surface circulation here is dominated by the flow of the South Equatorial Current (SEC) westward across the ocean, forced by the southeast trades of the South Pacific Ocean, a response well described by Wyrtki (1966). This westward stream extends across the equator to about 5°N and thus requires a ridge in the topography of the thermocline along the equator so that, in each hemisphere, it flows along the poleward slope of this ridge. At the equator, flow in the SEC is relatively shallow, extending only to as little as 20–50 m although flow is strongest, reaching 50 cm $sec^{-1}$. South of the equator, the SEC is interrupted by variable threads of eastward flow attributable to the SECC; that part of the SEC that lies equatorward of the SECC is designated SEC 01, while that lying south of the SECC is designated SEC 02 (Wyrtki, 1984). The latter, of course, forms the flow around the northern part of the South Pacific Subtropical Gyre, the SPSG discussed later. The flow of the SEC westward across the ocean represents about $50 \times 10^6$ $m^3$ $sec^{-1}$ but, of this, only about 25% flows within SEC 02. The narrow (<200 km), long (14,000 km) jet of the Equatorial Undercurrent (EUC) lies in the upper part of the equatorial thermocline ridge noted earlier. Townsend Cromwell was the first to describe this flow, shortly before his untimely death in an air crash on his way to join a Scripps ship on the Mexican coast. Here, in the PEQD province, the EUC extends to within 40–50 m of the surface, whereas further to the west it lies at 100–300 m.

The linearity of zonal flow in the equatorial Pacific, including both the NECC and the SEC, is testified to by remarkable images representing mesoscale sea-level anomalies obtained by TOPEX-POSEIDON sensors; the ridge-trough system at the sea surface observed subsequent to the 1997–1998 El Niño showed a ridge precisely aligned along the equator and another, stronger one along the poleward side of the NECC; a trough separated the two. During the event itself, the equatorial zone of positive sea-level anomalies was almost entirely absent, while that associated with the NECC was exceptionally strong (see Color plate 18). The multiple threads of the SECC are represented in the relatively poorly defined linear zones of sea surface elevation anomalies to the south of the equator.

Within this general circulation, linear divergence and upwelling occur along the equator forced by the change of sign of Coriolis parameter because, in each hemisphere, westward motion must generate a poleward force. In many transequatorial sections from 135°W to 160°W (Wyrtki and Kilonsky, 1984; Colin *et al.*, 1987; Carr *et al.*, 1992) the divergence—as indicated by the surfacing of isotherms and the nitrate distribution—is aligned almost precisely along the equator and extends to about 2°N–2°S. East of 120°W, however, the divergence lies somewhat to the south of the equator at all seasons; the EASTROPAC sections show that 25°C water is exposed at the surface from 2°N to 5°S in August and from 1°N to 3°S in February. Satellite thermal imagery shows that the boundary between upwelled and surface water carries 1000-km wavelength instability waves that propagate westward.

The ECF follows these waves, and this strongly convergent front may form a spectacular (the word is carefully chosen) feature at the sea surface, because the transition between cold, clear upwelled water within the SEC and the warm, greener water of the NECC may be only a few tens of meters wide. This phenomenon is associated with current shear and eddying and may also be marked by a field of whitecaps or an aggregation of floating *Thalassiosira* mats. Very high concentrations of chlorophyll (background × 3) and extremely high rates of autotrophic production (1.4–1.8 g C $m^2$ $d^{-1}$) have been observed along such a convergent front at 140°W (Barber, 1992). The ECF contains the surface isohalines for 34.4–34.9 ppt, whereas further south and sometimes at the edge of the divergence zone there is usually a second salinity front containing the isohalines for 35.0–35.4 ppt. These indicate progressive transitions in the relative influence of low-salinity water of the NECC and the denser water of the SEC.

In the eastern part of the province, the ECF passes close around the north of the Galapagos Islands so that important anomalies in circulation and hence in the surface chlorophyll field must occur here. Under normal conditions, the island wakes lie to the west or northwest of the islands and may exhibit plumes of high chlorophyll biomass, induced within the turbulent wake. Under El Niño conditions, when flow at the surface may even be temporarily reversed, such plumes may occur to the east and northeast of the islands.

This is, of course, one of the original high-*S* regions and fully to understand the nature of the nitrate enigma, it is necessary to consider the three-dimensional distribution of nitrate here (Thomas, 1972, 1978). I have already noted how sharp the boundary between nitrate-replete and nitrate-depleted water is at the ECF. We must examine both the eastern and western parts of the province: in the east, at 110–120°W, the EASTROPAC surveys found a zone of nitrate-replete water between about 4°N and 10–15°S, or far to the south of the divergence zone. North of this region, only in the CRD is 10-m nitrate higher than 0.1 μM. A nitrate section across the province shows that the mixed-layer nitrate from about 8°S is separated from the deep nitracline that lies at 150 m at 10°S and slopes down to 350 m at 29°S, by nitrate-depleted water (<0.1 μM). Therefore, the nitrate of the upper layer (6.0 μM near the equator and 1.0–2.0 μM near 12°S) has been transported poleward in the diverging surface layer from the equatorial upwelling, which passes above the more saline nitrate-depleted water, rather than originating in vertical mixing *in situ*. It will be appropriate to return to the high-*S* aspects of this province shortly.

## Regional Response of the Pelagic Ecosystem

The PEQD is one of the favorite study regions of the satellite image community because of the very spectacular consequences of the equatorial divergence that occurs here. One of the iconic images from the SeaWiFS sensors is that of the bloom that occurred along the equator at the termination of the unusually strong 1997–98 Niño episode in July–August 1998 already discussed in Chapter 8 (see Color plate 4). Other images of this bloom that may dominate the eastern equatorial Pacific show clearly how the narrow band of high chlorophyll values (<3.0 mg $m^{-3}$) reveals the form of planetary waves propagating toward the west in the eastward flow of the undercurrent (see Color plate 18). The transition event that terminated the Niño conditions in 1998 dominates the long-term, satellite-derived chlorophyll record for the entire period 1997–2002 shown in the regional synopsis; otherwise, seasonality is relatively weak—as was originally suggested by the CZCS images. Ryan *et al.* (2002) discuss the different mechanisms that forced the transition bloom in 1998 when the regional thermocline had become extremely shallow; toward the west, the bloom appears to have been forced by nutrients delivered by turbulent wind mixing and wind-driven upwelling, whereas further east it was planetary wave-induced shoaling of nutrient-rich source waters. Zonal advection within the equatorial undercurrent and meridional spreading by tropical instability waves were the major factors in determining the spatial extent of the transition bloom.

Otherwise, this province has attracted much attention for two reasons: not only because it is a high-*S* region, but also because it is very important in the global flux of carbon dioxide across the sea surface. I have dealt at some length with some of the high-*S* aspects of this province in earlier sections; unfortunately, the Fe-fertilization experiments done here did not develop significant new understanding of the structure and functioning of the pelagic ecosystem; nevertheless, I shall briefly discuss the results later.

But another initiative of the early 1990s, responding to concerns over increasing atmospheric $CO_2$ concentrations, did produce a major new body of significant research. The equatorial eastern Pacific attracted special attention from oceanographers at that time because of the critical role that this region plays in the global carbon cycle. It is the largest single natural source of atmospheric $CO_2$ (of order 1.0 Gt C $y^{-1}$) and is also the site of a major fraction of global oceanic primary production (0.8–1.9 Gt C $y^{-1}$)—and hence of the uptake of $CO_2$ from the atmosphere. It was thought urgent to understand the balance between these two fluxes, and to predict how each would respond to a changed global climate pattern; so, a primary objective of JGOFS was to understand more fully the nature of carbon flux in the PEQD province (Murray *et al.*, 1995). To this end, the JGOFS EqPac studies were undertaken in February–March and August–September, 1992, during which time both states of the ENSO regime were experienced. The U.S. NSF supported a time-series study at 140°W, between Hawaii and Papeete, and the U.S. NOAA mounted five meridional sections across the PEQD province between 140° and 95°W. Four special volumes of Deep-Sea Research II were devoted to the results of EqPac between 1995 and 2002.

The excess of macronutrients in PEQD is often presented in simple terms, but the reality of the ecological response to the imbalance is different, and more complex. The source of excess $NO_3$ is, as noted above, the undercurrent that runs along the equatorial geostrophic ridge, although only the upper part of the EUC is involved in surface enrichment. Moreover, as already noted, the enrichment of surface water in $NO_3$ is highly asymmetric about the equator and is very abruptly terminated northward at the EQF. The long waves that occur in the frontal systems here are tracked by chlorophyll biomass, readily observed both by satellite and also by moored EqPac instruments that lay within the field covered by the waves (Foley *et al.*, 1997). The onset of the ENSO warm event in September 1991 occurred as a series of wind reversals that induced eastward-propagating Kelvin waves, associated with downwelling pulses, zonally across the PEQD province (Kessler and McPhaden, 1995). The process peaked before the first survey of February–March 1992, so that the second survey, 6 months later, was undertaken during the recovery phase, when upwelling was renewed, and after the nutrients thus supplied had initiated a strong bloom (Barber *et al.*, 1996; McCarthy *et al.*, 1996).

The NOAA EqPac surveys also produced one of the few modern spatially extensive data sets of phytoplankton characteristics, so that the relations between these and the oceanic environment could be established (Chavez *et al.*, 1996). The area covered by this survey closely matched the boundaries of PEQD. The data from the five meridional transects suggest values for integrated chlorophyll for the year 1992 of 25–30 mg $m^{-2}$ for PEQD from 110°W to 170°W and for integrated primary production of 650–950 mg C $m^{-2}$ $d^{-1}$. The production values are rather higher than those calculated for PEQD for 1997–2001 from SeaWiFS data, although the chlorophyll values are very close. The biomass of fractions of autotrophic cells was relatively constant from 110°W to 170°W (each expressed as μg C $liter^{-1}$): *Synechococcus*, 0.9–2.9; eukaryotic picoplankton, 0.5–5.0; haptophytes, 2.1–3.0; dinoflagellates, 3.7–10.7; pennate diatoms, 0.5–11.7; centric diatoms, 0.2–2.7; and ciliates with sequestered chloroplasts, 0.001–0.01. Chlorophyll biomass and production rates of autotrophs were usually greatest close to the equator, whether or not El Niño conditions obtained; nevertheless, as Fig. 11.11 shows, productivity was significantly higher during the cool period later in the year than during the El Niño episode at the beginning of the year. The heterotrophic bacterial biomass ($6–8 \times 10^8$ cells $liter^{-1}$, 0–100 m) averaged 70% of autotrophic biomass at the equator at 140°W, and although relatively high ratios are characteristic of oligotrophic water, it was suggested by Ducklow *et al.* (1995)

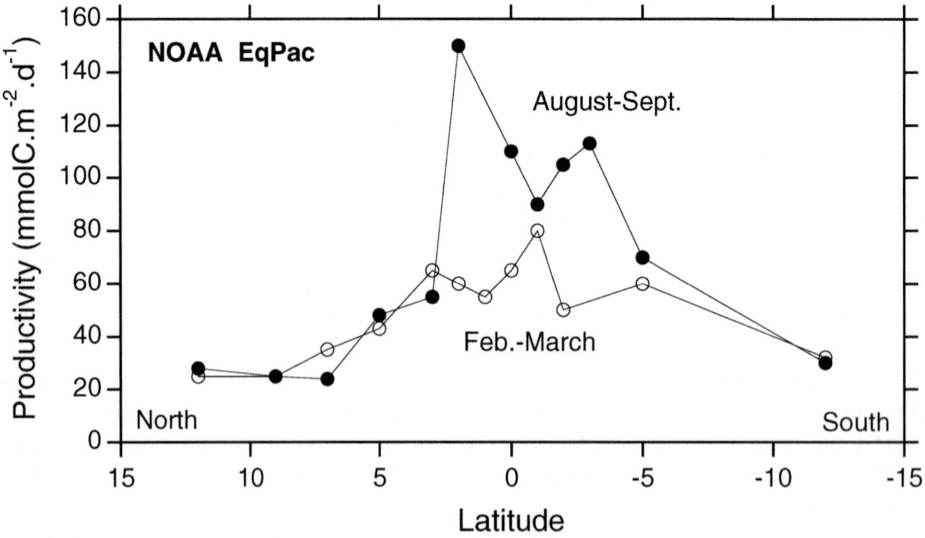


**Fig. 11.11** Transequatorial sections of surface productivity from the 1992 EqPac Surveys to illustrate the increase in productivity between mild Niño conditions early in the year, and 6 months later when upwelling had already recommenced. Source: redrawn from Barber *et al.*, 1996.

that bacteria respond relatively slowly to upwelling fields and that this causes persistent uncoupling with autotrophic biomass.

The distribution of picophytoplankton along the 140°W EqPac transect supported the reality of the ecological boundary between PEQD and PNEC; beyond 7°N the abundance and vertical distribution of small cells matched subtropical gyre profiles. Here, *Prochlorococcus* and heterotrophic bacteria occupied the upper euphotic zone, while maximum abundance of *Synechococcus* and small autotrophic eukaryotes occurred deeper. In the PEQD region, this vertical distinction broke down, and all cell-types of the pico-fraction were abundant throughout the euphotic zone (Landry *et al.*, 1996). Another EqPac study also provided support for the ecological reality of the boundaries suggested for this province. Optical studies of microorganisms along the section at 140°W revealed three "biohydrographic regimes" (Chung *et al.*, 1996). Characteristics of beam attenuation (Cp) due to particles, dominated by heterotrophic bacteria, prochlorophytes, cyanophytes, and small eukaryotes (<3.0 mm), showed that the 7°N–7°S zone could readily be distinguished from the more poleward parts of the section: this zone corresponds well enough with the region here defined here as the PEQD province. It was only in the central biohydrographic regime, corresponding to our PEQD, that the particle population responded significantly to the 1992 El Niño event; a 30% increase in beam Cp and depth-integrated Cp were noted only here.

There is much evidence that the major fluxes in the pelagic ecosystem are—as we have come to expect in recent years—dominated by flux through the small cells of the microbial loop. The basic composition of the autotrophs is approximately as follows (expressed as mg C liter$^{-1}$): *Synechococcus*, 2.0; prochlorophytes, 4.5; prymnesiophytes, 3.0; autotrophic dinoflagellates, 5.0; pennate diatoms, 2.0; and *Phaeocystis*, 0.5. We have become used to the apparent predominance of just two taxa of prokaryotes in the picophytoplankton, usually referred to simply as *Prochlorococcus* and *Synechococcus* without specific connotation. Here, during EqPac, analysis of chlorophyll and carotenins showed that *Prochlorococcus* occurs as at least three ecotypes having different pigment ratios and perhaps carbon-to-pigment ratios. *Synechococcus* may also possess similar heterogeneity (Mackey *et al.*, 2002a). However that may be, the ecology of these small cells is now becoming clearer and EqPac studies revealed a very marked diel cycle in abundance, in cellular light scattering, and in fluorescence of the prokaryotes in PEQD: cell division is synchronized and occurs in late afternoon or at night, whereas cell size, and hence light scattering, represents the balance between cell growth and division. Diel changes in cell fluorescence are due to changes in pigment content and light-dependent quenching (Binder and DuRand, 2002).

With the small autotrophic cells occur the protistan grazers, also expressed as mg C liter$^{-1}$: heterotrophic and dinoflagellates, 7.0; and mixotrophic ciliates, 0.25 (Coale *et al.*, 1996b). This is a recipe for a tightly coupled production/consumption system in which the population size of the grazers can respond as fast as that of the autotrophic organisms. Rate estimates for microzooplankton grazing at the EqPac stations and instantaneous autotrophic growth rates were highly variable, but quite similar. Bulk chlorophyll estimates for February–March 1991 suggested that 0.83 d$^{-1}$ (10–20 m), 0.34 d$^{-1}$ (40–50 m), and 0.22 d$^{-1}$ (70–80 m) would be useful average values. Corresponding rates for consumption by microzooplankton were 0.72, 0.22, and 0.21 d$^{-1}$, respectively (Landry *et al.*, 1995). Thus, grazing by these organisms imposed a daily mortality representing 83% of autotrophic growth at this season, which was reduced to only 55% in the second study period in August–September, when only the smallest autotrophs were controlled by protists. An independent analysis suggested that grazing by microzooplankton (especially microflagellates and dinoflagellates) in February–March removed the entire daily production of picophytoplankton. In this study, microzooplankton consumption only balanced the daily production of prymnesiophytes and cyanobacteria, while at least 50% of diatom loss was attributed to mesozooplankton grazing and to sinking (Verity *et al.*, 1996).

Turning now to mesozooplankton, Roman *et al.* (2002) used the EqPac transects to specify the relationship between biomass of the >200-μm fraction and the location of maximal autotrophic production in the upwelling zone. In accordance with theory, they found mesozooplankton biomass to be shifted "downstream" because of the delay involved in the response of mesozooplankton life cycles to enhanced food availability. Mesozooplankton biomass was generally ~25% of that of autotrophs in equivalent carbon units and ~30% of bacterial biomass. Mesozooplankton grazing was estimated in this study to be <5% daily of the standing stock of autotrophs, a figure that implies that mesozooplankton largely consume microzooplankton and detritus to support their growth ($0.58\,d^{-1}$ in the 64–200 μm fraction, $0.08\,d^{-1}$ in the 1–2 mm fraction). Independent estimates, also made during EqPac, put mesozooplankton grazing rates at <9% of chlorophyll and <12% of primary production daily (Dam *et al.*, 1996). This study suggested that >80% of the carbon ingested by mesozooplankton is not phytoplankton, yet may represent removal of >27% of the biomass of large (>2 μm) diatoms daily. But, as was noted earlier (see PNEC), although generalizations such as these concerning mesozooplankton feeding have become very familiar to us recently, they do tend to obscure the highly structured and highly differentiated pattern of consumption characteristic of individual species, and individual growth stages of species and of meso- and microzooplankton.

The mesozooplankton species and stages are arranged in relation to the stratification of the water column in a reasonably simple and well-known arrangement (Roman *et al.*, 1995). As the mixed layer deepens toward the west across the ocean and poleward from the equator, it carries with it the permanent DCM and associated vertical layering of zooplankton, of which the depth of greatest abundance lies near the depth of maximal primary production rate in the upper mixed layer and therefore significantly shoaler than the DCM. At 155°W the core of the DCM (0.2–0.3 mg chl $m^{-3}$) deepens from 50 m under the equator to 100 m at 15°S, whereas the rate of normalized primary production is maximal in the upper 10 m at about 3 or 4 g C $(g\ chl)^{-1}\ h^{-1}$. Diel migrants (euphausiids and metridiid copepods) are at 400–500 m by day and join the epiplankton in the upper 50 m at night.

Although, as I suggested earlier and in Chapter 5, the "enigma" of Fe limitation in PEQD may be resolved quite simply and without recourse to subaerial fluxes, it may nevertheless be useful to comment very briefly here on the IronEx I and II experiments that were done in PEQD. IronEx I, in 1993, led by John Martin, was the first such experiment in a now long series, and was only a partial success: Fe was released into the ship's wake and response of the autotrophic biota was instantaneous, with photosynthetic efficiency peaking after 2 or 3 days, when chlorophyll concentration and primary production rate had doubled. Unfortunately, after 5 days the patch was subducted below fast-moving low-salinity surface water and lost.

IronEx II, in 1995, was a resounding success (Coale *et al.*, 1996a,b). Repeated injections of Fe into the ship's wake during about 1 week produced a 70-$km^2$ patch with an Fe concentration of 2.0 μM. During this time, nitrate was drawn down from an initial >10 μM to <5.0 μM, chlorophyll increased from <0.2 mg $m^{-3}$ to >3.0 mg $m^{-3}$ in the patch center, and $pCO_2$ decreased as carbon was taken up for photosynthesis. When injection ceased, the patch began to weaken both through diffusion and by the loss of Fe. In control patches, there was no response to injection of biologically inert molecules.

The autotrophic biota showed differential responses that make perfect sense. The cells that responded most rapidly and most completely were diatoms, which increased in abundance by a factor of 85, whereas the picoautotrophs responded only by doubling their numbers. The micrograzers responded in step with the picoautotrophs, but mesozooplankton responded very little—as indeed their long generation time would ensure. The resulting imbalance between copepods and diatoms allowed a bloom to occur that was analogous to the imbalance during a high-latitude spring bloom.

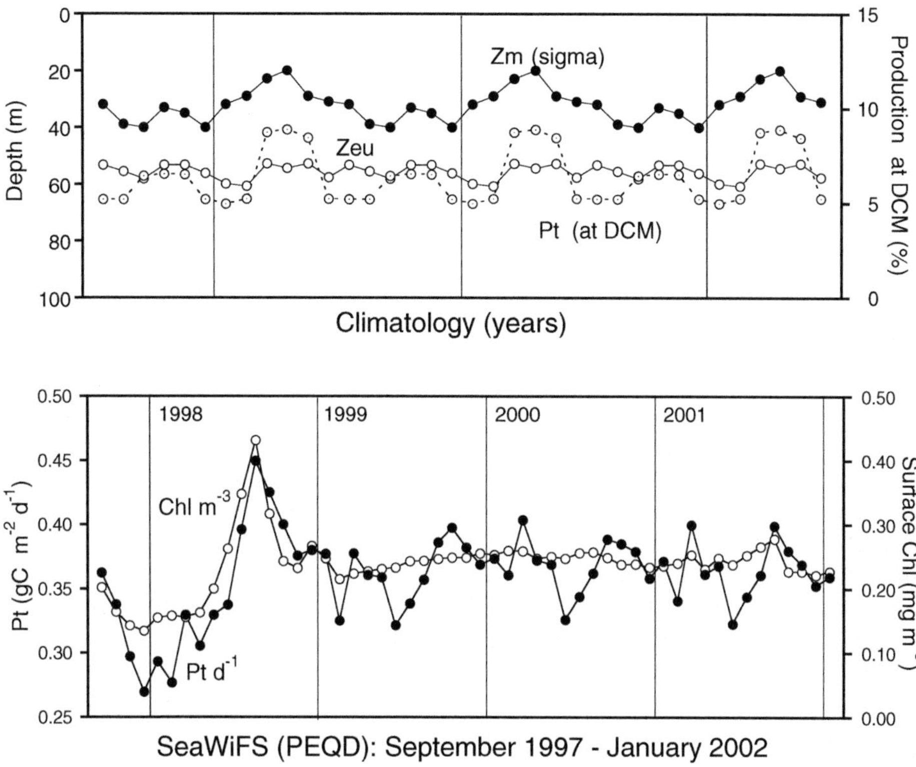


**Fig. 11.12** PEQD: seasonal cycles of monthly surface chlorophyll and depth-integrated autotrophic production for the years 1997–2002 from SeaWiFS data together with characteristic seasonal cycles of mixed-layer depths from Levitus climatological data and photic depths computed from characteristic irradiance and the archive of chlorophyll profiles discussed in Chapter 1.

## Synopsis

*Case 4—Small-amplitude response to trade-wind seasonality*—Mixed-layer depth has only a weak geostrophically forced seasonality when meaned over the whole area of the province with slight shoaling twice annually, in April–May and (weaker) in October–November. Photic depth is consistently deeper, so the thermocline is illuminated in all months (Fig. 11.12). Seasonal changes in P rate that follow the changes in $Z_m$ are so small (0.25–0.35 mg C $m^{-2}$ $d^{-1}$) as to be insignificant when averaged across the whole province, and chlorophyll biomass is seasonally almost invariant (0.12–0.15 mg $m^{-3}$); interannual differences are, however, significant and respond to the value of the SOI.

# Western Pacific Warm Pool Province (WARM)

## Extent of the Province

This province lies under the low-pressure region of the atmospheric Walker circulation—that is, from the date line to the western boundary current that flows NE-SW along the shelf edge on the eastern side of the Indo-Pacific archipelago. WARM therefore has a maximum meridional extent (at about 160°E) from 10°N to about 12°S. The southern boundary of the province passes along the northern coast of New Guinea and then around the arc of the Solomon Islands. These limits will be found to correspond approximately with the regional 29°C isotherm at the surface.

## Defining Characteristics of Regional Oceanography

Circulation in the western part of this province is cyclonic because when the NEC encounters the eastern coast of the Philippines, it bifurcates north and south into the Kuroshio and the Mindenau Currents. The latter branch flows toward the southeast along the shelf edge and then back across the ocean. This cyclonic flow, together with wind stress curl associated with the NE Monsoon, results in the generation of the Mindenau Dome at about 7°N 130°E. Further to the southeast, of course, the flow of the SEC around the southern subtropical gyre of the Pacific Ocean is encountered. The weakly flowing SECC at about 7–10°S (at 165°E) is scarcely observable as a slope in the thermocline topography (Godfrey *et al.*, 1993; Gouriou, 1993), which is otherwise dominated by the westward flow of the SEC along the equator in both hemispheres, requiring the existence of a pycnocline ridge aligned along the equator.

These diverse circulation features of the western tropical Pacific are included within a single province to respect the dominant role played there by regional meteorology in regulating stratification in the ocean that lies below the heavy convective cloud cover of the low-pressure cell of the Walker circulation. The zonal, near-equatorial cloud band of the ITCZ (between the northeast and southeast trades) and the cloud band of the South Pacific Convergence Zone (between the southwest and southeast trades) meet the west winds coming out of the Indo-Pacific archipelago in this region, and this conjunction forms the largest region of persistent cloudiness in the tropics. The resulting heavy rainfall leads to an excess of precipitation over evaporation of 50–150 cm $y^{-1}$, so that a lens of warm, brackish surface water is formed, to be further diluted by the eastward advection of low-salinity surface water from the Indonesian archipelago. Similar conditions occur elsewhere only in two much smaller regions, to the west of Central America and in the northern Caribbean. As defined by the 29°C surface isotherm, the Western Pacific Warm Pool varies in size on the decadal scale (Yan *et al.*, 1992).

The surface isohaline layer overlies a halocline at an average depth of 30 m within a deeper thermostad that reaches down to about 75 m (Lukas and Lindstrom, 1991; Sprintall and Tomzcak, 1992). The Ekman layer, therefore, corresponds to the low-salinity surface layer while the deeper part of the thermostad forms a "barrier layer" between halocline and thermocline. Since there is negligible vertical temperature gradient within the halocline, vertical heat flux does not occur across it (Lindstrom *et al.*, 1987) but, because of the relatively calm winds over the western Pacific, the diurnal temperature cycle in the near-surface layer may be quite large, exceeding 3°C (Soloviev and Lukas, 1997). The recent WEPOCS surveys have changed the classical view that the Pacific mixed layer deepens progressively from east to west; in fact, the mixed layer deepens westward only to the date line, then remains at about 100 m across the western Pacific. At 15–20% of the WEPOCS stations, however, recent wind bursts had mixed the brackish water down to the thermocline: under these circumstances the barrier layer is eroded, some nutrient flux is induced, and the consequent profile resembles the classic view of western Pacific stratification.

It is in this province that the westerly wind anomalies that mark the onset of an El Niño event first develop and, during strong negative anomalies in the Southern Oscillation, conditions characteristic of the WARM province may come to lie so far east that we may find it convenient to evoke the concept of a trans-Pacific WARM-ENSO province. During at least some ENSO events, as occurred during 1986 and 1987, the effect of the Rossby upwelling wave in the WARM province is to shoal the thermocline until it is coincident with the bottom of the surface mixed layer; when this occurs, the "barrier layer" no longer exists and the nutricline comes to lie at the bottom of the wind mixed layer so that vertical flux of nutrients into the euphotic zone must be enhanced (Radenac and Rodier, 1996).

Such westerly wind bursts, generated north of New Guinea, occur mostly in boreal winter, and their impulsive forcing generates Kelvin waves so that, if these straddle the equator, convergence and equatorial downwelling are induced in just a few days: upwelling is thus induced along the pycnocline ridges at 2–3° from the equator, especially that to the north. When the Southern Oscillation is relaxed and westward trade-wind stress is persistent, divergence occurs along the equator, just as in the eastern Pacific. But because the nitracline is so deep, this process enhances euphotic-zone nutrients only during periods of strong westward wind stress; obviously, during El Niño events when trade winds are reversed or absent here, equatorial upwelling is suppressed (Blanchot *et al.*, 1992).

## Regional Response of the Pelagic Ecosystem

There are scatterings of atolls and numerous high islands and island groups throughout this province, around which consequences for phytoplankton growth may be looked for, and perhaps observed, in satellite chlorophyll fields although the effect is multiple and complex (Dandonneau and Charpy, 1985).

Beyond such effects there are two regional blooms within the WARM province, each responding to a different process, and each only rather weakly seasonal; the most important of these responds to divergence at the equator and narrows progressively westward, toward the northern coast of New Guinea. The other, farther yet to the west and much smaller, is the narrow, meandering chlorophyll enhancement associated with the northern edge of the NECC, where this begins its flow eastward across the ocean. Christian *et al.* (2004) describe this as a "ribbon of dark water" in the western ocean and comment that it was especially well developed during El Niño periods, when it includes higher chlorophyll concentrations than at other times. It is associated with the formation of the Mindenau Dome in which nutrient flux is higher in the NECC that forms its southern boundary than in the Dome itself, despite the fact that it has a cyclonic circulation pattern. Christian *et al.* attribute the chlorophyll enhancement along the NECC variously (i) to upwelling associated with meandering, (ii) to seasonal Ekman pumping, and (iii) to the interannual differences in nutricline depth associated with ENSO events. There was a confluence of all these factors in 1997–98 that produced a stronger linear feature than in the previous 5 years of observations.

However, the major part of the WARM region lies well to the east of the Mindenau Dome, and to understand the nature of the seasonal and between-year variability of the biological response here, we may refer to a study by Le Borgne *et al.* (2002b) and to two sets of repeated meridional transects: the Australian JGOFS studies in 1992–93 along 10°S–10°N at 155°E (Mackey *et al.*, 1997), and the French sections from 20°S to 10°N along 165°E from 1985–89 (Radenac and Rodier, 1996). These studies encountered all conditions of the ENSO cycle and enabled useful comparisons to be made of processes typical of different ENSO states.

Because of the great depth of the nutricline, and the presence of the barrier layer above it, surface nitrate is depleted to $<0.008\,\mu M$ down to 70 m although silicate remains always $>2.0\,\mu M$; very little deeper, nitrate is typically $9.0\,\mu M$ at 100 m. There is, as LeBorgne *et al.* (2002a) remark, no significant zonal gradient in phytoplankton response across this province, and productivity is controlled entirely by flux across the nutricline. It might be expected, therefore, that the pelagic ecosystem would resemble that of the downwelling central gyres, but Le Borgne *et al.* suggest that there is—despite the barrier layer—sufficient upward flux of nutrients to ensure that depth-integrated chlorophyll biomass is greater than in the oligotrophic midocean gyral regions, and this suggestion is supported by observation.

An unusually large fraction (<75%) of total production occurs within the DCM, under low illumination, where the chl:C ratio of phytoplankton is very high; nevertheless, there is some evidence for midday photoinhibition of primary production in the extremely clear water. Serial observations demonstrate very close coincidence of the time-varying depths of DCM, of nutricline and of maximum stability (Brunt-Väisälä frequency $>0.02\,sec^{-1}$) at around 100 m over periods of several days.

During Niño events, the principal change is the progressive shoaling of both thermocline and nutricline with a consequent shift in the depth of greatest productivity, although this remains close to the depth of 10% surface irradiance; vertical nutrient flux is somewhat enhanced by the stress at the sea surface imposed by westerly wind bursts. However, during the 1986–87 El Niño, when the thermocline shallowed significantly, surface enrichment and a doubling of primary production rates was quite ephemeral (Radenac and Rodier, 1996). The Australian transects of 1992–92, done at the end of another Niño event, likewise found a relatively shallow thermocline, a relatively thin barrier layer, and a DCM that was everywhere deeper than the top of the thermocline. Surface water was undersaturated with $CO_2$ during the Niño event, although under what Radenac and Rodier (1996) call "reference" conditions, or in the absence of El Niño, $pCO_2$ is in equilibrium with the atmosphere.

The serial satellite images show that near-surface chlorophyll values are somewhat higher a few degrees on either side of the equator, this effect being clearer in the eastern and western parts of the province but weaker centrally. During those ENSO events that are sufficiently strong to bring the nutricline up to lie at the bottom of the wind-mixed layer, we may reasonably assume that this is associated with the 25–50% increase in average primary production that occurs in the WARM province in these years (Barber and Chavez, 1991).

Under reference conditions, relative productivity may also depend on the thickness of the barrier layer itself (Mackey *et al.*, 1997), which is very nonuniform; in 1990 at 155°E, it was 88 m deep at 2°S, compared with only 6 m at 1°N. Associated with this variance in barrier layer thickness are changes in chlorophyll biomass and productivity: chlorophyll in the DCM was $0.3\,\mu g/liter^{-1}$ at 2°S and almost $0.5\,\mu g/liter^{-1}$ at 1°N. The regions of thinner barrier layer have been modeled to have higher column productivity and higher numbers of cyanobacteria in the upper layer, both thought to reflect higher rates of intermittent mixing.

Because mixed-layer chlorophyll concentrations are so low, the deep euphotic zone generally comprises two ecologically distinct depth strata: an upper nitrate-limited, cyanobacteria-dominated zone and a deeper light-limited zone dominated by eukaryotic microalgae (Le Boutiller *et al.*, 1992). As in the PEQD, the picophytoplankton is dominated by *Prochlorococcus* and *Synechococcus* and, when transient nutrient flux induces increased productivity, it is the latter that shows a small increase in abundance along with diatoms and chlorophytes (Mackey *et al.*, 2002a). A recent novelty, *Bolidomonas*, comprises <4% of pico-fraction chlorophyll.

There is a strong zonal discontinuity in mesozooplankton biomass at around 170°W, or just to the east of the date line, chosen as the statutory boundary of the WARM province (Le Borgne and Rodier, 1997). Biomass to the west of this discontinuity, in the WARM province, is about one-third of that in PEQD to the east and its vertical distribution is different: in WARM, overall zooplankton biomass lies deeper, and there is a greater distinction in distribution between microzooplankton (here, principally shallow and associated with the cyanobacterial populations) and the deeper mesozooplankton. Biomass is concentrated in a broad zone from the DCM to the surface but is biased toward the depth of the DCM, especially at night. Under both El Niño and reference conditions, zooplankton biomass from 20°S to 6°N takes low values in the range 0.5–1.0 g dry weight $m^{-2}$; at the equator, when westward trade wind stress is strong and upwelling occurs, zooplankton biomass rises to 2.5 g dry weight $m^{-2}$.

There are therefore, as discussed by Le Borge and Rodier, significant differences in the functioning of the biological pump in the WARM and PEQD regimes. The active flux, contributed by diel migrants, represents 40% of the passive flux of sinking POM in the WARM province and only 9% in PEQD.

As would be anticipated in such clear water, several genera of euphausiids (e.g., *Thysanopoda*, *Euphausia*, and *Nematoscelis*) migrate between great depths of around 400–500 m by day, and the DCM at 100 m at night (Hirota, 1987). *Stylocheiron* remains within the DCM by day and disperses both up and down at night. These results, species for species, follow the same pattern as those described by Sameoto *et al.* (1986) and Brinton (1962) in the eastern tropical Pacific.

An enigma associated with this province is that it should be so productive of tunas of several species. The distribution of these open-ocean predatory fish is probably better known globally than that for any other pelagic organisms: modern Japanese, Korean, and U.S. long liners, bait boats, and purse seiners have efficiently explored all tropical and subtropical oceans and several international fishery commissions have meticulously recorded their catch rates. Much better than any comparable ocean basin-scale maps of the distribution of plankton organisms, the maps of the catch rates of skipjack, yellowfin, albacore, and bluefin tuna give confidence that the distributions they show are real. Even these must, of course, be read with caution—is the edge of a species distribution natural, or is it the effect of fishery regulations, as occurs in the eastern tropical Pacific, or is it yet the effect of a deepening thermocline on the efficiency of purse seines, as may occur toward the west of the ocean?

Nevertheless, quite consistently in all Pacific maps of tuna distribution the boundaries of the WARM province enclose the region of greatest Pacific abundance of skipjack (*Katsuwonus pelamis*) and yellowfin (*Thunnus albacares*), which are the characteristic species of the trade wind zone (e.g., Bayliff, 1980; Sund *et al.*, 1980). For both these species, and also for bigeye (*Thunnus obesus*), the greatest concentrations of larvae occur in WARM, though they are also widely distributed across the other trade wind biome provinces. Though adult yellowfin and skipjack are also widely distributed in PNEC, PEQD, and the warmer parts of South Pacific Subtropical Gyre Province (SPSG) and NPSG, it is in WARM Province that the most persistent high concentrations occur; for skipjack, this population center extends seasonally into KURO as far as the Japanese Islands.

There seems to be no simple explanation for the paradox that such anomalously high concentrations of tuna should occur in such an oligotrophic province. It has been noted that skipjack occur preferentially in very warm water, but it seems most unlikely that a difference of $<2°C$ in surface water temperature in this region compared with other Pacific tropical regions could be physiologically of such advantage as to select for this otherwise apparently unsuitable area, at least in terms of food supply. I suggest that the explanation may perhaps lie in the unique character of the province and its extraordinarily strongly stratified water column with a boundary layer. Such a feature may have two consequences for tuna: the multiple pycnoclines may serve to aggregate layers of food organisms, thus simplifying their location by foraging tuna (which are able to tolerate cold temperatures during very deep hunting forays during daylight hours to at least 300 m), or perhaps the very stable water column provides invariant and predictable conditions for first-feeding tuna larvae. The latter suggestion is made in the light of well-known studies of the differential survival of fish larvae when their prey abundance matches their hatching date and when concentrated layers of prey organisms are disrupted by wind-mixing episodes.

## Synopsis

*Case 4—Small-amplitude response to trade-wind seasonality*—The integrated climatological data illustrate this anomalous case in which the photic depth coincides rather closely with the thermocline, but the halocline lies shoaler; all are seasonally invariant (Fig. 11.13).

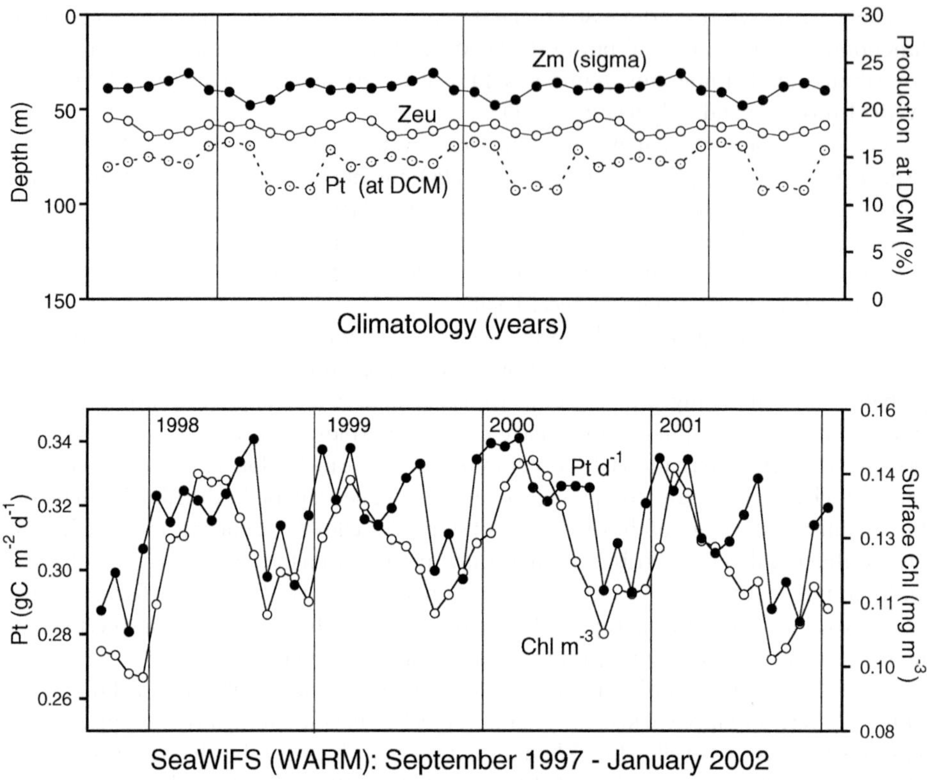


**Fig. 11.13** WARM: seasonal cycles of monthly surface chlorophyll and depth-integrated autotrophic production for the years 1997–2002 from SeaWiFS data together with characteristic seasonal cycles of mixed-layer depths from Levitus climatological data and photic depths computed from characteristic irradiance and the archive of chlorophyll profiles discussed in Chapter 1.

Slight shoaling of euphotic depth occurs in July–August, at the time of heaviest monsoon cloud cover. Both productivity and chlorophyll biomass exhibit very low seasonality with minimal values tending to occur in the boreal autumn (September–November). Once again, very significant between-year variability occurs, related to the value of the SOI.

## Archipelagic Deep Basins Province (ARCH)

### Extent of the Province

The Archipelagic Deep Basins Province (ARCH) is something of a grab bag, even though the arrangement does follow common oceanographic usage, because it coincides quite well with the concept of the Australo-Asiatic Mediterranean Seas of Dietrich *et al.* (1970).

This province comprises the deep basins of the Indonesian archipelago that lies between Indian and Pacific Oceans, together with the South China Sea and the Coral Sea to west and southeast, respectively, of the archipelago. The associated continental shelf areas are gathered into the Sunda-Arafura Shelves Province (SUND), named for the Sunda Islands. You may think that in this unique region of extraordinarily complex topography it would have been more convenient to consider both the deep basins and shelves as a unit, as was done for the Mediterranean and Black Sea. The reason that I did not take this route is because here the shelf areas are so very large that it is easy to forget that major open-ocean, deep-water conditions do exist within this region. A bibliographic search for studies pertaining to ARCH suggests that some regional investigators have not clearly differentiated shelf and deep-water conditions.

So, for those reasons, the ARCH province as used here comprises the deep basins of the South China Sea, of the centrally grouped Sulu, Celebes, Molucca, Flores, and Banda Seas, and, further to the east within the arcs of the Bismarck, Solomon, and Vanuatu Islands, of the Bismarck, Solomon, and Coral Seas. The last of these is by far the largest and could easily be considered separately as a unique province: it is bounded from the Pacific by the Solomon and New Hebrides and from the Tasman by the eastward passage across the north of that sea of the retroflected East Australian Current (see TASM). It therefore has a very large latitudinal extent, from about 22°N, south of Taiwan, to about 32–35°S just beyond Norfolk Island, a convenient marker for the southern border of the Coral Sea being the chlorophyll enhancement associated with the reflected flow.

## Continental Shelf Topography and Tidal and Shelf-Edge Fronts

In this complex region, it will not be useful to discuss in detail the relations between shelf areas and oceanic depths; these can be better understood simply by examining a simple bathymetric map of the province (Fig. 11.14). The differences between deep and shallow channels should be noted, as should the relative sill depths of deep basins, because these determine circulation patterns. It should also be noted at the outset that a common characteristic of all the subregions—shelf or deep sea—that are discussed is the presence of scattered groups of islands and atolls. The shelf region of the Coral Sea is, of course, the location of the Great Barrier Reef.

The individual deep basins are all open, in varying degrees, either to the Pacific or Indian Oceans or to both. The Sulu and Celebes Seas are more enclosed than the remainder, whereas, of the strictly archipelagic deep basins, the Banda and Molucca Seas represent a major deep-water connections between the two oceans. Of the two large marginal seas, the Coral Sea (together with the Bismarck and Solomon Seas) lies only

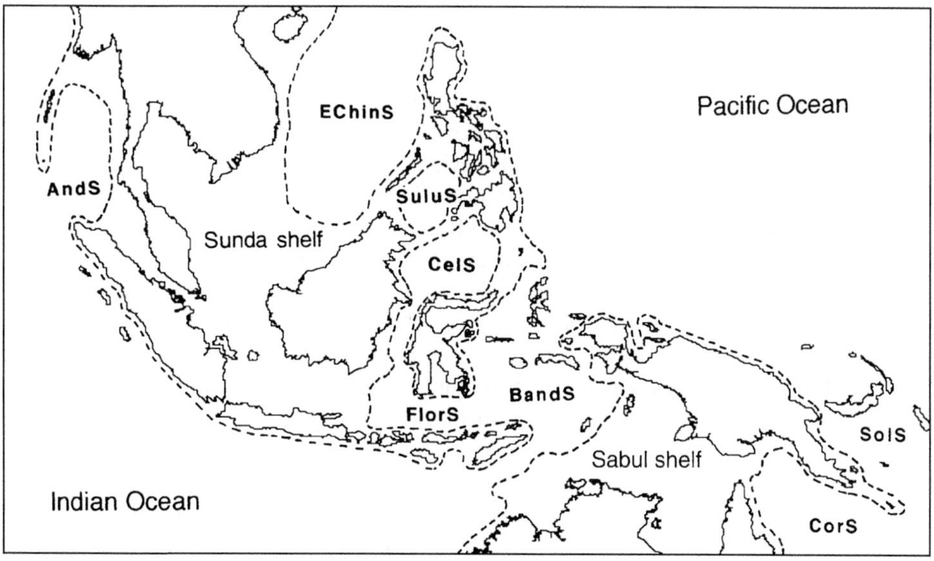


Fig. 11.14 This bathymetric map of the Indo-Pacific archipelago illustrates the characteristics of ARCH and SUND provinces. The dashed line is the 200-m isobath, enclosing SUND, whereas the deep basins of ARCH are indicated by abbreviations representing individually the Andaman, South China, Sulu, Celebes, Flores, Banda, Solomon and Coral Seas. Note that a central region of coral atolls inhabits the South China Sea and that these are visible in satellite chlorophyll fields.

partly isolated from the SW Pacific behind island arcs, whereas the South China Sea has a major deep-water connection to the Pacific between Taiwan and Luzon and extensive shallow connections to the southwest with the Indian Ocean.

## Defining Characteristics of Regional Oceanography

The circulation of the shelves and deep basins comprising the Indo-Pacific Archipelago and Coral Sea are treated as a single, complex hydrographic and atmospheric system, the Southeast Asian Waters, in the Naga Report of 1961. Nevertheless, it will be useful here to distinguish three components: the South China Sea, the archipelagic seas, and the Coral Sea. The discussion that follows of circulation through the archipelago and adjacent basins is intended to have relevance for the ecology of both the coastal (SUND) and oceanic (ARCH) provinces.

The regional oceanography of this region, of which the most complete description remains that of Wyrtki, is unique. Because of a small sea-level difference, in the absence of wind forcing, flow would be through the archipelago toward the Indian Ocean, but the near-surface flow, in which we shall be most interested, is in fact dominated by the effect of the reversing monsoon winds. Satellite imagery confirms the dynamic nature of the flows through passages between islands. Though apparently not described, it would not be surprising to find examples of barrier layers (shallow halocline above a deeper thermocline) here, as is typical of the WARM Province.

The East Asian monsoon system is the classical model of such seasonal patterns; the intertropical convergence zone (ITCZ) between the easterly winds of the northern and southern hemisphere lies seasonally on each side of the equator, following the sun between 23°N and S. In boreal winter, then, the Indonesian archipelago lies below the northerly trade winds that are first felt in October in the East China Sea, and then progressively strengthen as the ITCZ passes south of the equator; the northwest monsoon brings moist air, causing heavy rainfall and rapid discharge of freshwater from the short rivers in the archipelago. In May, the northerly monsoon collapses as the ITCZ moves north, so that southerly trades (Beaufort force 4 over the open sea) develop over the whole archipelago, and by August the monsoon of boreal summer is fully developed. This is a significantly drier season and in the archipelagic region the southeast trades are strongly zonal, so forcing a westward component in resultant flows. This seasonal near-reversal of wind stress induces reversing current systems through the channels of the archipelago and is associated with a great diversity of hydrographic processes likely to have biological consequences, usefully summarized in the review by Wyrtki (1961).

The strongest reversing flow is in the South China Sea, because of western intensification along the coast of Vietnam, and is usually associated with a countercurrent offshore. Thus, in February, at the peak of the Northeast Monsoon, a strong surface flow of cool water passes south as a western boundary current along the coasts of southern China and Vietnam, continuing on to the southeast around the southern side of the South China Sea, and then through the Java Sea to join Pacific water originating in the Molucca Sea. This combined stream then passes east through the Banda Sea to rejoin the Pacific Ocean around New Guinea. In boreal summer, this flow is reversed.

In August, the onset of the Southeast Monsoon reverses the circulation in the South China Sea, and flow is now forced to the northwest through the Java Sea and to the west through the Banda Sea by the westward component of the Southeast Monsoon winds north of Australia. Part of the water entrained from the Pacific around Mindanao returns to that ocean along the north coast of Celebes, while some passes through the Straits of Macassar. Here, flow is persistently to the south with maximum velocities in February–March into the Flores Sea and July–September into the Java Sea.

The deep basin of the Banda Sea, especially toward the east, experiences regional upwelling during the Southeast Monsoon, and downwelling during the northerly monsoon, to satisfy balance between the archipelagic flow-through and the SEC of the Indian Ocean. As noted earlier, the Banda Sea is a major pass for flow from the Pacific Ocean to the Indian Ocean past Halmahera and Ceram in the north, and by the Flores Sea and Timor to the south. Currents are weak and variable in intermonsoon periods.

Anticyclonic gyral flow persists in the Celebes Sea during all seasons, though it extends farther west during the Northeast Monsoon; strong flow passes south of Mindanao, returning to the Pacific (to enter the SECC) together with flow leaving the Molucca Sea around the Halmahera gyre. During boreal summer, when the gyre is displaced to the east, surface drift is received from the Sulu Sea; in winter, this drift through the Sulu archipelago is reversed.

The SEC of the Pacific Ocean flows westward into, and through, the Coral Sea diverging when it meets the Australian continent at 18–19°S during the dry season; during the Northeast Monsoon season of austral summer, the divergence migrates equatorward to at least 14°S. At this divergence, to the east of the 200-m topography of the Queensland Plateau, there is a feed into the East Australian Current to the south and—to the north—a cyclonic circuit around the Gulf of Papua and the Solomon Sea (Andrews and Clegg, 1989) which transports $10–15 \times 10^6\ m^{-3}\ sec^{-1}$ into the Indonesian archipelago. In the area of the divergence itself, currents are weak and include a persistent cyclonic eddy in which water is transported onto and over the Great Barrier Reef. The Coral Sea is a uniformly oligotrophic region, in which water-column stability is very strong and where the surface layers are strongly nitrate-deficient; at the end of the austral summer (wet) monsoon there is a strong difference between Solomon Sea (low) and Coral Sea (high) surface salinities.

Seasonal variability of mixed-layer depth is slight (Wyrtki, 1961) in the Sulu, Celebes, and Flores Seas, but elsewhere it responds as expected to changes in monsoon wind stress: in the central and northern parts of the South China Sea, mixed-layer depths are shallow (30–40 m) in the southeast monsoon, deepening (to 70–90 m) during the winter monsoon. By the end of boreal winter, the mixed layer has deepened here to 100 m. In the Banda Sea and also in the shelf areas of the Arafura Sea, the changes are even greater, upwelling under the influence of the southeast monsoon (May–August) forcing a 2°C temperature drop at the surface and reducing the mixed layer to $< 20$ m deep. Other monsoon-driven, persistent upwelling regions within the province were predicted by Wyrtki (1961) to be (i) off the Macassar peninsula; (ii) along the coast of Vietnam, where temperature drops of $>1$°C occur during the southerly monsoon; (iii) on the coast of Sarawak; and (iv) south of Hong Kong in the South China Sea. Satellite observations permit other locations of coastal upwelling to be located: given the geography of the region and the complexity of seasonal flows, it will be very surprising indeed if multiple locations are not identified when careful exploration is undertaken of the relevant high-precision AVHRR-derived SST images. Hendiarti *et al.* (2004) discuss such a situation on the southern coast of Java, and simple examination of SST and chlorophyll images confirms that the upwelling region off Vietnam that was predicted by Wyrtki does indeed exist.

The consequences of the unique meteorology (heavy rainfall and intense cloud cover) over the Indo-Pacific Archipelago are significant for biological oceanographic processes: the rivers of the archipelago discharge $3.0 \times 10^9$ tons of sediment annually into coastal water, or about twice the sediment discharge of the Amazon. The coastal rivers of Southeast Asia discharge another $4.1 \times 10^9$ tons. Together, this is more than twice the sediment discharged from all other rivers (Milliman and Meade, 1983).

## Regional Response of the Pelagic Ecosystem

Throughout this province, the complex topography both of the land and of the seabed is likely to produce many nonpersistent hydrographic instabilities, themselves likely to be associated with nutrient transport to the photic zone and biological enhancement. Such features are difficult to predict or map comprehensively, even with satellite imagery, because of the extensive and pervasive cloud cover of the region.

Despite such difficulties, it is clear that the pelagic ecosystem here responds to the strong seasonality of the reversing monsoon winds with winter phytoplankton blooms in each hemisphere: surface chlorophyll is maximum in the Coral Sea and adjacent basins in austral winter (July) and in the South China Sea in boreal winter (January). Although in the smaller archipelagic seas the seasonal signal is not so clear, the Coral Sea seasonal bloom does extend westward through the Banda and Flores Seas, while the Sulu and Celebes Seas follow—rather generally—the same pattern as the South China Sea. This pattern appears to be a simple response to the seasonal deepening of the mixed layer in the two hemispheres and the consequent flux of nutrients into the euphotic zone.

It is in this region that some of the earliest direct observations were made using satellite data of phytoplankton response to mesoscale physical processes, at least partly because here such responses are as direct and strong as anywhere in the oceans. Island wakes up to 300 km long were observed as high-contrast features in early CZCS images of the Bismarck Sea, representing the effects of turbulent plumes downwind of the mountains of New Britain and high islands such as Sakar to the east of the Vitiaz Straits (Wolanski *et al.*, 1986). These plumes represent cooler water surfacing in regions where the surface water is highly turbid, either with phytoplankton or suspended sediments; fields of large internal waves seen in the southern Bismarck Sea probably represent deeper, clearer water being brought to the surface. Some of these features have been examined at sea; thus, in the Coral Sea, temporary zonal ridges in the variable circulation have been observed at 12 and 17°S, with consequences for mixed-layer nutrient levels (Rougerie and Henin, 1977). We can anticipate that many such features will be revealed throughout the province now that examination of detailed images has become routine; we can also expect that their orientation and location will be found to depend on seasonal and shorter changes in wind direction and strength. Examination of almost any high-resolution chlorophyll or SST image of this province (including those shown here) will yield further examples, perhaps especially of plumes of high chlorophyll in the eddying flow through passes in the island arcs surrounding the Coral Sea. The consequences of eddying flow behind coral atolls has been investigated in the Coral Sea (Rissik *et al.*, 1997): uplift of the nutricline in the island wake induces enhanced production of phytoplankton and microzooplankton, although with little alteration of specific composition compared with the free stream. Ecosystem response occurs, principally in the organisms associated with the DCM, up to at least the micronekton level, with myctophid fish being observed to feed more successfully in the wake.

In the Solomon Sea, there are observations of a shade-adapted DCM (0.6 mg chl $m^{-3}$ and 0.15 mg C mg chl $hr^{-1}$) at about 75 m (near the 1.2% isolume), containing almost half of the total 0–200 m chlorophyll (Satoh *et al.*, 1992). Phytoplankton profiles in the Coral Sea have a somewhat deeper DCM, of order 100 m, where light-dependent maximum assimilation numbers of cyanobacteria are twice those for near surface populations, whereas for larger phytoplankton the relevant value is almost five times greater (Furnas and Mitchell, 1996b); this is a typical result for very oligotrophic open oceans. The southern part of the Solomon Sea, and the internal basin of the Louisiade Archipelago, which forms the southern margin of this body of water, has relatively high productivity. As in other oligotrophic regions, there is evidence in the stratified region that the cells at the DCM are more dependent on nitrate than those closer to the surface, and that the pico fraction generally dominates the assemblage of autotrophic cells. Most of the primary production occurs shoaler than the midday 20% isolume, and very little is lost to sinking, so we can assume that production and consumption of small cells are held closely in balance.

Under such conditions, it is small wonder that the Coral Sea and the other marginal seas share other ecological characteristics with the oligotrophic South Pacific, including the persistent occurrence of blooms of nitrogen-fixing tufts of cyanobacteria. Very large ($9 \times 10^4 km^2$) blooms of these organisms, identified as *Trichodesmium* by its chromatic signature, have been observed at 10°S 165°E New Caledonia in the surface chlorophyll field

(Dupouy *et al.*, 1988). Such occurrences have major significance for basin-scale budgets of inorganic nitrogen and are typical of the nitrate-limited oligotrophic ocean, a limitation that is only seasonally relieved in the southern Coral Sea by the deepening of the mixed layer, during the peak of the Southeast Monsoon of austral winter. Otherwise, the phytoplankton of the Coral Sea is recorded as being dominated by the pico fraction ($<2\,\mu m$), although quite high productivity (1–3 g C $m^{-2}$ $d^{-1}$) has been measured adjacent to the continental margins, as along the Papua Barrier Reef and in the Louisiade Archipelago (Furnas and Mitchell, 1996b); here, nutricline and associated DCM were deep (100 m and 60–125 m, respectively) and annual productivity was computed to be 100–200 g C $m^{-2}$.

In the Indonesian region, observations at sea confirm the general inferences made from satellite data; chlorophyll biomass at the surface is enhanced during the southeast monsoon from 0.25 to 2.5 mg $m^{-3}$ above levels occurring during the northeast monsoon. This seasonal buildup of biomass, associated with an increase in the rate of primary production, is greatest in the eastern seas (Banda, Flores, and Seram) and decreases westward through the archipelago, becoming a very slight effect in the Sulawesi Sea and the Makassar Strait (Kinkade *et al.*, 1997). The predicted downwelling-upwelling seasonal cycle in the Banda Sea has been observed, with predictable consequences for photic zone nitrate (Wetsteyn *et al.*, 1990) and seasonal primary production (Zevenboom and Wetsteyn, 1990; Gieskes *et al.*, 1990). In general, the degree of seasonal chlorophyll enhancement throughout the archipelago seems to be correlated with sea surface temperature and hence the relative strength of the seasonal upwelling effect; in the Banda Sea, chlorophyll integrated over 0–25 m is five times higher in austral winter, during the southeast monsoon (Gieskes *et al.*, 1988). In August, during the southeast monsoon (upwelling and low irradiance), the surface layer is homogenous down to 60 m and primary production rate is highest at the surface, whereas during February, under the influence of the (downwelling and high-irradiance) northerly monsoon, stratification is established and a DCM forms at 40–80 m. At these depths, phycoerythrin-containing *Synechococcus* occur at very high concentrations of $10^4$–$10^5$ cells $cm^{-3}$. Primary production rate is higher during the southerly than the northerly monsoon (1.85 and 0.9 g C $m^{-2}$ $d^{-1}$, respectively). Patches having even higher production rates ($<7$ g C $m^{-2}$ $d^{-1}$) occur to the south of New Guinea, where upwelling occurs over the outer slope of the Papuan Barrier Reef (Furnas and Mitchell, 1996a).

The mesozooplankton biomass of the Banda Sea is almost doubled at the onset of the upwelling process, when chlorophyll accumulation reaches 3–4 mg $m^{-3}$, by the rise from depth of large populations of *Calanoides philippinensis* together with *Rhincalanus nasutus*, which reach an abundance 4000 $m^{-2}$, grazing on phytoplankton and microzooplankton. Again as expected, vertical excursions during diel migration are reduced in the turbid water associated with the southerly monsoon. A consequence of this change in behavior is a stronger diel difference in gut chlorophyll in copepods between the two seasons; it should also be noted that, once again, tropical mesozooplankton herbivores account for only a small part of primary production (5–25% daily) or of standing stock (2–6% daily). The biomass of micronekton responds similarly but with smaller magnitude; there is a difference of only about 50% between the two monsoon seasons (Schalke *et al.* 1990).

The South China Sea is less oligotrophic than the adjacent waters of the western Pacific with which exchange occurs through the Straits of Luzon; the nutricline is shallower, and surface chlorophyll almost twice that of the adjacent Pacific. The physical processes, forced by the reversing monsoon winds, which control phytoplankton production in the South China Sea were recently modeled by Liu *et al.* (2002b). This reproduces quite closely the observed nutrient and biomass profiles, and suggests three areas where upwelling appears to induce exceptionally large seasonal blooms: off the east coast of Vietnam, to the east of Luzon, and along the northern edge of the Sunda Shelf. The SeaWiFS and AVHRR images confirm that coastal upwelling occurs on the east coast of Vietnam, especially during the southeasterly monsoon, and reveal major meandering plumes of high chlorophyll that extend out from the coast to the central region of the South China Sea.

Of course, the generally limiting nutrient in the South China Sea is again nitrate and so *Trichodesmium* is, as in the Coral Sea, a major component of the planktonic autotrophs; this organism is, of course, most frequently phosphate-limited and it may be that the effluents of Mekong and other great rivers draining into the South China Sea carry enough of this element to the surface waters to ensure massive and sustained blooms of *Trichodesmium* as occurs in the Guiana Current far downstream from the Amazon plume (Lenes *et al.*, 2005). The diatom symbiont *Richelia intracellularis* makes some contribution to nitrogen acquisition by the phytoplankton in the South China Sea, although probably < 5% of that which is based on nitrate flux across the nutricline during the winter monsoon (Chen *et al.*, 2004). The picoplankton of the DCM is increasingly dominated by *Prochlorococcus* toward the central regions, away from the shelf. The Celebes Sea appears to have a very similar pattern of phytoplankton ecology to the South China Sea with near-surface abundance of diatoms (with *Richelia*?) and *Trichodesmium* as well as a cyanobacteria-dominated DCM that is associated with a microzooplankton maximum.

## Synopsis

*Case 3—Winter-spring production with nutrient limitation*—The climatological seasonal cycles of chlorophyll and productivity offered for this province (Fig. 11.15) are somewhat misleading, because they are dominated by seasonality near the equator and further south. The pycnocline undergoes a weak seasonal excursion (20 m November–April in northerly

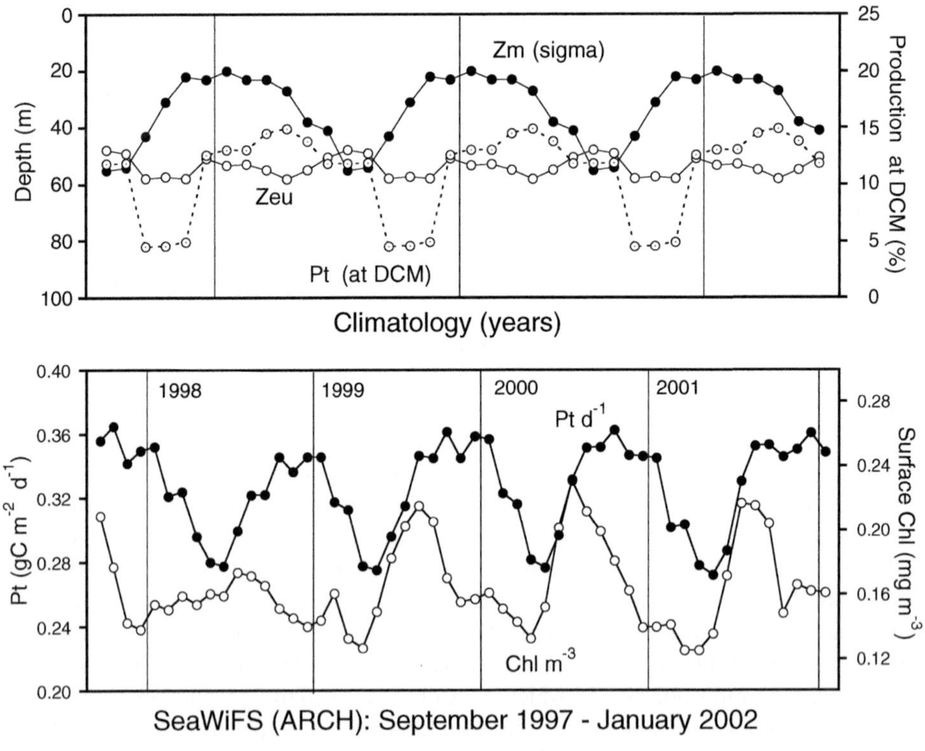


**Fig. 11.15** ARCH: seasonal cycles of monthly surface chlorophyll and depth-integrated autotrophic production for the years 1997–2002 from SeaWiFS data together with characteristic seasonal cycles of mixed-layer depths from Levitus climatological data and photic depths computed from characteristic irradiance and the archive of chlorophyll profiles discussed in Chapter 1.

monsoon, 60 m July–August in southerly monsoon) while photic depth is almost invariant at 50–55 m, so that the thermocline is permanently illuminated except briefly in austral winter (July–August). A very slight productivity rate increase occurs during onset of southerly monsoon winds, when the mixed layer also deepens through the photic depth. The dynamic range of chlorophyll biomass is very small, about a mean of about 0.15 mg $m^{-3}$. Accumulation and chlorophyll closely matches the seasonal increase in productivity, but decline sets in almost at once and continues until the next rate increase of the following year; this appears to be consistent with an overall very close match between consumer biomass and production rate. Overall, seasonality closely resembles that of MONS.

## South Pacific Subtropical Gyre Province, North and South (SPSG)

### Extent of the Province

The limits adopted for this province represent a pragmatic response to the fact that this is the least-well-known region of the oceans, about which it is very difficult to locate sufficient information for a reasoned description. For this reason I make no apologies for an arrangement that is contrary to the definition of the Trades and Westerlies biomes—obviously, this province includes both wind systems, as I shall discuss later, and therefore includes two oceanographic regimes that I believe are otherwise fundamentally different.

Thus, the SPSG province comprises the central and southern part of the subtropical gyre of the South Pacific Ocean, the northwestern regions are the WARM province just discussed. SPSG is bounded to the south by the far-field effects on chlorophyll enhancement of the Subtropical Convergence Zone (see SSTC) and to the north by the southern edge of the chlorophyll enhancement caused by equatorial divergence (see PEQD). To the east, SPSG is bounded by the offshore eddy field of the Humboldt Current (see Humboldt Current Coastal Province) and to the west by the 29°C surface isotherm at the edge of the western Pacific warm pool and the line of the New Hebrides (now Vanuatu) that enclose the Coral Sea.

This is therefore an entirely oceanic province, but the presence of the Polynesian Islands from Easter Island at 120°W to the Samoas at 170–180°W must be noted: the Marquesas, the Tuamotos, the Cook Islands, Pitcairn, Ducie, and many other isolated islands and atolls populate this otherwise empty ocean.

### Defining Characteristics of Regional Oceanography

This is the most data-poor region of the ocean, a situation that may finally be redeemed by the state-of-the-art drifting oceanographic floats now deployed globally in large numbers. A glance at any global maps showing where observations have been made of any variable will attest to the relative paucity of our knowledge and understanding of the South Pacific gyre.

The canonical single-gyre model has the SEC passing along its equatorward side (and flowing more strongly in boreal summer) and a South Pacific Current along its poleward side, associated with the oceanic Subtropical Convergence Zone. The western part of the gyre includes circulation into the Coral Sea and the Western Pacific Warm Pool (WARM and ARCH provinces). As noted in the discussion of the PEQD Province, an SECC, weaker and more variable than its counterpart of the Northern Hemisphere, flows across the northern part of the province, embedded within the westward flow of the SEC from 7°S to 14°S at 155°W, being farther south and stronger in austral winter (Wyrtki and Kilonsky, 1984; Eldin, 1983).

However, such anticyclonic surface circulation models that are based on a single gyre rotating around an axial point at about 25°S in the western part of the ocean are certainly a misleading oversimplification of the real pattern: flows within the southern subtropical gyre of the Pacific are structured just as in the North Pacific gyre, and that a South Subtropical Countercurrent traverses the center of the gyre at 20–25°S. This lies approximately below the zone of calms between the trades and westerlies, so that we should anticipate different seasonality to north and south of this line. If we had the same plenitude of studies here as we have for the North Pacific, I should certainly propose that we recognize two provinces here as there, one in the Westerlies and one in the Trades biomes.

Be that as it may, this southern subtropical gyre is, in its entirety, the most uniform and seasonally stable region of the open oceans and is (as noted by Tomczak and Godfrey) the origin of the name of the Pacific Ocean. Surface winds are as stable as the permanent atmospheric high pressure centered approximately above Easter Island (10°S, 110°W), and the trades are weak but remarkably constant year-round so that seasonal maps of wave height are almost invariant. The dry descending air mass that comprises the eastern termination of the Walker circulation cell (the wet, ascending air mass over the WARM province comprises the other pole of this atmospheric cell) maintains an evaporation-precipitation index of 40–80 cm $y^{-1}$ and thus warm, salty surface water.

The atmospheric convergence (the South Pacific Convergence Zone) between the two major wind systems lies NW-SE from New Guinea (15–20°S) to about 30°S as it approaches the American continent (Barry and Chorley, 1982). This atmospheric convergence is associated with an oceanwide line of relatively heavy cloud cover and is a feature in the distribution of wind divergence over the ocean (Lagler and O'Brien, 1980). It aligns with a thermocline ridge in winter, indicating some divergence of surface water, and it is aligned above a linear, zonal maximum in sea surface height variability that is associated with a South Subtropical Countercurrent (SSTC) that passes across the central gyre here like its counterpart of the northern hemisphere (Qiu and Chen, 2004). This flow, described almost 40 years ago as a "Tropical Countercurrent," has been lost sight of since, and does not appear in any map of general circulation that I have seen. The SSTC is most unstable in austral winter when cooling at the sea surface increases the vertical velocity shear between this and the underlying and opposing SEC flow.

The permanent pycnocline is bowl-shaped, lying at about 300 m at 124°W, at the center of the bowl, and rising to 150 m from 150°W to the western edge of the province near the date line; consequently, as in the South Atlantic Province, we should note the general baroclinic upslope of nitrate isopleths toward the edges of the gyre. A thermocline occurs at 25–40 m over the whole province during austral summer, but to the south of 25°S it deepens to 75–100 m during late austral winter. These meridional differences are forced by the different characteristics of the two wind systems—trades and westerlies—on either side of the atmospheric convergence zone; the former are less strong, but very much more predictable both as to strength and orientation, whereas the latter have high variability associated with the continual eastward passage of weather systems across the southern part of the gyre.

Photosynthetically active radiation at the surface in the central part of the gyre ranges seasonally from 75 W $m^{-2}$ in June to 135 W $m^{-2}$ in December–January even as SST varies from 23 to 19°C, the minimum being in September when MLD is deepest. Nitrate-replete surface water lies to the north (in PEQD) and to the south (in SSTC) of this province, whereas nitrate-depleted surface water occupies the whole of the center of the gyre, having an annual mean value of $<0.5\,\mu M$ from the surface to 150 m. The nitracline follows the pycnocline very closely so that the $0.3\,\mu M$ isopleth lies just deeper than the depth of sharpest gradients in the pycnocline. A recent analysis of the depth of the nitrate field shows that the isopleth for $2.0\,\mu M$ forms a trough below the intertropical convergence zone, lying NE-SW across the ocean and deepening toward the east: maximum depth of this isopleth is reached at about 25°S 120°W (McClain *et al.*, 2004). This feature lies below the ridge in TOPEX-derived dynamic height noted earlier.

## Biological Response and Regional Ecology

The generalized chlorophyll field clearly shows the consequences of the baroclinicity of the nitrate isopleths. McClain *et al.* have identified (for each of the subtropical gyres, let it be said) the location of minimum surface chlorophyll in each of the four seasons. The lowest values consistently occur very close to the location where the mixed layer is deepest, and hence where the nitracline lies furthest removed from wind-induced mixing at the surface. In the South Pacific gyre, this location is remarkably stable at about 25°S 120°W, as noted earlier, despite the significant seasonal change here in mixed-layer depth. Yet we must not suppose that the chlorophyll field of the subtropical gyre is featureless: the eddy field associated with STCC discussed previously is often located in the SeaWiFS and MODIS chlorophyll images, especially to the west of the date line, as a field of arcuate meanders of chlorophyll significantly higher than background. This is observed more commonly in austral winter and recalls the more elusive feature very occasionally seen in a comparable location in the South Atlantic subtropical gyre (see SANT). A more constant anomaly in the regional chlorophyll field is associated with the Marquesas archipelago at 10°N 140°W; why this group of islands should routinely produce a chlorophyll anomaly, while the Tuamotos not so far to the south do not, is something for which I can suggest no explanation.

Both the surface SeaWiFS and MODIS data used in this study, and those for the smaller box lying centrally in the gyre used by McLain *et al.*, yield the same result: a well-defined winter maximum occurs in August–September and the rate of seasonal change in chlorophyll values follows very closely the seasonal change in environmental forcing; here, equilibrium between irradiance, mixing, nutrient flux, and growth appears to be constantly maintained. The chlorophyll values attributed to the entire SPTG province are somewhat higher (0.06–0.13 mg $m^{-3}$) than those for the central box of McLain *et al.* (0.02–0.08 mg $m^{-3}$); this is to be expected because the peripheral regions of the province are everywhere adjacent to areas of higher productivity.

We have some sea-truth data obtained from ships of opportunity on the Panama to Auckland route to support the seasonal cycles inferred from SeaWiFS and similar data; for a box in the central part of this province, Dandonneau *et al.* (2004) confirm the permanent oligotrophy of SPSG (chlorophyll always $< 0.15$ mg $m^{-3}$) and report a weak winter maximum accompanied by unusually high numbers of picoeukaryotes (5000 cells $ml^{-1}$, compared with a few hundreds normally). These data also confirm the general dominance of cyanobacteria and the extensive occurrence of *Trichodesmium* blooms, especially in the southwest of the province and especially in summer.

An interesting meridional phytoplankton section along 115°W has been reported (Hardy *et al.*, 1996), which shows that phytoplankton taxonomic composition was remarkably invariant from about 11–12°S, a line coinciding with the southern edge of the PEQD province, right down to 36°S at the edge of the Subantarctic Convergence. Beyond these two boundaries, species composition differs while carbon biomass, chlorophyll, and primary productivity all take higher values. Principal component analysis of the whole section from 10°N to 60°S delivered five groupings of relative abundance of phytoplankton taxa in such a way that the meridional separation of these groups corresponds very well with the five provinces PNEC, PEQD, SPSG, SSTC, and SANT.

Apart from the EASTROPAC voyages of the 1960s in the northeast corner of the province, there are no comprehensive ecological studies of the open-ocean plankton ecosystem and the best that can be done is a brief extrapolation from other regions. It is probably safe to say that the observations discussed earlier for the CLIMAX and HOT site in the North Pacific subtropical gyre will obtain here—only in a more extreme sense. It is also safe to assume that a subsurface maximum chlorophyll layer lies across the province just shallower than the nitracline and on the upper part of the density gradient and that it acts to trap to utilize any nitrate mixed up across the upper pycnocline. Finally,

I also assume that maximum primary production rate will occur much shallower; that nitrogen-fixing organisms will be prominent; that the microbial loop, or regeneration of ammonium as a substrate for algal growth, will be very active and mediated by bacterioplankton, pico-fraction cyanobacteria, prochlorophytes, and microflagellates; and that most consumption will be by a complex protist community. Such mesozooplankton as occur will be small, except for diel migrant metridiids, which will rise at night, some to the DCM and some to graze on protists in the mixed layer.

For the EASTROPAC area, which extended only to 20°S at 90–125°W, some of these suggestions are generally confirmed. A DCM slopes downward in the upper thermocline, weakening to the south so that by 20°S it lies at about 120 m in February and March when the depth of maximum primary production rate is at about 20 m (Longhurst, 1976). The same pattern is repeated in August data, though the DCM is a little deeper. Whereas mesozooplankton profiles in PNEC and PEQD to the north are strongly structured with near-surface and near-DCM layers of maximum abundance, in SPSG their layering is relatively weak with a very broad zone of relative abundance progressively weakening downward to the midpycnocline.

Although coral reefs are not part of our discussion, a passing reference to their place in this ocean microcosm is in order because the myriad islands inhabiting the province owe their existence to the activities of reef-forming biota. If the growth of many of these reef-forming corals and mollusks is supported by photosynthesis of symbiotic algae within their tissues, a source of nitrogen still has to be found: this comes from three sources, the balance among which is complex to compute. In part, it is provided by the activity of nitrogen-fixing blue-green algae, in part by capture of the sparse oceanic plankton drifting over the reef, and in part—perhaps in some cases most important—by endo-upwelling within the fractured carbonate rock of the reef platform. Very slight geothermal heat flux from the earth's crust is sufficient to maintain a slow upward movement of water within fractured rock, drawing nitrate-replete water in at subpycnocline depths and releasing it at the surface. For this system to function another requisite is perfect clarity of the surface water to obtain maximum solar energy within the symbiotic tissues. These facts answer Darwin's paradox: why do we find the most exuberant growth of corals in the clearest water?

Therefore, where reef-forming organisms occur, we can be sure that seasonal blooms of planktonic algae and upwelling processes are not significant. This suggestion is confirmed by the observations of Rougerie and Rancher (1994), who found that in the immediate vicinity of numerous atolls the vertical structure of the oligotrophic profile is preserved intact. Right up to the coral front the nitracline (150–200 m), the pycnocline (100–175 m), and the DCM (125 m) remain undisturbed by the proximity of the reef platform. Weak, irregular flows past the atolls produce little turbulence or chlorophyll-enhanced wakes downstream of the atolls.

It is interesting to note that once again, McGowan's biogeography and Brinton's euphausiid maps support at least some of our conclusions. The distribution envelopes of 10 central water biota (*Stylocheiron suhmii, Euphausia brevis, Euphausia mutica, Sagitta serratodentata*, etc.) are a mirror image of the Northern Hemisphere pattern (see NPSG) but with the important difference that it is more diffuse in the eastern part of the ocean. This is exactly what we would expect from the hemispheric differences between the circulation patterns. However, the distribution of large pelagic fish, exemplified by tuna, is somewhat contrary, highest abundance being concentrated marginally along the northern and southern reaches of the province, adjacent to more productive regions.

## Synopsis

*Case 3—Winter-spring production with nutrient limitation*—Pycnocline depth undergoes a weak austral winter excursion (30–40 m February–March, 80 m September) while photic

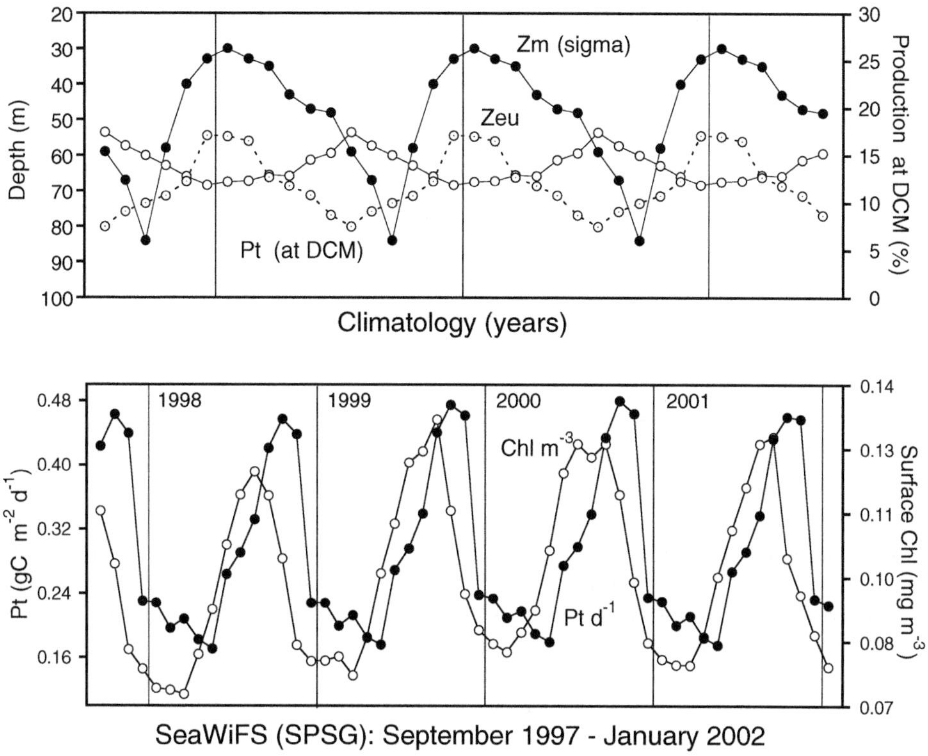


**Fig. 11.16** SPSG: seasonal cycles of monthly surface chlorophyll and depth-integrated autotrophic production for the years 1997–2002 from SeaWiFS data together with characteristic seasonal cycles of mixed-layer depths from Levitus climatological data and photic depths computed from characteristic irradiance and the archive of chlorophyll profiles discussed in Chapter 1.

depth remains at 50–60 m, so the pycnocline is dark only in the period July–September (Fig. 11.16). Productivity rate changes match this pattern: low during austral summer from which the rate increases in June through the austral winter to reach a spring maximum during September–November. Chlorophyll biomass tracks the seasonal cycle of productivity, although peaking a month or two earlier; this is consistent with closely balanced production and consumption processes.

# Pacific Coastal Biome

## Alaska Coastal Downwelling Province (ALSK)

### Extent of the Province

The ALSK province comprises the coastal boundary region from 53°N (Queen Charlotte Sound) to the end of the Aleutian chain of islands at about 180°W, and from the velocity maximum of the Alaska Current to the coastline.

### Continental Shelf Topography and Tidal and Shelf-Edge Fronts

The continental shelf is very narrow, or absent, along the north-south fjord coastline except behind the Queen Charlotte Islands. The northern and western coasts have a wider shelf, and at Kodiak it is about 200 km wide. This being a highly glaciated region, many deep canyons intersect the shelf, especially along the eastern side of this province; glaciation is the cause of the highly inflected nature of the coastline itself. About 20% of the adjacent continental drainage areas remain covered by glaciers that serve to modify the seasonality of the discharge of freshwater to the sea.

The shelf is also remarkable for the steepness of the topography that lies on either side: seaward, a very steep continental slope into abyssal depths and landward the very abrupt topography of high coastal mountain ranges. These mountains are essential elements in the forcing of the characteristic coastal oceanography of this province.

## Defining Characteristics of Regional Oceanography

This coastal boundary province comprises the rim current (the Alaska Current or Alaska Stream as it is termed off the Aleutians) of the subarctic gyre, from its origin at the divergence of the North Pacific Current when this encounters the American continent near Vancouver Island right around to the end of the Aleutian chain of islands; this province is approximately synonymous with the Alaska Current System of Favorite *et al.* (1976) and the Alaska Downwelling Province of Ware and McFarlane (1989).

Circulation is carried around the periphery of the Subarctic gyre in the Alaska Current that lies just beyond the shelf and, after it turns southwest to pass along the Aleutian chain as a western boundary current, is spoken of as the Alaska Stream. Close inshore, the Alaska Coastal Current is especially strongly developed west of 140°W and carries about two-thirds of the total along-shelf flow. Circulation and water column stability in this province are forced by wind stress and by buoyancy. As discussed recently by Royer (1998) the presence of the coastal ranges constrains the further passage to the east of the warm, wet, low-pressure systems that pass along the Aleutian storm track in winter. When these encounter the coastal mountains, they rise and lose a great deal of their moisture as rain or snow. Because the consequent precipitation rates are very high ($<800\,cm\ y^{-1}$), and the drainage basins rather small, there is a very high rate of freshwater discharge to the coastal ocean. This is maximal in late summer because the glaciers retain some winter precipitation until the ensuing summer; the seasonal salinity regime over the shelf (28–31‰ at 25 m) follows the seasonal pattern of discharge. Because the drainage area comprises high mountains, rather than forested flatlands, the river water is nutrient-poor; supply of nutrients to the mixed layer is a complex and unusual process (see later discussion).

The large discharge of freshwater controls shelf salinities and hence density gradients, both horizontal and vertical, create a salinity front between the Alaskan Coastal Current and the more variable midshelf flow; this is situated at only 10 km from the coast along the Cape Fairfield section in the northern part of this province. A second salinity front separates shelf water from the sluggish flow (c. $5\,m\ min^{-1}$) of the warm Alaska Current itself beyond the shelf edge. The velocity of the flow beyond the shelf edge increases to $<50\,m\ min^{-1}$ in the Alaska Stream that returns southwest down along the Aleutian Chain.

The inner shelf is dominated by semidiurnal tidal streams that may be very rapid (up to $6\,m\ sec^{-1}$ in the Straits of Georgia) and are associated with tidal fronts at the mouths of bays and rivers. The heavy clouds and rainfall of this region means that irradiance has a very strong seasonal cycle from $70\,W\ m^{-2}$ on a rainy winter day to $1400\,W\ m^{-2}$ under a sunny sky in summer—moreover, water clarity is reduced in winter by silt effluent from flooding rivers; Secchi depths in river plumes of 0.1 m in winter and 3.0 m in summer are not unusual.

The persistent cyclonic winds, especially in winter, impose onshore Ekman transport at the surface and downwelling at the coast itself. The Alaska Coastal Current is therefore coastally trapped so that, from Vancouver Island to the Unimak Pass in the Aleutians, it flows preferentially through the coastal passages before delivering low-salinity water to the shelf of the Bering Sea. In so doing, it becomes a major component in the freshwater budget of the Bering shelf and the entire Arctic Ocean. During the period of anticyclonic winds of summer, flow in the Coastal Current may be reversed for several months along the eastern coast.

Forced by the seasonal wind regime, the velocity of the Alaska Stream above the steep-to continental slope is greater in winter than in summer as a result of intensification of the Aleutian atmospheric low-pressure cell (Schumacher and Reed, 1983; Thompson, 1981). The downwelling tendency at the coast is maintained by the density distribution within the coastal flow, resulting from dilution of the surface water mass, from wind-curl stress, and from a longshore wind-induced slope in sea level. Flow is therefore strongest during early winter when river effluent is greatest; where flow becomes approximately zonal, along the Alaska peninsula, instability develops and the stream may separate from the continental edge or may regain stability by anticyclonic meandering and eddy shedding (Okkonen, 1992).

This situation obtains along most of this coast: only off the Alaska peninsula in summer (near Kodiak Island) do local winds force any significant upwelling of isopleths and weak upwelling lasts only from June to August. Along the Alaska peninsula and the Aleutian chain to the southwest, the Aleutian Stream normally flows close along the shelf-edge topography to the end of the Aleutian Islands and is the dominant source water for the Bering Sea gyre; anomalous separation of the current at about 170°W can occur, and this condition may persist for some months (Stabeno and Reed, 1992).

The fractal nature of the fjord coastline results in instabilities and in the recurrent formation of mesoscale eddies at certain localities (Ladd *et al.*, 2005). The best-studied are the anticyclonic Haida eddies that form off the Queen Charlotte Islands, while similar features form off Yakutat in the northern Gulf and off Sitka on the Alaska panhandle; these were compared by Royer to the warm-core eddies of lower latitudes, although they have very moderate thermal signatures. Haida eddies are initiated in buoyant plumes from Hecate Strait, usually in late winter, so that coastal water is retained at their cores as they propagate offshore across the NE Pacific; they are of order 150–300 km diameter, form at a rate of about one per year, and persist for about 2 years.

The permanent halocline of the Gulf of Alaska extends through this coastal province, though it is shoaler than out in the open gulf; it restrains winter mixing to about 50–60 m, though convective cooling extends much deeper. The stability thus induced constrains the vertical entrainment of nitrate during winter, and it is suggested that the spring bloom would be as brief as it is offshore (see Pacific Subarctic Gyres Province) if interaction between the Alaska Current and coastal topography, where the shelf is narrow and steep-to, did not entrain some nitrate throughout the summer.

This is, as we shall see, a highly productive coastal region for which, until very recently, we had no information at all concerning the nutrient regime: how, in the face of the permanent density stratification over the shelf, was productivity maintained? What was the basis for the rich fish ecosystem of the shelf regions of the Gulf of Alaska? These problems have now been resolved by Childers *et al.* (2005), who describe a very unusual mechanism. Repetitive cross-shelf sections of the northern shelf during summer reveal that the upper 20 m becomes totally depleted of macronutrients, but that below this the water column is nutrient-replete. This situation is maintained by an onshore flux of oceanic water that is induced by the cessation of winter downwelling at the coast and the slight upwelling tendency induced by summer winds. This subsurface reservoir of nutrients is thoroughly mixed to the surface in winter and thus available in spring. I know of no other region where this process is so clear although it has some of the characteristics of an estuarine circulation—offshore at the surface, inshore along the bottom: it appears, however, not to be forced in the same manner as an estuary.

There is, of course, an opposite flux of nutrients in the large anticyclonic eddies discussed earlier because, as Whitney *et al.* (2005) point out, much of the transported shelf water mass both has a near-coastal origin and lies deeper than the euphotic layer. The consequent transport of macronutrients to the subthermocline layer of the open subarctic gyre may not be significant, but the transport of micronutrients may well be. Whitney *et al.* suggest that iron transported in this manner, originating in the coastal discharges, may be a significant source of this element for the gyre as a whole. Anticyclonic eddies that propagate across the Gulf of Alaska form a major transport system from coastal waters to the open ocean, providing a steady flux from their iron-rich core water (Johnson *et al.*, 2005).

Circulation, buoyancy, and relative temperatures are clearly highly sensitive to short and long-term periodic changes in atmospheric forcing, as well as to the aperiodic changes that occur each year in the effects attributable to individual storm systems. As I shall discuss later, it is also a place of great fisheries, and concern for the rational management of these has created an unusual accumulation of observations on the interdecadal changes in the marine environment. The dominant determinant of long-term changes in conditions is the Pacific Decadal Oscillation (see Chapter 8) associated with the alternation of ENSO conditions. In recent decades, the warm conditions of the very strong 1997–98 Niño were followed by cold conditions in 1999, and the subsequent return to warmer inshore temperatures continued at least until 2003, leading to comments about "regime changes." Again, freshwater discharge has been rather higher than the 50-year average since the mid-1980s and a progressive shoaling of MLD in the open Gulf of Alaska in the recent decades must be reflected appropriately in these coastal regions.

## Regional Response of the Pelagic Ecosystem

Until recently, there has been very little information on the biological consequences of circulation and buoyancy within this province (Anderson *et al.*, 1977), save that production in shelf waters is thought to occur at about twice the open-ocean rate (200–300 compared to 70–100 g C $m^{-2}$ $y^{-1}$) and that this is concentrated in a spring bloom that occurs progressively northward. Here, of course, the required stability for a bloom to occur is given by the salinity profile, rather than by progressive warming of a surface layer. We now have more comprehensive information, based mostly on analysis of SeaWiFS data by Brickley and Thomas (2004), who present seasonal data for several compartments of this province showing a spring bloom that is sustained well into summer with higher chlorophyll concentrations on the shelf than in a 200-km strip beyond the shelf. Their global estimates of seasonal changes in surface chlorophyll for the shelf region correspond quite closely to the analysis presented later as a climatology. Relatively weak phytoplankton growth in spring 1998 is coincident with positive SST anomalies associated with the 1997–98 Niño. Brickley and Thomas suggest a general relationship between strong winter downwelling conditions and a subsequent weak spring bloom.

Early ecological studies in the Strait of Georgia, at 48°N between Vancouver Island and the Canadian mainland, have long served as a model for the other major sounds (which were given regal names by their navigating discoverers: Queen Charlotte, Prince of Wales, Alexander, and Prince William) up to 60°N in Alaska (Harrison *et al.*, 1983). A strong spring bloom is initiated in March (about 15 mg chl $m^{-3}$) and phytoplankton biomass then progressively declines throughout the whole summer to reach very low values (< 1 mg chl $m^{-3}$) in November. A succession of diatom species occupies the seasons: *Thalassiosira* and *Skeletonema* dominate the spring bloom, followed by *Chaetoceros, Ditylum, Nitzschia*, and *Leptocylindricus* in summer and *Coscinodiscus* in autumn. During the spring bloom, these large cells exhibit growth rates from 0.5 to >1.5 doublings per day. Nanoflagellates dominate during winter but form only 10% of the 2–4 μm phytoplankton in summer. During the summer, a very shallow DCM develops at 10–20 m, below a primary production peak at about 5 m. At other times primary production takes highest values just under the surface.

Observations of nutrient dynamics, especially rates of nitrate depletion, made along the Seward line across the wide shelf in the northern part of the province (Childers *et al.*, 2005) suggest that new production rates here are equivalent to 2.5–7.0 μM $NO_3$ $m^{-1}$ $d^{-1}$ with highest values occurring either in midshelf or at the edge, depending on conditions obtaining in individual years. Seasonally, productivity on the inner shelf leads productivity on the outer shelf by about 30 days: March–April compared with April–May, respectively, although satellite images suggest greater accumulation of surface chlorophyll within the Alaska Current at the shelf edge, especially in the northern and Aleutian sections of the shelf.

Recent studies of the comparative distribution of mesozooplankton species finally unravel some of the uncertainties about their maintenance in appropriate locations in such a dynamic environment: we are fortunate in having a recent study of the Seward line in the northern part of the province (Coyle and Pinchuk, 2005) and a more general regional analysis by Mackas and Coyle (2005). On the anomalously wide northern shelf, the preferential distribution of neritic and oceanic species of subarctic zooplankton can be determined perhaps more easily than elsewhere. A neritic community, dominated by *Pseudocalanus* spp., *Metridia pacifica*, *Calanus marshallae*, and *Limacina helicina* is most abundant in the Coastal Current and in the fjords in spring and summer at the period of maximum freshwater discharge. The oceanic community, beyond the shelf break, is dominated by *Eucalanus bungii*, *Neocalanus cristatus*, and *Oithona similis*. Midshelf, between the two salinity fronts, a mixed assemblage occurs in spring, although with lower biomass than within the Coastal Current assemblage. As the summer advances, and the low-salinity conditions of the inner shelf expand outward, the distinction between neritic and midshelf zooplankton assemblages is reduced. More generally, Mackas and Coyle point out, the major ecotone in this series occurs between oceanic and shelf communities at the salinity front near the shelf break. The distinctions between oceanic and neritic assemblages are profound: shelf zooplankton is much smaller, has more generations per year so that multiple cohorts are often present, may lay dormant eggs, obtains less of its food from protistan plankton, rarely exhibits ontogenetic vertical migration, and less consistently exhibits diel vertical migration.

The winter downwelling regime discussed earlier may induce shoreward advection of oceanic species after these have risen to the surface in early spring, prior to the reduction of wind stress associated with the breakdown of the Aleutian Low. On the shelf, beyond the salinity front of the Coastal Current, zooplankton biomass in the mixed assemblage is therefore dominated by large oceanic species, although numerically neritic species dominate. Nevertheless, mesozooplankton biomass is highest overall within the Coastal Current.

The basins and fjords of the continental shelf, like the Straits of Georgia, are able to support permanent populations of oceanic species provided sufficient depth is available for their overwintering stages. The winter mesozooplankton here is dominated by small copepods (*Pseudocalanus*, *Oithona*, and the rest), but their development in spring is swamped by a massive influx of larvae and early copepodite stages of the offshore *Neocalanus plumchrus* together with immigration, also from the ocean, of *Calanus pacificus*, *C. marshallae*, and *M. pacifica*. Predators also join the summer community: *Sagitta* spp. and amphipods. *Neocalanus plumchrus* dominates biomass in spring and *C. marshallae* dominates somewhat later in summer. Most copepods remain in the upper tens of meters of the water column, with smaller numbers at about 50 m, whereas *Sagitta*, amphipods, coelenterates, and euphausiids form deeper layers fueled largely by sinking organic material and diel migrants. In Prince William Sound at 60°N, *N. plumchrus* together with *N. cristatus* and *E. bungii* are the dominant copepods during those months when they are not overwintering in deep water beyond the shelf (Cooney, 1986). Appearing in spring as nauplii and copepodites over the shelf, many *N. cristatus* and *E. bungi* descend to overwintering depths beyond the shelf edge in October where maturation and reproduction take place to produce the early larvae of the new generation. However, many individuals of these oceanic species complete their life cycle within the Sound, which is sufficiently deep for this to occur, and in the case of *Neocalanus* at least, may be distinguished by the isotopically heavy carbon content; their proportion of the species population in the Sound varies between seasons and years.

Mackas and Coyle (2005) present an important discussion on the mechanisms by which behavior patterns of individual species are adapted to the rigors of remaining within a suitable habitat, in three-dimensional space; this, they point out, is no simple matter for an organism that has (i) very limited motility and (ii) no mechanism for locating its present position in relation to regions suitable for its continued existence, although (iii) it does know which way is "up," as they put it, and (iv) can detect microscale shear. For these reasons, Mackas and Coyle insist on the importance of vertical motion—a few tens of meters of relative motion being well within the capabilities of even the smaller copepod species—because this affords the individual a mechanism for significant horizontal motion in a layered physical system with differential motion between layers. They are quick to point out that this is no novel concept: it is usually attributed to Alister Hardy, who described such organisms as having "seven-league boots, to set them striding through the sea." They report differential behavior between *Pseudocalanus mimus* and *Calanus marshallae*, so that the former is unable to avoid offshore transport in anticyclonic eddies, whereas the latter occurs very rarely in such systems.

Despite these advances, Liu *et al.* (2005) suggest that we still have inadequate understanding of the nutritional dynamics of even the dominant species, such as *Neocalanus cristatus*, adequately to characterize the interaction between copepods and autotrophs. *N. cristatus* appears to reject $<5$-μm cells in favor of larger organisms, probably largely protests in sufficient quantities as to induce a cascade effect *in situ*: their intake of autotrophic carbon little more than doubles from ~30% under nonbloom conditions to ~70% during blooms. But the measured rates are too small to support observed growth, and observations of intake of even larger particles are to be anticipated.

The response of biota to the anticyclonic eddies discussed earlier requires separate consideration and now, fortunately, we have a series of sustained studies of the Heida eddies to inform us. Matching SeaWiFS-derived chlorophyll and TOPEX/ERS-2 SLA images over a 5-year period, Crawford *et al.* (2005) demonstrate that these eddies support central spring blooms in their "natal year" (as these authors put it) that are stronger than in surrounding offshore water. This is the anticipated consequence of the difference in nutrient regime between coastal water contained in the eddy core, and the offshore water surrounding it; the arcuate streamers of high chlorophyll often observed around eddies arise by offshore entrainment of eutrophic, chlorophyll-rich coastal water around the (slowly) whirling eddy. Deployment of expatriated CPRs on lines parallel to the coast obtained direct evidence of the presence in offshore eddies and streamers of species of diatoms and copepods characteristic of shelf regions. The zooplankton retained within a Haida eddy progressively changes its composition during the life of the eddy, so that abundance within the eddy is higher than outside or in source regions. This appears to result from aggregation and retention mechanisms favoring those species that preferentially occur relatively deep in the eddy, these being selected by slow divergence and upwelling within the eddy, by rapid but intermittent flushing of the upper layers by storm events, and by exchange across geostrophic streamlines at eddy margins by wind-driven inertial currents (Mackas *et al.*, 2005).

## Regional Benthic and Demersal Ecology

Systematic exploration of benthic species assemblages being no longer fashionable, I have been able to locate no modern studies for this province, except some concerned with environmental impact statements around oil platforms that will be of no use to us. We are therefore still dependent on Thorson, who quotes the findings of Shelford (1935) in his account of benthic communities here: these follow the familiar pattern of high-latitude shelves. The *Macoma baltica* community on shallow mixed bottoms with low salinity, an *Amphioplus* community on softer muddy deposits, the *Venus fluctuosa* community on midshelf sandy deposits, the *Maldane sarsi* community on soft inshore muds—and so on. I know of no modern ecological studies of benthic/pelagic interactions in this province.

Of course, we have much more precise knowledge of the fish faunas that depend to a large extent on these invertebrate communities: the importance of the regional fisheries ($\sim 6.0 \times 10^{-5}$ t $y^{-1}$) may be thanked for that. This is, zoogeographically, a marginal region so that there is a significant gradient in species composition from east to west and, of course, the largest concentrations of fish occur on the wider shelves in the northwestern part of the province (PICES, undated). The diversity of the fish fauna is not great for, even allowing for directed fishing, 10 species form >75% of the captured biomass: arrowtooth flounder (*Atheresthes stomias*), Pacific ocean perch (*Sebastes alutus*), and Pacific halibut (*Hippoglossus stenolopus*) dominate the catches.

The five species of migratory Pacific salmon (*Onchorhynchus* spp.) are clearly a special case. Although it has apparently not been quantified, their life history pattern suggests that they may represent a significant transfer of biomass from the open ocean to coastal regions. As is well known, sockeye, chum, and pink salmon pass almost their entire period of growth at sea in the subarctic gyre and die on returning to spawn in coastal rivers and streams. Only chinook and coho growth is dependent on feeding in coastal regions.

There are strong cross-shelf gradients in biomass and diversity of fish, with the highest concentrations occurring just below the break of shelf, while the neritic regions are the principal spawning areas and support large populations of juveniles of offshore species, both demersal and pelagic. What the fishery science community calls "forage fish," such as *Osmerus*, are very abundant, along with >20 species of small flatfish. Pacific herring are widespread over the shelf, with spawning areas in Prince William Sound and around Kodiak; their abundance is highly variable, the fishery having been closed in the former area since 1992. It is thought that this variability (at least in our modern seas) may be due to viral and fungal infection and consequent disease. Similar variability in demersal fish, particularly in such vulnerable species as *Sebastes alutus*, is largely fishery-driven; in these latitudes, recruitment is a rather variable process. Analysis of fisheries data suggests that a major regime shift occurred in the species composition after 1976–77, involving a general increase in biomass that persisted until 2002, since when a decrease has set in that is largely due to reduced abundance of walleye pollack (*Pollachius*). I believe that it is very difficult to attribute causation to such changes as these, which occur in a fish ecosystem already significantly modified from its pristine state by industrial fishing. For one thing, population structure and growth rates are rapidly modified, so that the naturally evolved characteristics of the life-history parameters of each species are modified so that they are no longer the best possible fit to the natural environment. But that's an old story, which I discussed elsewhere in 2002.

## Synopsis

*Case 1—Polar irradiance-mediated production peak*—The pycnocline undergoes a winter excursion to 50 m from late summer depths of around 10 m; photic depth similarly

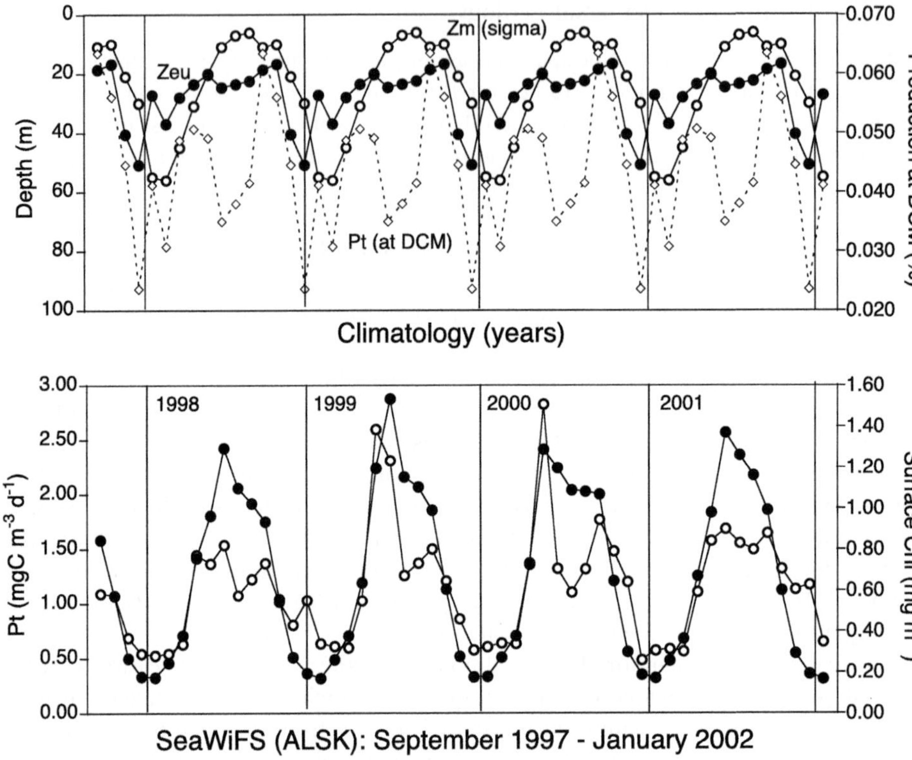


**Fig. 11.17** ALSK: seasonal cycles of monthly surface chlorophyll and depth-integrated autotrophic production for the years 2002–2005 from MODIS-Aqua data together with characteristic seasonal cycles of mixed-layer depths from Levitus climatological data and photic depths computed from characteristic irradiance and the archive of chlorophyll profiles discussed in Chapter 1.

undergoes excursion to >40 m in boreal winter, so that the thermocline is illuminated for 7 months (May–December). A consistent and linear productivity rate increase occurs from January to May, coincident with shoaling of the pycnocline (Fig. 11.17). Peak production rates are thereafter sustained through September, when a rapid decrease leads to a winter minimum in December. The seasonal cycle of chlorophyll biomass has a large dynamic range and rates of accumulation and consumption closely match rate changes in productivity; both have the shoulder in autumn that is observed in many high-latitude provinces.

## California Current Province (CALC)

### Extent of the Province

The CALC province comprises the California Current from the bifurcation of the eastward flow of the North Pacific Current near Vancouver Island at about 47–48°N, down to the convergent front that lies southwest of the tip of Baja California (22–23°N) at the root of the North Equatorial Current. The seaward limit is taken to be the thermal front between water characterized by the effect of coastal processes and the oceanic environment beyond, although this is progressively less well defined toward the south. In some previous iterations of this partition, two provinces were suggested, coastal and offshore: CCAL and OCAL. I now think that this was counterproductive.

## Continental Shelf Topography and Tidal and Shelf-Edge Fronts

The continental shelf is generally narrow, with the 200-m isobath occurring as little as 10 km from the shore in southern California (or even closer, where Scripps Canyon runs into the beach) to a distance of about 75 km off Oregon and Washington. A continental borderland with deep (approx. 200 m) basins, shallow banks, and islands occupies the bight south of Point Conception (34.5°N) that includes the Santa Barbara Basin. Tidal studies are reviewed by Hickey (1998), and these suggest that here the stratified water column over the shelf is rarely interrupted by tidally mixed regions; tidal currents do not appear to dominate over current variance associated with alongshore flow of the coastal currents.

## Defining Characteristics of Regional Oceanography

The California Current is the most intensively investigated of the eastern boundary currents, because of the marathon studies of the California Cooperative Fisheries Investigations (CalCOFI) jointly by federal, state, and university research groups. Much of what follows is drawn from reports, atlases, and research papers arising from CalCOFI; we are especially fortunate that the CalCOFI investigations recently celebrated their 50th anniversary of continuous study: a rich harvest of retrospective studies is therefore at our disposal. Many comments were made at this time concerning the importance of the CalCOFI series, and the information it presents about the manner in which long-term environmental changes are translated into ecological change in this region. Really, it does more than that: it should convince us that similar relationships probably occur in all seas, at all latitudes, but are presently concealed from us for lack of serial observations. It would be foolish to hope that similar studies can be undertaken comprehensively elsewhere, but CalCOFI enables us to lean more confidently on causal relationships that we infer from serial satellite observations of sea change and chlorophyll.

The California Current, which has been recently reviewed by Hickey (1998), takes its source in the divergence of the North Pacific Current where this approaches the American continent, within what Favorite *et al.* (1976) term the "dilution zone," where precipitation greatly exceeds evaporation. This eastward flow across the ocean diverges northward into the Alaska Current, and southward into the California Current where it forms an equatorward flow of cool, subarctic water. The characteristics of the present province derive from the conjunction between the southward geostrophic current, the coastal boundary, and an alternating wind regime—equatorward upwelling winds in summer and poleward downwelling winds in winter. Conditions here are much modified during El Niño events.

Equatorward, upwelling winds are maximal above the main jet of the California Current that lies 200–400 km offshore, where consequent wind stress is greatest. This zone coincides with the line of zero wind-stress curl, which is anticyclonic (convergent) to the west of this line and cyclonic (divergent) to the east (Bakun and Nelson, 1991). A narrow (10–40 km), jetlike poleward undercurrent underlies the California Current at all seasons over the continental slope and is continuous from Baja California to Vancouver Island.

Southward flow of the California Current past the prominence of Point Conception (35°N) creates a cyclonic eddy to the south and thus northward flow at the coast in the bight south of the cape. Upwelling wind stress creates an upward tilt of isopycnals toward the coast and an anomaly of the density surfaces that progressively moves offshore as summer advances. When the eddy is most strongly developed, coastal flow may escape past the cape and continue to the north as a short-lived California Coastal Countercurrent. In winter, however, from Point Conception northward to Vancouver Island, some poleward flow regularly occurs at the coast that may represent the surfacing of the undercurrent and generally goes by the name of the Davidson Current. In exceptional winters, this may originate as far south as northern Baja California (Wyllie, 1966).

The southward flow of the coastal current in summer is a meandering jet with persistent eddies and cool filaments that originate at prominent capes and may extend 200 km offshore. Some of these eddies, especially those at the Straits of Juan de Fuca, at Hecate Bank, and at Cape Blanco and Cape Mendocino, often form cold filaments that may extend several hundred kilometers offshore. These have a dipole termination, representing the development of a pair of counterrotating eddies that entrain upwelled water at the coast and advect it far offshore (Mooers and Robinson, 1984); they are now very well observed in AVHRR images. The actual development of these features is associated with variability of coastal winds.

Coastal winds are modified by the blocking of the zonal westerlies by the coastal mountain chain, creating a wind regime that is far from uniform: a local maximum of cyclonic curl is associated with the Southern California Bight, and a lobe of anticyclonic curl frequently reaches the coast at Punta Baja, where longshore equatorward wind stress is maximal. Off Oregon and Washington, wind-stress curl is variable, with frequent brief episodes of anticyclonic curl related to storm tracks (Bakun and Nelson, 1991) that are reflected in event-scale circulation changes. Here, during the summer coastal upwelling season, a front may occur 5–15 km offshore that is forced by cross-shelf circulation over the very narrow continental shelf (Peterson *et al.*, 1979).

Upwelling episodes generally last for periods of 1–3 weeks, including relaxation periods of 2 or 3 days, so that individual cold filaments may transport water from more than a single upwelling event (Lagerloef, 1992; Traganza *et al.*, 1981). These features are forced by along-shelf wind stress and curl, barotropic and baroclinic instabilities, and coastline irregularities (Hickey, 1998). At about 1000 km from the coast, mesoscale eddying gives way to less complex and slower flow to the south. This transition is often marked by a salinity front that is continuous with the oceanic Subarctic Front and is often termed the California Front: it forms a useful marker for the seaward boundary of this province. The regional characteristics of upwelling processes will be dealt described later, in the section dealing with the ecological response. One of the strengths of the 50-year CalCOFI series of observations is that it affords a unique opportunity of analyzing the relationship between processes over the shelf and in the Pacific Ocean as a whole. The most basic observation is perhaps that the overall strength of the California Current responds to the ENSO cycle. As Hickey (1998) comments, during El Niño events, flow in the California Current is anomalously weak while the Undercurrent is unusually strong; the upper 500 m of the water column is warmer than mean values. These effects extend to at least 200–300 km offshore. A set of standard observations from 1950 to 1980 show how changes in southward flow and temperature and salinity at 10 m are precisely coordinated until 1977, when the relationship changes; reduced southward flow is thereafter associated with increased, not reduced, temperature, although the relationship with salinity and zooplankton biomass is unchanged (Chelton *et al.*, 1982).

This simple observation serves to introduce the problem of the large-scale climate regime shift recently discussed by Bograd and Lynn (2003), who show that the period 1944–1976 was characterized by a weak atmospheric Aleutian Low so that surface Ekman processes were associated with relatively cool SSTs around the entire rim of the NE Pacific during this period. In 1976, the Aleutian Low strengthened very significantly, ushering in a period of warmer SSTs and a relatively greater frequency of El Niño events. Off California, after 1976, the upper 200–400 m of the water column warmed by as much as 1°C, especially nearshore, perhaps related to increased inshore penetration of subtropical gyre water. A decrease in surface salinity was associated with increased stratification, deeper density surfaces, and greater stability, so rendering upwelling less effective in the induction of phytoplankton growth. These effects were associated with shifts in position and strength of the flow structure; the California Current shifted offshore, and the eddy behind Point Conception was strongly modified. Such conditions persisted until at least

1998–1999, when a major El Niño event occurred, associated with another possible decadal climate shift, after which the character of the California returned to the pre-1976 condition. In 2002, another El Niño event was apparently associated with an anomalous intrusion of subarctic water.

## Regional Response of the Pelagic Ecosystem

The effects of wind-stress curl on nutrient dynamics are similar in each of the four principal eastern boundary currents (Bakun and Nelson, 1991). Though the paradigm of eastern boundary current nutrient dynamics is usually described simply as "coastal upwelling," the reality is far more complex and a variety of enrichment processes occur in (i) coastal upwelling cells, caused by Ekman divergence at the coast, usually of small dimension and locked to topography; (ii) continued upwelling in offshore-trending filaments of cool upwelled water; (iii) cyclonic eddies shed from meanders of the coastal current; (iv) and the offshore divergent front at the shear zone between inshore poleward and offshore equatorward flows.

Here, it is clear that the offshore jet of the California Current separates inshore, eutrophic water having a high concentration of chlorophyll from oligotrophic offshore water (Collins *et al.*, 2003). There is a striking discontinuity in properties at the sea surface so that, where it is parallel to the coast, the offshore jet runs along the junction between low oceanic chlorophyll values ($<0.5\,\text{mg m}^{-3}$) and higher values inshore. Along the same line, nitrate values are of order $<1\,\mu\text{M liter}^{-1}$ rising to around $20\,\mu\text{M liter}^{-1}$ near the coast. Offshore squirts and dipoles do not disrupt this pattern, for the jet current continues around their margins, enclosing high-productivity water. Such a pattern suggests that an amendment is required to the earlier model of Chelton *et al.* (1982), who found that nutrient input in the offshore California Current is principally advective in the cool, low-salinity, high-nutrient core of the offshore jet.

However this may be, off both northern California and Oregon–British Columbia, a winter-spring bloom occurs that appears to be unrelated to the effects of coastal upwelling plumes and filaments. Here, in March–April, chlorophyll values $>1.0\,\text{mg chl m}^{-3}$ occur over wide areas as far as 300 km from the coast. Between these two regions, chlorophyll values offshore are generally lower even at this season, and a "blue hole" region (covering several degrees of latitude) is persistent offshore, centered off the Columbia River. Satellite images clearly show that this is continuous with high surface chlorophyll at this season right across the North Pacific Ocean.

Because the processes leading to upwelling are not the same throughout the coastal area, it may be useful to review the processes that are characteristic of four compartments: Oregon–British Columbia, Point Conception to Cape Mendocino, the Southern California Bight, and Baja California.

*Oregon–British Columbia (42–48° N)*: Winter storms are strong and frequent and seasonal current reversal occurs regularly; primary production is strongly seasonal. Upwelling occurs in summer at a coastwise front about 10 km offshore as well as at the coast itself, and response to wind events typically results, after 4 or 5 days, in the development of an upwelling cell, whose seaward front moves progressively offshore during development and returns shoreward during the subsequent relaxation of upwelling. During this process, mean offshore flow is restricted to the surface 20–30 m (Brink, 1983; Smith, 1981a,b). Upwelling occurs in persistent, topographically locked gyres during summer, as at Juan de Fuca. During both summer and winter, frequent cold tongues of upwelled water on the scale of hundreds of kilometers extend westward from the continent across and beyond the relatively broad continental shelf. Relatively high levels of nutrients occur throughout this region offshore, entrained toward the south from the subarctic zone.

*Point Conception to Cape Mendocino (33–41°N)*: Upwelling is strongest here but primary production is markedly seasonal. Satellite images show persistent offshore meanders, shed eddies (anticyclonic to the north and cyclonic to the south), and offshore cool filaments entraining coastally upwelled water. These most frequently occur in summer and south of Cape Mendocino, Point Arena, and Point Sur. Upwelling also occurs at the shallow banks off Point Reyes. Off Monterey Bay (36°N), Wilkerson *et al.* (2000) showed clearly the formation of a plume of upwelled water and the active utilization of $NO_3$ within 20 km of the shore, which was transformed into a plume of $NO_3$-deplete water 90 km seaward. Above a threshold of 10–12 $\mu$M $NO_3$, the larger ($>5-\mu$m) fraction of the autotrophs dominated as they did at higher biomass concentrations, supporting the idea that to reach maximal biomass, the contribution of larger cells is essential.

*Southern California Bight (32–33°N)*: The upwelling plume from Point Conception is frequently observed in summer to pass to the south around the outer limb of the eddy that occupies this bight. This plume has consistently lower surface chlorophyll values than those north of Point Conception or off Baja California. Off Southern California, the winter wind regime is established only relatively briefly although, because the bight is wholly occupied by a quasipermanent cyclonic eddy, poleward flow along the coast is more continuous here than elsewhere. Upwelling occurs as small coastal cells in summer (Point Dume and Del Mar) and at offshore islands and banks during all seasons. The inshore flow around the southern limb of the eddy may be a prominent chlorophyll feature.

*Baja California (22–31°N)*: The coastal wind regime is weaker but apparently favorable for upwelling almost year-round, though some seasonality in upwelling is evident in the chlorophyll field. Upwelling cells south of prominent capes are persistent: Cape Colonet, Punta Baja, Cape San Quintin, Punta Eugenia, Punta Abreojos, and Cape Falso all may generate such cells. The seasonal surface pigment fields suggest that jets and filaments of upwelled water pass farther offshore from late summer (August) through early winter (November) and that south of Cape San Lazaro coastal upwelling ceases in autumn and winter (September–January).

Because of the large number of published studies of the California Current, in reviewing this province it is easier to concentrate on the individual processes rather than to see the whole. For instance, it is easy to lose sight of the fact that there is a seasonal cycle in the depth of the pycnocline that obtains over the whole area of the province (summer, 20–25 m; winter, ~75 m). At all seasons, if one ignores the effects of mesoscale features, the thermocline slopes downward to the west, offshore, as it must. In upwelling cells, the density profile may be relatively featureless, but wherever a significant mixed layer exists, a DCM occurs on the density gradient, usually with the depth of primary production a few meters shoaler. The offshore region has a seasonal cycle typical of subtropical oceans: chlorophyll accumulation begins as soon as the mixed layer begins to shoal in spring to peak in May–June, closely matching primary production rate. Analysis of chlorophyll fields, integrated for the entire province (see later discussion), show that this process—not summer upwelling at the coast of California—dominates the seasonal cycle of this province. This observation recalls earlier suggestions that between-year variability in biological properties was forced primarily by changes in the advection of nutrients from the source of the California Current rather than by variation in the nutrients brought to the surface by coastal upwelling (Chelton, 1982).

Although coastal upwelling cells—and other places where nutrients are brought into the euphotic zone intermittently—are generally thought of as diatom-dominated, recent studies have shown that growth of phycoerythrin-rich cyanobacteria may also be enhanced in such places. Collier and Palenik (2003) have shown that in the California Current, *Synechococcus* takes higher abundance on the coastal side of the offshore jet than beyond it, and also near the surface than deeper. As has been found elsewhere, cell types that may be distinguished by flow cytometry appear to be differentially distributed in relation to irradiance and nutrients. Here, the principal change in cell type occurs at the thermocline. A recent study (Bruland *et al.*, 2001) suggests why blooms in some upwelling centers on the Californian coast (and, by extension, in other similar systems) should be dominated by small cells, and others by larger diatoms: it is a question of the relative supply of Fe in the upwelled water. Where the shelf is relatively wide and Fe is regenerated in the benthic ecosystem, as from Monterey Bay to Point Reyes, upwelled water is Fe-replete (>10 nM), whereas in an upwelling center, such as the Big Sur coast, where the shelf is very narrow, upwelled water is generally Fe-deplete (<1 nM). Consequently, the response of the autotrophic cells is very different in the two regions: off Point Reyes, extensive blooms of large diatoms deplete the upwelled water of macronutrients ($NO_3$ and $SiO_3$), whereas off Big Sur the bloom that occurs after an upwelling event has a low abundance of diatoms, so that the macronutrients remain largely unused. This accounts for some of the differences in observed concentrations of macronutrients in upwelling cells.

The decadal-scale changes in the ocean conditions discussed earlier have been associated with appropriate changes in the response of the biota in this province; Hernández de la Torre *et al.* (2003) show that integrated nitrate-based production in the California Current 1970–2002 followed the environmental forcing pattern very closely. From 1970 to 1976, mean new production, integrated along CalCOFI lines 90, 107, and 120 (26–34°N), was 0.186 g C $m^{-2}$, but for 1977–1998 it was only 0.085 g C $m^{-2}$. Between 1999 and 2002, high values (0.148 g C $m^{-2}$) typical of the earlier period returned. Anomalies from the long-term mean values responded appropriately not only to these decadal changes of state but also to individual ENSO-scale events.

This result recalls the relationship between zooplankton biomass and 10-m temperature that was identified as a proxy for relative upwelling strength from 1950 to 1980; zooplankton biomass responds rapidly and positively to periods, and even individual years, of temperature anomalies of both signs. Thus, from 1958 to 1961 there was a positive temperature anomaly of >1°C, and a simultaneous negative anomaly for zooplankton biomass; the opposite condition occurred in 1955–1956. Seasonal patterns of change in zooplankton biomass also respond to temperature anomalies; in the post-1977 warm period, the spring biomass increase was earlier and sharper than was normal during the pre-1977 cool period (McGowan *et al.*, 2003).

Studies off Oregon, in the northern California Current, show that such responses in overall zooplankton biomass reflect changes in community composition (Petersen and Keister, 2003); analysis of 206 serial samples from a midshelf station here showed that strong upwelling was associated with the presence of *Centropages abdominalis*, *Acartia longiremus*, and *Microcalanus pusillus*, and El Niño conditions with *Calanus pacificus*, *Corycaeus anglicus*, and *Ctenocalanus vanus*. Comparable changes in distribution of euphausiids, tunicates, and fish larvae have also been recorded.

The upwelling cells themselves are inhabited by diatom-copepod assemblages of remarkably low diversity: in repeated net tows in one such cell a few kilometers off Baja California (25°N), I could find no more than 29 species of mesozooplankton of all groups (and no more than 20 in any one sample), or about one-quarter the number taken 25 km farther offshore and examined with equivalent attention. Large, filter-feeding copepods, especially *C. pacificus* (which you will find described as *C. helgolandicus* in much of the Californian literature) at 115 ind $m^{-3}$ comprised 77% of zooplankton dry weight at the coast. The seasonal ontogenetic migrations of this species cause the deep basins on the continental shelf to trap large concentrations of overwintering stage 5 copepodites; early in the winter these aggregate near the bottom, but the layers progressively shoal as oxygen concentration in the bottom water declines, eventually forcing them over the sill depth of the basin (Osgood and Checkley, 1997). This situation resembles that of *C. finmarchicus* in the deep basins of the Scotia Shelf in the Northwest Atlantic Shelves Province.

A few species of diatoms (*Coscinodiscus, Nitzschia*, and *Tripodonesis*) may form >80% of large algal cell volume in these upwelling cells; as well as by large copepods, these diatoms are utilized by a very unusual organism: the bright red, swimming, galatheid crab *Pleuroncodes planipes*. In their pelagic phase, these crowd into the surface layer off Baja California where they tail-flip up to the surface and then parachute down again with outstretched legs, filtering actively with their maxillipeds; this is a remarkable sight against the rich olive-green upwelled water. These crabs (at one per 3 $m^{-3}$, each capable of clearing diatoms from 3–4 $liter^{-1}$ $hr^{-1}$) may comprise 90% of the total zooplankton/nekton biomass in upwelling cells and contribute 85% of all zooplankton/nekton grazing pressure. *Pleuroncodes* is directly preyed on, and a preferred food of, yellowfin tuna in the same region, so this is a remarkably direct link from diatoms to your table.

The three-dimensional differential distribution of the five most abundant copepod species was used by Peterson *et al.* (1979) to clarify how these organisms exploit water movement during upwelling off Oregon. Over a very narrow continental shelf (the 100-m isobath is only 10 km from the coast) the pycnocline lies at 20–50 m, sloping up toward the beach: in an upwelling episode the pycnocline intersects the surface 5–10 km seaward and nitrate-replete deep water is brought to the surface. Within this system each copepod has a narrowly defined and specialized distribution, in which it is maintained by details of the circulation pattern and its reproductive behavior: *Acartia clausii* is restricted to the upper 5–10 m and within 5 km of the shore; *Acartia longiremis* occurs 10 km offshore and similarly near the surface; *Pseudocalanus* occurs out to 15 km from shore but also only within the pycnocline at 10–20 m; *Oithona similis* occurs at similar depths but not in the first 10 km offshore; and *C. pacificus* has wider ranges for both depth and distance offshore. Off California the endemic *Calanus* has a life history like that of *Calanoides* elsewhere; during winter and other periods when upwelling is not active, it descends to 400–600 m as a population of C5s and remains dormant in the oxygen minimum layer that underlies the California Current.

The California Current is, of course, home to what is perhaps the best known of all fluctuating fishery resources: sardines and anchovies. This is not the place for a major discourse on this classical case, the outlines of which have become clearer in recent years after decades of confusion. The collapse of the great California sardine (*Sardinops sagax*) fishery in 1950, and the subsequent building of an equally large stock of anchovies (*Engraulis mordax*) that likewise crashed in 1990, is too well known to require retelling. A similar pattern of relative abundance has occurred between sardines and anchovies off Japan, Peru, and South Africa, and competitive population dynamics, forced by environmental change, is the explanation that comes most readily to mind.

Unfortunately, it is not so simple as that and the relative consequences of fishery-induced changes in population structure and environmental forcing have not yet been satisfactorily untangled. The evidence of scales counted from unperturbed, varved cores in the Santa Barbara basin off Southern California shows that such changes in abundance have occurred naturally throughout the last 1700 years. But, and it's a large "but," there is no evidence whatever of the short-term alternation observed in modern data; instead, both species fluctuate in abundance at a 60-year period, with an additional 100-year frequency for anchovies—whose biomass, during these 1700 years, was approximately three times that of sardines (Baumgartner *et al.* 1992). Sardine biomass appears to have been significantly more variable than that of anchovies, and some spectacular episodes of very high abundance were recorded in the core samples. Thus, in the 16th century, over a period of around 25 years, sardine biomass reached 15 million tonnes, an order of magnitude greater than otherwise occurred in the 400 years from the 14th to the 18th centuries. At present, we have no way of knowing to what extent the pattern observed in recent decades, here and in other similar species pairs, is natural or is induced by the inevitable modification of the population characteristics of the target species by fishing. A scan of recent volumes of the CalCOFI Reports will convince you that we have a long way to go before consensus is achieved; the essay by Smith *et al.* (1992) would be a good entry point to these discussions.

## Regional Benthic and Demersal Ecology

The California Current is also home to a major population of Pacific hake (*Merluccius productus*) that dominates the demersal fish community of the outer shelf, although these gadoids behave as midwater pelagics during their migrations between feeding areas off Oregon-Washington and spawning areas off southern California; larvae and juveniles return to the northern region in the undercurrent.

The demersal fish fauna of the inner shelf, and especially of the coastal regions, is dominated by flatfish (*Citharichthys* spp. and *Microstoma* spp.) that form about 45% of all trawl-caught fish in the Southern California Bight; next in abundance are large sciaenids (*Genyonymus* spp.) and rockfish (*Sebastes* spp.). Large populations of decapods occur here also. There are major differences in the fish and benthic fauna with depth across the shelf; where the thermocline encounters the shelf (20–60 m) these changes are most profound and rapid, whereas from 60 m out to the shelf edge, changes are more gradual; this region is dominated by rockfish, hake, and grenadiers.

The dynamic circulation pattern of the California Current has major consequences for demersal—as well as for pelagic—fish; seasonal and episodic changes in the upwelling pattern cause shifts in the pattern of anoxic bottom water on the shelf and consequent shifts in the distribution of demersal fish (Mearns and Smith, 1975). But more than that, the specific patterns of reproduction of species of fish and invertebrates are matched to the exigencies of circulation so as to ensure closure of life cycles (Shanks and Eckert, 2005). Because drift in the California Current is unidirectional for long periods, organisms with planktonic larvae potentially risk becoming extinct at the population level: this is, as the authors say, "a marine equivalent of the drift paradox in streams." Probability of the closure of life cycles is enhanced by the evolution of suitable duration of planktonic larval drift, and by the abandoning of planktonic larvae in certain situations. Shanks and Eckert analyzed the reproductive stratagems of 154 fish and 50 benthic crustaceans and found three patterns: (i) long drift duration of c. 135 d and high fecundity in long-lived shelf/slope species, (ii) short drift duration of c. 45 d and lower fecundity in nearshore species having near-benthic larvae, and (iii) short drift duration of c. 48 d and high fecundity in coastal species having planktonic larvae.

These patterns appear to have evolved so as to maximize life-cycle closure: the long-lived larvae of offshore species below the mixed layer experience northward flow in winter and are returned southward in summer, while the short-lived planktonic larvae of coastal species maximize the probability of being retained within eddies, as in the Southern California Bight. The near-shore fishes may experience a reversal of coastwise flow during and after individual upwelling events. Furthermore, a higher proportion of nearshore species than others have evolved live bearing of larvae that directly enter the demersal habitat. These adaptations to the requirement for closure of life cycles, while exploiting the abundant and suitably scaled food offered to larval fish by the zooplankton, are probably of general occurrence in other regions influenced by a major coastal current system.

## Synopsis

*Case 6—Intermittent production at coastal divergences*—Seasonality of pycnocline depth, meaned over the entire province, is consistent with both boreal winter mixing (60 m,

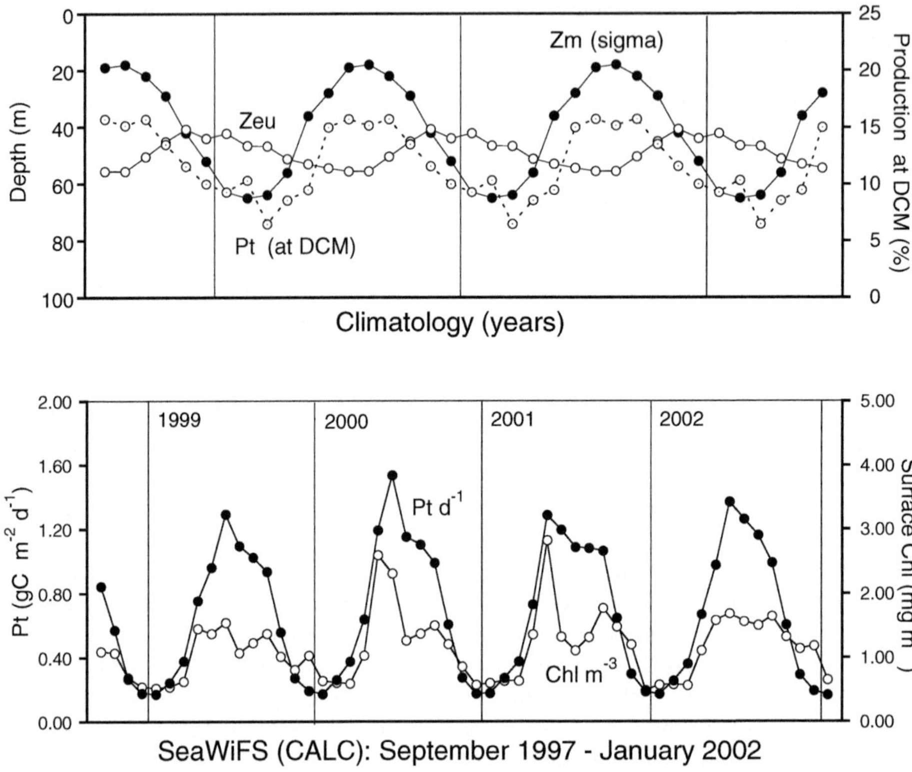


**Fig. 11.18** CALC: seasonal cycles of monthly surface chlorophyll and depth-integrated autotrophic production for the years 1997–2002 from SeaWiFS data together with characteristic seasonal cycles of mixed-layer depths from Levitus climatological data and photic depths computed from characteristic irradiance and the archive of chlorophyll profiles discussed in Chapter 1.

January) and coastal upwelling in summer (15–25 m, March–November). Photic depth lies consistently at 40–50 m, so the thermocline is illuminated except in boreal winter (December–April), and presumably also in strong upwelling cells, the effects of which are concealed in these mean values. Regional mean productivity rate increases from a January minimum (0.25 g C $m^{-2}$ $d^{-1}$) to an annual maximum in May–June that may reach >1.5 g C $m^{-2}$ $d^{-1}$, after which a decline is sustained until November or December (Fig. 11.18). Overall biomass accumulation appears to exhibit responses both to the boreal early summer bloom and to coastal upwelling. The relationship between productivity and chlorophyll changes is consistent with a close balance between consumption and production, even though seasonal vertical migrant herbivores are an important component of consumer biomass.

## Central American Coastal Province (CAMR)

### Extent of the Province

The Central American Coastal Province (CAMR) extends from the tip of Baja California at Cape San Lucas in Mexico to the Gulf of Guayaquil in Ecuador. For convenience, it also includes the long, narrow epicontinental sea of the Gulf of California that lies behind the peninsula of Baja California.

## Continental Shelf Topography and Tidal and Shelf-Edge Fronts

The west coast of America is an active continental margin in the geological sense, so the continental shelf is narrow, usually a few tens of kilometers wide. Only in the Gulfs of Panama and Guayaquil and at the head of the Gulf of California are there somewhat wider shelf areas. Tidal stirring has been invoked as a mechanism that contributes strongly to the productivity of the Gulf of California; here, the midbasin sill at the islands of Angel de la Guardia and Tiburón is the most important tidal mixing region augmented by breaking internal waves and hydraulic jumps.

## Defining Characteristics of Regional Oceanography

This province is limited by the northern and southern equatorial fronts where flow from the upwelling provinces that lie to north and south turns offshore. These fronts are sharp features, linear near the coast but increasingly meandering out into the open ocean (Griffiths, 1968). Circulation in CAMR is dominated by its location between the passage across the ocean of the North and South Equatorial Currents and by the arrival of the North Equatorial Countercurrent at the eastern margin of the ocean. Water transported in this current forms the characteristic water mass in the entire region between Baja California and the Gulf of Guayaquil. Seasonality is dominated by the meridional migration of the atmospheric ITCZ that in boreal winter lies across the southern part of the province and in boreal summer advances north to Nicaragua at about 15°N. Forced by the consequent shift of winds in winter, the North Pacific gyre is strengthened and the California Current extends farther south than in boreal summer (Badan-Dangon, 1998).

Except for a few months in boreal winter along the Mexican coast, circulation is predominantly toward the north, consistent with the poleward turn of the North Equatorial Countercurrent in the region of the Gulf of Panama: it is therefore continuous along the entire coast of the province and may indeed be continuous with the flow of the California Undercurrent although (as Badan-Dangon emphasizes) this is one of the least-investigated major coastal currents anywhere. Fiedler (2002), on the contrary, suggests that northward flow in the CAMR province eventually turns seaward to join the flow of the NEC. Various segments have been variously named, but I shall use the term Costa Rica Coastal Current for the whole. Recall that the eutrophic feature of the Costa Rica Dome (see PNEC) originates on a thermocline ridge at the coast in the Gulf of Papagaya in late winter, moving progressively offshore; the chlorophyll enhancement associated with the nascent CRD is indistinguishable from that proper to permanent coastal processes here.

Because drainage from the sierra mountain chains is mainly toward the east, few large rivers discharge onto the continental shelf, so that the Gulfs of Panama and Guayaquil are the sole recipients of significant freshwater and silt. Mixed-layer depth is generally shallow (30–40 m) and the permanent thermocline is sharp, although a very shallow brackish surface layer may occur during the wet season of boreal summer when southerly monsoon conditions occur. The coastal regime is modified by the passage of strong winds through the gaps in the mountains at head of the Gulfs of Tehuantapec, Papagayo, and Panama; these winds cause intense cooling of the sea surface due to upwelling but also to evaporative effects. Wind stress at such low latitude is translated into motion of offshore current jets that have warm-core anticyclonic eddies to the right and ephemeral cold-core cyclonic eddies to the left. Approximately five of these eddy pairs are produced annually and, as discussed later, they have singular effects on the ecology of the region as they propagate westward.

The Gulf of California, or Sea of Cortéz, is a long, narrow epicontinental sea that includes deep basins and has steep-to coasts and continental slopes and numerous islands.

It is convenient to include the bight lying between Cape San Lucas and Cape Corrientes within the gulf. The northern end of the gulf is desertic, whereas the southern end lies below the boreal summer southerly monsoon rainfall conditions; total rainfall depends on the incidence of tropical storms, or "chubascos." Circulation is complex and dynamic because of the interaction of the coastal wind regime with mountainous terrain of Baja California between the gulf and the open Pacific Ocean. Exchange with the open Pacific is minor and is largely driven by the fact that the upper gulf is an evaporative basin (Bray, 1988); the general circulation therefore resembles that of a small Red Sea, except there is no shallow sill across the mouth. In the northern gulf, the water column is fully mixed by the boreal northeasterly winds, assisted by convective overturning to a depth of 100 m (Roden, 1964). Over the shallow northern shelf, winter cooling of the highly saline water of the evaporative estuary of the Colorado River forces a complex surface pattern of turbidity gyres (Lepley *et al.*, 1975). This dense, cool water is the origin of an outflow of deep water from the gulf.

Residual, tidally forced circulation is strongly modified by local, temporary wind stress. Land-sea breeze effects are important and northwesterly winds dominate the region in October–May, whereas weaker southwesterlies dominate during boreal summer. This alternation of wind direction forces strong upwelling on east and west coasts alternately, according to season. Resultant plumes and dipoles of cool water may encompass the entire width of the gulf.

## Regional Response of the Pelagic Ecosystem

Perhaps because this region has a rather stable oceanographic regime, seasonality in productivity appears to be driven by an integration of characteristic local processes, rather than by changes common to the entire area of the province. Study of the SeaWiFS and MODIS images suggest that we should discuss separately the following individual processes: (i) possible winter upwelling on the coast from Cape Corrientes (21°N) to the Gulf of Tehuantapec (16°N), (ii) upwelling that is especially strong in winter, but not confined to that season, in the Gulfs of Tehuantapec (Mexico), Papagaya (Costa Rica), and Panama, and (iii) multiple processes leading to seasonal eutrophication in the Gulf of California.

*Gulf of California*: Tidal and drift currents can be strong through passes between islands, especially in the northern gulf around Angel de la Guardia Island, where the Ballenas Channel to the west of the island is isolated by relatively shallow sills (Alvarez-Borrego, 1983). There is a persistent pool of cool water that is associated with tidal fronts whose orientation depends on wind direction (Badon-Dongon *et al.*, 1985). Southwesterly winds in boreal autumn and northwesterly winds in spring force coastal upwelling plumes from opposite sides of the gulf: on the west coast in autumn and the east coast in spring. Wind-driven circulation changes are reflected in seasonal blooms, especially in autumn and winter when sea surface temperature cools under the influence of winter winds (from 30° down to 18°C) and SeaWiFS chlorophyll increases (from 0.25 to 2.5 mg chl $m^{-3}$), with diatom blooms initiating the sequence, followed by foraminifera and coccolithophores; the bloom ends with a population of silicoflagellates and other diatoms (Thunell *et al.*, 1996). These data were obtained from sequential sampling from sediment traps and must be presumed to integrate the individual local processes induced by topography referred to previously.

*Mexican west coast*: During boreal winter, as noted earlier, equatorward winds and flow of surface water along the west coast of Mexico, from Cape Corrientes southward, are associated with coastal chlorophyll enhancement and the formation of offshore filaments. This effect is strongest at Cape Corrientes, where it appears to be confluent with high chlorophyll at that season along the eastern coast of the Gulf of California. I have not seen a description of this effect but I suggest that the observation is consistent with the presumed slight upwelling tendency that may be expected along this coast at this season. It appears to be separable as a phenomenon from the Gulf of Tehuantapec enhancement immediately to the south.

*Gulfs of Tehuantapec, Papagaya, and Panama*: The Gulf of Tehuantapec is important oceanographically because it lies west of a major gap in the Central American sierra chain, the Chivela Pass, through which the transmontane winds from the Gulf of Mexico (only 150 km to the east) have easy passage (Blackburn, 1962). These winds, mostly from the northeast, force southward surface flow in the gulf. Interaction between this and the northward flow along the Mexican coast in the Costa Rica Coastal Current generates a meridional thermocline (pycnocline and nutricline) ridge. In boreal winter, the mixed layer above this feature may be completely eroded by wind stress, and surface temperature, salinity, and nutrient levels resemble values in the pycnocline. After extreme transmontane wind events, a plume having such characteristics may stretch several hundred kilometers to the southwest from the gulf. Surface productivity is high ($<150$ mg C $m^{-3}$ $day^{-1}$) and the plume, then lying above the Tehuantapec Ridge, becomes a feature of the surface chlorophyll field (see Color plate 19). Gonzales-Silvera *et al.* (2004) analyzed SeaWiFS and AVHRR data and found that surface chlorophyll ranged over two order of magnitude and SST by almost 10°C between upwelling events and stasis. They found that the Gulf of Tehuantapec produces a larger number of eddies, each being associated with several smaller cyclonic eddies; these rarely survive long, whereas the major anticyclonic eddies mentioned above were followed for up to 3 months and over trajectories of $<700$ km.

Upwelling also occurs, and for the same reason, in the Gulfs of Papagaya and Panama and at approximately the same season, though the occurrence of upwelling in one gulf does not necessarily denote upwelling in the others. In some years this process can produce a major feature in the sea-surface thermal field of the Gulf of Panama, which reaches seaward into the region of the Galapagos Islands. At such times, the northern thermal front off Panama carries 500-km waves that pass westward along the front at about 30 km $day^{-1}$ (Legeckis, 1988). Transmontane winds into the Gulf of Chiriqui appear not to cause upwelling.

Beyond these exceptional events the ecology of the coastal regions of Central America can best be regarded as comprising two systems: (i) those off arid coasts with little freshwater input (such as much of the Mexican coast and particularly the Gulf of California) where water clarity is high, and (ii) those within the area of tropical rainfall where coasts are invested with mangrove, there is significant river effluent, and water clarity is seasonally very low.

The inshore Gulf of Panama will serve as a model for the latter type of pelagic ecosystem: here the ecological balance and biota appear to closely resemble those of the mangrove coasts of the Gulf of Guinea (see GUIN). Here, there is an algal bloom, apparently dominated as in the Gulf of Guinea by large diatoms and at the same season (January–March); of course, wind-forced upwelling at the mouth of the Gulf of Panama is also implicated and perhaps the relaxation of river flow and consequent reduction in water clarity is less important here than in Gulf of Guinea. Less than 10% of daily production by phytoplankton can apparently be consumed by the standing stock of mesozooplankton; since these are nitrate and diatom-dominated blooms, it seems likely that modern investigations of the role of picoplankton and their protist grazers will modify the models based on older techniques less than they will in the open ocean. Another case of apparent surplus production, requiring an export process or a sink for organic matter, is the Gulf of Guayaquil, to the south of the Gulf of Panama.

## Regional Benthic and Demersal Ecology

Hermatypic corals are minor reef-forming agents in this province, in which benthic ecology strongly resembles that of the coast of intertropical West Africa (Longhurst and Pauly, 1987). Although we do not have for this province a survey of the quality of the GTS (see GUIN), it appears that the same genera of demersal fish, in similar relative proportions, occur here—or, at least, occurred in the pristine fish fauna—and that this is very distinct from that of the western Pacific at similar latitudes: some genera that are characteristic of that region are entirely lacking here, such as *Leignathus* and *Lactarius*.

## Synopsis

*Case 6—Intermittent production at coastal divergences*—The pycnocline is extremely shoal and shows very weak seasonality with some slight boreal winter mixing. Photic depth is consistently deeper (35–45 m), so the thermocline is illuminated year-round. Regional mean productivity rate increases rapidly to about 1.0 mg C $m^{-2}$ $d^{-1}$ in June, after which progressive reduction occurs during the second half of the year, perhaps reflecting events in the Gulfs of California, Tehuantapec, and Panama (Fig. 11.19). Chlorophyll biomass appears to be insensitive to changes in PP rate but shows significant accumulation during boreal spring. However, as noted earlier, it is the productivity of the Costa Rica Dome that dominates the regionally integrated data, rather than processes at the coast. So, this figure and the meaned data that represent this province may be somewhat misleading.

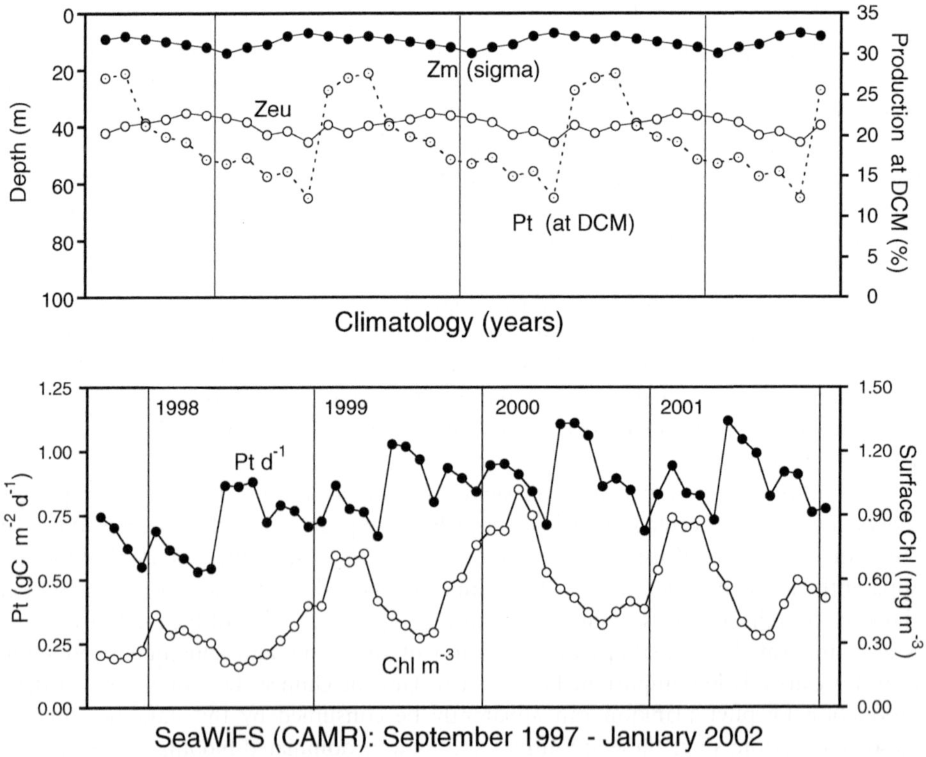


**Fig. 11.19** CAMR: seasonal cycles of monthly surface chlorophyll and depth-integrated autotrophic production for the years 1997–2002 from SeaWiFS data together with characteristic seasonal cycles of mixed-layer depths from Levitus climatological data and photic depths computed from characteristic irradiance and the archive of chlorophyll profiles discussed in Chapter 1.

## Humboldt Current Coastal Province (HUMB)

### Extent of the Province

This province has by far the greatest north-south extent of any of the coastal boundary current systems, and for that reason alone it would be advisable to subdivide it into Chilean and Peruvian provinces, separated at about the change in the orientation of the coast near 18°S, or (perhaps better) into tropical and subtropical provinces at about 40°S where the west wind drift encounters the coast and diverges north-south. The argument for this ideal arrangement is the same as for the SPSG province discussed earlier; the reason that I have not chosen the ideal partition is the same for the two provinces.

Because of its size, and because it covers both Peruvian and Chilean regions, I revert to the time-honored name of Humboldt for this province. In the north, the boundary is the equatorial front that separates the equatorward flow of the coastal current from tropical water at about 5°N, where the Peru Current turns offshore into the flow of the SEC. To the south, it is at Tierra del Fuego and, seaward, the edge of eutrophic conditions that lies at varying distances from the coast, depending on local eddy formation.

### Continental Shelf Topography and Tidal and Shelf-Edge Fronts

The continental shelf is generally very narrow along the whole west coast of South America because this is the leading edge of a drifting continent. The break of shelf occurs at 200 m, and only off northern Peru, from 6°S to 10°S (Trujillo and Chimbote), is the shelf somewhat wider, reaching a maximum of 125 km. From Callao (12°S) in Peru to Valparaiso (33°S) in Chile the 200-m isobath lies within 10–20 km of the shoreline. From Valparaiso southward, the shelf again widens and reaches 50 km off Concepción and Chiloe Islands, but over this region the continental edge at 200 m is as topographically complex as the coastline itself that here resembles the fjord coastline of Alaska or Norway. The coastal topography of the entire province is dominated by the range of the Andes mountains that everywhere follow the coastline, so that the 1000-m contour lies no more than 100–200 km inland. Coastal deserts are induced (see later discussion) by interaction between the wind regime and the mountains.

### Defining Characteristics of Regional Oceanography

This is a very complex and much-misunderstood region and we are fortunate in having a recent critical review of circulation by Strub *et al.* (1998); I lean heavily on it in what follows.

The regional winds are dominated by the presence of the South Pacific High, centered at about 30°S 95°W, which migrates meridionally with the seasons in step with the similar motion of the ITCZ. Winds at the coast are therefore equatorward everywhere north of 30°S, with some local exceptions, and poleward from 35°S down to Cape Horn. Two regions have very heavy precipitation: below the seasonally migrating ITCZ, as everywhere at low latitudes, and also off southern Chile. The southern wet region is—as in the Pacific Northwest (Oregon to Alaska)—the consequence of westerly winds which, on encountering a mountain barrier, generate sequential coastal low-pressure systems that propagate poleward, as off Alaska; these systems are most frequent and strongest in austral winter. Here, these winds induce downwelling at the coast in the same manner as in the Gulf of Alaska. However, winds are perennially upwelling favorable, from central Chile (35°S) to Pieta (5°N) where the coast changes orientation, and this effect is maximal in austral winter. For these reasons, the HUMB province, as pragmatically constituted here, comprises parts of two biomes: a northern region where upwelling is permanent or seasonal, and a southern region where winds are primarily downwelling and where heavy precipitation results in large surface buoyancy fluxes.

Thus, the pattern of wind stress and its curl over the province resembles the general pattern established for eastern boundary currents by Bakun and Nelson (1991), but with some special characteristics due to the alignment of the coast and coastal mountains. Cyclonic curl adjacent to the coast takes maximal values in two areas in austral winter: off Peru from 9°S to 15°S and off Chile at about 25°S. The Peru maximum is seasonally stable, whereas off Chile it moves poleward in summer to about 32°S and simultaneously strengthens so that the Peruvian and Chilean upwelling centers are seasonally out of phase. In the bight at 18–20°S (Arica to Iquique), there is anomalous and persistent anticyclonic curl, probably because the bight is open to wind from the south.

As a consequence of the specific characteristics of the wind field, the regional partitioning of positive and negative wind-stress curl is significantly less organized into inshore cyclonic and offshore anticyclonic areas, especially during austral winter (April–September). As Bakun and Nelson point out, wind stress is maximal not at the coast itself, but farther offshore, and this leads to offshore upwelling; right at the coast, local land-sea breezes may significantly reduce the potential effect at the sea surface of the regional wind systems. Such local effects are particularly strong at capes that enclose a bay open to the north, a conjunction (as at Punta Aguja) that may lead to the development of a local coastal desert region.

The interaction between the gyral flow, wind forcing at the coast, and coastal topography is very complex and involves both poleward and equatorward motion in the coastal currents; these have been very variously named, and I endeavor to use a conservative selection in what follows. Well offshore, the circulation around the eastern edge of the subtropical gyre from the divergence of the West Wind Drift (WWD) is weak, broad and continuously equatorward; I shall follow what appears to be common usage in using the term Humboldt Current for this flow. The offshore, poleward flow south of the divergence of the WWD is termed the Cape Horn Current by Strub *et al.* Neither of these flows are involved in the coastal effects in which we are interested.

Flow at the coast is equatorward during austral summer from Chiloe Island at 42°S right up to Cape Aguja at about 6°S, but, in winter, equatorward flow is initiated only at about 20°S as the Peru Coastal Current, while fluctuating flow obtains from here down to about 32°S and, even further south, flow becomes poleward in the Chile Coastal Current. A subsurface Poleward Undercurrent flows along the entire outer shelf and upper slope, transporting high-salinity, nutrient-rich water just as occurs also in the CALC province. Thus, off Peru, equatorward flow is shallow (~25 m), above poleward flow that has maximum velocities at about 100 m in the upper layers of an oxygen-depleted zone but shallower than the depth of maximum nitrate depletion. Because of the shallow depth of the equatorward flow, it is water of the Poleward Undercurrent that is upwelled at the coast (Brockman *et al.*, 1980); thus, at the upwelling near Concepción (37°S) high-salinity upwelling plumes are readily distinguished from low-salinity plumes of fjord water that emerge from the island passes to the south.

Upwelling water, and hence nutrient advection, differs in the Peruvian and Chilean sectors of the Humboldt Current system: off Peru, equatorial subsurface water flowing south in the subsurface countercurrent is upwelled, whereas off Chile it is subantarctic water of the equatorward coastal current that upwells, though countercurrent upwelling has been observed at Valparaiso (Johnson *et al.*, 1980). The boundary between these two source-water regimes lies at about 15°S according to Wyrtki (1963).

Though there are apparently regional differences in the depth of the source water in upwelling cells, the canonical view is that upwelling is restricted to relatively shallow depths and that there is an onshore flow of bottom water over the continental shelf (e.g., Codispoti *et al.*, 1982). Upwelling occurs as numerous small coastal upwelling cells, from which relatively small plumes (< 100 km) of cool water may be entrained offshore. These cluster in the following preferred locations on the Peruvian coast:

*Point Negro (5°S)*: Here, the trend of the coast changes abruptly and to the north lies the tropical surface water of the Gulf of Guayaquil of the CAMR province. This cape is a major and persistent upwelling center, and offshore filaments may be aligned with the convergent front that marks the separation of the Peru Current from the continent, so forming a major hydrographic discontinuity.

*Chimbote (9°S)* and *Callao (12°S)*: Though there is no major topographic feature here, this seems to be a consistent upwelling center whose plume proceeds seaward over the widest region of shelf on the Peruvian coast. The coastal prominence at Callao and Lima appears to focus significantly the upwelling processes.

*Pisco (14°S) and Cape Nazca-Punta Sta. Ana (15°S)*: In the bight north of Pisco, surface water is consistently cool and a major upwelling center appears to be topographically linked to the Paracas peninsula, whereas the upwelling center off San Juan is the most consistent and the strongest upwelling center on the Peruvian coast and was well known to early investigators (e.g., Gunther, 1936). Much of what has been learned about upwelling dynamics on the Peruvian coast, and about the ecological physiology of phytoplankton during upwelling, is due to work at this site during JOINT I and II of the Coastal Upwelling Ecosystem Analysis program (MacIsaac *et al.*, 1985).

Despite the existence of these persistent upwelling centers off Peru, as many as 15–20 small (<25–km) cool-water cells may occur simultaneously along the upwelling coast from San Juan to Paita; we now see these in AVHRR images, but they were first visualized by Peruvian scientists who, in the early 1970s, deployed over a coastwide grid that was occupied during a single day many tens of chartered fishing boats that were equipped solely with bucket thermometers: this produced the first temperature map that resembled what we now see routinely in AVHRR images.

Because appropriate wind stress may be imposed for long periods off Peru, but often not at the coast itself, the upwelling front may move far offshore and become indistinct (Brink, 1983). At such times the whole coast may present a single inshore cool zone from Arica to Paita (5–15°S) within a weak temperature front about 100–125 km offshore, though progressively farther offshore to the north (Bohle-Carbonell, 1989). This zone may even extend south into the corner of the Arica Bight at about 18°S.

It has been suggested that there is a relatively weak relationship between wind stress and surface nutrients in the Peru Current because the generally persistent, generally equatorward winds drive only a shallow surface layer northward along the coast and this flow is rather frequently interrupted, as already noted, by surfacing of the poleward subsurface current. However, because nutrient levels below the photic zone are relatively high in the Pacific, the nutrient content of upwelled water is high, at least in normal years (Codispoti *et al.*, 1982). Especially in the Peruvian sector, or wherever countercurrent water is upwelled, the oxygen content of upwelled water is relatively low. The existence of offshore flow both at the surface and on the bottom, with onshore flow toward the site of upwelling, has consequences for nutrient regeneration over the shelf. Sinking organic material is transported across the shelf and sequestered in the bottom layer of oxygen-deficient water. In these circumstances, this organic material is no longer available for regeneration (Codispoti *et al.*, 1982).

Off Chile, the upwelling processes are different from those off Peru. Here, especially in winter and spring, a wide zone is populated by meanders, eddies, and very prominent cool filaments within which vorticity occurs and may extend 200–300 km offshore. As shown earlier by Fonseca and Farias (1987), these mesoscale features resemble the better-known eddies and cool filaments of the northern part of the California Current off Oregon and Washington.

Upwelling-favorable winds are relatively lighter along the Chilean coast, and their maximum potential for upwelling occurs in the vicinity of Valparaiso. Upwelling fronts are numerous, complex, and highly variable: we may distinguish coastal fronts no more than 10–20 km offshore, outlining clear upwelled water, whereas plume fronts outlining tongues of green upwelled water may be encountered up to 30–50 km offshore. An oceanic front separating the coastal regime from warmer, blue oceanic water is often indicated by the 15°C isotherm, which may meander between 100 km and 300 km offshore.

Five major, semipermanent upwelling centers have been identified on the Chilean coast, from analysis of CZCS images by Fonseca and Farias (1987). The offshore mesoscale tongues of cool water that are a prominent feature of upwelling off Chile can be traced back to these upwelling centers. All are apparently topographically locked to the vicinity of capes and submarine topographic features, though their individual characteristics are at the present time little known:

*Point Patache (20° S) and Mejillones-Antofagasta (23° S)*: The coastal prominence between Punta Tetas and Punta Angamos is a major topographic feature and the site of persistent upwelling, whereas to the south of Iquique, major upwelling events frequently occur and are distinct from the southernmost upwelling on the Peruvian coast.

*Point Lengua de Vaca (30° S) and Point de Curaimilla (33° S)*: These, and the Gulf of Arauco south of Valparaiso, are the best known of the persistent upwelling centers, and the maximum potential for upwelling-favorable winds on the Chilean coast occurs here. This is the only location where the dynamics of upwelling on the Chilean coast have been studied in detail (Johnson *et al.*, 1980), and the data from these studies are consistent with the existence of a classical double-celled, cross-shelf upwelling circulation. Water that is upwelled originates in the high-salinity, low-oxygen coastal undercurrent, and the upwelling effect reaches to a maximum depth of about 250 m. Short-term variations in upwelling wind intensity result in variations in the location of the undercurrent on the shelf and in pulsations in upwelling intensity.

*Point Lavapie (37° S)*: This peninsula, to the south of Concepción, is the site of persistent upwelling events and also the most southerly location where they occur with regularity (Arcos *et al.*, 1987). Upwelling response to favorable winds may occur in response both to short-period (1 or 2 days) wind events and to those of longer duration (6 or 7 days).

To the south of these regions, in the high-precipitation and downwelling regime of the fjordland that extends down to Tierra del Fuego, coastal processes are very similar to those discussed earlier for the Alaska coastline, and here we can reach back to the 1970 round-America maiden voyage of C.S.S. *Hudson* for some information. The effect of glaciers is strongest in the central regions; runoff is of the same order as in Alaska and is more important than direct precipitation to the sea surface. The density profile and stratification, therefore, is forced rather by relative salinity than by temperature. Everywhere in this region, despite strong winds, tides are the dominant forcing for water movement.

## Regional Response of the Pelagic Ecosystem

The regional fields of pigment and SST derived from satellite sensors confirm the anticipated relationship between upwelled water and high chlorophyll values, and mesoscale features resemble those observed in other eastern boundary currents. Rapidly changing cool filaments and plumes, exhibiting vorticity, are normal features of the chlorophyll field; development and decay of such features occurs on a scale of a few days to a few tens of days. Only in the southern Chilean region, beyond Concepción, does the seasonal regime follow a different pattern: here, as off Alaska, a spring bloom is to be expected with filaments and arcuate features of chlorophyll that reach far offshore; these appear to be not unlike the Heida eddies discussed earlier but nevertheless, as off Alaska, coastal chlorophyll enhancement does propagate far seaward to the west off Chile.

But to return to the canonical upwelling system of this province: the biological seasons are dominated by processes that occur off Peru, where austral spring-summer upwelling is the principal time-dependent forcing. The biological evolution of a parcel of upwelled water, which requires about 10 days to complete, has been well worked out (see MacIsaac *et al.*, 1985; Harrison *et al.*, 1981) in the Humboldt Current, in a location (Cape Nazca-Cape Santa Ana) having consistent upwelling for several weeks. In the actual upwelling zone, < 7 km from the coast, nitrate remains relatively constant at about 20 μM in the upwelling water, whereas silicate supply varies according to its source. The evolution of phytoplankton growth passes progressively through several phases: a small inoculum of phytoplankton cells is upwelled from 50–60 m depth, the origin of the seed stock being cells sinking from farther seaward; these cells are physiologically conditioned to low light levels and therefore have low rates of nutrient uptake and growth. In addition, the upwelled water itself may require conditioning by exposure to light to permit active growth of algae. If this occurs, it consists of progressive modification of trace metal chemistry, exchanging available copper and manganese ions.

Progressively, the cells entrained in the upwelled water shift up to increased physiological rates so that as stratification is induced by solar heating in the upwelled parcel, the entrained cells are adapted to high irradiance and high nutrient levels. The now fast-growing cells, held in a very shallow (< 10–15 m) Ekman layer, reduce ambient nitrate levels quite rapidly and there is a massive increase in standing stock of cells. Rates of primary production are maximal at 0–10 m, whereas two depths of high concentration of chlorophyll may occur—the first near the depth of maximum production rate and the other, a DCM, near the bottom of the thermocline, although this may merely represent unconsumed, sinking cells. Subsequently, in response to nutrient depletion in the Ekman layer, the cells respond by limiting some of their cellular processes, sequentially slowing nutrient uptake, photosynthesis, and storage of carbohydrates and lipids. Because most of the growth can be attributed to diatoms, the limiting nutrient may be silicate rather than nitrate. In the maximum growth phase, a typical upwelling plume has a phytoplankton species assemblage dominated by no more than 10 species of diatoms (*Detonula, Chaetoceros, Hemiaulis, Rhizosolenia*, and *Thalassiosira*) together with dinoflagellates and < 10-mm flagellates. The total cell number of diatoms is usually about the same as the total of the other two groups.

Herbivorous mesozooplankton consume only a small fraction (10% is a typical finding) of the daily primary production. Three copepods dominate the biomass of mesozooplankton offshore (*Calanus chilensis, Centropages brachiatus*, and *Eucalanus inermis*), of which only the last continues to perform diel migrations while it is entrained into an upwelling feature. These species aggregate close to the surface under such conditions and their layer depth coincides with the lower part of the upper chlorophyll layer—perhaps near the depth of maximum production rate (Herman, 1984).

At midshelf depths (120 m and beyond), during the upwelling season off Peru, it was found that these large copepods had different feeding strategies and that the differences were strengthened during periods of low food availability: *Calanus* feeds continuously as a herbivore, but avoids the surface layers by day; *Centropages* is an omnivore, feeding only at night when offshore; and *Eucalanus* feeds intensively by day on POM in the anoxic layer, but near-surface at night (Boyd *et al.*, 1980). Inshore, near the foci of upwelling cells, the zooplankton is dominated by small copepods, just as it is off Alaska: species of *Paracalanus, Oithona*, and *Acartia* dominate this fauna within the cores of upwelling cells, their abundance responding negatively to SST (Escribano and Hidalgo, 2000).

It was the variable and sometimes disastrous changes in weather patterns on the coast of Peru that first attracted the attention of oceanographers and fisheries scientists to the ENSO phenomenon. Obviously, we cannot end the discussion of the Humboldt province without some consideration of how an ENSO event is manifested. There is a sufficient number of easily available accounts of the phenomenon not to require another full-blown description here: if you are not familiar with the issue, I recommend starting with Tomczak and Godfrey (1994) for the physics and with Pauly and Tsukuyama (1987) or Pauly *et al.* (1989) for the ecology and for the disastrous fisheries consequences that may ensue.

Briefly to recapitulate the effects of the ENSO cycle in the HUMB province, the onset of an ENSO event is heralded by the slackening of upwelling intensity and the reduction of the number and extent of upwelling centers as indicated by surface water temperatures. This most often occurs in the autumn or winter (see Tomczak and Godfrey, 1994, p. 364, for an interesting account of how this process was misnamed by oceanographers for El Niño, the Christ child) and accompanied by progressive warming of the surface water, in which Chile lags Peru. For the years 1948–1985, three major and three minor ENSO events were correlated by Chavez *et al.* (1989) with deepening of the 14°C isotherm (from 80 m to 100–175 m) by reduction in 60-m nitrate (from 25 to 14–22 μM) and by a decrease in new, nitrate-based primary production rate (by about 1.0 g C $m^{-2}$ $day^{-1}$). Curiously, the rate of upwelling at the coast may even increase during ENSO periods because the coastal wind system is enhanced by lessened cloud cover even as the offshore trades, which regulate the depth of the thermocline, are at their weakest: what is upwelled to the surface is, of course, warm tropical or subtropical surface water, already fully depleted of nutrients, rather than nutrient-replete subthermocline water.

A multivariate ENSO index has been devised recently by Wolff *et al.* (2003) that brings together six variables: sea-level pressure, components of wind direction, SST, air temperature, and cloudiness. Positive values of this index represent the warm ENSO phase. Comparing the seven strongest such events since 1950, it is found that three events between 1957 and 1973 each featured an early warming in the western Pacific and matured during their first year. Three events between 1982 and 1992 took longer to develop and matured only in spring of their second year. The unusual 1997–98 event developed very fast, and within only 50 days had major impact at depths down to 150 m, and in relative temperature increase in the HUMB province.

The consequences of the 20-fold decrease in productivity and phytoplankton biomass that may occur during El Niño events are catastrophic and cascade all the way from the phytoplankton through herbivores to the disappearance of pelagic fish stocks and great perturbations in the fishmeal trade and in soybean futures. The details of this collapse have been sufficiently described that I do not need to provide a litany of loss of reproduction in seabirds, starvation of marine mammals, and the disruption of the hake, jack mackerel, shrimp, and sardine fisheries because the changed distribution of the stocks put them beyond the traditional ambit of the fishery, whereas the Peruvian anchoveta stock, already stressed by heavy exploitation, may collapse.

This region lacks a CalCOFI-like data base so that relatively little certitude has emerged from studies of the pelagic ecosystem here; instead, what we must work with are fishery landings, conflicting inferences, best guesses, and insights. Some things have become clearer in recent years, however, despite these difficulties: the different manner in which sardine (*Sardinops sagax*) and anchoveta (*Engraulis ringers*) populations react to the same environmental forcing seems to hold the key for understanding changes in their stock abundance. Some interaction also between these and the pelagic predators *Trachurus trechae* (horse mackerel) and *Scomber japonicus* (mackerel) is now seen to contribute to interspecific balances.

It is clear that here, as off California, anchoveta is the dominant planktivorous species, and that the population is subject to natural fluctuations: the stock appears to be under stress in El Niño years when it has been observed to retreat to water >100 m deep. Although it has been assumed that anchoveta recruitment was poor in these years, the recruitment failure of 1971, which led to the subsequent population collapse, occurred prior to the 1972–73 El Niño. This occurred very shortly after the period 1965–1970 when the stock biomass was variously estimated at 12–20 million tons. Almost immediately, the very small biomass of the sardine stock began to increase so that catches of ~6 million tons were obtained in the early 1980s. Meanwhile, the anchoveta stock was reduced to less than 1 million tons by the end of the 1982–83 El Niño but subsequently exploded (the word is taken from Alheit and Bernal) by an as-yet unsatisfactorily explained recruitment process to reach well above 6 million tons by 1985.

What to make of all this? Certainly, as Garth Murphy pointed out long ago (to the chagrin of the Peruvian fishery managers) the truncation of the year-classes of the anchoveta stock by the fishery, so that each fish could count on only a single spawning event, must logically lead to population collapse. This simple truth, denied by some theoretical ecologists, reflects the fact that specific longevity in fish must evolve as a function of uncertainty of recruitment success. Industrial fisheries inevitably truncate longevity, with predictable results.

Nevertheless, over the relatively short period that the stocks have been observed, it is clear that while El Niño conditions reduce the survival of all life stages of anchoveta, the same conditions have a positive effect on sardines. This despite the fact that sardines perform extensive migrations during these events, apparently to avoid the intrusion of warm waters from the north.

## Regional Benthic and Demersal Ecology

The benthic phenomenon that has attracted most attention in this province is the generation of anoxic bottom water off Peru, with formation of dissolved $H_2S$. This is the consequence of the very heavy deposition of POM, including unconsumed phytoplankton cells, onto the continental shelf, there to undergo transformation by the filamentous sulfur bacteria *Thioplica* (Rosenberg *et al.*, 1983). Disturbance of bottom water, as by the 1828 earthquake at Callao, may release a high concentration of $H_2S$ into the atmosphere. This phenomenon is often confused with the red water caused by massive growth of *Gymnodinium splendens*, termed the "Callao painter" by mariners, thought to be stimulated by masses of decaying organic material on the seabed. This phenomenon occurs commonly enough here, and in similar places—as on the SW African coast.

There is some evidence, reviewed by Rosenberg *et al.* (1983), that the probability of the release of $H_2S$ is related to the biomass of planktivorous pelagic fish in the superjacent water column. After the 1971 collapse of the anchoveta stocks there was a decade-long increase in sediment organic matter and in sulfate reduction on the sea floor associated with declines in nitrate and oxygen content of near-bottom water. The relative distribution of anoxic conditions remained relatively unchanged, so that from the surface to 50 m, oxygen concentration was progressively reduced from 5.0 to 0.5 ml $liter^{-1}$ leading down to minimal concentrations (<0.3 ml $liter^{-1}$) at 400–600 m, below which concentrations again increase. There is little seasonality in this pattern, which persists from northern Peru to central Chile.

Both benthic invertebrates and demersal fish populations respond more directly to oxygen tension than to any other factor; a "classical" benthos survey with van Veen grabs at 65 stations on the Peruvian shelf by Rosenberg *et al.* clearly showed that here the Thorson-Petersen benthic communities are essentially absent, except in the surf zone along the beach and just below the beach, where a perfectly normal benthic assemblage was found: *Donax, Emerita, Nepthys, Callinassa*, Paguridae, and so on. This study confirmed the relative distribution of dissolved oxygen discussed earlier, showing that percentage

**Table 11.1. Benthic environmental factors and dominant benthic organisms off Peru.**

| Depth (m) | 3–20 | 20–80 | 80–700 |
|---|---|---|---|
| Oxygen tension (ml $l^{-1}$) | 0.2–5.0 | 0.1–2.0 | 0.0–1.0 |
| Sediment, org. content (%) | >4 | 2–8 | 2.5–11 |
| Biomass | $<40\,g\ m^{-2}$ | $\sim 6\,g\ m^{-2}$ | $1–3\,g\ m^{-2}$ |
| Animal size | all sizes | small | small |
| Dominant biota | *Sinum*<br>*Polynices*<br>*Nassarius*<br>*Owenia*<br>*Diopatra*<br>*Pectinaria*<br>*Hepatus*<br>*Ophiactis* | *Thioplica*<br>*Pitar*<br>*Nassarius*<br>*Magelona*<br>*Polinices* | *Thioplica*<br>*Polynices*<br>*Pitar*<br>*Thyasira*<br>*Notomastus*<br>Nematoda |

*Source: After Rosenberg,* et al., *1983.*

of organic carbon and of benthic biomass closely followed oxygen tension, and that characteristic groups of remnant benthos inhabited each of several zones specified by their depth and oxygen tension (Table 11.1). Benthic biomass becomes depressed at oxygen concentrations $<0.6$ ml liter$^{-1}$, and below 0.1 ml liter$^{-1}$ the deposits may lack macrobenthic organisms entirely. The remnant fauna in the regions of lowest oxygen tension is dominated by polychaetes and nematodes, and the surface of the deposits is often entirely covered by mats of the filamentous sulfur bacterium *Thioplica.* Rosenberg *et al.* also performed fish sampling during these surveys and found that demersal fish biomass was negatively correlated with the coverage of the deposits by *Thioplica.*

Of course, to the south of the upwelling regions in southern Chile, the situation is different so that, although the actors are different, the play resembles that of the Gulf of Alaska. Of the 40-odd demersal species that occur off the fjordland of southern Chile, 15 are gadoids. The overall demersal fish biomass increases with latitude, responding to relative abundance of *Micromesistius australis.* Greatest biomass also occurs rather deep on the shelf and upper slope.

## Synopsis

*Case 6—Intermittent production at coastal divergences*—The slight seasonality of pycnocline depth is consistent with both austral winter mixing (40 m August–September) and coastal upwelling in summer (<20 m November–April) in most sectors of the province (Fig. 11.20). Photic depth is consistently between 35 and 50 m, so the pycnocline is illuminated except in austral winter and early spring, and presumably also in strong upwelling cells, the effects of which are concealed in these mean values. The rate of productivity increases from an austral winter minimum in June to an annual austral summer maximum sustained from November to March during the extremes of trade-wind stress and coastal divergence. The seasonal dynamic range of biomass accumulation is weaker and does not appear to track P-rate changes; mean chlorophyll biomass is dominated by seasonal changes in the southern part of the province, off Chile, where during austral summer high values occur extensively, probably unrelated to upwelling events. There are significant between-year changes in values, responding to the value taken by the SOI.

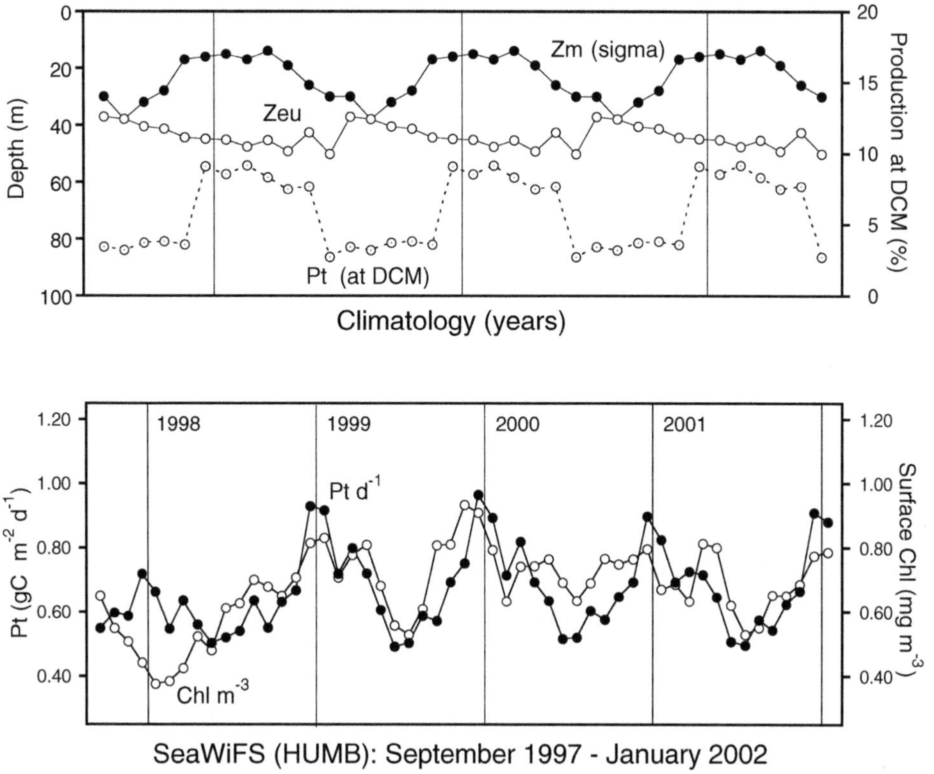


**Fig. 11.20** HUMB: seasonal cycles of monthly surface chlorophyll and depth-integrated autotrophic production for the years 1997–2002 from SeaWiFS data together with characteristic seasonal cycles of mixed-layer depths from Levitus climatological data and photic depths computed from characteristic irradiance and the archive of chlorophyll profiles discussed in Chapter 1.

# China Sea Coastal Province (CHIN)

*Extent of the Province* The CHIN coastal province extends from Sakhalin (45°N) down to Hainan (20°N) and so includes the continental shelves of the eastern part of the Sea of Japan, the Yellow Sea and its innermost gulf, the Bohai Sea, the East China Sea, and through the Taiwan Strait to the east coast of Hainan near 20°S. Because the Kuroshio passes along the edge of the continental shelf to the east of Taiwan, only the west coast of this island is included. The offshore limit of the province is taken to be the edge of the continental shelf, or else the edge of the eddy field of the Kuroshio Current if that is farther offshore; thus, where the Kuroshio makes brief incursions over the outer shelf, that process is not excluded. It should perhaps be noted here that for the northern part of the province, from eastern Korea to Sakhalin, I have been able to locate very little information: these waters lie off North Korea and the Russian Maritime Territories to the north of Vladivostok.

## Continental Shelf Topography and Tidal and Shelf-Edge Fronts

This is one of the largest areas of very shallow shelf anywhere; the Yellow Sea/Bohai Sea basin alone comprises 38,000 km$^2$ of an average depth of only 44 m. These shelves are strongly influenced by the discharge of freshwater and sediments from the great rivers draining the plains of China, principally the Yellow (Huangho) River opening into the Bohai Sea and the Yangtse (Changjiang) River entering at Shanghai into the northern East

China Sea. The resultant sediment regime in shoal water is a paradigm for domination by mobile, thixotropic mud banks: similar processes along western India or the Guiana coast of South America are weak in comparison. Sedimentary loads may be as much as 20 kg $m^{-3}$ at the mouths of the Yellow River at flood tide and, when transport slows down, this material sinks out and leads to the development of offshore mud banks that continually shift by tidal resuspension and transport (Wang, 1983). These mud banks form lunate features that lie normal to the coast and extend up to 75 km offshore; a feature termed the Great Sandbank formed in this way occupies much of the western part of the Yellow Sea. Further, recall that in 2009 the Three Gorges dam on the Yangtse should be completed, and that these works are expected to have major consequences for the ecology of the entire shelf region of eastern China: already, concerns are expressed for the potential loss of fishery resources.

Tidal fronts have been studied in three areas: in the Yellow Sea, on the western coast of Korea, and on the central part of the Chinese coast. Nearly vertical isotherms follow the familiar pattern in tidal mixing fronts that loop around groups of islands off SW Korea; the area of low stability within the loop is marked by cold water of high salinity (Lie, 1989).

## Defining Characteristics of Regional Oceanography

The East China and Yellow (Huanghai) Seas are large epicontinental shelf-depth seas, whereas the shelf of the eastern Sea of Japan is very narrow. The East China Sea is open to the Pacific Ocean except for the presence of Taiwan and the chain of Ryuku Islands, whereas the Yellow Sea is enclosed on three sides by the Asian continent. The Bohai Sea is a smaller, very shallow (< 20 m) basin at the head of the Yellow Sea. Nontidal circulation in these epicontinental seas is forced by the strongly seasonal regional wind system, and the invariant presence at the shelf edge of warm Kuroshio flow.

Southward from the Yellow Sea the wind regime is dominated by the seasonally reversing Asian monsoon: the Northeast Monsoon in winter and the Southwest in summer. Like New England and Atlantic Canada, in the northern part of this province there is an extreme contrast between winter and summer conditions. The cold, dry northerly winds strongly cool the surface water in winter; cooling is reversed with the advance of the southerly monsoon that begins in early May and by mid-July has reached the northern Yellow Sea so that the entire province lies under the influence of a southerly flow of tropical air (Barry and Chorley, 1982); this regime induces a seasonal temperature range of about 10°C between winter and summer, even at Hong Kong near the southern end of the province.

The thermocline is deepened in winter from 30–50 m to about 100 m at the shelf edge, but in the East China Sea intersects the bottom at only 20 km of the coast (Watts, 1972; Chan, 1970). Farther north, in the Yellow Sea, outbreaks of cold air from the Siberian High, accompanied by strong winds, force almost complete mixing in both Bohai and Yellow Seas. Thus, a very cold surface water mass develops there during winter, reaching 0°C in the inner Bohai Sea, where stratification is quite broken down. Subsequently, summer warming reaches 28°C in the central Yellow Sea but a cold bottom-water mass of 4–5°C lies below a remarkably steep thermocline (Wang and Zhu, 1991). The summer temperature of the cold subthermocline water mass is related to regional air temperatures of the previous winter.

Flow of modified coastal water from the western side of the Yellow Sea forms the southward China Coastal Current of winter that, under the influence of the Siberian northerly winds, passes through the Taiwan Strait (Shaw, 1992). This southward flow is reinforced by discharge of freshwater from the Yellow and Yangtse Rivers and is associated with a persistent cyclonic eddy, 100–200 km across, located south of Cheju Island; this eddy is well predicted by geostrophic calculations and has cool, high-salinity water at its center (Mao *et al.*, 1986). During summer, flow in the China Coastal Current is reversed from Hainan to Taiwan under the influence of the Southwest monsoon; at the Taiwan Straits this northward flow is reinforced by the northward flow of some Kuroshio water (see earlier discussion).

Beyond the eastern boundary of this province lies the warm flow of the Kuroshio, the velocity maximum of which is topographically locked to the continental slope (steep-to around Taiwan), which then continues northward through the Bashi Strait between Taiwan and the Ryuku chain and along the western slope of the Okinawa trough that here forms the edge of the continental shelf of the East China Sea. During its passage northward outside the shelf break, some Kuroshio surface water leaves the main axis and passes onto the continental shelf around Taiwan as the Taiwan Warm Current into the East China Sea, though mainly during winter (Fan, 1982), and into the Yellow Sea as the Yellow Sea Warm Current (Su *et al.*, 1990).

On the western side of the Sea of Japan, the flow of cold subarctic coastal water from the Sea of Okhotsk encounters Kuroshio water on the eastern side of the Korean peninsula, where a prominent thermal front is formed. Meanders of the Kuroshio propagate to the north along the edge of the shelf (Shibata, 1983) with a wavelength of about 300 km. A distinct shelf-edge front develops on the landward side of such meanders and of the cyclonic eddies that they shed (Chen *et al.*, 1992). Shelf-edge eddies appear in satellite images as warm Kuroshio water outlining cyclonic cores of cold shelf or slope water. Upwelling from 400–500 m occurs within the cold cores (Zheng *et al.*, 1992). Apart from the vorticity in shelf-edge meanders and cyclonic eddies, upwelling occurs along the Zhenjiang coast at about 26–30°N (Guan, 1984), on the northwest coast of Taiwan, and in the quasipermanent cyclonic eddy south of Cheju Island.

As the Kuroshio rounds Taiwan, it encounters the sharply curved and steep topography of the continental shelf of the East China Sea; a westward loop current across the shelf resembles the loop current of the Gulf of Mexico, but unlike that feature, it does not detach from the meandering jet (Hsueh *et al.*, 1992). The consequence of these processes is a regional upwelling in the vicinity of the Penghu Islands where a cold thermal anomaly is frequently observed at the shelf break; upwelling here is episodically enhanced by the onset of the northerly monsoon winds in boreal autumn. Nitrate-replete water reaches the surface from thermocline depths during these episodes (Fan, 1982).

During boreal summer, the wind stress of the Southwest Monsoon and the northward flow of the China Coastal Current causes upwelling at the coast in a water mass that is much modified by river water. Such water has unusual nutrient characteristics, being highly enriched in nitrogen, including nitrate, but relatively deficient in phosphate. Upwelled water has a phosphate concentration ($< 0.7\,\mu M$) that is an order of magnitude greater than in the overlying, river-modified surface water of the shelf so that after upwelling the resultant water mass is near the Redfield ratio (Chen *et al.*, 2001, 2004). A similar process may occur during winter mixing over the shelves of the East China Sea; in spring, the available phosphate in the euphotic zone may be rapidly depleted relative to nitrate, in excess because of the influence of river water in the shelf water mass.

This region is much subject to the passage of tropical cyclones, here called typhoons, during boreal summer (Shiah, 2000); when these pass over the normally oligotrophic conditions of the shelf of the East China Sea they induce deep mixing in the water column, together with resuspension of sediments and a brief but massive discharge of freshwater from flooded rivers. Nitrate and POM values at 40 m increase by about 175% and 75%, respectively, and a burst of productivity is thereby induced: autotrophic and bacterial production rates were each found to double after the passage of a typhoon in 1996.

## Regional Response of the Pelagic Ecosystem

This is one of the regions where we have to be very circumspect in our inferences concerning surface chlorophyll from satellite data. It was well understood that the CZCS algorithms were not able properly to distinguish CDOM from chlorophyll, although this instrument was intended originally for coastal analysis, and the user community progressively restricted its use to oceanic conditions once the limitations were understood. One of the first indications that the problem remained was the finding by Gong *et al.* (1998) that SeaWiFS data overestimated sea-truthed chlorophyll data on the Chinese shelf by a factor of 5. The recent demonstration that much of the signal from the Amazon plume derives from CDOM supports that finding (Hu *et al.*, 2004). Examination of the currently available images show a relatively invariant enhancement of "chlorophyll" the length of the coast of China from Hainan northward with a permanent penetration offshore that corresponds very well with the effluent from the Yangtse River. Only the eastern coast of the Sea of Japan lacks this coastwise signal. We assume that the climatology obtained from SeaWiFS data for the shelf of this province is dominated by nonliving Colored organic material (CDOM) from river water and await a resolution of this problem with impatience.

With the caution just expressed, examination of monthly and seasonal images shows that the Yellow and Bohai Seas and the coastal regions of the East China Sea are dominated by their sediment load almost year-round, whereas the Sea of Japan—having no major river effluents except the far-field effects of the Amur River late in the year—shows evidence of a spring bloom from February to May, after which oligotrophy appears to become established. A 50-km coastal band of high pigment, either CDOM or chlorophyll, occurs the length of the coast from the Yellow Sea to Hainan year-round, whereas off the Yangtse River a major high-pigment feature stretches far seaward toward the east before turning south along the coast.

Even if we cannot visualize them here as elsewhere, three principal processes must regulate primary production in this province: (i) the province-wide winter cooling and mixing of the water column alternating with summer stratification will generate a spring bloom whose timing will depend both on the seasonal irradiance cycle and on seasonal differences in water clarity; (ii) the coastal zone, especially in front of the major river mouths, will have a turbidity front, beyond which there will be a zone of enhanced primary production at the appropriate season; and (iii) there will be enhanced production in response to the various local eddying and upwelling processes.

From surface observations, a general boreal spring increase in chlorophyll of surface waters is noted in the East China Sea (Li and Fei, 1992), and enhancement of phytoplankton growth has also been observed in association with persistent eddies and upwelling. The river mouths, especially that of the Yangtse, and the cyclonic eddy south of Sheju Island were noted by Guo (1992) as supporting enhanced growth of neritic diatoms (*Coscinodiscus, Skeletonema, Melosira*, and *Chaetoceros*). The intrusions of Kuroshio water onto the shelf as shelf-edge eddies transport significant amounts of nitrate; one such eddy was computed to carry more nutrients than the combined annual river discharges (Chen *et al.*, 1992). Where the intrusion of these eddies over the shelf break results in locally strong vertical stratification, blooms and the development of strong subsurface chlorophyll maxima may occur at any season.

Fortunately, we now have access to the results of a systematic grid survey (four seasons, seven cross-shelf lines of stations) of the northern shelf of the East China Sea organized by the Ocean University of Taiwan (Gong *et al.*, 2003, and associated papers), and this is a useful model for the entire province south of the Yellow Sea. The northwest corner of the survey grid was well within the discharge plume from the Yangtse River, and this was reflected in nutrient enrichment there year-round; nevertheless, primary production in winter was light-constrained to $< 10\%$ of the peak rates of summer. Further seaward, away from the influence of the Yangtse plume, productivity was seasonally nutrient-limited but—somewhat curiously—seasonal changes in the production rate were rather small, so that the annual productivity of the two zones differed only by a small fraction: 155 g C $m^{-2}$ $y^{-1}$ in the plume and 144 g C $m^{-2}$ $y^{-1}$ on the open shelf. At coastal stations, *Skeletonema* dominated phytoplankton biomass, and at midshelf it was *Synechococcus* and *Pseudoselenia calcar-avis*, whereas at the outermost stations where biomass is very reduced, *Trichodesmium* dominated. This last is a very important contributor to productivity in the southern part of the East China Sea, where earlier investigations (Chang, 2000) showed N-fixation rates of up to 60 $\mu$M N $m^{-2}$ $d^{-1}$ from cell concentrations of $>10^7$ trichomes $m^{-2}$. In spring, highest concentrations occur near the shelf break, in autumn they are nearer the coastline. Here, the distribution of *Synechococcus* is strongly seasonal: in winter, these cells are evenly distributed on the shelf, whereas the highest concentrations in summer are at midshelf and in autumn at the shelf edge; distribution of heterotrophic ciliates closely matches that of cyanobacterial cells.

In coastal upwelling in the East China Sea (Chen *et al.*, 2004), nitrate-based new production formed about one-third of the daily production of $<4.5$ g C $m^{-2}$ $d^{-1}$, this being about four times the rate in adjacent stratified shelf water where nitrate was undetectable. This process is very dependent on the phosphate that is upwelled from deep shelf water into the relatively phosphate-deficient river plume water. In the southern part of the same region, Chen (2000) recorded a somewhat different relative distribution of picoplankton. The Kuroshio water at the shelf edge had the lowest productivity ($<0.61$ g C $m^{-2}$ $d^{-1}$) and the highest dominance of $<3$-$\mu$m picoplankton ($<85\%$ of total biomass and productivity). Although the larger fractions contributed more in coastal upwelling regions, picoplankton were still very important in the whole.

Uye (2000) asks why the large shelf-dominant *Calanus sinicus* should be restricted to the outer shelf waters in this region, and answers his own question in the following way: everywhere, the Kuroshio water mass that lies beyond the shelf edge is too warm for this temperate-zone species and suitable food particles are too dilute there. In the coastal regions, there is insufficient depth for adults to perform their habitual diel migration pattern and it is thought that developing eggs would sink to the bottom prior to hatching, there to be lost to the population. Additionally, as elsewhere, the inner shelf mesozooplankton is dominated by small competitors—*Paracalanus, Oithona, Acartia*, and so on. For such reasons, the habitat of *Calanus sinicus* is restricted to middle and outer shelf regions where, indeed, it is one of the dominant herbivores.

Mesozooplankton biomass increases by a factor of about 2 in summer, and *C. sinicus* and *E. pacifica* were the principal source of food for the large stock of Pacific herring that inhabited the Yellow Sea until it was fished out during the 1980s. A study of the long-term response of zooplankton to between-year environmental changes (Rebstock and Kang, 2003) on the three coasts of Korea found a common response in the abundance of the dominant zooplankton species to the climatic regime shift of 1989: temperature and zooplankton abundance increased subsequently in each of specific regions. Abundance of amphipods, chaetognaths, and euphausiids increased elsewhere, but not in the Yellow Sea. This study suggested that the ecosystem of the Yellow Sea differs strongly from the Japan and East China Seas, perhaps because of the partially closed nature of its circulation.

## Regional Benthic and Demersal Ecology

For higher trophic levels, it is difficult or impossible today to suggest what was the nature of the pristine ecosystem: the human population pressure is such here that the natural resource base is stretched very thin indeed. During the past three decades, fish stocks have become extremely stressed and the entire region subject to unrestrained pollution: effects of oil spills and heavy metal contamination are widespread in coastal areas. The species rank order, their diversity, and the abundance of demersal fish stocks changed significantly between surveys performed after 1967 even as the total landings progressively increased from 1970 ($127 \times 10^3$ t $y^{-1}$) to 1991 ($254 \times 10^3$ t $y^{-1}$). The 1967 surveys of the eastern Yellow Sea identified 134 species of 68 families of fish, but in 1980 only 51 species of 38 families figured in the listings (Zhang and Kim, 1999, and associated papers). Catch-per-unit-effort has progressively declined, and relative abundance shifted so that only 8 of the 20 most abundant species in 1980 had been among the top 20 in 1967. A now-classical effect of industrial exploitation of fish stocks is very extreme here: the population of chub mackerel, having a longevity of >10 years in the pristine state, are today represented by a stock that is highly dependent on 1- and 2-year old spawners. It may be confidently predicted that this stock will disappear after its next recruitment failure.

## Synopsis

*Case 3—Winter-spring production with nutrient limitation, much modified by the effects of seasonal river discharge.* Mixed-layer depth undergoes moderate winter deepening from 10 m in boreal summer to 40 m from December to February; since the photic depth lies at about 20–30 m in all seasons, the pycnocline is illuminated only in summer (April–September). Apparently significant chlorophyll accumulation in boreal winter (October–March) may be largely artifactual, due to river discharge of CDOM and winter resuspension of sediments, especially in the Yellow Sea (Fig. 11.21).

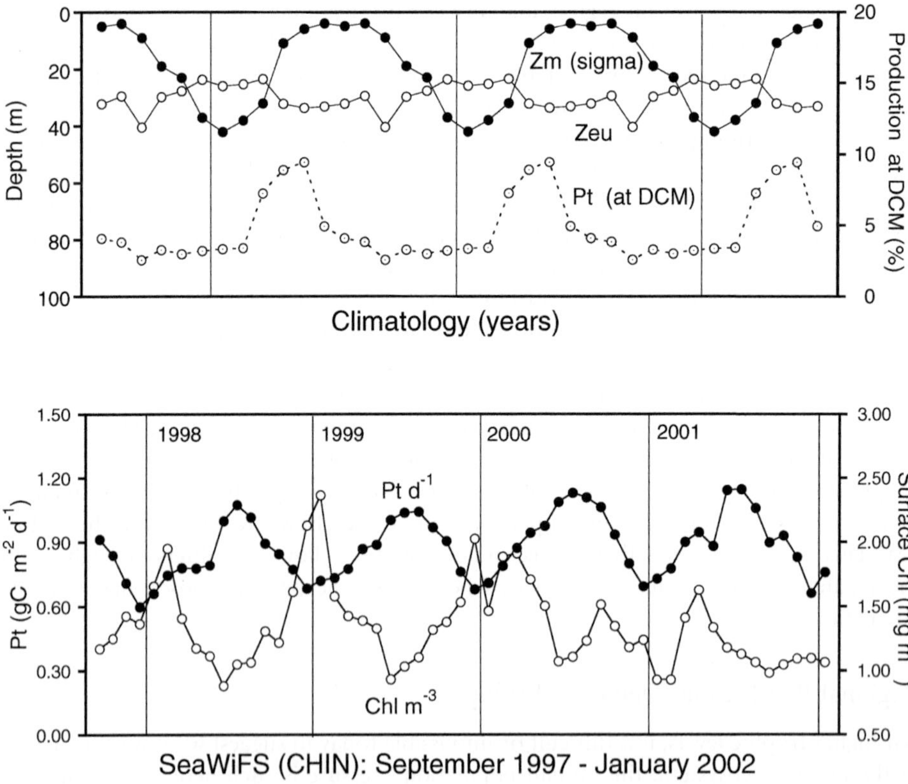


**Fig. 11.21** CHIN: seasonal cycles of monthly surface chlorophyll and depth-integrated autotrophic production for the years 1997–2002 from SeaWiFS data together with characteristic seasonal cycles of mixed-layer depths from Levitus climatological data and photic depths computed from characteristic irradiance and the archive of chlorophyll profiles discussed in Chapter 1.

# SUNDA-ARAFURA SHELVES PROVINCE (SUND)

## Extent of the Province

The SUND province comprises all the Asiatic continental shelves from Burma to Hainan and the archipelagic shelves south to the coast of Australia and New Guinea, at the Gulf of Papua. This is the shallow-water counterpart of the ARCH province that comprises the deep basins within the archipelagic region to the north of Australia; the logic of this partition has been discussed previously (see ARCH).

This province comprises both minor shelf areas and also large regions of continuous shallow water. The deep water of the Flores and Banda Seas separates the western from the eastern shelf areas, these seas being continuous with the deep basins lying farther north of the Celebes, Flores, and Sulu Seas. The many islands, large and small, and island arcs that inhabit these deep basins all have a coastal fringe of shallow water, and these shallow areas are included in this, the SUND province.

As will become very clear, this province could readily be subdivided into four or five logical subunits; unfortunately, we currently lack the information on which to base such a logical partitioning of this complex and beautiful region, which remains so poorly explored scientifically.

## Continental Shelf Topography and Tidal and Shelf-Edge Fronts

These are the largest continental shelves anywhere in the ocean, the scale of which is not often appreciated. The Arafura and Sunda shelves are together equivalent in area to about 10 Grand Banks of Newfoundland and the total extent of shelf area in this province (about $4.5 \times 10^6\,km^2$) is about 18–20% of the total shelf area of all oceans. There are three major blocks of shelf topography, each representing recently drowned coastlines (Ben-Avraham and Emery, 1973): (i) the shallow shelves fringing continental Southeast Asia (Gulf of Tonkin, Vietnam, Gulf of Thailand, eastern South China Sea, Malacca Straits, and the Andaman Sea) and (ii) the shelves of the western part of the archipelago (western South China Sea, Sunda Shelf, and the Java Sea), which are separated by deep water from (iii) the shelves of the eastern archipelago and Australia-New Guinea (Arafura Sea, Timor Sea, and the Gulf of Carpentaria). Obviously, this is a region ripe for further logical subdivision, for which these three sub-regions might be a useful start.

This shallow water province is unique in other ways than mere dimension: the unusually shallow shelf that nevertheless has great width, the complex tidal dynamics due to the interaction of the tides of two adjacent oceans, and the sediment input from rivers ($7.2 \times 10^9$ tons annually) all combine to make this a most unusual area. For instance, the central parts of the Sunda Shelf are less than 100 m, and the southern parts between Borneo and Sumatra are only 10–40 m deep; many submerged valleys of the rivers that previously drained the shelf area still exist, and some have been deepened by tidal flows. Some of the passes between the islands have opened only during the historical era and are still very shallow indeed.

Soft muddy deposits dominate the eastern complex of shelves more frequently than the western regions where strong tidal streams are the rule. The Torres Strait is such a case, because the Arafura and Coral Seas force different tides, such that there are fluctuations of up to 6 m in difference in sea level on each side of the strait between the two seas. Tidal streams (>5 knots) in and across the straits are controlled by sea-level slope, bottom friction, and acceleration (Wolanski *et al.*, 1988). Such tidal streams generate sufficient bottom friction in many places to overcome water-column stability and, in such places, we may expect to observe tidal fronts and associated biological enhancement. The Malacca Strait and the Arafura Sea/Gulf of Carpentaria region have been identified as potentially productive of such features; the area between Australia and the Lesser Sunda Islands ranks second among 50 shelf areas (for all seas) in dissipation of tidal energy per unit length of coastline (Hunter and Sharp, 1983; Forbes, 1984). Tidal mixing is especially strong around headlands in the Gulf of Carpentaria and also along the whole south coast of New Guinea. Fronts between tidally mixed and thermally stratified areas may develop only in austral summer, during the southerly monsoon, when wind stress is lowest in the eastern archipelago and when thermal stratification can occur over the shallow shelves.

## Defining Characteristics of Regional Oceanography

This province is wholly under the influence of the seasonally reversing monsoon winds and of the circulation thereby generated, which has been sufficiently noted already (see ARCH, MONS) and will not be reviewed again here in detail. It suffices to say that current flow between the Indian and Pacific Oceans generally reverses with the monsoon winds whose exact direction differs across the province: in the South China Sea it is seasonally NE-SW but in Indonesia closer to NW-SE. So one must be aware that both the relative meridional and zonal wind stress differs progressively across the province, because of the curvature of pressure contours around the main atmospheric systems. It must also be emphasized that in thinking about the oceanography of this province, you should also remember that it spans the equator and should expect the consequences of seasonality in the two hemispheres to be observable.

The source document for this region remains the splendid regional synthesis of Wyrtki (1961), from which I have drawn much of the following account, but I am indebted also to the recent reviews of Arief (1998) and of Church and Craig (1998) for additional information.

Over much of the province an Ekman layer of 30–50 m exists at all seasons, and only in the northern part of the South China Sea (north of Hainan) and in the seas north of Australia does this deepen in boreal winter to 70–90 m; in the former case, this is due to wind mixing (see China Sea Coastal Province), and in the latter caused by the seasonal alternation between upwelling and relaxation in the Banda and Flores Seas (see ARCH). A simple comparison between potential mixed-layer depths and the depth of water over the shelves shows that much of the southern Sunda Shelf between Borneo and Sumatra must be mixed to the bottom, and here we can expect active development of shelf sea fronts. The temperature of the mixed layer thus varies more at the northern and southern ends of the province, whereas salinity is more variable seasonally in the central regions, though actual seasonality is not uniform, following rainfall and river effluent patterns.

In the Gulf of Carpentaria, winter mixing is complete and in the austral summer season only the central gulf is stratified because tidal currents are sufficiently strong to mix the bottom 30 m of the water column (Church and Craig, 1998). There are significant consequences of this mixing for the benthic fauna (see later discussion). Circulation in the gulf changes with the monsoons, clockwise in the NE and counterclockwise in the SE trades, although the coastal jets do not readily mix with the higher-salinity central gulf water.

The mixed layer is almost universally nutrient-depleted, except in river plumes, and lies above a nutricline that is coincident with the thermocline; Wyrtki commented that the input of nutrients from rivers is, or was before serious clear-cut logging, remarkably low and that flux of nutrients into the mixed layer seems to occur mostly in tidally mixed areas or where seasonal upwelling of subthermocline water occurs. In such places, concern is currently expressed at the amount of nitrogen and phosphorus (and also, in some places, the biological oxygen demand of the discharge) entering some of the embayments. The Gulf of Thailand is the best example of this, or at least it is the most investigated; here, there are several episodic blooms during the year without a very clear seasonality, and the frequency of these is said to have significantly increased in recent decades and to be associated with the discharge of nitrate from the major rivers entering the gulf.

As already noted (see ARCH) some coastal upwelling occurs here, as we would expect from the complex and fractal geography of the coastlines; there must be many places where seasonal flow, seasonal wind stress, and coastal alignment combine to force intermittent or seasonal upwelling. Even with the precautions already discussed concerning the aliasing of the SeaWiFS data by CDOM, these images still provide the most efficient way of identifying these manifold locations. Some of these will be discussed later: the most obvious candidates for upwelling sites are the westward-facing gulf on the south coast of Papua–New Guinea, the edge of the Sunda shelf, and also along the east coast of Vietnam.

This region undergoes major change during ENSO events like that of 1982–83 (Chao *et al.*, 1996) that were initiated by increased evaporative cooling in late 1982, followed by weak NE and SW monsoons in succession. Increased heat content of surface water during this period is better explained by weakened circulation than by changes in latent heat flux. This resulted in seriously reduced upwelling on the Vietnam coast. Subsequently, the record NE monsoon that followed produced major downwelling at the coast around the marginal seas. This sequence was repeated during the 1986–87 event and are probably typical.

## Regional Response of the Pelagic Ecosystem

Although I know of no comprehensive surveys of regional primary productivity, it is evident that it is relatively high in the marginal gulfs compared with the open shelf, and this was already evident from the very first voyages deploying the $^{14}C$ production method in the 1950s. Platt and Subbha Rao (1975) list average figures for the Gulf of Thailand and the Vietnam coast: recall that the latter is an upwelling regime during the southerly monsoon (Liu *et al.*, 2002b) of 1.2–1.3 g C $m^{-2}day^{-1}$ compared to 0.3–0.6 g C $m^{-2}$ $day^{-1}$ for the remainder of the SUND province, though it is not easy to separate data from the deep basins and the continental shelves in their presentation. A glance at any SeaWiFS image of the entire region will convince you that, overall and at all seasons, chlorophyll values are higher over the shelf regions than over the intervening deep basins. A quick review of serial 7- or 30-day images also shows that the seasonal sequence differs in the eastern and western shelf regions: the western region responds to the Northeast Monsoon of boreal winter with a seasonal chlorophyll maximum, while the eastern region is greener during austral winter under the influence of the Southeast Monsoon. The coastal upwelling off Vietnam responds to the alongshore Southwest Monsoon winds of boreal summer; that in the coastal Aru Gulf is most active also at the same season, forced by the Southwest Monsoon winds of the southern hemisphere. We may assume that this is a general response to seasonal deepening of the mixed layer and consequent nutrient entrainment. The surface chlorophyll biomass in the SUND province is essentially invariant seasonally in the archived climatology.

The images also confirm what could be predicted from regional geography and circulation: that a coastal zone of high pigment concentration water occurs commonly: off Papua–New Guinea and in the adjacent Gulf of Carpentaria,; along the south coast of Borneo; in the Gulf of Tonkin and on the east coast of Vietnam; in the Straits of Malacca; and finally off the Mekong delta. To what extent this signal can be referred to biological activity, or is due to CDOM, is uncertain. However, one aspect of the ecology of this province can be observed, though with some technical reservations at these low latitudes, with satellite images: this is the occurrence in some tidally mixed shelf areas of coccolithophore blooms, presumably of *Emiliania huxleyi* (Brown, 1995b). The MODIS calcite images for 2001 suggest that such blooms are widespread in coastal regions to the east of the deep basins during the SE monsoon, and on the Sunda Shelf and Java Sea during the NW monsoon.

The locations of coastal upwelling features, forced by interaction between seabed and terrestrial topography and the local wind strength and direction, have been generalized in the discussion of the deep basins of the archipelago (see Archipelagic Deep Basins

Province, ARCH); this is a matter of convenience since these features may be sufficiently large as to occur over both a shelf and an adjacent deep basin. A prominent feature of this kind is the front aligned with the long shelf edge lying zonally across the South China Sea; flow approximately normal to this generates a meandering front with associated uplift of the nutricline by as much as 90 m (Lim, 1975).

Two of the most prominent and permanent areas of high pigment are on the southern coast of New Guinea: in the west-facing Aru Gulf where the Digoel River discharges and the eastward-facing Gulf of Papua, with the Fly River. Of these, the former is by far the more prominent, especially in austral winter, when the southeasterly monsoon winds sweep across this gulf. The high-pigment plume from the Aru Gulf often turns southwest and in some images appears to be traceable across the Gulf of Carpentaria.

What we know of these two features appears to be rather contradictory, and does not resolve the obvious question: are we looking at pigmented CDOM from the river effluents, or are we looking at a chlorophyll plume resulting from upwelling or riverine nutrient input? The outer Aru Gulf was the site of the Snellius II investigations in 1985, whereas the Gulf of Papua was investigated more recently by the TROPICS investigations that were intended to provide a model for "Tropical River-Ocean Processes in Coastal Settings."

The Snellius II investigations showed that in August, during the SE monsoon, upwelling occurs extensively over the continental shelf of the western Arafura Sea (Wetsteyn *et al.*, 1990). The mechanism appears to be that water from 100 to 150 m in the western part of the Aru Gulf spreads up over the shelf reaching several hundred kilometers distant the shelf edge; this phenomenon carries cool, saline, high-nutrient slope water (>10 μM $NO_3$) far up over the shelf both north and south of the Aru Islands. This transport is supported by a slow offshore Ekman drift of coastal, low-salinity water presumably, at least in part, forced by the easterly component of the monsoon winds. These processes are insufficient to upwell the slope water to the surface, the buoyancy imparted by river water maintaining stratification, but increased levels of primary production during boreal summer (Ilahude *et al.*, 1990) and higher cell counts of large phytoplankters (Adnan, 1990) suggest that upwelled nutrients, mixed into the buoyant layer, largely support the bloom: riverborne nutrient inputs, though utilized, are less important. Investigations during the NW monsoon in February showed, to the contrary, that productivity is then dependent on riverborne nutrients (Ilahude *et al.*, 1990). As we would expect, neritic diatoms dominate the macro-phytoplankton in the August blooms and are more abundant then by a factor of 3 than in February when nanoplankton account for 70–90% of chlorophyll except in *Trichodesmium* blooms.

The dynamics of the Gulf of Papua, facing east, appear to be different from this model (Robertson *et al.*, 1998). Here, upwelling is not advanced as a factor in nutrient supply, but rather this comes from the 30-m-deep, buoyant plume of warm river water that overlies cooler, saline water. The plume has a high sediment load and carries much floating plant debris with it. Nutrient dynamics are complex and whereas silicate is conservative, nitrate and phosphate are taken up locally by autotrophs and also released from particulates into the dissolved phase. Locally, phosphate dynamics corresponded with changes in chlorophyll concentration. Production rates were low near the coast ($<20$ mg C $m^{-2}$ $d^{-1}$) but increased to 225–350 mg C $m^{-2}$ $d^{-1}$ in the plume farther offshore. Bacterial production was highly variable but significantly correlated with autotrophic production, as also was zooplankton biomass, at rates very close to those for autotrophic production. All this suggested to Robertson *et al.* that bacteria in the Gulf of Papua use riverine DOC and POC and that Gulf waters are probably net heterotrophic with the consequence that benthic production in the Gulf of Papua is dependent largely on organic material derived from riverine sources rather than from local autotrophic production.

SeaWiFS images clearly confirm the spatial context of these observations in both gulfs, but probably overestimate the contribution of chlorophyll everywhere in this region. In the Gulf of Papua, observations by Robertson *et al.* indicated inshore chlorophyll concentrations of 0.25–5.07 $\mu$g chl liter$^{-1}$, whereas offshore, they found 0.32–0.73 mg chl m$^{-3}$. Values suggested by SeaWiFS sensors are not—subjectively—greatly different from these in either of the two gulfs, but Davies (2004) confirms that these data, as we would expect, overestimate chlorophyll because of interference from CDOM and suspended particulates.

Investigation of *Synechococcus* ecology around islands in the Philippines shows that populations of these cyanobacteria in shelf regions are very sensitive to disturbance of the coastal ecosystem whether by anthropogenic increases in local siltation or by the episodic passage of typhoons (Agawin *et al.*, 2003). High biomass is induced near nutrient sources in river effluents, but significant declines in growth rate and biomass occur in response to silt content of effluent river water or in coastal water after passage of a typhoon. Changes due to such factors were found to be greater than the overall seasonal abundance cycle. Where human population pressure is heavy, in the western part of the province around the Gulf of Thailand, there is already much evidence of regional eutrophication of coastal waters with increased frequency of phytoplankton blooms and knock-on consequences for the benthic ecosystem of the gulf that will be touched on in the next section.

## Regional Benthic and Demersal Ecology

Perhaps it is worth noting in passing that the biogeography of SUND does not respond to the presence of Wallace's Line, which passes across the province between the Lombok and Makassar Straits; marine organisms do not obey the same rules as terrestrial species for whom this line divides an Asiatic from an Australian fauna.

Because this is a region of very heavy monsoon rainfall, major discharge of freshwater to the shelf regions is to be expected everywhere; because it is also a region of intense recent expansion of the human population and deforestation, it is also to be expected that heavy deposition of organic-rich sediments will occur on the shelves. From the coastline of this province a total of around $3 \times 10^9$ tons of silt annually is discharged to the shelf: this amount is twice that of the discharges from the Amazon and Orinoco combined.

Where the topography of the shelf is such as to form isolated basins, these accumulate very soft muddy deposits, even if relatively far from any estuary. Such is the case in the fluid-mud bottoms of the central Java Sea that support only a very low biomass of macrofaunal benthic organisms ($< 1$ g m$^{-2}$ ash-free dry weight), mostly of small polychaetes. These fluid-mud bottoms are the habitat of very large numbers of meiofaunal nematodes that may reach half-a-million individuals per square meter. Annual productivity of such assemblages equals about one half of the autotrophic production in the water column above (De Wilde *et al.*, 1989). In the Gulf of Papua, the silt carried by the Fly River (see earlier discussion) is deposited as either laminated or bioturbated muds that are the trawling grounds for a major fishery for penaeid shrimps; here, oxic and suboxic diagenesis resemble those described for the Amazon shelf, with a very rapid release of nutrients that satisfies 71% and 35% of the daily N and P requirements of the superjacent phytoplankton.

Where the accumulation of silt does not occur, as over much of the Gulf of Carpentaria, sandy and shelly deposits with many coral heads, sand waves and an abundant epifauna of soft corals and cup sponges. Although there is abundant information on the demersal fish fauna, I have not located any satisfactory accounts of the invertebrate benthos of SUND: nevertheless, the surveys reported by Chatananthawej *et al.* (1987) in the Gulf of Thailand suggest nothing unusual in the relative composition of the major invertebrate groups in relation to bottom type.

The fishery potential of the very large shelf areas of this province was sufficiently self-evident that scientific trawling surveys were initiated during the colonial period, and the earliest of which I have a record was performed by the Singapore fisheries department in 1926. The well-documented survey by Ommaney (1961) of the Gulf of Thailand gives us an invaluable record of the nearly pristine structure of the demersal fish populations. More recent surveys record a dismal decline of the once-dominant species and of overall fish biomass that has attracted much attention internationally, so much so that the Gulf of Thailand served as the exemplar for Pauly's paradigm of "fishing down the food web" (e.g., Pauly and Chuenpagdee, 2003). However, the problems of maintaining sustainable fisheries are not what we are about here, and more to the point is that a massive alteration of the entire shelf ecosystem has occurred. Attempting to model how this ecosystem might function (e.g., Pauly and Christensen, 1993) has, as elsewhere, become like trying to hit a moving target: to understand its pristine state is now very difficult and perhaps impossible.

It will be sufficient here to draw attention to some features of the pristine state of the benthic ecosystem as far as it can now be discerned; the general arrangement of ecotypes apparently followed the same pattern as on other tropical shelves. The near-permanence of thermal stratification in the tropics at least doubles the number of characteristics ecotypes, by permitting the development of warm-water and cool-water faunas of both invertebrates and fish. Elsewhere, the cool-water fauna has many elements that are enabled to penetrate from higher latitudes into the tropical zone, and the same is undoubtedly true here. The overall pattern is very like that described for the eastern Atlantic (see GUIN) except that here the depth at which the thermocline intercepts the shelf is much deeper. The pristine fauna was dominated by the same families (Ariidae, Drepanidae, Polynemidae, Sciaenidae) as occur everywhere on tropical shelves that are not invested with coral reefs, together with some Indo-West Pacific endemics (Leiognathidae, Lactaridae). Between 1963 and 1982, this fauna in the Gulf of Thailand was reduced to 17% of its original biomass, so that squid (*Loligo* spp.) have replaced leiognathids as the dominant group in terms of biomass. A more recent review suggests that demersal biomass of fish in the Gulf of Thailand has progressively declined from around a standing stock of around 700,000 tons in 1960 to less than 50,000 in 1995; similar declines are recorded on the NW Borneo shelf and on the Phillipine coasts (Stobutzki, 2006).

The demersal fish fauna of the Timor Sea, the Gulf of Carpentaria, and the Arafura Sea were surveyed (266 standard trawl hauls) by Okera (1982), who used recurrent group analysis to specify the characteristic fauna of six habitats: neritic mud (15–20 m), inner shelf muddy sand (10–50 m), offshore sands (80–90 m), midshelf mixed deposits (60–110 m), shelf break mixed deposits (120–220 m), hard bottoms of boulders at all depths. With each of these habitats is associated a characteristic group of 10–20 species, such that the genera associated with each resemble genera on similar deposits elsewhere in the tropics. A very similar result was obtained in the Ragay Gulf, in the Philippines, by Jamir (1999), who applied statistical procedures, including cluster analysis, to describe three depth zones (warm $<90$ m, cool 90–150 m, and cold $>150$ m) depths of mixed layer, thermocline, and subthermocline water. There is some rearrangement of this distribution during major ENSO events, as in 1982–83. Within each zone, the important determinants of fish population structure is bottom type, benthic food biomass, and effects of river effluents. A very similar arrangement has been described for the shelf of the South China Sea, partitioned at 40, 100, and 200 m; here, the shallow partition is dominated by lizardfish (*Saurida*) and the middle zone by snappers and bream (*Pagrus, Dentex, Lutjanus, et al.*) and ponyfish (Leignathidae).

## Synopsis

*Case 4—Small-amplitude response to trade-wind seasonality*—The SUND province has a very shoal pycnocline because of the extent of the surface brackish layer of river water. The mixed layer is 15–25 m in all months, the photic depth being deeper at about 40 m;

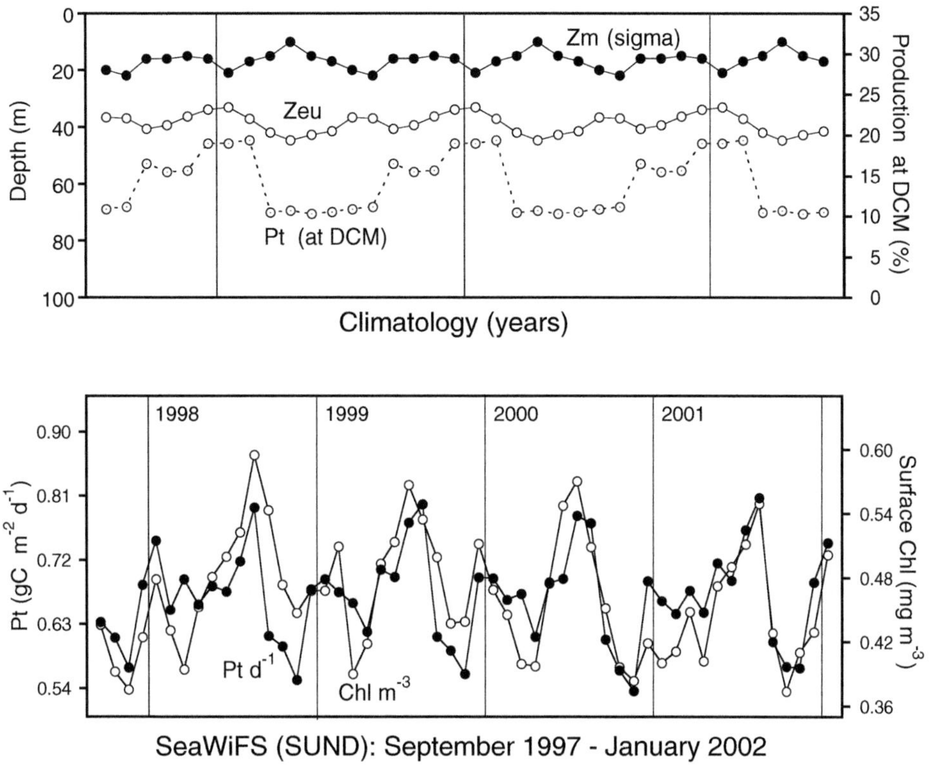


**Fig. 11.22** SUND: seasonal cycles of monthly surface chlorophyll and depth-integrated autotrophic production for the years 1997–2002 from SeaWiFS data together with characteristic seasonal cycles of mixed-layer depths from Levitus climatological data and photic depths computed from characteristic irradiance and the archive of chlorophyll profiles discussed in Chapter 1.

thus, the pycnocline is illuminated in all months except during the SW Monsoon period, July–August. The two productivity maxima in a seasonal cycle of weak dynamic range correspond with the seasons of greatest monsoon wind stress (0.5 mg C $m^{-2}$ $d^{-1}$ in July and 0.6 mg C $m^{-2}$ $d^{-1}$ in December–January), and the consequent effects of enhanced productivity and chlorophyll accumulation cannot be separated from resuspension of sediments (Fig. 11.22).

## East Australian Coastal Province (AUSE)

### Extent of the Province

For convenience, two ecologically different coasts are aggregated into the East Australian Coastal Province (AUSE) province, which thus extends the entire length of the eastern coast of Australia. For many purposes, and to follow strictly the logic of the Trades and Westerlies biomes discussed earlier, this province should be partitioned. One logical partition would be from the Papuan Barrier Reef down the Queensland coast to Sandy Cape (25°S) where the wide northern shelf narrows and changes orientation; a second partition would be from there to near Sydney (c. 35°S), where the East Australian Current is retroflected eastward across the Tasman Sea. The most southerly partition would extend down along the Tasman Sea coast to SE Cape at 45°S, thus including the eastern coast of Tasmania. It might, for some purposes, also be logical to include the reef-bound coast of New Caledonia within the northern partition.

## Continental Shelf Topography and Tidal and Shelf-Edge Fronts

The continental shelf north of Sandy Cape is almost wholly occupied by the Great Barrier Reef off Queensland that is both wide at its southern end (250 km at 22°S) and long, extending for about 2000 km from Torres Strait to Sandy Cape. It is broad and shallow, with no major canyons; the coral facies is not uniformly distributed upon it, as shall be discussed later, and a coastal lagoon extends almost the entire length of the Queensland shelf. The ribbon reef on the outer shelf is densest on the wider, southern part, whereas in the northern, narrower sections the fringing barrier is less complete and sandbanks and isolated reefs populate the inner shelf.

The southern coast of New Caledonia carries a similar fringing reef, and although it is much narrower (10–25 km), it does extend northwest of the island as far as the Entrecasteaux reefs, so the total length of the habitat is almost 600 km.

South of Sandy Cape, along the New South Wales and Tasmanian coasts, the shelf is rarely more than 10–20 km wide and is interrupted by the shallow Bass Strait between Tasmania and the mainland.

## Defining Characteristics of Regional Oceanography

The coastal circulation in this province is the western boundary current of the SW Pacific that has been reviewed recently by Church and Craig (1998). North of 18°S, where westward flow of the South Equatorial Current across the Coral Sea diverges north and south on encountering the continent, the flow at the shelf edge is highly variable and unpredictable, though generally equatorward. This northward flow rounds the Gulf of Papua and turns eastward along the Papuan Barrier Reef toward the coast of New Caledonia. The southward limb, as the East Australian Current, forms a typical western boundary current with predictable poleward flow from the central Great Barrier Reef south to near Sydney near 30°S, where retroflection occurs into an eastward jet along the Tasman Front (see Tasman Sea Province) that continues on into the western boundary current down the east coast of New Zealand.

The relatively slow flow of the East Australia Current along the Queensland coast lies above the continental slope, beyond the barrier reef, and induces similar flow along the outer part of the wide shelf. On the inner shelf, the Southeast Monsoon of austral winter forces a near-coastal equatorward flow, largely occupying and mixing the shelf area behind the ribbon reef. In austral summer, the rate of southward flow over the Queensland shelf is enhanced by the Northwest Monsoon winds and, at the same season, the flow of the offshore East Australian Current is strongest.

The mixed layer of the Coral Sea is deeper than typical shelf depths off Queensland, so that here the entire feature is typically covered in tropical surface water. Buoyancy is sporadically induced by freshwater discharge after austral winter monsoon rains on the inner shelf, while subsurface intrusions of cool, nutrient-rich slope water onto the outer shelf also occur. The mechanism of these intrusions is that in the slope region, alongshore winds excite barotropic motion that displaces the thermocline by as much as 50 m, and so induces Coral Sea water to flood up onto the shelf. This process induces the development of strong jet currents through the narrow gaps between individual reefs with which are associated eddy systems extending over the inner continental shelf behind the reef. Little flow occurs over the reefs themselves (Young *et al.*, 1994), but the eddies provide a mechanism to enhance the exchange of deep water with the inner shelf water of the lagoon.

In the central Barrier Reef region (17–20°S) this process occurs episodically several times during each austral summer while occasional large upwelling events may displace up to one third of the total outer shelf water mass (Furnas and Mitchell, 1996a); these intrusions penetrate the whole width of the reef facies and form the major source of nutrients to the central Barrier Reef, being of order 75% and 30% of gross N and P imports respectively. Similar processes have been described for the Papuan Barrier Reef, off which both mixed layer and nutricline are shallower (25–75 m). Here, upwelling at the reef edge occurs during both NE and SE winds (parallel and normal to the reef axis, respectively), although surface outcropping of upwelled water seems not to occur (Furnas and Mitchell, 1996a).

South of Sandy Cape, the East Australia Current is no longer a slow southerly stream, but its velocity is markedly increased, reaching a maximum flow rate at 25–30°S and in austral summer. This lies over the outer shelf and velocity is progressively reduced toward the coast; on the outer shelf, velocity fluctuates strongly at scales of 100 d and 100 km, inducing topographic Rossby waves that drift south at about 9 km $d^{-1}$. Where the East Australian Current spreads onto the shelf, it drives an Ekman boundary layer of nutrient-rich water shoreward that may be upwelled and advected to the south. This may induce a plume of nutrient-rich water over several hundreds of kilometers (Cresswell, 1994).

The offshore eddy field of the East Australian Current is relatively weak during austral winter (May–July) but at other seasons its intensive meandering induces enrichment of the coastal water mass. When a meander or eddy approaches the shelf break, nutrient-rich slope water is driven onto the shelf by the peripheral anticyclonic current associated with the shoreward side of the feature. Such flow creates an Ekman boundary layer that would flow northward, displacing the surface coastal waters. This process results in intrusions of slope water to the 60-m isobath.

Cyclonic frontal eddies spin off from the landward side of the velocity maximum of the coastal current and become entrained in the coastal water mass (Tranter *et al.*, 1987); similar elongated cyclonic features are termed "shingles" on the landward side of the Florida Current by Atkinson *et al.* (1984) and are an important source of nutrients over the shelf.

The major wind events on this coast ("Southerly Busters") generate equatorward stress and downwelling, but are of short duration. On the other hand, strong current events are of longer duration. Therefore, bottom Ekman-layer forcing produces upwelling events of long duration, whereas wind stress produces downwelling of shorter duration (Griffiths and Middleton, 1992). On the narrow shelf of eastern Australia there are several sites where coastal upwelling cells are persistent, notably at Evans Head (29°S), Laurieton (31°S), and Sydney (34°S). In part, the location of these cells is due to topographic influence, such as the narrowing of the shelf just north of Laurieton (Rochford, 1975) or where the flow of the East Australia Current separates from the coast (Tranter *et al.*, 1987). Nutrient-rich water may be uplifted at about 45-d intervals from 125 to 275 m (at 15–17°C) either to a bottom Ekman layer over the shelf, as is usually the case off Sydney, or to the surface, as occurs farther north. The primary source of nutrients on the shelf from 29°S to 38°S appears to be intrusions of slope water of this type onto shelf during the spring and summer that is subsequently mixed up into the surface layers by wind stress in autumn and winter.

South of the retroflection of the main transport of the East Australian Current, conditions are much influenced by the passage of water through the Bass Straits, which is then entrained northward toward the well-known "Bass Strait Cascade" off Victoria, where cold, dense water from the straits slips below lighter Tasman Sea water.

I have already mentioned (see TASM) the long time series of observations made in South Bay, on the Tasmanian coast at midshelf (50 m), that have relevance also for the processes in the oceanic regions (Harris *et al.*, 1991). Here, during a 4-year period (1985–1989) a 2.5°C warming of the regional SST was observed; this is apparently a not-unusual occurrence that is associated with the balance between intrusive subtropical and subantarctic influences. During this warming period, intrusions of subtropical water were detected by their high salinity and low nutrient content. The periods of increased subtropical intrusions are also periods of lower westerly wind stress and shallower penetration of winter mixing offshore.

## Regional Response of the Pelagic Ecosystem

The climatological sea surface chlorophyll archive reveals a strong overall seasonality. The 7-day and 30-day images of sea surface chlorophyll now available strikingly reveal that two processes dominate the response of phytoplankton to physical forcing in this province. First, during the late austral winter (April–June) the seasonal chlorophyll enhancement of the Tasman Sea reaches its most northerly extent, involving the southwestern part of the Coral Sea, beyond the Tasman Front, so that the Queensland shelf inshore of the Great Barrier Reef shows some seasonal chlorophyll enhancement, although this remains less strong than the chlorophyll biomass indicated for the reef itself. This phenomenon is attributed by Gabric *et al.* (1990) to the effects of higher nutrient loading from the coast, and wind mixing on the inner shelf, during the wet monsoon season.

It has long been understood that the winter-spring bloom of the Tasman Sea, stronger toward the south and reaching highest surface biomass in the period September–November, influences the western coast that is otherwise dominated by two processes that cause biological enhancement: coastal upwelling cells and vorticity in the offshore eddy field (Tranter *et al.*, 1987; Rahmstorf, 1992). But, once again, the enhancement of chlorophyll in features in the eddies is not in any way comparable with, for example, the offshore eddies in the North Pacific; they may be distinguished from background in the chlorophyll fields, but they are not prominent.

It is also clear from a general inspection of such images that the coastal upwelling locations, discussed earlier, are not prominent features in the chlorophyll field. The bight south of Ballina, including Evans Head (see earlier discussion), is perhaps the most persistent, but I have seen no images in which it forms more than a very narrow coastal strip in no way resembling chlorophyll plumes on classical upwelling coasts. But beyond that, it is not only at the localities listed earlier that coastal enhancement of chlorophyll is observed: satellite data for September 1999 suggest that a larger feature—though still minor in the global sense—had developed in the bight north of Moreton Bay (27°S) and also, although more diffuse, on the wider shelf south of Sandy Cape.

The various processes, discussed previously, that transport nutrients to the narrow shelf of New South Wales result in regional enhancement, as discussed by Cresswell (*op. cit.*), who found a strong DCM associated with a nutrient-rich coastal plume at Sydney. The more general annual spring blooms that occur on this coast annually, and are recorded in the SeaWiFS images, are due to short-lived diatom blooms that increase regional column chlorophyll by a factor of 10 and evolve in a predictable sequence from small chain-forming species to large centrics and finally to dinoflagellates (Hallegraef and Jeffrey, 1993). This regional-scale bloom is not forced by the same mechanism as classical spring blooms, but by the dynamic processes in the East Australian Current that lead to nutrient enrichment and were discussed earlier.

Much attention has been given to the response of the pelagic ecosystem to the warm-core eddies, and their associated cyclonic crescents, that originate in the meanders of the East Australian Current (e.g., Tranter *et al.*, 1983). Elevated nitrate levels (1.5 $\mu$g liter$^{-1}$ compared with 0.4 $\mu$g liter$^{-1}$) and chlorophyll enhancement over background occur in the cool water crescent on the margins of warm-core eddies; frontal curvature is, of course, usually supposed to generate shear and to destabilize the interface between the eddy and the surrounding water (Woods *et al.*, 1977) and is often associated with biological enhancement, up to higher trophic levels, such as tuna. The cool crescent studied by Tranter *et al.* off Sydney supported very large concentrations of *Calanoides carinatus* of 300–400 mg m$^{-3}$, an abundance similar to that occurring in coastal upwelling along the New South Wales coastline.

The most prominent feature of enhanced chlorophyll, by far, on this coast is that associated with the shelf break to the east of Bass Strait and along the Victoria coast as far as the protruding feature of Cape Howe (38°S): here, the shelf-break front forms the so-called Bass Strait Cascade where cold, dense straits water slips below lighter Tasman Sea water, producing a linear zone over the upper slope of enhanced chlorophyll that extends to the coast of Victoria where the bloom may continue later than to the east of the Strait itself. It is suggested by Gibbs *et al.* (1991) that phytoplankton growth continues actively in the plume in the northward flow along the shelf break after it has ceased at the entrance to Bass Strait. Eventually, the constituent cells will be carried to depth when the plume enters the cascade area itself off the coast of Victoria.

The 4-year study in South Bay on the Tasmanian shelf, referred to previously, demonstrated the ecosystem consequences of the proximity of the strongly meandering South Subtropical Convergence Zone: pelagic ecosystem structure, the timing of the spring bloom, and the recruitment success of fish stocks may all be correlated with incursions of water from the Convergence Zone at the coast (Harris *et al.*, 1991). The pelagic ecosystem responded strongly to conditions obtaining during a relatively warm, calm year after several cooler windy years, associated with changes in mixed-layer depth and hence in nutrient supply at the end of winter: cooler SST, deeper MLD, and more 10-m nutrients in windy years. The phytoplankton particle-size spectrum significantly changes between such periods, being less steep in windy years and dominated by large cells. Small cells were more dominant in years of greater stability. Hence, the flux of organic matter to the shelf sediments is much greater in windy, more productive years. However, nitrate is drawn down to limiting levels for several months every summer, but phosphate only briefly.

Net-caught plankton volumes were greater by an order of magnitude in the windy year 1986–87 than in other years, a result of larger fractions both of diatoms and of salps. The dominance of larger organisms in net-taken samples changed strikingly between years: *Nyctiphanes* dominated in 1985–86, salps in 1986–87, and salps and *Nyctiphanes* in 1987–88. In 1988–89, only small volumes of small copepods were taken. These last are the "background" plankton, on which are imposed *Nyctiphanes* and salps in years when production is dominated by external sources of nutrients, rather than on regeneration.

## Regional Benthic and Demersal Ecology

If only because the coral reefs along the Queensland coast are the largest example of this habitat anywhere, it will be appropriate to note some of the critical characteristics of this benthic system, so far ignored in this work. In what follows I lean heavily on the classical review of John Lewis (1981) and on Alongi (1998).

The first questions that come to mind when examining this region concern the relationship between the regional ocean circulation discussed earlier and the extraordinary accumulation of benthic biomass represented by the Great Barrier Reef. To what extent are the fluxes of slope water through the channels in the reef onto the inner shelf, carrying inorganic nutrients, particulate organic matter, and oceanic plankton, involved in maintaining the great biomass of the reef? The observation that the reef biota are most strongly developed where the shelf edge is steepest led Orr (1933) to suggest that the richness of the Great Barrier Reef might be a response to upwelling in the western boundary current.

So it may be, but in that case the mechanism is very complex and obscure: as Lewis emphasizes, there is generally little reduction in the nitrate content of water that flows across reefs, whereas near-bottom phosphate concentration in water over a reef appear to be higher than background. And although oceanic zooplankton are consumed during their passage across a reef, it is calculated that this can provide only a small part of what is required for growth of the reef ensemble.

Rather than looking for external sources of particulate and dissolved nitrogen, we should perhaps remember that fixation rates of $<50–250\,\mathrm{g\ N\ m^{-2}\ y^{-1}}$ are typical for the gamut of nitrogen-fixing organisms associated with coral and, according to some budgets, such rates are sufficient to satisfy the nitrogen needs of the reef community. Regeneration is active and rapid so that with uptake of P during the daytime and release at night, bottom water becomes enriched over background. Recycling of nutrients occurs between the gamut of symbionts inhabiting individual coral heads.

It is also a solid generalization that primary productivity of reefs is dominated by benthic plants, rather than by phytoplankton—usually comprising a large fraction of nano- and pico-fraction cells. Even during the period June–August, when a general chlorophyll enhancement occurs over the entire shelf, as discussed earlier, SeaWiFS images indicate a significantly higher chlorophyll biomass for the Barrier Reef (as well as for the Capricorn line of reefs north of Sandy Cape) than the clearer water of the inner shelf. During the dry season during austral summer (December–February) the very clear water of the lagoon inside the Barrier Reef separates a near-shore zone of high chlorophyll from that over the reef itself. The latter signal probably emanates at least in part from the benthic algal mats and other components of the reef flora: such a suggestion is reinforced by the clear image given by the carbonate shoals of such places as the Bahama Banks, identifiable in SeaWiFS images as a persistent feature surrounded by water of high clarity.

Perhaps we may assume that this signal represents the benthic autotrophs whose productivity is far greater than that of phytoplankton over the reef: not only benthic macroalgae (*Sargassum, Halimeda,* and *Penicillus*) and sea grasses (*Thalassia* and *Cymodecea*), but also the symbionts of corals contribute to the high productivity associated with this warm, shallow, and very well-lit environment. Characteristic carbon fixation rates for reefs are from 1 to 5 kg C $m^{-2}$ $y^{-1}$ and this is partitioned between different groups of autotrophs.

This is not the place for a discussion in any detail of the higher trophic-level biota of the shelf in this province, nor am I equipped to describe adequately the extraordinary diversity of the Queensland shelf benthic and demersal fauna; suffice it to emphasize that the long-term stability of the reef ecosystem follows the generalities that we now accept for the maintenance of dynamic balance of ecosystems: that the balance between production and consumption can be very easily disturbed by the removal of individual species or the addition of others. Whether the observed episodic population explosions of crown-of-thorns starfish (*Acanthaster*) are natural or induced by human disturbance, they are testament to the fragility of the ecological balance of reefs, as of other ecosystems.

This province, of course, because its great latitudinal extent, includes both tropical coral-shelf regions and others of the western and southwestern Tasman Sea that are characterized by subtropical demersal fish (e.g., Sparidae, Lutjanidae) and benthic communities that follow the Thorson-Petersen model and resemble the New Zealand faunas. There is evidence here that the low-density plumes of estuarine water that extend out across the shelf of New South Wales serve actively to aggregate the larvae of demersal fish spawned on the shelf; those, such as mullet, that pass their juvenile period preferentially within estuaries are cued to the presence of that habitat by an encounter with a buoyant plume having estuarine characteristics.

## Synopsis

*Case 3—Winter-spring production with nutrient limitation*—This is perhaps the most relevant case, but this province has a very large latitudinal extent and is thus very heterogeneous; for the more northerly parts of the coast, this model is dysfunctional. Mixed-layer depth undergoes a moderate seasonal excursion (January 20 m, July 80 m) that is consistent with austral winter mixing, at the season when the coastal eddy field is also weak. Photic depth responds to the irradiance cycle, varying between 20 and 50 m seasonally, so that the pycnocline is illuminated only in austral summer (October–April).

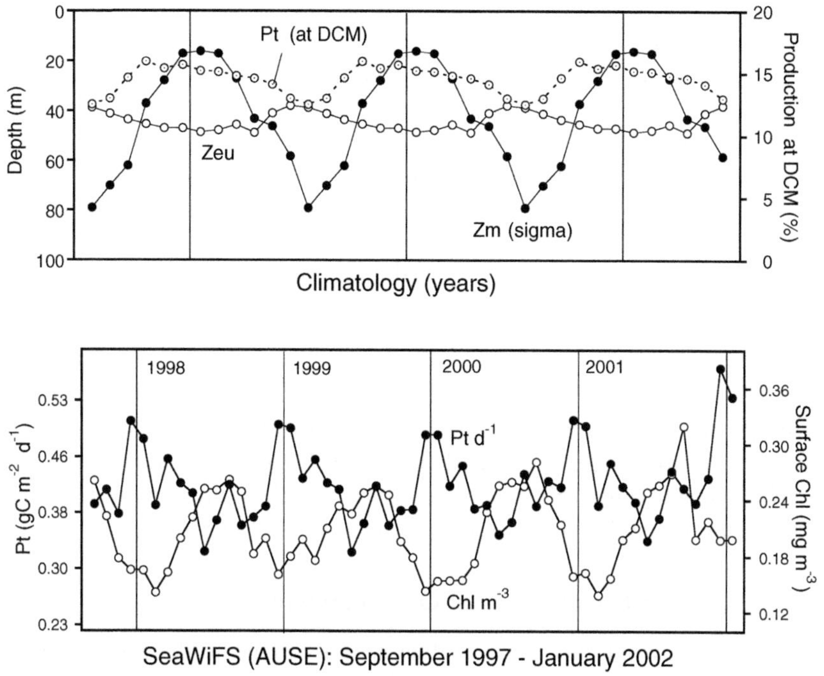


**Fig. 11.23** AUSE: seasonal cycles of monthly surface chlorophyll and depth-integrated autotrophic production for the years 1997–2002 from SeaWiFS data together with characteristic seasonal cycles of mixed-layer depths from Levitus climatological data and photic depths computed from characteristic irradiance and the archive of chlorophyll profiles discussed in Chapter 1.

A moderate productivity-rate increase is initiated 30 days after austral midwinter (July) from annual low of 0.35 g C $m^{-2}d^{-1}$, reaching maximal values in December–January, in austral midsummer, when coastal eddies and upwelling (from multiple causes) are most frequent. Chlorophyll biomass has a small dynamic range of 0.1–0.3 mg chl $m^{-3}$ taking maximal values in early austral spring (Fig. 11.23).

# New Zealand Coastal Province (NEWZ)

## Extent of the Province

The NEWZ province comprises the continental shelf around New Zealand. In earlier versions, the areas within the 1000-m isobath on the Campbell and Chatham plateaus to south and east were included, but this was not a good idea: these plateaus, large relative to the area of the continental shelf proper, are dominated by processes within the South Subtropical Convergence. So, the data used previously to represent this province were strongly aliased by events external to the shelf. In the version used here, the province is defined strictly by the shelf edge fronts and consequently, because of the scale (2° × 2°) of the gridded data in this work, it is not possible to derive adequate seasonal cycles of chlorophyll and productivity from satellite images.

## Continental Shelf Topography and Tidal and Shelf-Edge Fronts

New Zealand lies adjacent to the Chatham Rise that extends eastward and to the Campbell plateau that lies to the south; these two extensive deep shoals are everywhere more than 1000 m deep and lie beyond the relatively narrow shelf platform that covers about

$3 \times 10^5$ km$^2$. The southwest coast of the South Island is especially steep-to, the 1000-m contour lying only a few kilometers from the shore along the entire Fjordland coast. The only extensive areas of shelf within the 200-m isobath around New Zealand are to the west of Cook Strait, and the adjacent coasts of North and South Island, and also the Snares Shelf beyond Stewart Island, off the south coast of South Island. On the north coast of North Island, off the Hauraki Gulf and Bay of Plenty, the shelf width is significant, though the area involved is smaller than to the west of Cook Strait. The shelf on the east side of South Island widens somewhat off the Canterbury Bight (45°S), from where the deeper topography (>1000 m) of the Chatham Rise extends eastward.

Tidal currents and mixing are not as relatively important as in regions where the shelf is wider; the $M_2$ tide is a coastally trapped Kelvin wave that travels counterclockwise around New Zealand. It is only in Cook Strait that important tidal currents are generated, and here they may attain 4–5 m sec$^{-1}$. A recent study of internal tide dissipation on the northeast shelf showed that the energy associated with the internal tide was greater than that of barotropic tides or internal shear and that it generated significant vertical diffusivity at the nutricline. This flux, of $7 \times 10^{-4}$ m$^2$ sec$^{-1}$, is sufficient to support annual new production on the shelf of 100 g C m$^{-2}$ (Sharples *et al.*, 2001).

## Defining Characteristics of Regional Oceanography

New Zealand lies between subtropical weather patterns on the north coast of North Island and the frequent passage of cyclones generated in the Tasman across the high topography of South Island; this conjuncture generates extremely high rainfall on the Southern Alps and heavy discharge of freshwater along the west coast. The consequences of the location of New Zealand, lying across the wind and current regimes of the Southern Ocean, have been reviewed recently by Sharples and Greig (1998). As he points out, for this province we have access to only a relatively limited number of regional studies of coastal ocean circulation. Much of the following account of both the physical and the biological oceanography of New Zealand waters is drawn from Heath (1981, 1985), Heath and Bradford (1980), Vincent *et al.* (1991), and Butler *et al.* (1992).

The geographical location of New Zealand and the relatively shallow topography of the Campbell Plateau and the Chatham Rise strongly modify the zonal arrangement of the circumpolar frontal systems of the Southern Ocean, and determine circulation around New Zealand. Sharples comments that only here, off New Zealand, does shallow topography intercept a western boundary current in midocean.

The circulation around New Zealand over the shelf and just beyond is complex, reflecting the fact that two major currents encounter its western topography: (i) the flow of the retroflected East Australian Current along the Tasman Front meets the North Island at 33–35°S and (ii) that the gyral circulation of the southern Tasman Sea, together with the subtropical convergence zone beyond, is constrained by the Snares Plateau at about 45–47°S.

Each of these flows, on encountering the shelf edge, becomes locked to its topography and passes around New Zealand as the East Auckland and Southland Currents, respectively. Thus, the main flow along the Tasman Front passes as a high-velocity stream down the east coast of North Island to about 43°S, off the Banks Peninsula on South Island. Here it meets the frontal jet of the Subtropical Convergence that, after encountering the Snares Plateau, has passed around South Island, beyond the shelf edge but locked to its topography; this conjuncture—at the margin of the NEWZ Province—is very dynamic, with frequent cross-frontal exchange of subtropical and subantarctic surface water bodies. Here, the distinction between subtropical and subantarctic water masses is usually clear because the interleaving water parcels are typified by their origins: mixed-layer depth, chlorophyll profiles, and nitrate levels in the surface water are all clearly different. Finally, associated with the gyre of the Tasman Sea, flow to the north along the coast is induced over the shelf as the Westland Current.

Within these flows there are several sites of coastal enrichment. At the northern tip of the South Island and at the entrance to Cook Strait, bottom friction induces upwelling that is enhanced by the curvature of the coastline and curvature of the coastline at Cape Farewell (Shirtcliffe *et al.*, 1990). The very strong winds of Cook Strait then force the nutrient-rich, upwelled water through the straits, and the intensification of this flow also induces a sea-surface slope that accelerates water over the Kaharungi Shoals, thus creating a strong upwelling source near Kaharungi Point. South of Cape Providence (46°S) there is some uplift of water from the deep salinity maximum over the continental slope of the Puysegur Bank resulting in high nitrate concentrations inshore (Bradford, 1983). At the East Cape of North Island, which is rounded by the East Auckland Current, this is accompanied by the formation of a persistent warm-core eddy offshore and upwelling nearshore over Ranfurly Bank (Bradford *et al.*, 1982).

However, even more important is the persistent upwelling in the open bight between Cape Cascade (44°S) and Cape Foulwind (42°S). This simple wind-induced coastal divergence is limited by the shelf-edge front of the upwelled water that bears a series of wavelike, regularly spaced plumes of cooler water (about 13°C) in the warmer subtropical water mass of the Tasman Sea. Seaward of Abut Head the plume may reach 70 km offshore and carry much of the discharge from the Buller River. Inshore, both here and on the east coast, freshwater runoff from these rainy islands creates a sandwich of multiple, overlying surface-water masses identifiable in the density profiles of the upper 25 m. More generally, cross-shelf canyons on this coast may induce cross-shelf plumes of buoyant, inshore water that resemble the offshore squirts observed on the Californian coast: they stabilize the water column and have consequences for the pelagic production system on the outer shelf.

In the seas west of New Zealand, wherever frontal processes are weak, the mixed-layer depth can be modeled simply by local irradiance and wind stress. The seasonal range in mixed-layer depths is large—about 20 m in summer and 140–200 m in winter. This cycle is largely convective, being generated more by variance in irradiance (simply, sun angle and cloudiness) than by variance in wind stress. To the southwest, along the Fjordland coast, the exceedingly heavy rainfall on the Southern Alps creates a well-defined buoyant layer, 5–10 m deep, which is extremely opaque within the fjords because of a very heavy load of dissolved organic humics or "yellow substance." This is widely distributed, at different dilutions, over the shelves of the South Island. The deeper layers are oxygen-deficient because of very low rates of light-limited primary production.

## Regional Response of the Pelagic Ecosystem

This is a very difficult region in which to generalize the ecological processes, though a winter-spring bloom, forced by the seasonal cycle of irradiance and changes in mixed-layer depth, is the dominant signal in the chlorophyll field around the islands. However, the local dominance of a brief seasonal bloom in the confluence region east of the Banks Peninsula over the Chatham Rise must also be emphasized, even though as noted earlier, it occurs mostly on or beyond the logical boundary of the province. The brevity of this bloom is striking, as seen in satellite images and in sediment trap mass fluxes (King and Howard, 2001). More generally, chlorophyll accumulation is strongest during September–October over the entire shelf area of New Zealand with strongest biomass along the west and north coasts as well as the confluence region on the east coast from Cook Strait to Banks peninsula.

Also observable in the seasonal images are the consequences of the dynamic processes discussed above, especially in austral spring and summer. However, as Vincent *et al.*

(1991) note, a simple statistical correlation cannot be found between physical and nutrient variables at groups of individual stations and biological variables at the same stations; moreover, Chang *et al.* (1992) suggest that strong winds and unusually constant deep wind mixing around the New Zealand coasts may constrain both the accumulation of phytoplankton cells and the complete utilization of nitrate. This is too dynamic a region for simple relationships between nitrate and chlorophyll to be evident in survey data, and the relationships that might exist with phytoplankton growth rate are, as Vincent *et al.* point out, probably masked by the fact that phytoplankton biomass is a secondary variable representing, as I have noted several times, merely the resultant between accumulation and loss. Nevertheless, high F ratios (new/regenerated or nitrate/ammonium utilization ratios) reaching 0.9 are often encountered inshore in the Westland upwelling region.

Nitrate is often in excess of limiting concentration except in the shelf areas adjoining the TASM province, where a simple Sverdrup model is probably appropriate and where nitrate reaches limiting levels in surface water in summer. To the south and east of New Zealand, the surface water masses are distinguishable by their $NO_3$ levels: 4–7 $\mu$M in subtropical and 12 $\mu$M in subantarctic water. Elsewhere in the coastal regions, values of 0.5 to about 3.0 $\mu$M are normally encountered. Production processes have been studied extensively in Tasman Bay in terms of the relation between salinity stratification early in the year and thermal stratification later in the summer; in early summer, the photic zone is nutrient-deplete and has very low chlorophyll, but the turbid, well-mixed subpycnocline water has relatively high phytoflagellate biomass: this distribution has significance for filter-feeding benthic organisms. When the salinity stratification breaks down, and prior to thermal stratification, a diatom bloom ensues (MacKenzie and Adamson, 2004). More typical of shelf areas is the East Cape region studied by Bradford *et al.* (1982), where the phytoplankton response to the anticyclonicity of the eddying induced by the change in direction of flow around the cape. Breakdown of stratification and consequent high near-surface nutrient concentration caused by flow over Ranfurly Bank is associated with regional maxima in phytoplankton biomass and productivity. This, in turn, is closely reflected in the regional distribution of zooplankton biomass.

Off Otago, on the southeast coast, the Southland Current remains largely subsurface, below buoyant plumes of coastal water that sustain larger biomass of algal cells than does the Southland Current when it surfaces; the diatom-rich buoyant water is several kilometers wide along the coast, suggesting that riverborne nutrients are very important here. The same pattern has been observed on the west coast of South Island: diatom-rich coastal water, and phytoflagellates offshore. However, in summer (February) dinoflagellates may reach bloom concentrations both inshore and off, whereas diatoms occur only at the bottom of the photic zone (Bradford and Chang, 1982). In El Niño years, unusually cold SST around New Zealand may be associated with outbreak blooms of the coccolithophores *Gephyrocapsa oceanica* and *Emiliania huxleyi*.

Along the northwest coast of South Island, both upwelling from aphotic depths and river water enrich the euphotic zone in nutrients (MacKenzie *et al.*, 1983) with resulting elevated biomass of both heterotrophic bacteria and autotrophic cells; a DCM occurs near the nutricline here.

Here, off New Zealand, as in so many other sea areas in the past 10 years, we have confirmation of the paramount importance of the picophytoplankton, both prokaryotes and eukaryotes. In the upwelling region of the Westland coast, Hall and Vincent (1994) have demonstrated the important role these organisms play in the whole phytoplankton community. In summer the $<2-\mu$m fraction accounted for 73% of phytoplankton particulate nitrogen, whereas in winter 40–80% of particulate nitrogen, 55% of chlorophyll, and 45–70% of primary productivity are attributable to the pico fraction. At single stations (e.g., at the edge of the upwelling plumes) prokaryotes contributed 99% of all cells. These studies also reinforce the general observation of the high variability in abundance of the small photosynthetic cells: prokaryotes are normally within the range 5–7 × $10^7$ cells $liter^{-1}$, which is at least twice the relative variability normally observed for the small eukaryotes. In general, and as occurs elsewhere, maximum prokaryote cell numbers occur deep in the mixed layer and in association with a nitracline. In the inshore phytoplankton, biomass is more evenly distributed among the size classes. Chang *et al.* (1992) suggest that the 20- to 200-mm fraction contributes here as much as 55% of total phytoplankton nitrogen because high nitrate levels favor the growth of diatoms. Cell division rates among large cells are actually higher here than for the smaller fractions: at surface irradiance, the 20–200 mm fraction = 0.5 doublings $d^{-1}$, the 2–20 mm fraction = 0.22 doublings $d^{-1}$, and the <2 mm fraction = 0.18 doublings $d^{-1}$.

Both the Westland and the Cape Kahurangi (Cook Strait) upwelling blooms support populations of herbivorous macrozooplankton, as might be anticipated. Here, crustacean herbivores are dominated by small copepods (species of *Acartia, Paracalanus, Clausocalanus, Oithona,* and *Centropages*) and a euphausiid (*Nyctiphanes australis*), which is an important component in the pelagic food web both here and in the AUSE Province. At Cape Kahurangi, *Acartia* may form up to 60% of net zooplankton biomass, and on the Westland coast it dominates the inshore crustacean plankton; offshore in the upwelling water a mixture of oceanic and neritic species is normally encountered. Off Westland, zooplankton grazing represents 0.3–20% of algal production, with consumption being relatively higher inshore. *Nyctiphanes* manages its diel migration in the inshore-offshore transport associated with upwelling cells so as to place its eggs and larvae inshore, whereas the main feeding population of older larvae and adults remains offshore. In the Taranaki Bight, south of the upwelling at the Cape, *Nyctiphanes* may comprise more than half of all zooplankton biomass (James and Wilkinson, 1988; Bradford and Chapman, 1988).

A 3-year study in the Hauraki Gulf revealed between-year differences that were probably significant for the recruitment ecology of spring- and summer-spawned fish larvae (Zeldis *et al.*, 1995); during the first two summers the salps *Thalia democratica* and *Salpa fusiformis* dominated the later summer zooplankton biomass, whereas in the third year salps were very rare. The presence of salps induced a small phytoplankton biomass with a very deep DCM, well below the mixed layer. Without salps, mixed-layer chlorophyll biomass was much greater and the DCM lay at the base of the mixed layer. The between-year differences in stratification were minor, and the effects observed were due to the effect of salp grazing. Nevertheless, where salps were relatively sparse in the first 2 years, phytoplankton were not correspondingly abundant: modeling studies suggested that the incorporation of nutrients into salp biomass, and its sedimentation down through the pycnocline, sufficiently reduced mixed-layer nutrients as to inhibit phytoplankton growth even in those places where salp consumption rate was low. This situation is in stark contrast with that more commonly found, in which ammonium excretion by micro- and mesozooplankton maintains some balance with the requirements of the autotrophic community, as observed off Westland by James (1987).

On the east coast of North Island, where conditions are dominated by the effect of the subtropical water of the western boundary current, especially during ENSO events, there may be extensive coastal microalgal blooms of "nuisance" species: *Mesodinium* (a red tide organism), *Gonioceros*, and *Aulacodiscus* are the main microalgae in these blooms. In Tasman Bay, similar blooms occur after especially heavy coastal rainfall and runoff from the land during summer.

## Regional Benthic and Demersal Ecology

Benthic ecology, in the classical sense, seems to have survived longer off New Zealand than in many other places, so here we have several modern studies of the macrobenthos of the continental shelf and the relations between the fauna and deposit type. Early studies here, strongly influenced by the Petersen school of thought, located at least three coastal community types that could be accommodated in the global system of isocommunities—and even if this was a strictly subjective judgment, it was not lacking in useful information. Thus, Thorson identified a regional version of the *Tellina* isocommunity, dominated in this case by *T. liliacea*, on clean sandy bottoms in the sublittoral; deeper off the open coasts of the South Island, on soft muddy grounds, an *Amphiura* isocommunity was found to occur. This was dominated by *Amphiura rosea*, with *Echinocardium, Nucula, Dosinia*, and polychaetes—*Glycera, Lumriconeries*, and *Pectinaria*. In some other regions this assemblage is dominated by one of its components, *Turritella*; here, the *Turritella*-like *Maoricolpus* dominates in some regions, as in deep fjords.

Recent investigations have looked at larger areas and used less subjective techniques, but their results appear compatible. Studies that have examined the grouping of biota across the entire width of the shelf by numerical classificatory techniques have identified a small number of depth-specific communities of benthic invertebrates, of which the most remarkable is perhaps the midshelf community off the Otago peninsula dominated by bryozoans (*Filicea elegans*) on a substrate of pebbles, with little silt deposition. Otherwise, three or four depth-specific, substrate-specific assemblages are recognized with the use of numerical classification (Probert and Wilson, 1984; Probert and Grove, 1998); polychaetes dominate numerically at most stations, in most assemblages. Thus, off the west coast of South Island, the assemblages identified by such techniques were (i) on silty sand at 30–50 m, a grouping dominated by *Sthenelais, Nephthys, Ampharete*, and others, (ii) at 90–300 m on sandy mud characterized by bivalves (*Poroleda*) and amphipods (*Ampelisca*) and polychaetes (*Aglaophamus, Lumbrinereis* and others), (iii) on sand at 200–250 m characterized by bryozoans and *Chloeia*, another polychaete, and (iv) at >500 m on sandy mud characterized by tanaids and ophiuroids.

A trawling survey off southern New Zealand likewise specified demersal fish assemblages that could be related to characteristic depth zones, but not to sediment type because the range of types was too narrow. Shallow assemblages were dominated by Gempylidae, Squalidae, and Moridae, whereas deeper on the shelf Chimaeridae, Argentidae, Merluccidae, and Macrouridae were characteristic. Off the northeast coast and down to at least Cook Strait, subtropical families dominate: breams (Sparidae) are very characteristic. More generally, the overall diversity of demersal fish communities off New Zealand from many such surveys has been related to depth, latitude, and regional phytoplankton productivity (McClatchie *et al.*, 1997): species richness decreases with latitude and increases with depth and surface productivity. It is no surprise, therefore, that diversity of demersal fish is maximal between the 500- and 1000-m contours on Chatham Rise; it was thought that the relatively high degree of mesoscale activity over the northern slopes of this feature adjacent to the New Zealand shelf is responsible for the large number of species-rich locations found here on the steeply sloping slope. However, perhaps it is not as simple as it seems.

This is, in fact, a beautiful example with which to finish this chapter, to warn us of the dangers inherent in the subjective relationship that I have proposed, or reported on, throughout this book. For, having offered the relationship just expressed, the authors warn that perhaps a quite different relationship may actually determine relative diversity. The greatest diversity, found on steep slopes on the Chatham Rise, may, they suggest, have more to do with the fractal nature of this habitat than the overlying productivity that itself may be provoked by the same topographic roughness that is likely to induce mesoscale eddies and productivity above. They also point out that diversity may be low in the higher latitudes surveyed simply because of lack of rough habitat there. Although I think it is not their favored interpretation, they point out that the association between fish diversity and surface productivity may be correlative but not linked by causation. Would that more authors understood that possibility.

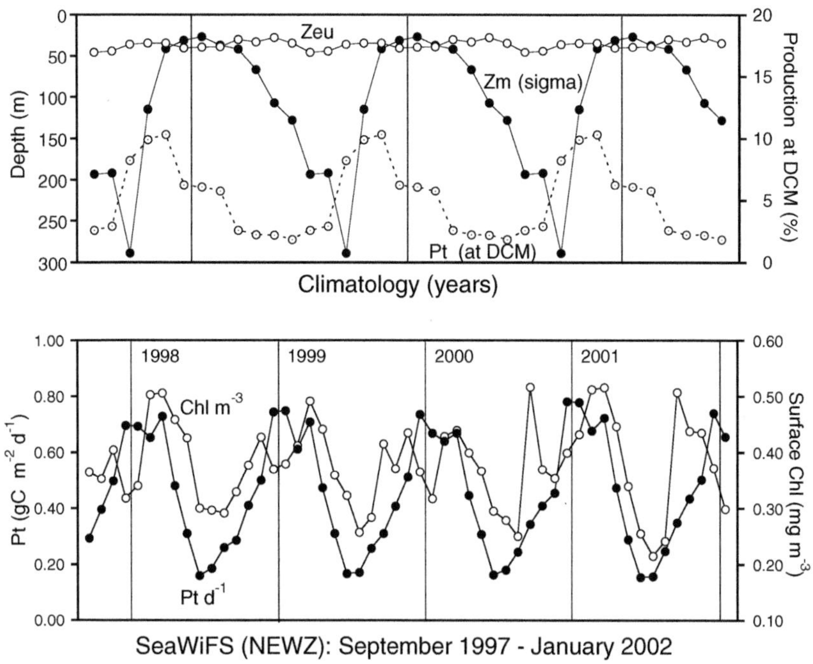


**Fig. 11.24** NEWZ: seasonal cycles of monthly surface chlorophyll and depth-integrated autotrophic production for the years 1997–2002 from SeaWiFS data together with characteristic seasonal cycles of mixed-layer depths from Levitus climatological data and photic depths computed from characteristic irradiance and the archive of chlorophyll profiles discussed in Chapter 1.

## Synopsis

*Case 2—Nutrient-limited spring production peak*—The mixed layer undergoes a strong austral winter excursion (25 m January, 290 m July–September) from which it shoals rapidly in October–November. The photic layer is consistently at 35–50 m, so the thermocline is illuminated only briefly in austral summer (December–February). The productivity rate increases progressively from winter (September) to autumn (April), after which it declines rapidly to low winter values, rate changes that match the long winter deepening of the mixed layer (Fig. 11.24). The dynamic range of chlorophyll biomass is less than 0.5 mg $m^{-3}$, and accumulation of chlorophyll matches changes in the production rates, reaching annual maximum biomass (0.5 mg $m^{-3}$) in fall.

Chapter 12

# The Southern Ocean

The Southern Ocean is unique in having no continental barrier to flow eastward around the globe; only the bottleneck of the Drake Passage between South America and the Antarctic Peninsula is a constraint, pinching together the components of the circumglobal current structure but not eliminating them. The lack of land barriers also permits the westerly winds of the Southern Ocean to build up a wave field having a mean height greater than anywhere else; this is the windiest and roughest of all oceans. It goes without saying that poleward of the Polar Front it is subject to seasonal ice cover, that the Antarctic coastline bears large permanent ice shelves, and also that here the sun is never very high in the sky. Winter darkness occurs south of the Antarctic Circle, the dark period increasing poleward at a rate of 40 additional days for each 5° of latitude. Thus, the region having some winter darkness includes the entire marginal ice zone and all three marginal seas: Ross, Amundsen, and Weddell.

The Southern Ocean is also unusual in the richness of the ecological studies that have been performed there. Although it is distant from all oceanographic centers, the lure of the rich whaling grounds early attracted the attention of biologists. The list of research ships that have worked the Southern Ocean is a roll call of early oceanography, strikingly evoked in a chapter by Deacon (1984). I personally relish the fact that Captain Thaddeus von Bellingshausen, German leader of Tsar Alexander I's expedition of 1819–1821 aboard *Vostok* (East) and *Myrni* (Peaceful), caused net tows to be made that first demonstrated the diel vertical migration of zooplankton: some people are still debating the "real" significance of that behavior today. From then on, there was a cascade of biological studies, including the *Discovery II* expeditions organized by the British Colonial Office a century after Bellingshausen; Alister Hardy's 1967 account gives the full flavor of these. The modern voyages of *Polarstern* show that the tradition of oceanographic research is alive and well in the Southern Ocean. JGOFS studies, carried by that ship and others in the 1990s, gave us new insight into these fascinating seas.

Circulation in the Southern Ocean is carried by the Antarctic Circumpolar Current (ACC), which is forced by exceptionally strong and stable winds; these build up an equally exceptionally steep wave field that has infinite fetch. The consequent mean flow of the ACC is sufficiently rapid that Rossby waves appear to propagate eastward (Killworth *et al.*, 2004); in all other zonal currents, as in the equatorial current system, they propagate westward against the baroclinic mean flow. The eastward flow of the ACC is tightly constrained by both deep and shallow topography, though the frontal systems maintain their integrity even when squeezed through Drake Passage. Sharp poleward turns in the ACC flow are induced through a gap in the deep Southwest Indian Ocean Ridge and through fracture zones in the Pacific-Antarctic Ridge, while the Scotia Arc induces a northward turn. The ACC is strongly sheared and contains three relatively narrow, meandering jets between which the flow is significantly slower. Each of these current cores is associated with one or more oceanic fronts, which form prominent, permanent, and almost continuous circumpolar features around the Southern Ocean. At sea, these may be relatively easily identified by the crowding of isotherms or isohalines that slope upward toward the surface of the Southern Ocean. The surface water masses of the ACC, and their ecological characteristics, are thus partitioned into zonal bands. Only far to the south of the most poleward of these fronts is there contrary, westward transport in the East Wind Drift, close to the Antarctic continent. I refrain from quoting characteristic latitudes for each feature, because of the extent to which coastline and deep bottom topography deform the annular flow; thus, the characteristic latitude of each feature differs by at least 10° of latitude as it passes around the ocean.

From the earliest days of Antarctic oceanography, this frontal pattern has suggested a partition into zones for which, unfortunately, there is much duplication of terminology. To make the reading of this chapter simpler, it may be helpful to summarize how these fronts, and the annular zones between, are arranged. What follows depends heavily on the fresh interpretation and revised terminology of Orsi *et al.* (1995), although the names used in this chapter for each individual feature are a compromise and do not conform precisely to those of Orsi *et al.* The sequence of fronts and zones across the Southern Ocean is shown in section and map format in Figs. 12.1 and 12.2. It is to be noted that these fronts lie much further poleward than their counterparts in the northern hemisphere; this, of course, is yet one more consequence of the irregular geography of the continents and the consequent distribution of the wind systems of our planet.

The partition discussed below forms the basis for the four biogeochemical provinces proposed in this chapter as the basis for the ecological geography of the Southern Ocean. This arrangement differs slightly from other recent suggestions, such as that of Tréguer and Jacques (1992), who divide the regions south of the SAF into a Polar Frontal Zone, a Permanently Open Ocean Zone, a Seasonal Ice Zone, and a Coastal and Continental Shelf Zone.

*South Subtropical Convergence Zone.* This lies below a permanent line of zero wind-stress curl, being the equatorward limit of northward Ekman drift of the subantarctic water and of southward drift of subtropical water within the three main oceanic gyres.

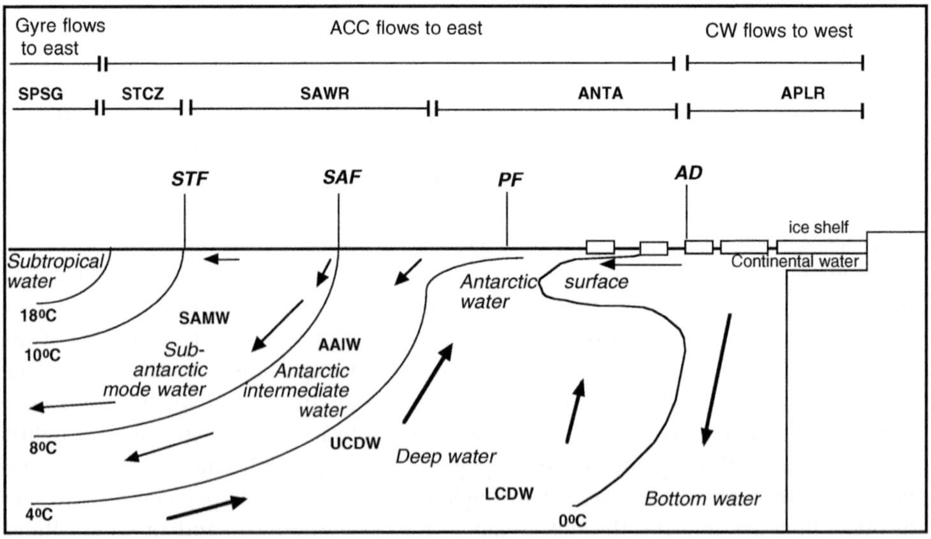


**Fig. 12.1** Idealized section across the Southern Ocean to illustrate the relationship between water masses, their motion and of the fronts between them: SPSG, S. Pacific Central Gyre; STCZ, Subtropical Convergence Zone; SAWR, Subantarctic Water Ring; ANTA, Antarctic Zone; APLR, Antarctic Polar Zone; STF, Subtropical Front; SAF, Subantarctic Front; PF, Polar Front; AD, Antarctic Divergence.

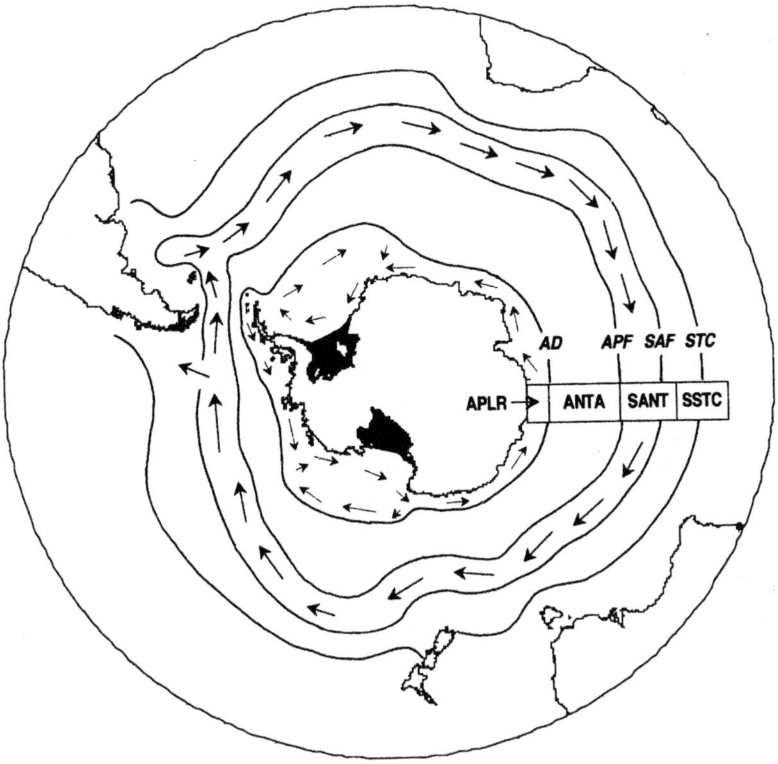


**Fig. 12.2** Regional circulation pattern of the Southern Ocean, as influenced by land mass distribution, to show the arrangement of the ecological provinces of austral regions, viz. APLR, ANTA, SANT, SSTC; (AD = Antarctic divergence, APF = Antarctic Polar Front, SAF = Subantarctic Front, STC = Subtropical Convergence).

This zone is associated with a zonal band of strong mesoscale meanders and eddies that occupy ~5° of latitude. Meander formation is strongest to the east of Patagonia, of South Africa, and of Tasmania and New Zealand, and also in the South Pacific at Crozet Island and the Kerguelen-Heard plateau. Motion of the detached eddies is complex and depends on interaction with the general field of potential vorticity.

The transitions to subtropical and to subantarctic surface waters on either side of the zone have frontal characteristic; the term *Subtropical Front (STF)* is now usually applied to the more poleward of the pair. Here, the entire convergence zone between these fronts shall be referred to as the South Subtropical Convergence Province (SSTC). The STF, therefore, lies on the poleward side of a fast zonal jet that occupies the upper 1000 m as it flows to the east (Stramma and Petersen, 1990; Stramma, 1992). On the cold side of the STF, low-salinity subantarctic surface water converges with the modified subtropical surface water of the frontal jet and passes beneath it. The STF forms an almost continuous and prominent feature around the entire Southern Ocean.

*Subantarctic Water Ring*. This is the most complex entity to grasp. It comprises the two current cores within the ACC and the slower-moving flow of the subantarctic surface water mass between these cores and the STF. The two relatively fast, constricted flows lie to the south of the slower flow and together form the *Polar Frontal Zone* (Anonymous, 1989). This zone, which is embedded in the Subantarctic Water Ring, is defined by two fronts at which the shoaling, upsloping isotherms and isohalines are more tightly clustered than elsewhere in the ACC: these fronts are the relatively weak *Subantarctic Front* (SAF) and the stronger *Polar Front* (PF). The location of the latter corresponds approximately with the limit of winter ice cover. The STF and the PF are used here as the limits of a Subantarctic Water Ring Province (SANT), integrating two entities: a Subantarctic Zone (SAZ) and a Polar Frontal Zone (PFZ).

Because both the PF and, to a lesser extent, the SAF are associated with jetlike eastward flow, each typically exists as a multiple front with active meandering and eddy shedding (Read *et al.*, 1995; Veth *et al.*, 1997). In fact, one merit of the collective term Polar Frontal Zone (PFZ) is that it recognizes the fact that—in the real ocean—PF and SAF often merge. The steering of the ACC by deep topography in several regions induces the SAF and PF to approach very close to each other and frequently to form a single dynamic feature: this occurs most commonly north of the Falkland and Kerguelen Plateaus, near 32°E, and also in the vicinity of the mid-Pacific Rise (Orsi *et al.*, 1995).

An additional deep front, the Southern ACC Front, occurs to the south of the PF (Orsi *et al.*, 1995). This does not separate distinct surface water masses, for it is expressed only in the deeper density structure and will not much concern us here. Nevertheless, you will frequently see references to the fact that the flow of the ACC is subdivided by three, rather than the habitual two, circumpolar fronts.

*Antarctic Region (ANTA* and *APLR).* From the PF southward to the continent, there is only a single superficial water mass, Antarctic Surface Water, and this is widely subject to winter ice and the effects of melting within the marginal ice zone. Orsi *et al.* (1995) suggest that this region is unitary, even though fronts do occur within it; nevertheless, as was done in the first edition of this work, I continue to believe it useful to partition the region into two provinces, of which one represents the Coastal biome. The principal discontinuity in the Antarctic region is the Antarctic divergence (AD) that lies below the change in direction of mean wind stress and marks the poleward limit of flow of ACC water to the east and the limit of the Coastal biome. Less useful is the boundary of diluted continental shelf water.

At Drake Passage, between Cape Horn and the South Shetland Islands, the West Wind Drift is restricted to a narrow passage of only about 300 km compared with the 1200 km between South Africa and the Antarctic continent. This requires that the annular zones should squeeze together as they pass through the straits, although there is some loss from the northern ACC into the southern part of the Humboldt Current. Nevertheless, the zonal bands retain their relative positions, and only the STF itself is not identifiable here. East of the Drake Passage, flow conforms to the continental margin and so sweeps to the north along the edge of the Falkland Plateau before again passing eastward on meeting the Brazil Current. Consequently, in the Atlantic sector of the Southern Ocean the zonal boundaries take their most northerly position. Further yet to the east, the relatively narrow passage south of Africa again somewhat constrains the northern parts of the banded flow of the ACC.

From analytical accounts of the physical forcing of the banded flow of the ACC, it is rather easy to obtain a false understanding of how this flow is actually manifest at the surface of the ocean. The best way of visualizing this is to consult a collection of satellite images of chlorophyll, which forms an excellent tracer for differential horizontal motion of the near-surface water mass (see Color plate 20). Such images demonstrate that textbook interpretations of the mean banded flow of the ACC are singularly misleading. Although the frontal regions discussed earlier are prominent features in all images, they exhibit far from an "ideal" pattern; linear but highly meandering and fragmented features dominate the flow field and it is usually not immediately obvious whether a frontal jet observed in this way represents the PF or the SAF, or both mingled—or even neither. The certainty with which each front may be identified from an observing ship, properly equipped, may suggest wrongly that these are simple linear features. It is also perhaps significant that so much of the work at sea that has required frontal identification has been done in the Atlantic sector, where the frontal regions are unusually discrete and are more widely separated than elsewhere in the Southern Ocean.

As has already been discussed in a general manner in Chapter 5, the Southern Ocean presents us with an apparently simple question, posed by the observation that it habitually exhibits regions of clear, blue water where phytoplankton biomass and primary production must take low values. Is this lack of plant growth caused by micronutrient deficiency, given the characteristic excess availability of macronutrients, or is it due to Sverdrupian dynamics appropriate to unusually deep wind-mixing and low solar irradiance? I have had these questions in the front of my mind when writing each section of the present chapter. I have also had in mind that the satellite chlorophyll fields now available suggest that the widely held view that the Southern Ocean does not accumulate chlorophyll is an overstatement: really, blooms are more widespread in this ocean than some accounts would have us believe.

In the first edition of this book, I rather confidently asserted that the then proposed partition was consistent with concepts concerning regional differences in the regulation of algal growth by iron in the annular zones of the Southern Ocean (de Baar *et al.*, 1995). I stated that Fe-deficient water occurred south of the STF, Fe-replete water in the PFZ, and Fe-limiting water in the ACC south of the PF. The Fe-replete water of the coastwise East Wind Drift completed the logic.

However, as you will realize if you have read Chapter 5, I am no longer so confident that this apparently simple model is useful because, for one thing, it is not Fe repletion alone that induces strong blooms in the PFZ. It should be enough to note that average chlorophyll biomass along 170°W does not correlate with Fe concentration in the 0–100 m layer (Buesseler *et al.*, 2003). Repeated cruises during austral spring and summer 1997–98 down this longitude by U.S. JGOFS ships revealed a more complex model of nutrient supply and limitation (Hiscock *et al.*, 2003) than I had in mind when writing earlier. The Hiscock *et al.* model involves the upsloping of Fe-rich, nutrient-rich Upper Circumpolar Deep Water (UCDW) to the surface between the PF and the southern boundary of the ACC, south of which nutrient-poor LCDW comes to the surface. As austral summer conditions move south along 170°W, and $SiO_3$ is progressively depleted poleward by diatom growth (Nelson *et al.*, 2001), a strong latitudinal gradient ($Si_{max}$) develops and shifts from 62–63°S near the PF in early spring down to 66°S at the southern boundary of the ACC by midsummer.

Evidently, the winter recharge of UCDW in the PFZ and to the south provides sufficient Fe to support a diatom bloom that propagates south only to the southern boundary of the ACC, where it encounters Fe-deficient LCDW. Productivity and phytoplankton biomass are maximal within the $Si_{max}$ as this gradient shifts poleward. In this progression, Hiscock *et al.* recognize three phases or zones of differential nutrient limitation that were revealed by the repeated sections down 170°W:

(i) Already in spring, the SAZ (north of the PF, see above) is $SiO_3$-limited and Fe-replete
(ii) Between the PFZ and CWB the surface layers are $SiO_3$-replete and Fe-limited in spring, but $SiO_3$-limited and Fe-replete in summer
(iii) Between the southern boundary of the ACC and the shelf edge around Antarctica, conditions are consistently $SiO_3$-replete, Fe-limited

You will have noted that nitrate is omitted from these comments because, in these seas, it is always in excess of requirement for reasons discussed earlier in this chapter, and in Chapter 5. During the open-water season, surface nitrate values of about 10–15 μM occur widely (Foster, 1984), although they are somewhat lower in the Indian than in the other sectors of the Southern Ocean. Note also that light limitation is not discussed but—once again if you have read Chapter 5—you will remember that in the deeply mixed regime south of the PFZ, this limitation occurs in the second half of summer.

This characterization of regional production processes represents the situation along 170°W, a meridian that was selected specifically because here the individual fronts tend to lie zonally and to be well separated meridionally. As far as can be managed, it therefore represents an ideal situation from which we may reasonably extrapolate to other regions. The complex evolution of differential nutrient relationships during the summer growth season suggested by these JGOFS observations provides a more satisfactory basis for further discussion in the sections that follow than was provided by the earlier, simpler model of de Baar *et al.* that was conceived from a region having greater local characterization.

A little to the east, down 150°W, several CPR tows have recently been reported that demonstrate the partition of zooplankton species distributions in relation to the zonal frontal structure of the Southern Ocean (Hunt and Hosie, 2005); the northern Subantarctic Front was the strongest biogeographic boundary at the southern limits of many subtropical and temperate zone species. Five zonally arranged species clusters were identified by Bray-Curtis techniques, their distributions clearly determined by the occurrence along the transect of the frontal zones discussed earlier.

# Antarctic Westerly Winds Biome

## South Subtropical Convergence Province (SSTC)

### Extent of the Province

This province covers the entire Subtropical Convergence Zone (STCZ) that is defined operationally by two fronts, or sharp gradients, in surface water properties that lie across the southern side of the southern subtropical gyres of each ocean basin. The location of these fronts is predictable in the general sense, but flow along them is meandering, so that their instantaneous location can be observed usefully only in satellite imagery. Although there is no coastline in this province, the STCZ does in some places pass close along the edge of shallow shelf topography, as around the Falkland and Campbell Plateaus.

### Defining Characteristics of Regional Oceanography

While heading south, George Deacon observed a steep temperature gradient at 40°S "accompanied by a sharp decrease in salinity . . . [that] . . . provides clear evidence of a sharp, though moveable and sometimes fragmented boundary between water of subtropical origin to the north and subantarctic to the south." I shall try to emphasize throughout this chapter that although it is convenient (and sometimes useful) to portray the annular fronts of the Southern Ocean as simple linear features, in reality they are, as I have already noted, meandering and fragmented.

As Deacon remarked, the STCZ divides the anticyclonic circulations of the southern Atlantic, Indian, and Pacific Oceans from the cyclonic circulation of the ACC and can be traced almost continuously around the globe at about 35–45°S. It turns equatorward when it encounters shallower water and poleward over deep water.

The STCZ occurs below a sharp decrease in the westerly winds of the Southern Ocean and a permanent line of zero wind-stress curl, and is therefore the equatorward limit of northward Ekman drift of the subantarctic water and of southward drift in the subtropical gyre: strong convergence and downwelling consequently occurs here. Sharp surface thermal gradients (14–18°C summer, 11–15°C winter) separate warm, salty subtropical water from the colder, fresher water to the south. These are the NSTF and SSTF of Belkin and Gordon (1996), usually separated by 4–5° in each ocean basin; the NSTF is a shallow (200–300 m) front at the southern boundary of the warm, salty subtropical surface water. The STF of many previous authors represents a concatenation of the two fronts of the STZ, but was most frequently identified by the conjunction here of cold, low-salinity subantarctic surface water with water having some residual subtropical characteristics.

The entire frontal zone is sufficiently dynamic to have a major eddy field associated with it, so that several surface discontinuity fronts may be encountered in the same transect (Deacon, 1984; Colborn, 1975; Lutjeharms, 1985). The STF on the cold side of the STCZ is most frequently observed because of the transition across it to subantarctic conditions. Because of seasonal changes that occur in surface temperatures, the associated surface salinity gradient is the best practical indicator of the presence of the STF.

A fast zonal jet in the upper 1000 m of the water column passes eastward across each ocean basin along the warm side of the STF (Stramma and Petersen, 1990; Stramma, 1992). Other names have been applied to it, but we may simply call this the STCZ flow, because its lateral extent corresponds closely to the region of the entire STCZ as otherwise determined. The STCZ flow is bounded to the north by the termination of the thermal and salinity gradients at the surface, a change that is sufficiently sharply marked in surface properties as to be recognized by some authors as a Northern Subtropical Front. Velocity in the near-surface core of the STCZ flow is of order 10–30 cm $sec^{-1}$, and transport is two to three times faster than the flow to the north and south.

Since the STCZ lies between nitrate-depleted subtropical and nitrate-replete subantarctic water, there is a major gradient in dissolved nitrogen across the feature: $NO_3 + NH_4$ may be undetectable in subtropical water to the north, but take values of ~2 μM in the STCZ itself, and 8–10 μM in the subantarctic water to the south. Silicate values remain undetectable across the STCZ and remain low (0.4 μM) to the south, not increasing (to >5 μM) until the PF is crossed at about 50°S. Available illumination at the surface (12–52 E $m^{-2}$ $d^{-1}$) and mean mixed-layer depth (30–125 m) each show a simple seasonal pattern having maxima and minima in December–January and June–July, as appropriate. Euphotic zone depth varies only rather slightly with the seasons (~85 ± 15 m). Surface temperature maxima and minima (12–15°C) lag by about 1 month.

The strength of the flow in the STCZ explains what is observed so clearly in satellite imagery. TOPEX-POSEIDON sea-surface elevation data reveal that across each ocean a zonal band of strong mesoscale meanders and eddies occupies a zone of approximately 5° latitude in extent. This is, at once, the flow of the STC current and an operational means of defining the SSTC province, although the concept of a simple linear STF lying zonally around the Southern Ocean is very far from reality. This had already been observed in section data; a meridional section down the southern Indian Ocean at 70°E cut twice across the STCZ, with a 100 km section of subantarctic water between, indicating very strong meandering (Stramma, 1992). Although, as we shall see, the STCZ is clearly represented in sea surface chlorophyll fields, these are somewhat misleading because, to eliminate cloud masking, they are usually examined as 30-day composites. There is so much motion within the meanders of the STCZ that it then appears in these images wider than instantaneous images would reveal. Only from cloud-piercing TOPEX-Poseidon radar maps of sea surface elevation can we obtain true dimensions and location of the STCZ. Both 7-day chlorophyll images and sea-level anomaly maps show that where the STCZ is most strongly developed, meanders take rather uniform scale: about 5–7° of latitude seems in the SW Indian Ocean, for example. Meander formation is strongest immediately downstream of where the zonal flow is deformed by topography to the east of Patagonia, of South Africa, and of Tasmania and New Zealand.

South of Africa, the location of the Subtropical Convergence is determined by the position of the retroflection zone of the Agulhas Current where the warm Agulhas water thrusts south of the Cape of Good Hope (Lutjeharms *et al.*, 1985; Lutjeharms and van

Ballegooyen, 1984); this retroflection is again associated with an active eddy field. Large meanders form between the warm Agulhas water and the cooler water to the south; on the coastal side of the Agulhas Current, eddies are to some extent topographically controlled by the Agulhas Plateau. To the east of Cape Horn, flow in the ACC turns northward (as the Falkland Current) across the Argentine Basin, and the STF is re-established between the eastward-turning subtropical water of the Brazil Current and the colder subantarctic water at about 40–45°S (Peterson and Whitworth, 1989; Peterson and Stramma, 1991). The STCZ then proceeds across the South Atlantic, reaching its most northerly position close to 35°S at the Greenwich meridian.

As noted in the previous chapter, the STCZ conforms to topography in the New Zealand sector, looping northward along the continental edge east of Tasmania (Harris *et al.*, 1987) and south and east of New Zealand. It is only across the Tasman Sea and eastward from the Chatham Rise that the convergence again becomes detached from the steep topography of a continental edge.

## Biological Response and Regional Ecology

Here we have to address, in a local sense, some of the questions discussed generally in Chapter 5: what exactly are the mechanisms that support phytoplankton growth and its accumulation in the Southern Ocean? Why should the seasonal progression of light, temperature, and stratification produce phytoplankton growth preferentially in this frontal zone? The nutrient distribution, discussed briefly earlier, offers no simple explanation; but since frontal processes everywhere tend to establish near-surface stability in mixed water columns, this may provide sufficient stability to initiate and maintain a bloom (Marra, 1982), especially where, as here, the density surfaces in the front slope downward. It is not necessary to postulate, as did Banse and English (1997), that chlorophyll patches in this habitat represent undemonstrated "patchy" dust fallout. The suggestion that dust deposition induces the high chlorophyll observed in the STCZ east of Patagonia (Erickson *et al.*, 2003) is based on an illogical argument, as has already been observed.

Similar situations have been observed elsewhere in the STCZ and have been generalized by Franks and Walstad (1997). Following their model, chlorophyll accumulation in the STCZ should be strongest where eddy kinetic energy is highest, downstream of topographic disturbances to the zonal flow; this, of course, is what is observed. The mingling of subtropical water (nitrate-depleted, Fe-sufficient) with subantarctic water (nitrate-sufficient, Fe-depleted) across the STCZ provides an unusually strong nutrient gradient that must result in local phytoplankton growth where other conditions permit. Blooms are proximally controlled by the dynamics of eddies and meanders as is clearly shown by the abundance of arcuate streamers and annular features in appropriate 7-day chlorophyll images. These may be matched—feature for feature—with appropriate sea-level anomalies (see Color plate 13, for example).

The STCZ is one of the most prominent features routinely seen in global sea-surface chlorophyll images for which several mechanisms have been invoked (Lutjeharms *et al.*, 1985; Furuya *et al.*, 1986). Although simple physical aggregation of biota may occur in the STCZ, as in all convergent fronts, there seems no reason to invoke this mechanism because the scales are not right nor are there obvious sources of biota to be transported. Rather, it is more likely that a variety of mechanisms, differing from place to place and from season to season, are involved. Seasonally, chlorophyll accumulates more rapidly than it is dissipated, again an unsurprising observation. A simple seasonal cycle occurs in the STCZ, so that phytoplankton biomass begins to accumulate in austral spring soon after MLD has started to shoal and takes its regional maximum in midsummer when MLD is at its shoalest.

There is, as I noted in Chapter 5, a general tendency for dynamic frontal processes to establish near-surface stability in mixed water columns and so initiate accumulation of plant biomass. Where density surface in fronts slope downward, as here, this may provide sufficient stability to initiate and maintain a bloom (Marra, 1982). Cross-frontal mixing of warm, nutrient-poor subtropical water and nutrient-rich subantarctic water will induce thermal stability and a sufficient nutrient supply to establish a bloom. We would anticipate that these processes would be least effective in austral winter when deeper convective mixing will occur—and when irradiance will be minimal. A return to net growth in spring in response to reduced mixing depth and increased irradiance is entirely in accordance with Sverdrup.

Such a process was observed along a meridional section down 150°E during austral summer 1983–83 that crossed the STF (Furuya *et al.*, 1986). Here, subtropical water was found to have penetrated (at 30–70 m) for about 16 km along an isopycnal surface down into the subantarctic water mass that therefore lay both above and below it. A sharp, shallow pycnocline resulted from the temperature inversion, the regional DCM at 75 m being accompanied by an additional near-surface (0–20 m) chlorophyll layer above the penetration. Phytoplankton of the subantarctic water mass (biomass was >90% diatom carbon, principally *Chaetoceros, Nitzschia, Thalassiothrix, Rhizosolenia*) lay at these two depths of high chlorophyll biomass, one above and the other below the subtropical phytoplankton that comprised only about 50% diatom biomass. Once incorporated in the front, the warm-water assemblage of unicellular cyanophytes and small flagellates were never more than 15% of total autotrophic carbon. Silicate depletion is likely to occur in these blooms and will result in an overall reduction in new production, and progressive evolution of the algal assemblage to one dominated by dinoflagellates and microalgae. Primary production was higher across the intrusion than either to north or to south. This process occurred under very moderate wind stress and is unlikely to function in a similar manner during the deeper convective mixing of austral winter.

A similar situation has been observed in the Indian Ocean sector, where Russian investigations concluded that the strongest enhancement of productivity occurred at intrusions of warm into cool water associated with the STCZ. The induction of phytoplankton blooms by eddy- and meander-induced stratification is perhaps a sufficient general explanation of the existence of a strong chlorophyll field in the meandering STCZ.

Obviously, there are special regional characteristics in some parts of the STCZ; Gregg and Conkright (2001) discuss decadal-scale changes in the retroflection to the east of the Falkland plateau. This is a peculiarly complex region and it will bear some additional discussion. Although the zonal currents of the Southern Ocean pass through the Drake Passage without mingling, real continuity of frontal features may be weak. Many authors suggest that the STF of the Pacific does not penetrate to the Atlantic, but rather that it is formed again *de novo* where the ACC turns east across the Atlantic. This appears to be supported by Color plate 13, representative of many such patterns. Arising from the chlorophyll feature representing the summer bloom on the Falklands shelf and plateau is a SW–NE linear feature of high chlorophyll that lies just outside the shelf edge as far north as the River Plate. This linear bloom is unlikely to be an upwelling feature, having no offshore plumes, and it lies in the jetlike Falklands Current, part of the ACC flow. This cold water turns sharply back on itself at its confluence with the warm Brazil Current and first flows southward, then turns east across the ocean. There, its zonal flow lies to the south of the newly formed STF and STCZ of the South Atlantic. These features may then be followed around the globe in TOPEX-Poseidon SLA maps and chlorophyll images, right back to South America again.

A size-fractionated study of the autotrophic cells in the STCZ east of New Zealand (Hall *et al.*, 1999) found chlorophyll-*a* concentrations of up to 3.4 $\mu$g liter$^{-1}$, but only as much as 0.2 $\mu$g liter$^{-1}$ in the subantarctic water beyond the STF. At all seasons, the subantarctic water was dominated by microbiota, as was the subtropical water to the north of the STCZ in spring: here, up to 98% of phytoplankton was in the <20-μm fraction, with values of 0.9–1.9 for the prokaryote/eukaryote carbon ratio. In comparison, the phytoplankton of the STCZ was at all seasons dominated by the <20-μm fraction and the prokaryote/eukaryote carbon ratio was of order 0.25. These observations are now supported at the basin-scale by the images shown by Alvain *et al.* (2005) of the global distribution of four phytoplankton assemblages: the South Subtropical Convergence appears at all seasons, except briefly in late winter, as a linear zone dominated by *Synechococcus*-like bacterial chlorophyll. It is only briefly in summer that zonal bands of high diatom concentration partly obscure the dominance of small cells.

In each of the three water masses investigated, microzooplankton were consuming 78–118% of daily primary production. Despite such observations, the STCZ does harbor large populations of macrozooplankton, of which *Neocalanus tonsus* is a dominant component that contributes significantly to the vertical flux of carbon. Although this copepod has a latitudinal range of about 20° in the southern oceans, maximum biomass of >100 ind $m^{-3}$ in the mixed layer was located within the STCZ on each of 11 meridional transects in all three oceans (Bradford-Grieve and Jillett, 1998): this distribution suggests that this is one of the most abundant large copepods in all seas. Stage V copepodites of *Neocalanus tonsus* descend to >500 m in austral autumn, there to moult into adults; reproduction is then fueled by wax esters that have accumulated in the surface layer during spring and summer. Eggs are laid at depth, and the cycle starts again with the rise of larvae and of the surviving adult females to near-surface layers in austral spring. The annual descent of this population in autumn represents a very large contribution to downward carbon flux, lost to the interior of the ocean. This has been calculated as 2.5–8.9 g C $m^{-2}$ $y^{-1}$ by Bradford-Grieve (pers. commun., 1995), which is an order of magnitude larger than computed for the seasonal vertical migration of large copepods in the North Atlantic (Longhurst and Williams, 1992). But, because small cells form a larger fraction of the subantarctic phytoplankton, the 10.2 Gt C $y^{-1}$ that *Neocalanus* contributes to the deep carbon pool is a smaller proportion of total phytoplankton production than is represented by the flux in the North Atlantic.

Since "*Discovery*" days, it has been known that zooplankton biomass is greater in the STCZ than farther north and that it undergoes seasonal vertical migration. This increase in the abundance of zooplankton coincides with the southern "Transition Zone" of McGowan's Pacific biogeography. The group of 12 species of zooplankton that we encountered in the North Pacific Transition Zone Province (NPPF) occurs also here in the transition between subtropical and subpolar gyres. Representative species are the euphausiid *Thysanoessa gregaria*, copepods *Eucalanus bungii*, *E. elongatus*, and *E. longiceps*, the chaetognath *Sagitta scrippsae*, and the mollusk *Limacina helicina*. However, this region affords a good example of the dubious support given to ecological geography by biogeography: consider the euphausiid species pair *Nematoscelis difficilis* and *N. megalops*. In his monumental work on Pacific euphausiids, Brinton (1962) considered the former to be characteristic of the North Pacific transitional zone (the NPPF province), and *N. megalops* to be characteristic of the southern transitional zone (the SSTC province). He goes on to tell us that *N. megalops* occurs widely in the North Atlantic, a statement supported by later writers, in the Mediterranean, and in the Benguela Current, whereas *N. difficilis* is isolated in the North Pacific.

One might infer, from such a distribution, that one ocean is exceptional in having ecological similarity between boreal and austral high-latitude provinces. On the other hand, we may take comfort from Brinton's global distribution of *Thysanoessa gregaria*, which has three components. First, it occurs circumglobally in the SSTC, following the pattern of the STCZ closely in each ocean, though with some drift into the eastern boundary currents. Then, it occurs in the transitional zone of the North Pacific, where it closely matches the distribution of the NPPF and northern California Current Province. Finally, in the North Atlantic, it inhabits the terminal part of the Gulf Stream Province and the whole of North Atlantic Drift Province across the ocean. Whether these distinct populations belong to the same species, or to several cognate species, is a conundrum for the taxonomists that will not delay us here.

The enhanced biomass of lower trophic levels in the STCZ supports concentrations of large pelagic fish, including the mackerel *Trachurus picturatus murphyi*, which maintains a large population from Chile out to at least 160°W (Parrish, 1989; Evseenko, 1987). The STCZ is also the home range of the warm-blooded southern bluefin tuna (*Thunnus maccoyi*), which exploits the relatively high biomass of small fish and squid that must occur here. This species leaves the STCZ only to enter warmer water, principally to the northwest of Australia, to spawn in the austral winter, when fish appear in Tasman Sea Province and New Zealand Coastal Province as far north as the tropical convergence that passes across the northern Tasman Sea. The region of the STCZ adjacent to the Agulhas Retroflection appears to carry an especially large population of southern bluefin.

In the pristine ocean, before whaling, this was the preferred habitat of sei whales (*Balaenoptra borealis*), whose slender baleen plates and specialized tongue grooves are adapted to a wider range of planktonic organisms than those of the euphausiid specialists, blue and fin whales. Although the feeding mechanism of sei whales is thus specialized rather for obtaining copepods than swarming euphausiids, they utilize both skimming and engulfing behavior according to the prey species actually present: these include herbivorous copepods (e.g., *Drepanopus*), but also amphipods, especially *Parathemisto gaudichauda*, one of the three main swarming nekton species of the Southern Ocean, and in the SW Atlantic, *Munida gregaria* that is another swarming species—obviously! In the Southern Ocean, copepods are also swarm-forming: calanoids here form swarms of order 100 m in horizontal extent, at concentrations sufficient to support "skimming" feeding of sei and right whales. Nemoto and Harrison (1981) suggest that the absence of *Calanoides acutus* and *Euphausia triacantha* in whale stomachs is due to their nonswarming habit. Sei whales have now, after decimation of blue whales by whalers, to some extent shifted their range southward into the range of *Euphausia superba.*

This is also the feeding range of the bulk of the population of southern sperm whales (*Physeter macrocephalus*), although the larger males penetrate farther south into cold water. Sperm whales feed almost exclusively on squid, obtained by deep diving to at least 1200 m, where it is presumed that prey must be found entirely by echolocation. Very complex social structure is revealed by the composition of individual schools: specialized groups of individuals form nursery, juvenile, bachelor, harem, and bull schools that have some permanence and may each include >100 individuals (Brown and Lockyer, 1984).

## Synopsis

*Case 3—Winter-spring production with nutrient limitation*—The archived climatology illustrated in synopsis form in Fig. 12.3 shows that the mixed layer undergoes a major winter excursion to 175 m while photic depth remains at 40–50 m, so that the thermocline is illuminated only briefly in austral summer. Productivity rate increases progressively during the period of increasing illumination from June to the November maximum; thereafter, productivity rapidly declines to winter minimum. There is a rather weak seasonal change in chlorophyll biomass (0.25–0.35 mg $m^{-3}$), with some accumulation occurring during the period of the productivity rate increase. It is not clear why the seasonal changes in production and accumulation should not be matched more closely; in some years, consumption appears to require a period of about 60 days to come into

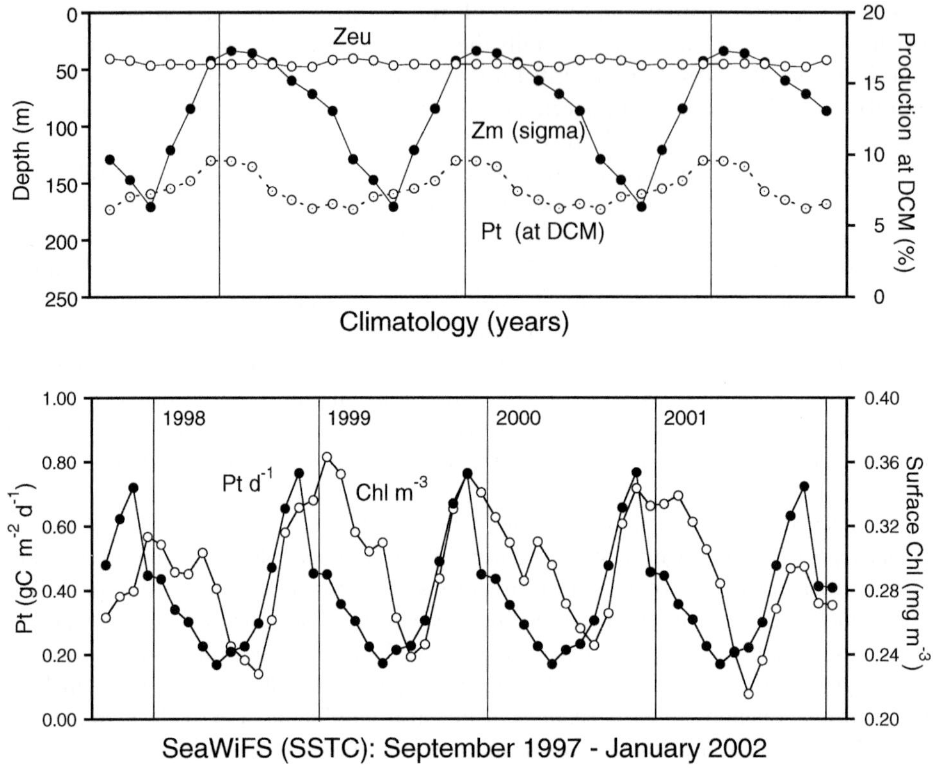


**Fig. 12.3** SSTC: seasonal cycles of monthly surface chlorophyll and depth-integrated autotrophic production for the years 1997–2002 from SeaWiFS data together with characteristic seasonal cycles of mixed-layer depths from Levitus climatological data and photic depths computed from characteristic irradiance and the archive of chlorophyll profiles discussed in Chapter 1.

balance with production. ISSG is an unreasonably large province, so perhaps this is a spurious result.

## Subantarctic Water Ring Province (SANT)

### Extent of the Province

This province is that part of the flow of the circumpolar ACC that lies between the Subtropical Front (STF) and the Polar Front (PF) at the southern limit of the Polar Frontal Zone. It comprises two distinct ecological zones that might be better separated for some purposes: the Polar Frontal Zone itself, in which algal growth is enhanced by a variety of processes, and the remainder of the Subantarctic Water Ring, which lies to the north and exhibits persistent oligotrophy. Banse (1996) refers to these as the northern and southern hydrographic regimes, but I prefer to retain the more conventional "Subantarctic Zone" (SAZ) and "Polar Frontal Zone" (PFZ) introduced earlier.

### Continental Shelf Topography and Tidal and Shelf-Edge Fronts

There is no coastline in this province, though the ACC does encounter shoal topography as it passes through the Drake Passage and the over Scotia Ridge on which stand South Georgia and the South Sandwich, Orkney, and Shetland Islands. This passage near shoal water has perhaps some consequences for the micronutrient concentration of surface water in the current jet along the PF (see also Chapter 7).

## Defining Characteristics of Regional Oceanography

The general characteristics of the northern flow of the ACC were noted briefly earlier, but it will be useful to reiterate that the two deep fronts with which we shall be concerned, the SAF and the PF, are each associated with frontal jet currents (Nowlin and Klinck, 1986). The PF jet is the stronger, but these two together comprise the main eastward transport within the northern part of the ACC. Flow in these jet currents is especially fast above the southwest Pacific midocean ridge, in the Drake passage, and south of the Agulhas Retroflection loop current. With each velocity maximum is associated a zone of intense baroclinicity (Carmack, 1990).

The Polar Frontal Zone (PFZ) occupies almost half of the latitudinal extent of the Subantarctic Water Ring (SWR), and is a complex feature bounded by two fronts, the SAF and the PF. At the PF, Antarctic Surface Water slips below and mixes with warmer water to the north, to form Atlantic Intermediate Water (AAIW). The strongest subsurface expression of the PF is usually 50 km north of the surface feature (Lutjeharms *et al.*, 1985). South of the PF, down to the southern boundary of the ACC, upsloping UCDW lies sufficiently close to the surface that deep winter mixing replenishes surface water with nutrients.

To the east of South America, the SAF and PF separate more widely than elsewhere (Tréguer and Jacques, 1992). Here, the SAF passes to the north close around the Falkland Islands, before looping around the Argentine Basin (see SSTC) to meet the southerly flow of the Brazil Current at about 40°N. Meanwhile, the PF passes directly north over deep water between the Ewing Bank and the Falkland Plateau to approach the SAF again, after which both pass eastward across the ocean, defining anew the PFZ lying between them (see Fig 12.2). This alignment will be referred to again later in relation to a claim that iron enrichment of the PF occurs by entrainment from shallow sediments during this passage.

These are the windiest seas in all oceans; only in some parts of the North Atlantic in winter does comparable wind stress occur at the sea surface. Even in December (austral midsummer) wind speed at 50–60°S is similar to almost anywhere in the winter northern hemisphere at similar latitudes, and the passage of depressions associated with force 6–7 winds must be expected even at that season. Winds are strongest in the region from the longitude of South Africa to the mid-Pacific, where *mean* wind speed over large regions exceeds 14 m $sec^{-1}$. Wind direction is consistently from the west or northwest, and the zonal component of velocity varies by a factor of about 2 between summer and winter. Wind-stress curl south of the PF imparts buoyancy gain through Ekman divergence while, north of the PF, downwelling and buoyancy loss is imparted.

The pattern of solar irradiance over the Southern Ocean involves winter darkness poleward of 66°S; the opposite side of this coin is that day length there is longer at midsummer than at lower latitudes. In some parts of this province, therefore, the total solar radiation received at the surface each summer day is greater than in the central subtropical gyres. Consequently, near-surface phytoplankton may become light-saturated at the surface so that the standardized photosynthetic rate may be maximal at 5–10 m depth.

The SAZ is characterized by deep winter mixing (April–October) and a shallow summer thermocline that has an unusually weak thermal gradient. Winter mixed-layer depths are greater here than anywhere else in the oceans; heat loss and deep convection may carry mixing sufficiently far down as to entrain deep high-salinity, high-nutrient water into a surface homogenous layer. Deepest winter mixing (>400 m) occurs at about 50°S and may prevent the formation of a permanent thermocline in an annulus around the entire

Southern Ocean, especially in the Indian and Pacific Ocean sectors, differing from regions to north and south where winter mixing is much shallower. Here, Subantarctic Mode Water (SAMW) is formed in winter, having nutrient characteristics imposed by biological processes further south. SAMW is a pycnostad and passes down and northward into the main thermocline of each ocean basin, which has consequences for the nutrient supply in other parts of the oceans (see later discussion). Within deep mixed layers formed in winter, near-surface stratification is induced as both sun angle and day length increase, and by midsummer the seasonal mixed layer (9–12°C) deepens in this way to 60–75 m. However, there is probably greater impermanence of stratification here than further to the south in APLR, where winds are quieter.

The Fe profile in the SWR in summer is that of a typical nutrient so that, after winter mixing, a ferricline is not observed. It is suggested that eastward jet flows, such as the PF, may transport water that has been enhanced in Fe from shelf deposits in the Drake Passage/Falkland Plateau region (Loscher *et al.*, 1997b). High chlorophyll accumulation in the Southern Ocean fronts have come to be attributed to "differences in iron availability" between the frontal systems and other regions (e.g., de Baar *et al.*, 1995). But it has never been demonstrated that the iron-rich character of water in the frontal jet currents can be sustained over great distances; in fact, the chemistry of iron in seawater suggests that this will not occur. More realistic, perhaps, is the alternative view that most iron utilized by phytoplankton in the Southern Ocean, including within frontal systems, is upwelled. Indeed, as Fennel *et al.* (2003) comment:

> In the contemporary Southern Ocean, deep mixing during winter and upwelling of iron-rich subsurface waters are the main sources of dissolved iron in surface water. Aeolian deposition plays only a minor role. Locally, significant amounts of iron can be added during ice melt, i.e. in the Ross Sea, but in the SW Pacific sector this input is negligible.

This is the model followed here, and I specifically reject that of Smetacek *et al.* (2004), who write: "Phytoplankton blooms are restricted in the modern Southern Ocean to waters receiving an adequate supply of iron from landmasses and sediments. Evidence is mounting that wind-blown dust is a major supply route of iron to the land-remote Southern Ocean."

There is a surface nitrate gradient across the SAZ, especially strong in summer that, at this season, increases poleward from $[NO_3^-]$ of 1–3 μM just south of the SAF to 10–15 μM in the PFZ. A similar gradient of $[Si(OH)_4]$ also occurs very strongly across the SAF; spring-summer values to the north of the front are < 1–15 μM but 40–60 μM on the southern side (Franck *et al.*, 2003), although these are but discontinuities in the general poleward progression of the surface nutrient gradient during summer. How the different nitrate-silicate balance evolves, both spatially and seasonally, north of the PF will be discussed in the following section.

## Biological Response and Regional Ecology

The Polar Frontal Zone is a major discontinuity in the Southern Ocean, well expressed in an evocative description (Deacon, 1984) of what it feels like to cross the convergence on returning from the polar regions, because the air smells again as air at sea *should* smell, but doesn't in the high Antarctic. It is also a significant border in species distributions and a major feeding zone of seabirds.

The principal observations of the planktonic ecosystem in this province that require explanation are (i) that a zonal band of high chlorophyll biomass should lie around the southern half of the province, (ii), that the seasonal surface chlorophyll cycle should apparently be a simple function of sun angle, and (iii) that the phytoplankton should be dominated everywhere by small cells. The explanations that can so far be offered are incomplete and complex, because these are interrelated and improperly understood problems. To begin to understand them it is necessary to invoke the special characteristics of this region discussed earlier in this chapter, and in Chapter 5.

Because it is here that subantarctic conditions are first encountered, it will perhaps be useful to note a few common characteristics of the ecology of this region and of the other antarctic provinces. I have found Trull *et al.* (2001) to be a particularly useful source for such an evaluation. The Southern Ocean shows two principal zonal bands of surface chlorophyll enhanced above background: the STCZ zone (discussed in the previous section) and another that lies to the south of it. Comparison with maps of SLA reveals that the southern zonal band corresponds both with the geographical position of the PFZ and also with a zonal band of strong eddying in the ACC that matches the line taken by the PF around the Southern Ocean. A zone of silica-rich sediment whose main component is the diatom *Fragilariopsis* lies on the deep-sea floor, aligned below the PFZ. Sediment traps and estimates of primary production suggest rates of $<80\,g\ C\ m^{-2}\ y^{-1}$, or about twice as high as the Antarctic average and an order of magnitude higher than in the oligotrophic zone of the Subantarctic Water Ring (Wefer and Fischer, 1991; Tréguer and Jacques, 1992).

We are therefore confronted by the same relationship between vorticity and phytoplankton blooms as we were in the STCZ. Once again, observations confirm that enhancement of chlorophyll biomass here is associated with mesoscale frontal dynamics. Strass *et al.* (2002) review the mechanisms that are involved in this process, especially the cross-frontal circulation related to baroclinic instability; confluence of surface water at the edges of meanders concentrates phytoplankton blooms that have grown in response to upwelling-supplied nutrients on the anticyclonic side of jets.

As shall be discussed later, these dynamic processes are imposed upon a general background of the winter recharge of essential molecules, including dissolved iron, by vertical transport into the APLR province from the subjacent and nutrient-rich UCDW. During the growth season, N, P, and Fe can be recycled only where organic matter is retained by a density discontinuity within the euphotic zone. Even in such places, of course, Si is preferentially depleted by loss of very slowly dissolving diatom frustules to the deeper layers. The findings of van Oijen (2004) concerning differential control of growth by Fe limitation and light limitation in the PF have already been discussed in Chapter 5. Here it is sufficient to remind ourselves of their finding that in the deep-mixing region of the ACC to the south of the PF in late summer, phytoplankton cells are light-limited, but in the shallower-mixing regime of the PF itself, they are Fe-limited. Nevertheless, the PF carries a larger standing stock of diatoms than the ANTA zone to the south.

The regional evolution of satellite-observed surface chlorophyll in the SWR closely tracks sun-angle, so that highest biomass occurs in midsummer, but local conditions exhibit rapid variability. Chlorophyll biomass across the PFZ at 170°W was followed by Abbott *et al.* (2000) at short time intervals throughout an austral summer. A few days of calm weather in November initiated weak stratification, and the phytoplankton came to photoadaptation with the resulting improved light environment within 1–2 days by modification of their fluorescence/chlorophyll (F/C) ratio. Chlorophyll concentrations rapidly increased, reaching maximum values ($1.25\,mg\ chl\ m^{-3}$) in about 14 days, or well before nutrient stress was revealed. A subsequent decrease of chlorophyll biomass was almost as rapid, so that by the end of December a low level of biomass ($0.25\,mg\ m^{-3}$) was reached and subsequently sustained. The initial rate of decrease, before the onset of nutrient stress, was consistent with grazing rates of $0.2\,d^{-1}$ measured nearby. For the remainder of the summer, steadily increasing F/C ratios indicated progressive nutrient stress and grazing-balanced accumulation. Again, there is a close relationship between vorticity and phytoplankton abundance in the frontal zones, and the mechanisms involved in this process, especially cross-frontal circulation related to baroclinic instability, are now quite clear. One of the principal mechanisms for cell accumulation is the confluence of surface water at the edges of meanders, thus serving to concentrate cells of phytoplankton blooms that themselves are responses to upwelling-supplied nutrients on the anticyclonic side of jets, especially where these exhibit sloping isopycnals (Strass *et al.*, 2002). High-chlorophyll anomalies in the PFZ along a cyclonic bend in flow are likely to be associated with nutrient flux and a sloping density surface; such enhancement is often sheared into 10- to 20-km-wide bands that are coherent at the scale of hundreds of kilometers.

Uptake of macronutrients in the developing bloom is co-regulated by levels of available iron, so that iron limitation prevents complete drawdown of both silicate and nitrate. The progressive excess during the summer of $NO_3$ over $Si(OH)_4$—normally required by diatoms in a 1:1 ratio—has generally been ascribed to the effect of the Si-pump model, which allows no regeneration of Si in the euphotic zone. This model reflects the fact that whereas $NO_3$ and $PO_4$ are regenerated in the upper kilometer of the water column from sinking organic particles, $SiO_3$ is regenerated from sinking silica particles only by solution. Because solution is a negative function of temperature, much silica in the form of partially dissolved diatom frustules reaches the deep sea floor. But this is not the only mechanism that produces a disproportionate drawdown of $Si(OH)_4$ relative to $NO_3$ as the growth season advances; this imbalance is accentuated by the anomalously high silica demand of diatoms under conditions of Fe stress. Southern Ocean JGOFS experiments found that, *in vitro*, Fe addition decreased $[Si(OH)_4]:[NO_3^-]$ ratios by three to five times over controls in high-Si water south of the nutrient gradient (Franck *et al.*, 2000, 2003).

Nutrient limitation of growth in the declining bloom is a complex interplay between iron and silicate. The meridional gradient in surface nitrate in summer is from 1–3 μM just south of the SAF to 10–15 μM at the PF itself, but the silicate gradient across the SAF is stronger: to the north of this front, silicate is <1–15 μM, whereas on the southern side it is 40–60 μM because of the influence of upwelled UCDW further to the south.

Deep convection of Subantarctic Mode Water (SAMW) having low concentrations of silicate occurs in the SAZ in winter and of Antarctic Intermediate Water (AAIW) in the PFZ. The descending SAMW water enters the main thermocline of the ocean and its Si deficiency is conserved, spreading almost globally through the main thermocline and so determining the nutrient environment of the global nutricline (Sarmiento *et al.*, 2004), excepting only in parts of the North Pacific. Thus the $NO_3/SiO_3$ ratio of the nutrient supply available to fuel algal blooms in much of the global ocean is consequent upon phytoplankton/nutrient dynamics within the SAWR province. Sarmiento *et al.* suggest that this may determine the surprisingly low production of diatoms, compared with other phytoplankton taxa, in much of the present-day ocean.

As I noted in the introduction to this chapter, I have previously attributed the high chlorophyll in the PFZ largely to relief from Fe limitation, as suggested by de Baar *et al.* (1995). An Fe anomaly had been located in the PF at 43–50°S, reaching >5 nM in near-surface water against a background of around 0.5 nM. Because this was associated with an Al anomaly, de Baar *et al.* invoked a terrestrial input from the shallow topography of the Scotia Arc. The most likely source, the Birdwood Bank, is far upstream, and it is very unlikely that the discrete patch that was observed could be conserved so far downstream in a dynamic, meandering frontal system. Similar results were obtained in the PF downstream of the Kerguelen plateau at 70°E during the JGOFS/ANTARES 3 investigations; $Fe_{td}$ was very high in coastal water (~10 nM), but complex mesoscale eddying made it difficult to trace this water progressively to the east.

I suggest, therefore, that de Baar *et al.*'s interpretation of its source may be incorrect. Nor do we need to suggest that the generally enhanced phytoplankton biomass in the circumglobal PFZ must result from an upstream terrestrial Fe input. It will also be well to remember, when considering sources of iron for surface water in the Southern Ocean that here, too, the model of Johnson *et al.* (1997) for the continuous regeneration of dissolved Fe from sinking particles must apply. So also does the model for the recycling of oligonutrients that goes back to Ramadhas and Venugopalam (1972). These authors demonstrated that "body-equivalent excretion time" of Fe in zooplankton is a negative function of body size, so we may anticipate very rapid turnover of iron through the microbial food web. Iron excreted by the macrozooplankton is largely in particulate form and this is available for recycling, according to the model of Johnson *et al.*

It was, of course, the oligotrophic regime of the SAZ in the northern part of the Subantarctic Water Ring that first attracted attention to the "HNLC" properties of the Southern Ocean. Essays such as Karl Banse's "Low seasonality of low chlorophyll concentrations in the Subantarctic Water Ring" leave no doubt that this is where the emphasis lay. The reader is left with the impression that this is the dominant condition of the open Southern Ocean, but the serial SeaWiFS images now give us a more realistic vision of that region. You will observe that the oligotrophic region of the SAZ is no wider than the two frontal zones of higher chlorophyll that bound it north and south. You will also note that some seasonality, much between-year difference, and many mesoscale patches of enhanced chlorophyll are typical of the SAZ, which is not a monotonous blue desert, as is abundantly confirmed by a viewing of any annual series of 8-day images of surface chlorophyll in movie format.

In the oligotrophic regime (seasonal in the PFZ, quasipermanent in the SAZ), the phytoplankton is dominated by very small cells despite the relative absence in these cold seas of the prokaryotic autotrophs. A relatively invariant background flora of pico- and nanoplankton contributes up to 90% in the SAZ, where overall chlorophyll concentrations are low (0.2–0.4 $\mu$g liter$^{-1}$), but $<50\%$ in the PFZ where chlorophyll biomass is higher ($>1.8\,\mu$g liter$^{-1}$) during the bloom period. As Detmer and Bathman (1997) put it, the pico and nano fractions "represent abundant and stable constituents of the phytoplankton assemblage of the Southern Ocean." However, their diversity appears to be somewhat greater in the PFZ (prymnesiophytes, green algae, dinoflagellates, cryptophytes, pelagophytes, and micromonads) than in the SAZ, where only the first two of these taxa are significant.

The principal consumers of these small autotrophs are protists and very small zooplankton, often the larval stages of larger species. Imposed upon the flora of the microbial loop, when nutrient, irradiance, and stratification conditions are suitable, is a flora of larger diatoms and the colonial cyanobacterium *Phaeocystis pouchetti.* The autotrophs of the microbial loop are here dominated not by cyanobacteria, but rather by small flagellates and diatoms that maintain biomass levels of the order of 0.2 mg chl m$^{-3}$, based on production rates of $\sim$200 mg C m$^{-2}$ d$^{-1}$. This biomass is maintained against the constant grazing pressure of abundant heterotrophic nanoflagellates, themselves consumed by larger ciliates and flagellates (Smetacek *et al.*, 2004).

In assessing the relative abundance of autotrophic flagellates in the Southern Ocean, there is a potential for misleading counts, because extracellular chloroplasts from damaged cells continue to fluoresce and may be mistaken for monads for some days at these very cold ambient temperatures. Estimates of the contribution of nanoflagellates and very small diatoms (3–20 $\mu$m) to total water column chlorophyll and primary production are on the order of 50–75% and 50–65%, respectively. Since small cells are able to utilize available irradiance more efficiently than large cells, it should be no surprise that, generally, nanoplankton (fueled by the microbial loop) contribute most in a well-mixed water column. Micro- or net plankton contribute more as soon as stability is introduced at periods longer than the potential time scale of bloom development (Weber and El-Sayed, 1987; Jochem *et al.*, 1995) or if trace nutrient deficiency should cease to limit nitrogen uptake.

It is the diatom flora that has, of course, attracted much attention in the Southern Ocean and is dominated by cells around 10–50 $\mu$m in dimension, frequently chain-forming and often of complex cell form. As has been proposed, the formation of chains and the development of long spines and bristles on the cell wall may reduce their rate of consumption by herbivores (Smetacek, 2001). It is also possible, of course, that in these seas where deep mixing rather than stratification is the rule, cell spines may function also to slow sinking rates as well as to inhibit ingestion. Iron-enrichment experiments in the PFZ have, nevertheless, demonstrated the usual fate of these large cells in episodic blooms. In EIFEX 2004 at 45°S in the Atlantic sector the summer mixed layer was very deep and "the rate at which the bloom sank through the water column exceeded our expectations"—to quote the Weekly Report from *Polarstern* for 22 March 2004. Within one week of the peak of the induced bloom, a mass of apparently healthy diatoms was found to be sinking at >500 m $d^{-1}$ and was subsequently followed clear to the bottom at 3500 m. No sustained natural bloom can occur here, simply because euphotic zone regeneration cannot occur: one or another element would be rapidly stripped down to limiting level by removal in sinking phytoplankton. It is not surprising, therefore, that a zone of silica-rich sediment lies on the deep-sea floor below the PFZ. Sedimentation rates appear to be of order 80 g C $m^{-2}$ $y^{-1}$, or about an order of magnitude higher than in the oligotrophic zone of the Subantarctic Water Ring (Wefer and Fischer, 1991; Tréguer and Jacques, 1992). Sporadic blooms in oligotrophic, perhaps Fe-limited places, a small number of unusually large diatom species are characteristic: *Fragilariopsis kerguelenensis, Thalassiothrix antarctica, T. lentiginosa, Corethron pennatulum.* Such species have a relatively high silica demand, because of their large size and extravagant cell architecture.

The high-resolution (15-km scale) distribution of diatom species in the PF at 10–11°W at austral midsummer of 1995–96 was resolved by the German JGOFS team, who observed three species complexes of diatoms biomass being dominated by the species *Fragilariopsis kerguelenensis, Corethron inerme* and *C. criophilum* (Bathman *et al.*, 1997; Peeken, 1997). Even in the absence of significant stratification in October–November, chlorophyll concentrations increased from 0.7 to 4.0 mg chl $m^{-3}$ to biomass stabilize at 2.0 mg chl $m^{-3}$. Maximum concentration of a *Thalassiothrix* species complex occurred north of the frontal jet, *Chaetoceros* biomass was highest within the jet itself, while *Pseudo-nitzschia* was more uniformly distributed. The dominant diatoms in each species complex were all very large and either needle-shaped or spinous (Smetacek *et al.*, 2002). Only euphausiids would have been capable of clearing blooms of these species and they were present only in small patches. Diatom biomass were therefore able to reach biomass of 2 mg chl $m^{-3}$ against a pico- and nanophytoplankton background of 0.5 mg chl $m^{-3}$, much more susceptible to herbivore consumption.

Historically, and until the quite recent past, attention has been focused on what were thought to be the dominant Antarctic herbivores, comprising copepods (*Calanoides acutus, Calanus propinquus, Metridia gerlachi,* and *Rhincalaus gigas*), the chaetognaths *Sagitta gazellae* and *Eukrohnia hamata*, and swarming tunicates (*Salpa thompsoni*). Though implicated in regulating diatom blooms (and providing food for whales and seabirds), the crustacea of this assemblage are also facultative predators and (in the case of some species) able to feed on detritus and algae from the lower surface of ice floes. The 1992 JGOFS investigations suggested that despite their abundance, the large herbivorous copepods (mostly *Calanoides acutus* and *Calanus propinquus*) consumed only 1% of the daily primary production in the diatom blooms at the Polar Front in spring and summer (Dubischar and Bathman, 1997). It was concluded that the massive swarms of salps (*Salpa thompsoni*) were the principal constraint on phytoplankton biomass: but these organisms, previously thought of as microherbivores, were found to consume important quantities also of small zooplankters.

This model was derived from results obtained with the large-mesh (200 μm) plankton nets habitually employed in the Southern Ocean, but the use of smaller-meshed nets by

Dubischar *et al.* (2002) showed that the large calanoids are—as it were—just the tip of the copepod iceberg. In the PFZ of the Atlantic sector, dense populations of small copepods occur within the frontal system, although diversity is not high: a few dominant species take abundances five to ten times higher than the "common" species. Concentrations of up to 50,000 ind $m^{-3}$, mostly copepodites (and mostly those of *Oithona similis, O. frigida,* and *Ctenocalanus citer*) were recorded in obliquely hauled nets. Even higher concentrations must occur at some depths, since profiles of copepod abundance are rarely uniform. The abundance of small copepods in the PFZ is therefore comparable with the greatest abundances reported for productive coastal ecosystems. Retention in the warm, downwelling side of meanders along the frontal jet assists in maintaining such concentrations, but the fivefold increase in numbers of small calanoids and cyclopoids in austral spring suggest that these may play an important role in the controlling diatom accumulation. Although the daily carbon requirement of the <2-mm copepods exceeds the 750–950 mg C $m^{-2}$ $d^{-1}$ typical of such blooms, at least *Oithona* is known to feed largely on fecal pellets of larger zooplankters. Whatever they consume in these seas, we cannot neglect the role that small copepods must play in regulating Southern Ocean plankton dynamics.

Despite the importance of smaller forms, it is the larger copepods that have attracted most attention in the Southern Ocean until now. The seasonal and ontogenetic vertical migrations, which allow them to feed near the surface and to overwinter deep, have been studied in detail since the early days of Antarctic oceanography. Two related species of calanoids have fundamentally different strategies: I deal with them together for comparison, though it must be noted that these species have somewhat different zonal distributions. *Calanus propinquus* occurs principally in the open water of the PFZ and immediately poleward, whereas *Calanoides acutus* is the more austral species, at home in the coastal water flow (Mackintosh, 1937).

In line with this observation, *C. propinquus* departs from the high-latitude norm and stores its lipid reserves as triglycerides rather than as wax esters as does *C. acutus.* Both species migrate deep in autumn, but differently. *Calanoides acutus* descends to great depths (500–1000 m) to overwinter in diapause so as to conserve lipid reserves that will be required for reproduction at the surface in spring. *Calanus propinquus* goes less deep (100–500 m) and does not enter physiological diapause; some feeding, often by predation on microzooplankton, may continue and reproduction occurs over a longer period than for *C. acutus.* Both species are concentrated near the surface in summer (usually at 0–50 m), and *C. acutus* both ascends and descends earlier than *C. propinquus.* The ascent of *C. acutus* females coincides with the onset of sea-ice melting, or with the very earliest spring bloom (October) in permanently open water: eggs are matured and shed quite rapidly. In the austral autumn (March), *C. acutus* descends (as copepodite 4s or 5s), although there is still sufficient phytoplankton in the surface waters for *C. propinquus* to continue feeding (Smith and Schnack-Schiel, 1990; Schnack-Schiel *et al.*, 1991).

In contrast, *Rhincalanus gigas* is equally at home in both SANT and ANTA provinces and may produce two generations per year, in spring and autumn. It lies somewhat deeper in summer (50–100 m) than does either *C. acutus* and *C. propinquus,* and in winter it descends to 500–1000 m. Several species perform regular diel migrations in this water body between the near-surface chlorophyll layer at night and the warmer subsurface layer by day at >200 m. These are familiar diel migrant genera: *Pleuromamma robusta* together with the small krill species *Euphausia triacantha* and *E. frigida* that preferentially inhabit the PFZ.

Here, sei whales feed on swarming zooplankton, largely copepods, that are present in sufficient concentrations to support their unique "swallowing and skimming" feeding technique that merges that of right whales and the blue whale. Sei whales are more omnivorous than right whales, taking not only herbivorous crustacea but also amphipods, euphausiids, and schooling small fish and squid. But only aggregated prey is taken by these as by other baleen whales.

Zooplankton aggregations attract also seabirds. Observed densities of 29 species, almost all Procelliformes (petrels, shearwaters, prions), in the PFZ in the austral summer of 1995–96, varied from very small numbers up to 400 ind $km^{-2}$ (van Franeker *et al.*, 2002). Where mesoscale circulation engendered high values for phytoplankton productivity and zooplankton abundance, aggregation of birds also occurred. This positive spatial correlation between plankton ecology and bird numbers was functional in that broad-billed prions, the numerically dominant bird species, foraged on very small copepods (*Oithona* spp.) that were aggregated in the frontal zones. Species that fed largely on squid and fish showed a much weaker correlation with the details of mesoscale features: these larger food organisms are much less passive tracers of water movement than are the zooplankton.

## Synopsis

*Case 3—Winter-spring production with nutrient limitation*—The mixed-layer depth undergoes strong austral winter excursion (60 m January–February, 300 m July–September) while photic depth remains about 40 m, so that the thermocline is not illuminated at any season. Productivity rate increases progressively during winter-spring starting in August to an annual peak rate in early austral summer (November); subsequently a rapid rate decline establishes the austral summer-fall period of low productivity rates (Fig. 12.4). There is no significant period of chlorophyll accumulation, so that biomass remains in the

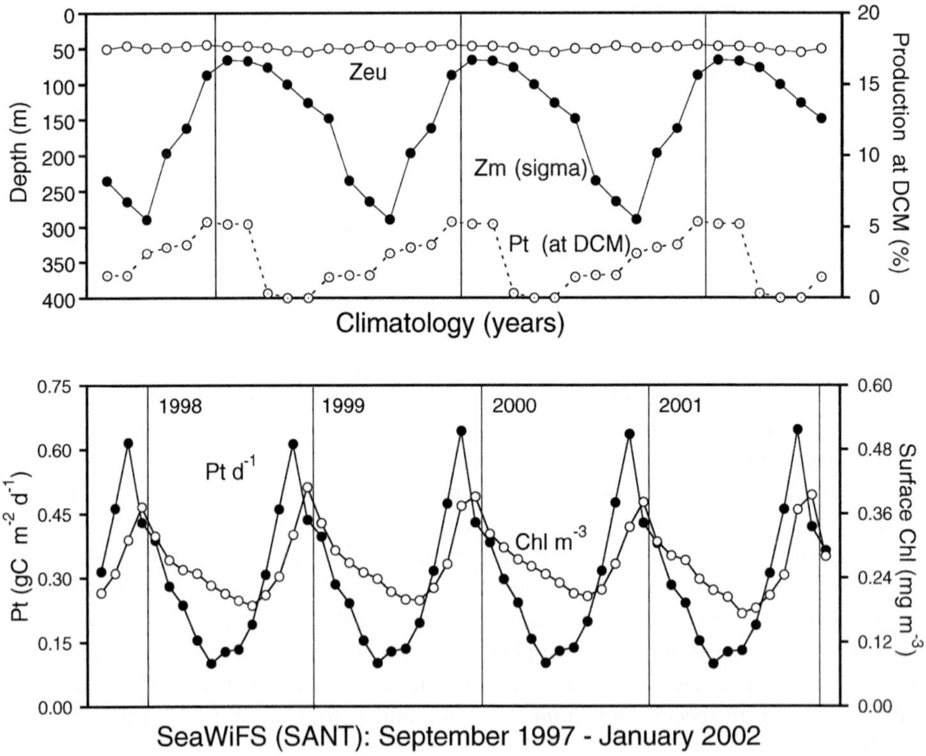


**Fig. 12.4** SANT: seasonal cycles of monthly surface chlorophyll and depth-integrated autotrophic production for the years 1997–2002 from SeaWiFSdata together with characteristic seasonal cycles of mixed-layer depths from Levitus climatological data and photic depths computed from characteristic irradiance and the archive of chlorophyll profiles discussed in Chapter 1.

range 0.2–0.4 mg C $m^{-3}$ year-round. Failure to achieve chlorophyll accumulation during the winter-spring period of P rate increase is consistent with closely matched production and consumption processes and also with extreme mixing-down of cells due to unusually strong wind stress at the surface.

# Antarctic Polar Biome

As already noted in the introduction to this chapter, two nutrient regimes were postulated by Hiscock *et al.* (2003) for what is treated here as the austral Polar biome, an approach to the question of regional nutrient distribution that has the great merit of being functional. The performance of phytoplankton cells induced by differential nutrients availability is what is really important, and not—as in some previous discussions of the Southern Ocean—simply the relative concentrations of nutrient molecules.

## Antarctic Province (ANTA)

### Extent of the Province

This is an annular province and is synonymous with the Southern Branch of the ACC of some authors. It lies between the Polar Front at about 50°S and the Antarctic Divergence at about 60–65°S.

### Continental Shelf Topography and Tidal and Shelf-Edge Fronts

No coastline is included, because the island arcs between the Antarctic Peninsula and South America are allocated to the APLR province. However, a few smaller oceanic islands do occur in this province: Bouvet, Kerguelen, and Heard of the Atlantic and Indian Ocean sectors.

### Defining Characteristics of Regional Oceanography

This province includes permanently ice-free conditions, regions that are ice-free in austral summer but covered by pack ice in winter, and also some regions that may be permanently ice-covered. Drifting, melting icebergs that have broken free from glacier fronts are, in some years and some places, very frequent and may have consequences that are ecologically significant. The area of seasonal pack ice approaches very close to the PF, especially in some years and at some longitudes. The 1997 US JGOFS section down 170°W in December found the rapidly retreating seasonal ice edge at 65°S, or only 2–3° poleward of the Polar Front even in austral summer. The 1992 German JGOFS sections at 6°W in October–November found the PF at 49°S with spun-off eddies to 52°S; the retreating ice edge was at 54–56°S at this time. Thus, what others have called the Permanently Open Ocean Zone (POOZ), equatorward of the zone across which the ice edge migrates seasonally, may be very limited in meridional extent.

A permanent halocline lies at about 150–200 m between the AD and the PF, sloping up toward the south. In summer, heating of the surface water mass induces a pycnocline at 50–100 m, below a shoal mixed layer. But, even at this season, the mixed layer is by no means immune to episodic deepening during the passage of a depression with force 6–7 winds. Because deep water below the halocline is warmer than surface water, a thermal inversion may occur if surface heating during a period of quiet winds has incompletely heated the entire surface water column above the halocline.

This is, of course, the region within which Atlantic Deep Water upwells, within and adjacent to the Antarctic Divergence. Lower Central Deep Water (LCDW) reaches the surface at the AD and Upper Central Deep Water (UCDW) closer to the PF. The LCDW water mass carries higher concentration of dissolved nutrients than UCDW.

## Biological Response and Regional Ecology

Nelson and Smith (1991) reformulated the Sverdrup model with parameters suitable for very high latitudes to estimate the maximum chlorophyll biomass permitted by the relationship between critical and mixed-layer depths in the APZ. Their predictions closely match observations, both for the marginal ice zone and for the open oceanic regions. At 60°S, where clear-sky irradiance in January would be ~60 mol $m^{-2}$ $d^{-1}$, this is reduced to <50% on most days by cloud cover to around 25 mol $m^{-2}$ $d^{-1}$. Critical depth ($Z_{cr}$) shoals in response to a very small increase in chlorophyll, but this relationship becomes much weaker at >2 μg $liter^{-1}$ chlorophyll. In the absence of chlorophyll, critical depth would exceed mixed depth, but $Z_{cr} = Z_m$ at 0.5–1.5 μg chlorophyll, depending on the value of $Z_m$. As Nelson and Smith comment, "This makes any bloom self-limiting" because above a critical chlorophyll biomass "an unfavourable relationship between $Z_{cr}$ and $Z_m$ would prevent further increase in phytoplankton biomass." In open water, this may restrain biomass to <1.0 μg $liter^{-1}$, although higher concentrations occur in the MIZ. Nelson and Smith predict that mixed-layer depth and integrated chlorophyll biomass will be inversely related, as indeed they appear to be now that we have adequate spatial information on chlorophyll and wind speed from satellite observations.

The regime in the permanently open water and of that part of the migrating marginal ice zone (MIZ) that lies north of the CWB is seasonally complex, as noted earlier. This region lies poleward of the maximum silicate gradient ($Si_{max}$) in spring, but equatorward in summer, because the strong uptake of silica by the diatom blooms that occur progressively further south during that period drives the silica gradient poleward. Progressive $SiO_3$ depletion induces a shift from Fe limitation in spring to Fe repletion in austral summer. This change is accompanied by the disappearance of diatoms from the phytoplankton assemblage, which comes to be dominated by small cells.

As suggested by such models of nutrient dynamics, the phytoplankton of the open-water areas in the ANTA province exhibits a complex response to changing seasons. During winter, the low sun angle effectively stops phytoplankton growth, but upwelling continues to bring UCDW water, relatively rich in both macro- and oligonutrients, toward the surface. In the Pacific sector, the surface water mass at the end of winter comprises almost entirely UCDW, with a small admixture of surface water from the austral polar regime that flows equatorward from the APLR province (Hiscock *et al.*, 2003). Thus, when sun angle increases and intermittent stratification occurs in surface waters (recall that the ice edge is usually not far distant), a diatom bloom develops in late November or December, propagating poleward and progressively depleting $SiO_3$ as it goes. Highest diatom biomass occurs at the silica gradient as the two features move poleward together, but the bloom is liable to be interrupted by local mixing events. Of course, where the zonal band of open water at the end of winter is relatively narrow, the open water and MIZ blooms are initiated almost simultaneously. Where the zone is relatively wide, the open-water bloom tends to be dominated by large diatoms (*Corethron, Chaetoceros*), whereas along the ice edge *Phaeocyctis* is often dominant. Local chlorophyll enhancement with values reaching about 4 mg chl $m^{-3}$ at a subsurface chlorophyll maximum (Smith and Nelson, 1985). Sediment trap data indicate that the very brief spring bloom is dominated by diatoms in December and January and is followed by a late-summer season of regenerative algal growth of small cells from which sedimentation rates are two orders of magnitude lower (Smetacek *et al.*, 1990).

As the blooms propagate poleward, a seasonal succession ensues with the most mature communities along the southern boundary of the PFZ and the youngest community near the receding ice edge. The mature pelagic ecology of the open-water zone is characteristic of a retention community in the sense of Peinert *et al.* (1989). Small cells dominate, together with an appropriate population of heterotrophic protists and bacteria; nevertheless, diatoms may remain abundant for extended periods far from the receding ice edge. This was observed during the AESOPS Process Study 1 voyage of 1997 along 107°W: here, 2° north of the ice edge, a residual bloom characterized by a strong DCM was dominated by bundle-forming *Thalassiothrix* and chains of *Fragilariopsis*. This bloom was quite distinct from the *Phaeocystis*-dominated bloom at the ice edge, although biomass of autotrophs was equivalent (Selph *et al.* 2001).

The JGOFS 1992 investigations confirmed that up to 90% of total chlorophyll in the open-water post-bloom phytoplankton resides in pico- and nanophytoplankton, with very small contributions from diatoms, dinoflagellates, prymnesiophytes and chlorophytes (Peeken, 1997). Low chlorophyll values under such conditions may result in water of exceptionally high clarity, and Secchi disc depths of >60 m have been recorded after the open-water bloom has collapsed. The lack of permanence of such extremely oligotrophic conditions in the Southern Ocean has been addressed earlier. The sedimentation rate from the summer algal growth here is at least two orders of magnitude lower than at the MIZ (Smetacek *et al.*, 1990). The seasonally migrating MIZ is associated near-surface stratification that we would expect to lead to the initiation of a bloom wherever irradiance is sufficient. But in the nutrient regime south of the CWB, beyond the influence of UCDW, Fe is always limiting until we reach the polar neritic regions further south. So the spring breakup and retreat of the MIZ may not always result in strong diatom blooms.

Although highest chlorophyll biomass in this province usually occurs at the ice edge, biomass at the receding MIZ bloom is not always significantly higher than in the adjacent open-water bloom. The repeated ADEOS sections along 170°W showed that chlorophyll biomass in open water at 65°S (far to the north of the MIZ) in late January was equivalent to biomass at the ice edge in mid-December (Buesseler *et al.*, 2003). Satellite images showed a band of chlorophyll (0.5–3.0 mg $m^{-3}$) advancing southward with the ice front, but with maximum chlorophyll always about 5–7° north of the 50% ice contour. During austral spring and summer 1997–98, observations suggested a complex model of nutrient supply and limitation (Hiscock *et al.*, 2003), based on the upsloping of Fe-rich, nutrient-rich Upper Circumpolar Deep Water (UCDW) to the surface between the PF and the southern boundary of the ACC, south of which nutrient-poor LCDW surfaces. Silicate is progressively depleted poleward by diatom growth (Nelson *et al.*, 2001) and a strong latitudinal gradient develops, the $Si_{max}$ shifting from 62–63°S near the PF in early spring down to 66°S at the southern boundary of the ACC by midsummer.

The dynamics of the consumption of Antarctic blooms has become clearer in recent years. The UK-JGOFS Southern Ocean studies showed that even here, in the APLR province, we cannot ignore the protist microplankton, as has been done in the past. Along the Bellingshausen Sea section mentioned previously, protistan microplankton were sparse below continuous ice cover but increased in numbers in the open water, their abundance being positively correlated with chlorophyll biomass. Where maximum numbers occurred (54 mg C or 17,000 organisms liter$^{-1}$( the microplankton was dominated by oligotrich ciliates (*Strombidium, Tontonia*, and *Lohmaniella*) and heterotrophic dinoflagellates (*Gyrodinium, Cochlodinium, Torodinium*, and *Pronoctiluca*) of biomass equivalent to about 25% of the phytoplankton. The feeding activity of these protists required from 3 to 40% of the phytoplankton biomass daily. Nano-protozoan biomass (especially heterotrophic dinoflagellates) was equivalent to about 70% of their total potential food biomass—that is, of bacterial and nanophytoplankton cells (Becquevort, 1997). There is heavy grazing pressure by small copepods (Smetacek *et al.*, 1990), though this is likely to remove no more than about 30% of the daily primary production: *Salpa thompsoni*, on the other hand, is present in sufficient numbers to remove all of the daily production (Dubischar and Bathman, 1997). It is not easy, here, to achieve a clear balance between production and consumption.

Some studies suggest that only an insignificant part of the daily production during diatom blooms is consumed by meso- and macro-zooplankton (e.g., 0.6% daily in the Drake Passage and 8% in the Bellingshausen Sea), but in other situations, as in the Ross Sea, it is suggested that larger herbivores are a major factor in constraining the accumulation of cells. For example, in the Drake Passage, where phytoplankton biomass is low (1 or 2 g C $m^{-2}$) and copepods are numerous (about 0.8 g C $m^{-2}$), the latter consume 50% of daily primary production (Schnack *et al.*, 1985). Whatever their relative importance, the mesozooplankters involved in the reduction of diatoms blooms are dominated by *Calanus propinquus*, *Calanoides acutus*, and *Rhincalanus gigas*, whereas *Metridia gerlachi* is the principal diel migrant. All these come up from overwintering depths to encounter the spring bloom, with *C. acutus* appearing a little later in the season than the others. This species has a life cycle resembling that of the arctic calanoids, though its life cycle appears to be completed within 1 year except where the summer season of open water is shortest, as in the eastern Weddell Sea, where a small proportion of the population takes 2 years to reach maturity (Atkinson *et al.*, 1997). Though these are the dominant species, it must be remembered that even here, in such high latitudes, lists of dominants conceal the real diversity of the mesoplankton that consume both phytoplankton and protists. In the western Weddell Sea in March, at the end of summer, a total of 113 mesoplankton species, of which 31 species were encountered in the upper 50 m, were listed by Hopkins and Torres (1989).

Antarctic krill are the keystone organisms for the biomass buildup in this province by baleen whales, crab-eater seals, and many seabirds that directly forage for it by a variety of feeding techniques. It is hard to imagine a more efficient organism to harvest the spring and under-ice blooms of this province and to transfer the harvest directly to mammals and seabirds, so enabling them to establish the truly astonishing populations that existed before whaling and other forms of exploitation were begun. Since salps generally occur in warmer water than krill, Atkinson *et al.* (2004) suggest that—since the western Atlantic sector of the Southern Ocean is one of the fastest-warming regions of the globe—they may come progressively to replace krill as the major large herbivore of the Southern Ocean: this, of course, would have very far-reaching consequences for the entire ecosystem.

Together with salps, krill (*Euphausia superba* and *S. thompson*) are the principal herbivores among the larger size classes of planktonic organisms; close to the coast, and especially in shallow water as in the inner Weddell Sea, these species are replaced largely by *E. cristallophoria*, whose life cycle is adapted to this environment. Krill occur everywhere in this province, but significant abundance does not extend equatorward of the PF; as Atkinson *et al.* (2004) showed, >50% of all krill present in the Southern Ocean occur in the SW Atlantic sector of this circumpolar province. Thus, they are more abundant than elsewhere by an order of magnitude eastward of a line joining the tip of the peninsula to South Georgia, passing by way of the island arcs. Here, of course, the Weddell Current (see earlier discussion), which is a component of the westward coastal current, passes back along the island arc to rejoin the general eastward flow of the ACC, carrying with it high concentrations of all planktonic biota. The consequent zone of high krill abundance may be traced eastward over deep water to about 30°W. Intermittent patches of high abundance may be encountered around the remainder of the southern part of the ACC, though none are so important as that in the Weddell Current. Despite exaggerated claims concerning their fishery potential, krill do represent an enormous biomass, although some regions have an enigmatic low abundance; the Ross Sea, for example, despite the very high concentrations of diatoms there, has rather few.

Krill occur in small to very large swarms that resemble those of clupeid fish, though fish shoals are more uniform in terms of size of individuals, swimming speeds, and maturity stage. Swarms may be aggregated into "superswarms" comprising more than a million tons of biomass spread over several hundred square kilometers, though these are mainly associated with coastal chlorophyll enrichment. During normal swimming, krill are oriented with their dorsal side uppermost, but when food is encountered and filter feeding begins, individuals may be oriented in any direction. Feeding is most active in darkness and swarms undergo diel vertical migration between the surface and about 200 m. One reason for the success of these organisms is their ability to feed on a very wide range of food particles from nanoflagellates to crustacean nauplii. Since krill have such a patchy distribution, attempts to balance their consumption against production in the blooms, or their browsing on under-ice algae, have been unsatisfactory on a regional basis. Nevertheless, it is calculated that krill consume up to 60–80% of daily production where they are abundant. The consumption of diatom blooms by krill can be very spectacular: a summer bloom in the Weddell Sea was grazed down in 20 hours from 4.0 to 0.5–1.0 mg chl $m^{-3}$, this remnant apparently representing the unconsumed nanophytoplankton component of the original bloom.

*Euphausia superba* has a life strategy that is more like that of a fish than of a planktonic crustacean, being long-lived (5–8 years), during which period reproduction occurs two or three times, usually late in the summer. Eggs of krill sink rapidly (out of the way of the grazing activities of their myriad parental stock) to as much as 1000 m prior to hatching. The nauplii and young larvae rise progressively toward the surface so that *E. superba* is typically much more abundant over deep than shallow water. The two species of krill exist in a dynamic balance with the relative abundance of one being associated with relative scarcity of the other (Loeb *et al.*, 1997). At Elephant Island, near the tip of the Antarctic Peninsula, this balance is determined each spring by the extent of sea-ice cover during the previous winter. The reestablishment of seasonal ice cover is progressive, starting with the establishment of frazil ice, honeycombed with cavities that occlude only when the frazil consolidates into pack ice (Smetacek *et al.*, 1990). The lower surface and cavities of frazil and lower surface of pack ice concentrate diatom cells that are readily "grazed" by krill that accumulate for this purpose in and around the rugosities and cavities on the lower surface of sea ice. During the remnant of the sunlit season and in the early spring, sufficient irradiance penetrates the ice to maintain a diatom population here. Leads in pack ice rapidly freeze over, and the undersurface of this new ice supports fresh growth of algae. Dense concentrations of krill are supported during the winter within the under-ice habitat, as many as 10–30 ind $m^{-2}$ having been observed within the interstices. Atkinson *et al.* suggest that strong recruitment to the summer population depends on ice conditions during the previous winter: extensive ice cover is associated with generous recruitment.

Large salp populations occur after winters of small ice coverage followed by an early open-sea spring bloom which then permits an early and rapid increase of the overwintering salp population. Krill have population dynamics resembling those of fish so that their stock size is determined by reproductive success the previous year. Heavy recruitment of first-year krill to the population occurs 1 year after a heavy ice winter that promotes early spawning by the adult population and, incidentally, inhibits a spring outburst of salps. Two consecutive heavy-ice years enhance the effect. In the seas around Antarctica, the consumption of phytoplankton production by both salps and krill is reported to be significantly lower than to the north of the Continental Water Boundary.

Salps are reported to consume only 19% of daily primary production in years of high abundance (e.g., 1994) and $< 1\%$ in poor salp years (e.g., 1995), but there is a correlation between computed salp/krill consumption and between-year variability in satellite-observed sea-surface chlorophyll in summer. The balance between krill and salps is not just a result of competition for food, because it has recently been shown that salps may be a preferred food of adult krill (Kawaguchi and Takahashi, 1996). During the period of routine observations, a trend to more frequent years with light ice cover and lower krill densities has been detected; this trend can be extrapolated securely back to 1947 when routine observations of relative winter ice cover were established. Using net-tow data, the trend has recently been extrapolated back to 1926 by Atkinson *et al.* (2004). The inferred progressive decrease in the abundance of krill during a period of 80 years is cause for concern for the direct fishery for *E. superba* of about $1.5 \times 10^6$ tons $y^{-1}$, managed by the Commission for the Conservation of Antarctic Marine Living Resources, and also for the marine mammals and birds dependent on krill for their livelihood. In fact, very recent evidence suggests that this trend can be extrapolated back to the early 1930s in the records of whaling ships (de la Mare, 1997).

### Synopsis

*Case 1—Polar irradiance-mediated production peak*—The low salinity mixed layer undergoes significant winter deepening to almost 150 m, while the photic depth remains shoaler, so that the thermocline is never illuminated. Seasonal evolution of productivity rate is symmetrical about a midsummer maximum that is significantly lower than in high boreal latitudes. Rates of increase and decline of productivity are consistent and linear, suggesting that vernal productivity increase is more closely related to seasonal changes in the light field than to mixed-layer depth. Chlorophyll biomass closely tracks the increase of P, peaking (but only at 0.5 mg chl $m^{-3}$) at midsummer, with some renewal of accumulation during March–May, until winter ice cover is again established. Consumption does not balance production during summer bloom and is inferred to decline further from balance when copepods descend to overwintering depths at the end of summer (March), this reduction being mitigated by continued presence of krill until the ice cover is renewed (Fig. 12.5).

## Austral Polar Province (APLR)

### Extent of the Province

This is the "real Antarctic" of oceanographers, whalers, early navigators, and 21st-century tourists. It comprises the seasonally ice-covered sea from the coasts of Antarctica (including the peninsula) north to the Antarctic Divergence, and consequently much of this area lies poleward of the Antarctic Circle and so has a period of total winter darkness. Because of their ecological similarity to the rest of APLR, we should include in this province the island arcs between the Antarctic Peninsula and South America (South Orkneys, South Sandwich, and South Georgia) because their ecology is dominated by coastal and ice-edge effects.

The Antarctic continental shelf is not extensive, and comprises only a few small patches where water depth is <200 m, mostly near 30°E: elsewhere, water depth of 200–500 m is found within a few miles of the coastline. Like that of Labrador, this deep continental platform is a submerged continental shelf, its submergence caused by the weight of the 4-km-deep ice cap resting on the Antarctic continent. A subsurface thermal front of about 2°C at 200 m (the CWB already noted) occurs close to the continent: this, of course, is not simply a typical shelf-edge front but rather a convergence of significance in the global thermohaline circulation.

The unique characteristic of the coast of Antarctica must be, of course, that each of the major shallow embayments on the coast, except the rather open Bellingshausen Sea,

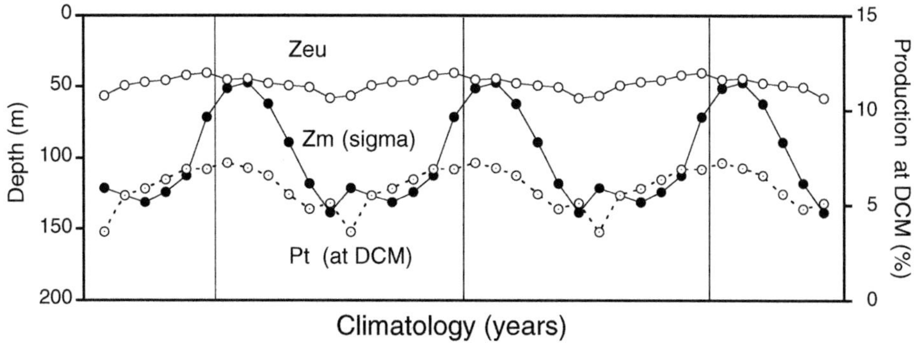


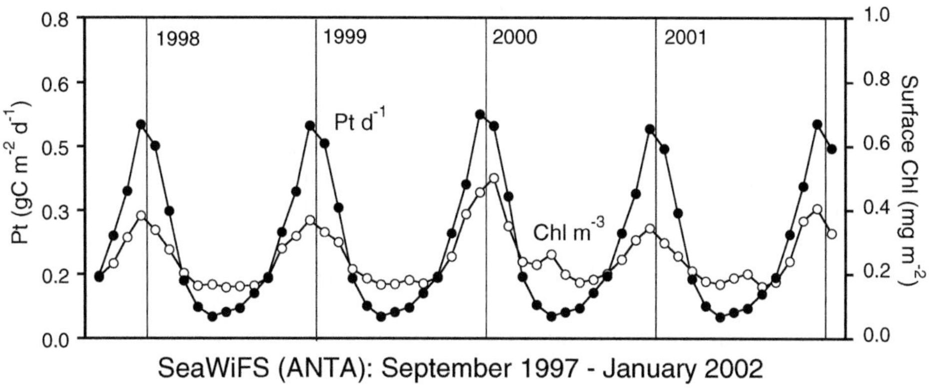


**Fig. 12.5** ANTA: seasonal cycles of monthly surface chlorophyll and depth-integrated autotrophic production for the years 1997–2002 from SeaWiFS data together with characteristic seasonal cycles of mixed-layer depths from Levitus climatological data and photic depths computed from characteristic irradiance and the archive of chlorophyll profiles discussed in Chapter 1.

is partially covered by a permanent ice shelf or ice barrier (Rϕnne, Shackleton, Filschner, etc.). These are rooted at the coast and extend seaward across the deep shelves and beyond. The Ross Sea ice shelf extends almost 800 km seaward from the termination of the Axel Heidberg glacier, whereas in the Weddell Sea the edge of the Ronne ice shelf lies only 300 km offshore. This ice shelf is perhaps in the initial stages of breaking up because of a warming climate.

## Defining Characteristics of Regional Oceanography

The Antarctic Divergence (AD) marks the transition between the oceanic west winds and the east-wind regime nearer the Antarctic continent. The mean position of the divergence is rather variable, responding to episodic weather patterns (Deacon, 1984); progressively to the north across the province toward the AD, wind stress becomes more persistent and periods of westerly stress occur. Poleward of the AD, the westward-moving Antarctic surface water is the East Wind Drift, up to 300 km wide, ice-covered and dark in winter but with some open-water areas in summer. Flow in the East Wind Drift is meandering, with eddies being generated by the topography of the coast and offshore shoals and islands. Recalling that since wind stress introduces kinetic energy to the Ekman layer at low latitudes, but potential energy preferentially at high latitudes, it will be no surprise to find that the field of mesoscale eddies here is unusually complex and active.

The area covered by ice is a direct function of sun angle, so that a simple sine wave represents the sea-ice coverage from a minimum ($5 \times 10^6\,km^{-2}$) in February to a maximum (19–20 × $10^6$ $km^{-2}$) in August–October when it extends to within a few degrees of the PF. It lies farthest from the continent in the central Atlantic and Pacific sectors. Especially in the western Pacific/Drake Passage sector the permanently open surface water is very narrow, and seasonal pack ice closely approaches the PF.

Water of the East Wind Drift is very cold: shoaler than 500 m it is always <2°C even in summer and is still colder in winter; it is also thermally unstratified. Where slight surface warming occurs, this results in a mixed-layer depth of no more than about 25 m. In austral winter, a combination of low sun angle, heat loss to the atmosphere, and deep convection cools the entire surface water mass so that mixing extends to halocline depth at 200–400 m (Trull *et al.*, 2001).

The convergent Continental Water Boundary (CB) that circles the Antarctic continent poleward of the AD (see Fig. 12.2) lies between shelf water and the warmer circumpolar deep water upwelled at the divergence that otherwise occupies the surface of this province. At the CB, deep convection occurs of the very dense (cold and salty) shelf water down into the interior of the ocean. It is only in places such as this, where surface evaporation from cold dry winds occurs simultaneously with ice formation and brine rejection, that sufficiently dense water is evolved to sink and form the deep water masses of the oceans. The Antarctic bottom water formed in this way along the coast of Antarctica is one of the two principal roots of the global thermohaline circulation of the oceans (Lutjeharms, 1985).

Because wind-stress curl is at its regional minimum here, Ekman transport diverges on either side of the front and density surface slopes downward both to the north and to the south; a convenient way of locating the axis of upwelling is the salinity maximum on the 200-m horizon. LCDW is upwelled at the AD and this—compared with UCDW (see earlier discussion)—has a relatively low content of macronutrients and iron: this fact has significance for the ecology of the province. It is only closer to the Antarctic continent, in the coastal region, that higher nutrient levels are encountered. To the south of the CWB (and consequently south of the $Si_{max}$) the surface water mass is permanently silica-replete. Although winter mixing of course occurs here, Central Deep Water is not reached and so the potentially strong input of dissolved Fe from this water mass is therefore not realized.

There are two major gyres in the westward flow around Antarctica, occupying the Weddell and Ross Seas, and each is confluent with the flow of the ACC to the north. The Weddell Sea gyre comprises very cold coastal water (−1.8°C in the center of the gyre even in summer), occupies the area enclosed by the Antarctic Peninsula, and extends eastward to about 10–15°W before rejoining the westward flow along the coast into the southern Weddell Sea. The eastward flow in this gyre from the tip of the peninsula forms a confluent front (at the AD) with the ACC. The smaller Ross Sea gyre is generated by flow past the coastal prominences on either side of this embayment.

Because so much of the research performed in Antarctic seas has concentrated on only a few places, it will be useful to discuss these in more detail. The Larsen ice shelf on the east coast of the Antarctic Peninsula and the Filchner-Rønne ice shelf at the head of the bight occupy a significant part of the surface of the Weddell Sea. The coastal current of the East Wind Drift enters the Weddell Sea at 20°W and contributes to its cyclonic circulation (clockwise in the Southern Hemisphere), whereas near Halley Bay (25°E, on the southeast coastline) the incoming coastal current passes through a divergence zone that somewhat separates the ecology of the sea into a northeastern and southern component. The gyral circulation passes toward the east at the tip of the Antarctic Peninsula and can be traced to about 15°W where it returns south toward the coast. Weddell Sea gyre water has anomalously high silicate and nitrate concentrations compared with the water beyond the gyre.

Although the Weddell Sea is within the zone usually mapped as "permanently ice-covered," much open water does occur seasonally within the confines of its gyral circulation. Along the southeast coast a persistent polynya is maintained by offshore winds; this may be 800 km long even in winter, and in summer it expands into a mosaic of shore leads and patches of open water that persist for several weeks. Persistent open water also occurs in the much-studied Bransfield Sound and in the passages between the islands of the South Orkneys.

The Ross Sea gyre is smaller and narrower than the Weddell gyre but has many of the same properties; a wide area of sea ice continues to invest the eastern coast and partly encloses the Ross Sea so that even in austral midsummer it is open to the ocean only toward the northwest, off Cape Adare in Victoria Land. The Bellingshausen Sea is hard to define, lying in the very open embayment west of the Antarctic Peninsula; most is ice-covered even in summer, and the southern edge of eastward flow in the ACC is unusually far south and lies quite close to the ice edge in summer (Read *et al.*, 1995).

## Biological Response and Regional Ecology

Because phytoplankton in this region always has a low photosynthetic performance as well as a positive response to Fe fertilization, it is likely that here the cells are largely Fe-limited. Nevertheless, it has been frequently suggested that, at least in the marginal ice zone, a near-surface input of Fe may commonly occur with the summer flow of meltwater from glacier mouths, from fleets of drifting icebergs having the same source, and from freshwater deposited as snow on the Antarctic continent. This process has been briefly discussed in Chapter 5 and it is, according to Smetacek *et al.* (2004), the principal factor in forcing the development of seasonal phytoplankton blooms; but we should remember that available Fe is only one among many relevant characteristics of the upper water column in early summer, adjacent to clusters of icebergs or near the retreating ice edge. Near-surface density stratification, which occurs in such regions and is so often the principal factor in initiating high-latitude blooms, is certainly equally critical here as elsewhere, as was suggested during the 1930s by the "*Discovery*" investigations.

Whatever the proximate factors may be in the neritic and MIZ regime, early summer blooms are initiated close to the coast and in polynyas such as those of the Ross Sea; these blooms may subsequently propagate seaward along with the effluent meltwater in the surface layer. It should be noted that sea ice is a more complex substance than it might appear to be at first sight. It is a two-phase material, containing seawater enriched in nutrients and autotrophic biota; these are distributed in a complex manner. A "surface slush community" extends down into crevices rather superficially, and within cavities in the ice a "freeboard community" may develop in partially rotted ice, mixed with seawater freshened by ice melt. Finally, a "bottom community" may form in the columnar or platelet ice that occurs below fast ice in spring or autumn, when light penetration is sufficient. The start of freezing in the autumn causes nutrient-depleted water within the ice to be replaced by nutrient-rich water from below the ice by "upwelling," and fluid motion in the ice cover can be driven by freezing-induced convection (Lytle and Ackley, 1996). Enrichment within second-year ice is due to brine rejection from the freezing process at the ice surface (as air temperature drops in autumn) and fuels significant blooms of algae.

Because seasonal ice is not very thick in these seas, there is often sufficient penetration of light for the growth of algal "superblooms" on the underside of the continuous ice cover, perhaps especially in the coastal regions of the Weddell Sea. These were investigated in very early spring (October and November) along the eastern coast of the Weddell Sea in a feature that left a brown wake behind the ship as it passed through continuous pack ice (Smetacek *et al.*, 1992). This ice was only 20 cm thick but was underlain by a low-salinity layer of nitrate-depleted water 1–5 m deep, rich in unconsolidated centimeter-scale ice platelets, where irradiance was significantly reduced. Within the stratification afforded by the platelets, a miniature chlorophyll profile had maxima in the range 7–36 mg chl $m^{-3}$ and with POC maxima of 250–1250 mg chl $m^{-3}$. Chlorophyll concentrations were highest below freshly frozen leads. The bloom was composed largely of centric diatoms, and the only herbivore of consequence was a large heterotrophic dinoflagellate. As nitrate became exhausted, the nutrient-depleted diatoms sank to form a miniature DCM at the bottom of the platelet layer and hence at a nitracline. Such bloom conditions, below continuous ice cover, extended over an area of 20, 000 $km^2$ and contrasted with the oligotrophic conditions then obtaining throughout all the open waters to the north. Other investigations of the below-the-ice autotrophs have revealed abundant pico- and nanophytoplankton along with the overwintering very early larval stages of copepods.

Blooms in the far south may be very extensive, but are not well seen by satellite sensors because of cloud cover. The Ross Sea is perhaps the site of the greatest and most consistent phytoplankton biomass in the Southern Ocean, marked by deep deposits of richly siliceous sediments. Coastal blooms along Victoria Land are consistently diatom-dominated, whereas offshore blooms are more likely to be dominated by *Phaeocystis antarctica* (DiTullio and Smith, 1996); an event in the coastal polynya of the western Ross Sea contained intense blooms of up to 10–40 mg chl $m^{-3}$ over an area of 106, 000 $km^2$ and occurred before the polynya opened up. Later, about 10 mg chl $m^{-3}$ was seen over an area of 126, 000 $km^2$ (Arrigo and McClain, 1994). Diatom blooms of this kind may be terminated rapidly by gales that destroy the stratification on which they form. Images obtained in December and January, later in the season, show the very great extent of the *Phaeocystis pouchetii* blooms that occur here and are advected into McMurdo Sound after the spring bloom of centric diatoms is over. This appears to be a characteristic sequence in many regions in this province. *Phaeocystis* also has a characteristic seasonal morphological cycle: in spring a very high percentage of cells are flagellated, solitary individuals, whereas late in the season almost all are in the colonial stage, when their mortality rate is much lowered.

These under-ice cell concentrations are the food resource for a variety of organisms, some having life cycles closely adapted to these singular conditions: the small copepod *Stephos longipes* is one such. This copepod manages the alternation between ice-covered and open-water conditions as follows. There are two concurrent generations, of which the younger overwinters within and below the mass of unconsolidated platelets below the fast ice as nauplii and early copepodites, while the older generation lies in the water column just below the ice as C4s that become adults in very early spring as the ice starts to break up. The summer population, in the upper 50 m of ice-free water, comprises mostly C1–C3s. In autumn, the adults from this generation enter the freezing ice to generate eggs and nauplii, whereas the slightly younger copepodites remain in the water column to overwinter as C4s, closing the cycle (Schnack-Schiel *et al.*, 1995). Using such an apparently unlikely stratagem, the equally improbable ice-platelet bloom is well utilized.

Around Elephant Island (61°S), $< 20$-μm cells were found to consistently comprise 55–100% of total chlorophyll. In spring, prokaryote contribution is negligible, but in summer these cells represent about 5–10%, though at anomalous stations with very low chlorophyll values they may contribute $< 50\%$ of total chlorophyll. In many areas, these small cells are probably comprised largely by motile cells of *Phaeocystis* ($< 3 \times 10^6$ cells $liter^{-1}$) released from macroscopic colonies of this prymnesiophyte. Several other investigations in the Weddell Sea have shown the existence of a rich protist fauna of heterotrophic flagellates, choanoflagellates, dinoflagellates, and ciliates whose abundance is correlated with autotrophic cells and is direct evidence of a well-developed microbial loop. We can expect this association to have greater biomass and flux near the ice edge and progressively less biomass out into the open water. In the Ross Sea, although abundant protists were correlated with chlorophyll, dilution experiments performed at three seasons all suggested very low rates of phytoplankton cell removal. Massive blooms observed here in extremely cold water are perhaps due to the low activity of microbial herbivores under such conditions (Caron *et al.*, 2000).

However, this is a very diverse region and a quite different balance between production and consumption also occurs (e.g., Brandini, 1993): in some coastal bays, such as Admiralty Bay in the South Shetlands, nanosized diatoms (*Nitzschia*) dominate the phytoplankton during open-water and stable conditions, and these are held at relatively low concentration by very intense grazing by ciliates that dominate the protistan herbivores. Production during the brief summer season is maintained entirely by regenerated nitrogen and micronutrients, and larger cells are very scarce, occurring deeper in the chlorophyll profile.

The benthos of the continental shelf responds both to the general constraints of a very cold environment, and to the special conditions imposed everywhere by topography and sediment type. The fauna of the sea floor below permanent ice shelves is sufficiently attenuated as to be effectively absent in some regions, but elsewhere it may be very dense though rapidly rarified with depth. Obviously, where the water depth is such as to permit scouring by drifting ice, the benthic community is sparse and is retained permanently in the early stages of recolonization of the substrate. As in the high arctic, the fauna is attenuated in the sense that some groups, abundant in the benthos of temperate shelves, are absent; the ecology of those organisms that are abundant on the Antarctic shelf is characterized by relatively slow growth compared with similar kinds of organisms in similar habitats in warmer seas, despite the effect of cold-adapted metabolism. Note that the seasonal temperature variation here is as small as anywhere and decreases with increasing latitude. There is also a characteristic avoidance of the planktonic larval stage in the life history in favor of brooding of eggs and larvae (as Thorson would have predicted), this preference leading to the relative dominance of groups that habitually brood, even in lower latitudes: thus, among the crustacea, amphipods and isopods are abundant, but decapods and cirripedes are not. Incubation duration is an excellent key to the consequences of cold temperatures on the physiology of benthic invertebrates: four species of Antarctic amphipods brood for 5–7 months, seven temperate species brood for only 8–49 days in summer. There seems no reason to quarrel with the long-accepted explanation for this peculiarity: although high-latitude seasonality is, in general, very predictable, what occurs in any year at any place is highly variable and, further, larval transformation for most species could not be completed within one season.

A thick ground cover of suspension- and particle-feeding sponges, bryozoa, brachiopods and coelenterates, together with asteroids and nudibranchs, may dominate the benthic communities on sand, gravel, and cobbles. These matlike communities appear to be restricted to water depths exceeding 50–60 m; presumably, in shallow water, ice scour would prevent the formation of this biotope. Otherwise, diverse bottom types support diverse benthic communities in a manner not very different from that in lower latitudes, the benthic community structure responding in the anticipated manner to the presence of rock, sand, mud, macroalgae, and tidal scour. Muddy sediments do not, of course, support mats of suspension-feeding organisms but rather a fauna typical of such a habitat: annelids, mollusks, nematodes, and so on. Fish are dominated by notothenids, a group that is rather poor in species: Sakurai *et al.* found only five: one generalist, two on rocky bottoms, and two species among kelp.

Of course, it was the large animals of the austral regions that first drew scientific attention to the Antarctic: various estimates, all very large, have been made of the pristine numbers and biomass of whales, seals and birds. Of 36 species in all seas, the crabeater seal (*Lobodon*) of the Antarctic comprised at one time half of the total world stock of pinnipeds (Laws, 1984). In this province, close to the continent, crabeaters, Ross's

(*Ommatophoca*), Weddell (*Leptonychotes*), fur (*Arctocephalus*), and leopard (*Hydrurga*) seals dominate, whereas elephant seals (*Mirounga*) occur mostly on the island groups in the Southern Ocean. Greatest numbers of all species occur in the pack-ice zone: fur and elephant seals breed on the ice, the others ashore. Their exact distributions at sea when not breeding are not well known. Krill dominate the diet of crabeaters and are important to fur seals. Diet otherwise largely comprises other seals and birds (leopard seal), cephalopods (all), and fish (all). Elephant seals eat fish and cephalopods almost exclusively.

It is in these seas that the large rorquals, the blue, fin, and Minke whales (*Balaenoptera* spp.) and the humpbacks (*Megaptera novazealandiae*) refuel themselves during austral summer, largely on krill. During austral winter, they pass north into tropical or subtropical breeding areas where very little feeding occurs. Fin, Minke, and humpback whales have deeply grooved throats that are prominent in photographs of whales on the slipways of the old whaling stations. These grooves enable the throat to open like a bag to enclose a very large volume of water engulfed by the mouth; this water is subsequently ejected, jetlike, by pressure from throat and tongue. Krill are strained from the ejected water as it passes through the filtering apparatus formed by the lateral baleen plates. The right whale (*Balaenoptera musculus*) has baleen plates that are rather finer, so instead of ingesting a large volume of water, subsequently to filter it, theses whales swim slowly forward with widely open mouth, filtering krill from the water as they go after the manner of a plankton net (Brown and Lockyer, 1984).

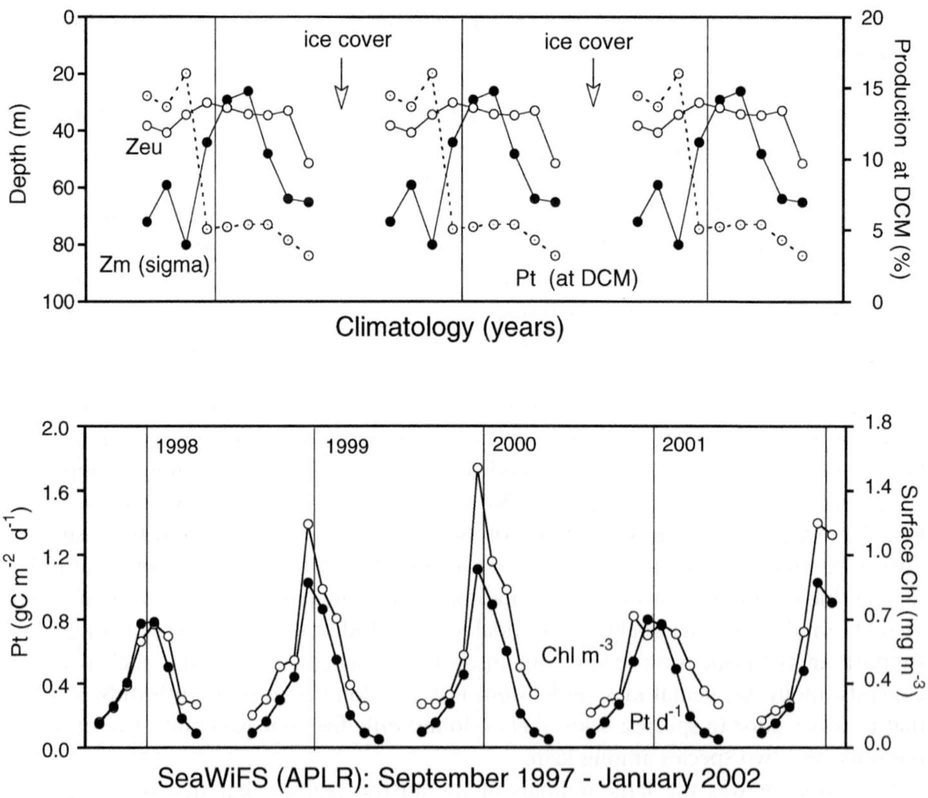


**Fig. 12.6** APLR: seasonal cycles of monthly surface chlorophyll and depth-integrated autotrophic production for the years 1997–2002 from SeaWiFS data together with characteristic seasonal cycles of mixed-layer depths from Levitus climatological data and photic depths computed from characteristic irradiance and the archive of chlorophyll profiles discussed in Chapter 1.

## Synopsis

*Case 1—Polar irradiance-mediated production peak*—In open water, the mixed layer is relatively deep at all seasons with a brief shoaling in January–February, so the pycnocline is illuminated only briefly in late austral summer. Darkness from May to July sets productivity to zero, and the return of light is immediately followed by a sustained linear increase to a midsummer maximum in December; subsequently, a sustained linear decline is initiated. Chlorophyll biomass tracks the linear increase and decrease very closely. Consumption does not balance production during the bloom and is inferred to decline even further from balance when copepods descend to overwintering depths at the end of summer (Fig. 12.6).

# REFERENCES


Abbott, M.R., *et al.* (2000) The spring bloom in the Antarctic Polar Frontal Zone as observed from a mesoscale array of bio-optical sensors. *Deep-Sea Res. II.* **47**, 3285–3314.

Abbott, M.R., *et al.* (2001) Meanders in the Antarctic Polar Frontal Zone and their impact on phytoplankton. *Deep-Sea Res. II.* **48**, 3891–3912.

Ackefors, H. (1966) Plankton and hydrography of the Landsort Deep. *Veroff. Inst. Meeresforsch. Bremerhaven.* **2**, 381–386.

Adnan, Q. (1990) Monsoonal differences in net-phytoplankton in the Arafura Sea. *Neth. J. Sea Res.* **25**, 523–526.

Agawin, N.S.R., *et al.* (2003) Abundance, biomass and growth rates of *Synechococcus* sp. in a tropical coastal ecosystem. *Estuar. Coast. Shelf Sci.* **56**, 493–502.

Agusti, S., and C.M. Duarte (1999) Phytoplankton chlorophyll-*a* distribution and water column stability in the central Atlantic Ocean. *Océanolog. Acta.* **22**, 193–203.

Aiken, J. (1981) The UOR Mk.2. *J. Plankton Res.* **3**, 551–560.

Aiken, J., and A.J. Bale (2000) An introduction to the Atlantic Meridional Transect Programme. *Prog. Oceanogr.* **25**, 251–256.

Ajao, E.A., and R.W. Houghton (1998) Coastal ocean of equatorial West Africa, from 10°N to 10°S. *In—The Sea*, Vol. 12, *The Global Coastal Ocean* (A.R. Robinson and K.H. Brink, Eds.), pp. 605–632. Wiley InterScience, New York.

Aller, J.Y., and I. Stupakoff (1996) The distribution and seasonal characteristics of benthic communities on the Amazon shelf as indicators of physical processes. *Cont. Shelf Res.* **16**, 717–751.

Aller, J.Y., and J.R. Todorov (1997) Seasonal and spatial patterns of deeply-buried copepods on the Amazon shelf: evidence for periodic erosional/depositional cycles. *Est. Coast. Shelf Sci.* **44**, 57–66.

Allison, S.K., and K.F. Wishner (1986) Spatial and temporal patterns of zooplankton biomass across the Gulf Stream. *Mar. Ecol. Prog. Ser.* **31**, 233–244.

Al-Mutairi, H., and M. Landry (2001) Active export of carbon and nitrogen at Station Aloha by diel migrant zooplankton. *Deep-Sea Res.* **48**, 2083–2103.

Alongi, D.M. (1998) Decomposition and recycling of organic matter in muds of the Gulf of Papua. *Cont. Shelf Res.* **15**, 1319–1337.

Alongi, D.M. (1999) *Coastal Ecosystem Processes*, p. 419. CRC Press, Boca Raton, FL.

Al-Saadi, H.A. (2001) Ecological status of phytoplankton in the NW Arabian Gulf during the seventies. *Mar. Mesop. Spec. Issue.* **16**, 157–168.

Alton, M.S. (1974) Bering Sea benthos as a food resource for demersal fish populations. *In—Oceanography of the Bering Sea* (D.W. Hood and E.J. Kelly, Eds.), pp. 257–278. University of Alaska Press, Fairbanks, AK.

Alvain, S., *et al.* (2005) Remote sensing of phytoplankton groups in Case 1 waters from global SeaWiFS imagery. *Deep-Sea Res. I.* **52**, 1989–2004.

Alvarez-Borrego, A. (1983) Gulf of California. *In—Estuaries and Enclosed Seas* (B.H. Ketchum, Ed.), pp. 427–450. Elsevier, New York.

Alvarez-Salgado, X.A., *et al.* (2001) Net ecosystem production of dissolved organic carbon in a coastal upwelling system: the Ria de Vigo, Iberian margin of the North Atlantic. *Limnol. Oceanogr.* **46**, 135–147.

Alvarino, A. (1965) Chaetognaths. *Oceanogr. Mar. Biol. Ann. Rev.* **3**, 115–194.

Alvarino, A. (1971) Siphonophores of the Pacific with a review of the world distribution. *Bull Scripps Inst. Oceanogr.* **16**, 1–432.

Ambler, J.W., and C.B. Miller (1987) Vertical habitat partitioning by copepodites and adults of subtropical oceanic copepods. *Mar. Biol.* **94**, 561–577.

Anda-Montañez, J.A. *et al.* (2004) Spatial analysis of yellowfin tuna catch rate and its relation to El Niño in the eastern tropical Pacific. *Deep-Sea Res. II.* **51**, 575–586.

Andersen, V., *et al.* (1997) Macroplankton and micronekton in the NE tropical Atlantic: abundance, community composition and vertical distribution in relation to different trophic environments. *Deep-Sea Res. I.* **44**, 193–222.

Anderson, G.C., *et al.* (1969) Nitrate distribution in the northeast Pacific Ocean. *Deep-Sea Res.* **16**, 329–334.

Anderson, G.C., *et al.* (1977) Description and numerical analysis of factors affecting the processes of production in the Gulf of Alaska. *Spec. Rep. Univ. Washington.* **76** (ref. M-77–40).

Anderson, L.A., and A.R. Robinson (2001) Physical and biological modeling in the Gulf Stream region. Part II. Physical and biological processes. *Deep-Sea Res. I.* **48**, 1139–1168.

Andrews, J.C. (1977) Eddy structure and the west Australian Current. *Deep-Sea Res. I.* **24**, 1133–1148.

Andrews, J.C., *et al.* (1980) Observations of the Tasman front. *J. Phys. Oceanogr.* **10**, 1854–1869.

Andrews, J.C., and S. Clegg (1989) Coral Sea circulation and transport deduced from modal information models. *Deep-Sea Res.* **36**, 957–974.

Anonymous (1973) Continuous plankton records: a plankton atlas of the North Atlantic and the North Sea. *Bull. Mar. Ecol.* **8**, 1–174.

Anonymous (1989) *Ocean Circulation.* Pergamon, Elmsford, NY.

Anonymous (1990) Confluence 1988–1990, an investigative study of the SW Atlantic Ocean. *Eos.* **71**, 1131–1133.

Antony, M.K., and S.C. Shenoi (1991) Current measurements over the continental shelf of India. *Contin. Shelf Res.* **11**, 81–93.

Archer, D.E., and K.S. Johnson (2000) A model of the iron cycle in the ocean. *Glob. Biogeochem. Cycles.* **14**, 269–279.

Arcos, E., *et al.* (1987) Variabilidad vertical de chlorofila a en un rea de surgencia frente a Chile. *Central Invest. Pesqu. Chile.* **34**, 47–55.

Arief, D. (1998) Outer southeast Asia: a region of deep straits including the Banda Sea. *In—The Sea*, Vol. 12, *The Global Coastal Ocean* (A.R. Robinson, and K.H. Brink. Eds.), pp. 507–522. Wiley InterScience, New York.

Aristegui, J., *et al.* (1997) The influence of island-generated eddies on chlorophyll distribution: a study of mesoscale variation around Gran Canaria. *Deep-Sea Res.* **44**, 71–96.

Arntz, W. E. (1986) The two faces of El Niño, 1982–83. *Meeresforschung.* **31**, 1–46.

Arrigo, K.R., and C.R. McClain (1994) Spring phytoplankton production in the western Ross Sea. *Science.* **266**, 261–263.

Arrigo, K.R., and G.L. van Dijken (2004) Annual cycles of sea ice and phytoplankton in Cape Bathurst polynya, SE Beaufort Sea. *Geophys. Res. Lett.* **31**(8) (unpaginated).

Arrigo, K.R., *et al.* (1999) Phytoplankton community structure and drawdown of nutrients and $CO_2$ in the Southern Ocean. *Science*, **283**, 365–367.

Ashjian, C.J., and K.F. Wishner (1993) Temporal persistence of copepod species groups in the Gulf Stream. *Deep-Sea Res. I.* **40**, 483–516.

Astthorsson, O.S., and A. Gislayson (2003) Seasonal variations in abundance, development and vertical distribution of *Calanus finmarchicus, C. hyperboreus* and *C. glacialis* in the East Icelandic Current. *J. Plankton Res.* **25**, 843–854.

Atjay, G.L., *et al.* (1979) Terrestrial primary production and phytomass. *In—The Global Carbon Cycle* (B. Bolin *et al.*, Eds.), pp. 129–181. John Wiley and Sons, New York.

Atkinson, A., *et al.* (1997) Regional differences in the life cycle of *Calanoides acutus* within the Atlantic sector of the Southern Ocean. *Mar. Ecol. Prog. Ser.* **150**, 99–111.

Atkinson, A., *et al.* (2004) Long-term decline in krill stock and increase in salps within the Southern Ocean. *Nature (Lond.).* **432**, 100–103.

Atkinson, D.B. (1997) Distribution changes and abundance of northern cod, 1981–1993. *Can. J. Fish. Aquat. Sci.* **54** (Suppl.), 132–138.

Atkinson, L.P., *et al.* (1984) The effect of summertime shelf break upwelling on nutrient flux in southeast USA continental shelf waters. *J. Mar. Res.* **42**, 9679–993.

Baars, M.A., *et al.* (1991) The ecology of the Frisian Front. Observations on a biologically-enriched zone in the North Sea between the Southern Bight and the Oyster Ground. *Int. Counc. Explor. Sea, Copenhagen.* **CM**-1991-/L:25.

Backus, RH 1986. Biogeographic boundaries in the open ocean. *In—Pelagic Biogeography, UNESCO Tech. Pap. Mar. Sc.*, **49**: 9–13.

Badan-Dangon, A. (1998) Coastal circulation from the Galápagos to the Gulf of California. *In—The Sea*, Vol. 12, *The Global Coastal Ocean* (A.R. Robinson, and K.H. Brink. Eds.), pp. 315–343. Wiley InterScience, New York.

Badan-Dongon, A., *et al.* (1985) Spring and summer in the Gulf of California: observations of surface thermal patterns. *Océanol. Acta.* **8**, 13–22.

Bailey, R.G. (1983) *Ecoregions—The Ecosystem Geography of the Oceans and Continents.* Springer, New York.

Bainbridge, V. (1960a) Occurrence of *Calanoides carinatus* in the plankton of the Gulf of Guinea. *Nature (Lond.).* **188**, 932–933.

Bainbridge, V. (1960b) The plankton of inshore waters off Sierra Leone. *Col. Off. Fish. Pubs. London.* **13**, 1–48.

Baker, A.R., *et al.* (2003) Atmospheric deposition of nutrients to the Atlantic Ocean. *Geophys. Res. Lett.* **30**(24) (unpaginated).

Bakun, A. (1996) *Patterns in the Ocean: Ocean Processes and Marine Population Dynamics.* California Sea Grant Publications.

Bakun, A., and Nelson, C.S. (1991) The seasonal cycle of wind-stress curl in subtropical eastern boundary current regions. *J. Phys. Oceanogr.* **21**, 1815–1834.

Banse, K. (1984) Overview of the hydrography and associated biological phenomena in the Arabian Sea, off Pakistan. *In—Marine Geology and Oceanography of the Arabian Sea and Coastal Pakistan* (B.U. Haq and J. Milliman, Eds.), pp. 271–303. van Nostrand Reinhold, New York.

Banse, K. (1987) Seasonality of phytoplankton chlorophyll in the central and northern Arabian Sea. *Deep-Sea Res.* **34**, 713–723.

Banse, K. (1990a) Does iron really limit phytoplankton production in the offshore subarctic Pacific? *Limnol. Oceanogr.* **35**, 772–775.

Banse, K. (1990b) Remarks on the oceanographic observations off the east coast of India. *Mahasagar—Bull. Natl. Inst. Oceanogr. India.* **23**, 75–84.

Banse, K. (1992) Grazing, temporal changes of phytoplankton concentrations, and the microbial loop in the open sea. *In—Primary Productivity and Biogeochemical Cycles in the Sea* (P.G. Falkowski and A.D. Woodhead, Eds.), pp. 409–440. Plenum, New York.

Banse, K. (1994) On the coupling of hydrography, phytoplankton, zooplankton and settling organic particles offshore in the Arabian Sea. *Proc. Indian Acad. Sci. (Earth Planet. Sci.).* **103**, 125–161.

Banse, K. (1996) Low seasonality of low concentrations of surface chlorophyll in the Subantarctic Water Ring: Underwater irradiance, iron, and grazing. *Prog. Oceanogr.* **37**, 241–291.

Banse, K., and D.C. English (1993) Revision of satellite-based phytoplankton pigment data from the Arabian Sea during the northeast monsoon. *Mar. Res.* **2**, 83–103.

Banse, K., and D.C. English (1994) Seasonality of CZCS phytoplankton pigment in the offshore oceans. *J. Geophys. Res.* **99**, 7323–7345.

Banse, K., and D.C. English (1997) Near-surface phytoplankton pigment from the CZCS in the Subantarctic region southeast of New Zealand. *Mar. Ecol. Prog. Ser.* **156**, 51–66.

Banse, K., and D.C. English (1999) Comparing phytoplankton seasonality in the eastern and western subarctic Pacific and the western Bering Sea. *Prog. Oceanogr.* **43**, 235–288.

Banse, K., and D.C. English (2000) Geographical differences in seasonality of CZCS-derived phytoplankton pigment in the Arabian Sea for 1978–1986. *Deep-Sea Res. II.* **47**, 1623–1677.

Banse, K., and C.R. McClain (1986) Winter blooms of phytoplankton in the Arabian Sea as observed by the Coastal Zone Colour Scanner. *Mar. Ecol. Prog. Ser.* **34**, 201–211.

Banse, K., *et al.* (1996) On the possible causes of the seasonal phytoplankton blooms along the southwest coast of India. *Ind. J. Mar. Sci.* **25**, 283–289.

Barber, R.T. (1988) Ocean basin ecosystems. *In—Concepts of Ecosystem Ecology* (L.R. Pomeroy and J.J. Alberts, Eds.), pp. 171–193. Springer, New York.
Barber, R.T. (1992) Introduction to the WEC88 cruise: an investigation into why the Equator is not greener. *J. Geophys. Res.* **97**, 609–610.
Barber, R.T., and R.L. Smith (1981) Coastal upwelling ecosystems. *In—Analysis of Marine Ecosystems* (A.R. Longhurst, Ed.), pp. 31–68. Academic Press, London.
Barber, R.T., and F.P. Chavez (1991) Regulation of primary productivity in the equatorial Pacific. *Limnol. Oceanogr.* **36**, 1803–1815.
Barber, R.T., *et al.* (1996) Primary productivity and its regulation in the equatorial Pacific during and following the 1991–1992 El Niño. *Deep-Sea Res. II.* **43**, 933–969.
Barry, R.G., and Chorley, R.J. (1982) *Atmosphere, Weather and Climate.* Methuen, New York.
Barton, E.D., *et al.* (2004) Variability in the Canary Islands area of filament-eddy exchanges. *Prog. Oceanogr.* **62**, 71–94.
Bathman, U.V., *et al.* (1997) Spring development of phytoplankton biomass and composition in major water masses of the Atlantic sector of the Southern Ocean. *Deep-Sea Res. II.* **44**, 51–68.
Bauer, S., *et al.* (1991) Influence of monsoonally-forced Ekman dynamics upon surface layer depth and plankton biomass distribution in the Arabian Sea. *Deep-Sea Res.* **38**, 531–553.
Baumgartner, T., *et al.* (1992) Reconstruction of the history of Pacific sardine and northern anchovy populations over two millennia from sediments of the Santa Barbara Basin, California. *Rpts. Cal Coop. Fish. Invest.* **33**, 24–40.
Bayliff, W.H. (1980) Species synopses of biological data on 8 species of scombroids. *Int. Am. Trop. Tuna Commun. Spec. Rep.* **2**, 1–530.
Beare, D., and E. McKenzie (1999a) The multinomial logit model: a new tool for exploring CPR data. *Fish. Oceanogr. (Suppl.)* **8**, 23–39.
Beare, D.J., and E. McKenzie (1999b) Temporal patterns in the surface abundance of *Calanus finmarchicus* and *C. helgolandicus* in the northern North Sea (1958–1996) inferred from the CPR data. *Mar. Ecol. Prog. Ser.* **190**, 241–251.
Beaugrand, G., *et al.* (2002a) Diversity of calanoid copepods in the North Atlantic and adjacent seas: species associations and biogeography. *Mar. Ecol. Prog. Ser.* **232**, 179–195.
Beaugrand G., *et al.* (2002b) Reorganisation of North Atlantic copepod diversity and climate. *Science.* **296**, 1692–1694.
Becquevort, S. (1997) Nanoprotozooplankton in the Atlantic sector of the Southern Ocean during early spring: Biomass and feeding activities. *Deep-Sea Res. II.* **44**, 355–373.
Behrenfeld, M.J., and P.G. Falkowski (1997a) Photosynthetic rates derived from satellite-based chlorophyll concentration. *Limnol. Oceanogr.* **42**, 1–10.
Behrenfeld, M.J., and P.G. Falkowski (1997b) A consumers guide to phytoplankton primary productivity models. *Limnol. Oceanogr.* **42**, 1479–1491.
Behrenfeld, M.J., and Z.S. Kolber (1999) Widespread limitation of phytoplankton in the South Pacific Ocean. *Science.* **283**, 840–843.
Beklemishev, K.V. (1969) *Ekologiya Biogeografiya Pelagiali (Ecology and Biogeography of the Open Ocean).* Nauka, Moscow.
Belkin, I.M., and A.L. Gordon (1996) Southern Ocean fronts from the Greenwich meridian to Tasmania. *J. Geophys. Res. (C Oceans).* **191**, **C2**, 3675–3696.
Ben-Avraham, Z., and Emery, K.O. (1973) Structural framework of the Sunda Shelf. *Bull. Am. Petrol. Geol.* **57**, 2323–2366.
Berger, W.M., *et al.* (1987) Ocean productivity and organic carbon flux. *Scripps Inst. Oceanogr.* **87-10**, 1–67.
Biastoch, A., and Krauss, W. (1999) The role of mesoscale eddies in the source regions of the Agulhas Current. *J. Phys. Oceanogr.* **29**, 2303–2317.
Binder, B.J., and M.D. DuRand (2002) Diel cycles in surface waters of the equatorial Pacific. *Deep-Sea Res. II.* **49**, 2601–2617.
Binet, D. (1983) Phytoplankton et production primaire des régions côtières et upwellings saisonières dans le Golfe de Guinée. *Océanogr. Trop.* **18**, 331–355.
Binet, D. (1997) Climate and pelagic fisheries in the Canary and Guinea currents 1964–1993: the role of trade winds and the Southern Oscillation. *Oceanolog. Acta.* **20**, 177–190.
Binet, D., *et al.* (2001) El Niño–like warm events in the eastern Atlantic and fish availability from Congo to Angola. *Aquat. Living Resour.* **14**, 99–113.

Bishop, J.K.B., *et al.* (2002) Robotic observations of dust storm enhancement of carbon biomass in the North Pacific. *Science* **298**, 817–821.

Blackburn, M. (1962) An oceanographic study of the Gulf of Tehuantapec. *Spec. Sci. Rep.-Fish. US Fish Wild. Serv.* **404**, 1–28.

Blackburn, M., *et al.* (1970) Seasonal and areal changes in standing stocks of phytoplankton, zooplankton and micronekton in the eastern tropical Pacific. *Mar. Biol.* **7**, 14–31.

Blacker, R.W. (1965) Recent changes in the benthos of the West Spitzbergen grounds. *ICNAF Spec. Pub.* **6**, 791–795.

Blain, S., *et al.* (2001) Quantification of algal iron requirements in the Subantarctic Southern Ocean (Indian sector). *Deep-Sea Res. II.* **49**, 3255–3273.

Blanchot, J., *et al.* (1992) Effect of ENSO events on the distribution and abundance of phytoplankton in the western tropical Pacific Ocean along 165°E. *J. Plankton Res.* **14**, 137–156.

Blanton, J.O., *et al.* (1981) The intrusion of Gulf Stream water across the continental shelf due to topographically induced upwelling. *Deep-Sea Res.* **28**, 393–405.

Blasco, D., *et al.* (1981) Short time variability of phytoplankton populations in upwelling regions—the example of northwest Africa. *In—Coastal Upwelling* (F.A. Richards, Ed.), pp. 339–347. American Geophysical Union, Washington, D.C.

Blindheim, J., *et al.* (2000) Upper layer cooling and freshening in the Norwegian Sea in relation to atmospheric forcing. *Deep-Sea Res.* **47**, 655–680.

Boelen, P., *et al.* (2002) Diel patterns of UVBR-induced DNA damage in picoplankton size fractions in the Gulf of Aqaba. *Microb. Ecol.* **44**, 164–174.

Bograd, S.J., and R.J. Lynn (2003) Long-term variability in the Southern California Current system. *Deep-Sea Res. II.* **50**, 2355–2370.

Bohle-Carbonell, M. (1989) On the variability of the Peruvian upwelling system. *In—The Peruvian Upwelling System: Dynamics and Interactions* (D. Pauly, Ed.), pp. 14–44. ICLARM, Makati, Phillipines.

Böhm, E. (1999) The Ras-al-Hadd jet: remotely sensed and acoustic Doppler current profiler observations in 1994–1995. *Deep-Sea Res. II.* **46**, 1531–1549.

Boicourt, W.C., *et al.* (1998) Continental shelf of the SE United States and Gulf of Mexico; in the shadow of the western boundary current. *In—The Sea*, Vol. 11, *The Global Coastal Ocean* (A.R. Robinson, and K.H. Brink. Eds.), pp. 135–182. Wiley, New York.

Boisvert, W.E. (1967) Major currents in the north and south Atlantic Oceans. *NavOceano. Trans.* **193**, 1–92.

Boltovsky, D. (1986) Biogeography of the SW Atlantic; overview, current problems and perspectives. *In—Pelagic Biogeography, UNESCO Tech Pap. Mar. Sci.* **49**, 14–24.

Bonilla, J., *et al.* (1993) Seasonal distribution of nutrients and primary productivity on the eastern continental shelf of Venezuela as influenced by the Orinoco River. *J. Geophys. Res.* **98**, 2245–2257.

Borstad, G.A. (1982) The influence of the meandering Guiana Current and Amazon River discharge on surface salinity near Barbados. *J. Mar. Res.* **40**, 421–434.

Bot, P.V.M., *et al.* (1996) Comparison of changes in the annual variability of seasonal cycles of chlorophyll, nutrients and zooplankton at eight locations on the NW European continental shelf (1960–1980). *Deutsch. Hydr.-Zeitsch.* **48**, 349–363.

Bowie, A.R., *et al.* (2002) Biochemistry of Fe and other trace elements (Al, Co, Ni) in the upper Atlantic Ocean. *Deep-Sea Res. I.* **49**, 605–636.

Boyd, C.M., *et al.* (1980) Grazing patterns of copepods in the upwelling system off Peru. *Limnol. Oceanogr.* **25**, 583–596.

Boyd, P., and P.J. Harrison (1999) Phytoplankton dynamics in the NE subarctic Pacific. *Deep-Sea Res. II.* **46**, 2405–2432.

Boyd, P.W. (2002) Environmental factors controlling phytoplankton processes in the Southern Ocean. *J. Phycol.* **38**, 844–861.

Boyd, P.W., and C.S. Law (2001) The Southern Ocean Iron Release Experiment (SOIREE)—Introduction and Summary. *Deep-Sea Res. II.* **48**, 2425–2438.

Bradford, J.M. (1983) Physical and chemical oceanography of the west coast of New Zealand. *N. Z. J. Mar. Freshwat. Res.* **17**, 71–81.

Bradford, J.M., and F.H. Chang (1982) Standing stocks and productivity of phytoplankton off Westland, New Zealand. *N. Z. J. Mar. Freshwat. Res.* **21**, 71–90.

Bradford, J.M., and B. Chapman (1988) *Nyctyphanes australis* and an upwelling plume in western Cook Strait, New Zealand. *N. Z. J. Mar. Freshwat. Res.* **22**, 237–247.

Bradford, J.M., *et al.* (1982) The effect of warm-core eddies on oceanic productivity off northeastern New Zealand. *Deep-Sea Res.* **29**, 1501–1516.

Bradford-Grieve, J.M., and J.B. Jillett (1998) Ecological constraints on horizontal patterns—with special reference to the copepod *Neocalanus tonsus* (Brady, 1883). In—*Pelagic Biogeography, Final Rpt. SCOR/IOC WG 93*, pp. 65–77.

Brandini, F.P. (1993) Phytoplankton biomass in an Antarctic coastal environment during stable water conditions—implications for the iron limitation theory. *Mar. Ecol. Prog. Ser.* **93**, 267–275.

Bray, N.A. (1988) Water mass formation in the Gulf of California. *J. Geophys. Res.* **93**, 9223–9240.

Brickley, P.J., and A.C. Thomas (2004) Satellite-measured seasonal and interannual chlorophyll variability in the NE Pacific and coastal Gulf of Alaska. *Deep-Sea Res. II* **51**, 229–245.

Briggs, J. (1974) *Marine Zoogeography*. McGraw-Hill, New York.

Briggs, J. (1994) Species diversity: land and sea compared. *System. Biol.* 43, 130–135.

Brink, K.H. (1983) The near-surface dynamics of coastal upwelling. *Prog. Oceanogr.* **12**, 223–257.

Brinton, E. (1962) The distribution of Pacific euphausiids. *Bull. Scripps Inst. Oceanogr.* **8**, 51–270.

Brock, J.C., and C.R. McLain (1992) Interannual variability in phytoplankton blooms observed in the northwestern Arabian Sea during the southwest monsoon. *J. Geophys. Res.* **97**, 733–750.

Brock, J.C., *et al.* (1991) The phytoplankton bloom in the northwestern Arabian Sea during the southwest monsoon of 1979. *J. Geophys. Res.* **96**, 20623–20642.

Brock, J.C., *et al.* (1992) A southwest monsoon hydrographic climatology for the northwest Arabian Sea. *J. Geophys. Res.* **97**, 9455–9465.

Brock, J.C., *et al.* (1993) Modelling the seasonality of subsurface light and primary production in the Arabian Sea. *Mar. Ecol.-Prog. Ser.* **101**, 209–221.

Brock, J.C., *et al.* (1998) Bio-hydro-optical classification of the NW Indian Ocean. *Mar. Ecol. Prog. Ser.* **165**, 1–15.

Brockman, C., *et al.* (1980) The poleward undercurrent along the Peru coast: 5–15°S. *Deep-Sea Res.* **27**, 847–856.

Broerse, A.T.C., *et al.* (2000) $CaCO_3$ flux in the Sea of Okhotsk. *Mar. Micropaleont.* **39**, 179–200.

Brown, C. W. (1995b) Global distribution of coccolithophore blooms. *Oceanography* **8**, 59–60.

Brown, C.W., and J.A. Yoder (1994) Distribution of coccolithophorid blooms in the western North Atlantic Ocean. *Cont. Shelf Res.* **14**, 175–197.

Brown, J.H. (1995a) *Macroecology*. Chicago University Press, Chicago.

Brown, S.G., and C.H. Lockyer (1984) Whales. *In—Antarctic Ecology*, Vol. 2 (R.M. Laws, Ed.), pp. 717–781. Academic Press, London.

Brown, S.L. (1999) Picophytoplankton dynamics and production in the Arabian Sea during the 1995 SW Monsoon. *Deep-Sea Res. II.* **46**, 1745–1768.

Brown, S.L., *et al.* (2002) Microbial community dynamics and taxon-specific phytoplankton production in the Arabian Sea during the 1995 monsoons. *Deep-Sea Res. II.* **49**, 2345–2376.

Bruce, J.G. (1979) Eddies off the Somali coast during the southwest monsoon. *J. Geophys. Res.* **84**, 7742–7748.

Bruce, J.G., *et al.* (1985) On the North Brazilian eddy field. *Prog. Oceanogr.* **14**, 57–63.

Bruland, K.W., *et al.* (1991) Interactive influences of bioactive trace metals on biological production in oceanic waters. *Limnol. Oceanogr.* **36**, 1555–1577.

Bruland, K.W., *et al.* (2001) Iron and macronutrients in California coastal upwelling regimes: implications for diatom blooms. *Limnol. Oceanogr.* **46**, 1661–1674.

Bubnov, V.A. (1972) Structure and characteristics of the oxygen minimum in the southeastern Atlantic. *Oceanology.* **12**, 193–201.

Buchanan, J.B., and R.M. Warwick (1974) An estimate of benthic macrofaunal production in the offshore mud of the Northumberland coast. J. *Mar. Biol. Assoc. U.K.* **54**, 197–222.

Bucklin, A., *et al.* (1995) Molecular systematics of six *Calanus* and three *Metridia* species (Calanoida: Copepoda). *Mar. Biol.* **121**, 655–664.

Buesseler, K.O., *et al.* (2003) The effect of marginal ice-edge dynamics on production and export in the Southern Ocean along 170°W. *Deep-Sea Res. II.* **50**, 579–603.

Burkill, P.H. (1999) ARABESQUE: an overview. *Deep-Sea Res. II.* **46**, 529–548.

Burkill, P.H., *et al.* (1993a) Biogeochemical cycling in the northwest Indian Ocean. *Deep-Sea Res. II.* **40**, 643–849.

Burkill, P.H., *et al.* (1993b) *Synechococcus* and its importance to the microbial food web of the north-western Arabian Sea. *Deep-Sea Res.* II. 40, 773–782.
Butler, E.C.V., *et al.* (1992) Oceanography of the subtropical convergence zone around southern New Zealand. *N.Z. J. Mar. Freshwat. Res.* **26**, 131–154.
Butler, J.N., *et al.* (1983) Studies of *Sargassum* and the *Sargassum* community. *Bermuda Biol. Station Spec. Pub.* **22**, 1–85.
Cabal, J., *et al.* (1997) Egg production rates of *Calanus finmarchicus* in the Northwest Atlantic (Labrador Sea). *Can. J. Fish. Aquat. Sci.* **54**, 1270–1279.
Cadee, G.C. (1979) Primary production and chlorophyll in the Zaire river, estuary and plume. *Neth. J. Sea Res.* **12**, 368–381.
Campbell, J.W., and Aarup, T. (1992) New production in the North Atlantic derived from seasonal patterns of surface chlorophyll. *Deep-Sea Res.* **39**, 1669–1694.
Campbell, L. (1998) Response of microbial community structure to environmental forcing in the Arabian Sea. *Deep-Sea Res. II.* **45**, 2301–225.
Campbell, R.G. (2001) Evidence for food limitation of *Calanus finmarchicus* production rates on the southern flank of Georges Bank during April 1997. *Deep-Sea Res. II.* **48**, 531–550.
Campillo-Campbell, C., and A. Gordoa (2004) Physical and biological variability in the Namibian upwelling system: October 1997–October 2001. *Deep-Sea Res. II.* **51**, 147–158.
Capone, D.G., *et al.* (1997) *Trichodesmium*, a globally significant marine cyanobacterium. *Science.* **276**, 1221–1229.
Cardon, K.L., *et al.* (1991) Determination of Saharan dust radiance and chlorophyll from CZCS imagery. *J. Geophys. Res.* **96**, 5369–5378.
Carmack, E.C. (1990) Large-scale physical oceanography of polar regions. *In—Polar Oceanography. Part A: Physical Science* (W. Smith, Jr., Ed.), pp. 171–221. Academic Press, San Diego.
Caron, D.A., *et al.* (2000) Microzooplankton herbivory in the Ross Sea, Antarctica. *Deep-Sea Res. II.* **47**, 3249–3272.
Carpenter, E.J. (1983) Nitrogen fixation by marine Oscillatoria. *In—Nitrogen in the Marine Environment* (E.J. Carpenter and D.G. Capone, Eds.), pp. 65–104. Academic Press, New York.
Carpenter, E.J., and J.L. Cox (1974) Production of pelagic *Sargassum* and a blue-green epiphyte in the western Sargasso Sea. *Limnol. Oceanogr.* **19**, 429–436.
Carr, M.-E., *et al.* (1992) Hydrographic patterns and vertical mixing in the equatorial Pacific along 150°W. *J. Geophys. Res.* **97**, 611–626.
Carton, J. A., and B. Huang (1994) Warm events in the tropical Atlantic. *J. Phys. Oceanogr.* **24**, 888–903.
Caspers, H. (1957) Black Sea and Sea of Azov. *Mem. Geol. Soc. Am.* **67**, 801–889.
Castro, B.M., and L.B.de Miranda (1998) Physical oceanography of the western Atlantic continental shelf between 4°N and 34°S. *In—The Sea*, Vol. 11, *The Global Coastal Ocean* (A.R. Robinson and K.H. Brink, Eds.), pp. 209–251. Wiley, New York.
Catabiano, A.C.V., *et al.* (2005) Multi-year satellite observations of tropical instability waves in the tropical Atlantic Ocean. *Eur. Geosci. Union, Ocean Sci. Discussions* **2**, 1–35.
Chan, K.M. (1970) Seasonal variation of hydrologic properties in the northern South China Sea. *In—The Kuroshio* (J.C. Marr, Ed.), pp. 143–162. University of Washington Press, Seattle, WA.
Chang, F.H., *et al.* (1992) Nitrogen utilisation by size-fractionated phytoplankton assemblages associated with an upwelling event off Westland, New Zealand. *N. Z. J. Mar. Freshwat. Res.* **26**, 287–301.
Chang, J. (2000) Seasonal variation and cross-shelf distribution of the nitrogen-fixing cyanobacterium *Trichodesmium* in the southern East China Sea. *Cont. Shelf Res.* **20**, 479–492.
Chao, S.-Y., *et al.* (1996) El Niño modulation of the South China Sea circulation. *Prog. Oceanogr.* **38**, 51–93.
Chapman, P., and L.V. Shannon (1985) The Benguela ecosystem. II. Chemistry and related processes. *Oceanogr. Mar. Biol. Ann. Rev.* **23**, 183–251.
Charria, G., *et al.* (2003) Rossby wave and ocean color: the cells uplifting hypothesis in the South Atlantic Subtropical Convergence Zone. *Geophys. Res. Lett.* **30**(3), 25–1.
Chatananthawej, B., *et al.* (1987) Quantitative survey of the macrobenthic fauna along the west coast of Thailand. *Res. Bull. Phuket Mar. Biol. Cent.* **47**, 1–23.
Chavez, F.P., *et al.* (1989) The potential primary production of the Peruvian upwelling ecosystem: 1953–1984. *ICLARM Stud. Rev.* **15**, 1–351.

Chavez, F.P., *et al.* (1996) Phytoplankton variability in the central and eastern tropical Pacific. *Deep-Sea Res. II.* **43**, 835–870.
Chavez, F.P., *et al.* (2003) From anchovies to sardines and back: multidecadal change in the Pacific Ocean. *Science.* **299**, 217–221.
Chelton, D.B. (1982) Large-scale response of the California Current to forcing by the wind stress curl. *Rept. Calif. Coop. Fish. Invest.* **23**, 130–148.
Chelton, D.B., and M.G. Schlax (1996) Global observations of oceanic Rossby waves. *Science.* **272**, 234–238.
Chelton, D.B., *et al.* (1982) Large-scale interannual physical and biological interaction in the California Current. *J. Mar. Res.* **40**, 1095–1125.
Chen, S., and B. Qiu (2004) Seasonal variability of the South Equatorial Countercurrent. *J. Geophys. Res.* **109**, C08003.
Chen, Y.-L. (2000) Comparisons of primary productivity and phytoplankton size structure in the marginal regions of southern East China Sea. *Cont. Shelf Res.* **20**, 437–458.
Chen, Y.-L., *et al.* (2001) New production in the East China Sea, comparison between well-mixed winter and stratified summer conditions. *Cont. Shelf Res.* **21**, 751–764.
Chen, Y.-L., *et al.* (2004) Phytoplankton production during a summer coastal upwelling in the East China Sea. *Cont. Shelf Res.* **24**, 1321–1338.
Chen, Z., *et al.* (1992) The effects of a Kuroshio frontal eddy on the biochemical structure of the East China Sea. *In—Essays on the Investigation of the Kuroshio 4*, pp. 115–124. China Ocean Press, Beijing.
Chester, R. (2000) *Marine Chemistry—Trace Elements in the Ocean*, p. 522. Blackwell, Oxford.
Childers, A.R., *et al.* (2005) Seasonal and interannual variability in the distribution of nutrients and chlorophyll across the Gulf of Alaska shelf: 1998–2000. *Deep-Sea Res. II.* **52**, 193–216.
Chisholm, S.W. (1992) Phytoplankton size. *In—Primary Productivity and Biogeochemical Cycles in the Sea* (P. G. Falkowski and A. D. Woodhead, Eds.), pp. 213–237. Plenum Press, New York.
Chisholm, S.W., *et al.* (1988) A novel free-living prochlorophyte abundant in the oceanic photic zone. *Nature, Lond.* **334**, 340–343.
Christian J.R., *et al.* (2002) Biogeochemical modelling of the tropical Pacific Ocean. II. Iron geochemistry. *Deep-Sea Res. II.* **49**, 545–556.
Christian, J.R., *et al.* (2004) A ribbon of dark water: phytoplankton blooms in the meanders of the Pacific North Equatorial Current. *Deep-Sea Res.* **51**, 209–228.
Christiansen, B., and A. Boetius (2000) Mass sedimentation of the swimming crab *Charybdis smithii* (Crustacea: Decapoda) in the deep Arabian Sea. *Deep-Sea Res. II.* **47**, 2673–2685.
Chung, S.P., *et al.* (1996) Beam attenuation and micro-organisms: spatial and temporal variations in small particles along 140°W during the 1992 JGOFS EqPac transects. *Deep-Sea Res. II.* **43**, 1205–1226.
Church, J.A. (1987) East Australian Current adjacent to the Great Barrier Reef. *Aust. J. Mar. Freshwat. Res.* **38**, 671–683.
Church, J.A., and P.D. Craig (1998) Australia's shelf seas: diversity and complexity. *In—The Sea*, Vol. 11, *The Global Coastal Ocean* (A.R. Robinson and K.H. Brink, Eds.), pp. 933–964. Wiley, New York.
Church, J.A., *et al.* (1989) The Leeuwin Current. *In—Poleward Flows along the Eastern Ocean Boundaries* (S.J. Neshyba, C.N.K. Mooers, R.L. Smith, and R.T. Barber, Eds.), pp. 230–253. Springer-Verlag, New York.
Citeau, J., *et al.* (1988) The march of ITCZ migrations. *Trop. Ocean-Atmos. Newslett.* **45**, 1–3.
Clark, P.U., *et al.* (2002) The role of the thermohaline circulation in abrupt climate change. *Nature (Lond.).* **415**, 863869.
Coachman, L.K. (1986) Circulation: water masses and fluxes on the southeastern Bering Sea shelf. *Cont. Shelf Res.* **5**, 22–108.
Coachman, L.K., *et al.* (1980) Frontal systems of the southeast Bering Sea shelf. *In—Second International Symposium on Stratified Flows. Trondheim. June 1980* (T. Carstens and T. McClimans, Eds.), pp. 917–933. 2nd IAHR Symposium, Trondheim.
Coale, K.H *et al.* (1996a) Control of community growth and export production by upwelled iron in the equatorial Pacific Ocean. *Nature (Lond).* **379**, 621–624.
Coale, K.H., *et al.* (1996b) A massive phytoplankton bloom induced by an ecosystem-scale iron fertilization experiment in the equatorial eastern Pacific Ocean. *Nature (Lond.).* **383**, 495–501.

Codispoti, L.A., *et al.* (1982) A comparison of the nutrient regimes off northwest Africa, Peru and Baja California. *Rapp. Proc. Verb. Réun. Int. Counc. Explor Mer.* **180**, 184–201.

Colborn, J.G. (1975) *The Thermal Structure of the Indian Ocean*. University of Hawaii Press, Honolulu, HI.

Colebrook, J.M. (1979) Continuous plankton records: seasonal cycles of phytoplankton and copepods in the North Atlantic Ocean and the North Sea. *Mar. Biol.* **51**, 23–32.

Colebrook, J.M. (1982) Continuous plankton records: seasonal variations in the distribution and abundance of plankton in the North Atlantic Ocean and the North Sea. *J. Mar. Res.* **4**, 435–462.

Colebrook, J.M. (1986) Environmental influences on long-term variability in marine plankton. *Hydrobiologia.* **142**, 309–325.

Colin, C., Gonella, J., and Merle, J. (1987) Equatorial upwelling at 4°W during the FOCAL programme. *Océanol. Acta.* **6**, 39–49.

Collier, J.L., and B. Palenik (2003) Phycoerythrin-containing picoplankton in the Southern California Bight. *Deep-Sea Res. II.* **50**, 2405–2422.

Collins, C.A., *et al.* (2003) The California Current system off Monterey, California: physical and biological coupling. *Deep-Sea Res. II.* **50**, 2389–2404.

Comiso, J.C., *et al.* (1993) Coastal zone color scanner pigment concentrations in the Southern Ocean and relationships to geophysical surface features. *J. Geophys. Res.* **98**, 2419–2451.

Concelman, S., *et al.* (2001) Distribution, abundance and benthic-pelagic coupling of suspended hydroids on Georges Bank. *Deep-Sea Res. II.* **48**, 645–658.

Conkright, M.E., *et al.* (1994) *NOAA World Ocean Atlas, NESDIS 1 (Nutrients)*. U.S. Department of Commerce, Washington, DC.

Conover, R.J. (1981) Nutritional strategies for feeding on small suspended particles. *In—Analysis of Marine Ecosystems* (A.R. Longhurst, Ed.), pp. 363–396. Academic Press, London.

Conover, R.J. (1988) Comparative life histories in the genera *Calanus* and *Neocalanus* in high latitudes of the northern hemisphere. *Hydrobiologia.* **167/168**, 127–142.

Conover, R.J., *et al.* (1991) Copepods in cold waters—how do they cope? *Bull. Plankton Soc. Japan, Spec. Vol.*, 177–199.

Conversi, A. (2001) Seasonal and interannual dynamics of *Calanus finmarchicus* in the Gulf of Maine, with reference to the NAO. *Deep-Sea Res. II.* **48**, 519–530.

Cooney, R.T. (1986) The seasonal occurrence of *Neocalanus cristatus, N. plumchrus* and *Eucalanus bungii* over the shelf of the northern Gulf of Alaska. *Cont. Shelf Res.* **5**, 541–553.

Cornillon, P. (1986) The effect of the New England seamounts on Gulf Stream meandering. *J. Phys. Oceanogr.* **16**, 386–389.

Coste, B., *et al.* (1988) Les élements nutritifs dans le bassin occidental de la Méditerranean. *Océanol. Acta Spec.* 87–94.

Cota, G.F., *et al.* (1990) Impact of ice algae on inorganic nutrients in seawater and sea ice in Barrow Strait, NWT, Canada. *Can. J. Fish. Aquat. Sci.* **47**, 1402–1415.

Coyle, K.O., and A.I. Pinchuck (2005) Seasonal cross-shelf distribution of major zooplankton taxa on the northern Gulf of Alaska shelf relative to water mass properties, species depth preferences and vertical migration behaviour. *Deep-Sea Res. II.* **52**, 217–246.

Cracraft, J. (1983) Species concepts and speciation analysis. *Curr. Ornithol.* **1**, 159–187.

Crawford, W.R., *et al.* (2005) Impact of Haida eddies on chlorophyll distribution in the eastern Gulf of Alaska. *Deep-Sea Res. II.* **52**, 975–989.

Cresswell, G. (1994) Nutrient enrichment of the Sydney continental shelf. *Aust. J. Mar. Freshwat. Res.* **45**, 677–691.

Cromwell, T. (1953) Circulation in a meridional plane in the central equatorial Pacific. *J. Mar. Res.* **12**, 196–213.

CSIRO (1969) Seasonal variations in the Indian Ocean at 110°E. *Aust. J. Mar. Freshwat. Res.* **20**, 1–90.

Cullen, J.J. (1991) Hypotheses to explain high nutrient low chlorophyll conditions in the sea. *Limnol. Oceanogr.* **36**, 1578–1599.

Cullen, J.J., *et al.* (2002) Physical forcing of marine ecosystem dynamics. *In—Biological-Physical Interactions in the Ocean* (A.R. Robinson, J.J. McCarthy and B.J. Rothschild, Eds.), pp. 245–336. Wiley InterScience, New York.

Cury, P., and Roy, C. (1987) Upwelling et pêche des éspèces pélagiques côtières de la Côte d'Ivoire. *Océanolog. Acta.* **10**, 347–357.

Cushing, D.H. (1973) Production in the Indian Ocean and the transfer from primary to the secondary level. *In—The Biology of the Indian Ocean* (B. Zeitzschel, Ed.), pp. 257–272. Springer, Berlin.

Cushing, D.H. (1982) *Climate and Fisheries*. Academic Press, San Diego.

Cushing, D.H. (1989) A difference in structure between ecosystems in strongly stratified waters and those that are rarely stratified. *J. Plankton Res.* **11**, 1–3.

Cushing, D.H. (1990) Plankton production and year-class strength in fish populations: an update of the match-mismatch hypothesis. *Adv. Mar. Biol.* **26**, 249–294.

Cushing, D.H., and R.R. Dickson (1966) The biological response in the sea to climatic change. *Adv. Mar. Biol.* **14**, 1–122.

Dadou, I., *et al.* (1996) Impact of the North Equatorial Counter Current meandering on a pelagic ecosystem: a modelling approach. *J. Mar. Res.* **54**, 311–342.

Dalpadado, P. (2000) Inter-specific variations in distribution, abundance and possible life-cycle patterns in *Themisto* spp. (Amphipoda) in the Barents Sea. *Polar Biol.* **25**, 656–666.

Dam, H.G., *et al.* (1993) The trophic role of mesozooplankton at 47°N 20°W during the North Atlantic Bloom Experiment. *Deep-Sea Res. II.* **40**, 197–212.

Dam, H.G., *et al.* (1996) Mesozooplankton grazing and metabolism at the equator in the central Pacific: implications for carbon and nitrogen fluxes. *Deep-Sea Res. II.* **42**, 735–756.

Dandonneau Y., *et al.* (2004) Seasonal and interannual variability of ocean colour and composition of phytoplankton communities in the North Atlantic, equatorial and South Pacific. *Deep-Sea Res. II.* **51**, 303–318.

Dandonneau, Y., and Charpy, L. (1985) An empirical approach to the island mass effect in the south tropical Pacific Ocean based on sea surface chlorophyll concentrations. *Deep-Sea Res.* **32**, 707–721.

Dandonneau, Y., and Gohin, F. (1984) Meridional and seasonal variations of the sea surface chlorophyll concentration in the southwestern tropical Pacific. *Deep-Sea Res.* **31**, 1377–1394.

Davies, P. (2004) Nutrient processes and chlorophyll in the estuaries and plume of the Gulf of Papua. *Cont. Shelf Res.* **24**, 2317–2341.

de Baar, H.W., *et al.* (1995) Importance of iron for plankton blooms and $CO_2$ drawdown in the Southern Ocean. *Nature (Lond.).* **373**, 412–415.

de Baar, H.W., *et al.* (1997) Nutrient anomalies in *Fragilariopsis* blooms, iron deficiency and the N/P ratio of the Antarctic Ocean. *Deep-Sea Res. II.* **44**, 229–260.

de Beaufort, L.F. (1951) *Zoogeography of the Land and Inland Waters*. Sidgewick & Jackson, London.

de Castro, B.M., and L.B. de Miranda (1998) Physical oceanography of the western Atlantic continental shelf located between 4°N and 34°S. *In—The Sea*, Vol. 11, *The Global Coastal Ocean* (A.R. Robinson and K.H. Brink, Eds.), pp. 209–252. Wiley, New York.

de la Mare, W.K. (1997) Abrupt mid-twentieth century decline in Antarctic sea-ice extent from whaling records. *Nature (Lond.).* **389**, 57–60.

De Oliveira, J.A.A., *et al.* (2005) Potential improvement in the management of the Bay of Biscay anchovy by incorporating environmental indices as recruitment predictors. *Fish. Res.* **75**, 2–14.

De Wilde P.A.W., *et al.* (1989) Structure and energy demand of the soft-bottom communities in the Java Sea. *Neth J. Sea Res.* **23**, 449–461.

Deacon, G. (1984) *The Antarctic Circumpolar Ocean*. Cambridge University Press, Cambridge, UK.

Deegan, L.A., *et al.* (1986) Relationships among physical characteristics, vegetation distribution and fisheries yield in Gulf of Mexico estuaries. *In—Estuarine Variability* (D.A. Wolfe, Ed.), pp. 83–100. Academic Press, New York.

Delacluse, P., *et al.* (1994) On the connection between the 1984 Atlantic warm event and the 1982–83 El Niño. *Tellus.* **46A**, 446–464.

Deming, J.W., *et al.* (2002) The International North Water Polynya Study (NOW): a brief overview. *Deep-Sea Res. II.* **449**, 4887–4892.

Denman, K.L., and M.A. Peña (1999) A coupled 1-D biological/physical model of the northeast subarctic Pacific with iron limitation. *Deep-Sea Res. II.* **46**, 2877–2908.

Desrosiers, G. (2000) Trophic structure of macrobenthos in the Gulf of St. Lawrence and on the Scotian Shelf. *Deep-Sea Res.* II. **47**, 663–697.

Dessier, A., and J.R. Donguy (1994) The sea surface salinity in the tropical Atlantic between 10°N and 10°S—seasonal and interannual variations 1977–1989. *Deep-Sea Res. I*, **41** 101–112.

Detmer, A.E., and U.V. Bathman (1997) Distribution patterns of autotrophic pico- and nanoplankton and their relative contribution to algal biomass during spring in the Atlantic sector of the Southern Ocean. *Deep-Sea Res. II.* **44**, 299–320.

Dickey, T., *et al.* (1994) Bio-optical and physical variability in the subarctic North Atlantic Ocean during the spring of 1989. *J. Geophys. Res.* **99**, 22541–22556.

Dickey, T., *et al.* (2001) Physical and biogeochemical variability from hours to years at the BATS site; June 1994–March 1998. *Deep-Sea Res. II.* **48**, 2105–2140.

Dickie, L., and Trites, R. (1983) The Gulf of St. Lawrence. *In—Estuaries and Enclosed Seas* (B.H. Ketchum, Ed.), pp. 403–425. Elsevier, Amsterdam.

Dickinson, W.R. (2003) The place and power of myth in geoscience. *Am. J. Sci.* **303**(9), 856–864.

Dickson, R. (1983) Global summaries and inter-comparisons; flow statistics from long-term current meter moorings. *In—Eddies in Marine Science* (A.R. Robinson, Ed.), pp. 278–328. Springer-Verlag, Berlin.

Dickson, R. (2003) Stirring times in the North Atlantic. *Nature (Lond.).* **424**, 152–156.

Dickson, R., *et al.* (1988) North winds and production in the eastern North Atlantic. *J. Plankton Res.* **10**, 151–169.

Dietrich, G. (1964) Oceanic polar front survey. *Res. Geophys.* **2**, 291–308.

Dietrich, G., *et al.* (1970) *General Oceanography.* Wiley, New York.

DiMarco, S.F., *et al.* (2000) Satellite observations of upwelling on the continental shelf south of Madagascar. *Geophys. Res. Lett.* **27**, 3965–3968.

DiTullio, G.R., and W.O. Smith (1996) Spatial patterns in phytoplankton biomass and pigment distributions in the Ross Sea. *J. Geophys. Res.* **101**, 18467–18477.

Dodimead, A.J., *et al.* (1967) Review of the oceanography of the North Pacific. *Bull. Int. North Pacific Fish. Comm.* **13**, 1–195.

Donaghey, P.L., *et al.* (1991) The role of episodic atmospheric nutrient inputs in the chemical and biological dynamics of oceanic ecosystems. *Oceanography.* **4**, 62–70.

Dorgham, M.M., and A. Moftah (1989) Environmental conditions and phytoplankton distribution in the Arabian Gulf and Gulf of Oman, September 1986. *J. Mar. Biol. Assoc. India.* **31**, 36–53.

Dorman, C.E., and D.P. Palmer (1981) Southern California summer coastal upwelling. *In—Coastal Upwelling* (F.A. Richards, Ed.), pp. 44–56. American Geophysical Union, Washington, D.C.

Dower, K., and M.I. Lucas (1993) Photosynthesis-irradiance relationships and production associated with a warm-core ring shed from the Agulhas Retroflection south of Africa. *Mar. Ecol. Prog. Ser.* **95**, 141–154.

Doyon, P., *et al.* (2000) Influence of wind mixing and upper-layer stratification on phytoplankton biomass in the Gulf of St. Lawrence. *Deep-Sea Res. II.* **47**, 415–433.

Dubischar, C.D., and U.V. Bathman (1997) Grazing impact of copepods and salps on phytoplankton in the Atlantic sector of the Southern Ocean. *Deep-Sea Res. II.* **44**, 415–433.

Dubischar, C.D., *et al.* (2002) High summer abundances of small pelagic copepods at the Antarctic Polar Front—implications for ecosystem dynamics. *Deep-Sea Res. II.* **49**, 3871–3887.

Duce, R.A., and N.W. Tindale (1991) Atmospheric transport of iron and its deposition in the ocean. *Limnol. Oceanogr.* **38**, 1715–1726.

Duce, R.A., *et al.* (1991) The atmospheric input of trace species to the world ocean. *Global Biogeochem. Cycles.* **5**, 193–259.

Ducklow, H.W., and R.P. Harris (1993) Introduction to the JGOFS North Atlantic study. *Deep-Sea Res. II.* **40**, 1–8.

Ducklow, H.W., *et al.* (1995) Bacterioplankton dynamics in the Equatorial Pacific during the 1992 El Niño. *Deep-Sea Res. II.* **42**, 621–638.

Dufour, P., and J.-M. Stretta (1973) Production primaire, biomasses du phytoplankton et du zooplankton dans l'Alantique tropical sud, le long de la Meridian 4°W. *Cah. ORSTOM, sér. Océanogr.* **11**, 419–429.

Dugdale, R.C., *et al.* (2002) Meridional asymmetry of source nutrients to the equatorial Pacific upwelling ecosystem and its potential impact on ocean-atmosphere $CO_2$ flux. *Deep-Sea Res. II.* **49**, 2513–2531.

Duineveld, G.C.A., *et al.* (1991) The macrobenthos of the North Sea. *Neth. J. Sea Res.* **28**, 53–65.

Dunbar, M.J. (1979) The relation between the oceans. *In—Zoogeography and Diversity of Plankton* (S. van der Spoel and A.C. Pierrot-Bults, Eds.), pp. 112–125. Bunge Scientific, Utrecht, The Netherlands.

Duncan, C.P., *et al.* (1982) Surface currents near the Greater and Lesser Antilles. *Int. Hydrogr. Rev.* **59**, 67–78.

Dupouy, C., *et al.* (1988) Satellite-detected cyanobacteria bloom in the southwest tropical Pacific: implications for oceanic nitrogen fixation. *Int. J. Remote Sensing.* **9**, 389–396.

DuRand, M.D., *et al.* (2001) Phytoplankton population dynamics at the BATS station in the Sargasso Sea. *Deep-Sea Res. II.* **48**, 1983–2003.

Dutkiewicz, S., *et al.* (2001) Interannual variability of phytoplankton abundances in the North Atlantic. *Deep-Sea Res. II.* **48**, 2323–2344.

Ebert, U., *et al.* (2001) Critical conditions for phytoplankton blooms. *Math. Biol.* **63**, 1095–1124.

Edwards, F. (1987) Climate and oceanography. *In—Key Environments—Red Sea* (A.J. Edwards and S.M. Head, Eds.), pp. 46–69. Pergamon Press, Oxford.

Edwards, M., *et al.* (2001) Case history and persistence of the non-indigenous diatom *Coscinodiscus wailesii* in the northeast Atlantic. *J. Mar. Biol. Assoc. UK.* **81**, 207–211.

Eilertsen, H.C., *et al.* (1989a) Potential of herbivorous copepods for regulating the spring phytoplankton bloom in the Barents Sea. *Rapp. Proc. Verb. Cons. Int. Explor. Mer.* **188**, 154–163.

Eilertsen, H.C., *et al.* (1989b) Vertical distributions of primary production and grazing in arctic waters (Barents Sea). *Polar Biol.* **9**, 253–260.

Eisma, D. (1988) An introduction to the geology of the continental shelves of the world. *In*—Continental Shelves (H. Postma and J.J. Zijlstra, Eds.). Elsevier, Amsterdam.

Ekman, S. (1953) *Zoogeography of the Sea.* Sidgewick and Jackson, London.

Eldin, G. (1983) Eastward flows of the South Equatorial Central Pacific. *J. Phys. Oceanogr.* **13**, 1461–1467.

Ellis, D.V. (1959) Arctic benthos of North America. *Nature (Lond.).* **184**, 79–80.

Elmoussaoui, A., *et al.* (2005) Model-inferred upper ocean circulation in the eastern tropics of the North Atlantic. *Deep-Sea Res. I.* **52**, 1093–1120.

Emery, W.J., *et al.* (1984) Geographic and seasonal distributions of Brunt-Väisälä frequency and Rossby radii in the North Pacific and North Atlantic. *J. Phys. Oceanogr.* **14**, 294–317.

Enfield, D.B., and Cid, L. (1990) Statistical analysis of ENSO over the last 500 years. *TOGA Notes.* **1**, 1–4.

English, T.S. (1961) Some biological oceanographic observations in the central Polar Sea. *Res. Pap. Arct. Inst. North Am.* **13**, 1–80.

Erickson III, D.J., *et al.* (2003) Atmospheric iron delivery and surface ocean biological activity in the Southern Ocean and Patagonian regions. *Geophys. Res. Lett.* **30**, 1–4.

Escribano, R., and P. Hidalgo (2000) Spatial distribution of copepods in the north of the Humboldt Current off Chile during coastal upwelling. J. *Mar. Biol. Assoc. UK* **80**, 283–290.

Estrada, M., and R. Margalef (1988) Supply of nutrients to the Mediterranean photic zone along a persistent front. *Océanol. Acta.* **9**, 133–142.

Estrada, M., *et al.* (1993) Variability of the deep chlorophyll maximum in the northwest Mediterranean. *Mar. Ecol.-Prog. Ser.* **92**, 289–300.

Evseenko. P. (1987) Reproduction of Peruvian jack mackerel in the southern Pacific. *J. Ichthyol.* **27**, 151–160.

Fager, E.W., and A.R. Longhurst (1968) Recurrent group analysis of species assemblages of demersal fish in the Gulf of Guinea. *J. Fish. Res. Bd. Canada* **25**, 1405–1421.

Falkowski, P.G., *et al.* (1991) Role of eddy pumping in enhancing primary production in the ocean. *Nature (Lond.).* 352, 55–88.

Fan, K.L. (1982) A study of the water masses of the Taiwan Strait. *Acta Oceanol. Taiwan* **13**, 140–153.

Fanning, K.A. (1991) Nutrient provinces in the sea: concentration ratios, reaction rate ratios and ideal covariation. *J. Geophys. Res.* **97**, 5693–5712.

Färber-Lorda, J., *et al.* (2004) Effects of wind forcing on the trophic conditions, zooplankton biomass and krill biochemical composition in the Gulf of Tehuantapec. *Deep-Sea Res. II.* **51**, 601–614.

Farstey, V., *et al.* (2002) Expansion and homogeneity of the vertical distribution of zooplankton in a very deep mixed layer. *Mar. Ecol. Prog. Ser.* **238**, 91–100.

Favorite, F. (1974) Flow into the Bering Sea through Aleutian Island passes. *In* Occas. Publ. (2), Inst. Mar. Sci., *Oceanography of the Bering Sea* (D.W. Hood and E.J. Kelly, Eds.), pp. 3–38. University of Alaska, Fairbanks.

Favorite, F., *et al.* (1976) Review of the oceanography of the North Pacific, 1960–1971. *Bull. Int. North Pacific Fish. Comm.* **33**, 1–187.
Feder, H.M., *et al.* (1994) The northeastern Chukchi Sea: benthos-environmental interactions. *Mar. Ecol. Prog. Ser.* **111**, 171–190.
Fedoseev, A. (1970) Geostrophic circulation of surface waters on the shelf of northwest Africa. *Rapp. Proc. Verb. Réun. Cons. Int. Explor. Mer.* **159**, 30–37.
Feldman, G.C., *et al.* (1989) Ocean color: availability of the Global Data Set. *Eos.* **70**, 634–641.
Fenchel, T. (1978) The ecology of micro- and meiobenthos. *Ann. Rev. Ecol. System.* **9**, 99–121.
Fennel, K., *et al.* (2003) Impacts of iron control on phytoplankton production in the modern and glacial Southern Ocean. *Deep-Sea Res. II.* **50**, 833–851.
Fenton, M., *et al.* (2000) Aplanktonic zones in the Red Sea. *Mar. Micropaleontology.* **40**, 277–294.
Fernandez, E., and R.D. Pingree (1996) Coupling between physical and biological fields in the North Atlantic subtropical front southeast of the Azores. *Deep-Sea Res. I.* **43**, 1369–1393.
Fiedler, P.C. (1994) Seasonal climatologies and variability of the eastern tropical Pacific surface waters. *Tech. Rep. U.S. Natl. Mar. Fish. Service.* **109**, 1–65.
Fiedler, P.C. (2002) The annual cycle and biological effects of the Costa Rica Dome. *Deep-Sea Res. I.* **49**, 321–338.
Fiedler, P.C., *et al.* (1991) Oceanic upwelling and productivity in the eastern tropical Pacific. *Limnol. Oceanogr.* **36**, 1834–1850.
Fiekas, V., *et al.* (1992) A view of the Canary Basin thermocline circulation in winter. *J. Geophys. Res.* **97**, 12495–12510.
Findlater, J. (1969) A major low-level air current near the Indian Ocean during the northern summer. *Q. J. R. Meteor. Soc.* **95**, 362–380.
Finenko, G.A., *et al.* (2003) Population dynamics and ingestion rates of *Beroe ovata* in the Black Sea. *J. Plankton Res.* **25**, 539–549.
Firth, A., *et al.* (2002) Variability of the Great Whirl from observations and models. *Deep-Sea Res. II.* **49**, 1279–1295.
Flagg, C.N., *et al.* (2002) Springtime hydrography of the southern Middle Atlantic Bight and the onset of seasonal stratification. *Deep-Sea Res. II.* **49**, 4297–4329.
Fleminger, A., and K. Hulseman. (1977) Geographic range and taxonomic divergence of *Calanus* in the North Atlantic. *Mar. Biol.* **40**, 233–248.
Flierl, G., and D.J. McGillicuddy (2002) Mesoscale and submesoscale physical-biological interactions. *In—The Sea*, Vol. 11, *The Global Coastal Ocean* (A.R. Robinson and K.H. Brink, Eds.), pp. 113–186. Wiley, New York.
Foley, D.G., *et al.* (1997) Longwaves and primary productivity variations in the equatorial Pacific at 140°W. *Deep-Sea Res. II.* **44**, 1801–1826.
Follows, M., and S. Dutkiewicz (2003) Meteorological modulation of the North Atlantic spring bloom. *Deep-Sea Res. II.* **49**, 321–344.
Fonseca, T.R., and M. Farias (1987) Estudio del proceso de surgencia en la costa chileña utilizando percepción remota. *Invest. Pesq.* **34**, 33–46.
Fontenau, A. (1997) *Atlas of Tropical Tuna Fisheries: World Catches and Environment.* ORSTOM Editions, Paris.
Forbes, A.M.G. (1984) The contributions of local processes to seasonal hydrography of the Gulf of Carpentaria. *Océanogr. Trop.* **19**, 193–201.
Foster, T.D. (1984) The marine environment. *In—Antarctic Ecology*, Vol. 2 (R.M. Laws, Ed.), pp. 345–372. Academic Press, London.
Fraga, F. (1981) Upwelling off the Galicia coast, Northwest Spain. *In—Coastal Upwelling* (F.A. Richards, Ed.), pp. 176–182. American Geophysical Union, Washington, DC.
Francis, R.C., *et al.* (1998) Effects of interdecadal climate variability on the oceanic ecosystems of the northeast Pacific. *Fish. Oceanogr.* **7**, 1–21.
Franck, V.M., *et al.* (2000) Iron and silicic acid availability regulate Si uptake in the Pacific sector of the Southern Ocean. *Deep-Sea Res. II.* **47**, 3315–3338.
Franck, VM., *et al.* (2003) Iron and zinc effects on silicic acid and nitrate uptake kinetics in three high-nutrient, low-chlorophyll (HNLC) regions. *Mar. Ecol. Prog. Ser.* **252**, 15–33.
Franks, P.J.S., and C. Chen (1996) Plankton production in tidal fronts: a model of Georges Bank in summer. *J. Mar. Res.* 54, 631–651.

Franks, P.J.S., and L.J. Walstad (1997) Phytoplankton at front: a model of formation and response to wind events. *J. Mar. Res.* **55**, 1–29.

Fratantoni, D.M., and D.A. Glickson (2002) North Brazil Current ring generation and evolution observed with SeaWiFS. *J. Phys. Oceanogr.* **32**, 1058–1074.

Frost, B., and A. Fleminger (1968) A revision of the genus *Clausocalanus* with remarks on distributional patterns in diagnostic characters. *Bull Scripps Inst. Oceanogr.* **12**, 1–235.

Fujii, M., *et al.* (2002) A one-dimensional ecosystem model applied to time-series Station KNOT. *Deep-Sea Res. II.* **49**, 5441–5461.

Fung, I.Y., *et al.*, (2000) Iron supply and demand in the upper ocean. *Glob. Biogeochem. Cyc.* **14**, 281–291.

Furnas, M.J., and A.W. Mitchell (1996a) Nutrient inputs into the central Great Barrier Reef from subsurface intrusions of Coral Sea waters: a 2-dimensional model. *Cont. Shelf Res.* **16**, 1127–1148.

Furnas, M.J., and A.W. Mitchell (1996b) Pelagic primary production in the Coral and southern Solomon Seas. *Mar. Freshwat. Res.* **47**, 695–706.

Furuya K., and R. Marumo (1983) Size distribution of phytoplankton in the Western Pacific Ocean and adjacent waters in summer. *Bull Plank. Soc. Japan.* **30**, 21–32.

Furuya, K., *et al.* (1986) Phytoplankton in the Subtropical Convergence during the austral summer: community structure and growth activity. *Deep-Sea Res.* **33**, 621–630.

Gabric, A.J., *et al.* (1990) Chlorophyll distribution in the Great Barrier Reef. *Aust. J. Mar. Freshwat. Res.* **41**, 313–324.

Gabric, A.J., *et al.* (1993) Offshore export of shelf production in the Cap Blanc giant filament as derived from CZCS imagery. *J. Geophys. Res.* **98**, 4697–4712.

Gallardo, C.S., and P.E. Pencheszardeh (2001) Hatching mode and latitude in marine gastropods: revisiting Thorson's paradigm in the southern hemisphere. *Mar. Biol.* **138**, 547–552.

Gammelsrφd, T., *et al.* (1998) Intrusion of warm surface water layer along the Angola-Namibian coast in February–March: the 1995 Benguela Niño. *S. Afr. J. Mar. Sci.* **19**, 41–56.

Ganssen, G., and D. Kroon (1991) Evidence for Red Sea surface circulation from oxygen isotopes of modern surface waters and planktonic foraminiferan tests. *Paleoceanography.* **6**, 73–82.

Garces, L.R., *et al.* (2006) Spatial structure of demersal fish assemlages in South and Southeast Asia, and implications for fisheries management. *Fish. Res.* 78, 143–157.

Garner, D.M. (1967) Hydrology of the south-east Tasman Sea. *Bull. N. Z. Dep. Sci. Ind. Res.* **181**, 1–40.

Garrett, C. (2003) Mixing with latitude. *Nature (Lond.).* **422**, 477.

Garrison, D.L. (1991) An overview of the abundance and role of protozooplankton in Antarctic waters. *J. Mar. Systems.* **2**, 317–331.

Garrod, D., and A. Schumacher (1994) The North Atlantic cod: the broad canvas. *ICES Mar. Sci. Symp.* **198**: 59–76.

Garzoli, S.L., and E.J. Katz (1983) The forced annual reversal of the Atlantic North Equatorial Countercurrent. *J. Phys. Oceanogr.* **13**, 2082–2090.

Garzoli, S., and P.L. Richardson, (1989) Low-frequency meandering of the Atlantic North Equatorial Countercurrent. *J. Geophys. Res.* **89**, 2079–2090.

Geider, R.J., and J. La Roche (1994) The role of iron in phytoplankton photosynthesis and the potential for iron-limitation of primary productivity in the sea. *J. Photosynth. Res.* **39**, 275–301.

Geyer, W.R., *et al.* (1996) Physical oceanography of the Amazon shelf. *Cont. Shelf Res.* **16**, 575–616.

Gibbons, M.J. (1997) Pelagic biogeography of the South Atlantic Ocean *Mar. Biol.* **129**, 757–768.

Gibbons, M.J., and L. Hutchings (1996) Zooplankton diversity and community structure around southern Africa with special attention to the Benguela upwelling system. *S. Afr. J. Sci.* **92**, 63–76.

Gibbs, C.F., *et al.* (1991) Nutrients and plankton distribution at a shelf-break front in the region of the Bass Strait Cascade. *Aust. J. Mar. Freshwat. Res.* **42**, 201–217.

Gibbs, R.K. (1980) Wind-controlled coastal upwelling in the western equatorial Atlantic. *Deep-Sea Res.* **27**, 857–866.

Gieskes, W.W.C., *et al.* (1988) Monsoonal alternation of a mixed and a layered structure in the phytoplankton of the euphotic zone of the Banda Sea. *Neth. J. Sea Res.* **22**, 123–137.

Gieskes, W.W.C., *et al.* (1990) Monsoonal differences in primary production in the eastern Banda Sea. *Neth. J. Sea Res.* **25**, 473–483.

Ginoux, P *et al.* (2001) Sources and distributions of dust aerosols simulated with the GOCART model. *J. Geophys. Res.* 106, **D17**, 20255–20274.
Gjostaeter, H., *et al.* (2002) Growth of Barents Sea capelin (*Mallotus villosus*) in relation to zooplankton abundance. *Journ. Mar. Sci.* 959–967.
Glémarec, M. (1973) The benthic communities of the European North Atlantic continental shelf. *Oceanogr. Mar. Biol. Ann. Rev.* **11**, 263–289.
Glorioso, P. (1987) Temperature distribution related to shelf-sea fronts on the Patagonian shelf. *Cont. Shelf Res.* 7, 27–34.
Glorioso, P.D., and J.H. Simpson (1994) Numerical modelling of the M2 tide on the northern Patagonian shelf. *Cont. Shelf Res.* **14**, 267–278.
Glover, D.M., *et al.* (1994) Dynamics of the transition zone in CZCS ocean colour in the North Pacific during oceanographic spring. *J. Geophys. Res.* **99**, 7501–7511.
Glover, D.M., and P.G. Brewer (1988) Estimates of wintertime mixed layer nutrient concentration in the North Atlantic. *Deep-Sea Res.* **35**, 1525–1546.
Godfrey, J.S., *et al.* (1986) Observations on the shelf-edge current south of Australia, winter 1982. *J. Phys. Oceanogr.* **16**, 668–679.
Godfrey, J.S., *et al.* (1993) Why does the Indonesian throughflow appear to originate from the North Pacific? *J. Phys. Oceanogr.* **23**, 1087–1098.
Goericke, R. (2002) Top-down controls of phytoplankton biomass and community structure in the monsoonal Arabian Sea. *Limnol. Oceanogr.* **47**, 1307–1323.
Goldblatt, R.H., *et al.* (1999) Mesozooplankton community characteristics in the northeast subarctic Pacific. *Deep-Sea Res. II.* **46**, 2619–2644.
Gomez, H.R., *et al.* (2000) Influence of physical processes and river discharge on the seasonality of phytoplankton regimes in the Bay of Bengal. *Cont. Shelf Res.* **20**, 313–330.
Gong G.-C., *et al.* (1998) Absorption coefficient of coloured dissolved organic matter and its effect on the SeaWiFS chlorophyll value in the East China Sea. *AGU Eos Transactions*, **79**, W42.
Gong G.-C., *et al.* (2003) Seasonal variation of chlorophyll-*a* concentration, primary production and environmental conditions in the subtropical East China Sea. *Deep-Sea Res. II.* **50**, 1219–1236.
Gonzales-Silvera, A., *et al.* (2004) Satellite observations of mesoscale eddies in the Gulfs of Tehuantepec and Papagayo. *Deep-Sea Res. II.* **51**, 587–600.
Gonzalez-Rodriguez, E. (1994) Yearly variation in primary productivity from Cabo Frio, Brazil. *Hydrobiology* **294**, 145–156.
Gonzalez-Rodriguez, E., *et al.* (1992) Upwelling and downwelling at Cabo Frio (Brazil): Comparing biomass and primary production responses. *J. Plankton Res.* **14**, 289–306.
Gordon, A.L. (1966) Potential temperature, oxygen and circulation of bottom water in the Southern Sea. *Deep-Sea Res.* **13**, 1125–1138.
Gordon, A.L., and C.L. Greengrove (1986) Geostrophic circulation of the Brazil-Falkland confluence. *Deep-Sea Res.* **33**, 573–585.
Gordon, A.L., and K.T. Bosley (1991) Cyclonic gyre in the tropical south Atlantic. *Deep-Sea Res.* **38**(Suppl. 1), S323-S343.
Gould, W.J. (1985) Physical oceanography of the Azores front. *Prog. Oceanogr.* **14**, 167–190.
Gouriou, Y. (1993) Mean circulation of the upper layers of the western equatorial Pacific Ocean. *J. Geophys. Res.* **98**, 22495–22520.
Gowing, M.M., and K.F. Wishner (1998) Feeding ecology of the copepod *Lucicutia aff. grandis* near the lower interface of the Arabian Sea oxygen minimum layer. *Deep-Sea Res.* **45**, 2433–2459.
Gradinger, R., and D. Nürenberg (1996) Snow algal communities in Arctic pack-ice flows dominated by *Chlamydomonas nivalis. NIPR Symp. Polar Biol.* **9**, 35–43.
Gradinger, R., *et al.* (1992) The significance of picoplankton in the Red Sea and the Gulf of Aden. *Bot. Mar.* **35**, 245–250.
Graham, N.E. (1994) Decadal-scale climate variability in the tropical and North Pacific during the 1970's and 1980's. *Clim. Dynam.* **10**, 135–162.
Grahame, J., *et al.* (1997) *Littorina neglecta* Bean: ecotype or species? *J. Nat. Hist.* **29**, 887–899.
Gran, H.H. (1931) On the conditions for the production of plankton in the sea. *Rapp. Proc. Verb. Réun. Cons. Int. Explor. Mer.* 7, 343–358.
Gran, H.H., and T. Braruud (1935) A quantitative study of the phytoplankton in the Bay of Fundy and the Gulf of Maine. *J. Biol. Bd. Can.* **1**, 279–467.

Grassle, J.F., and N.J. Maciolek (1993) Deep-sea species richness: regional and local diversity from quantitative bottom samples. *Am. Naturalist.* **139**, 313–341.

Gray, J.S. (2002) The species diversity of marine soft sediments. *Mar. Ecol. Prog. Ser.* **244**: 285–297.

Greene C.H., and A.J. Pershing (2000) Response of *Calanus finmarchicus* populations to climate variability in the Northeast Atlantic. *J. Mar. Sci.* **57**, 1536–1544.

Greene, C.H., *et al.* (2003) Trans-Atlantic responses of *Calanus finmarchicus* populations to basin-scale forcing associated with the North Atlantic Oscillation. *Prog. Oceanogr.* **58**, 301–312.

Gregg, W.W., and M.E. Conkright (2001) Global seasonal climatologies of ocean chlorophyll: blending in situ and satellite data for the CZCS era. *J. Geophys. Res.* **106**, 2499–2515.

Gregg, W.W., *et al.* (2003) Ocean primary production and climate: global decadal changes. *Geophys. Res. Lett.* **30**(15), unpaginated.

Gregg, W.W., *et al.* (2005) Recent trends in global ocean chlorophyll. *Geophys. Res. Lett.* **32**, L03606.

Grice, G.D., and A.D. Hart (1962) The abundance, seasonal occurrence and distribution of the epizooplankton between New York and Bermuda. *Ecol. Monogr.* **32**, 287–309.

Griffiths, D.A., and J.H. Middleton (1992) Upwelling and internal tides over the inner New South Wales continental shelf. *J. Geophys. Res.* **97**, 14389–14405.

Griffiths, R.C. (1968) Physical, chemical and biological oceanography of the entrance of the Gulf of California, spring of 1960. *U.S. Fish Wildlife Serv. Spec. Sci. Rep. Fish.* **573**, 1–47.

Griffiths, R.W., and A.F. Pearce (1985) Instability and eddy pairs on the Leeuwin Current south of Australia. *Deep-Sea Res.* **32**, 1511–1534.

Groendahl, F., and L. Hernroth (1986) Vertical distribution of copepods in the Eurasian part of the Nansen Basin. *Natl. Mus. Can. Syllogeus.* **58**, 311–320.

Gruber, N., and J.L. Sarmiento (2002) Large-scale biogeochemical-physical interactions in elemental cycles. *In—Biological-Physical interactions in the Sea* (A. Robinson *et al.*, Eds.), pp. 337–400. John Wiley and Sons, Inc.

Guan, B. (1984) Major features of the shallow water hydrography in the East China Sea and Huanghai Sea. *In—Ocean Hydrodynamics of the Japan and East China Seas* (T. Ichiye, Ed.), pp. 1–14 Elsevier, Amsterdam.

Gulland, J. (Ed.) (1971) *The Fish Resources of the Ocean.* Food and Agricultural Organization, Rome.

Gunther, E.R. (1936) A report on oceanographic investigations in the Peru Coastal Current. *Discovery Rep.* **13**, 109–276.

Guo, Y. (1992) Ecological studies on phytoplankton in the northern East China Sea. *Int. Symp. Mar. Plankton.* **37**, 756.

Guo, Y.J. (1990) The Kuroshio. Part II. Primary productivity and phytoplankton. *Oceanogr. Mar. Biol. Ann. Rev.* **28**, 155–189.

Haidar, A., and G. Thiersen (2001) Coccolithophore dynamics off Bermuda. *Deep-Sea Res.* II. **48**, 1925–1956.

Halim, Y. (1984) Plankton of the Red Sea and the Arabian Gulf. *Deep-Sea Res.* **31**, 969–982.

Hall, J.A., and W.F. Vincent (1994) Vertical and horizontal structure of the picophytoplankton community in a stratified coastal system off New Zealand. *N. Z. J. Mar. Freshwat. Res.* **28**, 299–308.

Hall, J.A., *et al.* (1999) Structure and dynamics of the pelagic microbial food web of the Subtropical Convergence region east of New Zealand. *Aquat. Microb. Ecol.* **20**, 95–105.

Hallegraef, G.M., and S.W. Jeffrey (1993) Annually recurrent diatom blooms along the New South Wales Coast. *Aust. J. Mar. Freshwat. Res.* **44**, 325–334.

Halliwell, G.R., *et al.* (1991) Westward-propagating SST anomalies and baroclinic eddies in the Sargasso Sea. *J. Phys. Oceanogr.* **21**, 635–649.

Hamakuaya, H. *et al.* (2001) The structure of demersal assemblages off Namibia in relation to abiotic factors. *S. Afr. J. Sci.* **23**, 397–417.

Hamilton, L.J. (1992) Surface circulation in the Tasman and Coral Seas: climatological features derived from bathythermograph data. *Aust. J. Mar. Freshwat. Res.* **43**, 793–822.

Hamilton, L. (1998) Demographic change and fisheries dependence in the northern Atlantic. *Res. Hum. Ecol.* **5**, 16–23.

Hansell, D.A., *et al.* (1989) Summer phytoplankton production and transport along the shelf break in the Bering Sea. *Cont. Shelf Res.* **9**, 1085–1104.

Hansen, B., *et al.* (2001) Decreasing overflow from the Nordic seas to the deep Atlantic Ocean since 1950. *Nature (Lond.).* **411**, 927–930.

Haq, S.M., *et al.* (1973) The distribution and abundance of zooplankton along the coast of Pakistan during post- and pre-monsoon periods. *In—The Biology of the Indian Ocean* (B. Zeitzschel, Ed.), pp. 475–486. Springer, Berlin.

Hardy, A. (1967) *Great Waters.* Collins, London.

Hardy, J., *et al.* (1996) Environmental biogeography of near-surface phytoplankton in the southeast Pacific Ocean. *Deep-Sea Res.* **43**, 1647–1660.

Harkantra, S.N., and A.H. Parukelar (1981) Ecology of benthic production in the coastal zone of Goa. *Mahasagar, Bull. Nat. Inst. Oceanogr.* **14**, 135–139.

Harkantra, S.N., *et al.* (1982) Macrobenthos of the shelf off NE Bay of Bengal. *Ind. J. Mar. Sci.* **11**, 115–121.

Harris, G.P., *et al.* (1987) The water masses of the East Coast of Tasmania. *Aust. J. Mar. Freshwat. Res.* **38**, 569–590.

Harris, G.P., *et al.* (1988) Interannual variability in climate and fisheries in Tasmania. *Nature (Lond.).* **333**, 754–757.

Harris, G.P., *et al.* (1991) Seasonal and interannual variability in physical processes, nutrient cycling and the structure of the food chain in Tasmanian shelf waters. *J. Plankton Res.* **13**(Suppl.), 109–131.

Harrison, P.J., *et al.* (1983) Review of the biological oceanography of the Straits of Georgia. *Can. J. Fish. Aquat. Sci.* **40**, 1064–1094.

Harrison, W.G., *et al.* (1981) Photosynthetic parameters and primary production of phytoplankton populations off the northern coast of Peru. *Coast. Estuar. Sci.* **1**, 303–311.

Harrison, W.G., *et al.* (1982) Primary production and nutrient assimilation by natural phytoplankton populations of the eastern Canadian Arctic. *Can. J. Fish. Aquat. Sci.* **39**, 335–345.

Harrison, W.G., *et al.* (1987) Depth profiles of plankton, particulate matter and microbial activity in the eastern Canadian arctic during summer. *Polar Biol.* **7**, 208–224.

Harrison, W.G., *et al.* (1993) The western North Atlantic Bloom Experiment. *Deep-Sea Res. II.* **40**, 279–306.

Harrison, W.G., *et al.* (1996) The kinetics of nitrogen utilisation in the oceanic mixed layer: nitrate and ammonium interactions at nanomolar concentrations. *Limnol. Oceanogr.* **41**, 16–23.

Harrison, W.G., *et al.* (1999) Variability in phytoplankton biomass on the Scotia Shelf as observed from space. *In—Microbial Systems: New Frontiers* (C.R. Bell *et al.*, Eds.). Proc. 8$^{th}$ Internat. Sympos. Microbial Ecology, Halifax, Canada.

Hastenrath, S. (1985) *Climate and Circulation of the Tropics.* Reidel, Dordrecht, The Netherlands.

Hastenrath, S. (1989) The monsoonal regimes of the upper hydrospheric structure of the tropical Indian Ocean. *Atmos.-Ocean.* **27**, 478–507.

Hastenrath, S., and J. Merle (1987) Annual cycle of subsurface thermal structure in the tropical Atlantic Ocean. *J. Phys. Oceanogr.* **17**, 1518–1538.

Hastenrath, S., and P.J. Lamb (1977) *Climatic Atlas of the Indian Ocean. Part 1: Surface Climate and Atmospheric Circulation*, pp. i-xv, 97 charts. University of Wisconsin Press, Madison, WI.

Hattori, H., and H. Saito (1995) Diel changes in the vertical distribution and feeding activity of copepods in ice-covered Resolute Passage, Canadian Arctic. *Proc. NIPR Symp. Pol. Biol.* **8**, 69–73.

Hayes, M.L., *et al.* (2001) How are marine climate and marine biological outbreaks functionally related? *Hydrobiologia.* 460, 213–220.

Hays, G.C. (1996) Large-scale patterns of diel vertical migration in the North Atlantic. *Deep-Sea Res.* **43**, 1601–1616.

Hayward, T.L. (1987) The nutrient distribution and primary production in the central North Pacific. *Deep-Sea Res.* **34**, 1593–1627.

Hayward, T.L., *et al.* (1983) Environmental heterogeneity and plankton community structure in the central North Pacific. *J. Mar. Res.* **41**, 711–729.

Head, E.J.H., and L.R. Harris (1985) Physiological and biochemical changes in *Calanus hyperboreus* from Jones Sound during the transition from summer feeding to overwintering conditions. *Polar Biol.* **4**, 99–106.

Head, E.J.H., *et al.* (2000) Investigations on the ecology of *Calanus* spp. in the Labrador Sea. I. Relationship between the phytoplankton bloom and reproduction and development of *Calanus finmarchicus* in spring. *Mar. Ecol. Prog. Ser.* **193**, 53–73.

Head, E.J.H., *et al.* (2003) Distributions of *Calanus* spp., and other mesozooplankton in the Labrador Sea in relation to hydrography in spring and summer (1995–2000). *Prog. Oceanogr.* **59**, 1–30.

Head, R.N., *et al.* (2002) Phytoplankton and mesozooplankton distribution and composition during transects of the Azores Subtropical Front. *Deep-Sea Res. II.* **49**, 4023–4034.

Head, S.M. (1987) Introduction to the Red Sea. *In—Key Environments—Red Sea* (A.J. Edwards and S.M. Head, Eds.), pp. 1–21. Pergamon Press, Oxford.

Heath, M.R., *et al.* (1999) Climatic fluctuations and the spring invasion of the North Sea by *Calanus finmarchicus. Fish. Oceanogr.* **8**(Suppl. 1), 163–176.

Heath, R.A. (1981) Oceanic fronts around New Zealand. *Deep-Sea Res.* **28**, 547–560.

Heath, R.A. (1985) A review of the physical oceanography of the seas around New Zealand—1982. *N. Z. J. Mar. Freshwat. Res.* **19**, 79–124.

Heath, R.A., and J.M. Bradford (1980) Factors affecting phytoplankton production over the Campbell Plateau. *J. Plankton Res.* **2**, 169–181.

Hebert, D., *et al.* (1991) Does ocean turbulence peak at the equator? *J. Phys. Oceanogr.* **21**, 1690–1698.

Hedgepeth, J.W. (Ed.) (1957) Treatise on marine ecology and palaeoecology. Vol 1: Ecology *Mem. Geol. Soc. Am.* **67**, 1–1296.

Heinrich, A.K. (1962a) The life histories of plankton animals and seasonal cycles of plankton communities in the oceans. *J. Cons. Int. Explor. Mer.* **27**, 15–24.

Heinrich, A.K. (1962b) On the production of copepods of the Bering Sea. *Int. Rev. Ges. Hydrobiol.* **47**, 465–469.

Hellerman, S. (1967) An updated estimate of the wind stress on the world ocean. *Mon. Weath. Rev.* **95**, 607–614. (See also correction in *Mon. Weath. Rev.* **96**, 63–74.)

Hendiarti, N., *et al.* (2004) Investigation of different coastal processes in Indonesian waters using SeaWiFS data. *Deep-Sea Res. II.* **51**, 85–97.

Herbland, A. (1983) Le maximum de chlorophylle dans l'Atlantique tropical orientale: déscription, ecologie, interpretation. *Océanog. Trop.* **18**, 295–318.

Herbland, A., and Voituriez, B. (1977) Production primaire, nitrate et nitrite dans l'Atlantique tropicale. *Cah. ORSTOM Sér. Océanogr.* **15**, 47–55.

Herbland, A., and Voituriez, B. (1979) Hydrological structure analysis for estimating primary production in the Atlantic. *J. Mar. Res.* **37**, 87–101.

Herbland, A., *et al.* (1985) Size structure of phytoplankton biomass in the equatorial Atlantic Ocean. *Deep-Sea Res.* **32**, 819–836.

Herbland, A., *et al.* (1987) Does the nutrient enrichment of the equatorial upwelling influence the size structure of phytoplankton in the Atlantic Ocean? *Océanol. Acta, Spec. Publ.* **6**, 115–120.

Herman, A.W. (1983) Vertical distribution patterns of copepods, chlorophyll and production in Baffin Bay. *Limnol. Oceanogr.* **28**, 709–719.

Herman, A.W. (1984) Vertical copepod aggregation and interactions with chlorophyll and production on the Peru Shelf. *Cont. Shelf Res.* **3**, 131–146.

Herman, A.W. (1989) Vertical relationships between chlorophyll, production and copepods in the eastern tropical Pacific Ocean. *J. Plankton Res.* **11**, 243–261.

Hernández de la Torre, B., *et al.* (2003) Interannual variability of new production in the southern region of the California Current. *Deep-Sea Res.* **50**, 2423–2430.

Hickey, B.M. (1998) Coastal oceanography of western North America. *In—The Sea*, Vol. 11, *The Global Coastal Ocean* (A.R. Robinson and K.H. Brink, Eds.), pp. 345–393. Wiley, New York.

Hirota, Y. (1987) Vertical distribution of euphausiids in the western Pacific Ocean. *Bull. Japan. Sea Regul. Fish. Res. Lab.* **37**, 175–224.

Hirsche, H.-J. (1991) Distribution of dominant calanoid species in the Greenland Sea during late fall. *Polar Biol.* **11**, 351–362.

Hisard, P. (1980) Observations de réponses de type "El Niño" dans l'Atlantique tropical orientale Golfe de Guinée. *Océanol. Acta.* **3**, 69–78.

Hiscock, M.R., *et al.* (2003) Primary productivity and its regulation in the Pacific Sector of the Southern Ocean. *Deep-Sea Res. II.* **50**, 533–558.

Hitchcock, G.L., *et al.* (2000) The fate of upwelled waters in the Great Whirl, August 1995. *Deep-Sea Res. II.* **47**, 1605–1621.

Hobson, L.A., and C.J. Lorenzen (1972) Relationship of chlorophyll maximum to density structure in the Atlantic Ocean and Gulf of Mexico. *Deep-Sea Res.* **19**, 297–306.

Holdridge, L.R. (1947) Determination of world plant formations from simple climatic data. *Science*. **105**, 267–268.
Holdridge L.R., *et al.* (1971) *Forest Elements in Tropical Life Zones*. Pergamon Press, Oxford.
Holligan, P.M. (1981) Biological implications of fronts on the northwest European shelf. *Phil. Trans. R. Soc. London (A)* **302**, 547–562.
Holligan, P.M., and D.S. Harbour (1977) The vertical distribution and succession of phytoplankton in the western English Channel. *J. Mar. Biol. Assoc. U.K.* **57**, 1075–1093.
Holligan, P.M., *et al.* (1983) Satellite studies on the distributions of chlorophyll and dinoflagellate blooms in the western English Channel. *Cont. Shelf Res.* **2**, 81–96.
Hooker, S.B., *et al.* (2000) An objective method for identifying oceanic provinces. *Prog. Oceanogr.* **45**, 313–338.
Hopkins, T.L. (1982) The vertical distribution of zooplankton in the Gulf of Mexico. *Deep-Sea Res.* **29**, 1069–1083.
Hopkins, T.L., and F. Torres (1989) Midwater foodweb in the vicinity of a marginal ice zone in the western Weddell Sea. *Deep-Sea Res.* **36**, 543–560.
Horne, E.P.W., and B. Petrie (1986) Mean position and variability of the sea surface temperature front east of the Grand Banks. *Atmos. Ocean.* **26**, 321–328.
Horne, E.P.W., *et al.* (1989) Nitrate supply and demand at the Georges Bank tidal front. *Sci. Mar.* **53**, 145–158.
Houghton, R.W. (1983) Seasonal variations of the subsurface thermal structure in the Gulf of Guinea. *J. Geophys. Res.* **13**, 2070–2080.
Houghton, R.W. (1989) Influence of local and remote wind forcing in the Gulf of Guinea. *J. Geophys. Res.* **94**, 4816–4828.
Houghton, R.W. (1991) The relationship of sea surface temperature to thermocline depth at annual and interannual time scale in the tropical Atlantic Ocean. *J. Geophys. Res.* **96**, 15173–15185.
Houghton, R.W., and C. Colin (1986) Thermal structure along 4°W in the Gulf of Guinea during 1983–1984. *J. Geophys. Res.* **91**, 11727–11739.
Houghton, R.W., and M.A. Mensah (1978) Physical aspects and biological consequences of Ghanaian coastal upwelling. *In—Upwelling Ecosystems* (R. Boje and M. Tomczak, Eds.), pp. 167–180. Springer, Berlin.
Houry, S., *et al.* (1987) Brunt-Väisälä frequency and Rossby radii in the South Atlantic. *J. Phys. Oceanogr.* **17**, 1619–1626.
Hsueh, Y., *et al.* (1992) The intrusion of the Kuroshio across the continental shelf northeast of Taiwan. *J. Geophys. Res.* **97**, 14323–14330.
Hu, C., *et al.* (2004) The dispersal of the Amazon and Orinoco River water in the tropical Atlantic and Caribbean: observation from space and S-PALACE floats. *Deep-Sea Res. II.* **51**, 1151–1172.
Huggett, J.A., and A.J. Richardson (2000) A review of the biology and ecology of *Calanus agulhensis* off South Africa. *ICES J. Mar. Sci.* **57**, 1834–1849.
Hughes, P., and A.F.G. Fiuza (1982) Observations of a bottom mixed layer in the coastal upwelling area off Northwest Africa. *Rapp. Proc. Verb. Réun. Cons. Int. Explor. Mer.* **180**, 75–82.
Huisman, J., *et al.* (1999) Critical depth and critical turbulence: two different mechanisms for the development of phytoplankton blooms. *Limnol. Oceanogr.* **44**, 1781–1787.
Hulseman, J., and F.J. Weissings (1999) Biodiversity of plankton by species oscillations and chaos. *Nature (Lond.)*. **402**, 407–410.
Humphrey, G.F., and J.D. Kerr (1969) Seasonal variation in the Indian Ocean along 110°E. (III. Chlorophylls-*a* and -*c*) *Aust. J. Mar. Freshwat. Res.* **20**, 55–64.
Hunt, B.P.V., and G.W. Hosie (2005) Zonal structure of zooplankton communities in the Southern Ocean south of Australia. *Deep-Sea Res. I.* **52**, 1241–1271.
Hunter, J.R., and G.D. Sharp (1983) Physics and fish populations: shelf sea fronts and fisheries. *Food Agric. Org. U.N. Fish. Rep.* **291**, 659–682.
Husar, R.B., *et al.* (1997) Characterisation of tropospheric aerosols over the ocean with the NOAA advanced very high resolution radiometer optical thickness operational product. *J. Geophys. Res.* **102**, D14, 16889–16910.
Huxley, J. (1942) *Evolution, the Modern Synthesis*. Allen and Unwin, London.
Huyer, A. (1976) A comparison of upwelling events in two locations: Oregon and Northwest Africa. *J. Mar. Res.* **34**, 532–545.

Hyrenbach, K.D. (2002) Oceanographic habitats of two sympatric North Pacific albatrosses during the breeding season. *Mar. Ecol. Prog. Ser.* **233**, 283–301.

Ichii, T., *et al.* (2002) Occurrence of jumbo flying squid aggregations associated with the countercurrent ridge off the Costa Rica Dome during 1997 El Niño and 1999 La Niña. *Mar. Ecol. Prog. Ser.* **231**, 151–166.

Ilahude, A.G., *et al.* (1990) On the hydrology and productivity of the northern Arafura Sea. *Neth. J. Sea Res.* **25**, 573–583.

Ingram, R.G., *et al.* (2002) An overview of physical processes in the North Water. *Deep-Sea Res. II.* **49**, 4893–4906.

Irigoien, X., *et al.* (2000) Feeding selectivity and egg production of *Calanus finmarchicus* in the English Channel. *Limnol. Oceanogr.* **45**, 44–54.

Irigoien, X., *et al.* (2002) Copepod hatching success in marine ecosystems with high diatom concentrations. *Nature (Lond.).* **419**, 387–389.

Iselin, C.O'D. (1936) A study of the circulation of the western North Atlantic. *Papers Phys. Oceanogr. Meteorol.* **4**, 1–104.

Isemer, H.J., and L. Hasse (1987) *The Bunker Climate Atlas of the North Atlantic Ocean. 2. Air-Sea Interactions.* Springer, Berlin.

Isobe, A., *et al.* (1994) Seasonal variability in the Tsushima Warm Current. *Cont. Shelf Res.* **14**, 23–35.

Jackson, J.B.C., *et al.* (2001) Historical overfishing and the recent collapse of coastal ecosystems. *Science.* **293**, 629–637.

Jacobs, G.A., *et al.* (1994) Decade-scale trans-Pacific propagation and warming effects of an El Niño anomaly. *Nature (Lond.).* **370**, 360–363.

James, M.R. (1987) Role of zooplankton in the nitrogen-cycle off the west coast of New Zealand, winter 1987. *N. Z. J. Mar. Freshwat. Res.* **23**, 507–518.

James, M.R., and V.H. Wilkinson (1988) Biomass, carbon ingestion and ammonium excretion by zooplankton in western Cook Strait, New Zealand. *N. Z. J. Mar. Freshwat. Res.* **22**, 249–257.

Jamir, T.V. (1999) Distribution, seasonal variation and community structure of demersal trawl fauna in the Gulf of Ragay. *Diss. Abstr. Int. B.* **59** (8) 39–43.

Jerlov, N.G. (1964) Optical classification of ocean water. *In—Physical Aspects of Light in the Sea*, pp. 45–49. University of Hawaii Press, Honolulu, HI.

Jickels, T.J. (1995) Atmospheric inputs of metals and nutrients to the oceans: their magnitude and effect. *Mar. Chem.* **48**, 199–214.

Jickels, T., and L. Spokes (in press) Atmospheric iron inputs to the oceans. *In—Biochemistry of Iron in Seawater* (Turner, D., and K. Hunter, Eds.), IUPAC Series Vol. 7. Wiley, New York.

Jochem, F.J., and B. Zeitzschel (1993) Productivity regime and phytoplankton size structure in the tropical and subtropical North Atlantic in spring 1989. *Deep-Sea Res. II.* **40**, 495–521.

Jochem, F.J., *et al.* (1995) Size-fractionated primary production in the open Southern Ocean in austral spring. *Polar Biol.* **15**, 381–392.

Johannessen, O.M. (1986) Brief overview of the physical oceanography. *In—The Nordic Seas* (B.G. Hurdle, Ed.), pp. 103–127. Tapir Academic Press, Trondheim, Norway.

Johns, B., *et al.* (1992) On the occurrence of upwelling along the east coast of India. *Est. Coast. Shelf Sci.* **35**, 75–90.

Johns, W.E., *et al.* (1990) The North Brazil Current retroflection: seasonal structure and eddy variability. *J. Geophys. Res.* **95**, 22103–22120.

Johns, W.E., *et al.* (2002) On the Atlantic inflow to the Caribbean Sea. *Deep-Sea Res.* I. **49**, 211–244.

Johnson, D.R., *et al.* (1980) Upwelling in the Humboldt Current near Valparaiso, Chile. *J. Mar. Res.* **38**, 1–16.

Johnson, K.S., *et al.* (1997) What controls dissolved iron conditions in the world ocean? *Mar. Chem.* **57**, *137–161.*

Johnson, P.W., and J.M. Sieburth (1979) Chroococcoid cyanobacteria in the sea: a ubiquitous and diverse phototrophic biomass. *Limnol. Oceanogr.* **24**, 928–935.

Johnson, W.K., *et al.* (2005) Iron transport by mesoscale Heida eddies in the Gulf of Alaska. *Deep-Sea Res. II.* **52**, 933–953.

Johnson Z., *et al.* (1999) Energetics and growth of a deep *Prochlorococcus* population in the Arabian Sea. *Deep-Sea Res. II.* **46**, 1719–1743.

Joint, I.R., and R. Williams (1985) Demands of the herbivore community on phytoplankton production in the Celtic Sea in August. *Mar. Biol.* **87**, 297–306.

Joint, I.R., *et al.* (2001a) Pelagic production at the Celtic Sea shelf break. *Deep-Sea Res. II.* **48**, 3049–3082.
Joint, I.R., *et al.* (2001b) Two lagrangian experiments in the Iberian upwelling system: tracking an upwelling event and an offshore filament. *Prog. Oceanogr.* **51**, 221–248.
Joint, I.R., *et al.* (2002) The response of phytoplankton production to periodic upwelling and relaxation events at the Iberian shelf break: estimates by the $^{14}C$ method and remote sensing. *J. Mar. Syst.* **32**, 219–238.
Jones, E.P., *et al.* (1990) Chemical oceanography. *In—Polar Oceanography Part B: Chemistry, Biology, and Geology* (W.O. Smith, Jr., Ed.), pp. 407–476. Elsevier.
Joyce, T.M., and P.H. Wiebe (Eds.) (1992) Warm core rings: Interdisciplinary studies of the Kuroshio and Gulf Stream Rings. *Deep-Sea Res.* **39**, S1–S417.
Kaartvedt, S. (2000) Life history of *Calanus finmarchicus* in the Norwegian Sea in relation to planktivorous fish. *ICES J. Mar. Sci.* **57**, 1819–1824.
Kabanova, J.G. (1968) Primary production in the Indian Ocean. *Oceanology.* **8**, 214.
Kachel, N.B., *et al.* (2002) Characteristics and variability of the inner front of the SE Bering Sea. *Deep-Sea Res. II.* **49**, 5889–5909.
Kaczmarska, I., and G.A. Fryxell (1995) Microphytoplankton of the equatorial Pacific: 140°W transect during the 1992 El Niño. *Deep-Sea Res. II.* **42**, 535–558.
Kamykowski., D., and S.-J. Zentara (1990) Hypoxia in the world ocean as recorded in the historical data set. *Deep-Sea Res.* **37**, 1861–1874.
Kamykowski, D., and S.-J. Zentura (2005) Changes in world ocean nitrate availability through the 20th century. *Deep-Sea Res. II.* **48**, 1719–1744.
Karl, D.M., and R. Lukas (1996) The Hawaiian Ocean Time-series (HOT) program: background, rationale and implementation. *Deep-Sea Res.* **43**, 129–155.
Karl, D.M., *et al.* (1995) Ecosystem changes in the North Pacific subtropical gyre attributed to the 1991–92 El Niño. *Nature (Lond.).* **373**, 230–234.
Karl, D.M., *et al.* (1996) Seasonal and annual variability in primary production and article flux at Station ALOHA. *Deep-Sea Res.* **43**, 539–568.
Karl, D.M., *et al.* (2001) Long-term changes in plankton community structure and productivity in the North Pacific gyre: the domain shift hypothesis. *Deep-Sea Res.* **48**, 1449–1470.
Katz, E.J. (1987) Seasonal response of the sea surface to the wind in the equatorial Atlantic. *J. Geophys. Res.* **92**, 1885–1893.
Kawaguchi, S., and Y. Takahashi (1996) Antarctic krill eat salps. *Polar Biol.* **16**, 479–481.
Kawai, H. (1972) Hydrography of the Kuroshio Extension. *In—Kuroshio* (H. Stommel and K. Yoshida, Eds.), pp. 235–341. University of Tokyo Press, Tokyo.
Kawai, H. (1979) Rings south of the Kuroshio and their possible roles in the transport of the intermediate salinity maximum and in the formation of tuna fishing grounds. *Proc. 4th CSK. Symp.*, 250–263.
Kawai, H., and S.I. Saitoh (1986) Secondary fronts, warm tongues and warm streamers of the Kuroshio Extension system. *Deep-Sea Res.* **33**, 1487–1507.
Kawamiya, M., and A. Oschlies (2001) Formation of basin-scale surface chlorophyll pattern by Rossby waves. *Geophys. Res. Lett.* **28**, 4139–4142.
Kawamiya, M., *et al.* (2000) An ecosystem model for the North Pacific embedded in a general circulation model. Part II: Mechanisms forming seasonal variations of chlorophyll. *J. Mar. Syst.* **25**, 159–178.
Kenchington, E.L.R., *et al.* (2001) Effects of experimental trawling on the macrofauna of a sandy-bottom ecosystem on the Grand Banks of Newfoundland. *Can. J. Fish. Aquat. Sci.* **58**, 1043–1057.
Kessler, W.S., and M.J. McPhaden (1995) The 1991–1993 El Niño in the central Pacific. *Deep-Sea Res. II.* **42**, 295–333.
Kielhorn, W.V. (1952) The biology of the surface zone zooplankton of a boreo-Arctic ocean area. *J. Fish. Res. Bd. Can.* **9**, 223–264.
Killworth, P.P., *et al.* (2004) Physical and biological mechanisms for planetary waves observed via satellite-derived chlorophyll. *J. Geophys. Res.* **109**, C7 (unpaginated).
Kinder, T.H. (1983) Shallow currents in the Caribbean Sea and Gulf of Mexico as observed with satellite-tracked drifters. *Bull. Mar. Sci.* **33**, 239–246.
King, A.L., and W.R. Howard (2001) Seasonality of foraminiferal flux in sediment traps at Chatham Rise. *Deep-Sea Res.* I. **48**, 1667–1708.

Kinkade, C., *et al.* (1997) Monsoonal differences in phytoplankton biomass and production in the Indonesian Seas: tracing vertical mixing using temperature. *Deep-Sea Res.* **44**, 581–592.

Kitano, K. (1979) Recent developments in the studies of the warm rings off Kuroshio, a review. *Proc. 4th CSK Symp.*, 243–251.

Knox, R.A., and Anderson, D. (1985) Recent advances in study of low latitude ocean circulation. *Prog. Oceanogr.* **14**, 259–317.

Koeve, W., *et al.* (2002) Storm-induced convective export of organic matter during spring in the NE Atlantic Ocean. *Deep-Sea Res.* I. **49**, 1431–1444.

Kolber, Z.S., *et al.* (2000) Bacterial photosynthesis in surface waters of the ocean. *Nature (Lond.).* **407**, 177–179.

Kolber, Z.S., *et al.* (2001) Contribution of aerobic photoheterotrophic bacteria to the carbon cycle of the ocean. *Science.* **292**, 2492–2495.

Kosobokova, K. and H.-J. Hirsche (2000) Zooplankton distributions across the Lomonosov Ridge, Arctic Ocean: species inventory, biomass and vertical structure. *Deep-Sea Res. I,* **47**, 2029–2060.

Krause, W. (1986) The North Atlantic Current. *J. Geophys. Res.* **91**, 5061–5074.

Krey, J., and B. Babenard (Eds.) (1976) *Atlas of the IIOE; Phytoplankton Production.* Institut für Meereskund an der Universität Kiel, Kiel, Germany.

Krupatkina, D.K., *et al.* (1991) Primary production and size-fractionated structure of the Black Sea in the winter-spring period. *Mar. Ecol. Prog. Ser.* **73**, 25–31.

Kullenberg, G. (1983) The Baltic Sea. *In—Ecosystems of the World, 26: Estuaries and Enclosed Seas* (B. H. Ketchum, Ed.), pp. 309–336. Elsevier, New York.

Kumar, S.P., and J. Narvekar (2005) Seasonal variability of the mixed layer in the central Arabian Sea and its implication on nutrients and primary productivity. *Deep-Sea Res. II.* **52**, 1848–1861.

Labiosa, R.G., *et al.* (2003) The interplay between upwelling and deep convective mixing in determining the seasonal phytoplankton dynamics in the Gulf of Aqaba: evidence from SeaWiFS and MODIS. *Limnol. Oceanogr.* **48**, 2355–2368.

Laborde, P., *et al.* (1999) Seasonal variability of primary production in the Cap Ferret canyon area, during the ECOFER cruises. *Deep-Sea Res. II.* **46**, 2057–2079.

Ladd, C., *et al.* (2005) A note on cross-shelf exchange in the northern Gulf of Alaska. *Deep-Sea Res. II.* **52**, 667–670.

Lagerloef, G.S.E. (1992) The Point Arena Eddy: a recurring summer anticyclone in the California Current. *J. Geophys. Res.* **97**, 12557–12568.

Lagler, D.M., and J.J. O'Brien (1980) *Atlas of Tropical Pacific Wind-Stress Climatology, 1971–1980.* Florida State University Press, Tallahassee, FL.

Landry M.R., *et al.* (1995) Microzooplankton grazing in the central equatorial Pacific during February and August 1992. *Deep-Sea Res. II.* **42**, 657–671.

Landry, M.R., *et al.* (1996) Abundances and distributions of picoplankton populations in the central equatorial Pacific at 140°W. *Deep-Sea Res. II.* **43**, 871–890.

Landry, M.R., *et al.* (1998) Spatial patterns in phytoplankton growth and microzooplankton grazing in the Arabian Sea during monsoon forcing. *Deep-Sea Res. II.* **45**, 2353–2368.

Landry. M.R., *et al.* (2001) Seasonal patterns of mesozooplankton abundance and biomass at Station ALOHA. *Deep-Sea Res.* **48**, 2037–2061.

Lavender, S.J., *et al.* (2005) Modification to the atmospheric correction of SeaWiFS ocean colour images over turbid waters. *Cont. Shelf. Res.* **25**, 539–555.

Laws, R.M. (1984) Seals. *In—Antarctic Ecology,* Vol. 2 (R.M. Laws, Ed.), pp. 621–716. Academic Press, London.

Lazier, J., *et al.* (2002) Convection and restratification in the Labrador Sea, 1990–2000. *Deep-Sea Res. I.* **49**, 1819–1835.

Le Borgne, R. (1981) Zooplankton production in the eastern tropical Atlantic: net growth and P:B in terms of carbon, nitrogen and phosphorus. *Limnol. Oceanogr.* **32**, 905–918.

Le Borgne, R., and M. Rodier (1997) Net zooplankton and the biological pump: a comparison between the oligotrophic and mesotrophic equatorial Pacific. *Deep-Sea Res. II.* **44**, 2003–2023.

Le Borgne, R., *et al.* (2002a) Pacific warm pool and divergence: temporal and zonal variations on the equator and the effects on the biological pump. *Deep-Sea Res. II.* **49**, 2471–2512.

Le Borgne, R., *et al.* (2002b) Carbon fluxes in the equatorial Pacific: a synthesis of the JGOFS programme. *Deep-Sea Res. II.* **49**, 2425–2442.

Le Boutiller, A., *et al.* (1992) Size distribution patterns of phytoplankton in the western Pacific: Towards a generalisation for the tropical Pacific ocean. *Deep-Sea Res.* **39**, 805–823.
Le Fèvre, J. (1986) Aspects of the biology of frontal systems. *Adv. Mar. Biol.* **23**, 164–299.
Lee, C.M. (2000) The upper-ocean response to monsoonal forcing in the Arabian Sea: seasonal and spatial variability. *Deep-Sea Res. II.* **47**, 1177–1226.
Leetma, A., and A.D. Voorhis (1978) Scales of motion in the Subtropical Convergence Zone. *J. Geophys. Res.* **83**, C9, 4589–4592.
Legeckis, R. (1987) Satellite observations of a western boundary current in the Bay of Bengal. *J. Geophys. Res.* **92**, 12974–12978.
Legeckis, R. (1988) Upwelling off the Gulfs of Panama and Papagaya in the tropical Pacific during March, 1988. *J. Geophys. Res.* **93**, 15845–15849.
Legendre, L., *et al.* (1992) Ecology of sea-ice biota. 2—Global significance. *Polar Biol.* **12**, 429–444.
Lenes, J.M., *et al.*, (2001) Iron fertilisation and the *Trichodesmium* response on the west Florida shelf. *Limnol. Oceanog.* **46**, 1261–1277.
Lepley, L.K., *et al.* (1975) Circulation in the Gulf of California from orbital photographs and ship investigations. *Cienc. Mar.* **2**, 86–93.
Letelier, R.M., *et al.* (1996) Seasonal and interannual variations in photosynthetic carbon assimilation at Station ALOHA. *Deep-Sea Res.* **43**, 467–490.
Levitus, S. (1982) Climatological atlas of the world ocean. *NOAA Professional Papers* **13**, 1–173. U.S. Government Printing Office, Washington, D.C.
Levitus, S., and T.P. Boyer (1994a) World ocean atlas 1994, I. Nutrients. *NOAA Atlas NESDIS* **1**, 1–150. U.S. Department of Commerce, Washington, D.C.
Levitus, S., and T.P. Boyer (1994b) World ocean atlas 1994, IV. Temperature. *NOAA Atlas NESDIS* **4**, 1–117. U.S. Department of Commerce, Washington, D.C.
Levitus, S., *et al.* (1994) World ocean atlas 1994, III. Salinity. *NOAA Atlas NESDIS* **3**, 1–99. U.S. Department of Commerce, Washington, D.C.
Lewis, J.B. (1981) Coral reef ecosystems. *In—Analysis of Marine Ecosystems* (A.R. Longhurst, Ed.), pp. 127–159. Academic Press, London.
Li, W.K.W. (1995) Composition of ultraphytoplankton of the central North Atlantic. *Mar. Ecol. Prog. Ser.* **122**, 1–8.
Li, W.K.W. (*2006*) Plankton populations and communities. *In—Marine Macroecology* (J Whitman and K Roy, Eds.). University of Chicago Press.
Li, W.K.W., and P. Dickie (1984) Rapid enhancement of heterotrophic but not photosynthetic activities in Arctic microbial plankton at mesobiotic temperatures. *Polar Biol.* **3**, 217–226.
Li, W.K.W., and Z. Fei (1992) Distribution and characteristics of chlorophyll in the northern East China Sea in 1989. *In—Essays on the Kuroshio, 4*, pp. 165–172. University of Washington Press, Seattle, WA.
Li, W.K.W., and W.G. Harrison (2001) Chlorophyll, bacteria and picophytoplankton in ecological provinces of the North Atlantic *Deep-Sea Res. II.* **48**, 2271–2294.
Li, W.K.W., and M. Wood (1988) Vertical distribution of North Atlantic ultraphytoplankton: analysis by flow cytometry and epifluorescent microscopy. *Deep-Sea Res.* **35**, 1615–1638.
Li, W.K.W., *et al.* (2002) Macroecological patterns of phytoplankton in the NW North Atlantic Ocean. *Nature (Lond.).* **419**, 154–157.
Li, W.K.W., *et al.* (2004) Macroecological limits of heterotrophic bacterial abundance in the ocean. *Deep-Sea Res.* I. **51**, 1529–1540.
Lie, H.-J. (1989) Tidal fronts in the Hwanghae. *Cont. Shelf Res.* **9**, 527–546.
Lierheimer, L.J., and K. Banse (2002) Seasonal and interannual variability of phytoplankton pigment in the Laccadive sea as observed by the CZCS. *Proc. Ind. Acad. Sci. (Earth Plan. Sci.)* **111**, 163–185.
Lighthill, M.J. (1969) Dynamic response of the Indian Ocean to the onset of the southwest monsoon. *Phil. Trans. R. Soc. London A* **265**, 45–93.
Lim, L.C. (1975) Record of an offshore upwelling in the Southern China Sea. *J. Prim. Ind.* **3**, 53–61.
Limsakul, A., *et al.* (2002) Seasonal variability in the lower trophic level environments of the western subtropical Pacific and Oyashio waters—a retrospective study. *Deep-Sea Res. II.* **49**, 5487–5512.
Lindstrom, E., *et al.* (1987) The Western Equatorial Pacific Ocean Circulation study. *Nature (Lond.).* **330**, 533–537.

Liu, H., *et al.* (2002a) Seasonal variability of picophytoplankton and bacteria in the western subarctic Pacific Ocean at station KNOT. *Deep-Sea Res. II.* **49**, 5409–5420.

Liu, H., *et al.* (2005) Grazing by the calanoid copepod *Neocalanus cristatus* on the microbial food web in the coastal Gulf of Alaska. *J. Plankton Res.* **27**, 647–662.

Liu K.-K., *et al.* (2002b) Monsoon-forced chlorophyll distribution and primary production in the South China Sea: observations and anumerical study. *Deep-Sea Res. II.* **49**, 1387–1412.

Livingston, P.A., *et al.*, (1999) Eastern Bering Sea ecosystem trends. *In—Large Marine Ecosystems of the Pacific Rim* (K. Sherman and Q. Teng, Eds.), pp. 140–162.

Loder, J.W., and T. Platt (1985) Physical controls on phytoplankton production at tidal fronts. *In—Proceedings of the 19th European Marine Biology Symposium* (P.E. Gibbs, Ed.), pp. 3–21, Blackwell Science.

Loder, J.W., and D.A. Greenberg (1986) Predicted positions of tidal fronts in the Gulf of Maine region. *Cont. Shelf Res.* **6**, 397–414.

Loder, J.W., *et al.* (1998) The coastal ocean off northeast North America. *In—The Sea*, Vol. 11, *The Global Coastal Ocean* (A.R. Robinson and K.H. Brink, Eds.), pp. 105–134. Wiley, New York.

Loeb, V., *et al.* (1997) Effects of sea-ice extent and krill or salp dominance on the Antarctic food web. *Nature (Lond.).* **387**, 897–900.

Loeng, H. (1991) Features of the physical oceanographic conditions of the Barents Sea. *Polar Biol.* **10**, 5–18.

Lohrenz, S.E., *et al.* (1993) Distribution of pigments and primary production in Gulf Stream meanders. *J. Geophys. Res.* **98**, 14545–14560.

Lohrenz, S.E., *et al.* (2002) Primary production on the continental shelf off Cape Hatteras. *Deep-Sea Res. II.* **49**, 4479–4510.

Longhurst, A.R. (1958) An ecological survey of the West African marine benthos. *Col. Off. Fish. Pubs., Lond.* **11**, 1–102.

Longhurst, A.R. (1964) The coastal oceanography of western Nigeria. *Bull. Inst. Fr. Afr. Noire. Ser. A.* **26**, 337–402.

Longhurst, A.R. (1966) The pelagic phase of *Pleuroncodes planipes* in the California Current. *Rep. California Coop. Fish. Invest.* **11**, 142–154.

Longhurst, A.R. (1967) Vertical distribution of zooplankton in relation to the eastern Pacific oxygen minimum. *Deep-Sea Res.* **14**, 1535–1570.

Longhurst, A.R. (1976) Interactions between zooplankton and phytoplankton profiles in the eastern tropical Pacific Ocean. *Deep-Sea Res.* **23**, 729–754.

Longhurst, A.R. (Ed.) (1981) *Analysis of Marine Ecosystems.* Academic Press, London.

Longhurst, A.R. (1983) Benthic-pelagic coupling and export of organic carbon from a tropical Atlantic continental shelf—Sierra Leone. *Estuar. Coast. Shelf Sci.* **17**, 261–285.

Longhurst, A.R. (1985a) Relationship between diversity and the vertical structure of the upper ocean. *Deep-Sea Res.* **32**, 1535–1570.

Longhurst, A.R. (1985b) The structure and evolution of plankton communities. *Prog. Oceanogr.* **15**, 1–35.

Longhurst, A.R. (1993) Seasonal cooling and blooming in tropical oceans. *Deep-Sea Res.* **40**, 2145–2165.

Longhurst, A.R. (1995) Seasonal cycles of pelagic production and consumption. *Prog. Oceanogr.* **36**, 77–167.

Longhurst, A.R. (2001) Pelagic Biogeography. *In—Encyclopedia of Ocean Sciences* (J.H. Steele, S.A. Turekian, and S.A. Thorpe, Eds.), pp. 2114–2122. Academic Press, London.

Longhurst, A.R. (2002) Murphy's Law revisited: longevity as a factor in recruitment to fish populations. *Fish. Res.* **56**, 125–131.

Longhurst, A.R., and W.G. Harrison (1989) The biological pump: profiles of plankton production and consumption in the open ocean. *Prog. Oceanogr.* **22**, 47–123.

Longhurst, A.R., and E.J.H. Head (1989) Algal production and variable herbivore demand in Jones Sound, Canadian High Arctic. *Polar Biol.* **9**, 281–286.

Longhurst, A.R., and D. Pauly (1987) *Ecology of Tropical Oceans.* Academic Press, San Diego.

Longhurst, A.R., and D.L. Seibert (1972) Oceanic distribution of *Evadne* in the eastern Pacific. *Crustaceana.* **22**, 239–248.

Longhurst, A.R., and R. Williams (1979) Materials for plankton modelling: vertical distribution of Atlantic zooplankton in summer. *J. Plankton Res.* **1**, 1–28.

Longhurst, A.R., and R. Williams (1992) Carbon flux by seasonal vertical migrant copepods is a small number. *J. Plankton Res.* **14**, 1495–1509.
Longhurst, A.R., and W.S. Wooster (1990) Abundance of oil sardine and upwelling on the southwest coast of India. *Can. J. Fish. Aquat. Sci.* **47**, 2407–2419.
Longhurst, A.R., *et al.* (1984) Vertical distribution of arctic zooplankton in summer: eastern Canadian archipelago. *J. Plankton Res.* **6**, 137–168.
Longhurst, A.R., *et al.* (1989) NFLUX: a test of vertical nitrogen flux by diel migrant biota. *Deep-Sea Res.* **36**, 1705–1719.
Longhurst, A.R., *et al.* (1992) Sub-micron particles in northwest Atlantic shelf water. *Deep-Sea Res.* **39**, 1–7.
Longhurst, A.R., *et al.* (1995) An estimate of global primary production in the ocean from satellite radiometer data. *J. Plankton Res.* **17**, 1245–1271.
Loo, L.-O., and R. Rosenberg (1989) Bivalve suspension-feeding dynamics and benthic-pelagic coupling in an eutrophicated marine bay. *J. Exp. Mar. Biol. Ecol.* **130**, 253–276.
Loscher, B.M., *et al.* (1997a) The global Cd/phosphate relationship in deep ocean waters and the need for accuracy. *Mar. Chem.* **59**, 87–93.
Loscher, B.M., *et al.* (1997b) The distribution of Fe in the Antarctic Circumpolar Current. *Deep-Sea Res. II.* **44**, 143–187.
Louanchi, F., and R.G. Najjar (2001) Annual cycles of nutrients and oxygen in the upper layers of the Atlantic Ocean. *Deep-Sea Res. II.* **48**, 2155–2171.
Lovejoy, C., *et al.* (2000) Growth and distribution of marine bacteria in relation to nanoplankton community structure. *Deep-Sea Res. II.* **47**, 461–487.
Lowe-McConnell, R.H. (1962) The fishes of the British Guiana continental shelf with notes on their natural history. *J. Linn. Soc. Lond. Zool.* **44**, 669–700.
Lukas, R., and E. Lindstrom (1991) The mixed layer of the western equatorial Pacific Ocean. *J. Geophys. Res.* **96**, 3343–3357.
Luo, J., *et al.* (2000) Diel vertical migration of zooplankton and mesopelagic fish in the Arabian Sea. *Deep-Sea Res. II.* **47**, 1451–1473.
Lutjeharms, J.R.E. (1985) Location of frontal systems between Africa and Antarctica: Some preliminary results. *Deep-Sea Res.* **32**, 1499–1509.
Lutjeharms, J.R.E. (1988) On the role of the East Madagascar Current as a source of the Agulhas Current. *S. Afr. J. Sci.* **84**, 236–238.
Lutjeharms, J.R.E., and A.D. Connell (1989) The Natal Pulse and inshore countercurrents off the South African east coast. *S. Afr. J. Sci.* **85**, 533–534.
Lutjeharms, J.R.E., and J.M. Meeuwis (1987) The extent and variability of south-east Atlantic upwelling. *S. Afr. J. Mar. Sci.* **5**, 51–62.
Lutjeharms, J.R.E., and R.C. van Ballegooyen (1984) Topographic control in the Agulhas Current system. *Deep-Sea Res.* **31**, 1321–1338.
Lutjeharms, J.R.E., and R.C. van Ballegooyen (1988) The retroflection of the Agulhas Current. *J. Phys. Oceanogr.* **18**, 1570–1583.
Lutjeharms, J.R.E., *et al.* (1981) Characteristics of currents east and south of Madagascar. *Deep-Sea Res.* **28**, 879–899.
Lutjeharms, J.R.E., *et al.* (1985) Oceanic frontal systems and biological enhancement. *In—Antarctic Nutrient Cycles and Food Webs* (W.R. Siegfried, P.R. Condy, and R.M. Laws, Eds.), pp. 11–21. Springer, Berlin.
Lutjeharms, J.R.E., *et al.* (1989) Eddies and other boundary phenomena of the Agulhas Current. *Cont. Shelf Res.* **9**, 597–616.
Lutz, V.A. (2002) Variability in pigment composition and optical characteristics of phytoplankton in Labrador Sea and central North Atlantic. *Mar. Ecol. Prog. Ser.* **260**, 1–10.
Lutz, V.A., *et al.* (2003) Variability in pigment composition and optical characteristics of phytoplankton in the Labrador Sea and the Central North Atlantic. *Mar. Ecol. Prog. Ser.* **260**, 1–18.
Lynch, D.R., *et al.* (2001) Can Georges Bank cod larvae survive on a calanoid diet? *Deep-Sea Res. II.* **48**, 609–630.
Lytle, V.I., and S.F. Ackley (1996) Heat flux through sea ice in the western Weddell Sea: convective and conductive transfer processes. *J. Geophys. Res. C. Oceans* **101**, 8853–8868.
MacIsaac, J.J., *et al.* (1985) Primary production in an upwelling center. *Deep-Sea Res.* **32**, 503–529.

Mackas, D.L., and K.O. Coyle (2005) Shelf-offshore exchange processes and their effects on mesozooplankton biomass and community patterns in the northeast Pacific. *Deep-Sea Res. II.* **52**, 707–726.

Mackas, D.L., *et al.* (2005) Zooplankton distribution and dynamics in a North Pacific eddy of coastal origin: II. Mechanisms of eddy colonisation by, and retention of, offshore species. *Deep-Sea Res. II.* **52**, 1011–1035.

MacKenzie L., and J. Adamson (2004) Water column stratification and the spatial and temporal distribution of phytoplankton in Tasman Bay. *N. Z. J. Mar. Freshwat. Res.* **38**, 705–728.

MacKenzie L., *et al.* (1983) Nutrients and microplankton biomass off the New Zealand northwest coast, January 1982. *N. Z. J. Mar. Freshwat.* Res. **22**, 551–564.

Mackey, D.J., *et al.* (1997) Phytoplankton productivity and the carbon cycle in the western Equatorial Pacific under El Niño and non-El Niño conditions. *Deep-Sea Res. II.* **44**, 1951–1978.

Mackey D.J., *et al.* (2002a) Phytoplankton abundances and community structure in the equatorial Pacific. *Deep-Sea Res. II.* **49**, 2561–2582.

Mackey, D.J., *et al.* (2002b) Iron in the western Pacific: a riverine or hydrothermal source for iron in the equatorial undercurrent? *Deep-Sea Res.* I. **49**, 877–893.

Mackintosh, N.A. (1937) The seasonal circulation of the Antarctic macroplankton. *Discovery Rep.* **16**, 365–412.

MacPherson, E., and A. Gordoa (1992) Trends in the demersal fish community off Namibia from 1983 to 1990. *S. Afr. J. Mar. Sci.* **11**, 12–17.

MacPherson, E., and A. Gordoa (1996) Biomass spectra in benthic fish assemblages in the Benguala system. *Mar. Ecol. Prog. Ser.* **138**, 27–32.

Madhupratap, M., and P. Haridas (1986) Epipelagic copepods of the northern Indian Ocean. *Océanolog. Acta.* **9**, 105–118.

Madhupratap, M., and P. Haridas (1990) Zooplankton, especially calanoid copepods, in the upper 1000 m of the south-east Arabian Sea. *J. Plankton Res.* **12**, 305–321.

Madhupratap, M., *et al.* (1994) Oil sardine and mackerel: their fishery, problems and coastal oceanography. *Curr. Sci. Bangalore.* **66**, 340–347.

Madhupratap, M., *et al.* (2003) Biogeochemistry of the Bay of Bengal during the summer monsoon of 2001. *Deep-Sea Res. II.* **50**, 881–896.

Madin, L.A., *et al.* (2001) Zooplankton at the BATS station: diel, seasonal and interannual variation in biomass 1994–98. *Deep-Sea Res. II.* **48**, 2063–2082.

Magazzu, G., *et al.* (1985) Picoplankton contribution to primary production in the NW coast of Madagascar. *Mem. Biol. Mar. Oceanogr.* **15**, 207–222.

Mahaffey, C., *et al.* (2004) Physical supply of nitrogen to phytoplankton in the Atlantic Ocean. *Glob. Biogeochem. Cycles.* **18**, GB1034, 12 pp.

Mahowald, N., *et al.* (1999) Dust sources and deposition during the last glacial and the present climate. *J. Geophys. Res.* **104**, 15895–15916.

Maldonado, M.T., *et al.* (1999) Co-limitation of phytoplankton growth by light and Fe during winter in the NE subarctic Pacific Ocean. *Deep-Sea Res. II.* **46**, 2475–2485.

Malej, A. (1997) Response of summer phytoplankton to episodic meteorological events. *Mar. Ecol.* **18**, 273–288.

Manghnani, V., *et al.* (1998) Advection of upwelled waters in the form of plumes off Oman during the Southwest Monsoon. *Deep-Sea Res. II.* **45**, 2027–2052.

Mann, C.R. (1967) The termination of the Gulf Stream and the beginning of the North Atlantic Current. *Deep-Sea Res.* **14**, 337.

Mann, K.H., and Lazier, J.R.N. (2006) *Dynamics of Marine Ecosystems*, 3rd ed. Blackwell, Boston.

Mantoura, F., *et al.* (1993) Nitrogen biogeochemical cycling in the northwestern Indian Ocean. *Deep-Sea Res. II.* **40**, 651–671.

Mao, H.-L., *et al.* (1986) A cyclonic eddy in the northeastern China Sea. *Stud. Mar. Sinica.* **27**, 21–31.

Maranger, R., *et al.* (1998) Iron acquisition by photosynthetic marine phytoplankton from ingested bacteria. *Nature (Lond.).* **396**, *248–251.*

Marañon, E., *et al.* (2000) Basin-scale variability of phytoplankton biomass, production and growth in the Atlantic Ocean. *Deep-Sea Res. I.* **47**, 825–857.

Margalef, R. (1994) Through the looking glass; how marine phytoplankton appears through the microscope when graded by size and taxonomically. *Sci. Mar.* **58**, 87–101.

Margalef, R. (1997) *Our Biosphere.* Inter-Research, Oldendorf, Germany.
Marra, J. (1982) Variability in surface chlorophyll-a at a shelf-break front (New York Bight). *J. Mar. Res.* **40**, 575–591.
Marra, J., and R.T. Barber (2005) Primary productivity in the Arabian Sea: a synthesis of the JGOFS data. *Prog. Oceanogr.* **65**, 159–175.
Martin, J.H., and Fitzwater, S. E. (1988) Iron deficiency limits phytoplankton growth in the north-east Pacific subarctic. *Nature (Lond.).* **331**, 341–343.
Martin, J.H., and R.M. Gordon (1988) NE Pacific iron distributions in relation to phytoplankton productivity *Deep-Sea Res.* **35**, 177–196.
Martin, J.H., *et al.* (1989) VERTEX: Phytoplankton studies in the Gulf of Alaska. *Deep-Sea Res.* **36**, 649–680.
Martin, J.H., *et al.* (1993) Iron, primary production and C-N flux studies during the JGOFS North Atlantic Bloom Experiment. *Deep-Sea Res. II.* **40**, 115–134.
Mascarenas, A.S., *et al.* (1971) A study of the oceanographic conditions in the region of Cabo Frio. *In—Fertility of the Sea* (J.D. Costlow, Ed.), pp. 285–297. Gordon and Breach, New York.
Mas-Riera, J.A. *et al.*, (1990) Influence of Benguela upwelling on the structure of demersal fish populations off Namibia. *Mar. Biol.* 104, 175–182.
Matano, R.P., *et al.* (1993) Seasonal variability in the South Atlantic. *J. Geophys. Res.* **98**, 18027–18035.
Mathew, B. (1982) Studies on upwelling and sinking in seas around India. Ph.D. thesis, University of Cochin.
Matishov, G.G., *et al.* (2003) Contemporary state and factors of stability of the Barents Sea Large Marine Ecosystem. *In—Large Marine Ecosystems of the World; Trends in Exploitation, Protection, Research.* (G. Hempel and K. Sherman, Eds.), pp. 41–74. Elsevier, New York.
McCarthy, J.J., and J.L. Nevins (1986) Utilisation of nitrogen and phosphorus by primary producers in warm core ring 82-B following deep convective nixing. *Deep-Sea Res.* **33**, 1773–1788.
McCarthy, J.J., *et al.* (1986) Global ocean flux. *Oceanus.* **30**, 16–26.
McCarthy, J.J., *et al.* (1996) New production along 140°W in the equatorial Pacific during and following the 1991–1992 El Niño. *Deep-Sea Res. II.* **43**, 1065–1093.
McClain, C.R., and Firestone, J. (1993) An investigation of Ekman upwelling in the North Atlantic. *J. Geophys. Res.* **98**, 12327–12339.
McClain, C.R., *et al.* (1990) Physical and biological processes in the North Atlantic during the first Global GARP Experiment. *J. Geophys. Res.* **95**(C10), 18027–18048.
McClain, C.R., *et al.* (1996) Observations and simulations of physical and biological processes at OWS P. 1951–1980. *J. Geophys. Res.* **101**, 3697–3713.
McClain, C.R., *et al.* (2004) Subtropical gyre variability observed by ocean-colour satellites. *Deep-Sea Res. II.* **51**, 281–302.
McClanahan, T.R. (1988) Seasonality in East Africa's coastal waters. *Mar. Ecol. Prog. Ser.* **44**, 191–199.
McClatchie, S., *et al.* (1997) Demersal fish community diversity off New Zealand: is it related to depth, latitude and regional surface phytoplankton? *Deep-Sea Res. I.* **44**, 647–668.
McCreary, J.P., *et al.* (1984) Effects of remote annual forcing in the eastern tropical Atlantic Ocean. *J. Mar. Res.* **42**, 45–81.
McCreary, J.P., *et al.* (1993) A numerical investigation of dynamics, thermodynamics and mixed-layer processes in the Indian Ocean. *Prog. Oceanogr.* **31**, 181–244.
McCreary, J.P., *et al.* (1996) A four component ecosystem model of biological activity in the Arabian Sea. *Prog. Oceanogr.* **37**, 193–240.
McDonald D., *et al.* (1999) Multiple late quaternary episodes of exceptional diatom production in the Gulf of Alaska. *Deep-Sea Res. II.* **46**, 2993–3018.
McGill, D.A. (1973) Light and nutrients in the Indian Ocean. *In—The Biology of the Indian Ocean* (B. Zeitzschel, Ed.), pp. 53–102. Springer, Berlin.
McGillicuddy, D.J., Jr., and A.R. Robinson (1997) Eddy-induced nutrient supply and new production in the Sargasso Sea. *Deep-Sea Res. II.* **44**, 1427–1450.
McGowan, J.A. (1971) Oceanic biogeography of the Pacific. *In—The Micropaleontology of the Oceans* (S.M. Funnell and W.R. Riedl, Eds.), pp. 3–74. Cambridge University Press, Cambridge.
McGowan, J.A., and P. Walker (1979) Structure in the copepod community of the North Pacific central gyre. *Ecol. Monogr.* **49**, 195–226.

McGowan, J.A., and P.M. Williams,(1973) Oceanic habitat differences in the North Pacific. *J. Exp. Mar. Biol. Ecol.* **12**, 187–212.

McGowan, J.A., *et al.* (2003) The biological response to the 1977 regime shift in the California Current. *Deep-Sea Res. II.* **50**, 2567–2582.

McMurray, H.F., *et al.* (1993) Size-fractionated phytoplankton production in western Agulhas Bank continental shelf waters. *Cont. Shelf Res.* **13**, 307–329.

McRoy, C.P., and J.J. Goering (1974) The influence of ice on the primary productivity of the Bering Sea. *In—Oceanography of the Bering Sea* (D.W. Hood and E.J. Kelly, Eds.), pp. 403–421. Institute of Marine Science, Fairbanks, AL.

McRoy, P.C., *et al.* (1985) Processes and resources of the Bering Sea shelf (PROBES): the development and accomplishments of the project. *Cont. Shelf Res.* **5**, 5–21.

Mearns, A.J., and L. Smith (1975) Benthic oceanography and the distribution of bottom fish in the Los Angeles Bight. *Rpts. Cal. Coop. Fish. Invest.* **18**, 118–124.

Measures, C.I., and S. Vink (2001) Dissolved Fe in the upper waters of the Pacific sector of the Southern Ocean. *Deep-Sea Res. II.* **48**, 3913–3941.

Melillo, J.M., *et al.* (1993) Global climate change and terrestrial net primary production. *Nature (Lond.).* **363**, 234–240.

Menkes, C.E. (2002) A whirling ecosystem in the equatorial Atlantic. *Geophys. Res. Lett.* **29** (48), 1–4.

Menon, M.D., and K.C. George (1977) The abundance of zooplankton along the west coast of India during the years 1971–75. *In—Proc. Symp. Warm-water Zooplank. (Spec. Publ. NIO, Goa)*, pp. 203–213. National Institute of Oceanography, Goa.

Mensah, M.A. (1974) The reproduction and feeding of the marine copepod *Calanoides carinatus* in Ghanaian waters. *Ghana J. Sci.* **14**, 167–191.

Menzel, D.W., and J.H. Ryther (1960) The annual cycle of primary production in the Sargasso Sea off Bermuda. *Deep-Sea Res.* **6**, 351–367.

Menzel, D.W., and J.H. Ryther (1961) Nutrients limiting the production of phytoplankton in the Sargasso Sea, with special reference to iron. *Deep-Sea Res.* **7**, 276–281.

Mercier, H., *et al.* (2003) Upper-layer circulation in the eastern Equatorial and South Atlantic Ocean in January–March 1995. *Deep-Sea Res. I.* **50**, 863–887.

Merle, J. (1983) Seasonal variability of subsurface thermal structure in the tropical Atlantic. *In—Hydrodynamics of the Equatorial Ocean* (J.C.J. Nihoul, Ed.). Elsevier Scientific, Amsterdam.

Merle, J., and S. Arnault (1985) Seasonal variability of the surface dynamic topography in the tropical Atlantic Ocean. *J. Mar. Res.* **43**, 267–288.

Metcalf, W.G., and M.C. Stalcup (1967) Origin of the Atlantic equatorial undercurrent. *J. Geophys. Res.* **72**, 4954–4975.

Metzler, P.M., *et al.* (1997) New and regenerated production in the South Atlantic off Brazil. *Deep-Sea Res.* **44**, 363–384.

Michaels, A.F., and A.H. Knap (1996) Overview of the U.S. JGOFS Bermuda Atlantic Time Series study. *Deep-Sea Res. II.* **43**, 157–198.

Michaels, A.F., *et al.* (1993) Episodic inputs of atmospheric nitrogen to the Sargasso Sea: contributions to new production and phytoplankton blooms. *Glob. Biogeochem. Cycles* **7**, 339–351.

Mihnea, P.E. (1997) Major shifts in the phytoplankton community (1980–1994) in the Romanian Black Sea. *Océanol. Acta.* **20**, 119–129.

Mileikovsky, S.A. (1971) Types of larval development in marine bottom invertebrates, their distribution and ecological significance: a reevaluation. *Mar. Biol.* **10**: 193–213.

Miller, A.R. (1983) Mediterranean Sea; Physical aspects. *In—Estuaries and Enclosed Seas* (B.H. Ketchum, Ed.), pp. 219–238. Elsevier Scientific, Amsterdam.

Miller, A., *et al.* (1994) The 1976–77 climate shift of the Pacific Ocean. *Oceanography* **7**, 21–26.

Miller, C.B. (1993) Pelagic processes in the Subarctic Pacific. *Prog. Oceanogr.* **32**, 1–15.

Miller, C.B. (2004) *Biological Oceanography.* Blackwell Publishing, Oxford.

Miller, C.B., *et al.* (1984) Life histories of large grazing copepods in a subarctic ocean gyre. *Prog. Oceanogr.* **13**, 201–243.

Miller, C.B., *et al.* (1991) Ecological dynamics in the subarctic ecosystem: a possibly iron-limited ecosystem. *Limnol. Oceanogr.* **36**, 1600–1615.

Milliman, J.D., and R.H. Meade (1983) World-wide delivery of river sediment to the oceans. *J. Geol.* **91**, 1–21.

Millot, C. (1987) Circulation in the Western Mediterranean Sea. *Océanol. Acta.* **10**, 143–150.

Millot, C. (1992) Are there differences between the largest Mediterranean Seas? *Bull. Inst. Océanogr. Monaco* **11**, 3–25.

Mills, E.L. (1989) *Biological Oceanography, an Early History, 1970–1960.* Cornell University Press, Ithaca, NY.

Minas, H.J., and M. Minas (1992) Net community production in "HNLC" waters of the tropical and Antarctic oceans: grazing v. iron hypothesis. *Océanol. Acta.* **15**, 145–162.

Minas, H.J., and P. Nival (1988) Pelagic Mediterranean oceanography. *Océanol. Acta.* **9**, 1–250.

Minas, H.J., *et al.* (1982) Nutrients and primary production in the upwelling region off northwest Africa. *Rapp. Proc. Verb. Réun. Cons. Int. Explor. Mer.* **180**, 148–183.

Minas, H.J., *et al.* (1991) Biological and geochemical signatures associated with the water circulation in the western Alboran Sea. *J. Geophys. Res.* 90, 8755–8771.

Mincks, S.L., *et al.* (2000) Distribution, abundance and feeding ecology of decapods in the Arabian Sea, with implications for vertical flux. *Deep-Sea Res.* **47**, 1475–1516.

Mitchell, B.G., *et al.* (1991a) Meridional zonation of the Barents Sea ecosystem. *Polar Biol.* **10**, 147–161.

Mitchell, B.G., *et al.* (1991b) Light limitation of phytoplankton biomass and macronutrient utilisation in the Southern Ocean. *Limnol. Oceanogr.* **36**, 1662–1677.

Mittelstaedt, E. (1991) The ocean boundary along the northwest African coast: circulation and oceanographic properties at the sea surface. *Prog. Oceanogr.* **26**, 307–355.

Miya, M., and M. Nishida (1997) Speciation in the open ocean. *Nature (Lond.).* **389**, 803–804.

Miyama, T., *et al.* (2003) Structure and dynamics of the Indian Ocean cross-equatorial cell. *Deep-Sea Res. II.* **50**, 2023–2047.

Mochizuki, M., *et al.* (2002) Seasonal changes in nutrients, chlorophyll and the phytoplankton assemblage in the W subarctic Pacific Ocean. *Deep-Sea Res. II.* **49**, 5421–5440.

Molinari, R.L., and J. Morrison (1988) The separation of the Yucatan Current from the Campeche Bank and the penetration of the Loop Current into the Gulf of Mexico. *J. Geophys. Res.* **93**, 10645–10654.

Monger, B., *et al.* (1997) Seasonal phytoplankton dynamics in the eastern tropical Atlantic. *J. Geophys. Res. Oceans* **102**, C6, 12,389.

Mooers, C.N.K., and G.A. Maul (1998) Intra-American Sea circulation. *In—The Sea*, Vol. 11, *The Global Coastal Ocean* (A.R. Robinson and K.H. Brink, Eds.), pp. 183–208. Wiley, New York.

Mooers, C.N.K., and A.R. Robinson (1984) Turbulent jets and eddies in the California Current and inferred cross-shore transports. *Science.* **223**, 51–53.

Mooers, C.N.K., *et al.* (1978) Prograde and retrograde fronts. In *Ocean Fronts in Coastal Processes* (M.J. Bowman and W.E. Esaias, Eds.), pp. 43–86. Springer-Verlag, New York.

Moon-van der Staay, S.Y., *et al.* (2001) Oceanic 18S rDNA sequences from picoplankton reveal unsuspected eukaryotic diversity. *Nature (Lond.).* **409**, 607–610.

Moore, J.K., *et al.* (2002) Iron cycling and nutrient-limitation patterns in surface waters of the open ocean. *Deep-Sea Res. II.* **49**, 463–507.

Moreira da Silva, P.C. (1971) Upwelling and its biological effects in southern Brazil. *In—Fertility of the Sea* (J.D. Costlow, Ed.), pp. 469–478. Proc. Internat. Symposium, Sao Paulo, Brazil. Gordon and Breach Science, London.

Morel, A., and J.-F. Berthon (1989) Surface pigments, algal biomass profiles and potential production of the euphotic layer: Relationships reinvestigated in view of remote sensing applications. *Limnol. Oceanogr.* **34**, 1545–1562.

Motoda, S. (1978) Differences in productivities between the Great Australian Bight and the Gulf of Carpentaria. *Mar. Biol.* **46**, 93–99.

Motoda, S., and T. Minoda (1974) Plankton of the Bering Sea. *In—Oceanography of the Bering Sea* (D.W. Hood and E.J. Kelly, Eds.), pp. 207–241. University of Alaska Press, Anchorage, AK.

Mueller, J.L., and R.E. Lang (1989) Bio-optical provinces of the Northeast Pacific Ocean: a provisional analysis. *Limnol. Oceanogr.* **34**, 1572–1586.

Muench, R.D. (1970) The physical oceanography of the northern Baffin Bay region. *Rep. Arch. Inst. North Am.* **1**, 1–150.

Muench, R.D. (1990) Mesoscale phenomena in the Polar Oceans. *In—Polar Oceanography. Part A. Physical Science*, (W.O. Smith, Jr., Ed.) pp. 223–285. Elsevier, New York.

Mulhearn, P.J. (1987) The Tasman front: a study using satellite infrared imagery. *J. Phys. Oceanogr.* **17**, 1148–1155.

Müller, P., and C. Garrett (2002) From stirring to mixing in a stratified ocean. *Oceanography* **15**, 12–19.

Müller-Karger, F.E., *et al.* (1988) The dispersal of the Amazon's water. *Nature (Lond.).* **333**, 56–59.

Müller-Karger, F.E., *et al.* (1989) Pigment observations in the Caribbean Sea: observations from space. *Prog. Oceanogr.* **35**, 23–64.

Muller-Karger, F.E., *et al.* (2004) Processes of coastal upwelling and carbon flux in the Carioco Basin. *Deep-Sea Res. II.* **51**, 927–943.

Murphy, R.J., *et al.* (2001) Phytoplankton distributions around New Zealand derived from SeaWiFS remotely-sensed ocean colour data. *N. Z. J. Mar. Freshwat. Res.* **35**, 343–362.

Murray, J.W., *et al.* (1995) A US JGOFS Process Study in the equatorial Pacific (EqPac). *Deep-Sea Res. II.* **42**, 275–293.

Murray, J.W., and E. Izdar (1989) The 1988 Black Sea oceanographic expedition: overview and new discoveries. *Oceanography.* **2**, 15–21.

Murray, J.W., Ed. (1991) Black Sea oceanography. *Deep-Sea Res.* **38**(Suppl. 2A), S655–S1266

Nagata, H.S.O. (1998) Seasonal changes and vertical distributions of chlorophyll and primary productivity on the Yamato Rise, central Japan Sea. *Plankt. Biol. Ecol.* **45**, 159–170.

Nagata, Y., *et al.* (1986) Detailed structure of the Kuroshio Front and the origin of the water in warm-core rings. *Deep-Sea Res.* **33**, 1509–1526.

Nair, S.R., *et al.* (1981) Zooplankton composition and diversity in western Bay of Bengal. *J. Plankton Res.* **3**, 493–508.

Naqui S.W.A., *et al.* (2000) Increased marine production of $N_2O$ due to intensifying anoxia on the Indian continental shelf. *Nature (Lond.).* **408**, 346–349.

Nathansohn, A. (1909) Beiträge zur Biologie des Planktons. II. Vertikalzirkulation und Plankton-maxima im Mittelmeer. *Int. Rev. Ges. Hydrobiol.* **2**, 580–632.

Nelson D.M., and W.O. Smith (1991) Sverdrup re-visited: critical depths, maximum chlorophyll levels and the control of the Southern Ocean productivity by the irradiance-mixing regime. *Limnol. Oceanogr.* **36**, 1650–1661.

Nelson, D.M., *et al.* (2001) A seasonal progression of Si limitation in the Pacific sector of the Southern Ocean. *Deep-Sea Res. II.* **48**, 3973–3995.

Nelson, D.M., *et al.* (2002) Vertical budgets for organic carbon and biogenic silica in the Pacific sector of the Southern Ocean, 1996–1998. *Deep-Sea Res. II.* **49**, 1645–1674.

Nelson, G., and L. Hutchings (1983) The Benguela upwelling area. *Prog. Oceanogr.* **12**, 333–356.

Nelson, N.B., *et al.* (2004) The spring bloom in the NW Sargasso Sea: spatial extent and relationship with winter mixing. *Deep-Sea Res. II.* **51**, 987–1000.

Nemoto, T., and W.G. Harrison (1981) High latitude ecosystems. *In—Analysis of Marine Ecosystems* (A.R. Longhurst, Ed.), pp. 95–126. Academic Press, London.

Neumann, G. (1969) The Equatorial Undercurrent in the Atlantic Ocean. *Proc. Symp. Oceanogr. Fish. Res. Trop. Atlantic UNESCO*, 33–44.

Newell, B.S. (1959) The hydrography of British East African coastal waters. *Col. Off. Fish. Pubs. Lond.* **12**, 1–118.

Niemann, H., *et al.* (2004) Red Sea gravity currents cascade near-reef phytoplankton to the twilight zone. *Mar. Ecol. Prog. Ser.* **269**, 91–99.

Niemann, R.G. (2003) The interplay between upwelling and deep convective mixing in the Gulf of Aqaba. *Limnol. Oceanogr.* **48**, 2355–2368.

Niermann, U. (1986) Distribution of *Sargassum natans* and some of its epibionts in the Sargasso Sea. *Helgol. Meeresunters.* **40**, 343–353.

Niiler, P.P. (1977) One-dimensional models of the seasonal thermocline. *In—The Sea* (E.D. Goldberg, Ed.), pp. 97–115. Wiley, New York.

Nishikawa, T., *et al.* (2000) Effects of temperature and salinity on the growth of the giant diatom *Coscinodiscus wailesii* isolated from Harima-Nada, Seto Inland Sea. *Nipp. Suis. Gak.* **66**, 993–998.

Nitani, H. (1972) The beginning of the Kuroshio. *In—Kuroshio: Physical Aspects of the Japan Current* (H. Stommel and K. Yoshida, Eds.), pp. 129–164. University of Washington Press, Seattle, WA.

Nittrouer, C.A., and D.J. DeMaster (1996) The Amazon shelf setting: tropical, energetic and influenced by a large river. *Cont. Shelf Res.* **16**, 553–573.

Nixon, K.C., and Q.D. Wheeler (1990) An amplification of the phylogenetic species concept. *Cladistics.* **6**, 211–233.

Nowlin, W.D., and Klinck, J.M. (1986) The physics of the Antarctic Circumpolar Current. *Rev. Geophys. Space Phys.* **24**, 469–491.

Nozaki, Y. (2001) Elemental distribution overview. *In—Encyclopedia of Ocean Sciences* (J.H. Steele, Ed.), p. 840. Academic Press, London.

Obata, A., *et al.* (1996) Global verification of critical depth theory for phytoplankton bloom with climatological in situ temperature and satellite ocean colour data. *J. Geophys. Res.* **101**, 20657–20667.

Odate, T., *et al.* (2002) Temporal and spatial patterns in the surface-water biomass of phytoplankton in the North Water. *Deep-Sea Res. II.* **49**, 4947–4958.

Odebrecht, C., and L. Djurfeldt (1989) The role of nearshore mixing on phytoplankton size structure of Cape Santa Maria Grande. *Arch. Fisch. Meeresforch.* **43**, 217–230.

Odum, E.P. (1971) *Fundamentals of Ecology.* Saunders, Philadelphia.

Oguz, T., *et al.* (1992) The upper layer circulation of the Black Sea: its variability as inferred from hydrographic and satellite observations. *J. Geophys. Res.* **97**, 12569–12584.

Okera, W. (1982) Organisation of fish assemblages on the northern Australian continental shelf (CSIRO, Cronulla, unpublished document).

Okkonen, S.R. (1992) Shedding of an anticyclonic eddy from the Alaskan Stream. *Geophys. Res. Lett.* **12**, 2397–2400.

Olson, D.B., *et al.* (1988) Temporal variations in the separation of Brazil and Malvinhas Currents. *Deep-Sea Res.* **35**, 1971–1990.

Olson, D.B., *et al.* (1993) Maintenance of the low-oxygen layer in the central Arabian Sea. *Deep-Sea Res. II.* **40**, 673–685.

Olson, M.B., and S.L. Strom (2002) Phytoplankton growth, microzooplankton herbivory and community structure in the SE Bering Sea: insight into the formation and temporal persistence of an *Emiliania huxleyi* bloom. *Deep-Sea Res. II.* **49**, 5969–5990.

Ommaney, F.D. (1961) Malayan offshore trawling grounds—experimental fishing of FRV *Manahine* in Malayan and Bornean waters. *Col. Off. Fish. Pubs. Lond.* **18**, 1–95.

Ono, S., *et al.* (2001) Shallow remineralisation in the Sargasso Sea estimated from seasonal variations in oxygen, dissolved inorganic carbon and nitrate. *Deep-Sea Res. II.* **48**, 1567–1582.

O'Reilly, J.E., and Busch, D.A. (1984) Phytoplankton primary production on the northwestern Atlantic shelf. *Rapp. Proc. Verb. Réun. Cons. Int. Explor. Mer.* **183**, 255–268.

Orr, A.P. (1933) Scientific report of the Great Barrier Reef Expedition, 1928–29. *Bull. Brit. Mus. (Nat. Hist.)* **2**, 37–86.

Orsi, A., *et al.* (1995) On the meridional extent and fronts of the Antarctic Circumpolar Current. *Deep-Sea Res. I.* **42**, 641–673.

Osborne, J., *et al.* (1992) Ocean Atlas for the Macintosh. SIO Rpt. 92–29. Scripps Institution of Oceanography, La Jolla, CA.

Osgood, K.E., and D.M. Checkley (1997) Seasonal variations in a deep aggregation of *Calanus pacificus* in the Santa Barbara Basin. *Mar. Ecol. Prog. Ser.* **148**, 59–69.

Østvedt, O.J. (1955) Zooplankton investigations from weather ship M in the Norwegian Sea, 1948–49. *Hvalradets Skrift.* **49**, 1–43.

Oszay, E. and U. Umluata (1998) The Black Sea. *In—The Sea*, Vol. 12, *The Global Coastal Ocean* (A.R. Robinson and K.H. Brink, Eds.), pp. 889–914. Wiley InterScience, New York.

Ott, J.A. (1992) Adriatic benthos; problems and perspectives. *Proc. 25th Eur. Mar. Biol. Symp.*, 367–378.

Ottens, J.J. (1991) Planktonic foraminifera as North Atlantic water mass indicators. *Océanol. Acta.* **14**, 123–140.

Oudot, C. (Ed.) (1987) *Observations physico-chimiques et biomasse vegetale dans l'océan Atlantique equatorial. Programme PIRAL.* Off. Res. Sci. Techn. Outre-Mer., Paris.

Oudot, C., and P. Morin (1987) The distribution of nutrients in the equatorial Atlantic: relation to physical processes and phytoplankton biomass. *Océanol. Acta.* **SP6**, 121–130.

Owen, R.W. (1981) Fronts and eddies in the ocean. *In—Analysis of Marine Ecosystems* (A.R. Longhurst, Ed.), pp. 197–233. Academic Press, London.

Owens, N.J.P., *et al.* (1993) Size-fractionated primary production and nitrogen assimilation in the N.W. Indian Ocean. *Deep-Sea Res. II.* **40**, 697–710.

Paffenhöfer, G.-A. (1983) Vertical zooplankton distribution on the northeast Florida shelf. *J. Plankton Res.* **5**, 15–34.

Paluskiewicz, T., *et al.* (1983) Observations of a Loop Current frontal eddy intrusion onto the West Florida continental shelf. *J. Geophys. Res.* **88**, 9639–9651.

Parekular, A.H., *et al.* (1982) Benthic production and assessment of demersal fishery resources of the Indian Seas. *Ind. J. Mar. Sci.* **11**, 107–114.

Parrish, R. (1989) The South Pacific oceanic horse mackerel (*Trachurus picturatus murphyi*) fishery. *ICLARM Stud. Rev.* **15**, 321–331.

Parsons, T.R., and C.M. Lalli (1988) Comparative oceanic biology of the plankton communities of the subarctic Atlantic and Pacific Oceans. *Oceanogr. Mar. Biol. Ann. Rev.* **26**, 317–359.

Partensky, F., *et al.* (1999) *Prochlorococcus*, a photosynthetic prokaryote of global significance. *Microbiol. Mol. Biol. Rev.* **63**, 106–127.

Partos, P., and M.C. Piccolo (1988) Hydrography of the Argentine continental shelf between 38° and 42°S. *Cont. Shelf Res.* **8**, 1043–1056.

Pauly, D., and V. Christensen (1993) Stratified models of large marine ecosystems: a general approach and an application to the South China Sea. *In—Large Marine Ecosystems of the World: Stress, Mitigation and Modelling of Large Marine Ecosystems* (K. Sherman *et al.* Eds.), pp. 148–174. AAAS, Washington, D.C.

Pauly, D., and R. Chuenpagdee (2003) Development of fisheries in the Gulf of Thailand LME: analysis of an unplanned experiment. *In—Large Marine Ecosystems of the World: Trends in Exploitation, Protection and Research* (G. Hempel and K. Sherman, Eds.), pp. 337–354. AAAS, Washington, D.C.

Pauly, D., and I. Tsukuyama (1987) The Peruvian anchoveta and its upwelling ecosystem: three decades of change. *ICLARM Stud. Rev.* **15**, 1–351.

Pauly, D., *et al.* (1989) The Peruvian upwelling system: dynamics and interactions. *ICLARM Conf. Proc.* **18**, 1–438.

Pauly, D., *et al.* (1998) Fishing down marine food webs. *Science.* **279**, 860–863.

Pauly, D., *et al.* (2000) Mapping fisheries onto marine ecosystems: a proposal for a consensus approach for regional, oceanic and global integration. *UBC Fisheries Cent. Res. Rep.* **8**.

Pearce, A.F., and M.L. Gruendlingh (1982) Is there a seasonal variation in the Agulhas Current? *J. Mar. Res.* **40**, 177–184.

Pedersen, O.P., *et al.* (2001) A model study of demography and spatial distribution of *C. finmarchicus* at the Norwegian coast. *Deep-Sea Res. II.* **48**, 567–587.

Peeken, I. (1997) Photosynthetic pigment fingerprints as indicators of phytoplankton biomass and development in different water masses of the Southern Ocean during austral spring. *Deep-Sea Res. II.* **44**, 261–282.

Peinert, R., *et al.* (1989) Impact of grazing on spring phytoplankton growth and sedimentation in the Norwegian Channel. *Mitt. Geol. Palaeontol. Inst. Hamburg.* **62**, 149–164.

Pérez, V., *et al.* (2004) Seasonal and interannual variability of chl-*a* and primary production in the Equatorial Atlantic: in situ and remote sensing observations. *J. Plankton Res.* **27**, 189–197.

Pérez, V., *et al.* (2005) Latitudinal distribution of microbial plankton abundance, production and respiration in the Equatorial Atlantic in autumn 2000. *Deep-Sea Res. I.* **52**, 861–880.

Petersen, C.J.G. (1918) The sea-bottom and the production of fish food. *Rep. Dan. Biol. Stat.* **25**, 1–62.

Petersen, W.T., and J.E. Keister (2003) Interannual variability in copepod community composition at a coastal station in the northern California Current. *Deep-Sea Res. II.* **50**, 2499–2517.

Peterson, R.G., and L. Stramma (1991) Upper-level circulation in the South Atlantic. *Prog. Oceanogr.* **26**, 1–73.

Peterson, R.G., and T. Whitworth (1989) The subantarctic and polar fronts in relation to deep water masses through the southwestern Atlantic. *J. Geophys. Res.* **94**, 10817–10838.

Peterson, W.T., *et al.* (1979) Zonation and maintenance of copepod populations in the Oregon upwelling zone. *Deep-Sea Res.* **26**, 467–494.

Petit, D., and Courties, C. (1976) *Calanoides carinatus* sur le plateau continentale congolaise II. *Cah. ORSTOM. Sér. Océanogr.* **14**, 177–199.

Philander, S.G.H. (1978) Variability of the tropical oceans. *Dynam. Atmos. Oceans.* **3**, 191–208.

Philander, S.G.H. (1979) Upwelling in the Gulf of Guinea. *J. Mar. Res.* **37**, 23–33.

Philander, S.G.H. (1985) Tropical oceanography. *Adv. Geophys.* **28**, 461–477.

Philander, S.G.H. (1990) *El Niño, La Niña and the Southern Oscillation.* Academic Press, San Diego.
Philander, S.G.H., and Y. Chao (1991) On the contrast between the seasonal cycles of the equatorial Atlantic and Pacific Oceans. *J. Phys. Oceanogr.* **21**, 1399–1406.
Philander, S.G.H., and R.C. Pacanowski (1981) The oceanic response to cross equatorial winds with application to coastal upwelling in low latitudes. *Tellus.* **33**, 201–210.
Philander, S.G.H., and E.M. Rasmusson (1985) The southern oscillation and El Niño. *Adv. Geophys.* **28A**, 197–215.
Philander, S.G.H., *et al.* (1996) Why the ITCZ is mostly north of the equator. *J. Climate.* **9**, 2958–2972.
Pianka, E.R. (1966) Latitudinal gradients in species diversity: a review of concepts. *Am. Naturalist.* **100**, 33–46.
Piccioni, A., *et al.* (1988) Wind-induced upwellings off the southern coast of Sicily. *Océanol. Acta.* **9**, 309–314.
Piccolo, M.C. (1998) Oceanography of the western South Atlantic continental shelf from 33–55°S. *In—The Sea,* Vol. 12, *The Global Coastal Ocean* (A.R. Robinson and K.H. Brink, Eds.), pp. 253–271. Wiley InterScience, New York.
PICES (undated) *Marine Ecosystems of the North Pacific* (278 pp.). Pac. Int. Counc. Explor. Sea.
Pingree, R.D. (1975) The advance and retreat of the thermocline on the continental shelf. *J. Mar. Biol. Assoc. U.K.* **55**, 965–974.
Pingree, R.D. (1997) The eastern subtropical gyre: flow rings, recirculations and subduction. *J. Mar. biol. Assoc. U.K.* **77**, 573–624.
Pingree, R.D., and D.K Griffiths (1978) Tidal fronts on the shelf seas around the British Isles. *J. Geophys. Res.* **83**, 4615–4622.
Pingree, R.D., and D.K. Griffiths (1980) A numerical model of the M2 tide in the Gulf of St. Lawrence. *Océanol. Acta.* **3**, 221–225.
Pingree, R.D., and G T. Mardell (1981) Slope turbulence, internal waves and phytoplankton growth at the Celtic Sea shelf edge. *Phil. Trans. R. Soc. London, A.* **302**, 663–682.
Pingree. R.D., and B. Sinha (2000) Westwards moving waves or eddies (storms) on the subtropical/Azores front. *J. Mar. Syst.* **29**, 239–276.
Pingree, R.D., *et al.* (1975) Summer phytoplankton blooms and red tides along tidal fronts in the approaches to the English Channel. *Nature (Lond.).* **258**, 672–677.
Pingree, R.D., *et al.* (1982) Celtic Sea and Armorican current structure and the vertical distributions of temperature and chlorophyll. *Cont. Shelf Res.* **1**, 99–116.
Pingree, R.D., *et al.* (1986) Propagation of internal tides from the upper slopes of the Bay of Biscay. *Nature (Lond.).* **321**, 154–158.
Pingree, R.D., *et al.* (1999) Position, structure of the Subtropical/Azores front region from Lagrangian and remote sensing. *J. Mar. Biol. Assoc. U.K.* **79**, 769–792.
Piontkowski, S.A., *et al.* (1997) Spatial heterogeneity of the planktonic fields in the upper mixed layer of the open ocean. *Mar. Ecol. Prog. Ser.* **148**, 145–154.
Piontkowski, S.A., *et al.* (2003) Plankton communities of the South Atlantic anticyclonic gyre. *Prog. Oceanogr.* **26**, 255–268.
Pitelka, F.A. (1941) Distribution of birds in relation to major biotic communities. *Am. Midl. Nat.* **25**, 113–157.
Planque, B., and A.H. Taylor (1998) Long-term changes in zooplankton and the climate of the North Atlantic. *J. Mar. Sci.* **55**, 644–654.
Planque, B., and P.C. Reid (1998) Predicting *Calanus finmarchicus* abundance from a climatic signal. *J. Mar. Biol. Assoc. U.K.* **78**, 1015–1018.
Planque, B., *et al.* (1997) Large scale spatial variations in the seasonal abundance of *Calanus finmarchicus. Deep-Sea Res.* I. **44**, 315–326.
Platt, T., and D.V. Subbha Rao (1975) Primary production of marine microphytes. *In—Photosynthesis and Productivity of in Different Environments* (J.P. Cooper, Ed.), Vol. 3, pp. 249–280. Cambridge University Press, Cambridge.
Platt, T., and S. Sathyendranath (1988) Oceanic primary production: estimation by remote sensing at local and regional scales. *Science.* **241**, 1613–1620.
Platt, T., and S. Sathyendranath (1999) Spatial structure of pelagic ecosystem processes in the global ocean. *Ecosystems.* **2**, 384–394.
Platt, T., *et al.* (1991a) Critical depth and marine production. *Proc. R. Soc. B,* **246**, 205–217.

Platt, T., *et al.* (1991b) Basin-scale estimates of oceanic primary production by remote sensing: the North Atlantic. *J. Geophys. Res.* **96**, C8, 15147–15159.

Platt, T., *et al.* (1994) Primary production, respiration and stratification in the ocean. *J. Geophys. Res.* **102**, 12765–12787.

Platt, T., *et al.* (2003) Nitrate supply and demand in the mixed layer of the ocean. *Mar. Ecol. Prog. Ser.* **254**, 3–9.

Platt, T., *et al.* (2005) Physical forcing and phytoplankton distributions. *Sci. Mar.* **69**(Supp. 1), 55–73.

POEM Group (1992) General circulation of the eastern Mediterranean. *Earth Sci. Rev. 32, 285–309.*

Polovina, J.J., *et al.* (2001) The transition zone chlorophyll front, a dynamic global feature defining migration and forage habitat for marine resources. *Prog. Oceanogr.* **49**, 469–483.

Popova, E.E. (2002) Coupled 3-D physical and biological modelling of the mesoscale variability observed in the NE Atlantic in spring 1997: biological processes. *Deep-Sea Res.* I. **49**, 1741–1768.

Prata, A.J., and J.B. Wells (1990) A satellite sensor image of the Leeuwin current, Western Australia. Int. *J. Remote Sensing.* **11**, *173–180.*

Prezelin B.B., and H.E. Glover (1991) Variability in phytoplankton, biomass and productivity in the Sargasso Sea. *J. Plankton Res.* **13**(Supp.), 45–67.

Price, N.M., *et al.* (1994) The equatorial Pacific Ocean: Grazer-controlled phytoplankton populations in an iron-limited ecosystem. *Limnol. Oceanogr.* **39**, *520–534.*

Probert, P.K., and J.B. Wilson (1984) Continental shelf benthos off Otago Peninsula, New Zealand. *Estuar. Coast. Shelf Sci.* **19**, 373–391.

Probert, P.K., and S.L. Grove (1998) Macrobenthic assemblages of the continental shelf and upper slope off the west coast of New Zealand. *J. R. Soc. N. Z.* **28**, 259–280.

Prospero, J.M. (1981) Eolian transport to the world ocean. *In—The Sea*, Vol. 7, *The Oceanic Lithosphere* (C. Emiliani, Ed.), pp. 801–874. Wiley InterScience, New York.

Prospero, J.M., *et al.* (2002) Environmental characteristics of global sources of atmospheric soil dust from the Nimbus 7 TOMS absorbing aerosol product. *Rev. Geophys.* **40**, 1, 1–31.

Purcell, R.M. (1977) Life at low Reynolds numbers. *Am. J. Phys.* **45**, *3–11.*

Qiu, B. (1999) Seasonal eddy field modulation of the NPSCC: TOPEX-Poseidon observations and theory. *J. Phys. Oceanogr.* **29**, 2471–2486.

Qiu, B., and S. Chen (1994) Seasonal modulations in the eddy field of the South Pacific Ocean. *J. Phys. Oceanogr.* **34**, 1515–1527.

Quaatey, S.K., and C.D. Maravelas (1999) Maturity and spawning pattern of *Sardinella aurita* in relation to water temperature and zooplankton abundance off Ghana, West Africa. *J. Appl. Ichthyol.* **15**, 63–69.

Quadfasel, D., and Cresswell, G. R. (1992) A note on the seasonal variability of the South Java Current. *J. Geophys. Res. 97, 3685–3688.*

Radenac, M.-H., and M. Rodier (1996) Nitrate and chlorophyll distributions in relation to thermohaline and current structures in the western tropical Pacific during 1985–1989. *Deep-Sea Res. II.* **43**, *725–752.*

Ragoonaden, V., *et al.* (1987) Physico-chemical characteristics and circulation of waters in the Mauritius-Seychelles ridge zone. *Ind. J. Mar. Sci.* **16**, 184–191.

Rahmstorf, S. (1992) Modelling ocean temperatures and mixed-layer depths in the Tasman Sea off the South Island, New Zealand. *N. Z. J. Mar. Freshwat. Res.* **26**, *37–51.*

Raja, B.T.A. (1969) The Indian oil sardine. *Bull. Cent. Mar. Fish. Res. Inst.* **16**, 1–128.

Ramadhas, V., and V.K. Venugopalam (1977) Iron excretion by some zooplankters. *In—Proc. Symp. Warm-water Zooplankt. (Spec. Publ. NIO, Goa)*, pp. 687–692. NIO, Goa, India.

Rao, D.P., and J.S. Sastry (1981) Circulation and distribution of some hydrographical properties during late winter in the Bay of Bengal. *Mahasagar* **14**, 1–15.

Rao, D.V.S., and D. Sameoto (1988) Relationship between phytoplankton and copepods in the deep tropical Pacific Ocean off Costa Rica. *Bull. Mar. Sci.* **42**, 85–100.

Rao, R.R. (1986) Cooling and deepening of the mixed layer in the central Arabian Sea during MONSOON-77: Observations and simulations. *Deep-Sea Res.* **33**, 1413–1424.

Rao, R.R., *et al.* (1989) Evolution of the climatological near-surface thermal structure of the tropical Indian Ocean. *J. Geophys. Res.* **94**, 10801–10815.

Rappé, M.S., *et al.* (2002) Cultivation of the ubiquitous SAR11 marine bacterioplankton clade. *Nature (Lond.).* **418**, 630–633.

Rea, D.K. (1994) The palaeoclimatic record provided by eolian deposition in the deep sea—the geologic history of wind. *Rev. Geophys.* **32**, 159–195.
Read, J.F., *et al.* (1995) On the southerly extent of the ACC in the southeast Pacific. *Deep-Sea Res. II.* **42**, 933–954.
Rebstock, G.A., and Y.S. Kang (2003) A comparison of three marine ecosystems surrounding the Korean peninsula: responses to climate change. *Prog. Oceanogr.* **59**, 357–379.
Reid, J.L. (1962) On the circulation, phosphate-phosphorus content and zooplankton volumes in the upper part of the North Pacific Ocean. *Limnol. Oceanogr.* **2**, 287–306.
Reid, J.L. (1964) A transequatorial Atlantic oceanographic section in July, 1963 compared with other Atlantic and Pacific sections. *J. Geophys. Res.* **69**, 5205–5215.
Reid, J.L., *et al.* (1978) Ocean circulation and marine life. *In—Advances in Oceanography* (H. Charnock and G. Deacon, Eds.), pp. 65–130. Plenum, New York.
Revelante, N., and M. Gilmartin (1994) Relative increase of larger phytoplankton in a subsurface maximum in the Adriatic Sea. *J. Plankton Res.* **17**, 1535–1562.
Reverdin, G., and M. Fieux (1987) Sections in the Indian Ocean—variability in the temperature structure. *Deep-Sea Res.* **34**, 601–626.
Rey, F. (1990) P/I relationship in natural phytoplankton populations in the Barents Sea. *Polar Res.* **10**, 105–116.
Richards, W.J., and J.A. Bohnsack (1990) The Caribbean Sea, a large marine ecosystem in crisis. *In—Large Marine Ecosystems: Patterns, Processes and Yields* (K. Sherman et al., Eds.), pp. 44–53. AAAS, Washington, D.C.
Richardson, A.J., *et al.* (1997) Assessment of the food available to Cape anchovy during their spawning season. *S. Afr. J. Mar. Sci.* **18**, 113–117.
Richardson, P.L. (1976) Gulf Stream rings. *Oceanus.* **19**, 65–68.
Richardson, P.L. (1980) Anticyclonic eddies generated near the Corner seamounts. *J. Mar. Res.* **38**, 673–686.
Richardson, P.L. (1983) Gulf Stream rings. *In—Eddies in Marine Science* (A.R. Robinson, Ed.), pp. 19–45. Springer, Berlin.
Richardson, P.L. (1985) Drifting derelicts in the North Atlantic, 1883—1902. *Prog. Oceanogr.* **14**, 463–483.
Richter, C. (1994) Regional and seasonal variability in the vertical distribution of mesozooplankton in the Greenland Sea. *Ber. Polarforsch.* **154**, 1–87.
Riley, G.A. (1942) The relationship between vertical turbulence and the spring flowering of diatoms. *J. Mar. Res.* **5**, 67–87.
Riley, G.A. (1957) Phytoplankton of the northern central Sargasso Sea. *Limnol. Oceanogr.* **2**, 252–267.
Ring Group (1981) Gulf stream cold core rings: their physics, chemistry and biology. *Science.* **212**, 1091–1100.
Ringuette, M., *et al.* (2002) Advanced recruitment and accelerated population development in Arctic copepods of the North Water. *Deep-Sea Res. II.* **49**, 5081–5099.
Rissik, D *et al.* (1997) Enhanced zooplankton abundance in the lee of an isolated reef in the southern Coral Sea. *J. Plankton Res.* **19**, 1347–1368.
Rivkin, R.B., *et al.* (1999) Microzooplankton bacterivory and herbivory in the NE subarctic Pacific. *Deep-Sea Res. II.* 46, 2579–2618.
Robertson, A.I., *et al.* (1998) The influence of fluvial discharge on pelagic production in the Gulf of Papua. *Estuar. Coast. Shelf Sci.* **46**, 319–331.
Robinson, A.R., and K.H. Brink (Eds.) (1998) *The Global Coastal Ocean, Regional Studies and Synthesis.* Wiley, New York.
Robinson, A.R., and P. Malanotte-Rizzoli (1993) Physical oceanography of the eastern Mediterranean. *Deep-Sea Res. II.* **40**, 1073–1332.
Robinson, C.L.K., *et al.* (1993) Simulated annual plankton production in the northeastern Pacific Coastal Upwelling Biome. *J. Plankton Res.* **15**, 161–183.
Robinson, M.K., *et al.* (1979) *Atlas of the North Atlantic Ocean Monthly Mean Temperatures and Salinities of the Surface Layer.* U.S. Naval Oceanographic Office, Washington, D.C.
Robles, F., *et al.* (1980) Water masses in the northern Chilean zone. In—*Proceedings of the Workshop on the Phenomenon Known as El Niño*, pp. 83–174. UNESCO publications, Paris.

Rochford, D.J. (1975) Nutrient enrichment of east Australian coastal waters. II. Laurieton upwelling. *Aust. J. Mar. Freshwat. Res.* **26**, 385–397.

Rochford, D.J. (1986) Seasonal changes in the distribution of Leeuwin Current waters off southern Australia. *Aust. J. Mar. Freshwat. Res.* **37**, 1–10.

Roden, G.I. (1964) Oceanographic aspects of the Gulf of California. *Mem. Am. Soc. Petr. Geol.* **3**, 30–58.

Roden, G.I. (1970) Aspects of the mid-Pacific Transition Zone. *J. Geophys. Res.* **75**, 1097–1109.

Roden, G.I. (1975) On North Pacific temperature, salinity, sound velocity and density fronts and their relation wind and energy flux fields. *J. Phys. Oceanogr.* **5**, 557–571.

Rodhe, J. (1998) The Baltic and North Seas: a process-oriented review of the physical oceanography. *In—The Sea*, Vol. 12, *The Global Coastal Ocean* (A.R. Robinson and K.H. Brink, Eds.), pp. 699–732. Wiley InterScience, New York.

Rodriguez, J., *et al.* (1987) Planktonic biomass spectra during a winter production pulse in Mediterranean waters. *J. Plankton Res.* **9**, 1183–1194.

Roemmich, D. (1981) Circulation of the Caribbean Sea: a well-resolved inverse problem. *J. Geophys. Res.* **86**, 7993–8005.

Roemmich, D., and J. McGowan (1995) Climatic warming and the decline of zooplankton in the California Current. *Science*. **267**, 1324–1326.

Roman, M., *et al.*, (2000) Mesozooplankton production and grazing in the Arabian Sea. *Deep-Sea Res. II.* **47**, 1423–1450.

Roman, M.R., *et al.* (2002) Latitudinal comparisons of equatorial Pacific zooplankton. *Deep-Sea Res. II.* **49**, 2695–2711.

Roman, M.R., *et al.* (1995) Zooplankton variability on the equator at 140°W during the JGOFS EqPac study. *Deep-Sea Res.* **42**, 673–693.

Rosa, H., and T. Laevastu (1959) Comparison of biological and ecological characteristics of sardines and related species—a preliminary study. *Proc. World Meeting Biol. Sardines Related Species (FAO)* **2**, 523–534.

Rosen, D.E. (1975) Doctrinal biogeography. *Q. Rev. Biol.* **50**, 69–70.

Rosenberg, R. (2001) Marine benthic faunal successional stages and related sedimentary activity. *Sci. Mar.* **65**(Suppl. 2), 107–119.

Rosenberg, R., *et al.* (1983) Benthic biomass and oxygen deficiency in the upwelling system off Peru. *J. Mar. Res.* **41**, 263–279.

Rougerie, F., and C. Henin (1977) Coral and Solomon Seas in the austral summer monsoon. *Cah. Orstom Oceanogr.* **15**, 261–278.

Rougerie, F., and J. Rancher (1994) The Polynesian South Ocean: features and circulation. *Mar. Pollut. Bull.* **29**, 14–25.

Roy, C. (1990) Les upwellings: Le cadre physique des pecheries cotières ouest-africains. *In*—"Pécheries Ouest-Africaines" (P. Curie and C. Roy, Eds.), pp. 38–66.

Roy, C., *et al.* (2001) The southern Benguela anchovy population reached an unprecedented level of abundance in 2000; another failure for fisheries oceanography? *GLOBEC Newslett.*, April.

Roy, K., *et al.* (2000c) Dissecting latitudinal diversity gradients: functional groups and clades of marine bivalves. *Proc. R. Soc. Ser. B.* **267**. 293–299.

Roy, S., *et al.* (2000a) A Canadian JGOFS process study in the Gulf of St. Lawrence. Introduction. *Deep-Sea Res. II.* **47**, 377–384.

Roy, S., *et al.* (2000b) Importance of mesozooplankton feeding for the downward flux of biogenic carbon in the Gulf of St. Lawrence. *Deep-Sea Res. II.* **47**, 519–544.

Royer, T.C. (1998) Coastal processes in the northern North Pacific. *In—The Sea*, Vol. 12, *The Global Coastal Ocean* (A.R. Robinson and K.H. Brink, Eds.), pp. 395–414. Wiley InterScience, New York.

Rudjakov, J.A. (1997) Quantifying seasonal phytoplankton oscillations in the global offshore ocean. *Mar. Ecol.-Prog. Ser.* **146**, 225–230.

Russell, F.S., *et al.* (1971) Changes in biological conditions in the English Channel off Plymouth during the last half-century. *Nature (Lond.).* **234**, 468–470.

Ryan, J.P., *et al.* (2002) Unusual large-scale phytoplankton blooms in the equatorial Pacific. *Prog. Oceanogr.* **55**, 263–285.

Ryther, J.H., *et al.* (1967) Influence of the Amazon River outflow on the ecology of the western tropical Atlantic. *J. Mar. Res.* **25**, 69–83.

Sabatini, M.E., and G.L.A. Colombo (2001) Seasonal pattern of zooplankton biomass in the Argentinean shelf off Southern Patagonia (45–55°S) *Sci. Mar.* **65**, 21–31.

Sabatini, M.E., *et al.* (2000) Distribution pattern and population structure of *Calanus australis* over the southern Patagonian Shelf in summer. *ICES J. Mar. Sci.* **57**, 1856–1866.

Saetre, R., and A. Jorge da Silva (1984) The circulation of the Mozambique Channel. *Deep-Sea Res.* **31**, 485–508.

Saijo, Y., *et al* (1970) Primary production in Kuroshio and its adjacent area. *Proc. 2nd CSK Symp., Tokyo.*, 169–175.

Sakurai, H., *et al.* (1996) Habitats of fish and epibenthic invertebrates in Fildes Bay, King George Isl., Antarctic Proc. NIPR Symp. *Polar Biol.* **9**, 231–242.

Salas de Leonand J. Morel-Gomez (1986) The role of the Loop Current in the Gulf of Mexico fronts. *In—Marine Interfaces Hydrodynamics* (J.C.J. Nihoul, Ed.), Elsevier Oceanography Series 42, pp. 295–311.

Saltzman, J., and K. Wishner (1997a) Zooplankton ecology in the eastern tropical Pacific oxygen minimum zone above a seamount. I. General trends. *Deep-Sea Res. II.* **44**, 907–930.

Saltzman, J., and K. Wishner (1997b) Zooplankton ecology in the eastern tropical Pacific oxygen minimum zone above a seamount. II. Vertical distribution of copepods. *Deep-Sea Res. II.* **44**, 931–954.

Sambrotto, R.N., *et al.* (1984) Large yearly production of phytoplankton in the western Bering Strait. *Science.* **225**, 1147–1150.

Sameoto, D., *et al.* (1986) Relations between the thermocline, meso- and microzooplankton, chlorophyll-*a* and primary production distributions in Lancaster Sound. *Polar Biol.* **6**, 53–61.

Sanders, H. (1968) Marine benthic diversity: a comparative study. *Am. Naturalist.* **102**, 243–282.

Santamaria-del-Angel, E., *et al.* (1994) Gulf of California biogeographic regions based on CZCS imagery. *J. Geophys. Res.* **99**, 7411–7421.

Sarma, V.V., and Aswanikumar, V. (1991) Subsurface chlorophyll maxima in the northwestern Bay of Bengal. *J. Plankton Res.* **13**, 339–352.

Sarmiento, J.L., *et al.* (2004) High latitude controls of thermocline nutrients and low latitude productivity. *Nature (Lond).* **427**, 56–60.

Sesama, S.K. (1990) Stability and mixing in the northern Bay of Bengal during the SW monsoon. *Mahasagar.* **23**, 19–28.

Sathyendranath, S., and Platt, T. (1994) New production and mixed-layer physics. *Proc. Ind. Acad. Sci. (Earth Planet. Sci.)* **103**, 177–188.

Sathyendranath, S., and T. Platt (2000) Mixed layers physics and primary production in the Arabian Sea. *In—The Changing Global Carbon Cycle* (R.B. Hanson, H.W. Ducklow, and J.G. Field, Eds.), pp. 285–299. Cambridge University Press, Cambridge.

Sathyendranath, S., *et al.* (1991) Biological control of surface temperature in the Arabian Sea. *Nature (Lond).* **349**, 54–56.

Sathyendranath, S., *et al.* (1995) Regionally and seasonally differentiated primary production in the North Atlantic. *Deep-Sea Res.* **42**, 1773–1802.

Sathyendranath, S., *et al.* (2004) Discrimination of diatoms from other phytoplankton using ocean-colour data. *Mar. Ecol. Prog. Ser.* **272**, 59–68.

Satoh, H.I., *et al.* (1992) Light conditions and photosynthetic characteristics of the deep chlorophyll maximum in Solomon Sea. *Jap. J. Phycol.* **40**, 135–142.

Saur, J.F.T. (1980) Surface salinity on the San Francisco-Honolulu route, June 1966—December 1975. *J. Phys. Oceanogr.* **10**, 1669–1680.

Saur, J.F., *et al.* (1994) Boundary current instabilities, upwelling, shelf mixing and eutrophication processes in the Black Sea. *Prog. Oceanogr.* **33**, 249–302.

Savenkoff, C., *et al.* (2000) Export of biogenic carbon and structure and dynamics of the pelagic food web in the Gulf of St. Lawrence. *Deep-Sea Res.* II. **47**, 585–607.

Savidge, G., and L. Gilpin (1999) Seasonal influences on size-fractionated chlorophyll concentrations and primary productivity in the NW Arabian Sea. *Deep-Sea Res. II.* **46**, 701–723.

Schalke, P.H., *et al.* (1990) Spatial and seasonal differences in acoustic recordings in the Banda Sea, obtained with a 30 kHz echosounder. Paper presented at Snellius II Symp. Jakarta, 23—29 Nov. 1987.

Scharek, R., *et al.* (1999) Temporal variations in diatom abundance and downward vertical flux in the oligotrophic North Pacific gyre. *Deep-Sea Res. I.* **46**, 1051–1075.

Scheffer, M., *et al.* (2001) Catastrophic shifts in ecosystems. *Nature (Lond).* **413**, 591–596.

Schnack, S.B., *et al.* (1985) Utilisation of phytoplankton by copepods in Antarctic waters during spring. *In—Marine Biology of Polar Regions* (J.S. Gray and M.E. Christiansen, Eds.), pp. 65–83. Wiley, Chichester, UK.

Schnack-Schiel, S.B., *et al.* (1991) Seasonal comparison of *Calanoides acutus* and *Calanus propinquus* (Copepoda: Calanoida) in the southeastern Weddell Sea, Antarctica. *Mar. Ecol. Prog. Ser.* **70**, 17–27.

Schnack-Schiel, S.B., *et al.* (1995) Life cycle strategy of the Antarctic copepod *Stephos longiceps*. *Prog. Oceanogr.* **36**, 45–75.

Schollaert, S.E., *et al.* (2004) Gulf Stream cross-frontal exchange: possible mechanisms to explain interannual variations in phytoplankton chlorophyll in the Slope Sea during the SeaWiFS years. *Deep-Sea Res. II.* **51**, 173–188.

Schott, F. (1983) Monsoon response of the Somali Current and associated upwelling. *Prog. Oceanogr.* **12**, 357–381.

Schott, F.A., and J.P McCreary (2000) The monsoon circulation of the Indian Ocean. *Prog. Oceanogr.* 51, 1–123.

Schouten, M.W., *et al.* (2003) Eddies and variability in the Mozambique Channel. *Deep-Sea Res. II.* 50, 1987–2003.

Schroeder, E.H. (1965) Average monthly temperatures in the North Atlantic. *Deep-Sea Res.* **12**, 323–343.

Schumacher, J., and P. Stabeno (1998) Continental shelf of the Bering Sea. *In—The Sea*, Vol. 12, *The Global Coastal Ocean* (A.R. Robinson and K.H. Brink, Eds.), pp. 789–822. Wiley InterScience, New York.

Schumacher, J.D., *et al.* (2003) Climate change in the southeast Bering Sea and some consequences for biota. *In—Large Marine Ecosystems of the World* (G. Hempel and K. Sherman, Eds.), pp. 17–40. AAAS, Washington, D.C.

Schumacher, J.D., and Reed, R. K. (1983) Interannual variability in abiotic environment of Bering Sea and Gulf of Alaska. *In—From Year to Year* (W.S. Wooster, Ed.), pp. 111–133. University of Washington, Seattle.

Schumann, E.H. (1982) Inshore circulation of the Agulhas Current off Natal. *J. Mar. Res.* **40**, 43–55.

Segerstrale, S. (1957) Baltic Sea. *Mem. Geol. Soc. Am.* **67**, 751–800.

Selph, K.E., *et al.* (2001) Microbial community composition and growth dynamics in the Antarctic Polar Front and seasonal ice zone during late spring 1997. *Deep-Sea Res. II.* **48**, 4059–4080.

Semina, H.J. (1997) An outline of the geographical distribution of oceanic phytoplankton. *Adv. Mar. Biol.* **32**, 528–563.

Servain, J., *et al.* (1982) Evidence of remote forcing in the equatorial Atlantic Ocean. *J. Phys. Oceanogr.* **12**, 457–463.

Seshappa, G. (1953) Observations on the physical and biological features of the inshore sea bottom along the Malabar coast. *Proc. Nat. Inst. Sci. India* **19**, 257–279.

Sevrin-Reyssac, J. (1993) Phytoplancton et production primaire dans les eaux marines ivoiriennes. *In—Environnement et Ressources Aquatiques de Côte d'Ivoire* (P. Le Loeuff, E. Marchal, and J.-B. Amon Kothias, Eds.), pp. 151–166. Editions de l'ORSTOM, Paris.

Shah, N.H. (1973) Seasonal variation of phytoplankton pigments in the Laccadive Sea off Cochin. *In—Biology of the Indian Ocean* (B. Zeitzschel, Ed.), pp. 175–185. Springer, Berlin.

Shanks, A.L., and G.L. Eckert (2005) Population persistence of California Current fishes and benthic crustaceans: a marine drift paradox. *Ecol. Monogr.* **75**, 505–524.

Shannon, L.V. (1985) The Benguela ecosystem. Part 1. Physical features and processes. *Oceanogr. Mar. Biol. Ann. Rev.* **23**, 105–182.

Shannon, L.V., and M. O'Toole (2003) The Benguela Current LME. *In—Large Marine Ecosystems of the World* (G. Hempel and K. Sherman, Eds.), pp. 227–253. AAAS, Washington, D.C.

Shannon, L.V., and S.C. Pillar (1986) The Benguela ecosystem. Part III. Plankton. *Oceanogr. Mar. Biol. Ann. Rev.* **24**, 65–170.

Shannon, L.V., *et al.* (1986) On the existence of an El Niño-type phenomenon in the Benguala system. *J. Mar. Res.* **44**, 495–520.

Shannon, L.V., *et al.* (1987) Large and mesoscale features of the Angola/Benguela Front. *S. Afr. J. Mar. Sci.* **5**, 11–34.

Sharma, V.V., and V. Aswanikumar (1991) Subsurface chlorophyll maxima in the northwestern Bay of Bengal. *J. Plankton Res.* **13**, 339–352.

Sharples, J., and M.J.N. Greig (1998) Tidal currents, mean flows and upwelling on the NE shelf of New Zealand. *N.Z. J. Mar. Freshwat. Res.* **32**, 215–231.

Sharples, J., *et al.* (2001) Internal tide dissipation, mixing, and vertical nitrate flux at the shelf edge of NE New Zealand. *J. Geophys. Res. (C Oceans).* **106**, 14069–14081.

Shaw, P.T. (1992) Shelf circulation off the southeast coast of China. *Rev. Aquat. Sci.* **6**, 1–28.

Schollaert, S.E. *et al.* (2004) Gulf Stream cross-frontal exchange. Deep-Sea Res. II 51, 173–188.

Schumann, E.H. (1998) The coastal ocean of southeastern Africa including Madagscar. *In—The Sea*, Vol. 12, *The Global Coastal Ocean* (A.R. Robinson and K.H. Brink, Eds.), pp. 557–582. Wiley InterScience, London.

Shelford, V.E. (1935) Some marine biotic communities of the Pacific Coast of North America. *Ecol. Monogr.* **5**, 251–292.

Shelford, V.E. (1963) *The Ecology of North America.* University of Illinois Press.

Sheppard, C.R.C., and D.J. Dixon (1998) Seas of the Arabian region. *In—The Sea*, Vol. 12, *The Global Coastal Ocean* (A.R. Robinson and K.H. Brink, Eds.), pp. 915–932. Wiley InterScience, New York.

Sherman, K., *et al.* (1996) *The Northeast Shelf Ecosystem: Assessment, Sustainability and Management.* Blackwell Scientific, Oxford.

Sherwin, T.J., *et al.* (1999) Eddies and a mesoscale deflection of the slope current in the Faeroe-Shetland Channel. *Deep-Sea Res. I.* **46**, 415–438.

Shetye, S.R. (1993) The western boundary current of the seasonal subtropical gyre in the Bay of Bengal. *J. Geophys. Res.* **98**, C1, 945–954.

Shetye, S.R., and A.D. Guveia (1998) Coastal circulation in the northern Indian Ocean. *In—The Sea*, Vol. 12, *The Global Coastal Ocean, Regional Studies and Synthesis* (A.R. Robinson and K.H. Brink, Eds.), pp. 523–556. Wiley InterScience, New York.

Shiah, F.-K. (2000) Biological and hydrographic responses to tropical cyclones on the continental shelf of the Taiwan Strait. *Cont. Shelf Res.* **15**, 2029–2044.

Shibata, A. (1983) Meander of the Kuroshio along the edge of continental shelf in the east China Sea. *Umi To Sora.* **58**, 113–120.

Shih, C.T. (1979) East-west diversity. *In—Zoogeography and Diversity in Plankton* (H. van der Spoel and I. Peirrot-Bults, Eds.), pp. 87–102.

Shillington. F.A. (1998) The Benguela upwelling system off Southwest Africa. *In—The Sea*, Vol. 12, *The Global Coastal Ocean* (A.R. Robinson and K.H. Brink, Eds.), pp. 583–604. Wiley InterScience, New York.

Shillington, F.A., *et al.* (1990) A cool upwelling filament off Namibia; preliminary measurements of physical and biological features. *Deep-Sea Res. I.* **37**, 1753–1772.

Shinn, E.A. (1987) Carbonate coastal accretion in an area of longshore transport. Northeast Qatar. *In—Key Environments—Red Sea* (A.J. Edwards and S.M. Head, Eds.), pp. 179–191. Pergamon Press, Oxford.

Shiomoto, A. (2000) Chlorophyll-*a* and primary productivity during spring in the oceanic region of the Oyahio Water. NW Pacific. *J. Mar. Biol. Assoc. U.K.* **80**, 343–354.

Shirtcliffe, T.G.L., *et al.* (1990) Dynamics of the Cape Farewell upwelling plume. *N. Z. J. Mar. Freshwat. Res.* **24**, 555–568.

Shukla, J. (1987) Interannual variability of monsoons. *In—Monsoons* (J.S. Fein and P.L. Stephens, Eds.), pp. 399–463. Wiley, New York.

Sieburth, J.M., and P.G. Davis (1984) The role of heterotrophic nanoplankton in nurturing and grazing planktonic bacteria. *Ann. Inst. Océanogr.* **58**, 285–296.

Siedler, G., *et al.* (1992) Seasonal changes in the tropical Atlantic circulation: observation and simulation of the Guinea Dome. *J. Geophys. Res.* **97**, 703–715.

Siegel, D.A. (2001) Oceanography—the Rossby rototiller. *Nature (Lond).* **409**, 576–577.

Siegel, D.A., *et al.* (2002) The North Atlantic spring bloom and Sverdrup's critical depth hypothesis. *Science.* **296**, 730–733.

Sievers, H.A., and N. Silva (1975) Water masses and circulation in the southeastern Pacific Ocean. *Cienc. Technol. Mar.* **1**, 7–67.

Signorini, S.R., *et al.* (1999) Biological and physical signatures in the tropical and subtropical Atlantic. *J. Geophys. Res.* **104** (C8) 18367–18382.

Silva Sandoval, N., and S. Neshyba (1977) Surface currents off the southern coast of Chile. *Cienc. Technol. Mar.* **3**, 37–42.

Simpson, J. H. (1981) The shelf sea fronts: Implications of their existence and behaviour. *Proc. R. Soc. London A* **302**, 531–546.

Simpson, J.H. (1998) The Celtic Seas *In—The Sea*, Vol. 12, *The Global Coastal Ocean* (A.R. Robinson and K.H. Brink, Eds.), pp. 659–698. Wiley InterScience, London.

Sinclair, M. (1988) *Marine Populations*. University of Washington Press, Seattle, WA.

Skjoldal, H.R., and F. Rey (1989) Pelagic production and variability of the Barents Sea Ecosystem. *In—Biomass Yields and Geography of Large Marine Ecosystems* (K. Sherman and L.M. Alexander, Eds.), AAAS Selected Symposium, pp. 173–186. AAAS, Washington.

Skjoldal, H.R., and R. Saetre (2004) Food webs and trophic interactions. *In—The Norwegian Sea Ecosystem* (H.R. Skjoldal, Ed.), pp. 507–534. Tapir Academic Press.

Smetacek, V.H. (2001) A watery arms race. *Nature (Lond)*. **411**, 745.

Smetacek, V.H., and U. Passow (1990) Spring bloom initiation and Sverdrup's critical depth model. *Limnol. Oceanogr.* **35**, 228–234.

Smetacek, V.H., *et al.* (1984) Seasonal stages characterizing the annual cycle of an inshore pelagic system. *Rapp. Proc. Verb. Réun. Cons. Int. Explor. Mer.* **183**, 126–135.

Smetacek, V.H., *et al.* (1990) Seasonal and regional variation in the pelagial and its relationship to the life history of krill. *In—Antarctic Ecosystems. Ecological Change and Conservation* (K.R. Kerry and G. Hempel, Eds.), pp. 103–114. Springer, Berlin.

Smetacek, V.H., *et al.* (1992) Early spring phytoplankton blooms in ice platelet layers of the southern Weddell Sea, Antarctica. *Deep-Sea Res.* **39**, 153–168.

Smetacek, V.H., *et al.* (1997) Ecology and biochemistry of the Antarctic Circumpolar Current during austral spring: A summary of SO JGOFS ANT X/6 of R.V. *Polarstern. Deep-Sea Res. II.* **44**, 1–22.

Smetacek, V.H., *et al.* (2002) Mesoscale distribution of dominant diatom species relative to the hydrographic field along the Antarctic Polar Front. *Deep-Sea Res. II.* **49**, 3835–3848.

Smetacek, V.H., *et al.* (2004) The role of grazing in structuring Southern Ocean pelagic ecosystems and biogeochemical cycles. *Antarct. Sci.* **16**, 541–558.

Smith, P.C., and B. Petrie (1982) Low-frequency circulation at the edge of the Scotian shelf. *J. Phys. Oceanogr.* **12**, 28–46.

Smith, P.C., and H. Sandstrom (1986) Shelf edge processes. *Bedford Inst. Oceanogr. Rev. 1986*, 40–46.

Smith, P.E., *et al.* (1992) Life-stage duration and survival parameters as related to interdecadal population variation in Pacific sardine. *Rep. Cal. Coop. Fish. Invest.* **33**, 41–49.

Smith, R.C. (1981) Remote sensing and depth distribution of chlorophyll. *Mar. Ecol. Prog. Ser.* **5**, 359–361.

Smith, R.E.H., *et al.* (1987) Intracellular photosynthate allocation and the control of Arctic marine ice algal production. *J. Phycol.* **23**, 124–132.

Smith, R.L. (1981a) Circulation patterns in upwelling regimes. *In—Coastal Upwelling* (F. W. Richards, Ed.), pp. 13–35. American Geophysical Union, Washington, D.C.

Smith, R.L. (1981b) A comparison of the structure and variability of the flow field in three coastal upwelling systems: Oregon, northwest Africa and Peru. *In—Coastal Upwelling* (F.W. Richards, Ed.), pp. 107–118. American Geophysical Union, Washington, D.C.

Smith, R.L., and J.S. Bottero (1977) On upwelling in the Arabian Sea. *In—A Voyage of Discovery* (M. Angel, Ed.), pp. 291–304. Pergamon, London.

Smith, R.L., *et al.* (1991b) The Leeuwin Current off Australia, 1986–1987. *J. Phys. Oceanogr.* **21**, 323–345.

Smith, S.L. (1982) The northwest Indian Ocean during the monsoons of 1979: distribution, abundance and feeding of zooplankton. *Deep-Sea Res.* **29**, 1331–1353.

Smith, S.L. (1984) Biological indications of active upwelling in the NW Indian Ocean in 1964 and 1979, and a comparison with Peru and NW Africa. *Deep-Sea Res. I.* **31**, 951–967.

Smith, S.L., and M. Madhupratap (2005) Mesozooplankton of the Arabian Sea: patterns influenced by seasons, upwelling and oxygen concentrations. *Prog. Oceanogr.* **65**, 214–239.

Smith, S.L., and S.B. Schnack-Schiel (1990) Polar zooplankton. *In—Polar Oceanography*, (Ed. W.O. Smith, Jr.), pp. 527–597. Academic Press, San Diego.

Smith, S.L., *et al.* (1998a) The 1994–96 Arabian Sea Expedition. *Deep-Sea Res. II.* **45**, 1905–1916.

Smith, S.L., *et al.* (1998b) Seasonal response of zooplankton to monsoonal reversals in the Arabian Sea. *Deep-Sea Res. II.* **45**, 2369–2403.

Smith, W.O. (1987) Phytoplankton dynamics in marginal ice zones. *Oceanogr. Mar. Biol. Ann. Rev.* 25, 11–38.
Smith, W.O., and R.I. Brightman (1991) Phytoplankton photosynthetic response during the winter-spring transition in the Fram Strait. *J. Geophys. Res.* **96**, 4549–4554.
Smith, W.O., and D. J. DeMaster (1996) Phytoplankton biomass and productivity in the Amazon River plume: Correlation with seasonal and river discharge. *Cont. Shelf Res.* **16**, 291–319.
Smith, W.O., and D. M. Nelson (1985) Phytoplankton biomass near a receding ice-edge in the Ross Sea. *In—Antarctic Nutrient Cycles and Food Webs* (W.R. Siegfried, Ed.), pp. 70–77. Springer, Berlin.
Smith, W.O., and E. Sakshaug (1990) Polar phytoplankton. *In—Polar Oceanography Part B: Chemistry, Biology, and Geology* (W.O. Smith, Ed.), pp. 477–525. Springer, Berlin.
Smith, W.O., *et al.* (1991a) Importance of *Phaeocystis* blooms in the high-latitude ocean carbon cycle. *Nature (Lond).* **352**, 514–516.
Soloviev, A., and R. Lukas (1997) Observation of large diurnal warming events in the near-surface layer of the western equatorial Pacific warm pool. *Deep-Sea Res.* **44**, 1055–1076.
Soltwedel, T. (1997) Meiobenthos distribution pattern in the tropical East Atlantic: indication for fractionated sedimentation of organic matter to the sea floor? *Mar. Biol.* **129**, 747–756.
Somayajulu, Y.K., *et al.* (2003) Seasonal and inter-annual variability of surface circulation in the Bay of Bengal from Topex/Poseidon altimetry. *Deep-Sea Res. II.* **50**, 867–880.
Sommer, U. (1998) Marine food webs under eutrophic conditions: desirable and undesirable forms of nutrient richness. *Dtsch. Hydrogr. Z. (Suppl.)* **6**, 167–176.
Sommer, U., *et al.* (2002) Grazing during the early spring in the Gulf of Aqaba and the northern Red Sea. *Mar. Ecol. Prog. Ser.* **239**, 251–261.
Sorokin, Y. (1983) The Black Sea. *In—Ecosystems of the World, 26: Estuaries and Enclosed Seas* (B.H. Ketchum, Ed.), pp. 253–292. Elsevier, Amsterdam.
Sorokin, Y., and P.Y. Sorokin (2002) Microplankton and primary production in the Sea of Okhotsk in summer 1994. *J. Plankton Res.* **24**, 453–470.
Sorokin, Y.I., and P.Y. Sorokin (1999) Production in the Sea of Okhotsk. *J. Plankton Res.* **21**, 201–230.
Sournia, A. (1994) Pelagic biogeography and fronts. *Prog. Oceanogr.* **34**, 109–120.
Springer, A., and McRoy, C. P. (1993) The paradox of pelagic food webs in the northern Bering Sea—III. Patterns of primary production. *Cont. Shelf Res.* **13**, 575–599.
Springer, A.M., *et al.* (1989) The paradox of pelagic food webs in the northern Bering Sea. II. Zooplankton communities. *Cont. Shelf Res.* **9**, 359–386.
Springer, A.M., *et al.* (1999) Marine birds and mammals of the Pacific subarctic gyres. *Prog. Oceanogr.* **43**, 443–487.
Sprintall, J., and M. Tomczak (1992) Evidence of the barrier layer in the surface layer of the tropics. *J. Geophys. Res.* **97**, 7305–7316.
Srokosz, M.A., *et al.* (2004) A possible plankton wave in the Indian Ocean. *Geophys. Res. Lett.* **31**, L13301.
Stabeno, P.J., and R.K. Reed (1992) A major circulation anomaly in the western Bering Sea. *Geophys. Res. Lett.* **19**, 1671–1674.
Stabeno, P.J., and R.K. Reed (1994) Circulation in the Bering Sea basin observed by satellite-tracked drifters: 1986–1993. *J. Phys. Oceanogr.* **24**, 848–864.
Steele, J.H. (1991) Marine functional diversity. *Bioscience.* **41**, 470–474.
Steele, J.H., and M. Schumacher (2000) Ecosystem structure before fishing, *Fish. Res.* **44**, 201–205.
Steeman Neilsen, E. (1952) The use of radioactive carbon (C-14) for measuring organic production in the sea. *J. Cons. Int. Explor. Mer.* **18**, 117–140.
Steftaris, N., *et al.* (2005) Globalisation in marine ecosystems: the story of non-indigenous marine species across European seas. *Oceanogr. Mar. Biol. Ann. Rev.* **43**, 419–453.
Steinberg, D.K., *et al.* (2001) Overview of the US JGOFS Bermuda Atlantic Time Series (BATS): a decade-scale look at ocean biology and biogeochemistry. *Deep-Sea Res. II.* **48**, 1405–1448.
Stelfox, C.E., *et al.* (1999) The structure of zooplankton communities in the Arabian Sea during and after the SW monsoon, 1994. *Deep-Sea Res. II.* **46**, 815–842.
Stergiou, K.I. (2002) Overfishing, tropicalisation of fish stocks, uncertainty and ecosystem management: resharpening Occam's razor. *Fish. Res.* **55**, 1–9.

Steuer, A. (1933) Zur planmassigen Erforschung der geographischen Verbreitung des Haliplankton, besonders der Copepoden. *Zoogeogr. Int. Rev. Comp. Caus. Anim. Geogr.* **1**, 269–302.

Stobutzski, I.C. (2006) Decline of demersal fisheries resources in three developing Asian countries. *Fish. Res.* 78, 130–142.

Stoecker, D. (1987) Large proportion of marine planktonic ciliates found to contain functional chloroplasts. *Nature (Lond).* **326**, 790–792.

Stommel, H., and A.B. Arons (1960) On the abyssal circulation of the world ocean. II. An idealised model of the circulation pattern and amplitude in ocean basins. *Deep-Sea Res.* **6**, 217–233.

Stoner, A.W. 1983. Pelagic sargassum: evidence for a major decrease in biomass. *Deep-Sea Res.* **30**, 469–474.

Stramma, L. (1992) The South Indian Ocean Current. *J. Phys. Oceanogr.* **22**, 421–430.

Stramma, L., and F. Schott (1999) The mean flow field of the tropical Atlantic Ocean. *Deep-Sea Res. II.* **46**, 279–303.

Stramma, L., and J.R.E. Lutjeharms (1997) The flow field of the subtropical gyre of the South Indian Ocean. *J. Geophys. Res* **102** (C3) 5513–5530.

Stramma, L., and R.G. Petersen (1990) The South Atlantic Current. *J. Phys. Oceanogr.* **20**, 846–859.

Strass, V.H. (1992) Chlorophyll patchiness caused by mesoscale upwelling at fronts. *Deep-Sea Res.* **39**, 75–97.

Strass, V.H., and J.D. Woods (1988) Horizontal and seasonal variation of density and chlorophyll profiles between the Azores and Greenland. *In—Towards a Theory of Biological-Physical Interactions in the Worlds Oceans* (B. J. Rothschild, Ed.), pp. 113–136. D. Reidel, Boston.

Strass, V.H., and J.D. Woods (1991) New production in the summer revealed by the meridional slope of the deep chlorophyll maximum. *Deep-Sea Res.* **38**, 35–56.

Strass, V.H., *et al.* (2002) A 3-D mesoscale map of primary production at the Antarctic Polar Front: results of a diagnostic model. *Deep-Sea Res. II.*, **49**, 3813–3834.

Streftaris, N. (2005) Globalisation in marine ecosystems: the story of non-indigenous species across European seas. *Oceanogr. Mar. Biol. Ann. Rev.* **43**, 419–453.

Strub, P.T., *et al.* (1998) Coastal ocean circulation off western South America. *In—The Sea*, Vol. 12, *The Global Coastal Ocean* (A.R. Robinson and K.H. Brink, Eds.), pp. 273–314. Wiley InterScience, London.

Su, J.L., *et al.* (1990) The Kuroshio. Part 1: Physical features. *Oceanogr. Mar. Biol. Ann. Rev.* **28**, 11–71.

Suárez-Morales, E. (1995) Pelagic copepod association during spring upwelling off the Yucatan peninsula. *In—Pelagic Biogeography, ICoPB II. IOC Wkshp. Rep. 142*, pp. 345–352.

Subba Rao, D. (1976) Temporal variations in primary production during upwelling season off Waltair. *Int. J. Ecol. Env. Sci.* **2**, 107–104.

Subbha Rao, D.V., and D. Sameoto (1988) Relationship between phytoplankton and copepods in the deep Tropical Pacific Ocean off Costa Rica. *Bull. Mar. Sci.* **42**, 85–100.

Suda, A. (1973) Tuna fisheries and their resources in the Indian Ocean. *In—Biology of the Indian Ocean* (B. Zeitzschel, Ed.), pp. 431–449. Springer, Berlin.

Sugimoto, T., and H. Tameishi (1992) Warm-core rings, streamers and their role on the fishing ground formation of Japan. *Deep-Sea Res.* **39**, 183–201.

Sund, P., *et al.* (1980) Tunas and their environment in the Pacific Ocean. *Oceanogr. Mar. Biol. Annu. Rev.* **18**, 23–56.

Sverdrup, H.U. (1947) Wind driven currents in a baroclinic ocean with application to the equatorial currents in the Pacific Ocean. *Proc. Natl. Acad. Sci. USA* **33**, 219–303.

Sverdrup, H.U. (1953) On the conditions for vernal blooming of the phytoplankton. *J. Cons. Perm. Int. Explor. Mer.* **18**, 287–295.

Swallow, J.C. (1984) Some aspects of the physical oceanography of the Indian Ocean. *Deep-Sea Res.* **31**, 639–650.

Swallow, J.C., and Fieux, M. (1982) Historical evidence for two gyres in the Somali Current. *J. Mar. Res.* **40**, 747–755.

Swart, V.P., and J.W. Gonzalves (1983) Episodic meanders in the Agulhas Current. *Rep 5th Natl. Oceanogr. Symp. S. Afr.* **79**, 161–172.

Swift, J.H. (1986) The Arctic waters. *In—The Nordic Seas* (B.G. Hurdle, Ed.), pp. 129–153. Springer, New York.

Szekielda, K.-H. (1978) Eolian dust into the northeast Atlantic. *Oceanogr. Mar. Biol. Ann. Rev.* **16**, 11–41.
Taft, B. (1972) Characteristics of the flow of the Kuroshio south of Japan. *In—Kuroshio* (H. Stommel and K. Yoshida, Eds.), pp. 165–216. University of Washington Press, Seattle, WA.
Takahashi, M., *et al.* (1985) Distribution of the SCM and its nutrient-light environment in and around the Kuroshio off Japan. *J. Oceanogr. Soc. Jpn.* **41**, 73–80.
Takenouti, A.Y., and K. Ohtani (1974) Currents and water masses in the Bering Sea: Japanese work. *In—Oceanography of the Bering Sea* (D.W. Hood and E.J. Kelly, Eds.), pp. 39–58. University of Alaska Press, Fairbanks, AK.
Taniguchi, A., and T. Kawamura (1972) Primary production in the western tropical and subtropical Pacific Ocean. *Proc. 2nd CSK Symp. Tokyo.*, 159–168.
Tarran, G.A., *et al.* (1999) Phytoplankton community structure in the Arabian Sea during and after the SW monsoon, 1994. *Deep-Sea Res. II.* **46**, 655–676.
Taylor, A.H., and J.A. Stephens (1993) Diurnal variations of convective mixing and the spring bloom of phytoplankton. *Deep-Sea Res. II.* **49**, 389–408.
Taylor, A.H., *et al.* (1998) Gulf Stream shifts following ENSO events. *Nature (Lond).* **393**, 638.
Tegen, I., and I. Fung (1994) Modelling of mineral dust in the atmosphere: sources, transport and optical thickness. *J. Geophys. Res.* **99**, 22897–22914.
Terazaki, M. (1999) The Sea of Japan Large Marine Ecosystem. *In—The Sea*, Vol. 12, *The Global Coastal Ocean* (A.R. Robinson and K.H. Brink, Eds.). Wiley InterScience, New York.
Tett, P. (1981) Modelling phytoplankton production at shelf sea fronts. *Phil. Trans. R. Soc. London A* **302**, 605–615.
Theocaris, A., *et al.* (1998) Physical and dynamical processes in the coastal and shelf areas of the Mediterranean. *In—The Sea*, Vol. 12, *The Global Coastal Ocean* (A.R. Robinson and K.H. Brink, Eds.), pp. 863–887. Wiley InterScience, New York.
Thibault, D., *et al.* (1999) Mesozooplankton in the Arctic Ocean in summer. *Deep-Sea Res. I.* **46**, 1391–1415.
Thomas, W.H. (1972) Nutrient inversion in the southeastern tropical Pacific Ocean. *U.S. Fish. Bull.* **70**, 929–932.
Thomas, W.H. (1978) Anomalous nutrient-chlorophyll relations in the offshore eastern tropical Pacific Ocean. *Deep-Sea Res.* **37**, 327–335.
Thompson, R.E. (1981) Oceanography of the British Columbia coast. *Can. Spec. Publ. Fish. Aquat. Sci.* **56**, 1–291.
Thompson, R.E. (1987) Continental shelf scale model of the Leeuwin Current. *J. Mar. Res.* **45**, 813–827.
Thomson, D.H. (1982) Marine benthos in the eastern Canadian high Arctic: multivariate analyses of standing crop and community structure. *Arctic* **35**, 61–74.
Thorson, G. (1957) Benthos. *In—A Treatise on Marine Ecology and Palaeoecology* (*Mem. Geol. Soc. Am.*), pp. 461–534. Geological Society of America, Washington, DC.
Thunell, R., *et al.* (1996) Plankton response to physical forcing in the Gulf of California. *J. Plankton Res.* **18**, 2017–2026.
Tianming, L., and S.G.H. Philander (1996) On the annual cycle of the eastern tropical Pacific. *J. Climate.* **9**, 2986–2998.
Tomczak, M., and J.S. Godfrey (1994) *Regional Oceanography.* Pergamon, Elmsford, NY.
Toole, J.M., and R.W.A.F. Schmidt (1987) Small-scale structure in the Northwest Atlantic subtropical front. *Nature (Lond).* **327**, 47–48.
Tortell, P.D., *et al.* (1999) Marine bacteria and biogeochemical cycling in the oceans. *Microbiol. Ecol.* **29**, 1–11.
Townsend, D.W. (1992) An overview of the oceanography and biological production. *In—The Gulf of Maine* (*U.S. NOAA, Regional Synthesis Series No. 1*), pp. 5–25. US Department of Commerce, Washington, DC.
Townsend, D.W., and A.C. Thomas (2001) Winter-spring transition of phytoplankton chlorophyll and inorganic nutrients on Georges Bank. *Deep-Sea Res. II.* **48**, 199–214.
Townsend, D.W., *et al.* (1988) Near-bottom chlorophyll maxima in southeastern Mediterranean shelf waters. *Océanol. Acta.* **9**, 235–244.
Townsend, D.W., *et al.* (1992) Spring phytoplankton blooms in the absence of vertical water column stratification. *Nature (Lond).* **360**, 59–62.

Traganza, E. D., *et al.* (1981) Satellite observations of a cyclonic upwelling system and giant plume in the California Current. *In—Coastal Upwelling* (F.A. Richards, Ed.), pp. 228–241. American Geophysical Union, Washington, DC.

Tranter, D.J. (1973) Seasonal studies of a pelagic ecosystem (110°E) *In—Biology of the Indian Ocean* (B. Zeitzschel, Ed.), pp. 487–520. Springer, Berlin.

Tranter, D.J., and G.S. Leech (1987) Factors influencing the standing crop of phytoplankton on the Australian Northwest Shelf seaward of the 40 m isobath. *Cont. Shelf Res.* 7, 115–133.

Tranter, D., *et al.* (1980) *In vivo* chlorophyll fluorescence in the vicinity of warm-core eddies of the coast of New South Wales. *Rep. Div. Fish. Oceanogr., Cronulla.*, 1–34.

Tranter, D., *et al.* (1983) Edge enrichment in an ocean eddy. *Aust. J. Mar. Freshwat. Res.* **34**, 665–680.

Tranter, D.J., *et al.* (1987) The coastal enrichment effect of the East Australia Current eddy field. *Deep-Sea Res.* **33**, 1705–1728.

Trees, C.C., and S.Z. El-Sayed (1986) Remote sensing of chlorophyll concentrations in the northern Gulf of Mexico. *Proc. Int. Soc. Optic. Eng.* **637**, 328–334.

Tréguer, P., and G. Jacques (1992) Dynamics of nutrients and phytoplankton and fluxes of carbon nitrogen and silicon in the Antarctic Ocean. *Polar Biol.* **12**, 149–162.

Trotte, J.R. (1985) Phytoplankton floristic composition and size-specific photosynthesis in the eastern Canadian arctic. M.Sc. thesis, Dalhousie University, Nova Scotia.

Trull, T., *et al.* (2001) Circulation and seasonal evolution of polar waters south of Australia: implications for iron fertilization of the Southern Ocean. *Deep-Sea Res. II.* **48**, 2439–2466.

Tsuchiya, M. (1985) Evidence of a double-cell subtropical circulation in the South Atlantic Ocean. *J. Mar. Res.* **43**, 57–65.

Tsuda, A., *et al.* (1989) Feeding of micro- and macro-zooplankton at the subsurface chlorophyll maximum in the subtropical North Pacific. *J. Exp. Mar. Biol. Ecol.* **132**, 41–52.

Turk, D., *et al.* (2001) Remotely-sensed biological production in the equatorial Pacific. *Science.* **293**, 471–474.

Uda, M. (1963) Oceanography of the subarctic North Pacific. *J. Fish. Res. Bd. Can.* **20**, 119–179.

Umatani, S., and T. Yamagata (1991) Response of the eastern tropical Pacific to meridional migration of the ITCZ: The generation of the Costa Rica Dome. *J. Phys. Oceanogr.* **21**, 346–363.

Uye, S.S.O. (2000) Why does *Calanus sinicus* prosper in the shelf ecosystem of the Northwest pacific Ocean? *ICES J. Mar. Sci.* **57**, 1850–1855.

Valentin, J.L. (1984) Spatial structure of the zooplankton community in the Cabo Frio region (Brazil) influenced by coastal upwelling. *Hydrobiologia.* **113**, 183–199.

van Aken, H.M., *et al.* (1991) The arctic front in the Greenland sea during February, 1989. *J. Geophys. Res.* **96**, 4739–4750.

van Camp, L., *et al.* (1991) Upwelling and boundary circulation off northwest Africa as depicted by infra-red and visible satellite observations. *Prog. Oceanogr.* **26**, 357–402.

Van der Spoel, S. (1994) A biosystematic basis for pelagic ecology. *Bijd. Dierkunde.* **64**, 3–31.

van der Spoel, S., and R.P. Heyman (1983) *A Comparative Atlas of Zooplankton.* Springer, Berlin.

van Franeker, J.A., *et al.* (2002) Responses of seabirds, in particular prions (*Pachyptila* sp.), to small-scale processes in the Antarctic Polar Front. *Deep-Sea Res. II.* **49**, 3931–3950.

van Oijen, T. (2004) Iron and light limitation of carbonate production by phytoplankton in the Southern Ocean. Ph.D. thesis, 172 pp., Univ. of Groningen, Netherlands.

Vedernikov, V.I., and A.B. Demidov (1991) Primary production and chlorophyll in deep regions of the Black Sea. *Okeanologia.* **33**, 193–199.

Veldhuis, M.J.W., *et al.* (1997) Seasonal and spatial variability in phytoplankton biomass, productivity and growth in the northeastern Indian Ocean; the southwest and northwest monsoon, 1992–1993. *Deep-Sea Res. II.* **44**, 431–449.

Venrick, E.L. (1974) Recurrent groups of diatoms species in the North Pacific. *Ecology.* **52**, 614–625.

Venrick. E.L. (1988) The vertical distribution of chlorophyll and phytoplankton species in the North Pacific central environment. *J. Plankton Res.* **10**, 987–988.

Venrick, E.L. (1989) The lateral extent and characteristics of the North Pacific Central environment at 35°N. *Deep-Sea Res.* **26**, 1153–1178.

Venrick, E.L. (1991) Mid-ocean ridges and their influence on the large-scale patterns of chlorophyll and production in the North Pacific. *Deep-Sea Res.* **38**, S83-S102.

Venrick, E.L., *et al.* (1973) Deep maxima of photosynthetic chlorophyll in the Pacific Ocean. *Fish. Bull.* **71**, 41–52.

Venrick, E.L., *et al.* (1994) Climate and chlorophyll: long-term trends in the central North Pacific Ocean. *Science* **238**, 70–72.

Verity, P.G., *et al.* (1996) Microzooplankton grazing of primary production at 140°W in the equatorial Pacific. *Deep-Sea Res. II.* **43**, 1227–1225.

Verity, P.G., *et al.* (2002) Coupling between primary production and pelagic consumption in temperate ocean margin pelagic ecosystems. *Deep-Sea Res. II.* **49**, 4553–4569.

Verstraete, J.-M. (1992) The seasonal upwellings in the Gulf of Guinea. *Prog. Oceanogr.* **29**, 1–60.

Veth, C., *et al.* (1997) Physical anatomy of fronts and surface waters in the ACC near the 6°W meridian during austral spring 1992. *Deep-Sea. Res. II.* **44**, 23–50.

Vethamony, P.V., *et al.* (1987) Thermal structure and flow patterns around Seychelles Islands during austral autumn. *Ind. J. Mar. Sci.* **16**, 179–183.

Vézina, A.F., and C. Savenkoff (1999) Inverse modelling of carbon and nitrogen flows in the pelagic food-web of the NE subarctic Pacific. *Deep-Sea Res. II.* **46**, 2909–2939.

Villareal, T., *et al.* (1993) Nitrogen transport by vertically migrating algal mats in the North Pacific Ocean. *Nature (Lond).* **363**, 709–712.

Vincent, W.F., *et al.* (1991) Distribution and biological properties of oceanic water masses around the South Island. *N. Z. J. Mar. Freshwat. Res.* **25**, 21–42.

Vinogradov, M.E., and N.M. Voronina (1961) Influence of the oxygen deficit on the distribution of plankton in the Arabian Sea. *Okeanologia.* **1**, 670–678.

Vinogradov, M.E., *et al.* (1985) Vertical distribution of mesozooplankton in the open area of the Black Sea. *Mar. Biol.* **89**, 95–107.

Vivekanandan, E., *et al.* (2005) Fishing down the food web along the Indian coast. *Fish Res.* **72**, 241–252.

Voituriez, B. (1981) The equatorial upwelling in the eastern Atlantic. *In—Recent Progress in Equatorial Oceanography: A Report of the Final Meeting of the SCOR WG 47* (J.P. McCreary, Ed.), pp. 229–247. Nova University Press, Dania, FL.

Voituriez, B., and A. Herbland (1981) Primary production in the tropical Atlantic Ocean mapped from oxygen values of EQUALANT 1 and 2. *Bull. Mar. Sci.* **31**, 853–863.

Voituriez, B., and A. Herbland (1982) Comparison des systèmes productifs de l'Atlantique tropical est: Domes thermiques, upwellings cotières et upwelling équatorial. *Rapp. Proc. Verb. Réun. Cons. Int. Explor. Mer.* **180**, 114–130.

Voorhuis, A.D. (1969) The horizontal extent and persistence of thermal fronts in the Sargasso Sea. *Deep-Sea Res.* **16**(Suppl.), 331–337.

Voorhuis, A.D., and J.G. Bruce (1982) Small-scale surface stirring and frontogenesis in the subtropical convergence of the North Atlantic. *J. Mar. Res.* **40**, 801–821.

Vordelwuebecke, N., *et al.* (2000) Phytoplankton biomass and activity in the northern Benguela. *Abstr. 10th S. Afr. Mar. Sci. Symp.*

Vukovich, F.M. (1988) Loop current boundary variations. *J. Geophys. Res.* **93**, 15585–15591.

Vukovich, F.M., and G.A. Maul (1985) Cyclonic eddies in the eastern Gulf of Mexico. *J. Phys. Oceanogr.* **15**, 105–107.

Vukovich, F.M., and E. Waddell (1991) Interaction of a warm core ring with the western slope in the Gulf of Mexico. *J. Phys. Oceanogr.* **21**, 1062–1074.

Walker, G.T., and E.W. Bliss (1932) World weather. *Mem. R. Meteorol. Soc.* **4**, 53–84.

Wallace, A.R. (1878) *Tropical Nature, and Other Essays.* Macmillan, London.

Walsh, J.J. (1998) *On the Nature of Continental Shelves.* Academic Press, San Diego.

Walsh, J.J., *et al.* (1989) Nitrogen exchange at the continental margin: a numerical study of the Gulf of Mexico. *Prog. Oceanogr.* **23**, 245–301.

Wang, X., and Zhu, B. (1991) Discussion of the subsurface chlorophyll maximum forming in the continental shelf and Kuroshio area of the South China Sea. *In—Essays on the Investigation of the Kuroshio* 3, pp. 297–304. University of Tokyo Press, Tokyo.

Wang, Y. (1983) The mudflat coast of China. *Can. J. Fish. Aquat. Sci.* **40**, 160–171.

Ware, D.M., and G.A. McFarlane (1989) Fisheries production biomes in the northeast Pacific Ocean. *Can. Spec. Publ. Fish. Aquat. Sci.* **108**, 359–379.

Warwick, R.M., and Ruswahyuni. (1987) A comparative study of the structure of some tropical and temperate marine soft-bottom macrobenthic communities. *Mar. Biol.* **95**, 641–649.

Warwick, R.M., and R.J. Uncles (1980) Distribution of benthic macrofaunal associations in the Bristol Channel in relation to tidal stress. *Mar. Ecol. Prog. Ser.* 3, 97–103.

Warwick, R.M., *et al.* (1978) Annual macrofauna production in a *Venus* community. *Est. Coast. Mar. Sci.* 7, 215–241.

Wassman, P., *et al* (1991) Patterns of production and sedimentation in the boreal and polar Northeast Atlantic. *Polar Biol.* **10**, 209–228.

Watson, A.J. (2001) Iron limitation in the oceans. *In—The Biogeochemistry of Iron in Seawater* (D.R. Turner and K.A. Hunter, Eds.), pp. 9–30. Wiley, New York.

Watts, J.C.D. (1972) The occurrence of the thermocline in the shelf areas south of Hong Kong. *In—Proceedings of the 2nd Symposium Kuroshio* (K. Sugawara, Ed.), pp. 43–46. University of Washington University Press, Seattle, WA.

Watts, L.J., and N.J.P. Owens (1999) Nitrogen assimilation and the f-ratio in the NW Indian Ocean during an intermonsoon period. *Deep-Sea Res. II.* **46**, 725–743.

Watts, L.J., S., *et al.* (1999) Modelling new production in the NW Indian Ocean region. *Mar. Ecol. Prog. Ser.* **183**, 1–12.

Weaks, M. (1984) Upwelling in the Gulf of Oman. *NOAA Oceanogr. Month. Sum.* **4**, 13.

Weber, L.H., and S. El-Sayed (1987) Contributions to the net, nano- and picoplankton standing crop and primary production in the Southern Ocean. *J. Plankton Res.* **9**, 973–994.

Wefer, G., and G. Fischer (1991) Annual primary production and export flux in the Southern Ocean from sediment trap data. *Mar. Chem.* **35**, 597–614.

Weikert, H. (1984) The vertical distribution of zooplankton in relation to habitat zones in the area of the Atlantis II Deep, Red Sea. *Mar. Ecol.-Prog. Ser.* **8**, 129–143.

Weikert, H. (1987) Plankton and pelagic environment. *In—Key Environments—Red Sea* (A.J. Edwards and S.M. Head, Eds.), pp. 90–111. Pergamon Press, Oxford.

Weller, R.A. (1991) Overview of the frontal air-sea interaction experiment, FASINEX. *J. Geophys. Res.* **96**, 8501–8516.

Wells, J.T. (1983) Dynamics of coastal fluid muds in low, moderate and high tide range environments. *Can. J. Fish. Aquat. Sci.* **40**, 130–142.

Wells, J.W. (1957) Coral reefs. *In—Treatise on Marine Ecology and Palaeontology* (*Mem. Geol. Soc. Am.*), pp. 609–632. Geological Society of America, Washington, DC.

Wells, M.L., *et al.* (1999) Tectonic processes in Papua New Guinea and past productivity in the eastern equatorial Pacific Ocean, *Nature (Lond).* **398**, 601–603.

Wells, M.L., *et al.* (1994) Iron limitation and the cyanobacterium *Synechococcus* in equatorial Pacific waters. *Limnol. Oceanogr. 39*, 1481–1486.

Welschmeyer, N.A., *et al.* (1993) Primary production in the subarctic Pacific Ocean: Project SUPER. *Prog. Oceanogr.* **32**, 101–135.

Westbrook, P. (2002) Biology on the global scale. *Nature (Lond).* **419**, 113–114.

Wetsteyn, F.J., *et al.* (1990) Nutrient distribution in the upper 300 m of the eastern Banda Sea during and after the upwelling season. *Neth. J. Sea Res.* **25**, 449–464.

Wheeler, P.A., and S.A. Kokkinakis (1990) Ammonium recycling limits nitrate use in the oceanic subarctic Pacific. *Limnol. Oceanogr.* **35**, 1267–1278.

White, J.R. (1996) Latitudinal gradients in zooplankton biomass in the tropical Pacific at 140°W during the JGOFS EqPac study: effects of El Niño. *Deep-Sea Res. II.* **42**, 715–733.

Whitney, F.A., and H.J. Freeland (1999) Variability in upper-ocean water properties in the NE Pacific Ocean. *Deep-Sea Res. II.* **46**, 2351–2370.

Whitney, F.A., *et al.* (2005) Physical processes that enhance nutrient transport and primary productivity in the coastal and open ocean of the subarctic North Pacific. *Deep-Sea Res. II.* **52**, 681–706.

Wiebe, P.H., *et al.* (1976) Gulf Stream cold-core rings: large interaction sites for open ocean plankton communities. *Deep-Sea Res.* **23**, 695–710.

Wiebe, P.H., and McDougall, T. J. (1986) Warm core rings. *Deep-Sea Res.* **33**, 1455–1922.

Wiggert, J.D., *et al.* (2000) The Northeast Monsoon's impact on mixing, phytoplankton biomass and nutrient cycling in the Arabian Sea. *Deep-Sea Res. II.* **47**, 1353–1385.

Wilkerson, F.P., *et al.* (2000) Biomass and productivity off Monterey Bay, California: contribution of the large phytoplankton. *Deep-Sea Res. II.* **47**, 1003–1022.

Williams, R. (1973) Vertical distribution and development of generations of copepods at OWS India. *Proc. Challenger Soc.* **4**(5) (unpaginated).

Williams, R., and D.V.P. Conway (1988) Vertical distribution and seasonal numerical abundance of the Calanidae in oceanic waters southwest of the British Isles. *Hydrobiology.* **167/168**, 259–266.

Williams, R., and C.C. Hopkins (1976) Biological sampling at OWS INDIA in 1974. *Ann. Biolog. Copenhag.* **31**, 60–62.

Williams, R., and J.A. Lindley (1982) Plankton of the Fladen ground during FLEX 76. III. Vertical distribution, population dynamics and production of *Calanus finmarchicus* (Crustacea: Copepoda). *Mar. Biol.* **60**, 47–56.

Williams, R., and A. Robinson (1973) Primary production at OWS "I" in the North Atlantic. *Bull. Mar. Ecol.* **8**, 115–121.

Wirth, A., *et al.* (2002) Variability of the Great Whirl from observations and models. *Deep-Sea Res. II.* **49**, 1279–1295.

Wolanski, E.M., *et al.* (1986) The CZCS views the Bismarck Sea. *Annu. Geophys. B Terr. Planet. Phys.* **4**, 55–62.

Wolanski, E.M., *et al.* (1988) Currents through Torres Strait. *J. Phys. Oceanogr.* **18**, 1535–1545.

Wolf, K.-U., and J.D. Woods (1988) Lagrangian simulation of primary production in the physical environment—the deep chlorophyll maximum and thermocline. *In—Towards a Theory of Biological-Physical Interactions in the World's Ocean* (B.J. Rothschild, Ed.), pp. 51–70. Kluwer Academic, Dordrecht, The Netherlands.

Wolff, M., *et al.* (2003) The Humboldt Current—trends in exploitation, protection and research. *In—Large Marine Ecosystems of the World* (G. Hempel and K. Sherman Eds.), pp. 179–309. AAAS, Washington, D.C.

Wong, C.S., *et al.* (2002) Seasonal cycles of nutrients and DIC at high and mid latitudes in the northern Pacific Ocean. *Deep-Sea Res. II.* **49**, 5317–5338.

Wong, C.S., and R.J. Matear (1999) Sporadic silicate limitation of phytoplankton production in the subarctic NE Pacific. *Deep-Sea Res. II.* **46**, 2539–2556.

Wood, A.M., *et al.* (1996) Mixing of chlorophyll from the Middle Atlantic Bight into the Gulf Stream at Cape Hatteras in July, 1993. *J. Geophys. Res.* **101**, 20579–20593.

Woods, J.D. (1984) The warmwatersphere of the northeast Atlantic: a miscellany. *Ber. Inst. Meeresk. Kiel.* **128**, 1–39.

Woods, J.D. (1988) Mesoscale upwelling and primary production. *In—Towards a Theory of Biological-Physical Interactions in the World's Ocean* (B.J. Rothschild, Ed.), pp. 1–29. Kluwer Academic, Dordrecht, The Netherlands.

Woods, J.D., *et al.* (1977) Vertical circulation at fronts in the upper ocean. *In—A Voyage of Discovery* (M. Angel, Ed.), pp. 253–276. Pergamon, London.

Woodward, E.M.S., *et al.* (1999) The influence of the southwest monsoon on the nutrient biogeochemistry of the Arabian Sea. *Deep-Sea Res. II.* **46**, 571–591.

Wooster, W.S., *et al.* (1967) Atlas of the Arabian Sea for fishery oceanography, *IMR Report No. 67–12*, pp. 1–39, 143 figs. Institute of Marine Resources, La Jolla, CA.

Wooster, W.S., *et al.* (1976) The seasonal upwelling cycle along the eastern boundary of the North Atlantic. *J. Mar. Res.* **34**, 131–141.

Worm, B., and R.A. Myers (2003) Meta-analysis of cod-shrimp interactions reveals top-down control in oceanic food webs. *Ecology* **84**, 162–173.

Worthington, L.V. (1986) On the North Atlantic circulation. *Johns Hopkins Oceanogr. Stud.* **6**, 1–110.

Wroblewski, J.S. (1989) A model of the spring bloom in the Atlantic and impact on ocean optics. *Limnol. Oceanogr.* **34**, 1563–1571.

Wroblewski, J.S., *et al.* (1988) An ocean basin scale model of plankton dynamics in the North Atlantic. *Global Biogeochem. Cycles.* **2**, 199–218.

Wyllie, J.G. (1966) Geostrophic flow of the California Current at the surface and at 200 meters. *CalCOFI. Atlas.* **4**, 1–288.

Wyrtki, K. (1961) The physical oceanography of the southeast Asian waters. *Naga Rept.* **2**, 1–195.

Wyrtki, K. (1962) The upwelling in the region between Java and Australia during the south-east monsoon. *Aust. J. Mar. Freshwat. Res.* **13**, 217–225.

Wyrtki, K. (1963) The horizontal and vertical field of motion in the Peru Current. *Bull. Scripps Inst.* **8**, 313–346.

Wyrtki, K. (1966) Oceanography of the eastern equatorial Pacific Ocean. Oceanogr. *Mar. Biol. Ann. Rev.* **4**, 33–68.

Wyrtki, K. (1971) *Oceanographic Atlas of the Indian Ocean Expedition*, p. 531. National Science Foundation, Washington, D.C.

Wyrtki, K. (1973a) An equatorial jet in the Indian Ocean. *Science.* **181**, 262–264.

Wyrtki, K. (1973b) Physical oceanography of the Indian Ocean. *In—Biology of the Indian Ocean* (B. Zeitzschel, Ed.), pp. 18–36. Springer, Berlin.

Wyrtki, K. (1982) Eddies in the Pacific North Equatorial Current. *J. Phys. Oceanogr.* **12**, 746–749.

Wyrtki, K. (1984) The slope of sea level along the equator during the 1982/1983 El Niño. *J. Geophys. Res.* **89**, 10419–10424.

Wyrtki, K., and Kilonsky, B. (1984) Mean water and current structure during the Hawaii-to-Tahiti Shuttle Experiment. *J. Phys. Oceanogr.* **14**, 242–252.

Yamamoto, T., and Nishizawa, S. (1986) Small-scale zooplankton aggregations at the front of a Kuroshio warm-core ring. *Deep-Sea Res.* **33**, 1729–1740.

Yan, X.-H., *et al.* (1992) Temperature and size variabilities of the Western Pacific warm pool. *Science.* **258**, 1643–1645.

Yebra, L., *et al.* (2004) The effect of upwelling filaments and island-induced eddies on indices of feeding, respiration and growth in copepods. *Prog. Oceanogr.* **62**, 151–169.

Yentsch, C.S. (1965) Distribution of chlorophyll and phaeophytin in the open ocean. *Limnol. Oceanogr.* **11**, 117–147.

Yentsch, C.S. (1974) The influence of geostrophy on primary production. *Tethys.* **6**, 111–118.

Yentsch, C.S. (1982) Satellite observation of phytoplankton distribution associated with large-scale oceanic circulation. *Sci. Stud. North Atlantic Fish. Org.* **4**, 53–59.

Yentsch, C.S. (1990) Estimates of "new" production in the mid-North Atlantic. *J. Plankton Res.* **12**, 717–734.

Yentsch, C.S., and D.A. Phinney (1985) Rotary motions and convection as a means of regulating primary production in warm core rings. *J. Geophys. Res.* **90**, 3237–3248.

Yentsch, C.S., and J.C. Garside (1986) Patterns of phytoplankton abundance and biogeography. *UNESCO Tech. Papers Mar. Sci.* **49**, 278–284.

Yilmaz, A. (1994) Phytoplankton fluorescence and DCM in the Mediterranean. *Océanol. Acta.* **17**, 69–77.

Yoder, J.A., *et al.* (1993) Annual cycles of phytoplankton chlorophyll concentrations in the global ocean: a satellite view. *Global Biogeochem. Cycles.* **7**, 181–193.

Young, I.R., *et al.* (1994) Circulation in the Ribbon Reef Region of the Great Barrier Reef. *Cont. Shelf Res.* **14**, 117–142.

Young, J., *et al.* (1999) A preliminary survey of the summer hydrography and plankton biomass of the Great Australian Bight. *In—Rep. 11th Workshop, Southern Bluefin Tuna Tagging Program*, pp. 1–15. CCAMLR, Hobart, Tasmania.

Young, R.W., *et al.* (1991) Atmospheric iron inputs and primary productivity: phytoplankton responses in the North Pacific. *Global Biogeochem. Cycles.* **5**, 119–134.

Zeitzschel, B.(Ed.) (1973) *The Biology of the Indian Ocean*, p. 549. Chapman and Hall, London.

Zeldis, J.R., *et al.* (1995) Salp grazing: effects on phytoplankton abundance, vertical distribution and taxonomic composition. *Mar. Ecol. Prog. Ser.* **126**, 267–283.

Zevenboom, W., and F.J. Wetsteyn (1990) Growth limitation and growth rates of pico-phytoplankton in the Banda sea during two different monsoons. *Neth. J. Sea Res.* **25**, 465–472.

Zhabin, I.A. *et al* (1990) Satellite-revealed surface cool patches in the northern part of the Sea of Okhotsk. *Issled. Zemli. Kosm.* **5**, 25–28.

Zhang, C.I., and S. Kim (1999) Living marine resources of the Yellow Sea ecosystem in Korean waters: status and perspectives. *In—Large Marine Ecosystems of the Pacific Rim* (K. Sherman and Q. Tang, Eds.), pp. 163–178. Blackwell Science.

Zheng, Y. *et al.* (1992) Observation of a Kuroshio frontal eddy in the East China Sea. *In—Essays on the Investigation of the Kuroshio 4*, pp. 23–32. University of Washington University Press, Seattle, WA.

Zhuang, G., and D.R. Kester (1990) The dissolution of atmospheric iron in the surface seawater of the open ocean. *J. Geophys. Res.* **95**, 16207–16216.

Zubkov, M.V., *et al.* (1998) Picoplanktonic community structure on an Atlantic transect from 50°N to 50°S. *Deep-Sea Res. I.* **45**, 1339–1356.

Zwanenburg, K.C.T., *et al.* (2003) Decadal changes in the Scotian Shelf Large Marine Ecosystem. *In—Large Marine Ecosystems of the North Atlantic* (K.H. Sherman and H.R. Skjodal, Eds.), pp. 105–150. AAAS, Washington, D.C.

# INDEX

## B

## C

## F

## L



## M

## N

## O

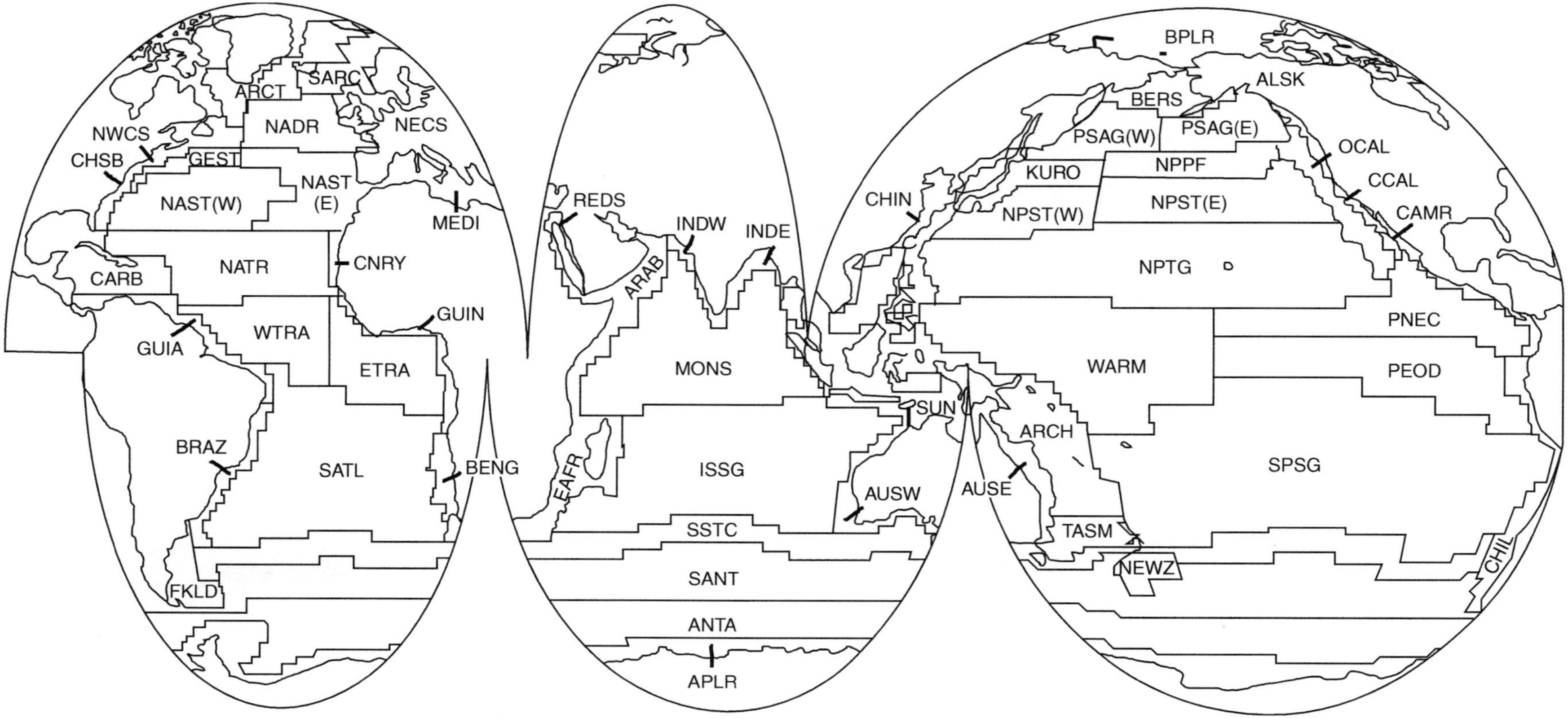

BPLR
ALSK
BERS
PSAG(W)
PSAG(E)
OCAL
CCAL
CAMR
KURO
NPPF
NPST(W)
NPST(E)
CHIN
NPTG
PNEC
PEOD
WARM
SUN
ARCH
SPSG
AUSW
AUSE
TASM
NEWZ
CHIL
SARC
ARCT
NADR
NECS
NWCS
CHSB
GEST
NAST(W)
NAST
(E)
MEDI
NATR
CNRY
CARB
GUIN
WTRA
GUIA
ETRA
BRAZ
SATL
BENG
FKLD
REDS
INDW
INDE
ARAB
MONS
EAFR
ISSG
SSTC
SANT
ANTA
APLR

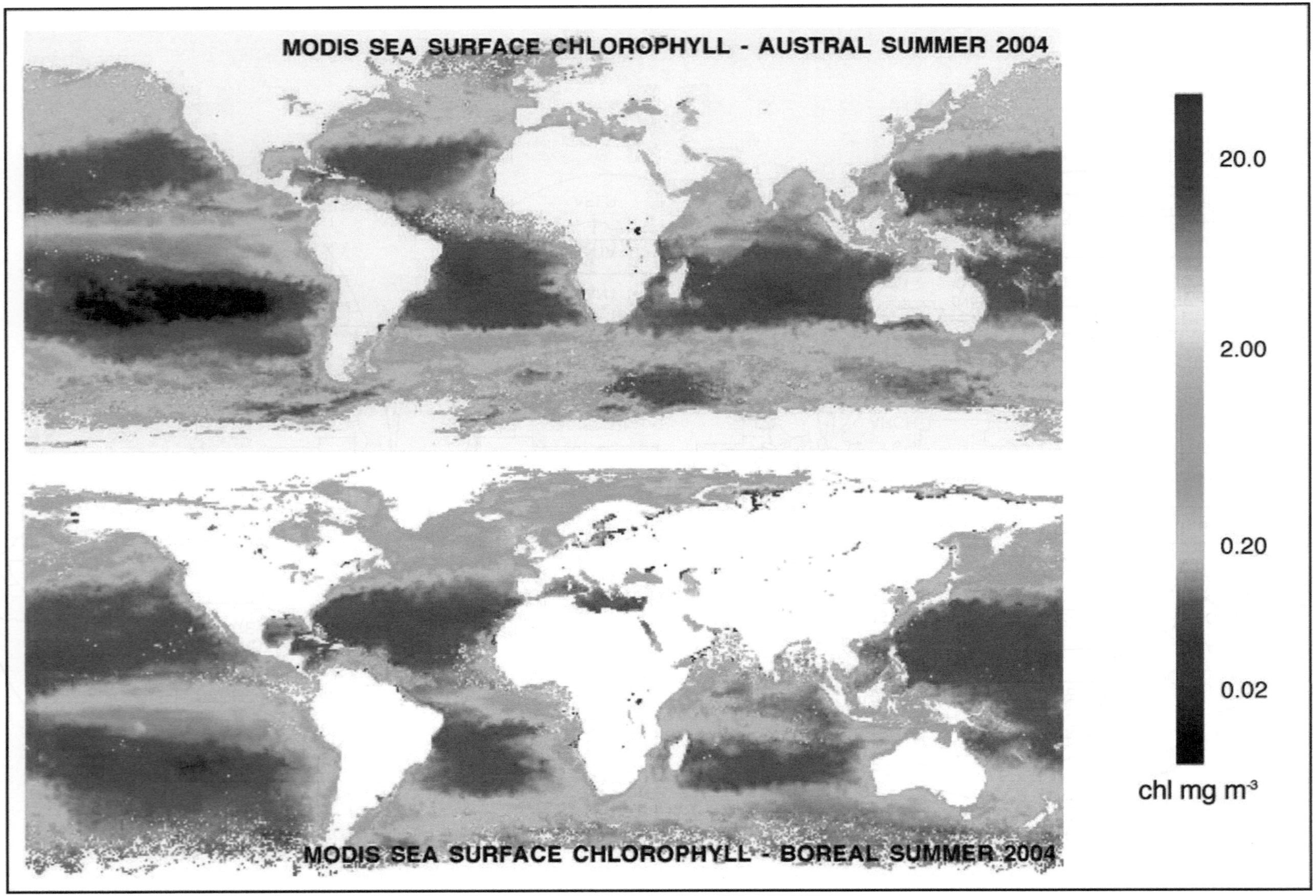
MODIS SEA SURFACE CHLOROPHYLL - AUSTRAL SUMMER 2004
20.0
2.00
0.20
0.02
chl mg m⁻³
MODIS SEA SURFACE CHLOROPHYLL - BOREAL SUMMER 2004

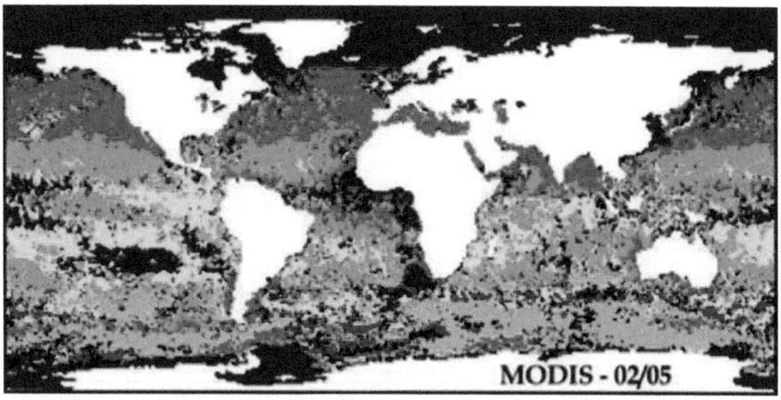


**Color plate 1** Analysis of multispectral radiometer data with the PHYSAT method of Alvain *et al.* (2005) to infer the relative global distribution of four autotrophic cell types; the image is derived from MODIS-Aqua data for February, 2005. Red, diatoms; green, *Prochlorococcus*; blue, *Synechococcus*; gray, haptophytes; black, no data.

*Source: NASA.*

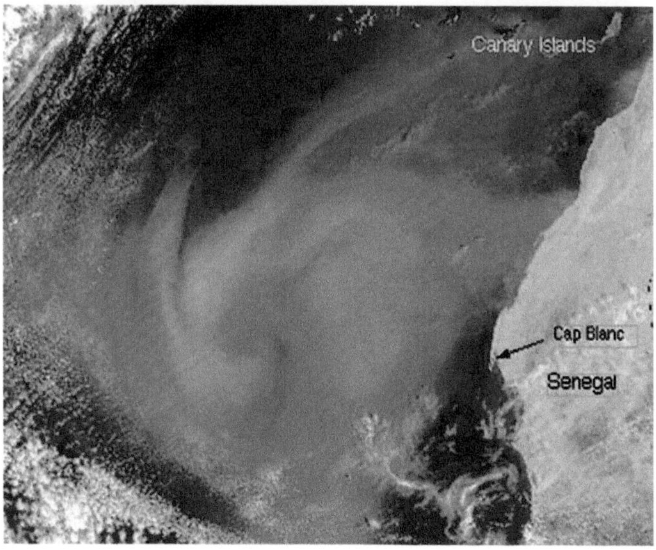


**Color plate 2** The instantaneous pattern over the sea surface formed by a dust storm off Senegal, derived from MODIS-Aqua data, to illustrate the difference between this and the pattern of sea surface chlorophyll forced by mesoscale eddy activity in the open ocean (see Fig. 9.14).

*Source: NASA.*

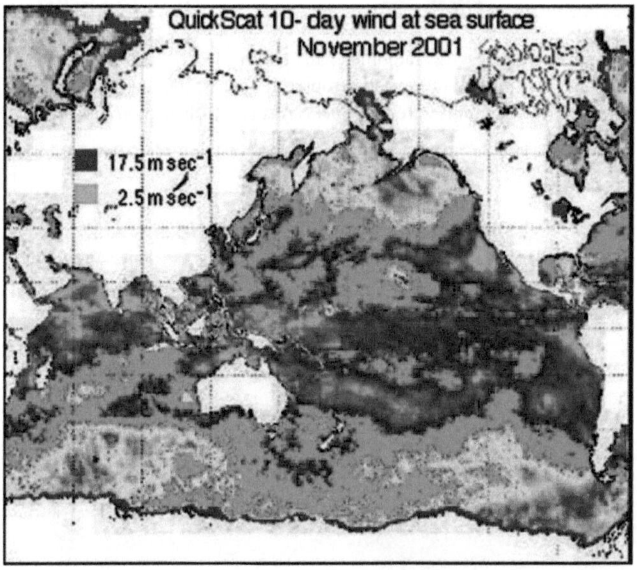


**Color plate 3** Average wind speed over the Indian, Pacific, and Southern Oceans, from satellite scatterometer (QuickScat) data for November 2001; this month was chosen to represent the midseason regime in both northern and southern hemispheres.

*Source: NASA.*

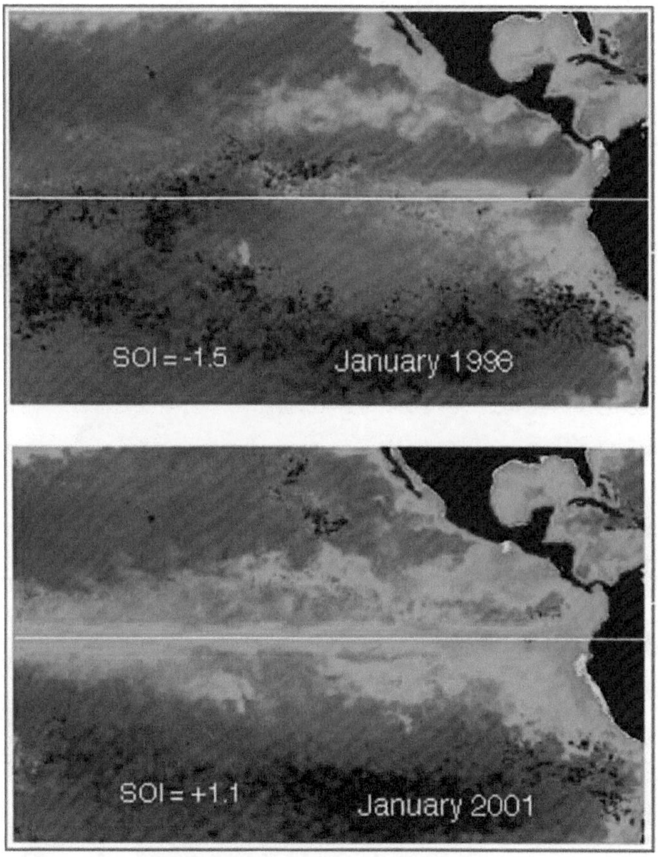


**Color plate 4** Characteristic patterns of sea surface chlorophyll over the eastern tropical Pacific Ocean during two phases of the ENSO cycle from SeaWiFS imagery: (top) during the 1997–98 El Niño episode and (bottom) during a later period of stronger upwelling.

*Source: SeaWiFS – OrbView 2 image from GeoEye Corporation.*

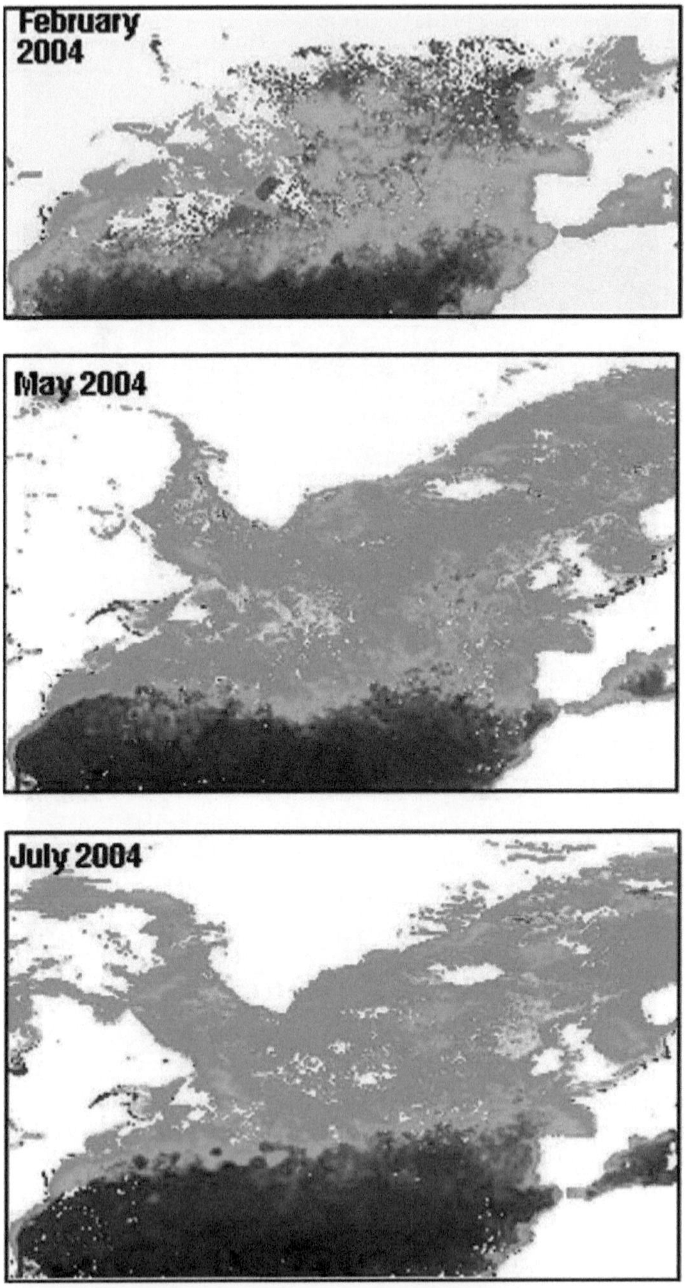


**Color plate 5** Seasonal progression of the North Atlantic bloom during February, May, and July of 2004 derived from MODIS-Aqua data. Compare these patterns of phytoplankton biomass with that of sea surface elevation, and hence of mesoscale circulation, shown in Fig. 3.1.

*Source: NASA.*

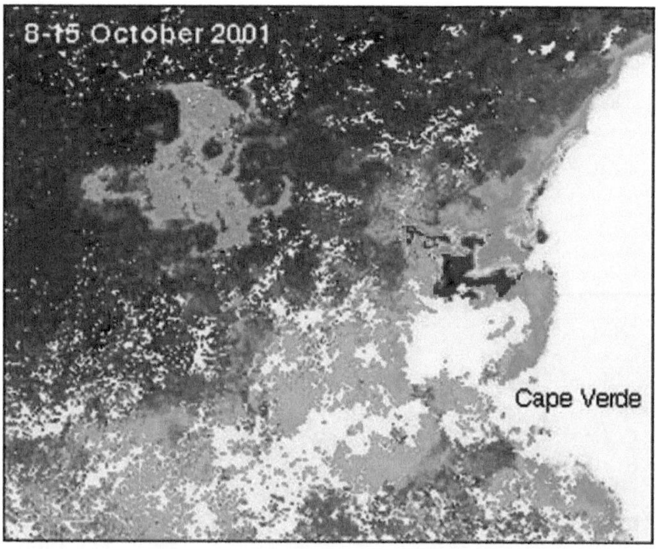


**Color plate 6** A very large, residual dendritic bloom lying far to the west of Mauretania shown in SeaWiFS imagery; this represents the final stages of a major upwelling event that occurred farther to the north than the small active bloom seen in the bight to the north of Cape Verde.

*Source: SeaWiFS – OrbView 2 image from GeoEye Corporation.*

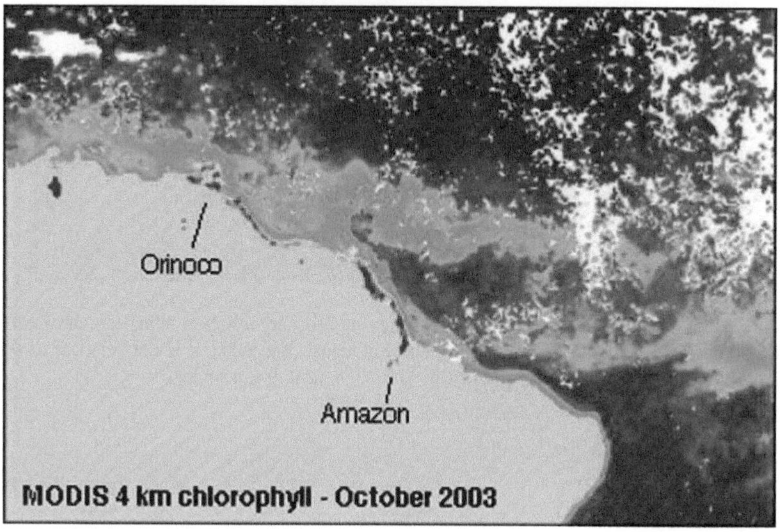


**Color plate 7** A typical case of the apparently high chlorophyll encircling the retroflection of shelf water to the north of the Amazon mouth derived from MODIS-Aqua data; it is yet to be established to what extent this represents CDOM entrained from the shelf water, or to what extent it represents production *in situ* related to eddy dynamics. Farther to the south, equatorial Rossby waves induce some degree of chlorophyll enhancement, terminating near the continent.

*Source: NASA.*

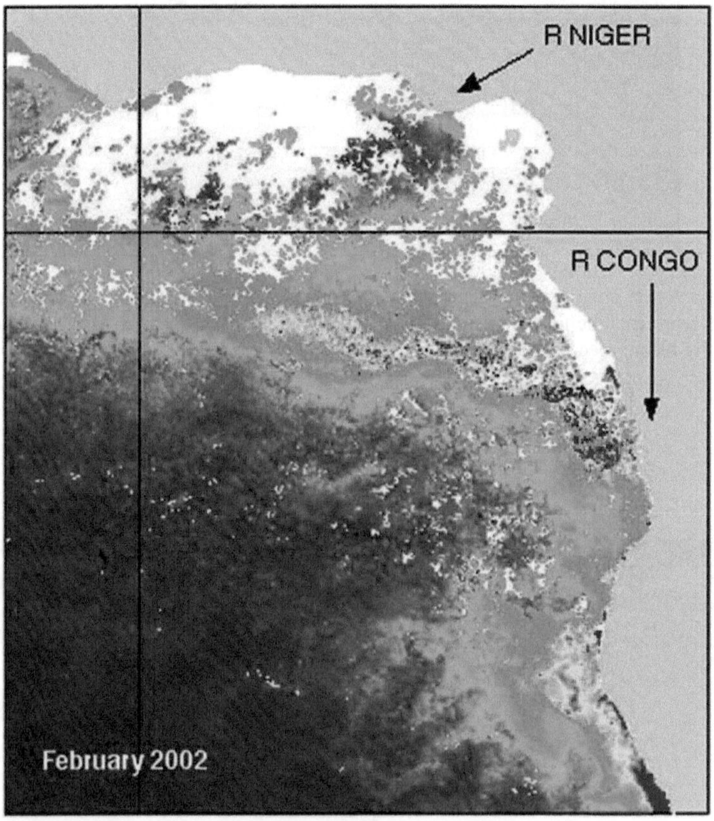


**Color plate 8** Sea-surface chlorophyll SeaWiFS image for February 2002, to show the zonal extent of the effects of the Congo River discharge; as with the Amazon signal (Fig. 9.17), it is not yet clear to what extent this is due to the persistence of riverborne CDOM (SeaWiFS 30-d, 9-km image).

*Source: SeaWiFS – OrbView 2 image from GeoEye Corporation.*

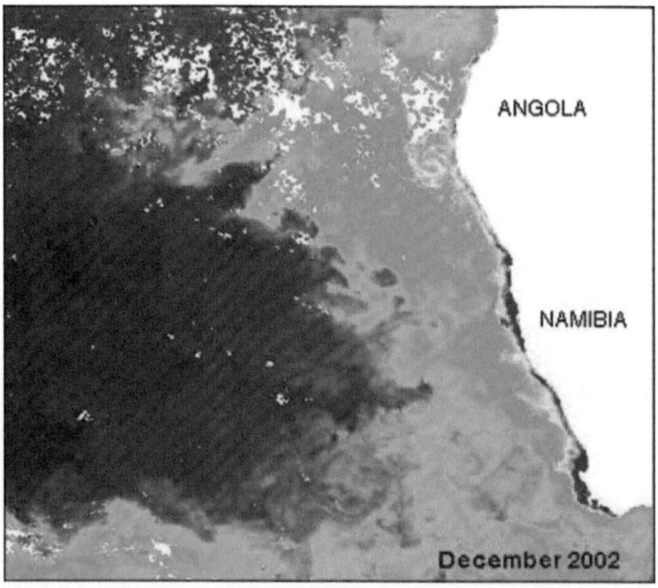


**Color plate 9** The extensive transport westward of upwelled water from the Benguela Current, carrying with it a high load of chlorophyll, seen here in MODIS-Aqua data. Because of the nature of the coastal regime, this plume is unlikely to be significantly aliased by the presence of CDOM.

*Source: NASA.*

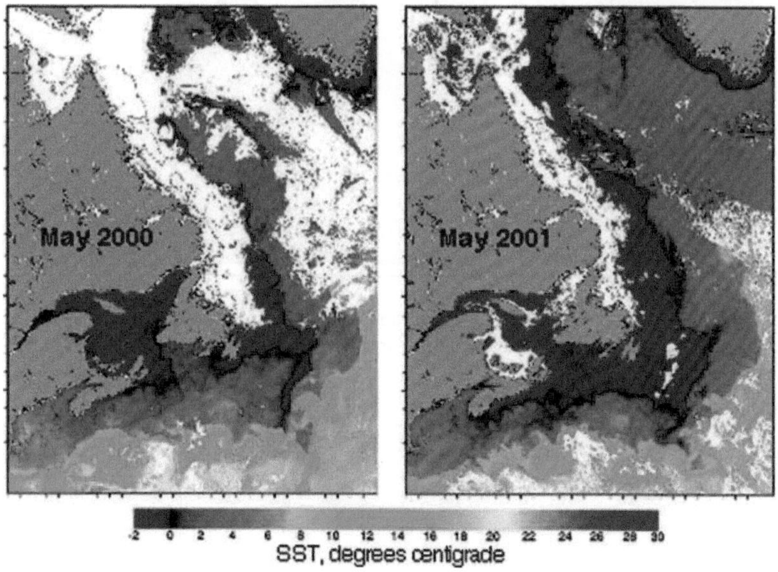


**Color plate 10** The differential distribution of sea surface temperature over the Grand Banks in years of high and low values of the North Atlantic Oscillation index; derived from NASA AVHRR data for the period 1–15 May of the years 2000 (NAO was +2.8) and 2001 (NAO was −1.89).

*Source: NASA.*

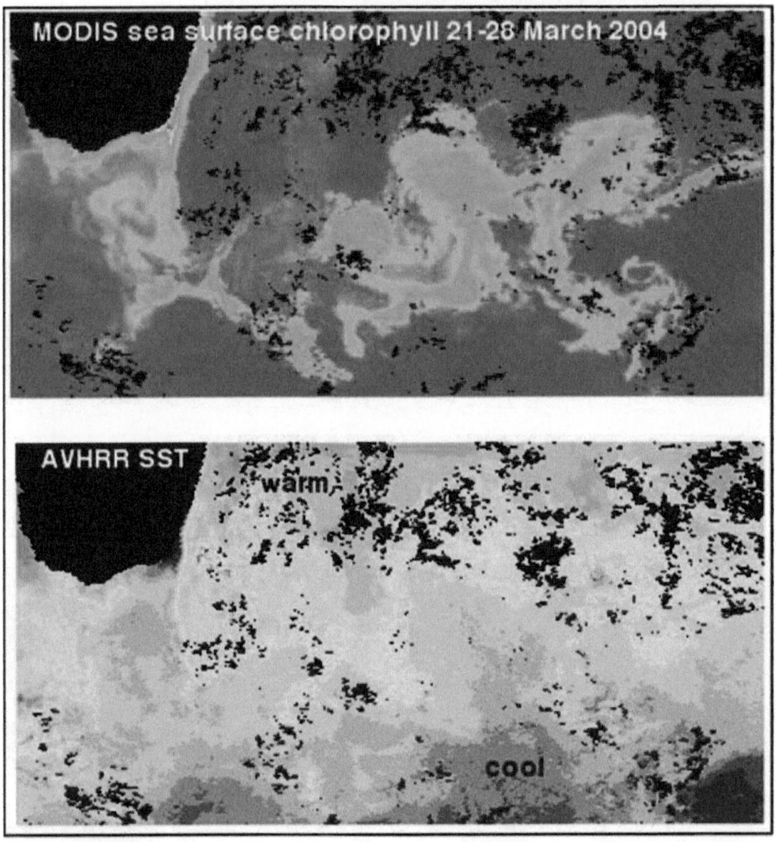


**Color plate 11** Simultaneous surface chlorophyll and surface temperature images to show the physical forcing of the East Madagascar dendroid bloom; these are matched MODIS-Aqua 4-km and TOPEX-POSEIDON images for 18–19 April 2004.

*Sources: NASA and the University of Colorado.*

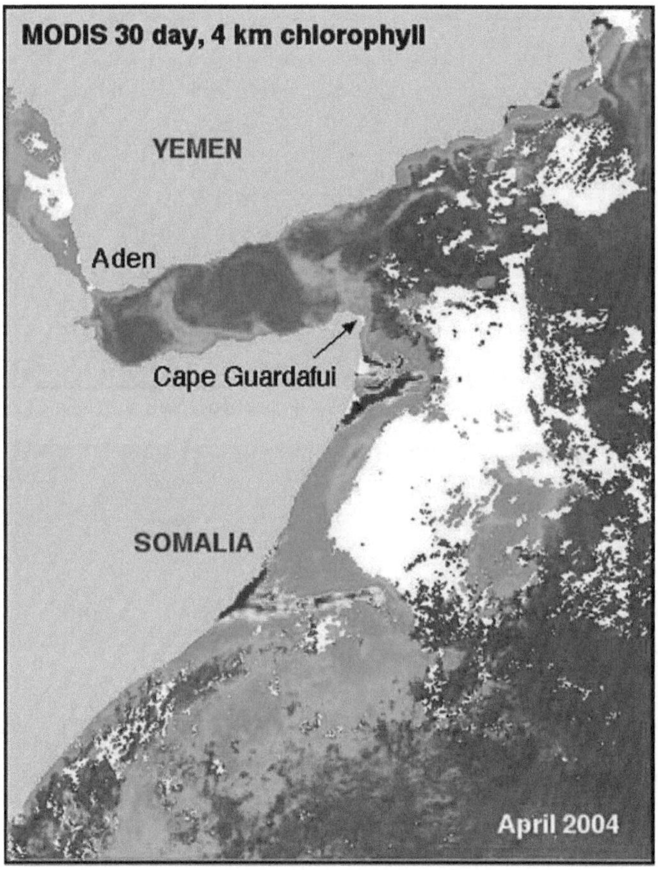


**Color plate 12** Sea surface chlorophyll distribution derived from MODIS-Aqua data in the Somali Current during the period of the SW Monsoon to show characteristic upwelling plumes and eddies.

*Source: NASA.*

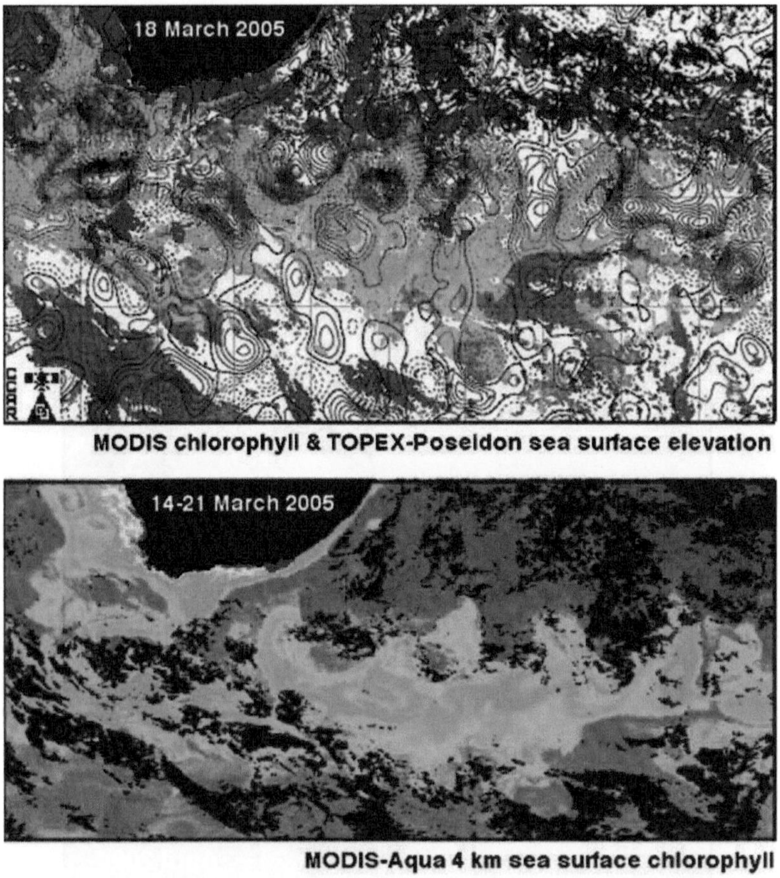


**Color plate 13** Analysis of the retroflection region of the Agulhas Current to the south of South Africa. The regional distribution of MODIS-Aqua surface chlorophyll and TOPEX-POSEIDON sea surface elevation overlaid, to show the congruence between mesoscale eddy pattern and chlorophyll concentrations. An 8-day, 4-km MODIS image is also shown to specify the surface chlorophyll field in more detail.

*Sources: NASA and the University of Colorado.*

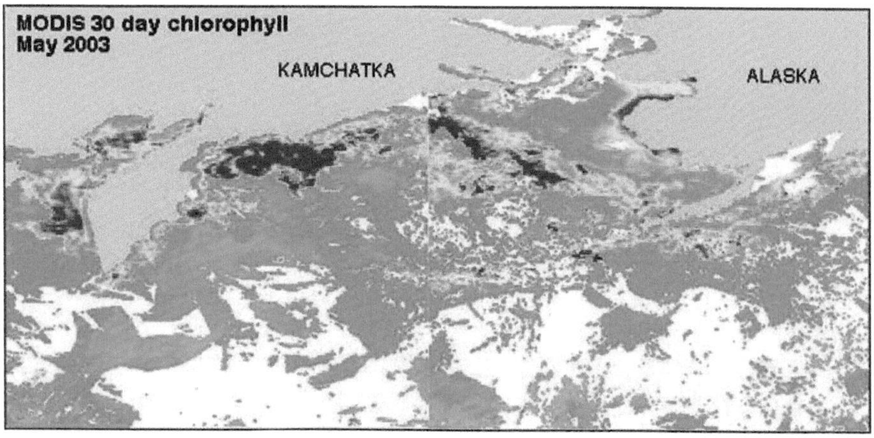


**Color plate 14** The distribution of surface chlorophyll during May 2003 over shelf regions of the extreme North Pacific derived from MODIS-Aqua data, to emphasize the almost permanent coastal blooms that are generated by the Kuril and Aleutian island arcs, and also the shelf-edge chlorophyll feature that is characteristic of the Bering Sea during the spring bloom period.

*Source: NASA.*

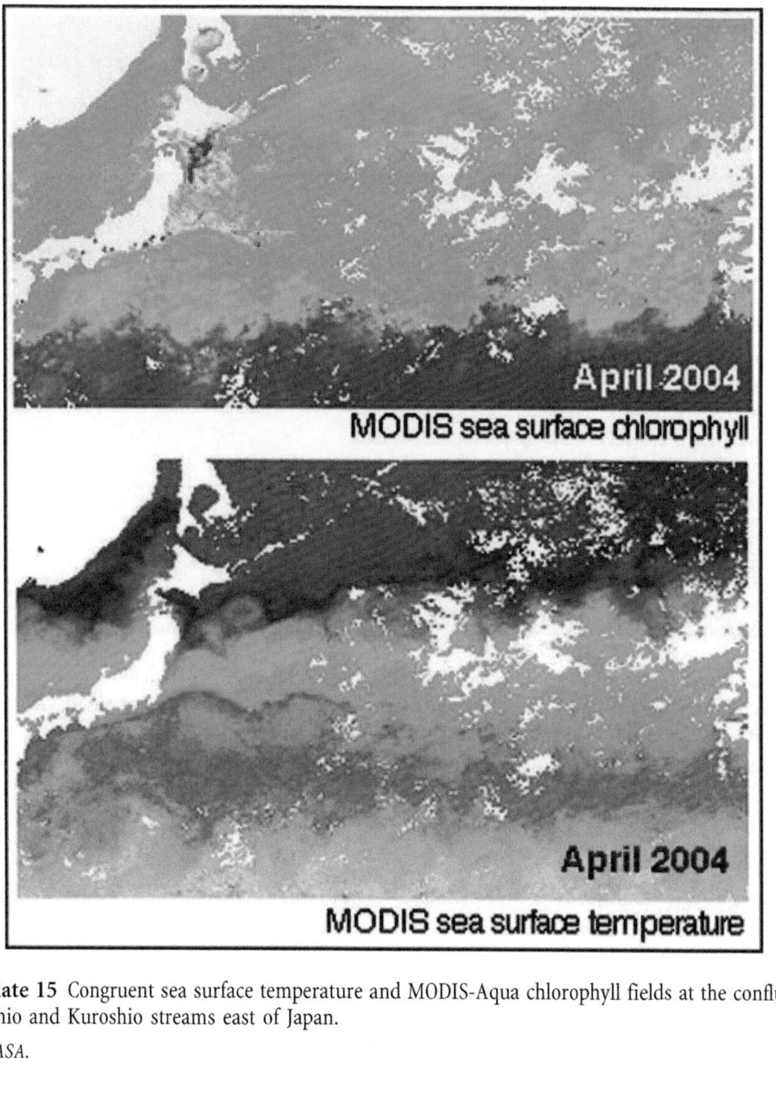


**Color plate 15** Congruent sea surface temperature and MODIS-Aqua chlorophyll fields at the confluence of the Oyashio and Kuroshio streams east of Japan.

*Source: NASA.*

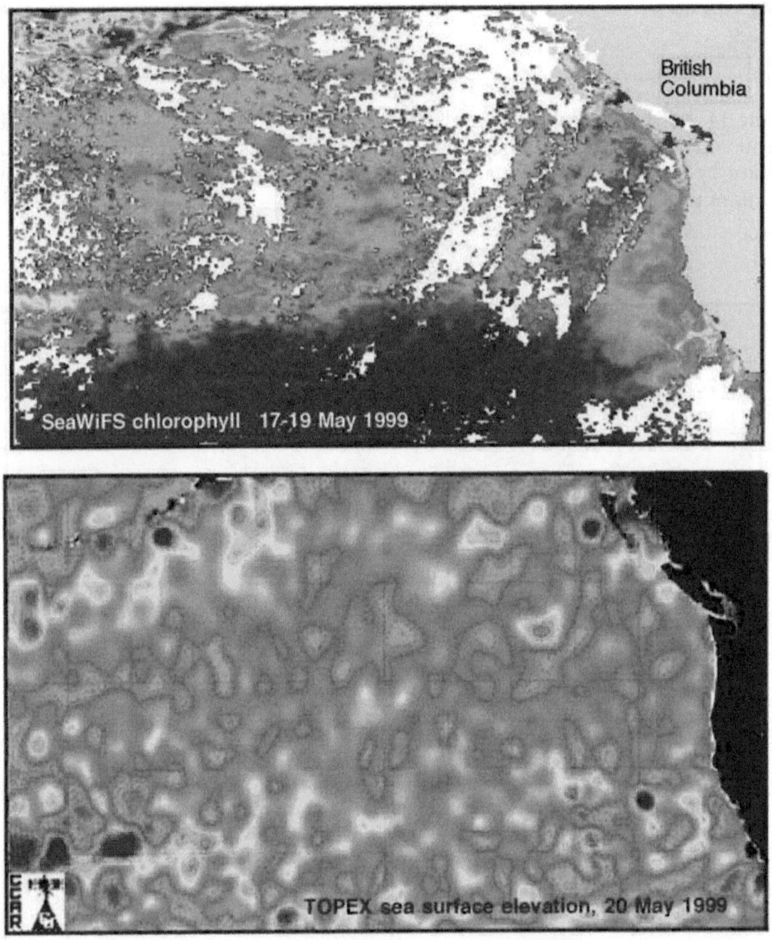


**Color plate 16** The North Pacific Front delimiting the subtropical recirculation gyre seen in near-simultaneous SeaWiFS chlorophyll and Topex-Poseiden sea surface elevation images; observe the congruence between the eddy field and relative production.

*Source: SeaWiFS – OrbView 2 image from GeoEye Corporation; Topex-Poseiden image from University of Colorado.*

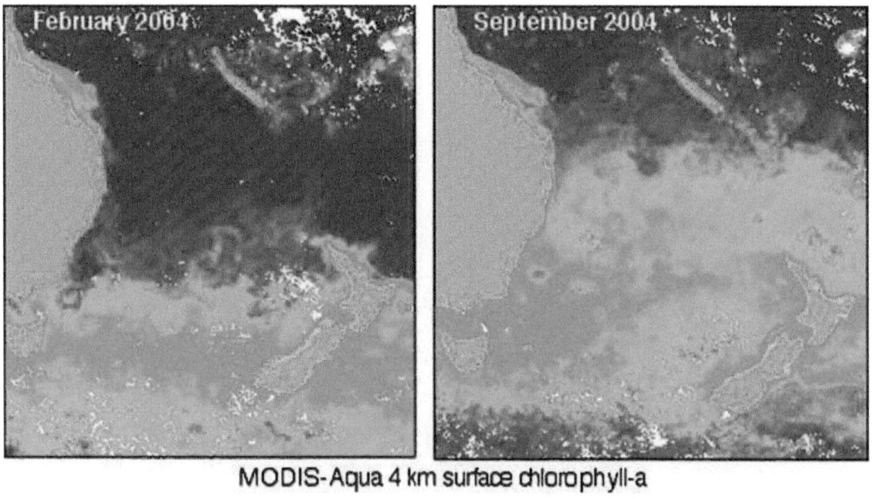


**Color plate 17** Stasis and bloom in Tasman Sea, to show the dominance of the effects of the South Subtropical Convergence Zone on the regional distribution of chlorophyll biomass derived from MODIS-Aqua data during the oligotrophic summer period, and the meridional extent of the winter-spring bloom.

*Source: NASA.*

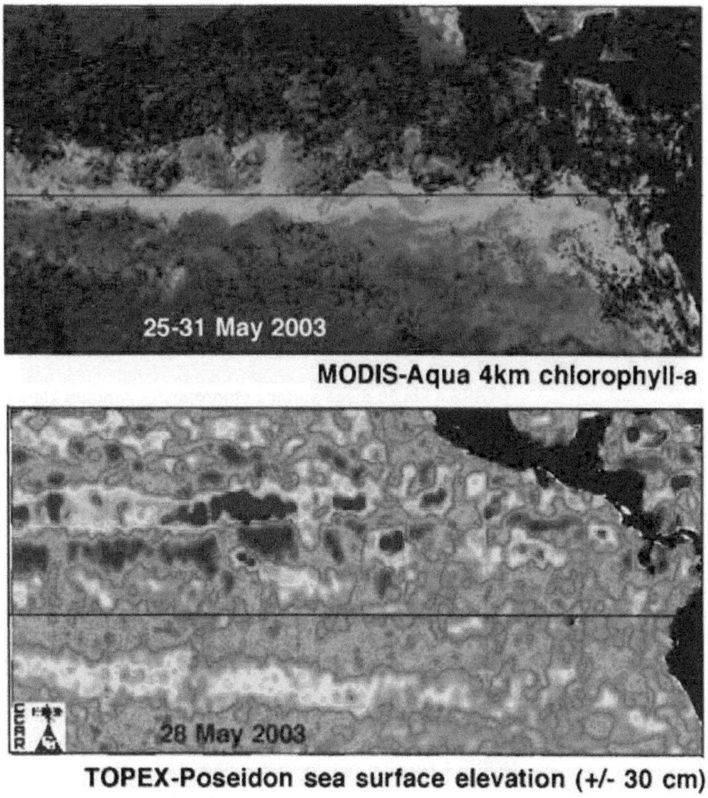


**Color plate 18** During this period of very moderate La Niña conditions, the patterns of TOPEX-Poeidon sea surface elevation and of MODIS-Aqua surface chlorophyll are congruent along the equatorial ridge. The 10°N ridge lies beyond the zonal region of the nitrate-replete superficial water mass (see Fig. 11.13).

*Source: NASA.*

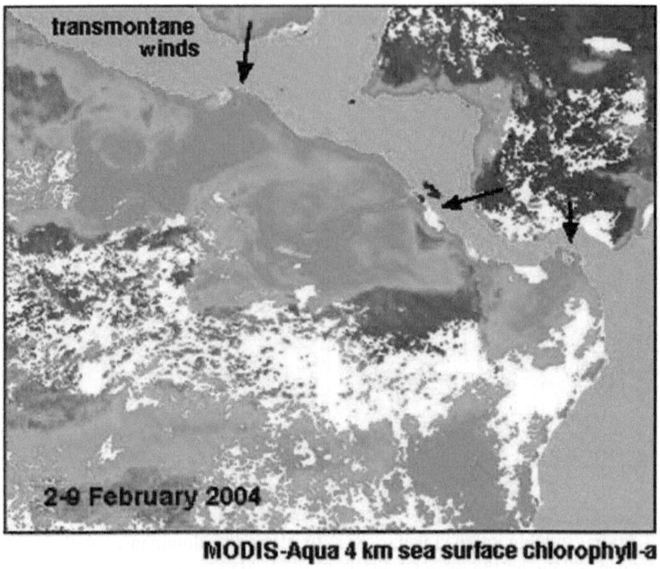


**Color plate 19** The effect of transmontane winds across the narrow regions of central America (at the Tehuantepec gap, over Lake Nicaragua and across the isthmus of Panama) on the chlorophyll field of the eastern tropical Pacific derived from MODIS-Aqua data.

*Source: NASA.*

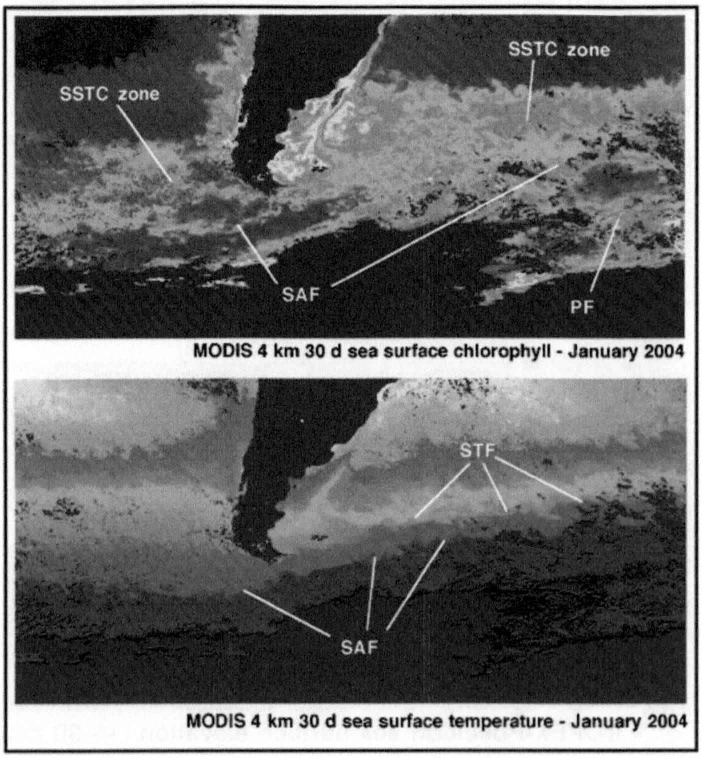


**Color plate 20** Distribution of relative phytoplankton biomass derived from MODIS-Aqua data and the sea surface temperature fronts shown diagrammatically in Fig. 12.01. Observe the prominence of the SSTC zone and the complexity of the chlorophyll biomass pattern over and around the Falkland Plateau (see FKLD).

*Source: NASA.*

Printed and bound by CPI Group (UK) Ltd, Croydon, CR0 4YY
03/10/2024
01040318-0009